T0181704

Lecture Notes in Computer Science   13844

Founding Editors

Gerhard Goos
Juris Hartmanis

The series Lecture Notes in Computer Science (LNCS), including its subseries Lecture Notes in Artificial Intelligence (LNAI) and Lecture Notes in Bioinformatics (LNBI), has established itself as a medium for the publication of new developments in computer science and information technology research, teaching, and education.

LNCS enjoys close cooperation with the computer science R & D community, the series counts many renowned academics among its volume editors and paper authors, and collaborates with prestigious societies. Its mission is to serve this international community by providing an invaluable service, mainly focused on the publication of conference and workshop proceedings and postproceedings. LNCS commenced publication in 1973.

Lei Wang · Juergen Gall · Tat-Jun Chin ·
Imari Sato · Rama Chellappa
Editors

Computer Vision – ACCV 2022

16th Asian Conference on Computer Vision
Macao, China, December 4–8, 2022
Proceedings, Part IV

 Springer

Editors
Lei Wang (iD)
University of Wollongong
Wollongong, NSW, Australia

Juergen Gall (iD)
University of Bonn
Bonn, Germany

Tat-Jun Chin (iD)
University of Adelaide
Adelaide, SA, Australia

Imari Sato
National Institute of Informatics
Tokyo, Japan

Rama Chellappa (iD)
Johns Hopkins University
Baltimore, MD, USA

ISSN 0302-9743   ISSN 1611-3349 (electronic)
Lecture Notes in Computer Science
ISBN 978-3-031-26315-6   ISBN 978-3-031-26316-3 (eBook)
https://doi.org/10.1007/978-3-031-26316-3

This Springer imprint is published by the registered company Springer Nature Switzerland AG
The registered company address is: Gewerbestrasse 11, 6330 Cham, Switzerland

Preface

The 16th Asian Conference on Computer Vision (ACCV) 2022 was held in a hybrid mode in Macau SAR, China during December 4–8, 2022. The conference featured novel research contributions from almost all sub-areas of computer vision.

For the main conference, 836 valid submissions entered the review stage after desk rejection. Sixty-three area chairs and 959 reviewers made great efforts to ensure that every submission received thorough and high-quality reviews. As in previous editions of ACCV, this conference adopted a double-blind review process. The identities of authors were not visible to the reviewers or area chairs; nor were the identities of the assigned reviewers and area chairs known to the authors. The program chairs did not submit papers to the conference.

After receiving the reviews, the authors had the option of submitting a rebuttal. Following that, the area chairs led the discussions and final recommendations were then made by the reviewers. Taking conflicts of interest into account, the area chairs formed 21 AC triplets to finalize the paper recommendations. With the confirmation of three area chairs for each paper, 277 papers were accepted. ACCV 2022 also included eight workshops, eight tutorials, and one grand challenge, covering various cutting-edge research topics related to computer vision. The proceedings of ACCV 2022 are open access at the Computer Vision Foundation website, by courtesy of Springer. The quality of the papers presented at ACCV 2022 demonstrates the research excellence of the international computer vision communities.

This conference is fortunate to receive support from many organizations and individuals. We would like to express our gratitude for the continued support of the Asian Federation of Computer Vision and our sponsors, the University of Macau, Springer, the Artificial Intelligence Journal, and OPPO. ACCV 2022 used the Conference Management Toolkit sponsored by Microsoft Research and received much help from its support team.

All the organizers, area chairs, reviewers, and authors made great contributions to ensure a successful ACCV 2022. For this, we owe them deep gratitude. Last but not least, we would like to thank the online and in-person attendees of ACCV 2022. Their presence showed strong commitment and appreciation towards this conference.

December 2022

Lei Wang
Juergen Gall
Tat-Jun Chin
Imari Sato
Rama Chellappa

Organization

General Chairs

Gérard Medioni   University of Southern California, USA
Shiguang Shan   Chinese Academy of Sciences, China
Bohyung Han   Seoul National University, South Korea
Hongdong Li   Australian National University, Australia

Program Chairs

Rama Chellappa   Johns Hopkins University, USA
Juergen Gall   University of Bonn, Germany
Imari Sato   National Institute of Informatics, Japan
Tat-Jun Chin   University of Adelaide, Australia
Lei Wang   University of Wollongong, Australia

Publication Chairs

Wenbin Li   Nanjing University, China
Wanqi Yang   Nanjing Normal University, China

Local Arrangements Chairs

Liming Zhang   University of Macau, China
Jianjia Zhang   Sun Yat-sen University, China

Web Chairs

Zongyuan Ge   Monash University, Australia
Deval Mehta   Monash University, Australia
Zhongyan Zhang   University of Wollongong, Australia

AC Meeting Chair

Chee Seng Chan   University of Malaya, Malaysia

Area Chairs

Aljosa Osep	Technical University of Munich, Germany
Angela Yao	National University of Singapore, Singapore
Anh T. Tran	VinAI Research, Vietnam
Anurag Mittal	Indian Institute of Technology Madras, India
Binh-Son Hua	VinAI Research, Vietnam
C. V. Jawahar	International Institute of Information Technology, Hyderabad, India
Dan Xu	The Hong Kong University of Science and Technology, China
Du Tran	Meta AI, USA
Frederic Jurie	University of Caen and Safran, France
Guangcan Liu	Southeast University, China
Guorong Li	University of Chinese Academy of Sciences, China
Guosheng Lin	Nanyang Technological University, Singapore
Gustavo Carneiro	University of Surrey, UK
Hyun Soo Park	University of Minnesota, USA
Hyunjung Shim	Korea Advanced Institute of Science and Technology, South Korea
Jiaying Liu	Peking University, China
Jun Zhou	Griffith University, Australia
Junseok Kwon	Chung-Ang University, South Korea
Kota Yamaguchi	CyberAgent, Japan
Li Liu	National University of Defense Technology, China
Liang Zheng	Australian National University, Australia
Mathieu Aubry	Ecole des Ponts ParisTech, France
Mehrtash Harandi	Monash University, Australia
Miaomiao Liu	Australian National University, Australia
Ming-Hsuan Yang	University of California at Merced, USA
Palaiahnakote Shivakumara	University of Malaya, Malaysia
Pau-Choo Chung	National Cheng Kung University, Taiwan

Yung-Yu Chuang   National Taiwan University, Taiwan
Zhaoxiang Zhang   Chinese Academy of Sciences, China
Ziad Al-Halah   University of Texas at Austin, USA
Zuzana Kukelova   Czech Technical University, Czech Republic

Additional Reviewers

Abanob E. N. Soliman   Atsushi Shimada   Chao Liu
Abdelbadie Belmouhcine   Attila Szabo   Chao Shi
Adrian Barbu   Aurelie Bugeau   Chaowei Tan
Agnibh Dasgupta   Avatharam Ganivada   Chaoyi Li
Akihiro Sugimoto   Ayan Kumar Bhunia   Chaoyu Dong
Akkarit Sangpetch   Azade Farshad   Chaoyu Zhao
Akrem Sellami   B. V. K. Vijaya Kumar   Chen He
Aleksandr Kim   Bach Tran   Chen Liu
Alexander Andreopoulos   Bailin Yang   Chen Yang
Alexander Fix   Baojiang Zhong   Chen Zhang
Alexander Kugele   Baoquan Zhang   Cheng Deng
Alexandre Morgand   Baoyao Yang   Cheng Guo
Alexis Lechervy   Basit O. Alawode   Cheng Yu
Alina E. Marcu   Beibei Lin   Cheng-Kun Yang
Alper Yilmaz   Benoit Guillard   Chenglong Li
Alvaro Parra   Beomgu Kang   Chengmei Yang
Amogh Subbakrishna   Bin He   Chengxin Liu
   Adishesha   Bin Li   Chengyao Qian
Andrea Giachetti   Bin Liu   Chen-Kuo Chiang
Andrea Lagorio   Bin Ren   Chenxu Luo
Andreu Girbau Xalabarder   Bin Yang   Che-Rung Lee
Andrey Kuehlkamp   Bin-Cheng Yang   Che-Tsung Lin
Anh Nguyen   BingLiang Jiao   Chi Xu
Anh T. Tran   Bo Liu   Chi Nhan Duong
Ankush Gupta   Bohan Li   Chia-Ching Lin
Anoop Cherian   Boyao Zhou   Chien-Cheng Lee
Anton Mitrokhin   Boyu Wang   Chien-Yi Wang
Antonio Agudo   Caoyun Fan   Chih-Chung Hsu
Antonio Robles-Kelly   Carlo Tomasi   Chih-Wei Lin
Ara Abigail Ambita   Carlos Torres   Ching-Chun Huang
Ardhendu Behera   Carvalho Micael   Chiou-Ting Hsu
Arjan Kuijper   Cees Snoek   Chippy M. Manu
Arren Matthew C.   Chang Kong   Chong Wang
   Antioquia   Changick Kim   Chongyang Wang
Arjun Ashok   Changkun Ye   Christian Siagian
Atsushi Hashimoto   Changsheng Lu   Christine Allen-Blanchette

Christoph Schorn
Christos Matsoukas
Chuan Guo
Chuang Yang
Chuanyi Zhang
Chunfeng Song
Chunhui Zhang
Chun-Rong Huang
Ci Lin
Ci-Siang Lin
Cong Fang
Cui Wang
Cui Yuan
Cyrill Stachniss
Dahai Yu
Daiki Ikami
Daisuke Miyazaki
Dandan Zhu
Daniel Barath
Daniel Lichy
Daniel Reich
Danyang Tu
David Picard
Davide Silvestri
Defang Chen
Dehuan Zhang
Deunsol Jung
Difei Gao
Dim P. Papadopoulos
Ding-Jie Chen
Dong Gong
Dong Hao
Dong Wook Shu
Dongdong Chen
Donghun Lee
Donghyeon Kwon
Donghyun Yoo
Dongkeun Kim
Dongliang Luo
Dongseob Kim
Dongsuk Kim
Dongwan Kim
Dongwon Kim
DongWook Yang
Dongze Lian

Dubing Chen
Edoardo Remelli
Emanuele Trucco
Erhan Gundogdu
Erh-Chung Chen
Rickson R. Nascimento
Erkang Chen
Eunbyung Park
Eunpil Park
Eun-Sol Kim
Fabio Cuzzolin
Fan Yang
Fan Zhang
Fangyu Zhou
Fani Deligianni
Fatemeh Karimi Nejadasl
Fei Liu
Feiyue Ni
Feng Su
Feng Xue
Fengchao Xiong
Fengji Ma
Fernando Díaz-del-Rio
Florian Bernard
Florian Kleber
Florin-Alexandru
    Vasluianu
Fok Hing Chi Tivive
Frank Neumann
Fu-En Yang
Fumio Okura
Gang Chen
Gang Liu
Gao Haoyuan
Gaoshuai Wang
Gaoyun An
Gen Li
Georgy Ponimatkin
Gianfranco Doretto
Gil Levi
Guang Yang
Guangfa Wang
Guangfeng Lin
Guillaume Jeanneret
Guisik Kim

Gunhee Kim
Guodong Wang
Ha Young Kim
Hadi Mohaghegh
    Dolatabadi
Haibo Ye
Haili Ye
Haithem Boussaid
Haixia Wang
Han Chen
Han Zou
Hang Cheng
Hang Du
Hang Guo
Hanlin Gu
Hannah H. Kim
Hao He
Hao Huang
Hao Quan
Hao Ren
Hao Tang
Hao Zeng
Hao Zhao
Haoji Hu
Haopeng Li
Haoqing Wang
Haoran Wen
Haoshuo Huang
Haotian Liu
Haozhao Ma
Hari Chandana K.
Haripriya Harikumar
Hehe Fan
Helder Araujo
Henok Ghebrechristos
Heunseung Lim
Hezhi Cao
Hideo Saito
Hieu Le
Hiroaki Santo
Hirokatsu Kataoka
Hiroshi Omori
Hitika Tiwari
Hojung Lee
Hong Cheng

Hong Liu
Hu Zhang
Huadong Tang
Huajie Jiang
Huang Ziqi
Huangying Zhan
Hui Kong
Hui Nie
Huiyu Duan
Huyen Thi Thanh Tran
Hyung-Jeong Yang
Hyunjin Park
Hyunsoo Kim
HyunWook Park
I-Chao Shen
Idil Esen Zulfikar
Ikuhisa Mitsugami
Inseop Chung
Ioannis Pavlidis
Isinsu Katircioglu
Jaeil Kim
Jaeyoon Park
Jae-Young Sim
James Clark
James Elder
James Pritts
Jan Zdenek
Janghoon Choi
Jeany Son
Jenny Seidenschwarz
Jesse Scott
Jia Wan
Jiadai Sun
JiaHuan Ji
Jiajiong Cao
Jian Zhang
Jianbo Jiao
Jianhui Wu
Jianjia Wang
Jianjia Zhang
Jianqiao Wangni
JiaQi Wang
Jiaqin Lin
Jiarui Liu
Jiawei Wang

Jiaxin Gu
Jiaxin Wei
Jiaxin Zhang
Jiaying Zhang
Jiayu Yang
Jidong Tian
Jie Hong
Jie Lin
Jie Liu
Jie Song
Jie Yang
Jiebo Luo
Jiejie Xu
Jin Fang
Jin Gao
Jin Tian
Jinbin Bai
Jing Bai
Jing Huo
Jing Tian
Jing Wu
Jing Zhang
Jingchen Xu
Jingchun Cheng
Jingjing Fu
Jingshuai Liu
JingWei Huang
Jingzhou Chen
JinHan Cui
Jinjie Song
Jinqiao Wang
Jinsun Park
Jinwoo Kim
Jinyu Chen
Jipeng Qiang
Jiri Sedlar
Jiseob Kim
Jiuxiang Gu
Jiwei Xiao
Jiyang Zheng
Jiyoung Lee
John Paisley
Joonki Paik
Joonseok Lee
Julien Mille

Julio C. Zamora
Jun Sato
Jun Tan
Jun Tang
Jun Xiao
Jun Xu
Junbao Zhuo
Jun-Cheng Chen
Junfen Chen
Jungeun Kim
Junhwa Hur
Junli Tao
Junlin Han
Junsik Kim
Junting Dong
Junwei Zhou
Junyu Gao
Kai Han
Kai Huang
Kai Katsumata
Kai Zhao
Kailun Yang
Kai-Po Chang
Kaixiang Wang
Kamal Nasrollahi
Kamil Kowol
Kan Chang
Kang-Jun Liu
Kanchana Vaishnavi
   Gandikota
Kanoksak Wattanachote
Karan Sikka
Kaushik Roy
Ke Xian
Keiji Yanai
Kha Gia Quach
Kibok Lee
Kira Maag
Kirill Gavrilyuk
Kohei Suenaga
Koichi Ito
Komei Sugiura
Kong Dehui
Konstantinos Batsos
Kotaro Kikuchi

Kouzou Ohara
Kuan-Wen Chen
Kun He
Kun Hu
Kun Zhan
Kunhee Kim
Kwan-Yee K. Wong
Kyong Hwan Jin
Kyuhong Shim
Kyung Ho Park
Kyungmin Kim
Kyungsu Lee
Lam Phan
Lanlan Liu
Le Hui
Lei Ke
Lei Qi
Lei Yang
Lei Yu
Lei Zhu
Leila Mahmoodi
Li Jiao
Li Su
Lianyu Hu
Licheng Jiao
Lichi Zhang
Lihong Zheng
Lijun Zhao
Like Xin
Lin Gu
Lin Xuhong
Lincheng Li
Linghua Tang
Lingzhi Kong
Linlin Yang
Linsen Li
Litao Yu
Liu Liu
Liujie Hua
Li-Yun Wang
Loren Schwiebert
Lujia Jin
Lujun Li
Luping Zhou
Luting Wang

Mansi Sharma
Mantini Pranav
Mahmoud Zidan
  Khairallah
Manuel Günther
Marcella Astrid
Marco Piccirilli
Martin Kampel
Marwan Torki
Masaaki Iiyama
Masanori Suganuma
Masayuki Tanaka
Matan Jacoby
Md Alimoor Reza
Md. Zasim Uddin
Meghshyam Prasad
Mei-Chen Yeh
Meng Tang
Mengde Xu
Mengyang Pu
Mevan B. Ekanayake
Michael Bi Mi
Michael Wray
Michaël Clément
Michel Antunes
Michele Sasdelli
Mikhail Sizintsev
Min Peng
Min Zhang
Minchul Shin
Minesh Mathew
Ming Li
Ming Meng
Ming Yin
Ming-Ching Chang
Mingfei Cheng
Minghui Wang
Mingjun Hu
MingKun Yang
Mingxing Tan
Mingzhi Yuan
Min-Hung Chen
Minhyun Lee
Minjung Kim
Min-Kook Suh

Minkyo Seo
Minyi Zhao
Mo Zhou
Mohammad Amin A.
  Shabani
Moein Sorkhei
Mohit Agarwal
Monish K. Keswani
Muhammad Sarmad
Muhammad Kashif Ali
Myung-Woo Woo
Naeemullah Khan
Naman Solanki
Namyup Kim
Nan Gao
Nan Xue
Naoki Chiba
Naoto Inoue
Naresh P. Cuntoor
Nati Daniel
Neelanjan Bhowmik
Niaz Ahmad
Nicholas I. Kuo
Nicholas E. Rosa
Nicola Fioraio
Nicolas Dufour
Nicolas Papadakis
Ning Liu
Nishan Khatri
Ole Johannsen
P. Real Jurado
Parikshit V. Sakurikar
Patrick Peursum
Pavan Turaga
Peijie Chen
Peizhi Yan
Peng Wang
Pengfei Fang
Penghui Du
Pengpeng Liu
Phi Le Nguyen
Philippe Chiberre
Pierre Gleize
Pinaki Nath Chowdhury
Ping Hu

Ping Li
Ping Zhao
Pingping Zhang
Pradyumna Narayana
Pritish Sahu
Qi Li
Qi Wang
Qi Zhang
Qian Li
Qian Wang
Qiang Fu
Qiang Wu
Qiangxi Zhu
Qianying Liu
Qiaosi Yi
Qier Meng
Qin Liu
Qing Liu
Qing Wang
Qingheng Zhang
Qingjie Liu
Qinglin Liu
Qingsen Yan
Qingwei Tang
Qingyao Wu
Qingzheng Wang
Qizao Wang
Quang Hieu Pham
Rabab Abdelfattah
Rabab Ward
Radu Tudor Ionescu
Rahul Mitra
Raül Pérez i Gonzalo
Raymond A. Yeh
Ren Li
Renán Rojas-Gómez
Renjie Wan
Renuka Sharma
Reyer Zwiggelaar
Robin Chan
Robin Courant
Rohit Saluja
Rongkai Ma
Ronny Hänsch
Rui Liu

Rui Wang
Rui Zhu
Ruibing Hou
Ruikui Wang
Ruiqi Zhao
Ruixing Wang
Ryo Furukawa
Ryusuke Sagawa
Saimunur Rahman
Samet Akcay
Samitha Herath
Sanath Narayan
Sandesh Kamath
Sanghoon Jeon
Sanghyun Son
Satoshi Suzuki
Saumik Bhattacharya
Sauradip Nag
Scott Wehrwein
Sebastien Lefevre
Sehyun Hwang
Seiya Ito
Selen Pehlivan
Sena Kiciroglu
Seok Bong Yoo
Seokjun Park
Seongwoong Cho
Seoungyoon Kang
Seth Nixon
Seunghwan Lee
Seung-Ik Lee
Seungyong Lee
Shaifali Parashar
Shan Cao
Shan Zhang
Shangfei Wang
Shaojian Qiu
Shaoru Wang
Shao-Yuan Lo
Shengjin Wang
Shengqi Huang
Shenjian Gong
Shi Qiu
Shiguang Liu
Shih-Yao Lin

Shin-Jye Lee
Shishi Qiao
Shivam Chandhok
Shohei Nobuhara
Shreya Ghosh
Shuai Yuan
Shuang Yang
Shuangping Huang
Shuigeng Zhou
Shuiwang Li
Shunli Zhang
Shuo Gu
Shuoxin Lin
Shuzhi Yu
Sida Peng
Siddhartha Chandra
Simon S. Woo
Siwei Wang
Sixiang Chen
Siyu Xia
Sohyun Lee
Song Guo
Soochahn Lee
Soumava Kumar Roy
Srinjay Soumitra Sarkar
Stanislav Pidhorskyi
Stefan Gumhold
Stefan Matcovici
Stefano Berretti
Stylianos Moschoglou
Sudhir Yarram
Sudong Cai
Suho Yang
Sumitra S. Malagi
Sungeun Hong
Sunggu Lee
Sunghyun Cho
Sunghyun Myung
Sungmin Cho
Sungyeon Kim
Suzhen Wang
Sven Sickert
Syed Zulqarnain Gilani
Tackgeun You
Taehun Kim

Takao Yamanaka
Takashi Shibata
Takayoshi Yamashita
Takeshi Endo
Takeshi Ikenaga
Tanvir Alam
Tao Hong
Tarun Kalluri
Tat-Jen Cham
Tatsuya Yatagawa
Teck Yian Lim
Tejas Indulal Dhamecha
Tengfei Shi
Thanh-Dat Truong
Thomas Probst
Thuan Hoang Nguyen
Tian Ye
Tianlei Jin
Tianwei Cao
Tianyi Shi
Tianyu Song
Tianyu Wang
Tien-Ju Yang
Tingting Fang
Tobias Baumgartner
Toby P. Breckon
Torsten Sattler
Trung Tuan Dao
Trung Le
Tsung-Hsuan Wu
Tuan-Anh Vu
Utkarsh Ojha
Utku Ozbulak
Vaasudev Narayanan
Venkata Siva Kumar
    Margapuri
Vandit J. Gajjar
Vi Thi Tuong Vo
Victor Fragoso
Vikas Desai
Vincent Lepetit
Vinh Tran
Viresh Ranjan
Wai-Kin Adams Kong
Wallace Michel Pinto Lira

Walter Liao
Wang Yan
Wang Yong
Wataru Shimoda
Wei Feng
Wei Mao
Wei Xu
Weibo Liu
Weichen Xu
Weide Liu
Weidong Chen
Weihong Deng
Wei-Jong Yang
Weikai Chen
Weishi Zhang
Weiwei Fang
Weixin Lu
Weixin Luo
Weiyao Wang
Wenbin Wang
Wenguan Wang
Wenhan Luo
Wenju Wang
Wenlei Liu
Wenqing Chen
Wenwen Yu
Wenxing Bao
Wenyu Liu
Wenzhao Zheng
Whie Jung
Williem Williem
Won Hwa Kim
Woohwan Jung
Wu Yirui
Wu Yufeng
Wu Yunjie
Wugen Zhou
Wujie Sun
Wuman Luo
Xi Wang
Xianfang Sun
Xiang Chen
Xiang Li
Xiangbo Shu
Xiangcheng Liu

Xiangyu Wang
Xiao Wang
Xiao Yan
Xiaobing Wang
Xiaodong Wang
Xiaofeng Wang
Xiaofeng Yang
Xiaogang Xu
Xiaogen Zhou
Xiaohan Yu
Xiaoheng Jiang
Xiaohua Huang
Xiaoke Shen
Xiaolong Liu
Xiaoqin Zhang
Xiaoqing Liu
Xiaosong Wang
Xiaowen Ma
Xiaoyi Zhang
Xiaoyu Wu
Xieyuanli Chen
Xin Chen
Xin Jin
Xin Wang
Xin Zhao
Xindong Zhang
Xingjian He
Xingqun Qi
Xinjie Li
Xinqi Fan
Xinwei He
Xinyan Liu
Xinyu He
Xinyue Zhang
Xiyuan Hu
Xu Cao
Xu Jia
Xu Yang
Xuan Luo
Xubo Yang
Xudong Lin
Xudong Xie
Xuefeng Liang
Xuehui Wang
Xuequan Lu

Xuesong Yang
Xueyan Zou
XuHu Lin
Xun Zhou
Xupeng Wang
Yali Zhang
Ya-Li Li
Yalin Zheng
Yan Di
Yan Luo
Yan Xu
Yang Cao
Yang Hu
Yang Song
Yang Zhang
Yang Zhao
Yangyang Shu
Yani A. Ioannou
Yaniv Nemcovsky
Yanjun Zhu
Yanling Hao
Yanling Tian
Yao Guo
Yao Lu
Yao Zhou
Yaping Zhao
Yasser Benigmim
Yasunori Ishii
Yasushi Yagi
Yawei Li
Ye Ding
Ye Zhu
Yeongnam Chae
Yeying Jin
Yi Cao
Yi Liu
Yi Rong
Yi Tang
Yi Wei
Yi Xu
Yichun Shi
Yifan Zhang
Yikai Wang
Yikang Ding
Yiming Liu

Yiming Qian
Yin Li
Yinghuan Shi
Yingjian Li
Yingkun Xu
Yingshu Chen
Yingwei Pan
Yiping Tang
Yiqing Shen
Yisheng Zhu
Yitian Li
Yizhou Yu
Yoichi Sato
Yong A.
Yongcai Wang
Yongheng Ren
Yonghuai Liu
Yongjun Zhang
Yongkang Luo
Yongkang Wong
Yongpei Zhu
Yongqiang Zhang
Yongrui Ma
Yoshimitsu Aoki
Yoshinori Konishi
Young Jun Heo
Young Min Shin
Youngmoon Lee
Youpeng Zhao
Yu Ding
Yu Feng
Yu Zhang
Yuanbin Wang
Yuang Wang
Yuanhong Chen
Yuanyuan Qiao
Yucong Shen
Yuda Song
Yue Huang
Yufan Liu
Yuguang Yan
Yuhan Xie
Yu-Hsuan Chen
Yu-Hui Wen
Yujiao Shi

Yujin Ren
Yuki Tatsunami
Yukuan Jia
Yukun Su
Yu-Lun Liu
Yun Liu
Yunan Liu
Yunce Zhao
Yun-Chun Chen
Yunhao Li
Yunlong Liu
Yunlong Meng
Yunlu Chen
Yunqian He
Yunzhong Hou
Yuqiu Kong
Yusuke Hosoya
Yusuke Matsui
Yusuke Morishita
Yusuke Sugano
Yuta Kudo
Yu-Ting Wu
Yutong Dai
Yuxi Hu
Yuxi Yang
Yuxuan Li
Yuxuan Zhang
Yuzhen Lin
Yuzhi Zhao
Yvain Queau
Zanwei Zhou
Zebin Guo
Ze-Feng Gao
Zejia Fan
Zekun Yang
Zelin Peng
Zelong Zeng
Zenglin Xu
Zewei Wu
Zhan Li
Zhan Shi
Zhe Li
Zhe Liu
Zhe Zhang
Zhedong Zheng

Zhenbo Xu
Zheng Gu
Zhenhua Tang
Zhenkun Wang
Zhenyu Weng
Zhi Zeng
Zhiguo Cao
Zhijie Rao
Zhijie Wang
Zhijun Zhang
Zhimin Gao
Zhipeng Yu
Zhiqiang Hu
Zhisong Liu
Zhiwei Hong
Zhiwei Xu

Zhiwu Lu
Zhixiang Wang
Zhixin Li
Zhiyong Dai
Zhiyong Huang
Zhiyuan Zhang
Zhonghua Wu
Zhongyan Zhang
Zhongzheng Yuan
Zhu Hu
Zhu Meng
Zhujun Li
Zhulun Yang
Zhuojun Zou
Ziang Cheng
Zichuan Liu

Zihan Ding
Zihao Zhang
Zijiang Song
Zijin Yin
Ziqiang Zheng
Zitian Wang
Ziwei Yao
Zixun Zhang
Ziyang Luo
Ziyi Bai
Ziyi Wang
Zongheng Tang
Zongsheng Cao
Zongwei Wu
Zoran Duric

Contents – Part IV

Pose and Action

Video Analysis and Event Recognition

Vision and Language

Biometrics

Face and Gesture

Confidence-Calibrated Face Image Forgery Detection with Contrastive Representation Distillation

Puning Yang[1,2], Huaibo Huang[1,2], Zhiyong Wang[3], Aijing Yu[1,2], and Ran He[1,2(✉)]

[1] Center for Research on Intelligent Perception and Computing, NLPR, CASIA, Beijing, China
{puning.yang,huaibo.huang,aijing.yu}@cripac.ia.ac.cn, rhe@nlpr.ia.ac.cn
[2] School of Artificial Intelligence, University of Chinese Academy of Sciences, Beijing, China
[3] Biomedical and Multimedia Information Technology (BMIT) Research Group, School of Information Technologies, University of Sydney, Camperdown, Australia
zhiyong.wang@sydney.edu.au

Abstract. Face forgery detection has been increasingly investigated due to the great success of various deepfake techniques. While most existing face forgery detection methods have achieved excellent results on the test split of the same dataset or the same type of manipulations, they often do not work well on unseen datasets or unseen manipulations due to the issue of model generalization. Therefore, in this paper, we propose a novel contrastive distillation calibration (CDC) framework, which distills the contrastive representations with confidence calibration to address this generalization issue. Different from previous methods that equally treat the two forgery types, Face Swapping and Face Reenactment, we devise a dual-teacher module where the knowledge is separately learned for each forgery type. A contrastive representation learning strategy is further presented to enhance the representations of diverse forgery artifacts. To prevent the proposed model from being overconfident, we propose a novel Kullback-Leibler divergence loss with dynamic weights to moderate the dual-teacher's outputs. In addition, we introduce label smoothing to calibrate the model confidence with the target outputs. Extensive experiments on three popular datasets show that our proposed method achieves the state-of-the-art performance for cross-dataset face forgery detection.

Keywords: Deepfake detection · Confidence calibration · Knowledge distillation

1 Introduction

Recent years have witnessed the rapid development of various deepfake techniques, such as face swapping and face reenactment [27,46–48]. As a result, face

---

Supplementary Information The online version contains supplementary material available at https://doi.org/10.1007/978-3-031-26316-3_1.

L. Wang et al. (Eds.): ACCV 2022, LNCS 13844, pp. 3–19, 2023.
https://doi.org/10.1007/978-3-031-26316-3_1

Fig. 1. Forgery samples from FaceForensic++ (FF++) [40]. Face Swapping wants to change the identity and keep the motion of the target face. On the contrary, Face Reenactment wants to preserve the identity and change the motion of the target face. Different forgery artifacts are generated during the different forgery processes. Based on this observation, we propose a feature-augmented contrastive representation approach.

forgery detection has been increasingly investigated to address the societal concerns on identity fraud, and many face forgery detection methods have been developed with promising progress [15,55,56].

Existing methods generally learn representations from spatial, temporal, and frequency domains to distinguish forged faces from genuine ones. Early studies focused on detecting face forgery in a single domain [37,38]. Although these methods achieved excellent performance on seen datasets and manipulations, they lack generalization to unseen datasets and manipulations. Then multi-domain methods [33,35] were proposed to improve generalization by fusing the features of multiple domains. However, the redundant representations introduced by multi-domain features can easily lead to over-fitting. As a result, additional constraints on inter-domain independence are required, which eventually increases computational costs. Recently, deep learning based methods [1,40] have been investigated to learn features common to different types of forgeries. However, none of the existing techniques consider the forgery artifact differences between different types of forgery. As illustrated in Fig. 1, two types of forgeries, Face Swapping and Face Reenactment, have distinct characteristics due to different objectives, which need to be exploited separately.

Therefore, in this paper, we propose a novel contrastive representation distillation model to address the issue of model generalization. Specifically, we devise a dual-teacher module in which each teacher is trained with one of two forgery types. That is, a Face Swap teacher and a Face Reenactment teacher are trained separately to obtain discriminative features for individual forgeries. Then, the knowledge from both teachers will be jointly distilled to a student model for improved model generalization.

To further improve model generalization, we refine the task of face forgery detection from binary classification (i.e., fake vs real) to fine-grained confidence scores in the range of $[0, 1]$. To this end, we propose to calibrate the model confidence with the scores. Therefore, we further devise a new loss function, namely Confidence Calibration Loss, to calibrate the output of our framework. We exploit label smoothing to quantify the target output at finer levels. The distribution of targets is changed into a uniform distribution between the ground-truth labels and the labels processed by label smoothing regularization. The new targets will guide the model to make more fine-grained predictions to alleviate the issue of being overconfident. In addition, according to the outputs from each teacher, two dynamic confidence weights are assigned to each input sample. A loss with dynamic weights expresses the prediction from two teachers to the current sample more accurately. Therefore, we name our proposed framework as *Contrastive Distillation Calibration* (CDC). Overall, the key contributions of this paper are summarized as follows:

- We propose a novel contrastive distillation calibration framework to enhance the generalization of face forgery detection by designing a dual-teacher module for knowledge distillation and utilizing contrastive representation learning without additional disentangling.
- We propose to calibrate the model confidence with the targets for the first time for face forgery detection and devise a new Kullback-Leibler divergence based loss function with label smoothing strategy.
- We perform comprehensive experiments on three widely used benchmark datasets: FaceForensics++ [40], Celeb-DF [31], and DFDC [8] to demonstrate that our proposed method outperforms the state-of-the-art face forgery detection methods on unseen datasets.

Code is publicly available at: https://github.com/Puning97/CDC_face_forgery_detection.

2 Related Work

2.1 Face Forgery Detection

Most of the existing face forgery detection methods identify forgery traces from the perspective of spatial [28,37,40,53], frequency [33,38], and temporal [29,35,56] domains. For example, Face x-ray [28] mainly paid attention to the mixing step existing in most face forgery cases and achieved state-of-the-art performance from the generalization perspective on raw videos. F^3-Net [38] exploited Discrete Cosine Transform (DCT) coefficients to extract the frequency features and achieved state-of-the-art performance on highly compressed videos. Various methods have also been proposed to exploit specific temporal incoherence in the temporal domain, such as eye blinking [29], lips motion [15], or expression [35].

However, these methods generally focus on learning low-level features from given datasets with a specific type of forgery, which could not be generalized across different datasets and different types of forgeries. In our work, we propose a dual-teacher module to enhance the learning of generalized representations with contrastive learning.

2.2   Knowledge Distillation

Knowledge distillation refers to a learning process of distilling knowledge from a big teacher model to a small student model. It has a history of more than ten years. Busira *et al.* [3] presented a method to compress a set of models into a single model without significant accuracy loss. Ba and Caruana [2] extended this idea to deep learning by using the logits of the teacher model. Hinton *et al.* [21] revived this idea under the name of knowledge distillation that distills class probabilities by minimizing the Kullback-Leibler (KL)-divergence between the softmax outputs of the teacher and student. In addition, the hidden representation from the teacher has been proved to hold additional knowledge that can contribute to improving the student's performance. Especially in computer vision, some recent knowledge distillation methods [24,39,42,54] were proposed to minimize the mean squared error (MSE) between the representation-level knowledge of two models. They addressed how to better extract more useful knowledge from the teacher model and transfer it to the student. In the case of vision tasks, the most common methods [10,14,20] of knowledge distillation focus on combining the ground truth and the teacher's predictions as the overall targets to train students.

On the one hand, we continue the idea of most knowledge distillation models: distilling the knowledge of the teacher model into the student model. On the other hand, existing distillation methods generally aim to transfer features from a big model to a small one. Such an operation often leads to an issue that the student lacks prominent local advantages and the global representation ability is weaker than their teachers. Therefore, we first train two small teachers with local salient feature representation, and then distill these two small teachers' knowledge into a big student model. Theoretically, the large student preserves the representations of both teachers and improve model generalization through the transfer process.

2.3   Confidence Calibration

Modifying model improves its robustness and generalization [17,18]. Confidence calibration is effective in enhancing the generalization of a model and improving its reliability to be deployed in realistic scenarios [13,43]. As there exists an overfitting issue for deep neural networks, Guo *et al.* [13] explained that existing neural networks often make overconfident predictions and proposed the concept of confidence calibration. Hein *et al.* [19] suggested that neural networks using the ReLU activation function are essentially piecewise linear functions, thus explaining why out-of-distribution data can easily cause softmax classifiers to generate highly confident but incorrect outputs: piecewise linear functions imply that the methods which operate on the output of classifiers cannot recognize an input as out-of-distribution inputs. Confidence calibration improves the generalization of a model from this perspective.

Many confidence calibration methods have been proposed in recent years, including temperature scaling [22,32], mixup [49], label smoothing [44], Monte

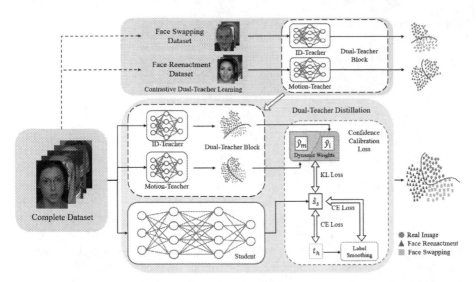

Fig. 2. Overview of our proposed CDC framework, we firstly perform the Contrastive Dual-teacher Learning to train the two teachers. Then, we undertake contrastive representation-based distillation with confidence calibration. Note that abbreviations have been used for some symbols. For instance, $\hat{y}_m = \hat{y}_{motion}$, $t_h = t_{hard}$, $\hat{y}_i = \hat{y}_{id}$, $\hat{s}_s = \hat{s}_{stu}$.

Carlo Dropout [12], and Deep Ensembles [26]. Label smoothing is a method of fusing the ground truth with a uniform distribution. It enforces a model to generate a smoother probability distribution to fit the soft targets, which is utilized in our distillation framework. In addition to label smoothing, we also propose to exploit dynamic confidence weights to better leverage the knowledge of the two teachers.

3   Methodology

As shown in Fig. 2, our proposed forgery detection method consists of three key components: a dual-teacher block, a student block, and a confidence calibration block. Firstly, we divide the FF++ dataset into two sub-datasets: Face Swapping Dataset (FSD) containing face swapped images and their corresponding genuine images and Face Reenactment Dataset (FRD) containing motion reenacted images and their corresponding genuine images, and obtain an id-teacher and a motion-teacher by training our backbone on FSD and FRD, respectively. Secondly, the knowledge of both teachers in the dual-teacher block is transferred to the student by training it on a multi-category forgery dataset. Finally, to achieve fine-grained classification of multiple types of forgeries, we devise confidence calibration-based loss functions with dynamic weights and introduce label smoothing to better supervise network training.

Table 1. Cross-evaluation results of the Dual-teacher block in terms of AUC (%) on FSD and FRD.

Teachers	Face swapping	Face reenactment
ID-Teacher	99.53	58.12
Motion-Teacher	76.58	99.25

3.1 Feature Representation

Existing forgery methods include two modes: face swapping and face reenactment. Face swapping aims to exchange identity information while keeping the motion information maximally. On the contrary, face reenactment aims to exchange the motion information and keep the identity information. Inspired by the principle of contrastive learning, we divide the FF++ dataset into two sub-datasets: Face Swapping Dataset, which includes face swapped forgeries and corresponding genuine images, and Face Reenactment Dataset, which provides face reenacted forgeries and corresponding genuine images. Two backbones are trained on these two sub-datasets, respectively. Now we obtain two teachers: ID-Teacher and Motion-Teacher. A cross-evaluation experiment was undertaken with results shown in Table 1, to verify the independence and validity between the ID-Teacher and the Motion-Teacher. The difference in cross-category evaluation indicates existing face reenactment forgery methods keep the identity information well. However, face swapping methods do not keep the motion information well.

We introduce the details of our backbone, which consists of two parts: the EfficientNet-B3 [45] and the Feature Transformer. These two parts are trained in an end-to-end manner for teacher training. Overall, given a suspect image $I \in \mathbb{R}^{H \times W \times 3}$, the first stage is the EfficientNet-B3 [45] pre-trained from TIMM [52]. It extracts identity or motion feature $F = \Phi_{EfficientNet-B3}(I)$, $F \in \mathbb{R}^{H' \times W' \times C}$.

The second stage is Feature Transformer based on Vision Transformer [9]. With the EfficientNet-B3, we obtain the features $F \in \mathbb{R}^{H' \times W' \times C}$. Note that a global feature F can be represented as a sequence of local features $F_l \in \mathbb{R}^{H' \times W'}$, $l \in 1, 2, ..., C$, which is a 2D sequence of token embeddings in the standard Vision Transformer [9]. Note that C denotes the input sequence length, H and W are as same as the input patch size p. Following the settings in ViT [9], we flatten the patches and map them to D dimensions with a trainable linear projection (Eq. 1). We add a learnable class embedding to the sequence of embedded patches ($z_0^0 = F_{class}$), learnable 1D position embeddings (E_{pos}) also have been included to retain positional information.

$$z_0 = [F_{class}; F_1 E; F_2 E; \cdots ; F_C E] + E_{pos}, \tag{1}$$

where F_n denotes the n-th part in feature F, $E \in \mathbb{R}^{(H' \cdot W') \times D}$, $E_{pos} \in \mathbb{R}^{(C+1) \times D}$.

The core block of Feature Transformer is L standard Transformer Encoder blocks and each block consists of a multi-head self-attention(MSA) (Eq. 2) block and an MLP block (Eq. 3). Layernorm (LN) [50] is applied before every block and residual connections after every block. The MLP contains two layers with a GELU activation function:

$$z'_\ell = \text{MSA}(\text{LN}(z_{\ell-1})) + z_{\ell-1}, \ell = 1 \cdots L, \tag{2}$$

$$z_\ell = \text{MLP}(\text{LN}(z'_\ell)) + z'_\ell, \ell = 1 \cdots L. \tag{3}$$

The output of the Transformer Encoder (z_L^0) is the representation of the image. We can apply an MLP head for the final prediction (Fig. 3):

$$y = \text{MLP}(\text{LN}(z_L^0)). \tag{4}$$

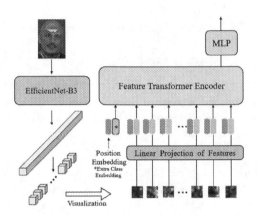

Fig. 3. Illustration of the backbone architecture which consists of EfficientNet-B3 and a variant of Vision Transformer. We split the features extracted from EfficientNet-B3 along the channel dimension and feed the sequences of features to a standard Transformer Encoder.

3.2 Dual-Teacher Knowledge Distillation (DKD)

Now we have two teachers to form the dual-teacher block. Specifically, the ID-Teacher maps face swapped samples I_{id} to the output y_{id}, and the Motion-Teacher maps face reenacted samples I_{motion} to the output y_{motion}. Our goal is to teach a student that inherits advantages from both teachers by distilling the knowledge from the dual-teacher block. Considering that the scales of the data between teachers and students are different, we fine-tuned the backbone structure. The difference between teachers and student is the number of layers and heads in the Feature Transformer encoder. A given image I_{train} is fed to three branches: the ID-Teacher, the Motion-Teacher, and the Student. On the one hand, we obtain the predictions $\hat{s}$ from the Student and the hard targets

t_{hard} from the groundtruth of the training dataset. On the other hand, the ID-Teacher and the Motion-Teacher respectively predict a score of the input, then we get $\hat{y}_{id}$ and $\hat{y}_{motion}$. We use the binary cross-entropy (BCE) loss to supervise the prediction on hard targets (Eq. 5). The learning process between teachers and student is supervised by the Kullback-Leibler (KL) divergence loss (Eq. 6).

$$\mathcal{L}_{hard} = \frac{1}{N}\sum_{1}^{N} \text{BCE}(\hat{s}_{stu}, \hat{t}_{hard}), \tag{5}$$

$$\mathcal{L}_{teacher} = \frac{1}{N}\sum_{1}^{N}(\text{KL}(\hat{s}_{stu}, \hat{y}_{id}) + \text{KL}(\hat{s}_{stu}, \hat{y}_{motion})). \tag{6}$$

The overall loss function of our distillation model is as follows:

$$\mathcal{L}_{distill} = \lambda\mathcal{L}_{hard} + \lambda\mathcal{L}_{teacher}. \tag{7}$$

3.3   Confidence Calibration Loss (CCL)

As shown in Fig. 4, our model also suffers from overconfidence. It is necessary to calibrate the confidence of our framework. We present two sub-functions to achieve this goal: Dynamic Confidence Weights and Label Smoothing Regularization.

Dynamic Confidence Weights (DCW). Given a training sample, we usually get different results from the two teachers, which means different levels of confidence. Therefore we propose the Dynamic Confidence Weight strategy, which is based on the outputs from the teachers. For instance, one teacher's output of a training sample is $\hat{y}_s$, the probability on the other category is $1 - \hat{y}_s$. We define the absolute value (Eq. 8) between these two probability scores as the model confidence index, which represents the teacher's confidence in its prediction.

$$\lambda_s = |1 - (\hat{y}_s) - \hat{y}_s| = |2\hat{y}_s - 1|. \tag{8}$$

For each sample, we get two dynamic confidence weights λ_{id} and λ_{motion}:

$$\lambda_{id} = |2\hat{y}_{id} - 1|, \lambda_{motion} = |2\hat{y}_{motion} - 1|. \tag{9}$$

we can calculate the loss of samples:

$$\mathcal{L}_{id} = \frac{1}{N}\sum_{1}^{N}(\lambda_{id}KL(\hat{s}_{stu}, \hat{y}_{id})), \tag{10}$$

$$\mathcal{L}_{motion} = \frac{1}{N}\sum_{1}^{N}(\lambda_{motion}\text{KL}(\hat{s}_{stu}, \hat{y}_{motion})), \tag{11}$$

and the final loss function can be written as:

$$\mathcal{L}'_{teacher} = \mathcal{L}_{id} + \mathcal{L}_{motion}. \tag{12}$$

Label Smoothing Regularization (LSR). To further improve the generalization ability, we exploit the label smoothing method in our task. Based on our observations, forged images often contain some genuine content. For instance, both face swapped samples and face reenacted samples have common real contents: the background in the images. Besides, face swapped samples also have part of the same motion information as the genuine samples. Coincidentally, face reenacted samples have the same identity information as the genuine samples. Thus, it is reasonable to set the targets from the combination between the ground-truth labels and its converted outputs. Considering a smoothing parameter ϵ, a sample of the ground-truth label y, we replace the label distribution:

$$P_i = \begin{cases} 1, if(i = y) \\ 0, if(i \neq y) \end{cases} \Rightarrow P_i = \begin{cases} (1 - \epsilon), if(i = y) \\ \epsilon, if(i \neq y) \end{cases} . \tag{13}$$

Now we get a new label distribution, which is a mixture of the original ground-truth distribution and the converted distribution with weights $1 - \epsilon$ and ϵ. We practically interpret LSR with the cross entropy:

$$LS(\hat{p}, t) = \begin{cases} (1 - \epsilon) * (-\sum_{i=0}^{1} t_i \log \hat{p}_i), if(i = y) \\ \epsilon * (-\sum_{i=0}^{1} t_i \log \hat{p}_i), if(i \neq y) \end{cases} . \tag{14}$$

The final label smoothing loss is:

$$\mathcal{L}_{smoothing} = \frac{1}{N} \sum_{1}^{N} LS(\hat{s}_{stu}.t_{hard}). \tag{15}$$

Finally, our total loss function is as follows:

$$\mathcal{L}_{total} = (1 - \lambda_t)\mathcal{L}_{hard} + \lambda_t \mathcal{L}'_{teacher} + \lambda_l \mathcal{L}_{smoothing}. \tag{16}$$

4   Experimental Results and Discussions

We evaluated the performance of our proposed CDC (i.e., DKD + CCL) against multiple state-of-the-art methods on three publicly available datasets. We show that our model achieves convincing performance under the in-dataset setting. To demonstrate the robust generalization ability of our model, we conducted the cross-dataset evaluation by training the model with only FF++ [40] datasets and testing on unseen datasets. Ablation studies explore the contribution of each component in our framework, such as the impact of DKD and CCL.

4.1   Experimental Settings

Datasets. Following recent related works on face forgery detection [1,30,33,55], we conducted our experiments on the three benchmark public deepfake datasets: FaceForensics++ (FF++) [40], Celeb-DF [31], and Deepfake Detection Challenge (DFDC) [8]. FaceForensics++ (FF++) is the most widely used dataset in many deepfake detection approaches. It contains 1,000 original real videos from the Internet, and each real video corresponds to 4 forgery ones, which are manipulated by Deepfakes [11], NeuralTextures [47], FaceSwap [25], and Face2Face [48], respectively. Celeb-DF consists of high-quality forged celebrity videos using an advanced synthesis process. Deepfake Detection Challenge (DFDC) public test set was released for the Deepfake Detection Challenge, which contains many low-quality videos and makes it exceptionally challenging.

Evaluation Metrics. We utilized the Accuracy rate (ACC) and the Area Under Receiver Characteristic Curve (AUC) as our evaluation metrics. (1) ACC. The accuracy rate is the most popular metric in the classification task. It is also applied to evaluate the performance of face forgery detection, and we used ACC as one of the evaluation metrics in the experiment. (2) AUC. Following the Celeb-DF [31] and DFDC [8], we used AUC as the other evaluation metric to evaluate the performance in the cross-dataset evaluation.

Pre-processing and Training Setting. For all video frames, we used Retinaface [7] to detect faces and saved the aligned facial images as inputs with a size of 224×224. We augmented our training data using Albumentations [4]. We set hyper-parameters $\lambda_t = 0.15$, $\lambda_l = 0.1$ in the Eq. 16. Optimizer was set to Adam [23] for end-to-end training of the complete model with a learning rate of $1e - 4$ and decay of $1e - 6$. We trained our models with a batch size of 32 for 100 epochs. All our results were obtained on four NVIDIA RTX3090 GPUs.

4.2   In-dataset Evaluation

Although our framework focuses on the generalization ability for the face forgery detection task, it still achieves competitive results in the in-dataset evaluation with FF++ [40] and DFDC [8]. Given a dataset, our model is trained on both genuine and deepfake data from train split, and its performance is evaluated with the corresponding test split. Different from existing works, we compare the performance in two sub-datasets: the Face Swapping Dataset and the Face Reenactment Dataset. As shown in Table 2, 3, and Fig. 4, our framework is on par with existing state-of-the-art methods. Selim [41] achieved a better performance in terms of AUC than our framework because it was devised for the setting of the

Table 2. Detection performance on unseen manipulations with our framework and others compared on the FF++ dataset.

Method	Face swapping		Face reenactment	
	ACC	AUC	ACC	AUC
Face X-ray [28]	98.97	99.18	**98.73**	**98.99**
I3D [5]	97.85	98.32	94.65	95.17
MIL [51]	97.54	97.21	98.19	98.27
Ours	**99.45**	**99.52**	98.12	98.56

known DFDC dataset and our model was not fine-tuned for the known dataset evaluation. Besides, model confidence index distribution demonstrates that the model's confidence is also better calibrated without a clear loss of detection accuracy.

4.3   Cross-Dataset Evaluation

In real-world scenarios, the target of the forgery detection task is often the outcome of images generated by a new model with an unknown source. Successfully detecting unseen images indicates the robustness of the model. Cross-dataset evaluation can reflect the generalization ability of the model well. Our experiments were designed to train our framework on the FF++ dataset and then test it on the Celeb-DF dataset to verify the model's generalization ability. As shown in Fig. 4 and Table 3, our model outperforms the state-of-the-art methods. Besides, the benefit of calibrating model confidence in the training phase has been reflected in a more reasonable confidence index distribution and a higher AUC in the cross-dataset evaluation.

Table 3. Detection performance on known dataset (DFDC, on the left) and unseen dataset (Celeb-DF, on the right) with our framework and others compared in terms of AUC (%). Our method's performance is comparable to that of the best model Selim result[48] in the DFDC competition and obtains the state-of- the-art performance in the cross-dataset evaluation. Results of some other methods are cited directly from [33].

Method	DFDC (AUC (%))	Methods	FF++	Celeb-DF
Capsule[37]	53.3	Two-stream[57]	70.1	53.8
Multi-task[36]	53.6	Multi-task[36]	76.3	54.3
HeadPose[53]	55.9	HeadPose[53]	47.3	54.6
Two-stream[57]	61.4	Meso4[1]	84.7	54.8
VA-MLP[34]	61.9	MesoInception4	83.0	53.6
VA-LogReg	66.2	VA-MLP[34]	66.4	55.0
MesoInception4	73.2	VA-LogReg	78.0	55.1
Meso4[1]	75.3	FWA[30]	80.1	56.9
Xception-raw[40]	49.9	Capsule[37]	96.6	57.5
Xception-c40	69.7	Xception-raw[40]	99.7	48.2
Xception-c23	72.2	Xception-c23	99.7	65.3
FWA[30]	72.7	Xception-c40	95.5	65.5
DSP-FWA[30]	75.5	DSP-FWA[30]	93.0	64.6
Emotion[35]	84.4	F^3-Net[38]	98.1	65.2
Selim[41]	**98.6**	Multi-attentional[55]	**99.8**	67.4
Ours	97.9	Two-branch[33]	93.2	73.4
		Face X-ray[28]	99.2	74.8
		Ours	99.1	**75.1**

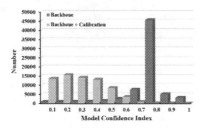 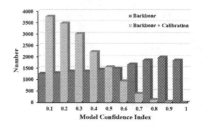

Fig. 4. Model Confidence Index (Eq. 8) on the FF++ dataset (left) and the Celeb-DF dataset (right), and we can conclude that CCL can effectively mitigate the model's overconfidence problem. On the unseen data, unlike the baseline lack of confidence, the model using CCL can confidently predict the authenticity of an image.

4.4   Ablation Study

We performed comprehensive studies on the FF++ [40] and the Celeb-DF [31] dataset to validate our design of the overall framework. In summary, compared with the ablation study that removes any components, the proposed framework shows higher detection accuracy in different face forgery models and general authenticity detection. Specifically, We have analyzed the performance of each component separately from the two major aspects of DKD and CCL. The results are as shown below:

Effect of DKD. We first validate the backbone of teachers and students. For example, for the Motion Teacher, we trained five variants of our framework: 1) ResNet-50; 2) Xception; 3) EfficientNet-B7; 4) EfficientNet-B4; 5) EfficientNet-B3.

Table 4. Abalation studies for backbone variants. Frame-level AUC (%) is reported.

Teacher backbone	FF++	Celeb-DF	Params
ResNet-50 [16]	98.62	66.83	24M
Xception [6]	98.85	67.42	23M
EfficientNet-B7 [45]	99.64	67.59	88M
EfficientNet-B4 [45]	99.42	67.55	19M
EfficientNet-B3 [45]	99.24	67.49	12M

Table 4 has shown the results of different backbones. We can conclude that different backbones have little influence on the detection results under the in-dataset and cross-dataset settings. Considering the computation cost, we finally chose EfficientNet-B3 as our backbone.

To validate the effectiveness of our Feature Transformer, we performed an ablation study on the framework. We trained four layer variants and four head variants of our framework. The results are presented in Table 5.

Table 5. Ablation study on the number of layers and heads of the Feature Transformer Encoder in our teacher framework. Frame-level AUC (%) is reported.

Teacher Model	FF++	Celeb-DF
B3+FTL × 12	99.32	67.27
B3+FTL × 3	99.12	67.58
B3+FTL × 2	98.99	67.62
B3+FTL × 1	98.89	67.72
B3+FTL × 1+HEAD × 12	98.89	67.72
B3+FTL × 1+HEAD × 16	98.95	67.70
B3+FTL × 1+HEAD × 8	98.84	68.02
B3+FTL × 1+HEAD × 6	98.78	68.98

We notice that 1) Feature Transformer can improve the generalization capability; 2) too many layers of the standard encoder in Feature Transformer cannot further improve the teacher's performance; and 3) an appropriate number of heads can achieve the best result.

With two teachers, we automatically choose the same structure for the student. However, the student with the same structure as teachers was mediocre. We speculated that the current structure is too small to accommodate the hypothetical space of two teachers simultaneously. Thus, we adjusted the structural parameters of the student model. As shown in Table 6, the student with two layers of the Feature Transformer encoder and twelve heads in each layer achieves the best performance.

Effect of CCL. After verifying and carefully adjusting the structure of teachers and students, we confirmed the necessity of the various components of the calibration section and adjusted them.

As shown in Table 7, we firstly validated the effectiveness of dynamic confidence weights and label smoothing. The result shows that while the model generalization ability is improved, the in-dataset evaluation is almost unaffected.

Table 6. Ablation study on the number of heads in each layer of the Feature Transformer Encoder in our student framework. Frame-level AUC (%) is reported.

Student model (model only)	FF++	Celeb-DF
B3+FTL × 1+HEAD × 6	98.97	68.78
B3+FTL × 1+HEAD × 8	98.92	69.84
B3+FTL × 1+HEAD × 10	98.88	69.82
B3+FTL × 1+HEAD × 12	98.94	70.68
B3+FTL × 1+HEAD × 14	99.01	70.32
B3+FTL × 2+HEAD × 12	99.13	71.62
B3+FTL × 4+HEAD × 12	99.17	68.73

Table 7. Ablation study of calibration components in our framework, Frame-level AUC (%) is reported.

Calibration components			Datasets	
Backbone	DCW	LSR	FF++	Celeb-DF
√			99.13	71.62
√	√		99.17	73.66
√		√	99.20	72.18
√	√	√	99.19	75.12

Secondly, we adjusted the proportion of each part in the loss function, namely the weight hyperparameters λ_t and λ_l. As shown in Table 8, the performance increases as the proportion of the two calibration components increase. However, after their ratio reaches a specific value, the performance decreases again. There are two possible reasons. On the one hand, we speculate that the excessively high proportion of label smoothing causes the difference between targets to become too small. On the other hand, over-proportioned dynamic confidence weights will cause students to learn in the wrong direction when both teachers make mistakes.

Table 8. Ablation study on different hyper-parameters λ_t and λ_l in our framework. Frame-level AUC (%) is reported.

λ_t	λ_l	FF++	Celeb-DF
0.1	0.1	98.84	70.98
0.2	0.1	98.95	71.72
0.15	0.1	99.04	72.09
0.15	0.2	99.17	75.12
0.15	0.4	99.25	73.47

Due to the limited space, more ablation studies are available in supplementary materials.

5   Conclusions

In this paper, we have presented the CDC framework for improved generalization on face forgery detection. The proposed framework consists of three components: contrastive representation learning, dual-teacher distillation, and confidence calibration. Instead of treating forgery detection as a simple binary classification task, we calibrate the labels and model confidence to refine the targets. Through extensive experiments with contrastive representation distillation and confidence calibration, we demonstrated the superiority of our method compared with the state-of-the-art method in terms of AUC.

References

1. Afchar, D., Nozick, V., Yamagishi, J., Echizen, I.: MesoNet: a compact facial video forgery detection network. In: WIFS (2018)
2. Ba, J., Caruana, R.: Do deep nets really need to be deep? In: NIPS (2014)
3. Busira, C., Caruana, R., Niculescu-Mizil, A.: Model compression. In: ACM KDD (2006)
4. Buslaev, A., Iglovikov, V.I., Khvedchenya, E., Parinov, A., Druzhinin, M., Kalinin, A.A.: Albumentations: fast and flexible image augmentations. Information **11**(2), 125 (2020)
5. Carreira, J., Zisserman, A.: Quo vadis, action recognition? A new model and the kinetics dataset. In: CVPR (2017)
6. Chollet, F.: Xception: deep learning with depthwise separable convolutions. In: CVPR (2017)
7. Deng, J., Guo, J., Zhou, Y., Yu, J., Kotsia, I., Zafeiriou, S.: RetinaFace: single-stage dense face localisation in the wild. arXiv preprint arXiv:1905.00641 (2019)
8. Dolhansky, B., et al.: The deepfake detection challenge (DFDC) dataset. arXiv preprint arXiv:2006.07397 (2020)
9. Dosovitskiy, A., et al.: An image is worth 16×16 words: transformers for image recognition at scale. arXiv preprint arXiv:2010.11929 (2020)
10. Du, S., et al.: Agree to disagree: adaptive ensemble knowledge distillation in gradient space. In: NIPS (2020)
11. FaceSwapDevs: Deepfakes (2019). https://github.com/deepfakes/faceswap. Accessed 7 Nov 2021
12. Gal, Y., Ghahramani, Z.: Dropout as a Bayesian approximation: representing model uncertainty in deep learning. In: ICML (2016)
13. Guo, C., Pleiss, G., Sun, Y., Weinberger, K.Q.: On calibration of modern neural networks. In: ICML (2017)
14. Guo, J., et al.: Distilling object detectors via decoupled features. In: CVPR (2021)
15. Haliassos, A., Vougioukas, K., Petridis, S., Pantic, M.: Lips don't lie: a generalisable and robust approach to face forgery detection. In: CVPR (2021)
16. He, K., Zhang, X., Ren, S., Sun, J.: Deep residual learning for image recognition. In: CVPR (2016)
17. He, R., Hu, B.-G., Yuan, X.-T.: Robust discriminant analysis based on nonparametric maximum entropy. In: Zhou, Z.-H., Washio, T. (eds.) ACML 2009. LNCS (LNAI), vol. 5828, pp. 120–134. Springer, Heidelberg (2009). https://doi.org/10.1007/978-3-642-05224-8_11
18. He, R., Hu, B., Yuan, X., Zheng, W.S.: Principal component analysis based on non-parametric maximum entropy. Neurocomputing **73**(10–12), 1840–1852 (2010)
19. Hein, M., Andriushchenko, M., Bitterwolf, J.: Why relu networks yield high-confidence predictions far away from the training data and how to mitigate the problem. In: CVPR (2019)
20. Heo, B., Kim, J., Yun, S., Park, H., Kwak, N., Choi, J.Y.: A comprehensive overhaul of feature distillation. In: ICCV (2019)
21. Hinton, G., Vinyals, O., Dean, J.: Distilling the knowledge in a neural network. In: NIPS Workshop (2015)
22. Hsu, Y.C., Shen, Y., Jin, H., Kira, Z.: Generalized ODIN: detecting out-of-distribution image without learning from out-of-distribution data. In: CVPR (2020)

23. Kingma, D.P., Ba, J.: Adam: a method for stochastic optimization. arXiv preprint arXiv:1412.6980 (2014)
24. Komodakis, N., Zagoruyko, S.: Paying more attention to attention: improving the performance of convolutional neural networks via attention transfer. In: ICLR (2017)
25. Kowalski, M.: Faceswap (2018). https://github.com/MarekKowalski/FaceSwap. Accessed 7 Nov 2021
26. Lakshminarayanan, B., Pritzel, A., Blundell, C.: Simple and scalable predictive uncertainty estimation using deep ensembles. In: NIPS (2017)
27. Li, L., Bao, J., Yang, H., Chen, D., Wen, F.: Advancing high fidelity identity swapping for forgery detection. In: CVPR (2020)
28. Li, L., et al.: Face X-ray for more general face forgery detection. In: CVPR (2020)
29. Li, Y., Chang, M.C., Lyu, S.: In Ictu Oculi: exposing AI created fake videos by detecting eye blinking. In: WIFS Workshop (2018)
30. Li, Y., Lyu, S.: Exposing deepfake videos by detecting face warping artifacts. In: CVPR Workshops (2019)
31. Li, Y., Yang, X., Sun, P., Qi, H., Lyu, S.: Celeb-DF: a large-scale challenging dataset for deepfake forensics. In: CVPR (2020)
32. Liang, S., Li, Y., Srikant, R.: Enhancing the reliability of out-of-distribution image detection in neural networks. In: ICLR (2018)
33. Masi, I., Killekar, A., Mascarenhas, R.M., Gurudatt, S.P., AbdAlmageed, W.: Two-branch recurrent network for isolating deepfakes in videos. In: Vedaldi, A., Bischof, H., Brox, T., Frahm, J.-M. (eds.) ECCV 2020. LNCS, vol. 12352, pp. 667–684. Springer, Cham (2020). https://doi.org/10.1007/978-3-030-58571-6_39
34. Matern, F., Riess, C., Stamminger, M.: Exploiting visual artifacts to expose deepfakes and face manipulations. In: WACV Workshops (2019)
35. Mittal, T., Bhattacharya, U., Chandra, R., Bera, A., Manocha, D.: Emotions don't lie: an audio-visual deepfake detection method using affective cues. In: ACM MM (2020)
36. Nguyen, H.H., Fang, F., Yamagishi, J., Echizen, I.: Multi-task learning for detecting and segmenting manipulated facial images and videos. In: BTAS (2019)
37. Nguyen, H.H., Yamagishi, J., Echizen, I.: Capsule-forensics: using capsule networks to detect forged images and videos. In: ICASSP (2019)
38. Qian, Y., Yin, G., Sheng, L., Chen, Z., Shao, J.: Thinking in frequency: face forgery detection by mining frequency-aware clues. In: Vedaldi, A., Bischof, H., Brox, T., Frahm, J.-M. (eds.) ECCV 2020. LNCS, vol. 12357, pp. 86–103. Springer, Cham (2020). https://doi.org/10.1007/978-3-030-58610-2_6
39. Romero, A., Ballas, N., Kahou, S.E., Chassang, A., Gatta, C., Bengio, Y.: Fitnets: hints for thin deep nets. In: ICLR (2015)
40. Rossler, A., Cozzolino, D., Verdoliva, L., Riess, C., Thies, J., Nießner, M.: Faceforensics++: learning to detect manipulated facial images. In: ICCV (2019)
41. Seferbekov, S.: (2019). https://github.com/selimsef/dfdc_deepfake_challenge
42. Srinivas, S., Fleuret, F.: Knowledge transfer with jacobian matching. In: ICML (2018)
43. Stutz, D., Hein, M., Schiele, B.: Confidence-calibrated adversarial training: generalizing to unseen attacks. In: ICML (2020)
44. Szegedy, C., Vanhoucke, V., Ioffe, S., Shlens, J., Wojna, Z.: Rethinking the inception architecture for computer vision. In: CVPR (2016)
45. Tan, M., Le, Q.: EfficientNet: rethinking model scaling for convolutional neural networks. In: ICML (2019)

46. Thies, J., Elgharib, M., Tewari, A., Theobalt, C., Nießner, M.: Neural voice pup-
    petry: audio-driven facial reenactment. In: Vedaldi, A., Bischof, H., Brox, T.,
    Frahm, J.-M. (eds.) ECCV 2020. LNCS, vol. 12361, pp. 716–731. Springer, Cham
    (2020). https://doi.org/10.1007/978-3-030-58517-4_42
47. Thies, J., Zollhöfer, M., Nießner, M.: Deferred neural rendering: image synthesis
    using neural textures. ACM TOG **38**(4), 1–12 (2019)
48. Thies, J., Zollhofer, M., Stamminger, M., Theobalt, C., Nießner, M.: Face2face:
    real-time face capture and reenactment of RGB videos. In: CVPR (2016)
49. Thulasidasan, S., Chennupati, G., Bilmes, J.A., Bhattacharya, T., Michalak, S.:
    On mixup training: improved calibration and predictive uncertainty for deep neural
    networks. In: NIPS (2019)
50. Wang, Q., et al.: Learning deep transformer models for machine translation. arXiv
    preprint arXiv:1906.01787 (2019)
51. Wang, X., Yan, Y., Tang, P., Bai, X., Liu, W.: Revisiting multiple instance neural
    networks. Pattern Recogn. **74**, 15–24 (2018)
52. Wightman, R.: Pytorch image models (2019). https://github.com/rwightman/
    pytorch-image-models
53. Yang, X., Li, Y., Lyu, S.: Exposing deep fakes using inconsistent head poses. In:
    ICASSP (2019)
54. Yim, J., Joo, D., Bae, J., Kim, J.: A gift from knowledge distillation: fast opti-
    mization, network minimization and transfer learning. In: CVPR (2017)
55. Zhao, H., Zhou, W., Chen, D., Wei, T., Zhang, W., Yu, N.: Multi-attentional
    deepfake detection. In: CVPR (2021)
56. Zheng, Y., Bao, J., Chen, D., Zeng, M., Wen, F.: Exploring temporal coherence
    for more general video face forgery detection. In: ICCV (2021)
57. Zhou, P., Han, X., Morariu, V.I., Davis, L.S.: Two-stream neural networks for
    tampered face detection. In: CVPRW (2017)

Exposing Face Forgery Clues
via Retinex-Based Image Enhancement

Han Chen[1,2,3,4], Yuzhen Lin[1,2,3,4], and Bin Li[1,2,3,4(✉)]

[1] Guangdong Key Laboratory of Intelligent Information Processing, Shenzhen, China
[2] Shenzhen Key Laboratory of Media Security, Shenzhen, China
[3] Guangdong Laboratory of Artificial Intelligence and Digital Economy (SZ), Shenzhen, China
[4] Shenzhen University, Shenzhen 518060, China
{2016130205,linyuzhen2020}@email.szu.edu.cn, libin@szu.edu.cn

Abstract. Public concerns about deepfake face forgery are continually rising in recent years. Existing deepfake detection approaches typically use convolutional neural networks (CNNs) to mine subtle artifacts under high-quality forged faces. However, most CNN-based deepfake detectors tend to over-fit the content-specific color textures, and thus fail to generalize across different data sources, forgery methods, and/or post-processing operations. It motivates us to develop a method to expose the subtle forgery clues in RGB space. Herein, we propose to utilize multi-scale retinex-based enhancement of RGB space and compose a novel modality, named MSR, to complementary capture the forgery traces. To take full advantage of the MSR information, we propose a two-stream network combined with salience-guided attention and feature re-weighted interaction modules. The salience-guided attention module guides the RGB feature extractor to concentrate more on forgery traces from an MSR perspective. The feature re-weighted interaction module implicitly learns the correlation between the two complementary modalities to promote feature learning for each other. Comprehensive experiments on several benchmarks show that our method outperforms the state-of-the-art face forgery detection methods in detecting severely compressed deepfakes. Besides, our method also shows superior performances on cross-datasets evaluation.

Keywords: Deepfake detection · Multi-scale retinex · Generalization

1 Introduction

Deepfake techniques [1–3] refer to a series of deep learning-based facial forgery techniques that can swap or reenact the face of one person in a video to another. While deepfake technology is very popular in the entertainment and film industries, it is also notorious for its unethical applications that can threaten politics,

---

H. Chen and Y. Lin—Contributed equally to this work.

L. Wang et al. (Eds.): ACCV 2022, LNCS 13844, pp. 20–34, 2023.
https://doi.org/10.1007/978-3-031-26316-3_2

economics, and personal privacy. Over the past few years, a large number of deepfake videos (called deepfakes) uploaded to the Internet with potential harms have been reported. Accordingly, the countermeasures being desired to identify deepfakes become an urgent topic in social security.

To prevent malicious deepfake media from threatening the credibility of human society, deepfake detection (i.e., face forgery detection) is becoming an urgent topic that has attracted widespread attention. Early works leverage hand-crated features (e.g., eyes-blinking [4] or visual artifacts [5]) or semantic features extracted by universal CNNs [6, 7] to identify real and fake images/videos. These methods achieve promising performance when the training and testing data are sampled from the same distribution. However, deepfakes in the real-world are different from those contained in a training set in terms of the data source, forgery method, and post-processing. Due to these mismatched domain gaps, most deepfake detection methods suffer from severe performance drops in practical applications. Therefore, generalization capability is one of the major concerns for existing deepfake detection systems.

In general, there are two typical manners for addressing the generalizing problem have been explored. On the one hand, some works train the deepfake detector with synthetic data that artificially simulates the forgery traces (e.g., visual resolution [8] or blending boundary [9]), which encourages models to learn generic features for face forgery detection. However, these methods suffer severe performance drop when facing post-process distortions (e.g. video compression). On the other hand, some works utilize two-stream networks that introduce information from other domains, such as DCT [10] and SRM [11] features. These methods either simply concatenate RGB and other features at the end of the network or fuse them with at a shallow layer, which rarely considers the relation and interaction between the additional information and regular color textures. This makes it difficult for them to fully utilize the additional information.

As pointed in [12, 13], the poor generalization in CNN-based deepfake detection can attribute to the fact that deep CNN models tend to easily capture the content-specific texture patterns in the RGB space. Thus, designing a deepfake detector with good generalization should consider suppressing the content-specific color textures and exposing discrepancies between forged and real regions. With this simple but powerful insight, herein, we utilize a multi-scale retinex enhancement inspired by the illumination-reflection model [14, 15] and compose a novel modality, named MSR, to complementary capture the forgery traces. To take full advantage of the MSR information, we propose a two-stream network combined with salience-guided attention and feature re-weighted interaction modules. The salience-guided attention module guides the RGB feature extractor to concentrate more on forgery traces from the MSR perspective at multi-scale level. The feature re-weighting module leverages the correlation between the two complementary modalities to promote feature learning for each other. Extensive experiments demonstrate that the proposed framework achieves consistent performance improvement compared to state-of-the-art methods. The main contributions of our work are summarized as follows.

- To expose the forgery traces, we perform retinex-based enhancement and propose a multi-scale retinex (MSR) feature as the complementary modality for RGB images.
- To take full advantage of MSR information, we devise a novel two-stream framework to collaboratively learn comprehensive representation. We design two functional modules to promote the correlation and interaction between the MSR and RGB components, i.e., the feature re-weighted interaction module and salience-guided attention module.
- Comprehensive experiments are presented to reveal the robustness and generalization of our proposed method compared to several state-of-the-art competitors.

2   Related Works

2.1   Face Forgery Detection

The past four years have witnessed a wide variety of methods proposed for defending against the malicious usage of deepfakes. Early works focus on hand-crafted features such as eyes-blinking [4] and visual artifacts [5]. Due to the tremendous success of deep learning, convolutional neural networks (CNNs) [6,7,16] is widely used to deepfake detection task and achieved better performance. As have been criticized, most of the methods suffer from severe over-fitting to the training data and cannot be effectively used in many practical scenarios. There are methods trying to cope with the over-fitting issue of deep-fake detectors. One of the effective approaches to address this problem is training models with synthetic data. For instance, Li *et al.* [8] noticed the quality gap between GAN-synthesized faces and natural faces, and proposed FWA (Face Warping Artifacts) to simulate the fake images by blurring the facial regions of real images. BI (Blending Image) [9] and I2G (Inconsistency Image Generator) [17] were introduced to generate blended faces which can simulate the blending artifacts from of some pristine image pairs with similar facial landmarks. In addition, it is also a common idea to use multi-modality (e.g., frequency domain) learning framework and auxiliary supervisions (e.g., forgery mask) to further mine the heuristic forgery clues and improve the robustness of the model. Qian *et al.* first employed the global and local frequency information for deepfake detection task. Luo *et al.* [11] employed SRM filter that extract the high-frequency noise to guide RGB features. Wang *et al.* [18] amplified implicit local discrepancies from RGB and frequency domain with a novel multi-modal contrastive learning framework. Kong *et al.* [19] introduced PRNU noise information to guide the RGB features, and proposed a novel two-stream network for not only identifying deepfakes but also localizing the forgery regions.

In this work, we utilize a novel MSR modality based on the Retinex theory [20] that exposes the forgery traces in RGB space. Furthermore, we devise a novel two-stream network that combines the MSR and RGB information to collaboratively learn comprehensive representation for detecting deepfakes.

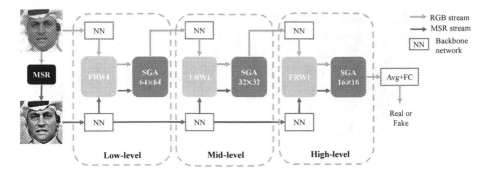

Fig. 1. Overall framework of our proposed method.

2.2 Retinex-Based Methods

Retinex theory [20] models the color perception of human vision on natural scenes. It assumes that the observed images can be decomposed into two components, i.e., reflectance and illumination, which can be mathematically formulated as:

$$S(x, y) = R(x, y) \otimes I(x, y) \qquad (1)$$

where x and y are image pixel coordinates. $R(x, y)$ represents reflectance, $I(x, y)$ represents illumination and $\otimes$ represents element-wise multiplication.

Retinex-based methods are widely accepted among image enhancement methodologies [14,21] due to their robustness. Besides, it can also be viewed as a fundamental theory for the intrinsic image decomposition problem, which aims at disentangling an image into two independent components, such as the structure and texture [22].

As for image forensics task, Chen *et al.* [15] proposed to use retinex-based information for face anti-spoofing task and achieve great generalization performances.

In this work, we apply the retinex-based information as the complementary of RGB modality, which aims to improve the generalization performance of deepfake detection.

3 Methodology

3.1 Overall Framework

In this work, we propose a novel two-stream framework that utilizes MSR information for face forgery detection. Specifically, the original RGB images are first converted to MSR images. Following that, to learn comprehensive feature representation, the RGB and MSR information are integrated at three semantic levels (low, mid and high) through a two-stream network combined with salience-guided attention and feature re-weighted interaction modules.

Real   Fake (F2F)   Forgery mask   MSR

(a) MSR modality can attenuate content-specific colors and enhance forgery clues

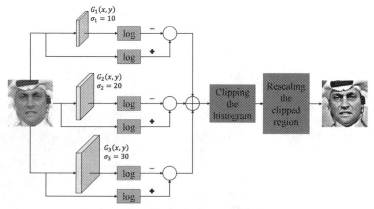

(b) Pipeline of MSR extraction in this work

Fig. 2. Illustration of multi-scale retinex (MSR) enhancement. (a) Inspired by [15], we adopt a MSR-based algorithm for deepfake detection task. (b) Red boxes mark blending traces that are hard to recognize in the RGB space but distinctive in the MSR space. (Color figure online)

According to the resolutions of output feature maps, we abstractly divide the whole network into three semantic layers. As for CNN, the low-resolution feature maps at the end of the network contain high semantic information, and vice versa. Thus, we denote these semantic layers as $l \in \{low, mid, high\}$ for simplicity. H^l, W^l, C^l are the height, width, and channel of the feature map of the corresponding layer. Formally, we define the feature map from the RGB and MSR stream at l-th layers of network as $F_R^l \in \mathbb{R}^{H^l \times W^l \times C^l}$ and $F_M^l \in \mathbb{R}^{H^l \times W^l \times C^l}$, respectively. The overall framework of our proposed approach is illustrated in Fig. 1, and several components are elaborated in more detail as follows.

3.2   Multi-scale Retinex Extraction

For the retinex theory, Eq. (1) is usually transformed into the logarithmic domain as:

$$\log[S(x,y)] = \log[R(x,y)] + \log[I(x,y)] \tag{2}$$

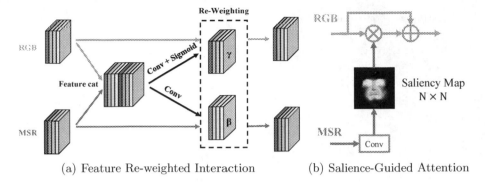

(a) Feature Re-weighted Interaction   (b) Salience-Guided Attention

Fig. 3. The pipeline of our proposed method. We design a novel two-stream architecture, which aims to collaboratively learn comprehensive representations from the RGB and MSR information with Feature Re-weighted Interaction and Salience-Guided Attention modules.

We respectively represent $\log[S(x,y)]$ and $\log[R(x,y)]$ by $s(x,y)$ and $r(x,y)$ for convenience. Summarizing the previous work, the illumination image can be generated from the source image using the center/surround etinex.

$$r(x,y) = s(x,y) - \log[S(x,y) * F(x,y)] \tag{3}$$

where $F(x,y)$ denotes the surround function, and $*$ is the convolution operation, and it is the so called Single Scale Retinex (SSR) model. To overcome the highly dependency on the parameter of $F(x,y)$, Jobson $et\ al.$ [14] proposed a Multiscale Retinex (MSR) model, which weights the outputs of several SSRs with different $F(x,y)$.

As shown in Fig. 2(a), the MSR modality can attenuate content-specific colors and enhance forgery clues. Thus, we utilize the multi-scale retinex enhancement and compose a novel modality, named MSR, for deepfake detection task. As shown in Fig. 2(b), the pipeline of MSR extraction in this work can be formulated as:

$$r_{MSR}(x,y) = \sum_{i=1}^{3} w_i \left\{ \log[S(x,y)] - \log[S(x,y) * G_i(x,y)] \right\} \tag{4}$$

where $G_i(x,y)$ denotes three Gaussian filters with $\sigma_i = 10, 20, 30$. We also add color restoration operations for eliminating color shifts after employing the above MSR pipeline.

3.3 Feature Re-weighted Interaction

To collaboratively align and integrate the feature maps from two domains, we proposed a novel Feature Re-weighted Interaction (FRWI) module inspired by the mechanisms of SPADE [23]. The computation block of FRWI is described in Fig. 3(a).

Firstly, we concatenate feature maps from RGB and MSR streams with $F_{Concat}^l = \mathcal{C}\left(F_R^l, F_M^l\right)$, where $\mathcal{C}\left(\cdot, \cdot\right)$ being feature concatenation in channel dimension. In order to align and aggregate the feature maps from two domains, we respectively generate the weight γ^l and bias β^l by utilizing F_{Concat}^l as:

$$\gamma^l = \delta\left(f_1\left(F_{Concat}^l\right)\right) \tag{5}$$

$$\beta^l = f_2\left(F_{Concat}^l\right) \tag{6}$$

Specifically, γ^l is learned through a 3×3 convolution layer (denoted as f_1) and a sigmoid activation (denoted as $\delta\left(\cdot\right)$) while β^l learned through another 3×3 convolutional layer (denoted as f_2). The outputs of the FRWI module can be formulated as:

$$\widetilde{F}_R^l = \gamma^l \otimes F_R^l + \beta^l \tag{7}$$

$$\widetilde{F}_M^l = \gamma^l \otimes F_M^l + \beta^l \tag{8}$$

where the $\widetilde{F}_R^l$ and $\widetilde{F}_M^l$ represents the aligned feature maps of RGB and MSR streams, respectively.

3.4   Salience-Guided Attention

Utilizing the forgery mask as auxiliary supervision is a universal trick to improve the performance of face forgery detection. Inspired by this, we further adopt the forgery mask as the saliency map to highlight the manipulation traces. In particular, we introduce the spatial attention mechanism and design a Salience-Guided Attention (SGA) which guides a feature learning in the RGB modality with MSR information at different semantic layer. The computation block of SGA is described in Fig. 3(b).

Specifically, we predict the saliency map (denoted as $\hat{\mathcal{M}}^l$) of l-th semantic layer as:

$$\hat{\mathcal{M}}^l = \delta\left(f_3^l\left(\widetilde{F}_M^l\right)\right) \tag{9}$$

where f_3^l represents a 1×1 convolution layer to transform the channels of $\widetilde{F}_M^l$ with 1. We respectively set $N = 64, 32, 16$ for predicting the $\hat{\mathcal{M}}^l$ in the low, mid and high level layer.

The output of SGA module is formulated as:

$$\widetilde{F}_{out}^l = \widetilde{F}_R^l + \hat{\mathcal{M}}^l \otimes \widetilde{F}_R^l \tag{10}$$

3.5   Training Details and Loss Functions

We employ the Efficient-B4 (EN-b4) as the backbone of our work. In order to capture more artifacts at higher resolutions, we change the stride of the first convolution layer at the backbone model from 2 to 1. The whole end-to-end training process involves the supervision of binary classification and salience prediction task, and the overall loss function consists of two components:

$$\mathcal{L} = \mathcal{L}_{cls} + \lambda\mathcal{L}_{SM} \tag{11}$$

where $\mathcal{L}_{cls}$ and $\mathcal{L}_{SM}$ represents the binary cross-entropy loss and saliency map loss, respectively. λ is the balance weight.

Specifically, the cross-entropy binary classification loss $\mathcal{L}_{cls}$ is formulated as:

$$\mathcal{L}_{cls} = \mathbf{y}_t \log \mathbf{y}_p + (1 - \mathbf{y}_t) \log (1 - \mathbf{y}_p) \tag{12}$$

where $\mathbf{y}_t$ and $\mathbf{y}_p$ represents ground-truth label and the prediction logits, respectively.

The saliency map loss $\mathcal{L}_{SM}$ consists of the l_2 loss in of three semantic layers, which can be formulated as:

$$\mathcal{L}_{SM} = \sum_{l=1}^{3} \frac{1}{\Omega(\mathcal{M}^l)} \left\| \mathcal{M}^l - \hat{\mathcal{M}}^l \right\|_2 \tag{13}$$

where $\Omega(\cdot)$ repents the total number of elements. $\mathcal{M}^l$ is the ground truth forgery mask of the l-th semantic layer. We employ DSSIM [24] algorithm, which compute the paired face manipulation images and their corresponding source face pristine images with threshold, to get the original forgery mask $\mathcal{M}_{gt}$ with size of 256×256. Besides, we use bi-linear interpolation to down-sample $\mathcal{M}_{gt}$ by $\{4\times, 8\times, 16\times\}$, respectively obtain ground-truth saliency maps for the low, mid and high semantic layer (i.e., $\mathcal{M}^l, i = 1, 2, 3$).

4   Experiments

4.1   Experimental Setup

Datasets and Pre-processing. In this paper, we mainly conducted experiments on the challenging *FaceForensics++ (FF++)* [7] dataset. FF++ contains 1000 Pristine (PT) videos (i.e., the real sample) and 4000 fake videos forged by five manipulation methods, i.e., Deepfakes (DF), Face2Face (F2F) [25], FaceSwap (FS), NeuralTextures (NT) [26]. Besides, FF++ provides three quality levels controlled by the constant rate quantization parameter (QP) in compression for these videos: raw (QP = 0), HQ (high-quality, QP = 23) and LQ (low-quality, QP = 40). Considering the deployment in real-world application scenarios, we conduct our experiments on both HQ videos and LQ videos. The samples were split into disjoint training, validation, and testing sets at the video level follows the official protocol [7].

As for pre-processing, we utilized MTCNN [27] to detect and crop the face regions (enlarged by a factor of 1.3) from each video frame, and resized the them to 256×256 as the input images.

Table 1. Detection performances (%) on FF++ dataset. HQ and LQ denote the high-quality and low-quality data. * indicate the model is trained by us implementing the official code. The best results are in bold.

Methods	HQ		LQ	
	ACC	AUC	ACC	AUC
MesoNet [6]	83.10	–	70.47	–
Xception [7]	95.73	–	81.00	–
PRRNet [29]	96.15	–	86.13	–
SPSL [30]	91.50	95.32	81.57	82.82
MTA [31]	97.60	99.29	**88.69**	90.40
MC-LCR [18]	97.89	99.65	88.07	90.28
D&L [19]	**98.40**	**99.77**	84.84	87.10
Xception*	96.06	98.89	86.35	90.25
RGB baseline	96.08	98.98	86.36	91.00
Ours	96.94	99.32	88.39	**92.98**

Implementation Details and Evaluation Metrics. The proposed framework is implemented by PyTorch on an NVIDIA Tesla A100 GPU (40 GB). We use Efficient-B4 (EN-b4) [28] as the backbone network and initialized with the weights pre-trained on ImageNet. We employed an Adam optimizer with a cosine learning rate scheduler and set the training hyper-parameters by: the mini-batch size as 12, the initial learning rate as 0.0002, the weight decay as 0.05, $\beta_1 = 0.9$, $\beta_2 = 0.999$. We implemented the training stage with 500 epochs. We set $\lambda = 10$ for the loss function.

Following most existing face forgery detection methods, we mainly utilize the Accuracy rate (ACC) and the Area Under Receiver Operating Characteristic Curve (AUC) as our evaluation metrics. We take AUC as the key evaluation metric and reports the frame-level performances.

4.2 Comparison with Previous Methods

In this part, we compare the proposed method with several stat-of-the-art face forgery detection methods. Since following the official data splitting settings [7], we directly cite the results of previous methods from the corresponding papers. We also report performance for the RGB baseline that removes the MSR stream and the proposed FRWI and SGA in our proposed framework.

Detection on Different Video Qualities. In the real-world situation, the videos spread in the social medias are always compressed by popular algorithms such as H.264. Therefore, we evaluate our models on two video qualities, i.e., FF++(HQ and LQ). Table 1 reports the comparison results with previous methods. For FF++(HQ), our method achieves comparable high performances (nearly 100%AUC) compared to state-of-the-art methods. Detecting low-quality manipulated face images is a challenging task as severe compression

Table 2. AUC (%) performance of binary detection on the FF++(LQ) dataset with each four manipulation methods. * indicate the model is trained by us implementing the official code. The best results are in bold.

Methods	DF	F2F	FS	NT
MesoNet [6]	89.52	84.44	83.56	75.74
Xception [7]	94.28	**91.56**	93.70	82.11
PRRNet [29]	95.63	90.15	94.93	80.01
SPSL [30]	93.48	86.02	92.26	76.78
MC-LCR [18]	**97.23**	91.08	94.44	82.13
Xception* [7]	95.60	89.76	93.33	78.87
RGB baseline	95.67	88.48	93.50	80.10
Ours	97.14	91.37	**94.94**	**82.54**

erases much detailed information from the original faces. For FF++(LQ), our method achieves the remarkable performance. Comparing the very recent work D&L [19], our method improves ACC and AUC in 3.55% and 5.88%, respectively. Furthermore, our method achieve better than comparing with the RGB baseline. It demonstrates that introducing the MSR information and joint learning with RGB features can effectively improve the detection performance (Table 2).

Table 3. Recall rate (%) of multi-class classification on the FF++(LQ) dataset with each four manipulation methods. * indicate the model is trained by us implementing the official code. The best results are in bold.

Methods	DF	F2F	FS	NT	PT	Avg
MesoNet [6]	62.45	40.37	28.89	63.35	40.93	47.20
Xception [7]	86.61	78.88	83.16	52.94	75.55	75.43
SPSL [30]	91.16	78.31	88.75	58.97	**77.49**	78.94
D&L [19]	**95.28**	**86.96**	**93.24**	71.66	63.80	82.19
Xception* [7]	93.13	79.24	85.93	66.74	67.25	78.46
RGB baseline	89.38	80.41	86.01	70.36	59.94	77.22
Ours	92.97	84.47	91.13	**74.61**	71.22	**82.88**

Detection on Specific Manipulation Methods. Although identifying the authenticity of input faces is of great importance, specifying the manipulation method is also a non-trivial problem. We evaluate the proposed method against different manipulation methods in FF++(LQ). The models were trained and tested on the FF++(LQ) for each manipulation method. Comparing with previous detection methods, the proposed model achieves the best detection accuracy on all four manipulation methods.

Furthermore, multi-classification is more challenging and significant than binary classification. We further evaluate the proposed model on this five-way (pristine and four respective manipulation methods) classification task.

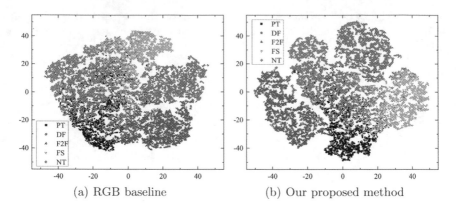

(a) RGB baseline   (b) Our proposed method

Fig. 4. A t-SNE [32] visual comparison of embedding spaces between RGB baseline (left) and our proposed method (right) on FF++(LQ) in the multi-classification task.

As reported in Table 3, our method achieve the best average recall performance. As for the recent work D&L [19], although the performance of identifying DF, F2F, and FS is slightly better than our method, the accuracy of distinguishing NT and real samples is very poor. This reveals that the D&L method tends to over-fit deepfake samples with large forgery artifacts, and perform a high false alarm rate.

We also show the t-SNE [32] feature spaces of data in FF++(LQ) with the multi-class classification task. As shown in Fig. 4, the RGB baseline is more likely to confuse pristine faces with NT-based fake faces because this manipulation method modifies very limited pixels in the spatial domain. In particular, NT-based images, which just slightly tampered with lip, are very similar to pristine images causing almost indistinguishable in the RGB domain. Conversely, our proposed method can split up all classes in the embedding feature spaces. These improvements may benefit from the introduction of the MSR information.

Cross-Datasets Evaluations. Most existing detection models always suffer a significant performance drop when applied to unseen datasets. To comprehensively evaluate the generalization ability of the proposed model, we conduct extensive cross-dataset experiments in this paper. We train our model on the FF++/DF and Pristine (HQ) data and test it on the unseen Deepfake-TIMIT(DT-HQ/LQ) [33], CelebDF [24], DFD-HQ[1] and DFDC-p [34] datasets.

As shown in Table 4, the proposed method achieves the best generalization performances under all cross-dataset settings. For trained on FF++/DF and tested on CelebDF, which is a common protocol that indicates the generalization performance, our method outperforms at least 2.8% at the AUC metric compared with other methods. The cross-dataset experiment demonstrates that the proposed model is capable of achieving high generalization capability.

---

[1] http://ai.googleblog.com/2019/09/contributing-data-to-deepfake-detection.html.

Table 4. AUC (%) performance of cross-dataset evaluations. * indicate the model is trained by us implementing the official code. The results for within-dataset settings are shown in gray and the best results are in bold.

Methods	FF++/DF	DT-LQ	DT-HQ	DFD-HQ	DFDC-P	CelebDF
Xception [7]	99.70	95.90	94.40	85.90	72.20	65.50
Nirkin *et al.* [35]	99.70	–	–	–	–	66.00
MC-LCR [18]	99.84	–	–	–	–	71.61
D&L [19]	**99.85**	56.08	47.20	76.23	–	70.65
Xception* [7]	99.74	92.83	90.49	88.21	71.66	59.09
Ours	99.67	**97.69**	**94.58**	**89.44**	**76.70**	**74.46**

Table 5. Ablation studies on the FF++(LQ) dataset with identifying specific manipulation methods.

RGB	MSR	FRWI	SGA	DF	FF	FS	NT
√	√	√	√	97.14	91.37	94.94	82.54
√	√		√	↓ 0.65	↓ 0.71	↓ 1.07	↓ 0.51
√	√	√		↓ 1.13	↓ 1.27	↓ 1.46	↓ 0.91
√				↓ 1.47	↓ 2.89	↓ 1.44	↓ 2.44
√				↓ 3.43	↓ 3.45	↓ 2.94	↓ 5.79

4.3   Ablation Studies and Visualizations

To explore the influence of each component, we evaluated the proposed model and its variants by identifying the specific manipulation method on FF++(LQ). The results are present in Table 5. From these experiments we get the following observations. In the single-stream setting, using only the RGB or MSR data as input leads to poor results. In the two-stream setting, combining the original two stream with the proposed FRWI or SGA can improve the performance, which verifies that the MSR input is distinct and complementary to the RGB data. The performance can be further improved by both adding the proposed FRWI and SGA, reaching the peak when using the overall proposed framework. This shows the effectiveness of each module: MSR exposes fake clues in the RGB space as supplementary information, and the FRWI and SGA enhance the above information by integrating them.

Furthermore, we presented the visualization of MSR and predicted saliency maps with SGA in Fig. 5 and qualitatively analyzed the results. We enlarged the predicted saliency maps $\hat{\mathcal{M}}^l$ to the same size as the ground-truth forgery mask $\mathcal{M}_{gt}$. We can observe that the MSR image can attenuate content-specific colors and enhance forgery clues. As for predicted saliency maps, the predicted saliency map can accurately localize the forgery region at all three semantic levels. It demonstrates that SGA can promote the RGB feature to capture the subtle forgery clues with the help of multi-scale spatial guidance.

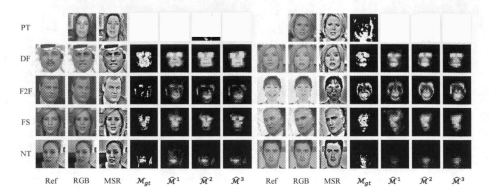

Fig. 5. Qualitative results. For the fake faces, we also display the corresponding real faces for reference.

5   Conclusions

In this work, we present a novel two-stream framework for deepfake detection. In particular, we utilize a multi-scale retinex enhancement inspired by the illumination-reflection model and compose a novel modality, named MSR, to complementary capture the forgery traces. To take full advantage of the MSR information, we propose a two-stream network combined with multi-scale salience-guided and feature re-weighting modules. The multi-scale salience-guided attention module guides the RGB feature extractor to concentrate more on forgery traces from the MSR perspective at multi-scale level. The feature re-weighting module leverages the correlation between the two complementary modalities to promote feature learning for each other. Extensive experiments demonstrate that the proposed framework achieves consistent performance improvement compared to state-of-the-art methods. Future studies can focus on extending this work at a video level so that multiple types of manipulated facial videos can be identified by using a general model.

Acknowledgments. This work was supported in part by NSFC (Grant 61872244), Guangdong Basic and Applied Basic Research Foundation (Grant 2019B151502001), Shenzhen R&D Program (Grant JCYJ20200109105008228).

References

1. Li, L., Bao, J., Yang, H., Chen, D., Wen, F.: Advancing high fidelity identity swapping for forgery detection. In: Proceedings of the IEEE/CVF Conference on Computer Vision and Pattern Recognition, pp. 5074–5083 (2020)
2. Ngo, L.M., aan de Wiel, C., Karaoglu, S., Gevers, T.: Unified application of style transfer for face swapping and reenactment. In: Proceedings of the Asian Conference on Computer Vision (2020)
3. Wang, Y., et al.: HifiFace: 3D shape and semantic prior guided high fidelity face swapping. In: Twenty-Ninth International Joint Conference on Artificial Intelligence, vol. 2, pp. 1136–1142 (2021)

4. Li, Y., Chang, M., Lyu, S.: In Ictu Oculi: exposing AI created fake videos by detecting eye blinking. In: 2018 IEEE International Workshop on Information Forensics and Security (WIFS), pp. 1–7 (2018)
5. Matern, F., Riess, C., Stamminger, M.: Exploiting visual artifacts to expose deepfakes and face manipulations. In: 2019 IEEE Winter Applications of Computer Vision Workshops (WACVW), pp. 83–92 (2019)
6. Afchar, D., Nozick, V., Yamagishi, J., Echizen, I.: MesoNet: a compact facial video forgery detection network. In: 2018 IEEE International Workshop on Information Forensics and Security (WIFS), pp. 1–7 (2018)
7. Rossler, A., Cozzolino, D., Verdoliva, L., Riess, C., Thies, J., Niessner, M.: FaceForensics++: learning to detect manipulated facial images. In: Proceedings of the IEEE/CVF International Conference on Computer Vision, pp. 1–11 (2019)
8. Li, Y., Lyu, S.: Exposing deepfake videos by detecting face warping artifacts. In: Proceedings of the IEEE/CVF Conference on Computer Vision and Pattern Recognition Workshops, pp. 46–52 (2019)
9. Li, L., et al.: Face X-ray for more general face forgery detection. In: Proceedings of the IEEE/CVF Conference on Computer Vision and Pattern Recognition, pp. 5001–5010 (2020)
10. Qian, Y., Yin, G., Sheng, L., Chen, Z., Shao, J.: Thinking in frequency: face forgery detection by mining frequency-aware clues. In: Vedaldi, A., Bischof, H., Brox, T., Frahm, J.-M. (eds.) ECCV 2020. LNCS, vol. 12357, pp. 86–103. Springer, Cham (2020). https://doi.org/10.1007/978-3-030-58610-2_6
11. Luo, Y., Zhang, Y., Yan, J., Liu, W.: Generalizing face forgery detection with high-frequency features. In: Proceedings of the IEEE/CVF Conference on Computer Vision and Pattern Recognition, pp. 16317–16326 (2021)
12. Geirhos, R., Rubisch, P., Michaelis, C., Bethge, M., Wichmann, F.A., Brendel, W.: ImageNet-trained CNNs are biased towards texture; increasing shape bias improves accuracy and robustness. In: International Conference on Learning Representations (2019)
13. Liu, Z., Qi, X., Torr, P.H.S.: Global texture enhancement for fake face detection in the wild. In: Proceedings of the IEEE/CVF Conference on Computer Vision and Pattern Recognition, pp. 8060–8069 (2020)
14. Jobson, D.J., Rahman, Z., Woodell, G.A.: A multiscale retinex for bridging the gap between color images and the human observation of scenes. IEEE Trans. Image Process. 6(7), 965–976 (1997)
15. Chen, H., Hu, G., Lei, Z., Chen, Y., Robertson, N.M., Li, S.Z.: Attention-based two-stream convolutional networks for face spoofing detection. IEEE Trans. Inf. Forensics Secur. 15, 578–593 (2020)
16. Han, J., Gevers, T.: MMD based discriminative learning for face forgery detection. In: Proceedings of the Asian Conference on Computer Vision (2020)
17. Zhao, T., Xu, X., Xu, M., Ding, H., Xiong, Y., Xia, W.: Learning self-consistency for deepfake detection. In: Proceedings of the IEEE/CVF International Conference on Computer Vision, pp. 15023–15033 (2021)
18. Wang, G., Jiang, Q., Jin, X., Li, W., Cui, X.: MC-LCR: multimodal contrastive classification by locally correlated representations for effective face forgery detection. Knowl.-Based Syst. 250, 109114 (2022)
19. Kong, C., Chen, B., Li, H., Wang, S., Rocha, A., Kwong, S.: Detect and locate: exposing face manipulation by semantic- and noise-level telltales. IEEE Trans. Inf. Forensics Secur. 17, 1741–1756 (2022)
20. Land, E.H., McCann, J.J.: Lightness and retinex theory. J. Opt. Soc. Am. 61(1), 1–11 (1971)

21. Ma, Q., Wang, Y., Zeng, T.: Retinex-based variational framework for low-light image enhancement and denoising. IEEE Trans. Multimed. 1–9 (2022)
22. Jun, X., et al.: STAR: a structure and texture aware retinex model. IEEE Trans. Image Process. **29**, 5022–5037 (2020)
23. Park, T., Liu, M.-Y., Wang, T.-C., Zhu, J.-Y.: Semantic image synthesis with spatially-adaptive normalization. In: Proceedings of the IEEE/CVF Conference on Computer Vision and Pattern Recognition, pp. 2337–2346 (2019)
24. Li, Y., Yang, X., Sun, P., Qi, H., Lyu, S.: Celeb-DF: a large-scale challenging dataset for deepfake forensics. In: Proceedings of the IEEE/CVF Conference on Computer Vision and Pattern Recognition, pp. 3207–3216 (2020)
25. Thies, J., Zollhofer, M., Stamminger, M., Theobalt, C., Niessner, M.: Face2Face: real-time face capture and reenactment of RGB videos. In: Proceedings of the IEEE Conference on Computer Vision and Pattern Recognition, pp. 2387–2395 (2016)
26. Thies, J., Zollhöfer, M., Nießner, M.: Deferred neural rendering: image synthesis using neural textures. ACM Trans. Graph. **38**(4), 1–12 (2019)
27. Zhang, K., Zhang, Z., Li, Z., Qiao, Y.: Joint face detection and alignment using multitask cascaded convolutional networks. IEEE Signal Process. Lett. **23**(10), 1499–1503 (2016)
28. Tan, M., Le, Q.: EfficientNet: rethinking model scaling for convolutional neural networks. In: International Conference on Machine Learning, pp. 6105–6114. PMLR (2019)
29. Shang, Z., Xie, H., Zha, Z., Lingyun, Yu., Li, Y., Zhang, Y.: PRRNet: pixel-region relation network for face forgery detection. Pattern Recogn. **116**, 107950 (2021)
30. Liu, H., et al.: Spatial-phase shallow learning: rethinking face forgery detection in frequency domain. In: Proceedings of the IEEE/CVF Conference on Computer Vision and Pattern Recognition, pp. 772–781 (2021)
31. Zhao, H., Zhou, W., Chen, D., Wei, T., Zhang, W., Yu, N.: Multi-attentional deepfake detection. In: Proceedings of the IEEE/CVF Conference on Computer Vision and Pattern Recognition, pp. 2185–2194 (2021)
32. Van der Maaten, L., Hinton, G.: Visualizing data using t-SNE. J. Mach. Learn. Res. **9**(11), 1–27 (2008)
33. Korshunov, P., Marcel, S.: Vulnerability assessment and detection of deepfake videos. In: 2019 International Conference on Biometrics (ICB), pp. 1–6 (2019)
34. Dolhansky, B., Howes, R., Pflaum, B., Baram, N., Ferrer, C.C.: The Deepfake Detection Challenge (DFDC) Preview Dataset. arXiv:1910.08854 (2019)
35. Nirkin, Y., Wolf, L., Keller, Y., Hassner, T.: DeepFake detection based on discrepancies between faces and their context. IEEE Trans. Pattern Anal. Mach. Intell. **44**(10), 6111–6121 (2021)

GB-CosFace: Rethinking Softmax-Based Face Recognition from the Perspective of Open Set Classification

Mingqiang Chen[✉], Lizhe Liu, Xiaohao Chen, and Siyu Zhu

Alibaba, Hangzhou, China
{mimingqiang.cmq,lizhe.llz,xiaohao.cxh,siting.zsy}@alibaba-inc.com

Abstract. State-of-the-art face recognition methods typically take the multi-classification pipeline and adopt the softmax-based loss for optimization. Although these methods have achieved great success, the softmax-based loss has its limitation from the perspective of open set classification: the multi-classification objective in the training phase does not strictly match the objective of open set classification testing. In this paper, we derive a new loss named global boundary CosFace (GB-CosFace). Our GB-CosFace introduces an adaptive global boundary to determine whether two face samples belong to the same identity so that the optimization objective is aligned with the testing process from the perspective of open set classification. Meanwhile, since the loss formulation is derived from the softmax-based loss, our GB-CosFace retains the excellent properties of the softmax-based loss, and CosFace is proved to be a special case of the proposed loss. We analyze and explain the proposed GB-CosFace geometrically. Comprehensive experiments on multiple face recognition benchmarks indicate that the proposed GB-CosFace outperforms current state-of-the-art face recognition losses in mainstream face recognition tasks. Compared to CosFace, our GB-CosFace improves 5.30%, 0.70%, and 0.36% at TAR@FAR=1e-6, 1e-5, 1e-4 on IJB-C benchmark.

1 Introduction

Research on the training objectives of face recognition (FR) has effectively improved the performance of deep-learning-based face recognition [1–4]. According to whether a proxy is used to represent a person's identity or a set of training samples, face recognition methods can be divided into proxy-free methods [5–16] and proxy-based methods [17–29]. The proxy-free methods directly compress the intra-class distance and expand the inter-class distance based on pair-wise learning [5–8] or triplet learning [9–13,15,16]. However, when dealing with a large amount of training data, the hard-mining operation which is crucial for

---

M. Chen and L. Liu—These authors contributed equally to this work.

Supplementary Information The online version contains supplementary material available at https://doi.org/10.1007/978-3-031-26316-3_3.

L. Wang et al. (Eds.): ACCV 2022, LNCS 13844, pp. 35–51, 2023.
https://doi.org/10.1007/978-3-031-26316-3_3

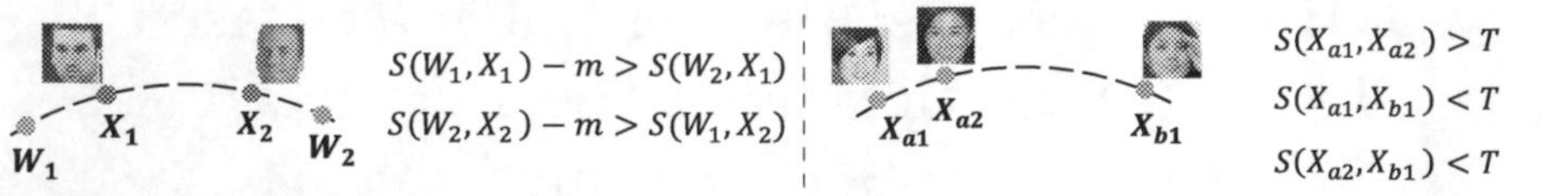

(a) Softmax Training Objective   (b) Open-set Testing Objective

Fig. 1. The difference of the objective between softmax-based training and the open set classification testing, where $S(\cdot)$ is the function to measure the distance between two samples, W_1 and W_2 are the prototypes of two identities respectively. In **(a)**, X_1 and X_2 is the given training sample, m is the margin parameter. In **(b)**, X_{a1} and X_{a2} are two testing samples of ID "a", and X_{b1} is a testing sample of ID "b". ID "a" and "b" are not included in the training data.

proxy-free methods becomes extremely difficult. Recently, proxy-based method have achieved great success and shown advantages in big data training. Most of them take a softmax-based multi-classification pipeline and use cross-entropy loss as the optimization objective. In these methods, each identity in the training set is represented by a prototype, which is the weight vector of the final fully connected layer. We refer to this type of method as the softmax-based face recognition method in this paper.

Despite the great success of softmax-based face recognition, this strategy has its limitation from the perspective of the open set classification [30–33]. As is shown in Fig. 1(a), the training objective of softmax-based multi-classification is to make the predicted probability of the target category larger than other categories. However, face recognition is an open set classification problem where the test category generally does not exist in the training category [1]. A typical requirement for a face recognition model is to determine whether two samples belong to the same identity by comparing the similarity between them with a global threshold T, as is shown in Fig. 1(b). The inconsistency of the objective of training and testing limits the performance.

To reduce the impact of this inconsistency, current softmax-based face recognition methods have made various improvements to the training objective. One of the most vital improvements is to normalize the face features to the hypersphere for unified comparison [18,19]. Typically, the similarity between two samples is represented by the cosine similarity of their corresponding feature vectors. Large-margin-based methods [18,20,21,23] are proposed to further compress the intra-class distance and expand the inter-class distance. Recently, the dynamic schemes for the scale parameter [34] and the margin parameter [26,35] have been studied and further improved the model performance.

From the perspective of training strategy, Lu et al. [36] proposed an optimal sampling strategy to address the inconsistency between the direction of gradient descent and optimizing the concerned evaluation metric. For face feature alignment, DAM [37] proposed a Discrepancy Alignment Metric, which introduces local inter-class differences for each face feature obtained from a pre-trained model, in the face verification stage. However, none of these methods consider introducing the global boundary in the testing process into the training objective.

In this paper, we propose a novel face recognition loss named global boundary CosFace (GB-CosFace), which resolve the above-mentioned inconsistencies well and can be easily applied for end-to-end training on face recognition task. In our GB-CosFace loss, the training objective is aligned with the testing process by introducing a global boundary determined by the proposed adaptive boundary strategy. First, we compare the objective difference between the softmax-based loss and the face recognition testing process. Then, we abstract the reasonable training objective from the perspective of open set classification and derive a antetype of the proposed loss. Furthermore, we combine the excellent properties of softmax-based losses with the proposed antetype loss and derive the final GB-CosFace formulation. We further prove that CosFace [20,21] is a special case of the proposed GB-CosFace. Finally, we analyze and explain the proposed GB-CosFace geometrically. The contributions of this paper are summarized as follows.

- We propose GB-CosFace loss for face recognition, which matches the testing objective of the open set classification while inheriting the advantages of the softmax-based loss. To the best of our knowledge, we are the first work which introduces a global boundary into the training objective for face recognition.
- We analyze the difference and connection between GB-CosFace and general softmax-based losses, and give a reasonable geometric explanation.
- Our GB-CosFace obviously improve the performance of softmax-based face recognition (e.g., improves 5.30%, 0.70%, and 0.36% at TAR@FAR=1e-6, 1e-5, 1e-4 on IJB-C benchmark compared to CosFace).

2   Softmax-Based Face Recognition

To better understand the proposed GB-CosFace, this section review the general softmax-based face recognition.

2.1   Framework

The training framework of the general softmax-based face recognition is shown in Fig. 2. In this framework, each identity in the training set has its corresponding prototype. The prototypes are represented by the weight vectors of the final fully connected layer. Given a training sample, we call the prototype representing the identity of this sample "target prototype", and call other prototypes "non-target prototypes". After extracting face features from the backbone, the predicted scores which represent the similarity between the feature vector and each prototype are calculated through the final fully connected layer (FC layer). The similarity between the feature vector and the target prototype is called "target score", and the other predicted scores are called "non-target scores". Generally, the output feature vector and the prototypes are normalized to the unit hyper-sphere. Therefore, the predicted scores are usually represented by the cosine of the feature vector and the prototype. In training, the softmax-based loss is adopted to optimize the backbone and the final FC layer through backpropagation.

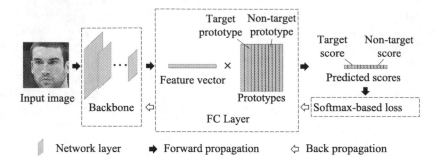

Fig. 2. The training framework of the general softmax-based face recognition.

2.2   Objective

For each iteration in n-class face recognition training, given a training sample and its label y, the general softmax-based loss is as follows:

$$\mathcal{L}_S = -log\frac{e^{s(cos(\theta_y+m_\theta)-m_p)}}{e^{s(cos(\theta_y+m_\theta)-m_p)} + \sum_i e^{scos\theta_i}} \tag{1}$$

where θ_y is the arc between the predicted feature vector and the target prototype, θ_i is the arc between the predicted feature vector and the non-target prototype, y is the index of the target identity, i is the index of the non-target identities, $i \in [1, n]$ and $i \neq y$. There are three hyper-parameters: the scale parameter "s", and the two margin parameters "m_θ" and "m_p".

We can reach several common softmax-based losses from Eq. 1. E.g., normalized softmax loss will be reached if both m_θ and m_p are set as zero. ArcFace and CosFace will be reached if we respectively set m_p and m_θ as 0.

Softmax-based losses can be regarded as the smooth form of the following optimization objective $\mathcal{O}_S$.

$$\begin{aligned}
\mathcal{O}_S &= ReLU(max(cos\theta_i) - (cos(\theta_y + m_\theta) - m_p)) \\
&= \lim_{s\to+\infty} -\frac{1}{s}log\frac{e^{s(cos(\theta_y+m_\theta)-m_p)}}{e^{s(cos(\theta_y+m_\theta)-m_p)} + \sum_{i=1,i\neq y}^{n} e^{scos\theta_i}} \\
&= \lim_{s\to+\infty} -\frac{1}{s}\mathcal{L}_S
\end{aligned} \tag{2}$$

where the SoftPlus function is used as a smooth form of ReLU operator and $log\sum exp(\cdot)$ is used as a smooth form of $max(\cdot)$ operator. More detailed derivation is included in the supplementary material.

From this perspective, we can find that the training objective $\mathcal{O}_S$ constrains the target score to be larger than the maximum non-target score. The margin is introduced for a stricter constraints. However, this constraint is not completely consistent with the objective of the testing process. Based on Eq. 2, we can visualize the decision boundaries of normalized softmax loss [19], CosFace [20,21], and ArcFace

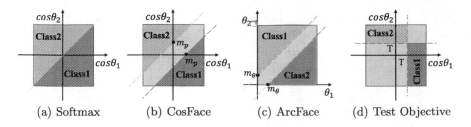

(a) Softmax   (b) CosFace   (c) ArcFace   (d) Test Objective

Fig. 3. Decision boundaries of different loss functions under binary classification case. Figure (d) shows the expected decision boundary in the testing phase.

[23] under binary classification case, as is shown in Fig. 3 (a)–(c). In the testing phase, a global threshold T of the cosine similarity needs to be fixed to determine whether two samples belong to the same person, as is shown in Fig. 3(d). We can see that, even if a margin is added, the decision boundaries of softmax-based losses do not completely match the expected boundary for testing.

2.3 Properties

Current face recognition models do not directly apply $\mathcal{O}_S$ as the training objective. On the one hand, $max(\cdot)$ operator only focuses on the maximum value and the gradients will only be backpropagated to the target score and maximum non-target score. On the other hand, if the argument of the RELU function is less than 0, no gradient will be backpropagated. As a smooth form of $\mathcal{O}_S$, the softmax-based loss can avoid the above problems. The success of softmax-based loss is due to its excellent properties.

Property 1. The gradients of the non-target scores are proportional to their softmax value

For softmax-based loss, the backpropagated gradients will be assigned to all non-target scores according to their softmax value. This property ensures that each non-target prototype can play a role in training, and hard non-target prototypes get more attention.

Property 2. The gradient of the target score and the sum of the gradients of all non-target scores have the same absolute value and opposite signs.

$$\frac{\partial \mathcal{L}_S}{\partial cos(\theta_y + m_\theta)} = -\sum_i \frac{\partial \mathcal{L}_S}{\partial cos(\theta_i)} \tag{3}$$

Softmax-based loss has balanced gradients for the target score and the non-target scores. This property can maintain the stability of training and prevent the training process from falling into a local minimum.

Considering the key role that these two properties play in face recognition training, we expect to inherit them in the loss design. In this paper, we add the consistency of training and testing to the loss design by introducing an adaptive global boundary. From the expected training objective, we derive our

GB-CosFace framework and prove compatibility with CosFace. This compatibility allows the proposed loss to inherit the excellent properties of the general softmax-based loss while solving the inconsistency between the training and testing objective.

3   GB-CosFace Framework

3.1   Antetype Formulation

Based on the face recognition testing process which is shown in Fig. 3(d), we propose to introduce a global threshold p_v as the boundary between target score and non-target scores. The target score is required to be larger than p_v while the maximum of the non-target scores is required to be less than p_v. Following this idea, we improve Eq. 2 as follows:

$$\begin{cases} \mathcal{O}_T = ReLU(p_v - (p_y - m)) \\ \mathcal{O}_N = ReLU(max(p_i) - (p_v - m)) \end{cases} \tag{4}$$

where we divide the training objective into the target score $\mathcal{O}_T$ and the non-target scores $\mathcal{O}_N$ respectively. p_y is the target score, where $p_y = cos\theta_y$. p_i is the non-target score, where $p_i = cos\theta_i$. m is the margin parameter introduced for stricter constraints. The training objective is to minimize $\mathcal{O}_T$ and $\mathcal{O}_N$.

Inspired by the success of the softmax-based loss, similar to Eq. 2, we take the smooth form of $\mathcal{O}_T$ and $\mathcal{O}_N$ as the antetype of the proposed loss.

$$\begin{cases} \mathcal{L}_{T1} = -log\frac{e^{s(p_y-m)}}{e^{s(p_y-m)}+e^{sp_v}} \\ \mathcal{L}_{N1} = -log\frac{e^{s(p_v-m)}}{e^{s(p_v-m)}+\sum_i e^{sp_i}} \end{cases} \tag{5}$$

The loss for target score and non-target scores are represented as $\mathcal{L}_{T1}$ and $\mathcal{L}_{N1}$ respectively. p_v is the global boundary hyper-parameter, which also means "virtual score". For $\mathcal{L}_{T1}$, p_v is a virtual non-target score. For $\mathcal{L}_{N1}$, p_v is a virtual target score. Since we take $log\sum exp(\cdot)$ as the smooth form of $max(\cdot)$, the distribution of the gradients of non-target scores inherits Property 1. (stated in Sect. 2.3) of the softmax-based loss.

However, the proposed antetype introduces another problem: the setting of hyper-parameter p_v. First, the inappropriate setting of p_v may cause a serious gradient imbalance problem. Since we separate the constraints on the target score and the non-target scores, the gradient balance for target score and non-target scores is broken and the antetype loss no longer retains Property 2. (stated in Sect. 2.3). Second, considering the rapid rise of the exponential function and the amplification effect of the hyper-parameter "s", the model is extremely sensitive to the choice of the hyper-parameter p_v. As can be seen in Fig. 4, a slight change in p_v can cause an order of magnitude difference between the gradients for target score p_y and non-target scores p_i. Therefore, an adaptive scheme for the global boundary is necessary.

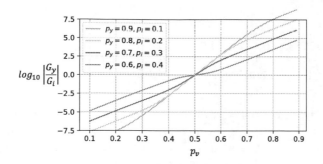

Fig. 4. The ratio of the target gradient to the non-target gradient varies with p_v under binary classification case using different p_y and p_i. Hyper-parameter s and m are set to 32 and 0.15 respectively. Note that the ordinate is the base 10 logarithm of the ratio.

3.2 Adaptive Global Boundary

To control the gradient balance and adapt the global boundary to different training stages, we propose an adaptive global boundary method. We believe that an ideal global boundary should meet the following conditions: **a)** Under this boundary setting, the gradients of the target score and the non-target scores should be roughly balanced from a global perspective; **b)** The global boundary should change slowly during the training process to keep the training stable while adapting to different training stages. Based on these two conditions, we make the following design.

Gradient Balance Control. We define $\hat{p}_v$ as the balanced threshold of the target score and the non-target scores which satisfies $\frac{\partial \mathcal{L}_{T1}}{\partial p_y} = -\sum_i \frac{\partial \mathcal{L}_{N1}}{\partial p_i}$. Based on this condition, we reach the following form of $\hat{p}_v$:

$$\hat{p}_v = (p_y + \frac{1}{s}log\sum_i e^{sp_i})/2 \tag{6}$$

Ideally, for each iteration, to satisfy the above condition **a)**, we expect to calculate $\hat{p}_v$ for each sample in the data set and get the mean value as the threshold p_v. Considering the efficiency, we calculate the mean of $\hat{p}_v$ for each batch and update it by the momentum update strategy.

$$p_{vg} = (1 - \gamma)p_{vg} + \gamma p_{vb} \tag{7}$$

where $\gamma \in [0, 1]$ is the update rate, p_{vb} is the mean of p_v in a batch. A small γ can keep the stability of p_v. We empirically set γ to 0.01.

This dynamic threshold strategy makes the gradient balanced globally. However, for each sample, the problem of gradient imbalance can be very serious. Therefore, we modify the value of p_v to be the weighted sum of p_{vg} and $\hat{p}_v$ as follows.

$$p_v = \alpha p_{vg} + (1 - \alpha)\hat{p}_v \tag{8}$$

where α is a hyper-parameter and $\alpha \in [0,1]$. When $\alpha = 0$, the gradients for the target score and the non-target scores are completely balanced. We can control the degree of the gradient imbalance by adjusting α.

Compatible with CosFace. In Eq. 8, if we take α as 0, the proposed loss will fully conform to Property 1 and Property 2 (stated in Sect. 2.3) of softmax-based loss. Through the following analysis, we can further find that Cosface [20,21] is a special case of the proposed loss when $\alpha = 0$.

The gradients based on CosFace is calculated as follows.

$$\mathcal{G}_{T-CosFace} = -\mathcal{G}_{N-CosFace} = -\frac{s \cdot \sum_i e^{sp_i}}{e^{s(p_y-m)} + \sum_i e^{sp_i}} = -\frac{s \cdot e^{sp_n}}{e^{s(p_y-m)} + e^{sp_n}} \tag{9}$$

where the gradient for the target score is represented as $\mathcal{G}_{T-CosFace}$, the sum of the gradients of the non-target scores is represented as $\mathcal{G}_{N-CosFace}$, and $p_n = \frac{1}{s}log\sum_i e^{sp_i}$.

For the proposed loss, based on Eq. 5, we can get the gradient for target score p_y ($\mathcal{G}_{T1}$) and the sum of the gradients for non-target scores p_i ($\mathcal{G}_{N1}$) when α is set to 0.

$$\mathcal{G}_{T1} = -\mathcal{G}_{N1} = -\frac{s \cdot e^{\frac{1}{2}sp_n}}{e^{\frac{1}{2}s(p_y-2m)} + e^{\frac{1}{2}sp_n}} \tag{10}$$

As the above equation shows, if we take p_v as $\hat{p}_v$ (Eq. 6), the difference of the proposed loss (Eq. 5) and CosFace only lies on the margin and the scale. The more detailed proof is included in the supplementary material.

Final Loss. For formal unity with CosFace, we rewrite the proposed loss into the following form.

$$\mathcal{L}_{GB-CosFace} = -\frac{1}{2}log\frac{e^{2s(p_y-m)}}{e^{2s(p_y-m)} + e^{2sp_v}} - \frac{1}{2}log\frac{e^{2s(p_v-m)}}{e^{2s(p_v-m)} + e^{2sp_n}} \tag{11}$$

where $p_n = \frac{1}{s}log\sum_i e^{sp_i}$. The value of p_v is in accordance with Eq. 8. In training, p_v is a detached parameter which does not require gradients.

This is the final form of the proposed GB-CosFace. Under this formulation, the hyper-parameter α controls the degree of gradient imbalance. If we set α as 0, the gradients for the target score and the non-target scores are balanced, and the proposed GB-CosFace is equivalent to CosFace which has the margin of $2m$ and the scale of s.

3.3   Geometric Analysis

To analyze the properties of the proposed loss and compare it with other softmax-based losses, we analyze the loss boundaries in the binary classification case. The boundaries of ArcFace [23] and CosFace [20,21] are determined by the following Eq. 12 and Eq. 13 respectively.

$$|arccos(P \cdot P_1) - arccos(P \cdot P_2)| = m \tag{12}$$

$$|P \cdot P_1 - P \cdot P_2| = m \tag{13}$$

where P is the predicted normalized n-dimensional feature vector and n is the face feature dimension, P_1 and P_2 are the feature vectors of ID1 and ID2 respectively.

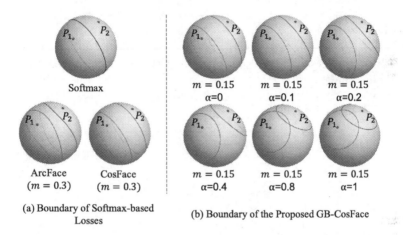

(a) Boundary of Softmax-based Losses

(b) Boundary of the Proposed GB-CosFace

Fig. 5. Boundaries of the softmax-based losses and the proposed GB-CosFace loss. P_1 and P_2 are two points at a distance of $60°$. The red line and blue line are the target boundaries for P_1 and P_2 respectively. For the normalized softmax loss, the boundaries for P_1 and P_2 are coincident and represented in black color.(Color figure online)

For normalized softmax loss, the boundary is determined by Eq. 12 or Eq. 13 with a zero margin. We set the angle between vector P_1 and P_2 as $60°$ and show the boundaries of normalized softmax, ArcFace, and CosFace in the 3D spherical feature space in Fig. 5(a).

The boundaries of the proposed GB-CosFace can be determined according to Eq. 14.

$$\begin{cases} |P \cdot P_1| = p_v + m \\ |P \cdot P_2| = p_v + m \end{cases} \tag{14}$$

According to Eq. 6 and Eq. 8, in the binary classification case, p_v can be represented as follows.

$$p_v = \alpha p_{vg} + (1 - \alpha)(P \cdot P_1 + P \cdot P_2)/2 \qquad (15)$$

We show the boundaries of the proposed GB-CosFace loss in Fig. 5(b), where p_{vg} is fixed to 0.62 (a reasonable value according to the experiments in Sect. 4) and m is fixed to 0.15.

In the face recognition problem, feature vectors of the same identity are expected to cluster together. However, by observing Fig. 5(a), we can find that the boundaries in the case of binary classification do not meet this expectation. Only the positions near the line from point P_1 to point P_2 on the sphere can be effectively constrained. Fortunately, the training set has far more than two identities. Ideally, the prototypes of different identities will be evenly distributed on the sphere. The feature vectors of the same identity will be constrained in all directions. But actually, it cannot be guaranteed that in the sparse high-dimensional spherical feature space, there are enough non-target prototypes evenly distributed around each training sample.

The proposed loss $\mathcal{L}_{GB}$ alleviates this problem by introducing a global boundary. As is shown in Fig. 5(b), when $\alpha = 0$, the boundary is the same as CosFace. When $\alpha = 1$, the boundary is a circle on the sphere centered on P_1 or P_2 with a fixed radius completely determined by p_{vg} and the margin m. With the increase of α, the boundary is closer to the ideal open set classification objective. However, an excessively large α will cause blurring or even crossing of the boundaries between different identities.

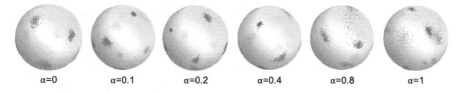

α=0   α=0.1   α=0.2   α=0.4   α=0.8   α=1

Fig. 6. Visualization of the toy experiments on the proposed GB-CosFace. Different colors represent different identities.

To study the appropriate range of α, we conduct a toy experiment based on a seven-layer convolutional neural network on a small face recognition dataset containing ten identities. We set the feature dimension as three, and visualize the distribution of the feature vectors on the unit sphere under different α settings, as is shown in Fig. 6. The margin is fixed to 0.15 and α is adjusted from 0 to 1. When $\alpha = 0$, our GB-CosFace is exactly the same as CosFace with the margin of 0.3, as indicated in Sect. 3.2. As α increases, e.g., $\alpha = 0.2$, the feature vectors of the same identity are more concentrated as expected. The model performance will deteriorate if α is further increased, e.g., $\alpha = 0.8$ or $\alpha = 1$. The setting of α will be studied in detail in the Sect. 4.2.

4   Experiments

In this section, we verify our GB-CosFace on two important face tasks: face recognition and face clustering. Furthermore, we conduct ablation experiments to verify the proposed strategies and the settings of the hyper-parameters.

Dataset. We employ MS1MV3 [38], a refined version of MS1M [39] as our training set for all the following experiments. This is a large-scale face recognition dataset containing 5.1M face images of 93K celebrities. We use several popular benchmarks as the validation set, including LFW [40], CFP-FP [41], CPLFW [42], AgeDB-30 [43], and CALFW [44]. And we use IJB-B [45] and IJB-C [46] as the testing sets.

Implementation Details. We use ResNet50 [47] and ResNet100 as the backbone for the following experiments. The BN-FC-BN structure is added after the last convolution layer to output 512-dimensional face feature vectors. For data pre-possessing, all face images are set to 112×112 and normalized by utilizing five facial points following recent papers [23,26]. Each RGB pixel is normalized to $[-1, 1]$. Random horizontal flip is the only data augmentation method employed in the training process. For optimization, we adopt the stochastic gradient descent (SGD) optimizer with a momentum of 0.9 and weight decay of 1e-4. We adopt the step learning rate decay strategy with an initial learning rate of 0.1. We train 24 epochs and divide the learning rate by 10 at 5, 10, 15, and 20 epochs. The training batch size is fixed to 512. Eight NVIDIA GPUS are employed for training. We fix the hyper-parameters s, m, α and γ as 32, 0.16, 0.15 and 0.01 respectively if not specified.

4.1   Face Recognition

Analysis of Gradient Blance. Figure 7 shows the gradients and the global boundary p_v in the training process. Throughout the training process, the gradient of the target score G_T and the gradients sum of the non-target scores G_N maintain a same convergence trend, and the values of G_T and G_N are approximately equal after 50k iterations. The change trend of the global boundary parameter p_v during the training process is consistent with the gradients G_T and G_N, and eventually converges to 0.62. This result shows that our adaptive global boundary strategy can guarantee the stability of model training and keep G_T and G_N balanced throughout the training process, which is consistent with our discussion in Sect. 3.2.

Results on Validation Datasets. To compare with recent state-of-the-art competitors, we compare the results on several popular face recognition benchmarks, including LFW, CFP-FP, AgeDB-30, CALFW, and CPLFW. LFW focuses on unconstrained fa ce verification.

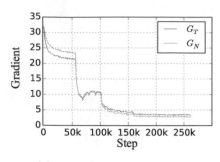
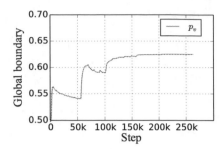

(a) Gradients during training.   (b) Global boundary p_v during training.

Fig. 7. G_T is the gradient of the target score, G_N is the gradients sum of the non-target scores, and p_v is the global boundary in Eq. 11.

The results are shown in Table 1. We achieve the best results on two of the five benchmarks. Even though both datasets are highly-saturated, our GB-CosFace surpasses the recent methods on CFP-FP and CPLFW, and achieves comparable results on other three datasets.

Results on IJB-B and IJB-C. IJB is one of the largest and most challenging benchmarks to evaluate unconstrained face recognition. IJB-B contains 1845 identities with 55025 frames and 7011 videos. IJB-C is an extension of IJB-B which contains about 3.5K identities from 138K face images and 11K face videos.

The results are shown in Table 2. We achieve SOTA results on IJB-B and IJB-C. Compared to CosFace, our GB-CosFace improves 6.07%, 4.07% and 0.41% at TAR@FAR=1e-6, 1e-5, 1e-4 on IJB-B, and improves 5.30%, 0.70% and 0.36% at TAR@FAR=1e-6, 1e-5, 1e-4 on IJB-C.

Table 1. 1:1 verification accuracy is reported on the LFW, CFP-FP, AgeDB-30, CALFW, CPLFW datasets. Backbone network: ResNet100.

Method	Validation dataset				
	LFW	CFP-FP	AgeDB-30	CALFW	CPLFW
CosFace [21] (CVPR18)	99.81	98.12	98.11	95.76	92.28
ArcFace [23] (CVPR19)	**99.83**	98.27	98.28	95.45	92.08
Sub-center ArcFace [48] (ECCV20)	**99.83**	98.80	**98.45**	–	–
BroadFace [28] (ECCV20)	**99.83**	98.63	98.38	**96.20**	93.17
CurricularFace [25] (CVPR20)	99.80	98.37	98.32	**96.20**	93.13
URFace [49] (CVPR20)	99.78	98.64	–	–	–
CosFace+SCF [50] (CVPR21)	99.80	98.59	98.26	96.18	93.26
MagFace [26] (CVPR21)	**99.83**	98.46	98.17	96.15	92.87
GB-CosFace	99.80	**98.84**	98.31	96.15	**93.55**

Table 2. The face verification accuracy on IJB-B and IJB-C. We evaluated the TAR@FAR from 1e-4 to 1e-6. Backbone network: ResNet100.

Method	IJB-B(TAR)			IJB-C(TAR)		
	1e-6	1e-5	1e-4	1e-6	1e-5	1e-4
CosFace [21] (CVPR18)	36.49	88.11	94.80	85.91	94.10	96.37
ArcFace [23] (CVPR19)	38.28	89.33	94.25	89.06	93.94	96.03
Sub-center ArcFace [48] (ECCV20)	-	-	**95.25**	-	-	96.61
BroadFace [28] (ECCV20)	46.53	90.81	94.61	90.41	94.11	96.03
CurricularFace [25] (CVPR20)	-	-	94.80	-	-	96.10
GroupFace [51] (CVPR20)	**52.12**	91.24	94.93	89.28	94.53	96.26
CosFace+DAM [37] (ICCV21)	-	-	94.97	-	-	96.45
CosFace+SCF [50] (CVPR21)	-	91.02	94.95	-	94.78	96.22
MagFace [26] (CVPR21)	40.91	89.88	94.33	89.26	93.67	95.81
GB-CosFace	42.56	**92.18**	95.21	**91.21**	**94.80**	**96.73**

4.2 Ablation Study

To analyze the effect of the adaptive boundary strategy and the setting of hyper-parameter α, we train ResNet-50 networks on MS1MV3 with different settings and evaluated the TAR@FAR=1e-4 on IJB-C.

Hyperparameter Setting. Compared to CosFace, we introduce another hyper-parameter α in Eq. 8. Since the settings of the scale parameter s and the margin parameter m have been studied in detail in the previous works [20,21,23], we empirically set $s = 32$ and $m = 0.16$ (equivalent to $m = 0.32$ in CosFace), and focus on the setting of α. For more detailed theoretical analysis, please refer to Sect. 3.3.

Table 3. The results of the proposed GB-CosFace under different settings of α.

Settings	IJB-C(TAR)
FAR=1e-4, R50, adaptive p_v, $\alpha = 0$	96.10
FAR=1e-4, R50, adaptive p_v, $\alpha = 0.05$	96.15
FAR=1e-4, R50, adaptive p_v, $\alpha = 0.15$	96.24
FAR=1e-4, R50, adaptive p_v, $\alpha = 0.25$	**96.35**
FAR=1e-4, R50, adaptive p_v, $\alpha = 0.35$	96.33
FAR=1e-4, R50, adaptive p_v, $\alpha = 0.60$	96.08

We conduct the controlled experiment where the value of α is set from 0 to 0.6 and other parameters are fixed. The results are shown in Table 3. When the value of α gradually increases from 0 to 0.25, the performance of the model gradually improves and the model performs best with $\alpha = 0.25$. After the value of α exceeds 0.35, the model performance obviously degenerates with the increase of α. Overall, the performance of the model can maintain relatively good results as the value of α is between 0.15 and 0.35. This result is consistent with the previous discussion and the toy experiments in Sect. 3.3.

Effect of the Adaptive Boundary Strategy. To evaluate the effectiveness of the adaptive boundary strategy, we compare the fixed boundary strategy and the proposed adaptive boundary strategy in Sect. 3.2. We fix the p_v in our GB-CosFace(Eq. 11) to different values and keep the other experimental settings the same as Sect. 4.1. Since p_v converges to 0.62 in the experiment in Sect. 4.1, we choose $p_v = 0.62$ and additionally choose values near 0.62.

Table 4. Comparison of the results of the proposed adaptive global boundary strategy and the fixed global boundary strategy.

Settings	IJB-C(TAR@FAR $= 1$e-4)
FAR=1e-4, R50, $\alpha = 0.15$, $p_v = 0.50$	91.19
FAR=1e-4, R50, $\alpha = 0.15$, $p_v = 0.58$	**96.27**
FAR=1e-4, R50, $\alpha = 0.15$, $p_v = 0.62$	96.19
FAR=1e-4, R50, $\alpha = 0.15$, $p_v = 0.66$	96.17
FAR=1e-4, R50, $\alpha = 0.15$, $p_v = 0.74$	95.09
FAR=1e-4, R50, $\alpha = 0.15$, adaptive p_v	96.24

The results are shown in Table 4. For the fixed boundaries, the model performs best when $p_v = 0.58$ and gets worse rapidly when the p_v changes, eg. the TAR decreases to 91.19% when $p_v = 0.50$. What's more, if we reduce p_v to 0.42 or increase it to 0.82, the training will not converge. For the adaptive boundary strategy, the performance is very close to the best fixed boundary strategy result. This indicates that for the fixed boundary strategy, the model performance is sensitive to the value of p_v, a very careful setting of p_v is required to obtain good results. While the adaptive global boundary strategy does not require careful tuning of hyper-parameters to achieve a similar performance. This result is consistent with the previous discussion and the toy experiments in Sect. 3.3.

5   Conclusion

In this work, we discuss the inconsistency between the training objective of the softmax-based loss and the testing process of face recognition, and derive a new loss from the perspective of open set classification, called the global boundary CosFace(GB-CosFace). Our GB-CosFace aligns the training objective with the face recognition testing process while inheriting the good properties of the softmax-based loss. Comprehensive experiments indicate that our GB-CosFace has an obvious improvement over general softmax-based losses.

References

1. Wang, M., Deng, W.: Deep face recognition: a survey. Neurocomputing **429**, 215–244 (2021)
2. Taigman, Y., Yang, M., Ranzato, M., Wolf, L.: Deepface: closing the gap to human-level performance in face verification. In: Conference on Computer Vision and Pattern Recognition (CVPR) (2014)
3. Sun, Y., Liang, D., Wang, X., Tang, X.: DeepID3: face recognition with very deep neural networks. arXiv preprint arXiv:1502.00873 (2015)
4. Wen, Y., Zhang, K., Li, Z., Qiao, Yu.: A discriminative feature learning approach for deep face recognition. In: Leibe, B., Matas, J., Sebe, N., Welling, M. (eds.) ECCV 2016. LNCS, vol. 9911, pp. 499–515. Springer, Cham (2016). https://doi.org/10.1007/978-3-319-46478-7_31
5. Chopra, S., Hadsell, R., LeCun, Y.: Learning a similarity metric discriminatively, with application to face verification. In: Conference on Computer Vision and Pattern Recognition (CVPR) (2005)
6. Sun, Y.: Deep learning face representation by joint identification-verification. In: Advances in Neural Information Processing Systems (NIPS) (2014)
7. Ustinova, E., Lempitsky, V.: Learning deep embeddings with histogram loss. In: Advances in Neural Information Processing Systems (NIPS) (2016)
8. Han, C., Shan, S., Kan, M., Wu, S., Chen, X.: Face recognition with contrastive convolution. In: European Conference on Computer Vision (ECCV) (2018)
9. Schroff, F., Kalenichenko, D., Philbin, J.: Facenet: a unified embedding for face recognition and clustering. In: Conference on Computer Vision and Pattern Recognition (CVPR) (2015)
10. Parkhi, O.M., Vedaldi, A., Zisserman, A.: Deep face recognition. In: British Machine Vision Association (BMVC) (2015)
11. Ge, W.: Deep metric learning with hierarchical triplet loss. In: European Conference on Computer Vision (ECCV) (2018)
12. Zhong, Y., Deng, W.: Adversarial learning with margin-based triplet embedding regularization. In: International Conference on Computer Vision (ICCV) (2019)
13. Oh Song, H., Xiang, Y., Jegelka, S., Savarese, S.: Deep metric learning via lifted structured feature embedding. In: Conference on Computer Vision and Pattern Recognition (CVPR) (2016)
14. Rippel, O., Paluri, M., Dollar, P., Bourdev, L.: Metric learning with adaptive density discrimination. In: International Conference on Learning Representations (ICLR) (2015)
15. Sohn, K.: Improved deep metric learning with multi-class n-pair loss objective. In: Advances in Neural Information Processing Systems (NIPS) (2016)
16. Wu, C.Y., Manmatha, R., Smola, A.J., Krahenbuhl, P.: Sampling matters in deep embedding learning. In: International Conference on Computer Vision (ICCV) (2017)
17. Sun, Y., Wang, X., Tang, X.: Deep learning face representation from predicting 10,000 classes. In: Conference on Computer Vision and Pattern Recognition (CVPR) (2014)

18. Liu, W., Wen, Y., Yu, Z., Li, M., Raj, B., Song, L.: Sphereface: deep hypersphere embedding for face recognition. In: Conference on Computer Vision and Pattern Recognition (CVPR) (2017)
19. Wang, F., Xiang, X., Cheng, J., Yuille, A.L.: Normface: L2 hypersphere embedding for face verification. In: Proceedings of the 25th ACM International Conference on Multimedia (ACM) (2017)
20. Wang, F., Cheng, J., Liu, W., Liu, H.: Additive margin softmax for face verification. IEEE Signal Process. Lett. **25**(7), 926–930 (2018)
21. Wang, H., et al.: Cosface: large margin cosine loss for deep face recognition. In: Conference on Computer Vision and Pattern Recognition (CVPR) (2018)
22. Sun, Y., et al.: Circle loss: a unified perspective of pair similarity optimization. In: Conference on Computer Vision and Pattern Recognition (CVPR) (2020)
23. Deng, J., Guo, J., Xue, N., Zafeiriou, S.: Arcface: additive angular margin loss for deep face recognition. In: Conference on Computer Vision and Pattern Recognition (CVPR) (2019)
24. Zheng, Y., Pal, D.K., Savvides, M.: Ring loss: convex feature normalization for face recognition. In: Conference on Computer Vision and Pattern Recognition (CVPR) (2018)
25. Huang, Y., et al.: Curricularface: adaptive curriculum learning loss for deep face recognition. In: Conference on Computer Vision and Pattern Recognition (CVPR) (2020)
26. Meng, Q., Zhao, S., Huang, Z., Zhou, F.: Magface: a universal representation for face recognition and quality assessment. In: Conference on Computer Vision and Pattern Recognition (CVPR) (2021)
27. Chen, B., Deng, W., Du, J.: Noisy softmax: improving the generalization ability of DCNN via postponing the early softmax saturation. In: Conference on Computer Vision and Pattern Recognition (CVPR) (2017)
28. Kim, Y., Park, W., Shin, J.: BroadFace: looking at tens of thousands of people at once for face recognition. In: Vedaldi, A., Bischof, H., Brox, T., Frahm, J.-M. (eds.) ECCV 2020. LNCS, vol. 12354, pp. 536–552. Springer, Cham (2020). https://doi.org/10.1007/978-3-030-58545-7_31
29. Deng, J., Guo, J., Yang, J., Lattas, A., Zafeiriou, S.: Variational prototype learning for deep face recognition. In: Proceedings of the IEEE/CVF Conference on Computer Vision and Pattern Recognition (2021)
30. Scheirer, W.J., de Rezende Rocha, A., Sapkota, A., Boult, T.E.: Toward open set recognition. IEEE Trans. Pattern Anal. Mach. Intell. **35**(7), 1757–1772 (2012)
31. Geng, C., Huang, S.J., Chen, S.: Recent advances in open set recognition: a survey. IEEE Trans. Pattern Anal. Mach. Intell. **43**(10), 3614–3631 (2020)
32. Ge, Z., Demyanov, S., Chen, Z., Garnavi, R.: Generative openmax for multi-class open set classification. arXiv preprint arXiv:1707.07418 (2017)
33. Yoshihashi, R., Shao, W., Kawakami, R., You, S., Iida, M., Naemura, T.: Classification-reconstruction learning for open-set recognition. In: Conference on Computer Vision and Pattern Recognition (CVPR) (2019)
34. Zhang, X., Zhao, R., Qiao, Y., Wang, X., Li, H.: Adacos: adaptively scaling cosine logits for effectively learning deep face representations. In: Conference on Computer Vision and Pattern Recognition (CVPR) (2019)
35. Liu, H., Zhu, X., Lei, Z., Li, S.Z.: Adaptiveface: adaptive margin and sampling for face recognition. In: Conference on Computer Vision and Pattern Recognition (CVPR) (2019)

36. Lu, J., Xu, C., Zhang, W., Duan, L.Y., Mei, T.: Sampling wisely: deep image embedding by top-k precision optimization. In: Proceedings of the IEEE/CVF International Conference on Computer Vision (2019)

37. Liu, J., et al.: DAM: discrepancy alignment metric for face recognition. In: Proceedings of the IEEE/CVF International Conference on Computer Vision (2021)

38. Deng, J., Guo, J., Zhang, D., Deng, Y., Lu, X., Shi, S.: Lightweight face recognition challenge. In: International Conference on Computer Vision Workshops (ICCVW) (2019)

39. Guo, Y., Zhang, L., Hu, Y., He, X., Gao, J.: MS-celeb-1M: a dataset and benchmark for large-scale face recognition. In: Leibe, B., Matas, J., Sebe, N., Welling, M. (eds.) ECCV 2016. LNCS, vol. 9907, pp. 87–102. Springer, Cham (2016). https://doi.org/10.1007/978-3-319-46487-9_6

40. Huang, G.B., Mattar, M., Berg, T., Learned-Miller, E.: Labeled faces in the wild: a database for studying face recognition in unconstrained environments. In: Workshop on Faces in 'Real-Life' Images: Detection, Alignment, and Recognition (2008)

41. Sengupta, S., Chen, J.C., Castillo, C., Patel, V.M., Chellappa, R., Jacobs, D.W.: Frontal to profile face verification in the wild. In: IEEE Winter Conference on Applications of Computer Vision (WACV) (2016)

42. Zheng, T., Deng, W.: Cross-pose LFW: a database for studying cross-pose face recognition in unconstrained environments. Beijing University of Posts and Telecommunications, Technical report (2018)

43. Moschoglou, S., Papaioannou, A., Sagonas, C., Deng, J., Kotsia, I., Zafeiriou, S.: AgeDB: the first manually collected, in-the-wild age database. In: Conference on Computer Vision and Pattern Recognition Workshops (CVPRW) (2017)

44. Zheng, T., Deng, W., Hu, J.: Cross-age LFW: a database for studying cross-age face recognition in unconstrained environments. arXiv preprint arXiv:1708.08197 (2017)

45. Whitelam, C., et al.: IARPA janus benchmark-b face dataset. In: Conference on Computer Vision and Pattern Recognition Workshops (CVPRW) (2017)

46. Maze, B., et al.: IARPA janus benchmark-C: face dataset and protocol. In: International Conference on Biometrics (ICB) (2018)

47. He, K., Zhang, X., Ren, S., Sun, J.: Deep residual learning for image recognition. In: Conference on Computer Vision and Pattern Recognition (CVPR) (2016)

48. Deng, J., Guo, J., Liu, T., Gong, M., Zafeiriou, S.: Sub-center ArcFace: boosting face recognition by large-scale noisy web faces. In: Vedaldi, A., Bischof, H., Brox, T., Frahm, J.-M. (eds.) ECCV 2020. LNCS, vol. 12356, pp. 741–757. Springer, Cham (2020). https://doi.org/10.1007/978-3-030-58621-8_43

49. Shi, Y., Yu, X., Sohn, K., Chandraker, M., Jain, A.K.: Towards universal representation learning for deep face recognition. In: Proceedings of the IEEE/CVF Conference on Computer Vision and Pattern Recognition, pp. 6817–6826 (2020)

50. Li, S., Xu, J., Xu, X., Shen, P., Li, S., Hooi, B.: Spherical confidence learning for face recognition. In: Proceedings of the IEEE/CVF Conference on Computer Vision and Pattern Recognition (2021)

51. Kim, Y., Park, W., Roh, M.C., Shin, J.: Groupface: learning latent groups and constructing group-based representations for face recognition. In: Proceedings of the IEEE/CVF Conference on Computer Vision and Pattern Recognition (2020)

Learning Video-Independent Eye Contact Segmentation from In-the-Wild Videos

Tianyi Wu$^{(\boxtimes)}$ ⬥ and Yusuke Sugano ⬥

Institute of Industrial Science, The University of Tokyo, Tokyo, Japan
{twu223,sugano}@iis.u-tokyo.ac.jp

Abstract. Human eye contact is a form of non-verbal communication and can have a great influence on social behavior. Since the location and size of the eye contact targets vary across different videos, learning a generic video-independent eye contact detector is still a challenging task. In this work, we address the task of one-way eye contact detection for videos in the wild. Our goal is to build a unified model that can identify when a person is looking at his gaze targets in an arbitrary input video. Considering that this requires time-series relative eye movement information, we propose to formulate the task as a temporal segmentation. Due to the scarcity of labeled training data, we further propose a gaze target discovery method to generate pseudo-labels for unlabeled videos, which allows us to train a generic eye contact segmentation model in an unsupervised way using in-the-wild videos. To evaluate our proposed approach, we manually annotated a test dataset consisting of 52 videos of human conversations. Experimental results show that our eye contact segmentation model outperforms the previous video-dependent eye contact detector and can achieve 71.88% framewise accuracy on our annotated test set. Our code and evaluation dataset are available at https://github.com/ut-vision/Video-Independent-ECS.

Keywords: Human gaze · Eye contact · Video segmentation

1 Introduction

Human gaze and eye contact have strong social meaning and are considered key to understanding human dyadic interactions. Studies have shown that eye contact functions as a signaling mechanism [6,19], indicates interest and attention [2,22], and is related to certain psychiatric conditions [3,35,37]. The importance of human gazes in general has also been well recognized in the computer vision community, leading to a series of related research work on vision-based gaze estimation techniques [7,8,38,39,47,56,57,59–62]. Recent advances in vision-based gaze estimation have the potential to enable robust analyses of gaze behavior, including one-way eye contact. However, gaze estimation is still challenging in images with extreme head poses and lighting conditions, and it is not a trivial task to robustly detect eye contacts in in-the-wild situations.

L. Wang et al. (Eds.): ACCV 2022, LNCS 13844, pp. 52–70, 2023.
https://doi.org/10.1007/978-3-031-26316-3_4

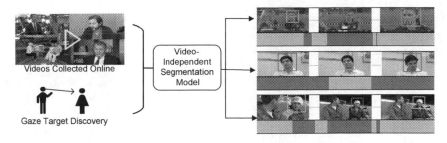

Fig. 1. Illustration of our proposed task of video-independent eye contact segmentation. Given a video sequence and a target person, the goal is to segment the video into fragments of the target person having and not having eye contact with his potential gaze targets.

Several previous studies have attempted to directly address the task of detecting one-way eye contact [9,36,46,54,58]. Given its binary classification nature, one-way eye contact detection can be a simpler task than regressing gaze directions. However, unconstrained eye contact detection remains a challenge. Fundamentally speaking, one-way eye contact detection is an ill-posed task if the gaze targets are not identified beforehand. Fully supervised approaches [9,46,54] necessarily result in environment-dependent models that cannot be applied to eye contact targets with different positions and sizes. Although there have been some work that address this task using unsupervised approaches that automatically detect the position of gaze targets relative to the camera [36,58], they still require a sufficient amount of unlabeled training data from the target environment. Learning a model that can detect one-way eye contact from arbitrary inputs independently of the environment is still a challenging task.

This work aims to address the task of unconstrained video-independent one-way eye contact detection. We aim to train a unified model that can be applied to arbitrary videos in the wild to obtain one-way eye contact moments of the target person without knowing his gaze targets beforehand. Since the position and size of the eye contact targets vary from video to video, it is nearly impossible to approach this task frame by frame. However, we humans can recognize when eye contact occurs from temporal eye movements, even when the target object is not visible in the scene. Inspired by this observation, we instead formulate the problem as a segmentation task utilizing the target person's temporal face appearance information from the input video (Fig. 1). The remaining challenge here is that this approach requires a large amount of eye contact training data. It is undoubtedly difficult to manually annotate training videos covering a wide variety of environmental and lighting conditions.

To train the eye contact segmentation model, we propose an unsupervised gaze target discovery method to generate eye contact pseudo-labels from noisy appearance-based gaze estimation results. Since online videos often contain camera movements and artificial edits, it is not a trivial task to locate eye contact targets relative to the camera. Instead of making a strong assumption about a stationary camera, we assume only that the relative positions of the eye contact target and the person are fixed. Our method analyzes human gazes in the body coordinate system and treats high-density gaze point regions as positive samples. By applying our gaze target discovery method to the VoxCeleb2 dataset [12], we obtain a large-scale pseudo-labeled training dataset. Based on the initial pseudo-labels, our segmentation model is trained iteratively using the original facial features as input. We also manually annotated 52 videos with eye contact segmentation labels for evaluation, and experiments show that our approach can achieve 71.88% framewise accuracy on our test set and outperforms video-dependent baselines.

Our contributions are threefold. First, to the best of our knowledge, we are the first to formulate one-way eye contact detection as a segmentation task. This formulation allows us to naturally leverage the target person's face and gaze features temporally, leading to a video-independent eye contact detector that can be applied to arbitrary videos. Second, we propose a novel gaze target discovery method robust to videos in the wild. This leads to high-quality eye contact pseudo-labels that can be further used for both video-dependent eye contact detection and video-independent eye contact segmentation. Finally, we create and release a manually annotated evaluation dataset for eye contact segmentation based on the VoxCeleb2 dataset.

2   Related Work

2.1   Gaze Estimation and Analysis

Appearance-based Gaze Estimation. Appearance-based gaze estimation directly regresses the input image into the gaze direction and only requires an off-the-shelf camera. Although most of the work take the eye region as input [7,8,38,47,56, 61], some demonstrated the advantage of using the full face as input [39,57,59,60, 62,63]. If the eye region is hardly visible, possibly due to low resolution, extreme head poses, and poor lighting conditions, the full-face gaze model can be expected to infer the direction of the human gaze from the rest of the face. Since most gaze estimation datasets are collected in controlled laboratory settings [16,17,57], in-the-wild appearance-based gaze estimation remains a challenge. Some recent efforts have been made to address this issue by domain adaptation [29,44] or using synthetic data [39,55,64]. Note that eye contact detection is a different task from gaze estimation and is still difficult even with a perfect gaze estimator due to the unknown gaze target locations. The goal of this work is to improve the accuracy of eye contact detection on top of the state-of-the-art appearance-based gaze estimation method.

Gaze Following and Mutual Gaze Detection. First proposed by Recasens *et al.* [40], gaze following aims to estimate the object where the person gazes in an image [10,11,14,41,49,51,52]. Another line of work is mutual gaze detection, which aims to locate moments when two people are looking at each other. Mutual gaze is an even stronger signal than one-way eye contact in reflecting the relationship between two people [31–33]. The problem of mutual gaze detection was first proposed by Marin-Jimenez *et al.* [30]. Our target differs from these tasks in two ways. First, we are interested in finding the moments in which one-way eye contact occurs to gaze targets, rather than determining the location of gaze targets on a frame-by-frame basis or detecting mutual gazes. Second, since our proposed method performs eye contact detection by segmenting the video based on the person's facial appearance, it can handle the cases where the gaze targets are not visible from the scene. Although some gaze following work [10, 11] can tell when gaze targets are outside the image, most of them are designed with the implicit assumption that the gaze target is included in the image.

2.2   Eye Contact Detection

Several previous works address the task of detecting eye contact specifically with the camera [9, 46, 54]. However, such pre-trained models cannot be applied to videos with the target person attending to gaze targets of different sizes and positions. Recent progress in appearance-based gaze estimation allows unsupervised detection of one-way eye contacts in third-person videos using an off-the-shelf camera [36, 58]. Zhang *et al.* assume a setting in which the camera is placed next to the gaze target and propose an unsupervised gaze target discovery method to locate the gaze target region relative to the camera [58]. They first run the appearance-based gaze estimator on all input sequences of human faces to get 3D gaze directions and then compute gaze points in the camera plane. This is followed by density-based clustering, which identifies high-density gaze point regions as the locations of gaze targets. Based on this idea, Müller *et al.* studies eye contact detection in a group of 3–4 people having conversations [36]. Based on the assumption that all listeners would look at the speaker most of the time, they use audio clues to more accurately locate gaze targets in the camera plane.

There are two major limitations that make these two approaches inapplicable to videos in the wild. First, in many online videos, camera movements and jump cuts are common, making the camera coordinate system inconsistent throughout the video. Meanwhile, since gaze points are essentially the intersection between the gaze ray and the plane $z = 0$ in the camera coordinate system, the gaze points corresponding to gaze targets far from the camera will naturally be more scattered than those corresponding to gaze targets close to the camera when receiving the same amount of eye gazes. Consequently, density-based clustering would fail to identify potential gaze targets far from the camera on the camera plane. Second, both works only explored video-dependent eye contact detection, *i.e.*, training one model for each test video. Instead, we study the feasibility of training a video-independent eye contact segmentation model that can be applied to different videos in the wild.

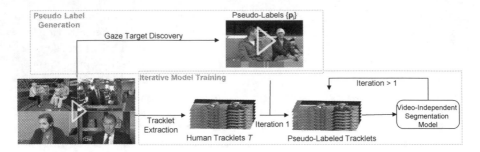

Fig. 2. An overview of our proposed unsupervised training pipeline for video-independent eye contact segmentation.

2.3 Action Segmentation

Action segmentation is the task of detecting and segmenting actions in a given video. Various research works have focused on designing the network architecture for the task. Singh *et al.* [45] propose to feed spatial-temporal video representations learned by a two-stream network to a bi-directional LSTM to capture dependencies between temporal fragments. Lea *et al.* [24] propose a network that performs temporal convolutions with the encoder-decoder architecture (ED-TCN). Recently, many works have tried to modify ED-TCN by introducing deformable convolutions [25], dilated residual layers [15], and dual dilated layers [28]. In this work, we adopt MS-TCN++ [28] as our segmentation model.

Since labeling action classes and defining their temporal boundaries to create annotations for action segmentation can be difficult and costly, some work explored unsupervised action segmentation [23,27,43,48,50]. Based on the observation that similar actions tend to appear in a similar temporal position in a video, most of these works rely on learning framewise representations through the self-supervised task of time stamp prediction [23,27,48,50]. However, it is difficult to apply these methods directly to our scenario because eye contact is a sporadic activity that can occur randomly over time. We instead leverage human gaze information and deduce the gaze target position from gaze point statistics.

3 Proposed Method

Our proposed eye contact segmentation network takes input a *tracklet*, *i.e.*, a sequence of video frames in which the target person is tracked and outputs framewise eye contact predictions. Formally, given a tracklet with I frames $T = \{I_i\}_{i=1}^{I}$, our objective is to train a model to produce framewise binary predictions of one-way eye contacts $Y = \{y_i\}_{i=1}^{I}$ of the person, where $y_i \in [0,1]^2$ is a two-dimensional one-hot vector. We define gaze targets as physical targets with which the person interacts, such as the camera and another person in the conversation. These gaze targets do not have to be visible in the video.

Figure 2 shows an overview of the proposed unsupervised approach to train the segmentation network. Our method consists of two stages, *i.e.*, *pseudo-label*

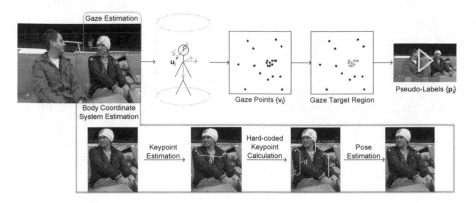

Fig. 3. An overview of the pseudo-label generation stage using our proposed method of gaze target discovery.

generation and *iterative model training.* We start by collecting a large set of M unlabeled conversation videos $\mathcal{V} = \{V_m\}_{m=1}^{M}$ from online. For each video V_m with J_m frames, we first generate framewise pseudo-labels $\{p_j\}_{j=1}^{J_m}$, where $p_j \in [0,1]^2$, using our proposed method of gaze target discovery. We also track the target person to extract a set of tracklets $\{T_k\}_{k=1}^{K_m}$ from each video V_m. The pseudo-labels $\{p_j\}$ are also split and assigned to each corresponding tracklet as a tracklet-wise set of pseudo-labels P_k. The collection of all tracklets $\mathcal{T} = \{T_n\}_{n=1}^{N}$ obtained from $\mathcal{V}$, where $N = \sum_m K_m$, and their corresponding collection of pseudo-labels $\mathcal{P} = \{P_n\}_{n=1}^{N}$ are then used to train our eye contact segmentation model. Since our proposed gaze target discovery does not leverage temporal information, we propose an iterative training strategy that iteratively updates the pseudo-labels using the trained segmentation model that has learned rich temporal information. In the following sections, we describe details of our pseudo-label generation and iterative model training processes.

3.1   Pseudo Label Generation

We generate framewise pseudo-labels $\{p_j\}$ for each training video V_m using our proposed gaze target discovery, which automatically locates the position of the gaze targets in the body coordinate system. An overview is illustrated in Fig. 3. In a nutshell, our proposed gaze target discovery obtains the target person's 3D gaze direction in the body coordinate system and identifies the high-density gaze regions. Since eye contact targets tend to form dense gaze clusters, these gaze regions are treated as potential gaze target locations. In the following sections, we give details of our proposed gaze target discovery and body pose estimation.

Gaze Target Discovery. For each frame I_j of the video V_m, we run an appearance-based gaze estimator to obtain the gaze vector g_j^c of the person of our interest in the camera coordinate system. Through the data normalization process for gaze estimation [59], we also obtain the center of the face o_j^c as the origin of the gaze vector. We also perform body pose estimation to obtain the translation vector t_j and the rotation matrix R_j from the body coordinate system to the camera coordinate system. g_j^c and o_j^c are then transformed from the camera coordinate system to the body coordinate system through $g_j^b = R_j^{-1} g_j^c$ and $o_j^b = R_j^{-1}(o_j^c - t_j)$, where o_j^b indicates the face center in the body coordinate system. Therefore, $o_j^b + \beta g_j^b$ defines the gaze ray in the body coordinate system.

For each video V_m, we then compute a set of intersection points $\{u_j\}$ between the gaze rays and a cylinder centered at the origin of the body coordinate system with a radius of r and convert these 3D intersection points to 2D gaze points $\{v_j\}$ on the cylinder plane by cutting the cylinder along the line $(0, y, r)$ parameterized with y. We apply OPTICS clustering [1] on the set of 2D gaze points $\{v_j\}$ and treat the identified clusters as eye contact regions for the m-th video V_m. We generate pseudo-labels $\{p_j\}$ for V_m by treating all identified eye contact regions as positive samples and others as negative samples.

Body Pose Estimation. We estimate the 3D body pose R and t of the target person based on the 3D model fitting. We define our six-point 3D generic body model according to the average human body statistics [34], consisting of nose, neck, left and right shoulder, and left and right waist keypoints. This 3D body model is in a right-handed coordinate system, which means that the chest-facing direction is the negative Z-axis direction. Given the corresponding 2D keypoint locations from the target image, we can fit the 3D model using the P6P algorithm [26] assuming (virtual) camera parameters. This gives us the translation vector t and the rotation matrix R from the body coordinate system to the camera coordinate system.

To locate the six 2D body keypoints, we rely on a pre-trained 2d keypoint-based pose estimator. For each frame, the pose estimator is expected to take the whole frame as input, and output body keypoints including the ones corresponding to our six-point body model. However, directly using the six 2D keypoints from the pose estimator can lead to inconsistent results throughout frames. This inconsistency arises from the fact that there normally exist at least two near-optimal solutions with similar reprojection errors due to the symmetric nature of the human body. Therefore, we introduce some subtle asymmetry in the 3D body model and stabilize the pose by introducing hard-coded keypoints, as illustrated in the lower part of Fig. 3. Specifically, we only use three keypoints that correspond to the left shoulder s^l, the right shoulder s^r, and the neck n from the pose estimator. We calculate the other three keypoints, the left waist w^l, the right waist w^r, and the nose e, assuming that the target person is standing straight. The keypoints of the waist are defined as $w^l = s^l + (0, d)^T$ and $w^r = s^r + (0, d)^T$, where $d = |s^r - s^l|_2$ indicates the length of the shoulder, and the nose is defined as $e = n - (0, \alpha d)^T$. We also set the ratio $\alpha = 1.632$ according to the same statistics of the human body [34].

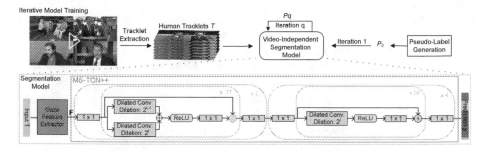

Fig. 4. An overview of the iterative model training stage with an illustration for our network structure.

3.2 Iterative Model Training

By applying gaze target discovery to all videos in $\mathcal{V}$ and extracting tracklets, we obtain an initial training dataset consisting of $\mathcal{T}$ and $\mathcal{P}$. However, our proposed gaze target discovery process generates noisy pseudo-labels. Consequently, the initial labels $\mathcal{P}$ obtained from appearance-based gaze estimation results can be oversegmented, violating the nature of eye contact segmentation. To address this problem, we propose to iteratively train new segmentation models supervised by the pseudo-labels $\mathcal{P}_q$ generated from the model trained in the previous iteration. Our segmentation model is trained to take the low-level gaze CNN feature as input and is expected to auto-correct the initial noisy pseudo-labels by attending to temporal information through the iterative process.

Figure 4 shows an overview of the iterative model training stage. Our segmentation network is based on the MS-TCN++ architecture [28] that takes as input a tracklet T_n and outputs its frame-wise eye contact predictions $\boldsymbol{y}_n$. At iteration 1, the model is supervised with $\mathcal{P}_1 = \mathcal{P}$ generated from the pseudo-label generator. In every subsequent iteration $q > 1$, the model will be supervised with better pseudo-labels $\mathcal{P}_1$ than the models trained before and could learn richer temporal information. We repeat this process to max Q iterations.

Network Architecture. The structure of the segmentation network is illustrated in the lower part of Fig. 4. For each frame in the input tracklet T of length I, we extract and normalize the face image of the target person according to [59]. It is then fed to a pre-trained gaze estimation network based on the ResNet architecture [18] with 50 layers followed by a fully connected regression head. We extract the gaze feature vectors $\boldsymbol{f}_i \in \mathbb{R}^{2048}$ from the last layer and concatenate all gaze features $\{\boldsymbol{f}_i\}$ collected from the tracklet along the temporal dimension to form the gaze feature matrix $\boldsymbol{F} \in \mathbb{R}^{2048 \times I}$, which will be used as input to the segmentation block.

Based on the MS-TCN++ architecture [28], the segmentation block consists of a prediction stage and several refinement stages stacked upon the prediction stage. The prediction stage has 11 dual dilated layers. At the l-th dual dilated layer, the network performs two dilated convolutions with dilation rates l and $11 - l$. The features after the two dilated convolutions are first concatenated, so that the network is able to attend to long-range temporal information even at the early stage. This is then followed by ReLU activation, pointwise convolution and skip connections. The refinement stages are similar to the prediction stage, except that the 11 dual dilated layers are replaced with 10 dilated residual layers.

For the loss function, we follow the original paper and use a combination of cross-entropy loss $\mathcal{L}_{\text{cls}}$ and truncated mean squared loss $\mathcal{L}_{\text{t-mse}}$ [28]. $\mathcal{L}_{\text{t-mse}}$ is defined based on the difference in the prediction between adjacent frames and encourages smoother model prediction. The loss for a single stage is defined as $\mathcal{L}_s = \frac{1}{|B|} \sum_{b \in B} (\mathcal{L}_{\text{cls}}(\boldsymbol{p}_b, \boldsymbol{y}_b) + \lambda \mathcal{L}_{\text{t-mse}}(\boldsymbol{y}_b))$, where B indicates the training batch, and λ is a hyperparameter controlling the extent of $\mathcal{L}_{\text{t-mse}}$. Finally, the overall loss for all stages is the sum of $\mathcal{L}_s$ at each stage.

Tracklet Extraction. To extract tracklets, we run face detection and face recognition on each video in $\mathcal{V}$ in the training dataset. A tracklet T_k is formed only if the IoU between the bounding boxes of the faces of consecutive frames is greater than τ_{IoU}. The tracklet is also disconnected if the neck and shoulder keypoints cannot be detected. We also discard short tracklets that do not exceed 4 s.

Since pseudo-labels extracted from the gaze target discovery are person-specific, we also need to make sure that the training tracklets are extracted from the specific target person. To this end, we add the cosine similarity threshold τ_c to construct training tracklets. Assuming that a set of reference face images is given, we compute the cosine similarity of the detected face with each of the reference faces. As long as one of them is greater than τ_c, the tracklet continues. Note that this threshold is not required during inference.

3.3   Implementation Details

The gaze estimation network is pre-trained on the ETH-XGaze dataset [57], and we follow their work to perform face detection, face normalization, and gaze estimation. We use OpenPose [5] as our 2d keypoint pose estimator. For face recognition, we use ArcFace [13], and set the cosine similarity threshold $\tau_c = 0.4$ and the IoU threshold $\tau_{\text{IoU}} = 0.4$.

During the pseudo-label generation stage, we set the radius of the cylinder $r = 1000$ mm. We also noticed that the hyperparameter of OPTICS can significantly affect the pseudo-label quality. In particular, we found that smaller max epsilon values should be given to longer videos, as the clustering space of long videos is much denser than that of short videos. To address this issue, we first set the max epsilon to 8 and perform OPTICS clustering. If no clusters are found, we continue to increment the max epsilon until at least one cluster is found.

During training, we split the pseudo-labeled dataset into training and validation splits with a ratio of 8 : 2. To train MS-TCN++, we follow the suggested hyperparameters for the network architecture. We use Adam [21] optimizer and

Fig. 5. Some example video frames randomly selected from our test dataset. We show the target celebrities in green and red bounding boxes, with green and red indicating positive and negative ground-truth labels. If visible, their gaze targets are also shown in orange bounding boxes. (Color figure online)

set the learning rate to 0.0005. We did not use dropout layers, and for the loss function, we set $\lambda = 0.15$. We empirically set the maximum number of iterations $Q = 4$ and trained 50 epochs for each iteration.

4 Experiments

4.1 Dataset

We build our dataset upon the VoxCeleb2 dataset [12], which consists of celebrities being interviewed or speaking to the public. The trimmed VoxCeleb2 dataset includes short video clips only containing the head crops of celebrities and has been used in various tasks including talking head generation [20] and speaker identification [12]. Since our method also requires body keypoints detection, we opt to process the raw videos used in the VoxCeleb2 dataset.

During training, we use randomly selected videos from celebrities id00501 to id09272. Due to the online availability and the computational cost of tracklet extraction, we used 5% of the entire dataset for training. After downloading the raw videos, we converted them to 25 fps. If the video is too short, there will not be enough gaze points to reliably identify high-density regions. On the other hand, if the video is too long, the possibility of the video containing multiple conversation sessions becomes high. Therefore, we only used videos of a duration between 2 and 12 min for training. In total, we pre-processed 4926 raw videos. During tracklet extraction, we obtain the sets of reference images from the VGGFace2 dataset [4]. For each celebrity, we only used the first 30 face images. This results in 49826 tracklets, which is equivalent to roughly 177.7 h.

To evaluate our method, we manually annotate 52 videos (summing up to 3.6 h) from celebrity id00016 to id00500 using ELAN [53]. Each video is selected from a different celebrity to ensure identity diversity. We treat the host, camera, and other interviewees who interacted with the target celebrity as eye contact targets, meaning that there can be multiple eye contact targets in each video. Note that these gaze targets do not necessarily need to be visible in the scene.

Table 1. Comparison between our video-independent eye contact segmentation model and video-dependent baseline approaches.

Method	Accuracy	Edit	F1@0.1	F1@0.25	F1@0.5
CameraPlane + SVM [58]	57.72%	46.83	38.37	30.08	20.48
CylinderPlane + SVM	68.52%	46.84	43.3	38.16	29.49
Ours	**71.88%**	**57.27**	**61.59**	**54.67**	**41.03**

In addition, some test videos have poor lighting conditions, extreme head poses, and low resolution, making them difficult to segment. Figure 5 shows some randomly sampled frames from the test videos with the celebrity and gaze targets that we identified highlighted in the boxes. In total, 48.14% of the frames are labeled positive. We further examine the quality of the labels by visualizing gaze positions on the cylinder plane. If the gaze target looks too scattered on the cylinder plane, we re-annotate the video. After forming tracklets from the test videos, we get 510 tracklets (1.9 h in total) with 48.12% of the frames labeled positive. These test tracklets are used as our test set to evaluate our proposed eye contact segmentation model.

4.2 Evaluation

We compare our method with video-dependent baselines and show the effectiveness of our design choices through ablation studies. Following previous work on action segmentation, we evaluate our proposed method using framewise accuracy, segmental edit score, and F1 scores at the overlap threshold 10%, 25%, and 50%, denoted by $F1@\{10, 25, 50\}$. Framewise accuracy measures the performance in the most straightforward way, but it favors longer tracklets and does not punish oversegmentation. The segmented F1 score reflects the degree of oversegmentation, and the F1 scores measure the quality of the prediction.

Performance Comparison. Table 1 shows the comparison of our proposed unsupervised video-independent eye contact segmentation model with unsupervised video-dependent eye contact detection baselines. The first row (*Camera-Plane + SVM*) is the re-implementation of the unsupervised method of Zhang *et al.* [58] that extracts framewise pseudo-labels in the camera plane and trains an SVM-based eye contact classifier based on a single frame input. We did not set a safe margin around the positive cluster to filter out unconfident negative gaze points because we observed that it does not work well in in-the-wild videos, especially when there exist multiple gaze targets. The second row (*CylinderPlane + SVM*) applies our proposed cylinder plane gaze target discovery method, and an SVM is trained on the resulting pseudo-labels. Note that these two methods are video-dependent approaches, *i.e.*, the models are trained specifically on the target video. Our proposed cylinder plane achieves 68.52% accuracy in video-dependent eye contact detection, outperforming the camera plane baseline by 10.8%, indicating the advantage of our pseudo-label generator in in-the-wild videos. The last

Table 2. Ablation results on our design choices.

Method	Accuracy	Edit	F1@0.1	F1@0.25	F1@0.5
CameraPlane + Generic SVM	61.28%	44.34	39.32	32.44	22.46
CylinderPlane + Generic SVM	58.83%	39.86	34.26	27.19	18.08
CameraPlane + Ours	65.55%	46.27	44.88	38.56	27.85
Ours (1 iteration)	70.04%	42.21	43.97	38.55	27.42
Ours (2 iteration)	71.15%	50.60	55.09	48.80	36.54
Ours (3 iteration)	71.41%	52.24	57.26	50.72	37.45
Ours (4 iteration)	**71.88%**	**57.27**	**61.59**	**54.67**	**41.03**

row (*Ours*) corresponds to the proposed unsupervised eye contact segmentation approach with an iterative training strategy. Our method achieves 71.88% framewise accuracy, outperforming the video-dependent counterpart (*CylinderPlane + SVM*) by 3.36% and the camera-plane baseline by 14.16%. It also achieves the highest segmented edit scores and F1 scores, indicating better segmentation qualities.

Ablation Studies. We also conduct ablation studies to show the effectiveness of our design choices, *i.e*, our problem formulation, our proposed gaze target discovery method and iterative training. The result is shown in Table 2. *CameraPlane + Generic SVM* is the baseline method that obtains pseudo-labels for tracklets using the gaze target discovery method of Zhang *et al.* [58] and trains an SVM-based generic video-independent eye contact detector. We choose to use SVM as the classifier simply for comparison with video-dependent baseline approaches, and SVM is trained through online learning optimized by SGD. *CylinderPlane + Generic SVM* replaces the gaze target discovery method of Zhang *et al.* [58] with our proposed gaze target discovery method and, therefore, is also a video-independent eye contact detection approach. Although the baseline camera-plane approach outperforms our proposed cylinder-plane approach by 2.45%, both SVM-based detection models achieve framewise accuracy only slightly better than chance. *CameraPlane + Ours* obtains pseudo-labels on the camera plane but replaces the SVM with our segmentation model, making it a video-independent eye contact segmentation method. It outperforms its detection counterpart by 4.27%, showing the superiority of our problem formulation.

Our proposed method without iterative training (*Ours (1 iteration)*) achieves 70.04% accuracy, showing the effectiveness of our proposed gaze target discovery when applied in the segmentation task setting. The last three rows of Table 2 shows the effectiveness of the proposed iterative training strategy. From iteration 1 to iteration 2, we observed a decent improvement in model performance in terms of both accuracy and segmentation quality. There are also gradual model improvements even in subsequent iterations, indicating its effectiveness. Our proposed method in the last iteration outperforms that of the first iteration by 1.84% in accuracy, and there is also a great improvement in the edit score and the F1 scores.

Table 3. Evaluation of the model performance with different input length.

Length [s]	Accuracy	Edit	F1@0.1	F1@0.25	F1@0.5
> 4	71.88%	57.27	61.59	54.67	41.03
> 10	74.27%	62.31	63.93	57.45	43.76
> 30	77.31%	63.76	64.86	58.60	45.96
> 60	85.14%	68.96	76.65	73.57	65.01

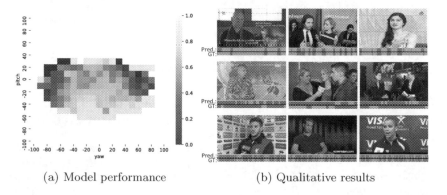

(a) Model performance   (b) Qualitative results

Fig. 6. Accuracy of our segmentation model on different head pose ranges. We (a) visualize segmentation accuracy for each yaw-pitch interval bin, and (b) show examples of the model performance on test tracklets with a thumbnail image followed by model predictions and groundtruth annotations.

Detailed Analysis and Discussions. During both training and testing time, we use tracklets for more than 4 s as input to the model. In Table 3, we vary the length of the input tracklet during the test time and analyze its effect on the performance of the model. As can be seen, the model performance improves as we increase the input tracklet length. In particular, if the input length is more than one minute, our proposed method reaches an accuracy of 85.14%.

The higher performance observed in longer tracklets can also be attributed to the performance of the model in different head poses. During the tracklet formation stage, we use face recognition to filter celebrity faces. Since face recognition performance is limited on profile faces, tracklets containing extreme head poses tend to be much shorter than those containing only frontal faces. During training, the lack of long tracklets with profile faces prevents the model from modeling long-term eye contact dependencies, leading to degraded performance on tracklets that contain mainly profile faces. The longer tracklets in the test set also contain mainly frontal faces, resulting in higher accuracy.

In Fig. 6a, we visualize the accuracy of our trained model conditioned on different head poses. We use HopeNet [42] to extract head poses of celebrity faces in each frame of the test tracklets and compute framewise accuracy for all yaw-pitch intervals. We can observe that our model works the best when the pitch is between -40 and $-60°$, *i.e.*, looking downward. It can also achieve decent performance when both yaw and pitch are around $0°$. However, when the yaw is lower than $-60°$ or higher than $60°$, our model is even worse than random chance.

In Fig. 6b, we show some qualitative visualizations. We randomly present test tracklets with frontal faces in the first row and test tracklets with profile faces in the second. Although our model has decent performance on frontal-face tracklets, the predictions on profile-face tracklets seem almost random. We also present test tracklets longer than one minute in the third row, and all of these long tracks contain frontal faces. Consequently, we can observe mainly fine-grained eye contact predictions in these examples.

Another limitation of our proposed approach lies in our full-face appearance-based estimator. Fundamentally, the line of work on full-face appearance-based gaze estimation regresses the face images into gaze directions in the normalized camera coordinate system. Consequently, the gaze features used as input to the segmentation model only have semantic meaning in the normalized space, but not in the real camera space. In addition, the full-face appearance-based estimator trained on the gaze dataset collected in controlled settings tends to have limited performance in unconstrained images, especially when the person has extreme head poses unseen in the training dataset. This may also be a reason for the low performance of the model on the face of the profile.

Finally, our approach cannot handle the cases of moving gaze targets and humans. Our gaze target discovery assumed fixed relative positions between the eye contact target and the person and would consequently give incorrect pseudo-labels in such cases. Eye contact detection with moving gaze targets is a more challenging task than that with stable gaze targets. We argue that in this case gaze information will not be sufficient for the model. Information about the spatial relationship between the person and gaze targets should be introduced.

5   Conclusion

In this paper, we proposed and challenged the task of video-independent one-way eye contact segmentation for videos in the wild. We proposed a novel method of gaze target discovery to obtain frame-wise eye contact labels in unconstrained videos, which allows us to train the segmentation model in an unsupervised way. By manually annotating a test dataset consisting of 52 videos for evaluation, we showed that our proposed method can lead to a video-independent eye contact detector that can outperform previous video-dependent approaches and is especially robust for non-profile face tracklets.

Acknowledgement. This work was supported by JST CREST Grant Number JPMJCR1781.

References

1. Ankerst, M., Breunig, M.M., Kriegel, H.P., Sander, J.: Optics: ordering points to identify the clustering structure. SIGMOD Rec. **28**(2), 49–60 (1999)
2. Argyle, M., Dean, J.E.: Eye-contact, distance and affiliation. Sociometry **28**, 289–304 (1965)
3. Broz, F., Lehmann, H., Nehaniv, C.L., Dautenhahn, K.: Mutual gaze, personality, and familiarity: dual eye-tracking during conversation. In: IEEE International Symposium on Robot and Human Interactive Communication, pp. 858–864 (2012)
4. Cao, Q., Shen, L., Xie, W., Parkhi, O.M., Zisserman, A.: VGGFace2: a dataset for recognising faces across pose and age. In: IEEE International Conference on Automatic Face & Gesture Recognition, pp. 67 74 (2018). https://doi.org/10.1109/FG.2018.00020
5. Cao, Z., Simon, T., Wei, S.E., Sheikh, Y.: Realtime multi-person 2D pose estimation using part affinity fields. In: IEEE Conference on Computer Vision and Pattern Recognition (CVPR), pp. 1302–1310 (2017). https://doi.org/10.1109/CVPR.2017.143
6. Cañigueral, R., de C. Hamilton, A.F.: The role of eye gaze during natural social interactions in typical and autistic people. Front. Psychol. **10**, 560 (2019). https://doi.org/10.3389/fpsyg.2019.00560
7. Cheng, Y., Lu, F., Zhang, X.: Appearance-based gaze estimation via evaluation-guided asymmetric regression. In: Ferrari, V., Hebert, M., Sminchisescu, C., Weiss, Y. (eds.) Computer Vision – ECCV 2018. LNCS, vol. 11218, pp. 105–121. Springer, Cham (2018). https://doi.org/10.1007/978-3-030-01264-9_7
8. Cheng, Y., Zhang, X., Lu, F., Sato, Y.: Gaze estimation by exploring two-eye asymmetry. IEEE Trans. Image Process. **29**, 5259–5272 (2020)
9. Chong, E., et al.: Detecting gaze towards eyes in natural social interactions and its use in child assessment. Proc. ACM Interact. Mob. Wearable Ubiquitous Technol. **1**(3), 1–20 (2017)
10. Chong, E., Ruiz, N., Wang, Y., Zhang, Y., Rozga, A., Rehg, J.M.: Connecting gaze, scene, and attention: generalized attention estimation via joint modeling of gaze and scene saliency. In: Ferrari, V., Hebert, M., Sminchisescu, C., Weiss, Y. (eds.) ECCV 2018. LNCS, vol. 11209, pp. 397–412. Springer, Cham (2018). https://doi.org/10.1007/978-3-030-01228-1_24
11. Chong, E., Wang, Y., Ruiz, N., Rehg, J.M.: Detecting attended visual targets in video. In: Proceedings of the IEEE/CVF Conference on Computer Vision and Pattern Recognition (CVPR), pp. 5396–5406 (2020)
12. Chung, J.S., Nagrani, A., Zisserman, A.: VoxCeleb2: deep speaker recognition. In: Proceedings of Interspeech, pp. 1086–1090 (2018)
13. Deng, J., Guo, J., Xue, N., Zafeiriou, S.: ArcFace: additive angular margin loss for deep face recognition. In: IEEE/CVF Conference on Computer Vision and Pattern Recognition (CVPR), pp. 4685–4694 (2019)
14. Fang, Y., et al.: Dual attention guided gaze target detection in the wild. In: Proceedings of the IEEE/CVF Conference on Computer Vision and Pattern Recognition (CVPR), pp. 11390–11399 (2021)
15. Farha, Y.A., Gall, J.: MS-TCN: multi-stage temporal convolutional network for action segmentation. In: Proceedings of the IEEE Conference on Computer Vision and Pattern Recognition, pp. 3575–3584 (2019)

16. Fischer, T., Chang, H.J., Demiris, Y.: RT-GENE: real-time eye gaze estimation in natural environments. In: Ferrari, V., Hebert, M., Sminchisescu, C., Weiss, Y. (eds.) ECCV 2018. LNCS, vol. 11214, pp. 339–357. Springer, Cham (2018). https://doi.org/10.1007/978-3-030-01249-6_21

17. Funes Mora, K.A., Monay, F., Odobez, J.M.: EYEDIAP: a database for the development and evaluation of gaze estimation algorithms from RGB and RGB-D cameras. In: Proceedings of the Symposium on Eye Tracking Research and Applications, pp. 255–258 (2014)

18. He, K., Zhang, X., Ren, S., Sun, J.: Deep residual learning for image recognition. In: Proceedings of the IEEE/CVF Conference on Computer Vision and Pattern Recognition (CVPR), pp. 770–778 (2016). https://doi.org/10.1109/CVPR.2016.90

19. Ho, S., Foulsham, T., Kingstone, A.: Speaking and listening with the eyes: gaze signaling during dyadic interactions. PloS One **10**(8), e0136905 (2015)

20. Joon Son Son, A.J., Zisserman, A.: You said that? In: Proceedings of the British Machine Vision Conference (BMVC), pp. 109.1–109.12 (2017)

21. Kingma, D.P., Ba, J.: Adam: a method for stochastic optimization. In: International Conference on Learning Representations (ICLR) (2015)

22. Kleinke, C.L.: Gaze and eye contact: a research review. Psychol. Bull. **100**(1), 78–100 (1986)

23. Kukleva, A., Kuehne, H., Sener, F., Gall, J.: Unsupervised learning of action classes with continuous temporal embedding. In: Proceedings of the IEEE Conference on Computer Vision and Pattern Recognition (CVPR), pp. 12066–12074 (2019)

24. Lea, C., Flynn, M.D., Vidal, R., Reiter, A., Hager, G.D.: Temporal convolutional networks for action segmentation and detection. In: Proceedings of the IEEE Conference on Computer Vision and Pattern Recognition (CVPR), pp. 156–165 (2017)

25. Lei, P., Todorovic, S.: Temporal deformable residual networks for action segmentation in videos. In: Proceedings of the IEEE Conference on Computer Vision and Pattern Recognition (CVPR), pp. 6742–6751 (2018)

26. Lepetit, V., Moreno-Noguer, F., Fua, P.: EPNP: an accurate O(n) solution to the PnP problem. Int. J. Comput. Vision (IJCV) **81**(2), 155–166 (2009)

27. Li, J., Todorovic, S.: Action shuffle alternating learning for unsupervised action segmentation. In: Proceedings of the IEEE/CVF Conference on Computer Vision and Pattern Recognition (CVPR), pp. 12628–12636, June 2021

28. Li, S.J., AbuFarha, Y., Liu, Y., Cheng, M.M., Gall, J.: MS-TCN++: multi-stage temporal convolutional network for action segmentation. IEEE Trans. Pattern Anal. Mach. Intell. 1 (2020)

29. Liu, Y., Liu, R., Wang, H., Lu, F.: Generalizing gaze estimation with outlier-guided collaborative adaptation. In: International Conference on Computer Vision (ICCV), pp. 3835–3844 (2021)

30. Manuel Marin-Jimenez, A.Z., Ferrari, V.: "Here's looking at you, kid". Detecting people looking at each other in videos. In: Proceedings of the British Machine Vision Conference (BMVC), pp. 22.1–22.12 (2011)

31. Marin-Jimenez, M.J., Zisserman, A., Eichner, M., Ferrari, V.: Detecting people looking at each other in videos. Int. J. Comput. Vision (IJCV) **106**(3), 282–296 (2014)

32. Marin-Jimenez, M.J., Kalogeiton, V., Medina-Suarez, P., Zisserman, A.: LAEO-Net: revisiting people looking at each other in videos. In: Proceedings of the IEEE/CVF Conference on Computer Vision and Pattern Recognition (CVPR), pp. 3477–3485, June 2019

33. Marin-Jimenez, M.J., Kalogeiton, V., Medina-Suarez, P., Zisserman, A.: LAEO-Net++: revisiting people looking at each other in videos. IEEE Trans. Pattern Anal. Mach. Intell. **44**(6), 3069–3081 (2022)

34. Marshall, R., Summerskill, S.: Chapter 25 - posture and anthropometry. In: DHM and Posturography, pp. 333–350. Academic Press (2019)

35. Miller, S.R., Miller, C.J., Bloom, J.S., Hynd, G.W., Craggs, J.G.: Right hemisphere brain morphology, attention-deficit hyperactivity disorder (ADHD) subtype, and social comprehension. J. Child Neurol. **21**(2), 139–144 (2006). https://doi.org/10.1177/08830738060210021901

36. Müller, P., Huang, M.X., Zhang, X., Bulling, A.: Robust eye contact detection in natural multi-person interactions using gaze and speaking behaviour. In: Proceedings of the ACM Symposium on Eye Tracking Research & Applications, pp. 1–10 (2018)

37. Mundy, P.C., Sigman, M.D., Ungerer, J.A., Sherman, T.: Defining the social deficits of autism: the contribution of non-verbal communication measures. J. Child Psychol. Psychiatry **27**(5), 657–69 (1986)

38. Park, S., Mello, S.D., Molchanov, P., Iqbal, U., Hilliges, O., Kautz, J.: Few-shot adaptive gaze estimation. In: International Conference on Computer Vision (ICCV), pp. 9368–9377 (2019)

39. Qin, J., Shimoyama, T., Sugano, Y.: Learning-by-novel-view-synthesis for full-face appearance-based 3D gaze estimation. In: Proceedings of the IEEE/CVF Conference on Computer Vision and Pattern Recognition (CVPR) Workshops, pp. 4981–4991 (2022)

40. Recasens, A., Khosla, A., Vondrick, C., Torralba, A.: Where are they looking? In: International Conference on Neural Information Processing Systems, pp. 199–207 (2015)

41. Recasens, A., Vondrick, C., Khosla, A., Torralba, A.: Following gaze in video. In: IEEE International Conference on Computer Vision (ICCV), pp. 1444–1452 (2017)

42. Ruiz, N., Chong, E., Rehg, J.M.: Fine-grained head pose estimation without keypoints. In: Proceedings of the IEEE/CVF Conference on Computer Vision and Pattern Recognition (CVPR) Workshops, pp. 2155–215509 (2018)

43. Sener, F., Yao, A.: Unsupervised learning and segmentation of complex activities from video. In: Proceedings of the IEEE Conference on Computer Vision and Pattern Recognition (CVPR), pp. 8368–8376 (2018)

44. Shrivastava, A., Pfister, T., Tuzel, O., Susskind, J., Wang, W., Webb, R.: Learning from simulated and unsupervised images through adversarial training. In: Proceedings of the IEEE Conference on Computer Vision and Pattern Recognition (CVPR), pp. 2107–2116 (2017)

45. Singh, B., Marks, T.K., Jones, M., Tuzel, O., Shao, M.: A multi-stream bi-directional recurrent neural network for fine-grained action detection. In: Proceedings of the IEEE Conference on Computer Vision and Pattern Recognition (CVPR), pp. 1961–1970 (2016)

46. Smith, B.A., Yin, Q., Feiner, S.K., Nayar, S.K.: Gaze locking: passive eye contact detection for human-object interaction. In: Proceedings of the Annual ACM Symposium on User Interface Software and Technology, pp. 271–280 (2013)

47. Sugano, Y., Matsushita, Y., Sato, Y.: Learning-by-synthesis for appearance-based 3D gaze estimation. In: Proceedings of the IEEE Conference on Computer Vision and Pattern Recognition (CVPR), pp. 1821–1828 (2014)

48. Swetha, S., Kuehne, H., Rawat, Y.S., Shah, M.: Unsupervised discriminative embedding for sub-action learning in complex activities. In: 2021 IEEE International Conference on Image Processing (ICIP), pp. 2588–2592 (2021)
49. Tu, D., Min, X., Duan, H., Guo, G., Zhai, G., Shen, W.: End-to-end human-gaze-target detection with transformers. In: Proceedings of the IEEE/CVF Conference on Computer Vision and Pattern Recognition (CVPR), pp. 2202–2210 (2022)
50. VidalMata, R.G., Scheirer, W.J., Kukleva, A., Cox, D., Kuehne, H.: Joint visual-temporal embedding for unsupervised learning of actions in untrimmed sequences. In: IEEE Winter Conference on Applications of Computer Vision (WACV), pp. 1237–1246 (2021)
51. Wang, B., Hu, T., Li, B., Chen, X., Zhang, Z.: GaTector: a unified framework for gaze object prediction. In: Proceedings of the IEEE/CVF Conference on Computer Vision and Pattern Recognition (CVPR), pp. 19588–19597 (2022)
52. Wei, P., Liu, Y., Shu, T., Zheng, N., Zhu, S.C.: Where and why are they looking? Jointly inferring human attention and intentions in complex tasks. In: Proceedings of the IEEE/CVF Conference on Computer Vision and Pattern Recognition (CVPR), pp. 6801–6809 (2018)
53. Wittenburg, P., Brugman, H., Russel, A., Klassmann, A., Sloetjes, H.: ELAN: a professional framework for multimodality research. In: Proceedings of the Fifth International Conference on Language Resources and Evaluation, pp. 1556–1559 (2006)
54. Ye, Z., Li, Y., Liu, Y., Bridges, C., Rozga, A., Rehg, J.M.: Detecting bids for eye contact using a wearable camera. In: IEEE International Conference and Workshops on Automatic Face and Gesture Recognition (FG), vol. 1, pp. 1–8 (2015)
55. Yu, Y., Liu, G., Odobez, J.M.: Improving few-shot user-specific gaze adaptation via gaze redirection synthesis. In: Proceedings of the IEEE Conference on Computer Vision and Pattern Recognition (CVPR), pp. 11937–11946 (2019)
56. Yu, Y., Odobez, J.M.: Unsupervised representation learning for gaze estimation. In: IEEE/CVF Conference on Computer Vision and Pattern Recognition (CVPR), pp. 7314–7324 (2020)
57. Zhang, X., Park, S., Beeler, T., Bradley, D., Tang, S., Hilliges, O.: ETH-XGaze: a large scale dataset for gaze estimation under extreme head pose and gaze variation. In: Vedaldi, A., Bischof, H., Brox, T., Frahm, J. M. (eds.) ECCV 2020. LNCS, vol. 12350, pp. 365–381. Springer, Cham (2020). https://doi.org/10.1007/978-3-030-58558-7_22
58. Zhang, X., Sugano, Y., Bulling, A.: Everyday eye contact detection using unsupervised gaze target discovery. In: Proceedings of the Annual ACM Symposium on User Interface Software and Technology, pp. 193–203 (2017)
59. Zhang, X., Sugano, Y., Bulling, A.: Revisiting data normalization for appearance-based gaze estimation. In: Proceedings of the ACM Symposium on Eye Tracking Research & Applications, pp. 1–9 (2018)
60. Zhang, X., Sugano, Y., Bulling, A.: Evaluation of appearance-based methods and implications for gaze-based applications. In: Proceedings of the CHI Conference on Human Factors in Computing Systems, pp. 1–13 (2019)
61. Zhang, X., Sugano, Y., Fritz, M., Bulling, A.: Appearance-based gaze estimation in the wild. In: Proceedings of the IEEE/CVF Conference on Computer Vision and Pattern Recognition (CVPR), pp. 4511–4520 (2015)
62. Zhang, X., Sugano, Y., Fritz, M., Bulling, A.: It's written all over your face: full-face appearance-based gaze estimation. In: Proceedings of the IEEE/CVF Conference on Computer Vision and Pattern Recognition (CVPR) Workshops, pp. 2299–2308 (2017)

63. Zhang, X., Sugano, Y., Fritz, M., Bulling, A.: Mpiigaze: real-world dataset and deep appearance-based gaze estimation. IEEE Trans. Pattern Anal. Mach. Intell. **41**(1), 162–175 (2019)
64. Zheng, Y., Park, S., Zhang, X., Mello, S.D., Hilliges, O.: Self-learning transformations for improving gaze and head redirection. In: International Conference on Neural Information Processing Systems, pp. 13127–13138 (2020)

Exemplar Free Class Agnostic Counting

Viresh Ranjan[1(✉)] and Minh Hoai Nguyen[1,2]

[1] Stony Brook University, Stony Brook, USA
vireshranjan@gmail.com
[2] VinAI Research, Hanoi, Vietnam

Abstract. We tackle the task of Class Agnostic Counting, which aims to count objects in a novel object category at test time without any access to labeled training data for that category. All previous class agnostic counting methods cannot work in a fully automated setting, and require computationally expensive test time adaptation. To address these challenges, we propose a visual counter which operates in a fully automated setting and does not require any test time adaptation. Our proposed approach first identifies exemplars from repeating objects in an image, and then counts the repeating objects. We propose a novel region proposal network for identifying the exemplars. After identifying the exemplars, we obtain the corresponding count by using a density estimation based Visual Counter. We evaluate our proposed approach on FSC-147 dataset, and show that it achieves superior performance compared to the existing approaches. Our code and models are available at: https://github.com/Viresh-R/ExemplarFreeCounting.git.

1 Introduction

In recent years, visual counters have become more and more accurate at counting objects from specialized categories such as human crowd [11,12,25,49], cars [26], animals [3], and cells [2,13,47]. Most of these visual counters treat counting as a class-specific regression task, where a class-specific mapping is learned to map from an input image to the corresponding object density map, and the count is obtained by summing over the density map. However, this approach does not provide a scalable solution for counting objects from a large number of object categories because these visual counters can count only a single category at a time, and it also requires hundreds of thousands [49] to millions of annotated training objects [38,44] to achieve reasonably accurate performance for each category. A more scalable approach for counting objects from many categories is to use class-agnostic visual counters [24,30], which can count objects from many categories. But the downside of not having a predefined object category is that these counters require a human user to specify what they want to count by providing several exemplars for the object category of interest. As a result, these class-agnostic visual counters cannot be used in any fully automated systems. Furthermore, these visual counters need to be adapted to each new visual category [24] or each test image [30], leading to slower inference.

L. Wang et al. (Eds.): ACCV 2022, LNCS 13844, pp. 71–87, 2023.
https://doi.org/10.1007/978-3-031-26316-3_5

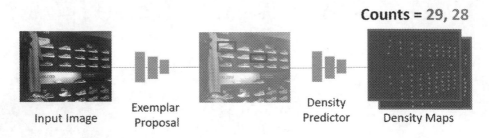

Fig. 1. Exemplar free class agnostic counter. Given an image containing instances of objects from unseen object categories, our proposed approach first generates exemplars from the repeating classes in the image using a novel region proposal network. Subsequently, a density predictor network predicts separate density maps for each of the exemplars. The total count for any exemplar, i.e. the number of times the object within the exemplar occurs in the image, is obtained by summing all the values in the density map corresponding to that exemplar.

In this paper, we present the first exemplar-free class-agnostic visual counter that is capable of counting objects from many categories, even for novel categories that have neither annotated objects at training time nor exemplar objects at testing time. Our visual counter does not require any human user in its counting process, and this will be very crucial for building fully automated systems in various applications in wildlife monitoring, healthcare and visual anomaly detection. For example, this visual counter can be used to alert environmentalists when a herd of animals with significant size pass by an area monitored by a wildlife camera. Another example is to use this visual counter to monitor for critical health conditions when any certain type of cells outgrows the other types. Unlike existing class-agnostic counters [24,30], our approach does not use any test time adaptation or finetuning.

At this point, a reader might wonder if it is possible to identify all possible exemplars in an image automatically by using a class-agnostic object detector such as a Region Proposal Network (RPN) [33], and run an existing class-agnostic visual counter using the detected exemplars to count all objects in all categories. Although this approach does not require a human's input during the counting process, it can be computationally expensive. This is because the RPNs usually produce a thousand or more of object proposals. And this in turn requires executing the class-agnostic visual counter at least a thousand times, a time-consuming and computationally demanding process.

To avoid this expensive procedure, we develop in this paper a novel convolutional network architecture called **Re**petitive **R**egion **P**roposal **N**etwork (RepRPN), which can be used to automatically identify few exemplars from the most frequent classes in the image. RepRPN is used at the first stage of our proposed two-stage visual counting algorithm named RepRPN-Counter. We use a density estimation based Visual Counter as the second stage of the RepRPN-Counter, which predicts a separate high resolution density map for each exemplar. Given an input image, RepRPN considers multiple region proposals, and compute the objectness and repetition scores for each proposal. The repetition score of a proposal is defined as the number of times the object contained within the proposal occurs in the image. The proposals with the highest repetition scores are chosen as the exemplars, and the second stage density predictor estimates the density maps only for the chosen exemplars with high repetition scores. This exemplar generation procedure relies on the underlying assumption that in an image containing different classes with varying counts, the classes of interest are the ones having larger counts. Compared to the traditional RPN [33], RepRPN is better suited for visual counting task, since it can significantly reduce the training and inference time for any two-stage counter. Furthermore, RepRPN can serve as a fast visual counter for applications which can tolerate some margin of error and do not require the localization information conveyed by density maps. Note that the second stage predictor of our visual counter estimates a separate density map for each of the chosen exemplars.

While training RepRPN-Counter, another technical challenge that we need to overcome is the lack of proper annotated data. The only dataset suitable for training class-agnostic visual counters is FSC-147 [30], which contains annotation for a single object category in each image, and may contain unannotated objects from other categories. To obtain annotation for unannotated objects in the FSC-147 dataset, we propose a novel knowledge transfer strategy where we use a RepRPN trained on a large scale object detection dataset [19] and a density prediction network [30] trained on FSC-147 as teacher networks.

In short, the contributions of this paper are threefold: (1) we develop the first exemplar free class agnostic visual counter for novel categories that have neither annotated objects at training time nor exemplar objects at testing time; (2) we develop a novel architecture to simultaneously estimate the objectness and repetition scores of each proposal; (3) we propose a knowledge transfer strategy to handle unannotated objects in the FSC-147 dataset.

2   Related Work

Visual Counting. Most previous methods for visual counting focus on specific categories [1,4,6,17,20,22,25,28,29,31,35,37,39,41–43,45,48,49]. These visual counters can count a single category at a time, and require training data with hundreds of thousands [49] to millions of annotated instances [44] for every visual category, which are expensive to collect. These visual counters cannot generalize to new categories at test time, and hence, cannot handle our class agnostic counting task. To reduce the expensive annotation cost, some of these methods focus on designing unsupervised [22] and semi-supervised tasks [23] for visual counting. However, these methods still require a significant amount of annotations and training time for each new category.

Class Agnostic Counting. Most related to ours is the previous works on class agnostic counting [24,30], which build counters that can be trained to count novel classes using relatively small number of examples from the novel classes. Lu and

Zisserman [24] proposed a Generic Matching Network (GMN) for class-agnostic counting, which follows a two-stage training framework where the first stage is trained on a large-scale video object tracking data, and the second stage consists of adapting GMN to a novel object class. GMN uses labeled data from the novel object class during the second stage, and only works well if several dozens to hundreds of examples are available for the adaptation. Few-shot Adaptation and Matching Network (FamNet) [30] is a recently proposed class agnostic few-shot visual counter which generalizes to a novel category at test time given only a few exemplars from the category. However, FamNet is an interactive visual counter which requires an user to provide the exemplars from the test image. Both GMN and FamNet require test time adaptation for each new class or test image, leading to slower counting procedures.

Zero-Shot Object Detection. Also related to ours is the previous work on zero-shot object detection [5,27,50]. Most of these approaches [5,27] use a region proposal network to generate class-agnostic proposals, and map the features from the proposals to a semantic space where they can be directly compared with semantic word embeddings of novel object classes. However, all of these zero-shot detection approaches require access to the semantic word embeddings for the test classes, and cannot work for our class agnostic counting task where the test classes are not known a priori.

Few-Shot Learning. Also related to ours is the previous works on few-shot learning [8,15,16,32,36], which aim to adapt classifiers to novel categories based on only a few labeled examples. One of the meta learning based few-shot approaches, Model Agnostic Meta Learning (MAML) [8], has been adapted for class-agnostic counting [30]. MAML focuses on learning parameters which can adapt to novel classes at test time by doing only a few gradient descent steps. Although these few-shot methods reduce the labeled data needed to generalize to new domains, most of these approaches cannot be used for our class agnostic counting task due to the unavailability of labeled data from the novel test class.

3   Proposed Approach

We propose an exemplar-free class-agnostic visual counter called RepRPN-Counter. Given an image containing one or more repetitive object categories, RepRPN-Counter predicts a separate density map for each of the repetitive categories. The object count for the repetitive categories can be obtained by simply summing up the corresponding density map. For a category that is counted, RepRPN-Counter also provides the bounding box for an example from the category.

RepRPN-Counter consists of two key components: 1) a Repetitive Region Proposal Network (RepRPN) for identifying exemplars from repetitive objects in an image, along with their approximate count; and 2) a Density Prediction Network (DPN) that predicts a density map corresponding to any exemplar produced by the RepRPN.

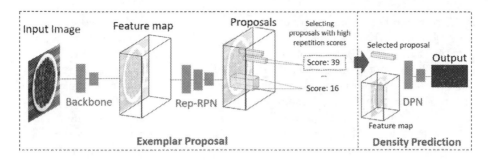

Fig. 2. RepRPN-counter is a two-stage Exemplar Free Class Agnostic Counter. RepRPN-Counter has two key components: 1) Repetitive Region Proposal Network (RepRPN) and 2) Density Prediction Network (DPN). RepRPN predicts repetition score and objectness score for every proposal. Repetition score is used to select few proposals, called exemplars, from the repeating classes in the image. The DPN predicts a separate density map for any proposal selected by the RepRPN. The total count for any proposal, i.e. the number of times the object within the proposal occurs in the image, is obtained by summing all the values in the density map corresponding to that proposal. The DPN ignores proposals which are less likely to contain repetitive objects, so as to reduce the time required for training and evaluation. To keep things simple, we have shown the density prediction step for a single proposal. In reality, several density maps are predicted by the DPN, one for every selected proposal.

Fig. 3. Missing labels in the FSC-147 dataset. Each image in the dataset comes with bounding box annotations for the exemplar objects (shown in blue), and dot annotations for all objects belonging to the same category as the exemplar. For each image, objects of only a single class are annotated. We present a knowledge transfer strategy to deal with incomplete annotation. (Color figure online)

For the rest of this section, we will describe the architecture of RepRPN in Sect. 3.1, the architecture of RepRPN-Counter in Sect. 3.2, the knowledge transfer approach for handling incomplete annotation in Sect. 3.3, and the overall training strategy in Sect. 3.4.

3.1 Repetitive Region Proposal Networks

Repetitive Region Proposal Network (RepRPN) proposes exemplars from the repetitive object classes in an image. RepRPN takes as input convolutional feature representation of an image computed by the Resnet-50 backbone [10], and predicts proposal bounding boxes along with objectness and repetition scores for every proposal at every anchor location. The objectness score is the probability of the proposal belonging to any object class and not the background class. The repetition score refers to the number of times the object within the proposal occurs in the image. For example, consider an image with m cats and n oranges. The RepRPN should predict m as the repetition score for any cat proposal, and n as the repetition score for any orange proposal. The repetition score is used to select exemplars from the repetitive classes in the image, i.e. the proposals with the highest repetition score are chosen as the exemplars. The original RPN formulation [33] uses a fixed window around an anchor location to predict the proposal boxes and objectness scores. However, this fixed sized window does not cover the entire image, and it does not contain sufficient information to predict the repetition score. This has been verified in our experiment where an RPN using only fixed-size window over convolutional features was unable to predict the repetition score. Predicting repetition score would require access to information from the entire image. To obtain this global information efficiently, we make use of the Encoder Self-Attention layers [40]. Given a feature vector at any location in the convolutional feature map, self-attention layers can pool information from *similar* vectors from the entire image, and can be used to estimate repetition score at any anchor location. To apply self-attention, we first transform the convolutional features into a sequence of length n: $S \in R^{n \times d}$. To preserve positional information, we concatenate appropriate $\frac{d}{2}$-dimensional row and column embeddings, resulting in d dimensional positional embeddings which are added with the sequence S. We refer to the resulting embeddings as $X \in R^{n \times d}$.

Given the sequence X, the self-attention layer first transforms X into query (X_Q), key (X_K), and value (X_V) matrices by multiplying X with matrices W_Q, W_K, and W_V:

$$X_Q = XW_Q, \quad X_K = XW_K, \quad X_V = XW_V. \tag{1}$$

The self-attention layer outputs a new sequence U where the ith element in the output sequence is obtained as a weighted average of the value sequence, and the weights are decided based on the similarity between the ith query element and the key sequence. The output sequence U is computed as follows:

$$U = softmax(X_Q X_K^T) X_V. \tag{2}$$

Tensor U will be reshaped into a tensor U' that has the same spatial dimensions as the input convolutional feature map. Tensor U' will be forwarded to the bounding box regression, objectness prediction, and repetition score prediction heads. Each prediction head consists of a single 1×1 convolutional layer. At each anchor location in the image, we consider k anchors boxes. For each anchor box, we predict an objectness score, a repetition score, and bounding box coordinates. The repetition score is used to identify proposals containing the repetitive objects in the image, i.e. proposals with a large repetition score contain repetitive objects.

3.2   RepRPN-Counter

As shown in Fig. 2, RepRPN-Counter consists of a Resnet-50 feature backbone, a RepRPN proposal network, and a Density Prediction Network (DPN). The RepRPN and the DPN share the same feature backbone. The RepRPN provides the DPN with the bounding box locations of the proposals with large repetition scores, also called exemplars, and the DPN predicts a separate density map for each exemplar. DPN is trained and evaluated on only the chosen exemplars, and not all the proposals, so as to reduce the training and inference time. Similar to the previous works on class-agnostic counting [24,30], DPN combines the convolutional features of an exemplar, with the convolutional features of the entire image to predict the density map for the exemplar. The exemplar features are obtained by performing ROI pooling on the convolutional features computed by the backbone, at the locations defined by the exemplar bounding boxes. The exemplar features are correlated with the image features, and the resulting correlation map is propagated through the DPN. The DPN is a fully convolutional network consisting of five convolutional layers and three upsampling layers (more architecture details are provided in the Supplementary submission), and the predicted density maps have the same spatial dimensions as the input image. Note that the DPN predicts several density maps, one for each exemplar. The overall count for an object class pertaining to an exemplar is obtained by simply summing all the values in the density map corresponding to the exemplar. The DPN is not evaluated on the proposals with a low repetition score. For such proposals, the repetition score can be used as the final count.

3.3   Knowledge Transfer for Handling Missing Labels

The only existing dataset consisting of images of densely populated objects from many visual categories that can be used for training class agnostic visual counters is FSC-147 [30]. However, it is not trivial to train RepRPN-Counter on FSC-147 because of the missing labels in the dataset. FSC-147 comes with two types of annotations for each image: a few exemplar bounding boxes to specify the object category to be counted, and dot annotations for all of the objects belonging to the same category as the specified exemplars. However, an image may contain objects from another category that has not been annotated, as shown in Fig. 3. Given the missing labels, forcing RepRPN-Counter to predict zero count for the unannotated objects may degrade the performance of the counter.

We use knowledge transfer from teacher networks to address the incomplete annotation issue. We first train a RepRPN on the MSCOCO object detection dataset [19]. The MSCOCO training set consists of over 82K natural images from 80 visual categories, and the RPNs trained on this large dataset have been shown to generalize to previously unseen classes, thereby proving useful for tasks like zero-shot object detection [27]. We use the RepRPN trained on MSCOCO as a teacher network for generating the target labels for the objectness scores and the repetition scores for those proposals not intersecting with the annotated objects in the FSC-147 dataset. To get the target density maps corresponding to the unannotated proposals, we use the pretrained class-agnostic visual counter FamNet [30], which can predict the density map for a novel object class given only a single exemplar. When needed, an unannotated proposal is fed into FamNet, and the output of FamNet is used as the target density map for training the proposed network RepRPN-Counter.

3.4   Training Objective

RepRPN-Counter is trained in two stages. The first stage consists of training the RepRPN. Once trained, the RepRPN is kept frozen and used to generate exemplars for the density estimation network DPN. The second stage of training consists of training the DPN to predict the density map for every exemplar.

Training Objective for RepRPN. For the ith anchor box, the outputs of the RepRPN are the objectness score y_i, the bounding box coordinates b_i, and the repetition score c_i. Let the corresponding ground truth labels be y_i^*, b_i^*, c_i^*. We follow the same protocol as used in Faster RCNN [33] for obtaining the binary objectness label y_i^*, and the same parameterization for the bounding box coordinates b_i. c_i^* is the number of times the object within the anchor box, if any, occurs in the image. Since predicting c_i requires access to global information about the image, RepRPN makes use of self-attentional features as described in Sect. 3.1. The training loss for the ith anchor box is:

$$\mathcal{L}_{RepRPN} = \lambda\mathcal{L}_{cls}(y_i, y_i^*) + \lambda\mathcal{L}_{reg}(b_i, b_i^*) + \mathcal{L}_{reg}(c_i, c_i^*), \qquad (3)$$

where $\mathcal{L}_{cls}$ is the binary cross entropy loss, and $\mathcal{L}_{reg}$ is the smooth L_1 loss. When training RepRPN on the FSC-147 dataset, the labels y_i^* and c_i^* for the positive anchors are obtained using the ground-truth annotation of FSC-147. Note that in the FSC-147 dataset, only three exemplars per image are annotated with bounding boxes, while the rest of the objects are annotated with a dot around their center. We obtain the bounding boxes for all the dot annotated objects by placing a bounding box of the average exemplar size around each of the dots. For anchors not intersecting with any of the annotated bounding boxes in FSC-147, y_i^* and c_i^* labels are obtained using a teacher RepRPN, which has been pre-trained on the MSCOCO dataset [19].

Training Objective for DPN. Given an exemplar bounding box b_i and the feature map U for an input image I of size $H \times W$, the density prediction network DPN predicts a density map $Z_{b_i} = f(U, b_i)$ of size $H \times W$. The training objective for the DPN is based on the mean square error:

$$\mathcal{L}_{mse}(Z_{b_i}, Z^*) = \frac{1}{HW}\sum_{r=1}^{H}\sum_{c=1}^{W}(Z_{b_i}(r, c) - Z^*(r, c))^2,$$

where Z^* is the target density map corresponding to Z_{b_i}. If the exemplar b_i intersects with any annotated object, Z^* is obtained by convolving a Gaussian kernel with the corresponding dot annotation map. Note that Gaussian blurred dot annotation maps are commonly used for training density estimation based visual counters [12, 21, 24, 28, 49]. For cases where b_i does not intersect with any annotated object, we use the pretrained FamNet [30] as a teacher network for obtaining Z^*. The FamNet teacher can predict a density map, given an exemplar b_i and an input image I.

3.5   Implementation Details

For training, we use Adam optimizer [14] with a learning rate of 10^{-5} and batch size of one. We use the first four convolutional blocks from the ImageNet pre-trained ResNet-50 [10] as the backbone. We keep the backbone frozen during training, since finetuning the backbone would yield poor results. This is because the backbone has feature maps suitable for detecting a large number of classes, and finetuning the backbone leads to specialization towards FSC-147 training classes, resulting in poor performance on the novel test classes.

The weights of the RepRPN and DPN are initialized from a zero mean univariate Gaussian with standard deviation of 10^{-3}. RepRPN uses five self-attention transformer layers, each with eight heads. For training the RepRPN, we use four anchors sizes of $32, 64, 128, 256$ and three aspect ratios of $0.5, 1, 2$. We sample a batch of 96 anchors from each image during training. Training is done for 1000 epochs.

4   Experiments

4.1   Dataset

We perform experiments on the recently proposed FSC147 dataset [30], which was originally proposed for the exemplar based class-agnostic counting task. The FSC147 dataset consists of 6135 images from 147 visual categories, which are split into train, val, and test splits comprising of 89, 29, and 29 classes respectively. There are no common categories between the train, val, and test sets. The mean and maximum counts for images in the dataset are 56 and 3701, respectively. We train our model on the train set, and evaluate it on the test and val sets. Each image comes with annotations for a single object category of interest only, which consists of several exemplar bounding boxes and complete dot annotation for the objects of interest in the image. Since our goal is to build an exemplar free counter, unlike previous methods [24, 30], we do not use human annotated exemplars as an input to our counter.

4.2   Evaluation Metrics

We use the Top-k version of Mean Absolute Error (MAE) and Root Mean Squared Error (RMSE) to compare the performance of the different visual counters. MAE and RMSE are defined as follows. $MAE = \frac{1}{n}\sum_{i=1}^{n}|y_i - \hat{y}_i|$; $RMSE = \sqrt{\frac{1}{n}\sum_{i=1}^{n}(y_i - \hat{y}_i)^2}$, where n is the number of test images, and y_i and $\hat{y}_i$ are the

Table 1. Comparing RepRPN-Counter to class-agnostic counters. FamNet, GMN and MAML are exemplar based class-agnostic counters which have been adapted and trained for the exemplar-free setting, where a RPN is used for generating exemplars. We report the Top-1, Top-3 and Top-5 MAE and RMSE metrics on the val and test sets of FSC-147 dataset. RepRPN-Counter consistently outperforms the competing approaches.

Method	MAE (Val set)			RMSE (Val set)			MAE (Test set)			RMSE (Test Set)		
	Top1	Top3	Top5	Top1	Top3	Top5	Top1	Top3	Top5	Top1	Top3	Top5
GMN	43.25	40.96	39.02	114.52	108.47	106.06	43.35	39.72	37.86	145.34	142.81	141.39
MAML	34.96	33.16	32.44	**98.83**	101.80	101.08	37.38	33.27	31.47	133.89	131.00	129.31
FamNet (pretrained)	47.66	42.85	39.52	125.54	121.59	116.08	50.89	42.70	39.38	150.52	146.08	143.51
FamNet	34.51	33.17	32.15	99.87	99.31	98.75	35.81	33.32	32.27	133.57	132.52	131.46
RepRPN-Counter	**31.69**	**30.40**	**29.24**	100.31	**98.73**	**98.11**	**28.32**	**27.45**	**26.66**	**128.76**	**129.69**	**129.11**

ground truth and predicted counts. MAE and RMSE are the most commonly used metrics for counting task [24,25,30,49]. However, RepRPN-Counter predicts several density maps and corresponding counts, one for each selected proposal. Given k predicted counts from k proposals, we compute Top-k MAE and RMSE by first selecting those proposals from the top k proposals which have an IoU ratio of at least 0.3 with any ground truth boxes, and average the counts corresponding to the selected proposals to get the predicted count $\hat{y}_i$. In case none of the k proposals intersect with any ground truth boxes, we simply average all of the k counts to get $\hat{y}_i$.

4.3 Comparison with Class-Agnostic Visual Counters

We compare our proposed RepRPN-Counter with the previous class-agnostic counting methods [8,24,30] on the task of counting objects from novel classes. We do not compare with class-specific counters [25,49] because such counters cannot handle novel classes at test time. Furthermore, these counters require hundreds [49] or thousands [38,44] of images per category during training, while FSC-147 dataset contains an average of only 41 images per category.

GMN [24], FamNet [30], and MAML [8,30] are exemplar based counters which can predict density map for any unseen object category based on few exemplars of the object category from the same image. These counters were originally proposed to work with human provided exemplars as an input to the counter. In order to make these exemplar based counters work in our exemplar free setup, we modify GMN, FamNet, and MAML based visual counters by replacing human provided exemplars with RPN [33] generated exemplars. We use the RPN of Faster RCNN [33] to generate the proposals for the competing approaches, and use the top k proposals with the highest objectness score as the exemplars. For fair comparison, both the RPNs used with the competing approaches as well as the RepRPN are pre-trained on the MSCOCO dataset [19]. We do not use MSCOCO to train the DPN. We train the competing approaches and our proposed approach on the train set of FSC-147, and report the results on the val and

Table 2. Comparing RepRPN-Counter with pre-trained object detectors, on Val-COCO and Test-COCO subsets of FSC-147, which only contain COCO classes. Pre-trained object detectors are available for these COCO classes. For RepRPN-Counter, we use the density map corresponding to the proposal with the highest repetition score. Without access to any labeled data from these COCO classes, our proposed approach outperforms all of the object detectors which are trained using the entire COCO train set containing a large number of images from these COCO classes

Method	Val-COCO Set		Test-COCO Set	
	MAE	RMSE	MAE	RMSE
Faster R-CNN	52.79	172.46	36.20	79.59
RetinaNet	63.57	174.36	52.67	85.86
Mask R-CNN	52.51	172.21	35.56	80.00
Detr	58.35	175.97	45.51	96.57
RepRPN-Counter (Ours)	**50.72**	**160.95**	**25.29**	**56.98**

test sets of FSC-147. We also compare our method with a pre-trained version of FamNet originally trained on the few-shot counting task. Following [30], all of the methods are trained with three proposals, and evaluated with 1, 3, and 5 proposals. We report the Top-1, Top-3 and Top-5 MAE and RMSE values in Table 1.

As can be seen from Table 1, our method RepRPN-Counter outperforms all of the competing methods. The pre-trained FamNet performs the worst, even though it was trained on the same FSC-147 training set. This shows that simply combining pre-trained exemplar based class agnostic counters with RPN-based exemplars does not provide a reasonable solution for the exemplar-free setting. When retrained specifically for the exemplar-free setting, the performance of FamNet significantly improves when compared to its pre-trained version. GMN performs worse than the other baselines, possibly due to the need for more examples for the adaptation process. This observation was earlier reported for exemplar based class-agnostic counting task as well [30].

4.4   Comparison with Object Detectors

One approach to counting is to use a detector and count the number of detections in an image. However, it requires thousands of examples to train an object detector, and the detector-based counters cannot be used for novel object classes. That being said, we compare RepRPN-Counter with object detectors on a subset of COCO categories from the validation and test sets of FSC-147. These subsets are called Val-COCO and Test-COCO, containing 277 and 282 images respectively. We compare our approach with the official implementations [46] of MaskRCNN [9], FasterRCNN [34], RetinaNet [18], and Detr [7]. The results are shown in Table 2. Without any access to labeled data from these COCO classes, our proposed method still outperforms the object detectors that have

Table 3. Comparing RepRPN with RPN, on the test set of FSC-147. Using RepRPN instead of RPN leads to significant boost in performance

| | RPN | | RepRPN | |
Method	MAE	RMSE	MAE	RMSE
GMN	43.35	145.34	32.17	137.29
MAML	37.38	133.89	32.09	141.03
FamNet (pre-trained)	50.89	150.52	38.64	144.27
FamNet	35.81	133.57	32.94	132.82

been trained using the entire COCO train set containing thousands of images from these COCO classes. Detr [7] performs worse than some of the earlier object detectors because Detr uses a fixed number of query slots (usually 100), which limits the maximum number of objects it can detect, while FSC-147 has images containing thousands of objects.

4.5   Comparing RepRPN with RPN

We are also interested in checking if RepRPN can boost the performance of class-agnostic visual counters other than RepRPN-Counter. For this experiment, we replace RPN [33] with RepRPN for GMN, FamNet, and MAML, and report the Top-1 MAE and RMSE scores on the FSC-147 test set in Table 3. Using RepRPN instead of RPN leads to significant boost in the performance for all class-agnostic visual counters. This suggests that RepRPN is much better suited for the exemplar proposal for exemplar free counting task in comparison to RPN. Also, RepRPN works well with different types of class agnostic counters, including the proposed RepRPN-Counter, GMN, FamNet, and MAML.

4.6   Ablation Studies

Our proposed RepRPN-Counter consists of two primary components: the RepRPN for exemplar proposal and the DPN for density prediction. Furthermore, our proposed knowledge transfer approach allows us to deal with unannotated objects in the FSC-147 dataset. In Table 4, we analyze the contribution of these components on the overall performance. The RepRPN baseline uses the repetition score as the final count. We propose to use RepRPN with DPN, but one can replace RepRPN by RPN [33] to get the method RPN+DPN. One can assume there are no unannotated objects in the FSC-147 dataset, and train our proposed RepRPN-Counter on FSC-147 without any knowledge transfer. As can be seen from Table 4, all the components of RepRPN-Counter are useful, and the best results are obtained when all the components are present. RPN+DPN performs much worse than RepRPN+DPN, which shows that RepRPN is better suited for our counting task than RPN.

Table 4. Analyzing individual components of RepRPN-Counter on the overall performance on the test set of FSC-147. RPN + DPN refers to the case where we replace RepRPN from our proposed approach with the RPN from Faster RCNN. As can be seen, RepRPN is a critical component of our proposed approach, and replacing it with RPN decreases the performance significantly. RepRPN+DPN-NoKT refers to the method when we do not use any knowledge transfer, which leads to a drop in performance. This shows the usefulness of the proposed knowledge transfer strategy.

Method	MAE			RMSE		
	Top1	Top3	Top5	Top1	Top3	Top5
RepRPN+DPN (proposed)	28.32	27.45	26.66	128.76	129.69	129.11
RepRPN+DPN-NoKT (no knowledge transfer)	29.52	28.80	28.42	132.76	131.03	130.82
RPN+DPN	35.81	33.32	32.27	133.57	132.52	131.46
RepRPN (without DPN)	29.60	29.18	28.95	136.25	136.21	136.26

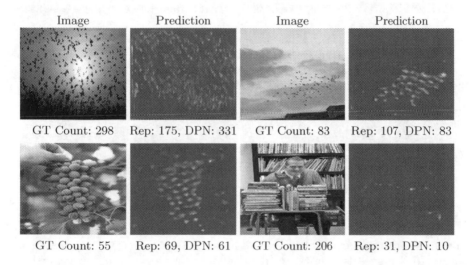

Fig. 4. Input images and the density maps predicted by RepRPN-Counter. Also shown in red are the selected proposals. Rep is the repetition score predicted by RepRPN, while DPN is the count obtained by summing the final density map.

4.7 Qualitative Results

In Fig. 4, we present a few input images, the proposal with the highest repetition score generated by RepRPN for each image, and the corresponding density map generated by the density prediction network. RepRPN-Counter performs well on the first three test cases. But it fails on the last one, because the aspect ratio of the chosen proposal is very different from the majority of the objects of interest.

In Fig. 5, we show the RepRPN proposal with the highest repetition score for several images from the Val and Test set of FSC-147. First three examples in each row contain test cases where the repetition score is close to the groundtruth count. The last example in each row shows test case which proved to be harder

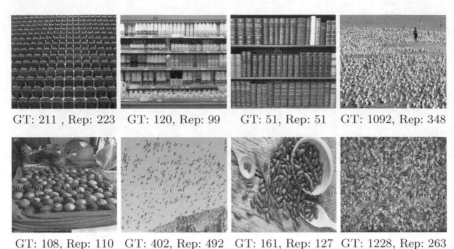

GT: 211 , Rep: 223   GT: 120, Rep: 99   GT: 51, Rep: 51   GT: 1092, Rep: 348

GT: 108, Rep: 110   GT: 402, Rep: 492   GT: 161, Rep: 127   GT: 1228, Rep: 263

Fig. 5. Selected proposal (shown in red) and corresponding repetition score (Rep) predicted by RepRPN (Color figure online). The first three examples in each row are success cases for RepRPN, and the predicted repetition score is close to the ground truth count. The last example in each row shows a failure case.

C_{red}: 12, C_{Blue}: 134 C_{red}: 38, C_{Blue}: 7 C_{red}: 10, C_{Blue}: 5 C_{red}: 14, C_{Blue}: 9

Fig. 6. Counting multiple classes in an image using RepRPN-Counter. Shown are RepRPN proposals from two of the most frequent classes in an image, and the corresponding counts predicted by RepRPN-Counter. C_{red}/C_{blue} is the predicted count for the proposal shown in red/blue. (Color figure online)

for RepRPN. RepRPN does not perform well in some cases when the objects are extremely small in size. Since RepRPN, and RPN in general, uses a fixed set of anchor sizes and aspect ratios, they may fail at detecting extremely small objects. It is also difficult for RepRPN to handle extreme variation in scale within the image, as evident from the failure cases.

RepRPN-Counter can be used for multi-class Class Agnostic counting task, i.e. counting multiple object classes in an image. In Fig. 6, we show few images from the Val and Test set of FSC-147 having at least two object classes, and the counts predicted by RepRPN-Counter for the two most frequent classes in the image. RepRPN-Counter provides a reasonable count estimate for both classes.

5   Conclusions

In this paper, we tackled the task of Exemplar Free Class Agnostic Counting. We proposed RepRPN-Counter, the first exemplar free class agnostic counter capable of handling previously unseen categories at test time. Our two-stage counter consists of a novel region proposal network for finding the exemplars from repetitive object classes, and a density estimation network to estimate the density map corresponding to each exemplar. We also showed that our region proposal network can significantly improve the performance of the previous state-of-the-art class-agnostic visual counters.

References

1. Abousamra, S., Hoai, M., Samaras, D., Chen, C.: Localization in the crowd with topological constraints. In: AAAI (2021)
2. Arteta, C., Lempitsky, V., Noble, J.A., Zisserman, A.: Detecting overlapping instances in microscopy images using extremal region trees. Med. Image Anal. **27**, 3–16 (2016)
3. Arteta, C., Lempitsky, V., Zisserman, A.: Counting in the wild. In: Leibe, B., Matas, J., Sebe, N., Welling, M. (eds.) ECCV 2016. LNCS, vol. 9911, pp. 483–498. Springer, Cham (2016). https://doi.org/10.1007/978-3-319-46478-7_30
4. Babu Sam, D., Sajjan, N.N., Venkatesh Babu, R., Srinivasan, M.: Divide and grow: Capturing huge diversity in crowd images with incrementally growing CNN. In: CVPR (2018)
5. Bansal, A., Sikka, K., Sharma, G., Chellappa, R., Divakaran, A.: Zero-shot object detection. In: Ferrari, V., Hebert, M., Sminchisescu, C., Weiss, Y. (eds.) ECCV 2018. LNCS, vol. 11205, pp. 397–414. Springer, Cham (2018). https://doi.org/10.1007/978-3-030-01246-5_24
6. Cao, X., Wang, Z., Zhao, Y., Su, F.: Scale aggregation network for accurate and efficient crowd counting. In: Ferrari, V., Hebert, M., Sminchisescu, C., Weiss, Y. (eds.) ECCV 2018. LNCS, vol. 11209, pp. 757–773. Springer, Cham (2018). https://doi.org/10.1007/978-3-030-01228-1_45
7. Carion, N., Massa, F., Synnaeve, G., Usunier, N., Kirillov, A., Zagoruyko, S.: End-to-end object detection with transformers. In: Vedaldi, A., Bischof, H., Brox, T., Frahm, J.-M. (eds.) ECCV 2020. LNCS, vol. 12346, pp. 213–229. Springer, Cham (2020). https://doi.org/10.1007/978-3-030-58452-8_13
8. Finn, C., Abbeel, P., Levine, S.: Model-agnostic meta-learning for fast adaptation of deep networks. In: Proceedings of the34thInternational Conference on Machine-Learning, Sydney, Australia, PMLR 70, 2017 (2017)
9. He, K., Gkioxari, G., Dollár, P., Girshick, R.: Mask R-CNN. In: ICCV (2017)
10. He, K., Zhang, X., Ren, S., Sun, J.: Deep residual learning for image recognition. In: CVPR (2016)
11. Idrees, H., Saleemi, I., Seibert, C., Shah, M.: Multi-source multi-scale counting in extremely dense crowd images. In: CVPR (2013)
12. Idrees, H., Tayyab, M., Athrey, K., Zhang, D., Al-Maadeed, S., Rajpoot, N., Shah, M.: Composition loss for counting, density map estimation and localization in dense crowds. In: Ferrari, V., Hebert, M., Sminchisescu, C., Weiss, Y. (eds.) ECCV 2018. LNCS, vol. 11206, pp. 544–559. Springer, Cham (2018). https://doi.org/10.1007/978-3-030-01216-8_33

13. Khan, A., Gould, S., Salzmann, M.: Deep convolutional neural networks for human embryonic cell counting. In: Hua, G., Jégou, H. (eds.) ECCV 2016. LNCS, vol. 9913, pp. 339–348. Springer, Cham (2016). https://doi.org/10.1007/978-3-319-46604-0_25

14. Kingma, D.P., Ba, J.: Adam: A method for stochastic optimization. arXiv preprint arXiv:1412.6980 (2014)

15. Koch, G., Zemel, R., Salakhutdinov, R.: Siamese neural networks for one-shot image recognition. In: ICML Deep Learning Workshop (2015)

16. Lake, B.M., Salakhutdinov, R., Tenenbaum, J.B.: Human-level concept learning through probabilistic program induction. Science **350**(6266), 1332–1338 (2015)

17. Li, Y., Zhang, X., Chen, D.: CsrNet: dilated convolutional neural networks for understanding the highly congested scenes. In: CVPR (2018)

18. Lin, T.Y., Goyal, P., Girshick, R., He, K., Dollár, P.: Focal loss for dense object detection. In: ICCV (2017)

19. Lin, T., et al.: Microsoft COCO: common objects in context. In: Fleet, D., Pajdla, T., Schiele, B., Tuytelaars, T. (eds.) ECCV 2014. LNCS, vol. 8693, pp. 740–755. Springer, Cham (2014). https://doi.org/10.1007/978-3-319-10602-1_48

20. Liu, W., Salzmann, M., Fua, P.: Context-aware crowd counting. In: CVPR (2019)

21. Liu, X., van de Weijer, J., Bagdanov, A.D.: Leveraging unlabeled data for crowd counting by learning to rank. In: CVPR (2018)

22. Liu, X., Van De Weijer, J., Bagdanov, A.D.: Leveraging unlabeled data for crowd counting by learning to rank. In: CVPR (2018)

23. Liu, Y., Liu, L., Wang, P., Zhang, P., Lei, Y.: Semi-supervised crowd counting via self-training on surrogate tasks. In: Vedaldi, A., Bischof, H., Brox, T., Frahm, J.-M. (eds.) ECCV 2020. LNCS, vol. 12360, pp. 242–259. Springer, Cham (2020). https://doi.org/10.1007/978-3-030-58555-6_15

24. Lu, E., Xie, W., Zisserman, A.: Class-agnostic counting. In: ACCV (2018)

25. Ma, Z., Wei, X., Hong, X., Gong, Y.: Bayesian loss for crowd count estimation with point supervision. In: ICCV (2019)

26. Mundhenk, T.N., Konjevod, G., Sakla, W.A., Boakye, K.: A Large contextual dataset for classification, detection and counting of cars with deep learning. In: Leibe, B., Matas, J., Sebe, N., Welling, M. (eds.) ECCV 2016. LNCS, vol. 9907, pp. 785–800. Springer, Cham (2016). https://doi.org/10.1007/978-3-319-46487-9_48

27. Rahman, S., Khan, S., Porikli, F.: Zero-Shot object detection: learning to simultaneously recognize and localize novel concepts. In: Jawahar, C.V., Li, H., Mori, G., Schindler, K. (eds.) ACCV 2018. LNCS, vol. 11361, pp. 547–563. Springer, Cham (2019). https://doi.org/10.1007/978-3-030-20887-5_34

28. Ranjan, V., Le, H., Hoai, M.: Iterative crowd counting. In: Ferrari, V., Hebert, M., Sminchisescu, C., Weiss, Y. (eds.) ECCV 2018. LNCS, vol. 11211, pp. 278–293. Springer, Cham (2018). https://doi.org/10.1007/978-3-030-01234-2_17

29. Ranjan, V., Shah, M., Nguyen, M.H.: Crowd transformer network. arXiv preprint arXiv:1904.02774 (2019)

30. Ranjan, V., Sharma, U., Nguyen, T., Hoai, M.: Learning to count everything. In: Proceedings of the IEEE/CVF Conference on Computer Vision and Pattern Recognition, pp. 3394–3403 (2021)

31. Ranjan, V., Wang, B., Shah, M., Hoai, M.: Uncertainty estimation and sample selection for crowd counting. In: ACCV (2020)

32. Ravi, S., Larochelle, H.: Optimization as a model for few-shot learning (2016)

33. Ren, S., He, K., Girshick, R., Sun, J.: Faster r-CNN: towards real-time object detection with region proposal networks. In: NeurIPS (2015)

34. Ren, S., He, K., Girshick, R., Sun, J.: Faster R-CNN: Towards real-time object detection with region proposal networks. In: NeurIPS (2015)

35. Sam, D.B., Surya, S., Babu, R.V.: Switching convolutional neural network for crowd counting. In: CVPR (2017)

36. Santoro, A., Bartunov, S., Botvinick, M., Wierstra, D., Lillicrap, T.: One-shot learning with memory-augmented neural networks (2016)

37. Shi, M., Yang, Z., Xu, C., Chen, Q.: Revisiting perspective information for efficient crowd counting. In: CVPR (2019)

38. Sindagi, V.A., Yasarla, R., Patel, V.M.: JHU-crowd++: Large-scale crowd counting dataset and a benchmark method. arXiv preprint arXiv:2004.03597 (2020)

39. Song, Q., et al.: Rethinking counting and localization in crowds: a purely point-based framework. In: Proceedings of the IEEE/CVF International Conference on Computer Vision, pp. 3365–3374 (2021)

40. Vaswani, A., et al.: Attention is all you need. In: Advances in neural information processing systems, pp. 5998–6008 (2017)

41. Wan, J., Chan, A.: Adaptive density map generation for crowd counting. In: Proceedings of the IEEE International Conference on Computer Vision, pp. 1130–1139 (2019)

42. Wan, J., Liu, Z., Chan, A.B.: A generalized loss function for crowd counting and localization. In: Proceedings of the IEEE/CVF Conference on Computer Vision and Pattern Recognition, pp. 1974–1983 (2021)

43. Wang, C., et al.: Uniformity in heterogeneity: Diving deep into count interval partition for crowd counting. In: Proceedings of the IEEE/CVF International Conference on Computer Vision, pp. 3234–3242 (2021)

44. Wang, Q., Gao, J., Lin, W., Li, X.: Nwpu-crowd: A large-scale benchmark for crowd counting. arXiv preprint arXiv:2001.03360 (2020)

45. Wang, Q., Gao, J., Lin, W., Yuan, Y.: Learning from synthetic data for crowd counting in the wild. In: CVPR (2019)

46. Wu, Y., Kirillov, A., Massa, F., Lo, W.Y., Girshick, R.: Detectron2 (2019)

47. Xie, W., Noble, J.A., Zisserman, A.: Microscopy cell counting and detection with fully convolutional regression networks. Comput. Methods Biomech. Biomed. Eng. Imaging Visual. **6**(3), 283–292 (2018)

48. Zhang, A., Yue, L., Shen, J., Zhu, F., Zhen, X., Cao, X., Shao, L.: Attentional neural fields for crowd counting. In: ICCV (2019)

49. Zhang, Y., Zhou, D., Chen, S., Gao, S., Ma, Y.: Single-image crowd counting via multi-column convolutional neural network. In: CVPR (2016)

50. Zhu, P., Wang, H., Saligrama, V.: Zero shot detection. IEEE Trans. Circuits Syst. Video Technol. **30**(4), 998–1010 (2019)

Emphasizing Closeness and Diversity Simultaneously for Deep Face Representation

Chaoyu Zhao[1], Jianjun Qian[1(✉)], Shumin Zhu[2], Jin Xie[1], and Jian Yang[1]

[1] PCA Lab, Key Lab of Intelligent Perception and Systems for High-Dimensional Information of Ministry of Education, and Jiangsu Key Lab of Image and Video Understanding for Social Security, School of Computer Science and Engineering, Nanjing University of Science and Technology, Nanjing, China
{cyzhao,csjqian}@njust.edu.cn

[2] AiDLab, Laboratory for Artificial Intelligence in Design, School of Fashion and Textiles, The Hong Kong Polytechnic University, Kowloon, Hong Kong

Abstract. Recent years have witnessed remarkable progress in deep face recognition due to the advancement of softmax-based methods. In this work, we first provide the analysis to reveal the working mechanism of softmax-based methods from the geometry view. Margin-based softmax methods enhance the feature discrimination by the extra margin. Mining-based softmax methods pay more attention to hard samples and try to enlarge their diversity during training. Both closeness and diversity are essential for discriminative features learning; however, we observe that most previous works dealing with hard samples fail to balance the relationship between closeness and diversity. Therefore, we propose a novel approach to tackle the above issue. Specifically, we design a two-branch cooperative network: the Elementary Representation Branch (ERB) and the Refined Representation Branch (RRB). ERB employs the margin-based softmax to guide the network to learn elementary features and measure the difficulty of training samples. RRB employs the proposed sampling strategy in conjunction with two loss terms to enhance closeness and diversity simultaneously. Extensive experimental results on popular benchmarks demonstrate the superiority of our proposed method over state-of-the-art methods.

Keywords: Deep face representation · Closeness and diversity · Difficulty measure · Two-branch cooperative network

1 Introduction

Deep face recognition (FR) has witnessed tremendous progress during recent years, mainly attributed to the growing scale of publicly available datasets, the

Supplementary Information The online version contains supplementary material available at https://doi.org/10.1007/978-3-031-26316-3_6.

development of convolutional neural network architectures, and the improvement of loss functions. In 2014, DeepFace [30] closely reached the human-level performance in unconstrained face recognition based on a nine-layer deep neural network. Subsequently, several successful FR systems such as DeepID2 [26], DeepID3 [27], VGGFace [20], and FaceNet [23] demonstrate that well-designed deep architectures can obtain promising performance.

Besides, the major advance comes from the evolution of loss functions for training deep convolutional neural networks. Early FR works often adopt methods based on metric learning, such as Contrastive Loss [3] and Triplet Loss [23]. However, most of them suffer from the combinatorial explosion, especially on large-scale datasets. Current deep FR approaches are typically based on margin-based softmax loss functions, such as L-Softmax [16], SphereFace [15], CosFace [32], AM-Softmax [31], and ArcFace [4]. These methods share a common idea of introducing a margin penalty between different classes to encourage feature discrimination. Subsequently, mining-based methods such as MV-Softmax [34] and CurricularFace [10] demonstrate that margin-based softmax methods fail to make good use of hard samples. They introduce the hard sample mining strategy and enlarge the distance between a misclassified sample and its negative class centers. By contrast, MagFace [18] learns the well-structured within-class feature distribution by loosening the margin constraint for uncertain samples.

As is well analyzed in several works [8,29], softmax aims to optimize $(s_n - s_p)$ to achieve the decision boundary $s_n - s_p = -m$ (m is the margin), where s_p is the intra-class similarity and s_n is the inter-class similarity. Moreover, the optimization of s_p and s_n are highly coupled. When mining-based methods enlarge the optimization strength of s_n for hard samples, their s_p will also get an extra tendency to be maximized (see Sect. 3.2). This naturally poses a problem: mining-based methods can indirectly enhance the within-class closeness while ignoring that hard samples usually contain much uncertainty and thus should lie on the edge of the intra-class distribution as is claimed in [18]. Consequently, the intra-class distribution structure in high dimensional space will be vulnerable, and the model will tend to overfit on noisy samples.

Based on the observation above, this paper first analyzes the working mechanism of softmax-based methods from the view of closeness and diversity in geometry space. Then we introduce the embedding feature constraint to enhance the closeness s_p for easy samples to establish robust class centers quickly, and meanwhile increase the diversity for hard samples to shift the optimization emphasis from their s_p to s_n and thus prevent overfitting.

To summarize, our key contributions are as follows:

- We analyze the working mechanism of softmax-based methods from the geometry view and claim that both closeness and diversity should be simultaneously emphasized for discriminative feature learning.
- We propose a two-branch cooperative network to simultaneously learn closeness and diversity so that the model can improve both discrimination and generalization ability.

– We conduct extensive experiments on several publicly available benchmarks. Experimental results demonstrate the superiority of our proposed method over state-of-the-art methods.

2   Related Work

2.1   Margin-Based Methods

In deep face recognition, softmax is widely applied to supervise the network for promoting features' separability. However, it exceeds softmax's ability when facing challenging tasks where intra-class variations get larger. Several margin-based methods [4,15,16,31,32] are then proposed. L-Softmax [16] first introduces margin penalty into traditional softmax. SphereFace [15] further normalizes weight vectors by $l2$-normalization to learn face representations on a hypersphere. Subsequently, CosFace [32], AM-Softmax [31], and ArcFace [4] introduce an additive margin penalty on cosine/angle space to further improve the discriminative power of learned face representations. AdaCos [41] employs the adaptive scale parameter to promote the training supervision in dealing with various facial samples. MagFace [18] improves the performance of previous margin-based methods by keeping ambiguous samples away from class centers.

2.2   Mining-Based Methods

Hard sample mining strategy is also a critical step to enhance the feature representation ability [1,25]. OHEM [25] automatically indicates and emphasizes hard samples within a mini-batch according to their loss values. Focal Loss [13] reduces the weight for easy samples during training by introducing the re-weighting factor into the standard cross-entropy loss. MV-Softmax [34] emphasizes hard samples to guide the networks for learning discriminative features by introducing an extra margin penalty when a sample is misclassified. CurricularFace [10] employs the Curriculum Learning (CL) strategy to focus on easy samples in the early training stage and concentrate on hard ones later.

2.3   Contrastive Learning

Traditional contrastive loss functions [6,35] are generally found in early FR works. However, most of them suffer from the combinatorial explosion when dealing with large-scale datasets [23,26,28]. Center Loss [36] proposes a joint supervision signal based on softmax to penalize the distances between the samples and their corresponding class centers. Range Loss [40] is proposed to address the long-tail problem by reducing within-class variances and enlarging inter-class differences in each mini-batch. Modern contrastive approaches [2,7,33] show promising results in the field of unsupervised representation learning. SimCLR [2] learns deep representations by minimizing the distance between multiple augmented views of the same image in the latent space. MoCo [7] proposes a dynamic

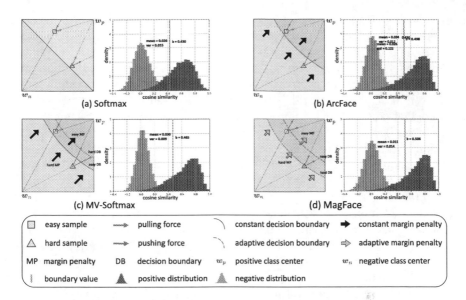

Fig. 1. The comparison between softmax and its variants from the geometry view (left) and the distribution view (right). The explanations of the components are listed under the plot. The distribution view is obtained by training a ResNet18 and evaluated on IJB-B. Closeness is higher when the boundary value ('b' in the figure) is larger. The boundary value asides the top 80% of positive scores on its right to indicate closeness. The diversity is higher when the mean of the red distribution is closer to zero and the variance is smaller. (Color figure online)

dictionary and a moving-averaged encoder to learn visual representations. Wang et al. [33] analyze alignment and uniformity on the feature hypersphere to guide the unsupervised representation learning. Supervised contrastive learning [12] significantly outperforms the traditional contrastive approaches by incorporating label information to construct genuine and imposter pairs.

3 Preliminary

3.1 Understanding Softmax-Based Methods

The softmax cross-entropy loss function (denoted as 'softmax' for short) can be formulated as follows:

$$\mathcal{L}_{CE} = -\log \frac{e^{f_y}}{e^{f_y} + \sum_{k=1,k\neq y}^{C} e^{f_k}} \tag{1}$$

The development of softmax variants in FR is mainly attributed to the advancement of the positive logit f_y and the negative logit f_k.

Let $\boldsymbol{x}$, $\boldsymbol{w}_i$, and b_i denote the feature vector of the input sample, the i-th class center, and the bias term, respectively. Then the traditional logit is calculated as $f_i = \boldsymbol{w}_i^T \boldsymbol{x} + b_i$. It is a common practice to ignore the bias term b_i in deep FR works [4,15,22,32]. The logit can be transformed as $\boldsymbol{w}_i^T \boldsymbol{x} = \|\boldsymbol{w}_i\| \|\boldsymbol{x}\| \cos \theta_i$, where θ_i is the angle distance between the feature $\boldsymbol{x}$ and the class center $\boldsymbol{w}_i$. Several works [4,32] further fix $\|\boldsymbol{w}_i\| = \|\boldsymbol{x}\| = 1$ and scale $\|\boldsymbol{x}\|$ to s. Then the logit can be reformulated as $f_i = s \cos \theta_i$.

As well-discussed in several works [8,29,38], the softmax's constraint can be decoupled into the pulling force from the same class center and the pushing forces from the negative ones. Softmax can thus increase intra-class similarity s_p with the pulling force while reducing the inter-class similarity s_n with the pushing forces. In addition, if the pulling force is enlarged, then the pushing forces will be amplified as well, and vice versa. To clarify this phenomenon, we provide a toy example in Sect. 3.2.

Original softmax simply optimizes $(s_n - s_p)$ to the decision boundary $s_n - s_p = 0$. However, the discriminative ability of facial features learned by the original softmax is limited. Therefore, several works introduce various margin penalties into softmax to improve feature discrimination. Generally speaking, they employ the decision boundary $s_n - s_p = -m$, and m is called margin.

Compared with original softmax, the positive logit is reformulated as $f_y = s \cos(\theta_y + m)$ in ArcFace [4]. As shown on the left side in Fig. 1(b), ArcFace shrinks the decision boundary towards the positive direction with a constant m. In this way, it can learn more discriminative face representations, as shown on the right side in Fig. 1(b). In MV-Softmax, the misclassified sample's negative logit is further reformulated as $f_k = s(t \cos \theta_k + t - 1)$, where $t > 1$. Based on ArcFace, MV-Softmax makes the decision boundary more rigorous by introducing an extra margin penalty on the negative logit for handling hard samples, as shown on the left side in Fig. 1(c). The right of Fig. 1(c) shows that (1) the negative distribution of MV-Softmax is more compact than ArcFace, which illustrates that MV-Softmax improves the diversity with the extra margin; (2) MV-Softmax has an inferior boundary value, indicating its insufficient closeness. A possible explanation is that MV-Softmax intends to improve the diversity for misclassified samples by the extra margin, however, it will indirectly enforce hard samples containing large uncertainty and noise to get closer to their positive class centers (see analysis in Sect. 3.2), leading to the overfitting and the inferior generalization ability.

Besides, another line of works exists, e.g., AdaCos [41], AdaptiveFace [14], and MagFace [18]. They substitute the constant margin penalty with the adaptive one to generate more effective supervision during training. MagFace learns well-structured intra-class features by dynamically adjusting the decision boundaries based on the feature magnitude. As shown on the left side in Fig. 1(d), Mag-Face relaxes the decision boundary for hard samples with large uncertainty and tightens the decision boundary for easy ones with high quality. The right side of Fig. 1(d) shows that the negative distribution in MagFace is not so compact as that in ArcFace. A probable reason is that MagFace prevents hard samples from obtaining excessive s_p during training by reducing the margin. Considering that the optimization of s_n and s_p are highly coupled in softmax, a suitable s_n is not well-learned.

In summary, MV-Softmax and MagFace adopt different strategies to deal with hard samples. MV-Softmax enhances the constraint strength for hard samples. Because MV-Softmax indirectly emphasizes s_p for uncertain samples, it fails to generalize well on challenging tasks during testing. By contrast, Mag-Face relaxes the constraint for hard samples. Although MagFace can prevent uncertain samples from obtaining excessive s_p, it can not ensure a desirable s_n for discriminative features learning. Therefore, both of the above methods fail to emphasize closeness and diversity simultaneously.

3.2 Derivative Analysis

In this subsection, we investigate how the margin penalty affects the pulling force and the pushing forces of softmax. Specifically, we demonstrate that if the pulling force is enlarged, the pushing forces will also get increased, and vice versa. A toy example is additionally provided to explain this phenomenon.

Let us start with the gradient of softmax with respect to the logit f_i, which is calculated as:

$$\frac{\partial \mathcal{L}_{CE}}{\partial f_i} = \underbrace{\mathbb{1}(i = y) \cdot (p_i - 1)}_{\text{pull}} + \underbrace{\mathbb{1}(i \neq y) \cdot p_i}_{\text{push}} \tag{2}$$

where $p_i = \frac{e^{f_i}}{\sum_{k=1}^{C} e^{f_k}}$ is the predicted probability of the i-th class, and satisfies $\sum_{i=1}^{C} p_i = 1$. Equation 2 contains two parts: (1) the first part aims to pull a sample towards its positive class center; (2) the second part aims to push a sample away from its negative class centers. The differences between the two parts lie in that the pulling force can quickly establish the class center; however, it can hurt the generalization ability by pulling a noise sample near its class center. The pushing forces can help to enhance the discrimination ability via encouraging diversity, but they can not be directly employed to establish the class centers.

Additionally, the pulling and pushing forces are equipped with the opposite signs due to the different relative directions. The gradient summation with respect to each class always equals to the constant zero:

$$\sum_{i=1}^{C} \frac{\partial \mathcal{L}_{CE}}{\partial f_i} = \underbrace{p_y - 1}_{\text{pull}} + \underbrace{\sum_{i=1, i \neq y}^{C} p_i}_{\text{push}} = \sum_{i=1}^{C} p_i - 1 = 0 \tag{3}$$

Therefore, if we enlarge either of the two forces by a margin, the other one will inevitably get increased at the same time.

Figure 2 exhibits a simplified example to depict the above issue. We assume the feature vector x and the classifier W are identical among all four cases. Therefore, we only need to care about the relative changes of the logits and

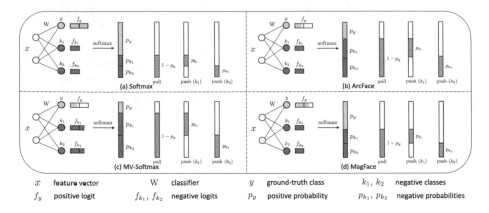

Fig. 2. The comparison between softmax and its variants on a toy example. The blue bars denote the magnitude of gradients generated by predictions. (Color figure online)

the gradients. In Fig. 2(a), we assume the positive logit f_y takes two units and the negative logits f_{k_1} and f_{k_2} take one unit. The classifier W makes the right classification during training with softmax and thus generates limited gradients. Figure 2(b) shows that the positive logit f_y gets smaller in ArcFace with the margin penalty. The pulling force is directly enhanced, and the pushing forces are indirectly enlarged according to Eq. 3.

Based on ArcFace, MV-Softmax introduces a margin penalty on the negative logits f_{k_1} and f_{k_2} for misclassified samples. The margin penalty directly enlarges f_{k_1} and f_{k_2}, leading to the enhanced pushing forces. Although MV-Softmax intrinsically enhance the diversity for hard samples, it also indirectly magnifies the pulling force, as shown in Fig. 2(c). By contrast, MagFace adopts a smaller penalty on f_y than ArcFace, leading to both kinds of forces getting smaller, as shown in Fig. 2(d).

Therefore, the pulling and pushing forces are highly coupled in softmax-based loss functions. Due to this phenomenon, softmax-based methods dealing with hard samples tend to get overfitting or underdiscriminative as discussed in Sect. 3.1.

4   Proposed Method

Based on the analysis in Sect. 3, we propose a two-branch cooperative framework to enhance closeness and diversity simultaneously. The proposed framework contains three parts: (1) the Hard Sample Mining Scheme for dynamically computing difficulty scores of training samples; (2) the Elementary Representation Branch (ERB) for learning initial face representations; (3) the Refined Representation Branch (RRB) for simultaneous closeness and diversity learning.[1]

---

[1] Code is available at: https://github.com/Zacharynjust/FR-closeness-and-diversity.

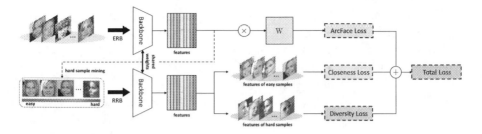

Fig. 3. The overview of the proposed framework. Overall, it contains three parts: (1) the Hard Sample Mining Scheme to maintain difficulty scores; (2) the Elementary Representation Branch (ERB) to learn basic face representations; (3) the Refined Representation Branch (RRB) to learn closeness and diversity.

4.1 Hard Sample Mining

Hard samples play an important role in guiding DCNNs to learn discriminative features. The previous works [10,34] indicate the misclassified samples as hard ones, but they can not measure the hardness quantitatively. MagFace [18] employs feature magnitude to determine the difficulty degree, but it lacks intuitive interpretability. Different from the above works, we employ cosine similarity to characterize the difficulty degree for its simplicity and effectiveness (Fig. 3).

To smooth out short-term fluctuations caused by the sampling sequence, we calculate the moving average of cosine similarity to characterize the difficulty degree for each sample:

$$d^{(t)} = \alpha d^{(t-1)} + (1 - \alpha) \cos \theta^{(t)} \tag{4}$$

where d represents the moving averaged similarity, when d is smaller the difficulty degree is higher. t stands for the t-th iteration. θ is the angle between a feature vector and its positive class center. α is the weight factor.

Based on Eq. 4, we propose two schemes to indicate hard samples: the Hard Mining Scheme (HMS) and the Soft Mining Scheme (SMS). HMS explicitly divides all training samples into easy/hard groups. SMS does not need the explicit division; by contrast, it adopts the difficulty degree to balance the weights of closeness and diversity for each sample in the training stage. In this way, easy samples can help the network establish robust class centers quickly. By contrast, hard samples are beneficial for the network to further improve feature discriminations.

4.2 Loss Design

Both closeness and diversity are indispensable for achieving better results; therefore, they should be simultaneously and properly emphasized for specific samples. Overall, we need to ensure the following conditions: (1) Easy samples should get enough closeness to ensure robust class centers [18]; (2) Hard samples should keep a suitable distance from their positive class centers to ensure generalization ability [18]; (3) Hard samples should also gain enough distance away from their negative class centers to ensure discrimination ability [34].

Equipped with the above three conditions, we design the following two loss functions in RRB to emphasize closeness and diversity simultaneously during training.

Loss for Closeness. To enhance the closeness, we directly minimize the feature differences of within-class samples. The loss term with HMS can be formulated as follows:

$$\mathcal{L}_{closeness}^{H} = \mathop{\mathbb{E}}_{(\boldsymbol{x},\boldsymbol{y}) \sim P_{pos}} \|y(\boldsymbol{x}) - y(\boldsymbol{y})\|^2 \tag{5}$$

where P_{pos} stands for the distribution of positive pairs constructed from the mini-batch. $g(\cdot)$ is the feature encoder with $l2$-normalization in its output layer. The loss term with SMS is actually the re-weighted version of Eq. 5:

$$\mathcal{L}_{closeness}^{S} = \mathop{\mathbb{E}}_{(\boldsymbol{x},\boldsymbol{y}) \sim P_{pos}} \phi\left(d_x, d_y\right) \|g(\boldsymbol{x}) - g(\boldsymbol{y})\|^2 \tag{6}$$

where d_x and d_y are the difficulty degrees of $\boldsymbol{x}$ and $\boldsymbol{y}$. $\phi(\cdot)$ is a monotonically non-decreasing function of d_x and d_y.

Loss for Diversity. In this part, we design diversity loss to enhance the discrimination and generalization ability. Inspired by the uniform loss format used in [33], we design loss for enhancing diversity which enlarges the distance between a sample and its negative class centers. The proposed diversity loss for a single sample can be formulated as follows:

$$\mathcal{L}_{diversity}^{i} = \log \mathop{\mathbb{E}}_{\boldsymbol{w}_j \sim W_{sub}^{(i)}} \left[e^{\max(0, \text{sgn}(\cos\theta_j)) \cdot s(\cos\theta_j)^2} \right] \tag{7}$$

where $W_{sub}^{(i)}$ stands for the subset of the negative class centers for i-th sample. s is the scale parameter. $\text{sgn}(\cdot)$ is the sign function and $\max(0, \text{sgn}(\cos\theta_j))$ is used to truncate the gradient when $\cos\theta_j$ is smaller than 0. $\cos\theta_j = \boldsymbol{w}_j^T \boldsymbol{x}_i$ is the similarity between a sample and its negative class j. For a mini-batch hard samples, we unite all their subsets of negative class centers and then exclude their positive labels to construct the final negative class centers for a mini-batch W, which can be formulated as follows:

$$W = \bigcup_{i=1}^{N} W_{sub}^{(i)} - \overline{W} \tag{8}$$

where N stands for the total number of hard samples within a mini-batch. $\overline{W}$ represents the positive class centers of hard samples in a mini-batch. Then the diversity loss function within a mini-batch using HMS can be further calculated as follows:

$$\mathcal{L}_{diversity}^{H} = \mathop{\mathbb{E}}_{x \sim X} \left[\log \mathop{\mathbb{E}}_{w_j \sim W} \left[e^{\max(0, \text{sgn}(\cos \theta_j)) \cdot s(\cos \theta_j)^2} \right] \right] \tag{9}$$

where X stands for the sample set in a mini-batch. W represents the final negative class centers for a mini-batch. The proposed diversity loss in conjunction with SMS is formulated as follows:

$$\mathcal{L}_{diversity}^{S} = \mathop{\mathbb{E}}_{x \sim X} \left[\psi(d_x) \log \mathop{\mathbb{E}}_{w_j \sim W} \left[e^{\max(0, \text{sgn}(\cos \theta_j)) \cdot s(\cos \theta_j)^2} \right] \right] \tag{10}$$

where d_x is the difficulty degree of x. $\psi(\cdot)$ is a monotonically non-increasing function of d_x.

Because the proposed diversity loss minimizes the cosine similarity between a sample feature and its negative class centers, it can directly push hard samples away from their negative class centers. Margin-based softmax actually seeks to minimize $(s_n - s_p)$ to achieve the decision boundary $s_n - s_p = -m$, when we reduce s_n via diversity loss, we actually shift the optimization emphasis of hard samples on reducing s_n. Therefore, diversity loss can somewhat prevent hard samples from achieving excessive s_p.

4.3   Sampling Strategy

In this subsection, we introduce the proposed sampling strategy for RRB in detail. We construct a mini-batch depending on class labels and difficulty degrees. The sampling strategy can be formulated as follows:

$$X = \mathcal{S}(\boldsymbol{y}, \boldsymbol{d}) \tag{11}$$

where $\boldsymbol{y}$ denotes the ground truth labels. $\boldsymbol{d}$ stands for the difficulty degrees of the whole dataset as introduced in Sect. 4.1.

For HMS, we further select a specific number of easy samples X_e and hard samples X_h according to the divided easy/hard groups within each class. Then, the total loss can be formulated as follows:

$$\mathcal{L}_{total}^{H}(X) = \mathcal{L}_{CE}(X) + \lambda_1 \mathcal{L}_{closeness}^{H}(X_e) + \lambda_2 \mathcal{L}_{diversity}^{H}(X_h) \tag{12}$$

where $\mathcal{L}_{CE}$ is ArcFace in our model. We can also choose other margin-based softmax loss functions. λ_1 and λ_2 are weight parameters for closeness and diversity respectively. For SMS, we randomly select N samples within each class and use their difficulty degrees to calculate the weight functions. Then, the total loss can be formulated as follows:

$$\mathcal{L}_{total}^{S}(X, \boldsymbol{d}) = \mathcal{L}_{CE}(X) + \lambda_1 \mathcal{L}_{closeness}^{S}(X, \boldsymbol{d}) + \lambda_2 \mathcal{L}_{diversity}^{S}(X, \boldsymbol{d}) \tag{13}$$

where $\boldsymbol{d}$ is the difficulty degrees of samples in a mini-batch.

5 Experiment

5.1 Implementation Details

Datasets. We employ MS1MV2 [4] as our training data for a fair comparison with other methods. MS1MV2 is a semi-automatic refined version of the MS1M [5], containing about 5.8M images of 85K different identities. For testing, we extensively evaluate our proposed method and the competed methods on several popular benchmarks, including LFW [9], CFP-FP [24], CPLFW [42], AgeDB [19], CALFW [43], IJB-B [37], IJB-C [17], and MegaFace [11].

Experimental Setting. We follow the setting in ArcFace [4] to align the images with five facial key points [39] and normalize the face images to 112×112. ResNet100 is used as the backbone network in our model. We implement our framework with PyTorch [21]. The models are trained by stochastic gradient descent. The batchsize is set to 512. The weight decay is set to $5e-4$, and the momentum is 0.9.

To obtain the difficulty degree of samples, the backbone is firstly trained with ERB for 4 epochs with the learning rate 0.1. The backbone is then trained with the joint supervisions of ERB and RRB for extra 21 epochs. The learning rate is set to 0.1 initially and divided by 10 when the extra epoch is 6, 12 and 18. For the sampling strategy, we choose 64 unique classes for a mini-batch. In addition, 500 negative class centers of each hard sample are selected to construct the imposter pairs.

For HMS, λ_1 and λ_2 are set to 0.5 and 1.0. We divide the top 20% of training samples into hard groups and collect seven easy samples and one hard sample within each class. For SMS, λ_1 and λ_2 are set to 0.5 and 2.0. In addition, we conduct experiments on both linear and non-linear weight functions. For the linear functions, we set $\phi(d_x, d_y) = \frac{d_x + d_y}{2}$ and $\psi(d_x) = 1 - d_x$. For the non-linear functions, we employ the sigmoid-like function $\sigma(x; \mu, \gamma) = \frac{1}{1+e^{-\gamma(x-\mu)}}$ to conduct non-linear transformation, where we fix $\mu = 0.5$ and $\gamma = 10$. We set $\phi(d_x, d_y) = \sigma(\frac{d_x + d_y}{2})$, and $\psi(d_x) = 1 - \sigma(d_x)$. We set α to 0.9 to calculate the difficulty degree.

5.2 Comparisons with SOTA Methods

Results on Small Benchmarks. In this subsection, we conducted experiments on various benchmarks, including LFW [9] for unconstrained face verification, CFP-FP [24] and CPLFW [42] for cross-pose variations, AgeDB [19] and CALFW [43] for cross-age variations.

The 1:1 verification accuracy among different methods on the above five benchmarks is listed in Table 1. According to Table 1, the proposed models achieve promising results, especially when they are integrated with HMS or SMS (non-linear). Note that our models give slight improvements on LFW since the performance of LFW is nearly saturated. Besides, for age/pose invariant face verification, our proposed method can achieve better results than our competitors.

Table 1. Performance comparisons between the proposed method and state-of-the-art methods on various benchmarks. * denotes our re-implement results on ResNet100. [**Best**, Second Best]

Methods	Verification accuracy					IJB		MegaFace	
	LFW	CFP-FP	CPLFW	AgeDB	CALFW	IJB-B	IJB-C	Id	Ver
Focal Loss* (CVPR16)	99.73	98.19	92.80	98.13	96.01	93.60	95.19	98.09	98.60
SphereFace (CVPR17)	99.42	–	–	97.16	94.55	–	–	–	–
CosFace* (CVPR18)	99.78	98.12	92.28	98.11	95.76	94.10	95.51	98.20	98.32
ArcFace* (CVPR19)	99.80	98.27	92.75	98.00	95.96	94.26	95.73	98.34	98.55
MV-Softmax* (AAAI20)	99.80	98.30	92.93	97.98	96.10	94.01	95.59	98.22	98.28
Circle Loss (CVPR20)	99.73	96.02	–	–	–	–	93.95	98.50	98.73
CurricularFace* (CVPR20)	_99.82_	98.30	_93.05_	_98.32_	96.05	94.75	96.04	_98.65_	98.70
MagFace* (CVPR21)	**99.83**	98.23	92.93	98.27	_96.12_	94.42	95.81	98.51	98.64
Ours, HMS	**99.83**	**98.44**	_93.05_	98.20	96.05	_94.86_	_96.25_	98.60	_98.75_
Ours, SMS, Linear	_99.82_	98.27	93.03	98.18	_96.12_	94.72	96.03	98.58	98.73
Ours, SMS, Non-Linear	_99.82_	_98.40_	**93.12**	**98.37**	**96.15**	**95.02**	**96.35**	**98.72**	**98.84**

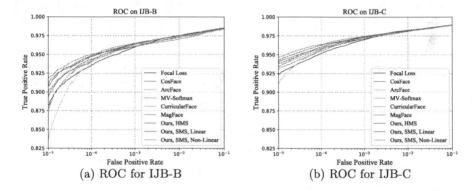

(a) ROC for IJB-B   (b) ROC for IJB-C

Fig. 4. ROC of 1:1 verification protocol on IJB-B/C.

Results on IJB-B and IJB-C. In this part, we compare our method with the state-of-the-art methods on IJB. Both IJB-B/C datasets are challenging tasks containing a considerable number of face images clipped from videos.

Table 1 lists the comparisons on TAR@FAR $= 1e - 4$ between our models and the competed methods. Without bells and whistles, our models achieve the leading results among all methods and improve the performance of IJB-B/C clearly. Among our three models, SMS with non-linearity achieves the top results on both IJB-B/C. The reason for the improvements is that both closeness and diversity should be simultaneously emphasized. As well analyzed in Sect. 3, both MV-Softmax and MagFace fail to emphasize closeness and diversity in a proper way. In addition, our models outperform the other competitors under most FPR variations, as shown in Fig. 4.

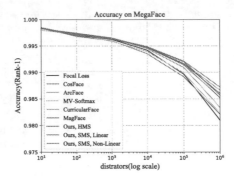

Fig. 5. The rank-1 face identification accuracy with different distractors on MegaFace.

Results on MegaFace. In this subsection, we evaluate the proposed method and the competed methods in terms of the identification and verification on MegaFace [11]. In our experiment, MegaFace is used as the gallery set, and FaceScrub is employed as the probe set.

The results of compared methods are listed in Table 1. "Id" refers to the rank-1 face identification accuracy with 1M distractors, and "Ver" refers to the face verification on TAR@FPR $= 1e - 6$. Table 1 shows that our models obtain the overall best results on both identification and verification tasks. Specifically, our model with SMS (non-linear) obtains the highest identification/verification results among all methods. In addition, although the performance will degrade with the increasing number of distractors, our model can achieve overall superiority over other methods, as shown in Fig. 5.

5.3   Ablation Study

In this part, we conduct experiments under different settings to investigate the effectiveness of the proposed components.

Loss Terms. In Table 2, Model 2, 3, 7, and 8 illustrate that it is difficult to obtain promising results if either closeness or diversity is absent. Model 2 and 7 pay more attention to closeness and achieve desirable results on the simple benchmark (e.g., LFW). However, their performances are degraded when dealing with the challenging datasets (e.g., IJB-B/C). Model 3 and 8 enforce diversity and perform better results on IJB-B/C. However, they have no improvements compared with Model 1 on LFW.

Division Thresholds for HMS. Here, we evaluate the proposed hard sample division scheme on several benchmarks. The experimental results of different division percentages for hard samples are listed in Table 2 (Model 4–6). Table 2 shows that our model achieves the best overall performance by taking 20% of

Table 2. Verification comparisons on several benchmarks under different loss terms and mining schemes. We conduct ablation experiments on the MS1MV2's subset containing 10K unique identites with ResNet34.[**Best**, Second Best]

Model	Closeness	Diversity	Mining scheme	LFW	CFP-FP	AgeDB	IJB-B	IJB-C
1			–	99.20	90.94	94.08	82.61	86.32
2	✓		HMS, $T = 20\%$	99.30	90.79	93.96	82.14	85.91
3		✓	HMS, $T = 20\%$	99.20	**91.68**	94.46	**83.64**	**86.82**
4	✓	✓	HMS, $T = 10\%$	99.25	91.55	94.38	83.50	86.71
5	✓	✓	HMS, $T = 20\%$	**99.32**	91.62	**94.63**	83.62	86.78
6	✓	✓	HMS, $T = 30\%$	99.26	91.65	94.60	83.53	86.87
7	✓		SMS, $\gamma = 10$	99.28	90.92	94.01	82.33	86.10
8		✓	SMS, $\gamma = 10$	99.13	91.46	94.25	**83.98**	**87.20**
9	✓	✓	SMS, Linear	99.25	91.01	94.10	82.88	86.53
10	✓	✓	SMS, $\gamma = 5$	99.27	91.55	94.23	83.80	86.95
11	✓	✓	SMS, $\gamma = 10$	**99.30**	**91.80**	94.55	83.92	87.04
12	✓	✓	SMS, $\gamma = 15$	99.25	91.37	**94.61**	83.85	87.10

the training data as hard samples. Besides, our model can also give competitive results when the hard samples occupy 10% and 30% in training samples.

Non-linearity Magnitudes γ for SMS. Model 9–12 in Table 2 provide the experimental results with different magnitudes of non-linearity. Model 9 achieves limited improvements compared with Model 1, indicating that it is difficult to achieve closeness and diversity by using the proposed SMS with linear function. Model 10–12 demonstrate that a suitable non-linearity by adjusting γ (i.e., 10) is helpful to achieve closeness and diversity. In addition, SMS with a properly γ can achieve better results on IJB datasets than HMS. The possible reason is that non-linear function can adjust the weight of samples adaptively based on different difficulty degrees.

6   Conclusion

This paper has proposed a two-branch cooperative network to learn discriminative features according to the understanding of margin-based softmax methods from the geometry view. Softmax-based methods can be considered as the pulling force from the corresponding class center and the pushing force from the negative class centers. Based on this, our model further enlarges the pulling force to enhance closeness and employ pushing force to enforce diversity. Several experimental results demonstrate the superiority of our proposed method over other competitors.

Acknowledgements. This work was supported by the National Science Fund of China under Grant Nos. 61876083, 62176124.

References

1. Chen, B., et al.: Angular visual hardness. In: International Conference on Machine Learning, pp. 1637–1648. PMLR (2020)
2. Chen, T., Kornblith, S., Norouzi, M., Hinton, G.: A simple framework for contrastive learning of visual representations. In: International Conference on Machine Learning, pp. 1597–1607. PMLR (2020)
3. Chopra, S., Hadsell, R., LeCun, Y.: Learning a similarity metric discriminatively, with application to face verification. In: 2005 IEEE Computer Society Conference on Computer Vision and Pattern Recognition (CVPR 2005), vol. 1, pp. 539–546. IEEE (2005)
4. Deng, J., Guo, J., Xue, N., Zafeiriou, S.: ArcFace: additive angular margin loss for deep face recognition. In: Proceedings of the IEEE/CVF Conference on Computer Vision and Pattern Recognition, pp. 4690–4699 (2019)
5. Guo, Y., Zhang, L., Hu, Y., He, X., Gao, J.: MS-Celeb-1M: a dataset and benchmark for large-scale face recognition. In: Leibe, B., Matas, J., Sebe, N., Welling, M. (eds.) ECCV 2016. LNCS, vol. 9907, pp. 87–102. Springer, Cham (2016). https://doi.org/10.1007/978-3-319-46487-9_6
6. Hadsell, R., Chopra, S., LeCun, Y.: Dimensionality reduction by learning an invariant mapping. In: 2006 IEEE Computer Society Conference on Computer Vision and Pattern Recognition (CVPR 2006), vol. 2, pp. 1735–1742. IEEE (2006)
7. He, K., Fan, H., Wu, Y., Xie, S., Girshick, R.: Momentum contrast for unsupervised visual representation learning. In: Proceedings of the IEEE/CVF Conference on Computer Vision and Pattern Recognition, pp. 9729–9738 (2020)
8. He, L., Wang, Z., Li, Y., Wang, S.: Softmax dissection: towards understanding intra-and inter-class objective for embedding learning. In: Proceedings of the AAAI Conference on Artificial Intelligence, vol. 34, pp. 10957–10964 (2020)
9. Huang, G.B., Mattar, M., Berg, T., Learned-Miller, E.: Labeled faces in the wild: a database for studying face recognition in unconstrained environments. In: Workshop on Faces in 'Real-Life' Images: Detection, Alignment, and Recognition (2008)
10. Huang, Y., et al.: CurricularFace: adaptive curriculum learning loss for deep face recognition. In: Proceedings of the IEEE/CVF Conference on Computer Vision and Pattern Recognition, pp. 5901–5910 (2020)
11. Kemelmacher-Shlizerman, I., Seitz, S.M., Miller, D., Brossard, E.: The megaface benchmark: 1 million faces for recognition at scale. In: Proceedings of the IEEE Conference on Computer Vision and Pattern Recognition, pp. 4873–4882 (2016)
12. Khosla, P., et al.: Supervised contrastive learning. arXiv preprint arXiv:2004.11362 (2020)
13. Lin, T.Y., Goyal, P., Girshick, R., He, K., Dollár, P.: Focal loss for dense object detection. In: Proceedings of the IEEE International Conference on Computer Vision, pp. 2980–2988 (2017)
14. Liu, H., Zhu, X., Lei, Z., Li, S.Z.: AdaptiveFace: adaptive margin and sampling for face recognition. In: Proceedings of the IEEE/CVF Conference on Computer Vision and Pattern Recognition, pp. 11947–11956 (2019)
15. Liu, W., Wen, Y., Yu, Z., Li, M., Raj, B., Song, L.: SphereFace: deep hypersphere embedding for face recognition. In: Proceedings of the IEEE Conference on Computer Vision and Pattern Recognition, pp. 212–220 (2017)
16. Liu, W., Wen, Y., Yu, Z., Yang, M.: Large-margin softmax loss for convolutional neural networks. In: ICML, vol. 2, p. 7 (2016)

17. Maze, B., et al.: IARPA Janus benchmark-C: face dataset and protocol. In: 2018 International Conference on Biometrics (ICB), pp. 158–165. IEEE (2018)
18. Meng, Q., Zhao, S., Huang, Z., Zhou, F.: MagFace: a universal representation for face recognition and quality assessment. In: Proceedings of the IEEE/CVF Conference on Computer Vision and Pattern Recognition, pp. 14225–14234 (2021)
19. Moschoglou, S., Papaioannou, A., Sagonas, C., Deng, J., Kotsia, I., Zafeiriou, S.: AgeDB: the first manually collected, in-the-wild age database. In: Proceedings of the IEEE Conference on Computer Vision and Pattern Recognition Workshops, pp. 51–59 (2017)
20. Parkhi, O.M., Vedaldi, A., Zisserman, A.: Deep face recognition (2015)
21. Paszke, A., et al.: Automatic differentiation in PyTorch (2017)
22. Ranjan, R., Castillo, C.D., Chellappa, R.: L2-constrained softmax loss for discriminative face verification. arXiv preprint arXiv:1703.09507 (2017)
23. Schroff, F., Kalenichenko, D., Philbin, J.: FaceNet: a unified embedding for face recognition and clustering. In: Proceedings of the IEEE Conference on Computer Vision and Pattern Recognition, pp. 815–823 (2015)
24. Sengupta, S., Chen, J.C., Castillo, C., Patel, V.M., Chellappa, R., Jacobs, D.W.: Frontal to profile face verification in the wild. In: 2016 IEEE Winter Conference on Applications of Computer Vision (WACV), pp. 1–9. IEEE (2016)
25. Shrivastava, A., Gupta, A., Girshick, R.: Training region-based object detectors with online hard example mining. In: Proceedings of the IEEE Conference on Computer Vision and Pattern Recognition, pp. 761–769 (2016)
26. Sun, Y.: Deep learning face representation by joint identification-verification. The Chinese University of Hong Kong, Hong Kong (2015)
27. Sun, Y., Liang, D., Wang, X., Tang, X.: DeepID3: face recognition with very deep neural networks. arXiv preprint arXiv:1502.00873 (2015)
28. Sun, Y., Wang, X., Tang, X.: Hybrid deep learning for face verification. In: Proceedings of the IEEE International Conference on Computer Vision, pp. 1489–1496 (2013)
29. Sun, Y., et al.: Circle loss: a unified perspective of pair similarity optimization. In: Proceedings of the IEEE/CVF Conference on Computer Vision and Pattern Recognition, pp. 6398–6407 (2020)
30. Taigman, Y., Yang, M., Ranzato, M., Wolf, L.: DeepFace: closing the gap to human-level performance in face verification. In: Proceedings of the IEEE Conference on Computer Vision and Pattern Recognition, pp. 1701–1708 (2014)
31. Wang, F., Cheng, J., Liu, W., Liu, H.: Additive margin softmax for face verification. IEEE Sig. Process. Lett. **25**(7), 926–930 (2018)
32. Wang, H., et al.: CosFace: large margin cosine loss for deep face recognition. In: Proceedings of the IEEE Conference on Computer Vision and Pattern Recognition, pp. 5265–5274 (2018)
33. Wang, T., Isola, P.: Understanding contrastive representation learning through alignment and uniformity on the hypersphere. In: International Conference on Machine Learning, pp. 9929–9939. PMLR (2020)
34. Wang, X., Zhang, S., Wang, S., Fu, T., Shi, H., Mei, T.: Mis-classified vector guided softmax loss for face recognition. In: Proceedings of the AAAI Conference on Artificial Intelligence, vol. 34, pp. 12241–12248 (2020)
35. Weinberger, K.Q., Saul, L.K.: Distance metric learning for large margin nearest neighbor classification. J. Mach. Learn. Res. **10**(2), 207–244 (2009)
36. Wen, Y., Zhang, K., Li, Z., Qiao, Yu.: A discriminative feature learning approach for deep face recognition. In: Leibe, B., Matas, J., Sebe, N., Welling, M. (eds.)

ECCV 2016. LNCS, vol. 9911, pp. 499–515. Springer, Cham (2016). https://doi.org/10.1007/978-3-319-46478-7_31

37. Whitelam, C., et al.: IARPA Janus benchmark-B face dataset. In: Proceedings of the IEEE Conference on Computer Vision and Pattern Recognition Workshops, pp. 90–98 (2017)

38. Zeng, D., Shi, H., Du, H., Wang, J., Lei, Z., Mei, T.: NPCFace: negative-positive collaborative training for large-scale face recognition. arXiv preprint arXiv:2007.10172 (2020)

39. Zhang, K., Zhang, Z., Li, Z., Qiao, Y.: Joint face detection and alignment using multitask cascaded convolutional networks. IEEE Sig. Process. Lett. **23**(10), 1499–1503 (2016)

40. Zhang, X., Fang, Z., Wen, Y., Li, Z., Qiao, Y.: Range loss for deep face recognition with long-tailed training data. In: Proceedings of the IEEE International Conference on Computer Vision, pp. 5409–5418 (2017)

41. Zhang, X., Zhao, R., Qiao, Y., Wang, X., Li, H.: AdaCos: adaptively scaling cosine logits for effectively learning deep face representations. In: Proceedings of the IEEE/CVF Conference on Computer Vision and Pattern Recognition, pp. 10823–10832 (2019)

42. Zheng, T., Deng, W.: Cross-Pose LFW: a database for studying cross-pose face recognition in unconstrained environments. Technical report **5**, 7. Beijing University of Posts and Telecommunications (2018)

43. Zheng, T., Deng, W., Hu, J.: Cross-Age LFW: a database for studying cross-age face recognition in unconstrained environments. arXiv preprint arXiv:1708.08197 (2017)

KinStyle: A Strong Baseline Photorealistic Kinship Face Synthesis with an Optimized StyleGAN Encoder

Li-Chen Cheng[1], Shu-Chuan Hsu[1], Pin-Hua Lee[1(✉)], Hsiu-Chieh Lee[1], Che-Hsien Lin[2], Jun-Cheng Chen[2], and Chih-Yu Wang[2]

[1] National Taiwan University, Taipei City, China
{b06902128,b06502152,b07303024,b07902030}@ntu.edu.tw
[2] Academia Sinica, Taipei City, China
{ypps920080,pullpull,cywang}@citi.sinica.edu.tw

Abstract. High-fidelity kinship face synthesis is a challenging task due to the limited amount of kinship data available for training and low-quality images. In addition, it is also hard to trace the genetic traits between parents and children from those low-quality training images. To address these issues, we leverage the pre-trained state-of-the-art face synthesis model, StyleGAN2, for kinship face synthesis. To handle large age, gender and other attribute variations between the parents and their children, we conduct a thorough study of its rich latent spaces and different encoder architectures for an optimized encoder design to repurpose StyleGAN2 for kinship face synthesis. The obtained latent representation from our developed encoder pipeline with stage-wise training strikes a better balance of editability and synthesis fidelity for identity preserving and attribute manipulations than other compared approaches. With extensive subjective, quantitative, and qualitative evaluations, the proposed approach consistently achieves better performance in terms of facial attribute heredity and image generation fidelity than other compared state-of-the-art methods. This demonstrates the effectiveness of the proposed approach which can yield promising and satisfactory kinship face synthesis using only a single and straightforward encoder architecture.

Keywords: Kinship face synthesis · StyleGAN Encoder

This work was supported by the National Science and Technology Council under Grant 108-2628-E-001-003-MY3, 111-2628-E-001 -002 -MY3, 111-3114-E-194-001 -, 110-2221-E-001 -009 -MY2, 110-2634-F-002-051-, 111-2221-E-001-002-, and the Academia Sinica under Thematic Research Grant AS-TP-110-M07-2.

Supplementary Information The online version contains supplementary material available at https://doi.org/10.1007/978-3-031-26316-3_7.

L. Wang et al. (Eds.): ACCV 2022, LNCS 13844, pp. 105–120, 2023.
https://doi.org/10.1007/978-3-031-26316-3_7

1   Introduction

With the recent popularity of deep image and face synthesis, kinship face synthesis gets increasing attention in the research community of facial analysis. The goal of kinship face synthesis is to render the possible children faces given a pair of parental face images. This facilitates plenty of kinship-related applications, including producing visual effects, delineating the possible facial appearances of a lost child after a long period of time, analyzing the facial traits of a family, etc. However, kinship face synthesis is still a challenging and ongoing research problem as compared with other face synthesis tasks due to a lack of large-scale training data, severe label noise, and poor image quality. In addition, it is also hard to trace the genetic traits between parents and children from those low-quality training images, especially when there is interference caused by the facial variations in illumination, pose, and other factors.

To synthesize a child face, one can use the image from a single reference parental image or images from both parents. For the former, because the information from the other parent is unavailable, there exist ambiguities of mapping a parental face to its child face. Ertuğrul et al. [1] propose the first work in this category, but the one-versus-one relation fails to capture enough information to yield promising and satisfactory results. For the latter, although two-versus-one relation between the parents and their children provides a good constraint for better kinship face synthesis, the limited amount of kinship data and data noise still restrict the generative models [2,3], to synthesize high fidelity child faces. This also usually results in the situations of overfitting or lacking diversity for synthesized faces that are close to the average face when no further regularizations are applied. Lin et al. [4] leverage the pre-trained state-of-the-art face synthesis model upon the FFHQ dataset, StyleGAN2 [5], and train an additional encoder to extract latent representations encoding rich parental appearance information from the face images of parents encoding rich for kinship face synthesis to mitigate the training data issues. Their method can effectively utilize the data manifold of the pre-trained StyleGAN2 as a regularization to effectively restrict the possible kinship face distributions and to synthesize meaningful and good child faces. Nevertheless, without considering the issues of attribute data imbalance and feature entanglement, the method still lacks the capability to perform further smooth and effective manipulation over specific facial traits towards the parents, such as face component-wise manipulation over eyes, nose, mouth, etc. Zhang et al. [6] proposed to use multiple encoders for each component to realize component-wise manipulation, but this introduces more efforts of training and computational costs than others.

In general, the pipeline for kinship face synthesis can be divided into three stages: parental feature extraction, parental feature fusion, and face rendering. To address above issues of kinship face synthesis, we also leverage the pre-trained StyleGAN2 for rendering due to its encoded rich face prior and superior face synthesis capability. However, due to the complex nature of StyleGAN2 model and large appearance variations (i.e., age, gender, and other facial attributes.) between parents and their children, it requires us to conduct a careful study of

its various latent spaces (i.e., Z, W, W+, S spaces.) and encoder architectures
for an optimized encoder to repurpose StyleGAN2 for kinship face synthesis. To
our knowledge, these have not been well studied for kinship face synthesis in
the literature. With thorough evaluations of different design choices as shown
in Table 1 (i.e., more details are presented in Sect. 4.), we propose an encoder
design consisting of an image encoder and a fusion block. The image encoder
is further composed of an ID-preserved block with the design of *Encoder for
Editing* (e4e) by Tov *et al.* [7] for better disentangled latent representation and
editability in addition to an attribute block for normalizing age and gender variations of the parental representations for better synthesis fidelity. The fusion block
fuses the latent representations of the parents for the final child representation.
The obtained representation from the proposed encoder pipeline with stage-wise
training strikes a better balance of editability and synthesis fidelity for identity
preserving and attribute manipulations than other compared approaches. With
extensive subjective, quantitative, and qualitative evaluations, the proposed approach consistently achieves better synthesis results using the Family-In-the-Wild
(FIW) [8] and TSKinFace [9] datasets in terms of facial attribute heredity and
image generation fidelity than other compared state-of-the-art methods. Furthermore, with our representation, we can also easily realize a component-wise
parental trait manipulation (CW-PTM) through a method proposed by Chong *et
al.* [10] to flexibly manipulate any desired face parts or regions of the synthesized
face towards the parents through latent interpolation while ensuring the transition is smooth and continuous. Surprisingly, the manipulation results are competitive with other approaches employing multiple facial component encoders
for the purpose. This also demonstrates the proposed method can not only yield
promising and satisfactory kinship face synthesis but also enable the fine control
of facial attributes using only a single and straightforward encoder architecture
without the complex multi-encoder structure. This reduces training difficulties
such as tuning the hyperparameters of multiple encoders simultaneously.

2   Related Work

In this section, we briefly review the relevant research works for kinship image
synthesis using deep generative models.

Deep Image Synthesis: Many studies of image synthesis using a deep neural
network rely heavily on Generative Adversarial Network (GAN) [11–14]. Based
on GAN models, StyleGAN [15] was proposed to generate high-level attributes
for synthesized images and preserve linearity in the latent space of generative
models. StyleGAN encodes the latent z from the latent space Z into the feature
space W, then w is chosen from feature space and input to multiple layers of convolution layers in order to control various styles of the output image through the
adaptive instance normalization (AdaIN) module. Abdal *et al.* [16] proposed an
efficient algorithm to perform inversion of the input image into an extended feature space W^+ instead of W space of a pre-trained StyleGAN for better image

reconstruction. StyleGAN2 [5] further improves the details of the synthesized image, such as removing the blob-like artifacts by redesigning the network of StyleGAN. Moreover, StyleGAN2 simplifies the instance normalization process with a weight demodulation operation. The latent space of GANs has been studied carefully in recent years especially in the field of computer vision [17–20]. For facial images, it is ideal to map the source into latent space in an effort to reduce dimension as well as provide image editing in latent space. Tov *et al.* proposed Encoder for Editing [7] that allows manipulation of inverted images.

Kinship Face Synthesis: Kinship face synthesis is a recently commenced problem that aims to generate the child image given the images of parents [21]. Some works studied generating a child image using the image of father or mother as reference [2,22,23]. Nevertheless, these approaches suffer from the problem of either low resolution or mode collapse, and thus the generated results are unsatisfactory. Some works use both parents' images as input, such as the methods proposed by Ghatas *et al.* [24] and Zaman *et al.* [25]. Still, the artifacts from the first work are sometimes corrupted, and the second work does not take the child image as guidance. Some recent works have further improved the synthesis results in the kinship face synthesis problem using more advanced architectures and loss designs. Gao *et al.* [3] introduce DNA-Net that leverages conditional adversarial autoencoder to generate the child images. Zhang *et al.* [6] generate child images by assigning inheritance control vector of a facial part so as to let the child inherit the facial region from the mother or father. Lin *et al.* [4] concatenate latent space embedding of a child with age and gender vector to render child images with pre-trained StyleGAN2. ChildGAN [26] extracts the representative semantic vectors and synthesizes the child image by macro fusion and micro fusion. Our proposed method utilizes a novel designed encoder and the method of attribute alignment, our model is capable of synthesizing the child images that inherit designated facial regions, which leads to an outstanding performance as well as diversity in the synthesized child images.

3   Methodology

Our goal is to build a framework to synthesize a high-fidelity child face from a pair of parental face images, I_F and I_M, while being able to smoothly control the age, gender, and specific facial features of the synthesized face. The overview of the proposed framework is shown in Fig. 1, which comprises three phases. The first phase is to encode face images of the parents into the optimal latent codes, s_F and s_M, while having a good compromise between fidelity and editability for StyleGAN2. The identity information of parental face images is preserved through the ID-Preserved block, and the age and gender attributes are further normalized by the attribute block. In the second phase, we performs weighted average upon the transformed latent representations of the parents into the final representation, s_C, for the child. The weights can be either manually assigned or learned by several multilayer perceptrons (MLPs) layers. In the third phase, with

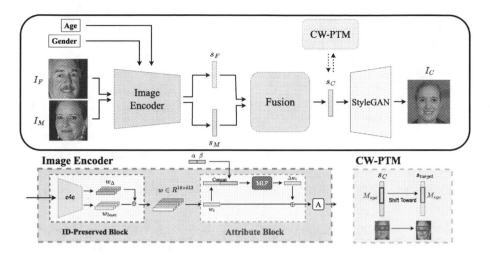

Fig. 1. The overview of the proposed encoder-decoder framework based on StyleGAN2 for kinship face synthesis.

the representation, a flexible and fine manipulation of the selected region towards the parents can be achieved through latent interpolation. In the following, we will describe the details of each component.

3.1 Phase 1: Image Encoder

Due to the immense diversity of age, gender, and identity for the face images of the parents and children, it is difficult to train a single encoder at a time to acquire a latent representation that not only preserves the identity information but also is age and gender invariant. Thus, we divide the image encoder into two blocks, one for preserving identity information and the other for normalizing the age and gender of the latent codes of the parents. Meanwhile, we perform stage-wise training, which provides more flexible training strategies and requires less computational resources. The details of each block are described as follows.

Identity-Preserved Block. Let $E(\cdot)$ denote identity-preserved encoder block (ID-preserved block), and $G_{W+}(\cdot)$ refers to StyleGAN2 taking the latent code in W^+ space as input. The objective of the ID-preserved block is to learn the mapping $E : \mathcal{I} \to W^+$, preserving the identity of the input image. As shown in Table 1, to preserve the identity information in the resultant latent representation while maximizing its editability, we adopt the *Encoder for Editing* (e4e) architecture [7] with the identity loss for the purpose. Given an input image, the encoder returns a base latent and a residual, each having the dimension $\mathbb{R}^{18 \times 512}$.

The final output latent code is obtained by adding them together. This design can minimize the variance between the latent codes from 18 layers of a W^+ space and make the latent code more editable. In addition, this also facilitates the age and gender normalization and other manipulations of the latent codes of the parents for the second and later stages. Let w denote the latent code encoded from the input image I, w_{base} denote the base latent code, w_Δ denote the residual, and I_{syn} denotes synthesized image. That is,

$$w = E(I) = w_{base} + w_\Delta \ , \tag{1}$$

$$I_{syn} = G_{W+}(w) \ . \tag{2}$$

For the loss function, to preserve the identity of the input image, we impose a common identity constraint.

$$\mathcal{L}_{ID_1} = 1- < R(I_{syn}), R(I) > \ , \tag{3}$$

where $< \cdot , \cdot >$ is cosine similarity and $R(\cdot)$ is a pre-trained ArcFace model that extracts the feature map of a face from the penultimate layer.

Moreover, we add an L2 facial landmark loss to further enhance the alignment of the synthesized face I_{syn} to be centered at the image. The landmarks on the center line of the face between the nose and mouth are used for the landmark loss, which is defined as

$$\mathcal{L}_{land} = \| E_{land}^x(I_{syn}) - C \|_2 \ , \tag{4}$$

where $E_{land}^x(\cdot)$ is a pre-trained landmark predictor and C is a vector with the values of the x coordinate of the image center which is 512 in the rest of experiments for a 1024×1024 image. Lastly, we also add two additional losses as in [7] for image editability and quality. The first one is the an L2 regularization loss on the residual w_Δ:

$$\mathcal{L}_{reg_1} = \| w_\Delta \|_2. \tag{5}$$

The second loss is the non-saturating GAN loss with R1 regularization, which is used for forcing w not to deviate from the W^+ space. Let $D(\cdot)$ denote a latent discriminator.

$$
\begin{aligned}
\mathcal{L}_{adv}^D = - & \underset{w_r \sim \mathcal{W}^+}{\mathbb{E}} [\log D(w_r)] - \underset{I \sim \mathcal{I}}{\mathbb{E}} [\log(1 - D(E(I)))] \\
& + \frac{\gamma}{2} \underset{w_r \sim \mathcal{W}^+}{\mathbb{E}} [\| \nabla_{w_r} D(w_r) \|_2^2] \ , \\
\mathcal{L}_{adv}^E = - & \underset{I \sim \mathcal{I}}{\mathbb{E}} [\log D(E(I))] \ .
\end{aligned}
\tag{6}
$$

To sum up, the total loss function of ID-preserved block is

$$\mathcal{L}_{enc} = \lambda_{ID_1} \mathcal{L}_{ID_1} + \lambda_{land} \mathcal{L}_{land} + \lambda_{reg_1} \mathcal{L}_{reg_1} + \lambda_{adv} \mathcal{L}_{adv}. \tag{7}$$

Attribute Block. After the ID-preserved block, the attribute block is further used to align the age and gender of input parental latent codes, $w = \{w_i\}_{i=1}^{18}$, where $w_i \in \mathbb{R}^{512}$ in W^+. As mentioned in Sect. 2, the attribute manipulation can be achieved by shifting the latent code along specific latent directions. Thus, we learn an offset vector for desired modification by employing MLPs with leaky ReLUs followed by each MLP, $M(\cdot)$, which take the concatenation of the latent codes of the parents w, desired age α, and gender β values as input. For α, the input value ranges from 0 to 1, corresponding to 0 years old to 100 years old. For β, 1 represents male, and 0 represents female. Then, we can obtain the modified latent code w' by adding the original w and offsets $\Delta w = M(w, \alpha, \beta)$.

In the training process, α is sampled from $Uniform[0, 1]$, and β is sampled from $Bernoulli(0.5)$. Besides arbitrary attributes, we can also use ground truth attributes to construct a reconstruction loss and a cycle consistency loss on w. We can further obtain three modified latent codes for each w.

$$
\begin{aligned}
w'_{syn} &= w + M(w, \alpha, \beta) \\
w'_{rec} &= w + M(w, \alpha_{gt}, \beta_{gt}) \\
w'_{cyc} &= w'_{syn} + M(w'_{syn}, \alpha_{gt}, \beta_{gt})
\end{aligned}
\tag{8}
$$

where α_{gt}, β_{gt} denote the ground truth age and gender labels. For each w' vector, we can obtain the corresponding synthesized images $I_{syn}, I_{rec}, I_{cyc}$ by passing it into $G_{W+}(\cdot)$. For the loss function of the attribute block, we first employ the following age and gender losses.

$$
\begin{aligned}
\mathcal{L}_{age} &= ||\alpha - C_a(I_{syn})||_2 + ||\alpha_{gt} - C_a(I_{rec})||_2 + ||\alpha_{gt} - C_a(I_{cyc})||_2 \\
\mathcal{L}_{gen} &= H(\beta, C_b(I_{syn})) + H(\beta_{gt}, C_b(I_{rec})) + H(\beta_{gt}, C_b(I_{cyc})) ,
\end{aligned}
\tag{9}
$$

where $C_a(\cdot)$ denotes a pre-trained age classifier, $C_b(\cdot)$ denotes a pre-trained gender classifier, and $H(\cdot)$ denotes the cross-entropy loss. Similarly, we also use identity loss for attribute block training. However, since the identity of a person may become obscure as the person ages, Alaluf et al. [27] proposed an identity loss decayed with the age difference between the prediction and the ground truth. We further extend the idea to both age and gender. The identity loss can be formulated as

$$
\begin{aligned}
\mathcal{L}_{ID_2} = \xi \cdot (1- <R(I_{syn}), R(I)>)+ \\
(1- <R(I_{rec}), R(I)>) + (1- <R(I_{cyc}), R(I)>),
\end{aligned}
\tag{10}
$$

where ξ is the decay coefficient, and we set $\xi = 0.45 + 0.35 \cdot cos(|\alpha - \alpha_{gt}| \cdot \pi) + 0.2 \cdot cos(|\beta - \beta_{gt}| \cdot \pi)$. Moreover, to make training faster and prevent latent codes from deviating from the original W^+ space, we utilize perceptual similarity losses on the images and use L2 regularization on the offsets.

$$
\begin{aligned}
\mathcal{L}_{reg_2} =&||M(w, \alpha, \beta)||_2 + ||M(w, \alpha_{gt}, \beta_{gt})||_2+ \\
&||M(w'_{syn}, \alpha_{gt}, \beta_{gt})||_2, \\
\mathcal{L}_{per} =&||P(I_{enc}) - P(I_{syn})||_2 + ||P(I_{enc}) - P(I_{rec})||_2+ \\
&||P(I_{enc}) - P(I_{cyc})||_2 ,
\end{aligned}
\tag{11}
$$

where $I_{enc} = G_{W+}(E(I))$ is the reconstructed image by passing the latent after ID-preserved block to StyleGAN2, and $P(\cdot)$ is a pre-trained AlexNet feature extractor upon the ImageNet dataset. Lastly, the total loss function is as follows.

$$\mathcal{L}_{attr} = \lambda_{ID_2}\mathcal{L}_{ID_2} + \lambda_{age}\mathcal{L}_{age} + \lambda_{gen}\mathcal{L}_{gen} + \lambda_{reg_2}\mathcal{L}_{reg_2} + \lambda_{per}\mathcal{L}_{per}. \tag{12}$$

As suggested in [28], the latent codes in S space of StyleGAN2 result in a better style mixing performance. We follow the idea and apply the affine transform layers, $A(\cdot)$, available in the StyleGAN2 model to convert the latent representation of each parent, w' vector, into s for the next stage as $s = A(w')$.

3.2 Phase 2: Fusion

Once the transformed latent representations of the parents in S space are obtained, we perform weighted average to blend them into one final child latent code as follows:

$$s_C = \gamma \circ s_M + (1 - \gamma) \circ s_F, \tag{13}$$

where s_F and $s_M \in \mathbb{R}^{9088}$ denote two latent codes in S space for the parents, s_C denotes the resultant child code, $\circ$ is the element-wise product, and $\gamma \in [0, 1]^{9088}$ is the blending coefficient which can be either manually specified or learned by a fusion network where we employ an MLP layer that takes the concatenated vector of s_F and s_M as the input trained with the $L2$ reconstruction and ID losses similar to the ID-preserved block. Besides S space, we also perform the blending in W^+ space for comparison, and more details can be found in Sect. 4.1.

3.3 Phase 3: Component-wise Parental Trait Manipulation (CW-PTM)

With our optimized encoder pipeline of kinship face synthesis in Phase 1 and 2, the resultant latent code for the child is suitable for editing through latent interpolation as compared with other methods, like StyleDNA. We can easily apply a similar approach as in [10] to realize component-wise parental trait manipulation (CW-PTM) in a single encoder-decoder framework. The relation between each dimension of the latent code and specific facial features, such as eyes, nose, and mouth, are obtained with K-means clustering. Then, a mask corresponding to the facial features over the latent code can be derived accordingly. A specific facial attribute can be transferred from one image to another by shifting the original latent code towards the target one based on the mask. For example, suppose the mask for the eyes is denoted by M_{eye}, where $M_{eye} \in \{0, 1\}^{9088}$ and one stands for the position of the latent vector controlling the synthesized face's eyes, as shown in Fig. 1. Therefore, we can shift the part of s_C within M_{eye} toward s_{target} by a coefficient ϵ. That is,

$$s'_C = s_C + \epsilon \cdot M_{eye} \circ (s_C - s_{target}), \tag{14}$$

where s'_C is the modified latent code, and $\circ$ denotes the element-wise product. The step size ϵ will determine how similar the eye is to the target. Leveraging this method, we can manipulate the latent code to make the synthetic child

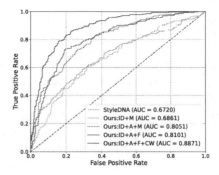

Fig. 2. It illustrates the ROC curves for ablation studies and the comparisons between the proposed approach and the baseline, StyleDNA. ID stands for ID-preserved block, A for attribute block, F for learned fusion, and CW for component-wise parental trait manipulation. M represents directly averaging parent latent codes without fusing them with a learned network.

inherit specific parental traits. The proposed approach is not only more memory and computation efficient but also allows more flexible and smooth manipulation towards any selected regions of the parents by latent interpolation as shown in Fig. 6b than other similar works, such as [6] which employs fixed multiple component-wise encoders. In addition, the proposed approach avoids using multiple encoders, which increases the training difficulties due to plenty of model parameters and hyperparameter tuning. The multiple region manipulations can be achieved through recursively applying the same procedure.

4 Experiment

In this section, we show the results of the proposed approach with several recent most representative state-of-the-art methods for quantitative, qualitative, and subjective comparisons.

Implementation Details: We use a pre-trained and fixed StyleGAN2 model upon the FFHQ dataset as our decoder. For the encoder, we train both ID-preserved and attribute blocks using the FFHQ-Aging dataset [29], which contains images with age and gender label information. We also perform weighted sampling based on the number of training instances per age and gender. Then, the learned fusion block is trained with the FIW dataset [8], which contains approximately 2,000 tri-pairs of kinship images. Since images in the FIW dataset are low-resolution, we preprocess them with GFP-GAN for super-resolution [30] before training. The facial landmarks are extracted using the MobileFaceNets [31]. We pre-train the age and gender classifiers following the same setting as [4]. Instead of one-hot vectors, age and gender conditionals are transformed to $[0, 1]$ and $\{0, 1\}$ respectively and then duplicate 50 times each in order to facilitate the stable training. For hyperparameters of training, we set batch size as 6, use a standard Ranger optimizer with a learning rate 0.0001, and set the loss weights as follows. We set $\lambda_{ID_1} = 1$, $\lambda_{land} = 0.0008$, $\lambda_{reg} = 0.0002$, $\lambda_{adv} = 0.5$ for ID-preserved block; $\lambda_{ID_2} = 0.5$, $\lambda_{gen} = 1$, $\lambda_{age} = 5$, $\lambda_{per} = 0.5$, $\lambda_{reg} = 0.05$ for attribute block; $\lambda_{ID_3} = 1$, $\lambda_2 = 1$ for the learned fusion block.

Training Pipeline: To facilitate the training process and handle the data imbalance issues, we perform stage-wise training to train a component at a time while freezing the model weights in prior stages. This enables us not only to apply relevant losses for the optimal training for each encoder block according to their task characteristics but also to train the model using much less GPU resources than other approaches.

4.1   Quantitative Evaluations of Different Encoder Configurations

Due to the complex nature of the StyleGAN2 model, for an optimized encoder design, we first perform an investigation of the encoder design choices of different latent spaces and network architectures for kinship face synthesis in terms of AUC of the ROC curve for face verification and FID scores for image synthesis where we measure the facial similarity between the synthesized child face and its corresponding ground truth child face. The pre-trained ArcFace model for face verification is used to extract the latent representations of both the prediction and ground truth followed by cosine similarity computation. We randomly sample 100 positive and negative pairs from the test set of the FIW dataset for the similarity computation. From Table 1, we find that encoding images to W^+ space or S space can generate offspring faces that are more realistic and more similar to ground truths. In addition, the editability of encoded latent codes needs to be considered in our styleGAN pipeline. The results show that the methods with another popular pSp backbone [32] which does not account for editability attained lower AUC and higher FID scores compared to the ones with e4e. Lastly, although the learning-based fusion can achieve slightly better performance improvement than the manual one, there is not much difference between them if the encoder is well designed. Then, performing fusion in W^+ space or S space attained similar AUC and FID scores. In the rest of the experiments, we select the configuration of (6) to further conduct qualitative and subjective evaluation, since it has the highest AUC for the best identity preservation and also yields good perceptual quality image. For the configuration of (6), we further compare the ROC curves of the proposed method with StyleDNA due to the public availability of its source code. As shown in Fig. 2, we can see the proposed approach achieves the best performance with AUC 0.8101. We also show the blue curve for the most promising result with AUC 0.8871 after applying CW-PTM using ID loss from the ground truth as the guidance. The improved number also shows the strength of CW-PTM to explore a more resembled face to the ground truth child face. Also, the learning-based fusion can achieve slightly better improvement than the manual one. In the rest of the experiments, we select the configuration of (6) to further conduct qualitative and subjective evaluation, since it has the highest AUC and yields good perceptual quality image.

Table 1. The quantitative results of pipelines with different combinations of the encoder and the fusion method.

	Encoder		Fusion		AUC (↑)	FID (↓)
	Space	Type	Space	Type		
(1)	W	Resnet	W	Learned	0.6720	197.9197
(2)	W^+	e4e	W^+	Mean	0.8050	133.3718
(3)	W^+	pSp	W^+	Mean	0.7738	176.4417
(4)	W^+	e4e	S	Mean	0.8051	133.3681
(5)	W^+	pSp	S	Mean	0.7738	176.4349
(6)	W^+	e4e	S	Learned	0.8101	138.2808
(7)	W^+	pSp	S	Learned	0.7783	173.5681

Table 2. (a) The weighted average rank for the resemblance between the synthesized child faces using different approaches and a pair of parental face images. (b) The weighted average rank of naturalness and photo-realism for the synthesized child faces of different approaches.

	GT	styleDNA [4]	ChildGAN [26]	Ours
Session I	2.83	2.92	2.33	**1.92**
Session II	2.27	2.82	2.18	**1.55** (CW-PTM)

(a)

	styleDNA [4]	ChildGAN [26]	Ours
Avg rank	**1.57**	2.59	1.84

(b)

4.2 Subjective Evaluation

To further compare the generation quality of different methods, we conduct the subjective evaluation in two independent online sessions, 186 participants for the first and 131 for the second. Each session contains 9 and 13 questions in total respectively. For each session, we ask the participants to rank the synthesized child faces by different methods, ground truth and other state-of-the-art methods, StyleDNA [4] and ChildGAN [26] along with the ground truth according to their perceived resemblance to the given reference face images of the parents, where a lower rank represents a higher score (*i.e.*, one is the most likely and four is the least likely.). Since the code of [6] which employs multiple facial component encoders is not publicly available and the image quality of the pdf file is low, we do not use it for subjective evaluation. In addition, we also asked participants to answer the extent of naturalness and photo-realism among different synthesized child faces. As shown in Table 2, we compute the weighted average ranks of rank for the resemblance between the synthesized child and the reference parental faces. The proposed method consistently achieve the best subjective

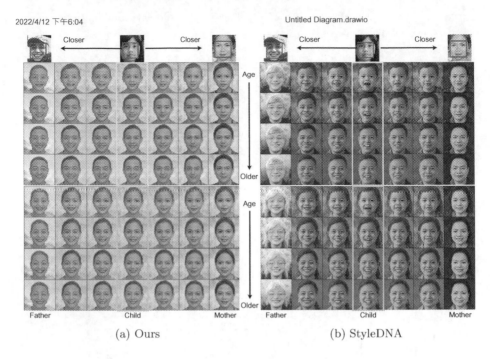

Fig. 3. The qualitative comparisons between the proposed approach and StyleDNA after performing fine attribute manipulation towards the parents using CW-PTM upon the respective extracted latent representations according to the selected facial component or region of the parent. The top row shows the ground truth faces of the parents and their child. The proposed approach achieves more continuous and better manipulation results than the compared baseline, keeping the age and gender intact while performing the manipulation.

performances as compared with other methods. For the photo-realism test in Table 2 for the first session, the average weighted rank of the proposed approach is close to StyleDNA. Instead of ranking, we further conducted the mean opinion score to compare the photo-realism of the proposed approach with StyleDNA in the second session, where the score ranges from 1 to 5, and a higher value is better. The proposed approach achieves better average score of **3.53** than StyleDNA, **3.23**. It is worth noting that the proposed approach even outperforms the ground truth. These results further demonstrate the strength of the proposed approach. For more details of the questions and rank scores, we refer interested readers to the supplementary materials.

4.3   Qualitative Evaluation

Finally, we also show various visual samples to compare the proposed approach with other methods in terms of the capability of various attribute manipulations, including age, gender, and parental traits. From Fig. 3, we find the synthesized

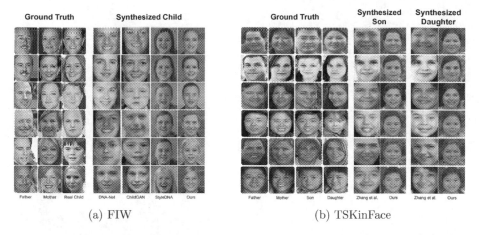

(a) FIW   (b) TSKinFace

Fig. 4. The results of ground truth face images along with the synthesized child images from the proposed and other compared methods. Left most three or four columns of (a) for FIW and (b) for TSKinFace respectively depict the ground truth images and right four columns depict the synthesized child images, where the compared methods include DNA-Net [3], ChildGAN [26], StyleDNA [4], and Zhang *et al.* [6].

faces by the proposed approach change more smoothly when adjusting the values of the corresponding latent codes as compared with StyleDNA (*i.e.*, the inference code of StyleDNA is publicly available, and StyleDNA is thus chosen as the main comparison target for the qualitative analysis.). This further demonstrates the advantages of the proposed approach to synthesize high fidelity kin faces while allowing smooth component-wise manipulation towards any selected facial components or regions of the parents as shown in Fig. 6b. Employing multiple encoders usually introduces additional computational costs and training difficulties. We also show the synthesis results in Fig. 4 using the face images of the FIW and TSKinFace datasets. The proposed method can synthesize the faces with better fidelities than other approaches.

4.4 Ablation Studies

In this section, we perform ablation studies to understand the effectiveness of each component where we indicate ID for ID-preserved block, A for attribute block, F for learned fusion, M represents directly averaging parent latent codes without fusing them with a learned network, and CW means the proposed component-wise parental trait manipulation. As shown in Fig. 6a, the fidelity of the synthesized child faces gets improved with applying more components of the proposed framework. With the assistance of CW-PTM, we can further flexibly make the synthesized child faces closer towards either the father or the mother or both with different facial components or regions at the same time. Surprisingly, from Fig. 2, the AUC scores of all the ablated results are better

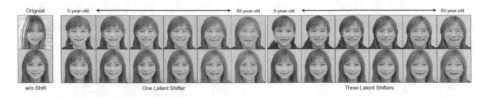

Fig. 5. This illustrates the effects using different numbers of MLP layers in the attribute block. For the left, it uses a single MLP and three MLPs for the right. The results using three MLPs show more prominent facial manipulation than the one using a single MLP.

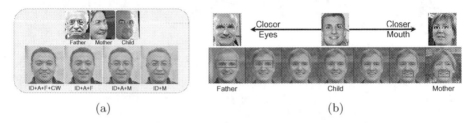

(a)   (b)

Fig. 6. (a) It shows qualitative results using different combinations of our network components for the ablation study. (b) It illustrates the proposed approach allows flexible parental trait manipulation with any selected facial components or regions towards parents.

than StyleDNA. This shows the strength of the proposed optimized encoder architecture and the stage-wise training.

On the other hand, for attribute block, besides a single MLP, we further divide the encoded latent code from the eighteen layers of the StyleGAN2 model into three groups with a corresponding MLP for transformation where the first three for coarse-grained, fourth to seventh layers for middle-grained, and the rest for fine-grained detail control. Although the differences in the quantitative results on FID of two settings are small, which are 37.056 and 36.663. However, we can see the three MLPs results in face images with better attribute manipulation than a single MLP one, as shown in Fig. 5.

5   Conclusion

The main contribution of our work is to conduct a thorough study of different encoder choices of different latent spaces and encoder architectures to repurpose StyleGAN2 for kinship face synthesis. The proposed optimized encoder striking a better balance of editability and synthesis fidelity for kinship face synthesis than other compared methods while allowing smooth and continuous face trait manipulation. With extensive subjective, quantitative, and qualitative evaluations, the proposed approach consistently achieves better performance in terms of facial attribute heredity and image generation fidelity than other state-of-the-art methods. This demonstrates the effectiveness of the proposed approach, which can yield promising and satisfactory kinship face synthesis using only a single straightforward encoder architecture.

References

1. Ertugrul, I.Ö., Dibeklioglu, H.: What will your future child look like? modeling and synthesis of hereditary patterns of facial dynamics. In: IEEE International Conference on Automatic Face and Gesture Recognition (FG) (2017)
2. Ozkan, S., Ozkan, A.: KinshipGAN: synthesizing of kinship faces from family photos by regularizing a deep face network. In: IEEE International Conference on Image Processing (ICIP) (2018)
3. Gao, P., Robinson, J., Zhu, J., Xia, C., Shao, M., Xia, S.: DNA-Net: age and gender aware kin face synthesizer. In: IEEE International Conference on Multimedia and Expo (ICME) (2021)
4. Lin, C.H., Chen, H.C., Cheng, L.C., Hsu, S.C., Chen, J.C., Wang, C.Y.: StyleDNA: a high-fidelity age and gender aware kinship face synthesizer. In: IEEE International Conference on Automatic Face and Gesture Recognition (FG) (2021)
5. Karras, T., Laine, S., Aittala, M., Hellsten, J., Lehtinen, J., Aila, T.: Analyzing and improving the image quality of styleGAN. In: Proceedings of the IEEE/CVF Conference on Computer Vision and Pattern Recognition (CVPR) (2020)
6. Zhang, Y., Li, L., Liu, Z., Wu, B., Fan, Y., Li, Z.: Controllable descendant face synthesis. arXiv preprint arXiv:2002.11376 (2020)
7. Tov, O., Alaluf, Y., Nitzan, Y., Patashnik, O., Cohen-Or, D.: Designing an encoder for styleGAN image manipulation. ACM Trans. Graph. (TOG) **40**, 1–14 (2021)
8. Robinson, J.P., Shao, M., Wu, Y., Liu, H., Gillis, T., Fu, Y.: Visual kinship recognition of families in the wild. IEEE Trans. Pattern Anal. Mach. Intell. **40**, 2624–2637 (2018)
9. Qin, X., Tan, X., Chen, S.: Tri-subjects kinship verification: understanding the core of a family. In: 2015 14th IAPR International Conference on Machine Vision Applications (MVA), pp. 580–583 (2015)
10. Chong, M.J., Chu, W.S., Kumar, A., Forsyth, D.: Retrieve in style: unsupervised facial feature transfer and retrieval. In: Proceedings of the IEEE/CVF International Conference on Computer Vision (ICCV) (2021)
11. Liu, S., et al.: Face aging with contextual generative adversarial nets. In: Proceedings of the 25th ACM International Conference on Multimedia (2017)
12. Tang, H., Bai, S., Sebe, N.: Dual attention GANs for semantic image synthesis. In: Proceedings of the 28th ACM International Conference on Multimedia (2020)
13. Brock, A., Donahue, J., Simonyan, K.: Large scale GAN training for high fidelity natural image synthesis. In: International Conference on Learning Representations (ICLR) (2019)
14. Zhao, J., et al.: Dual-agent GANs for photorealistic and identity preserving profile face synthesis. In: Advances in Neural Information Processing Systems 30 (2017)
15. Karras, T., Laine, S., Aila, T.: A style-based generator architecture for generative adversarial networks. In: Proceedings of the IEEE/CVF Conference on Computer Vision and Pattern Recognition (CVPR) (2019)
16. Abdal, R., Qin, Y., Wonka, P.: Image2styleGAN: how to embed images into the stylegan latent space? In: Proceedings of the IEEE/CVF International Conference on Computer Vision (2019)

17. Zhang, L., Bai, X., Gao, Y.: SalS-GAN: spatially-adaptive latent space in style-GAN for real image embedding. In: Proceedings of the 29th ACM International Conference on Multimedia, pp. 5176–5184 (2021)
18. Sainburg, T., Thielk, M., Theilman, B., Migliori, B., Gentner, T.: Generative adversarial interpolative autoencoding: adversarial training on latent space interpolations encourage convex latent distributions. arXiv preprint arXiv:1807.06650 (2018)
19. Mukherjee, S., Asnani, H., Lin, E., Kannan, S.: ClusterGAN: latent space clustering in generative adversarial networks. Proceed. AAAI Conf. Artif. Intell. **33**, 4610–4617 (2019)
20. Bojanowski, P., Joulin, A., Lopez-Paz, D., Szlam, A.: Optimizing the latent space of generative networks. arXiv preprint arXiv:1707.05776 (2017)
21. Robinson, J.P., Shao, M., Fu, Y.: Survey on the analysis and modeling of visual kinship: a decade in the making. In: IEEE Transactions on Pattern Analysis and Machine Intelligence (2021)
22. Ertuğrul, I.Ö., Jeni, L.A., Dibeklioğlu, H.: Modeling and synthesis of kinship patterns of facial expressions. Image Vis. Comput. **79**, 133–143 (2018)
23. Sinha, R., Vatsa, M., Singh, R.: FamilyGAN: generating kin face images using generative adversarial networks. In: Bartoli, A., Fusiello, A. (eds.) ECCV 2020. LNCS, vol. 12537, pp. 297–311. Springer, Cham (2020). https://doi.org/10.1007/978-3-030-67070-2_18
24. Ghatas, F.S., Hemayed, E.E.: GANKIN: generating kin faces using disentangled GAN. SN Appl. Sci. **2**, 166 (2020)
25. Zaman, I., Crandall, D.: Genetic-GAN: synthesizing images between two domains by genetic crossover. In: European Conference on Computer Vision Workshops (ECCVW) (2020)
26. Cui, X., Zhou, W., Hu, Y., Wang, W., Li, H.: Heredity-aware child face image generation with latent space disentanglement. arXiv preprint arXiv:2108.11080 (2021)
27. Alaluf, Y., Patashnik, O., Cohen-Or, D.: Only a matter of style: age transformation using a style-based regression model. ACM Trans. Graph. (TOG) **40**, 1–12 (2021)
28. Kafri, O., Patashnik, O., Alaluf, Y., Cohen-Or, D.: StyleFusion: a generative model for disentangling spatial segments. arXiv preprint arXiv:2107.07437 (2021)
29. Or-El, R., Sengupta, S., Fried, O., Shechtman, E., Kemelmacher-Shlizerman, I.: Lifespan age transformation synthesis. In: Vedaldi, A., Bischof, H., Brox, T., Frahm, J.-M. (eds.) ECCV 2020. LNCS, vol. 12351, pp. 739–755. Springer, Cham (2020). https://doi.org/10.1007/978-3-030-58539-6_44
30. Wang, X., Li, Y., Zhang, H., Shan, Y.: Towards real-world blind face restoration with generative facial prior. In: The IEEE Conference on Computer Vision and Pattern Recognition (CVPR) (2021)
31. Chen, S., Liu, Y., Gao, X., Han, Z.: MobileFaceNets: Efficient CNNs for accurate real-time face verification on mobile devices. CoRR abs/1804.07573 (2018)
32. Richardson, E., et al.: Encoding in style: a styleGAN encoder for image-to-image translation. In: Proceedings of the IEEE/CVF Conference on Computer Vision and Pattern Recognition (CVPR) (2021)

Occluded Facial Expression Recognition Using Self-supervised Learning

Jiahe Wang[1], Heyan Ding[1], and Shangfei Wang[1,2]($\boxtimes$)

[1] Key Lab of Computing and Communication Software of Anhui Province, University of Science and Technology of China, Hefei, China
{pia317,dhy0513}@mail.ustc.edu.cn
[2] Anhui Robot Technology Standard Innovation Base, University of Science and Technology of China, Hefei, China
sfwang@ustc.edu.cn

Abstract. Recent studies on occluded facial expression recognition typically required fully expression-annotated facial images for training. However, it is time consuming and expensive to collect a large number of facial images with various occlusions and expression annotations. To address this problem, we propose an occluded facial expression recognition method through self-supervised learning, which leverages the profusion of available unlabeled facial images to explore robust facial representations. Specifically, we generate a variety of occluded facial images by randomly adding occlusions to unlabeled facial images. Then we define occlusion prediction as the pretext task for representation learning. We also adopt contrastive learning to make facial representation of a facial image and those of its variations with synthesized occlusions close. Finally, we train an expression classifier as the downstream task. The experimental results on several databases containing both synthesized and realistic occluded facial images demonstrate the superiority of the proposed method over state-of-the-art methods.

Keywords: Occluded facial expression recognition · Self-supervised learning · Representation learning

1 Introduction

Facial expressions play an important role in our daily communication. In recent years, facial expression recognition has attracted increasing attention and achieved great progress [4,8,18,21] because of its application in many fields, such as psychological treatment, security and service robots. However, it is still challenging to recognize facial expressions from occluded facial images.

Current methods of occluded facial expression recognition can be divided into four categories: robust facial representation, non-occluded facial image reconstruction, sub-region analysis, and non-occluded facial image help. Robust facial representation methods aim to locate the representation that is insensitive to occlusion but discriminative for expression recognition. It is very difficult to find a robust representation because the types of occlusion are diverse and positions of occlusion are infinite. Non-occluded facial image reconstruction methods aim to construct non-occluded facial images using a generative model and train the facial expression classifier from the reconstructed facial images. However, the generated facial images are typically not as realistic as the real images, and this effects the performance of facial expression recognition. Sub-region analysis methods divide the image into several regions and recognize expressions from these regions and the entire image. Dividing facial images typically requires facial landmarks, and the attention mechanism is used to select important regions. However, facial landmark detection from the occluded facial image remains challenging. Non-occluded image help methods adopt non-occluded facial images as privileged information to assist occluded facial expression recognition. During training, these methods typically construct two networks: one for non-occluded facial expression recognition and the other for occluded facial expression recognition. During testing, these methods assume that all facial images are occluded and only the network for occluded expressions is used, whereas in a realistic scenario, we do not know whether the facial image is occluded or not. Furthermore, all the above methods require fully expression-annotated images for training. Because the types and positions of occlusion are infinite, collecting a large-scale dataset with various facial expressions and occlusions is difficult.

To address this, we propose an occluded facial expression recognition method through self-supervised learning [6]. We use a large number of unlabeled facial images in the pretext task to learn a robust and occlusion-insensitive facial representation. First, we synthesize many occluded images by randomly adding different occlusions to a large number of images. We apply occlusion detection as the pretext task to learn the representation. We also use contrastive learning to make the representation of facial image and those of its variations with synthesized occlusions similar. Finally, we set occluded expression recognition as the downstream task.

Our contributions are as follows: We are the fisrt to introduce a large number of unlabeled facial images for occluded expression recognition through self-supervised learning. We design an occluded detection and similarity constraint between the occluded and non-occluded facial images as the pretext task.

2   Related Work

Because of the variability of occlusion, occluded facial expression recognition is still a big challenge. Current work can be classified into four categories: sub-region analysis [3,10,11,18], robust facial representation [2,20], non-occluded facial image reconstruction [12,17] and non-occluded facial image help [16,19].

Sub-region analysis methods typically divide the image into several regions and obtain results from the regions and the entire image. These methods often apply the attention mechanism. Wang *et al.* [18] proposed a region attention network that adjusted the importance of facial parts and designed a region biased loss function to obtain a high attention weight for the important region. Li *et al.* [11] presented a patch-gated convolutional neural networks (PG-CNN) for facial expression recognition under occlusion. The PG-CNN chose 24 interest patches that were fed into an attention network to extract local features and learn an unobstructed score. The final classifier was constructed based on the weighted concatenated local features of all regions. Then, Li *et al.* [10] further proposed the global gated unit to add the global information of facial images for facial expression recognition. Dapogny *et al.* [3] used random forests to train partially defined local subspaces of the face and adapted local expression predictions as high-level representations. Then they weighted confidence scores provided by an autoencoder network. However, to divide the images or obtain regions, these methods typically require facial landmarks. It is difficult to detect facial landmarks from occluded facial images, which greatly affects the results of sub-region analysis methods. The detected error may be propagated to the classification task.

Robust facial representation methods aim to find a visual representation that is robust to occlusion. Zhang *et al.* [20] used a Monte Carlo algorithm to extract a set of Gabor templates from images and converted these templates into template match distance features. Cornejo and Pedrini [2] used robust principal component analysis to reconstruct occluded facial regions and then extracted census transform histogram features. However these methods do not have good generalization ability. It is difficult to find features that are insensitive to occlusion, because the positions of occlusion are unlimited.

Non-occluded facial image reconstruction methods exploit a deep generative model to construct non-occluded facial images. Lu *et al.* [12] exploited a generator to complement the non-occluded image and then the generated complementation image was used to predict the expression. They used reconstruction loss, triplet loss and adversarial loss to implement occluded facial image complementation. Ranzato *et al.* [17] used a deep belief network to construct a non-occluded face from the occluded face and then predicted the expression from the complete face. However, their visualization of occluded images was not good because of the unlimited positions and types of occlusion. Errors caused by an inaccurate reconstruction facial image may be propagated to the final task.

Non-occluded facial image help methods train facial expression classifiers from occluded facial images with the assistance of non-occluded facial images. Generally, non-occluded facial images have more useful information than occluded facial images. Pan *et al.* [16] used non-occluded images as privileged information to enhance the occluded classifier. Pan *et al.* trained two deep neural networks for occluded and non-occluded images separately and then used the non-occluded network to guide the occluded network. Xia *et al.* [19] proposed a stepwise learning strategy to obtain a robust network. They divided occluded and non-occluded images into three subsets from simple to difficult. Then they input the three subsets into the network to learn parameters in stages. They also used least squares generative adversarial networks [14] to reduce the feature gap between occluded facial images and non-occluded facial images. However, non-occluded facial image help methods trained two distinct networks for occluded and non-occluded images. During the tests, all facial images were assumed to be occluded, which may have affected the results of non-occluded facial images.

To summarize, existing methods require a large number of expression-labeled occluded images. However, occluded facial images with expression labels are difficult to collect. Thus, we generate a large number of occluded facial images from unlabeled facial images to simulate real occluded images, where the positions and types of occlusion vary. Therefore, in this study, we propose an occluded facial expression recognition method that uses self-supervised learning. Specifically, we design a pretext task related to occluded facial images. We apply contrastive learning to maximize the similarity between the facial image and its variations with synthesized occlusions. We also design an occlusion detection task as the pretext task. Then we set expression recognition as the downstream task and add a classifier to fine-tune the model.

3   Method

The framework of the proposed occluded facial expression recognition through self-supervised learning is shown in Fig. 1. The training process is mainly composed of two tasks. In the pretext task, we use the pre-training set to obtain an initial feature extractor F. In the downstream task, we add a classifier C after feature extractor F, and use the training set to fine-tune the parameters of extractor F and classifier C.

3.1   Problem Statement

Let $\mathcal{D}_{pre\_train} = \left\{ x_{pc}^{(i)} \right\}_{i=1}^{N_p}$ denote the pre-training set of N_p training samples, where $x_{pc}^{(i)} \in \mathbb{R}^{H \times W \times 3}$ is a non-occluded facial image, which has no expression label. We generate occluded facial image $x_{po}^{(i)}$ from non-occluded facial image $x_{pc}^{(i)}$. We randomly select the type of occlusions from N_c types of occlusions, and the positions of occlusions are random. $x_{po}^{(i)}$ has a corresponding occlusion mask $\mathbf{M} \in \{0,1\}^{H \times W}$, where H and W denote the height and width of the input image respectively. $\mathcal{D}_{fine\_train} = \left\{ x_{fo}^{(i)}, y^{(i)} \right\}_{i=1}^{N_{fo}} \cup \left\{ x_{fc}^{(i)}, y^{(i)} \right\}_{i=1}^{N_{fc}}$ denotes the training set of $N_{fo} + N_{fc}$ training samples, which have expression labels. We obtain $\left\{ x_{fo}^{(i)} \right\}_{i=1}^{N_{fo}}$ by adding occlusion to $\left\{ x_{fc}^{(i)} \right\}_{i=1}^{N_{fc}}$. $y^{(i)} \in \{0, 1, ..., N_e - 1\}$ represents the expression label of the ith sample. $\mathcal{D}_{test} = \{x_o^{(i)}\}_{i=1}^{N_1} \cup \{x_c^{(i)}\}_{i=1}^{N_2}$ denotes the N_1 occluded and N_2 non-occluded testing samples. Given $\mathcal{D}_{pre\_train}$ and $\mathcal{D}_{fine\_train}$, we first use self-supervised learning to obtain the initial parameters of extractor F on the pre-training set, and then fine-tune extractor F and classifier C on the training set. Our goal is to learn a network $f : \mathbb{R}^{H \times W \times 3} \rightarrow \{0, 1, \cdots, N_e - 1\}$, which improves prediction for occluded expression images.

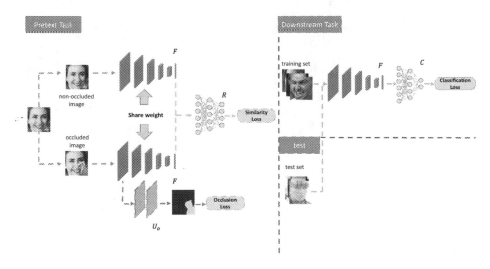

Fig. 1. Framework of the method. In the pretext task, extractor F extracts features; mask recognition network U_o predicts the location of the occlusion; and projection head R outputs the facial representation. In the downstream task, classifier C predicts the expression.

3.2 Pretext Task

Because of the infinite types and positions of occlusion, we always lack expression-annotated facial images with occlusions for learning robust facial representation. Fortunately, we may synthesize a large number of facial images with various occlusions. Therefore, we design pretext tasks to learn representations from facial images with synthesized occlusions.

To extract distinguishing and occlusion-insensitive features, we use many unlabeled facial images to obtain good initialization parameters of extractor F in the pretext task.

Specifically, we add occlusion to a non-occluded facial image to obtain an occluded facial image. Then we introduce an occlusion detector after a certain layer of the feature extractor to predict the occlusion position. Through occlusion detection, the learned features from a certain layer of the feature extractor may be aware of the importance of facial areas. Then, we adopt contrastive learning to to make the learned representation of facial image and those of its variations with synthesized occlusions similar. Therefore, the learned features are robust to occlusion.

Similarity Loss. After we generate occluded facial images from non-occluded facial images, we expect that these representation will be similar. Contrastive learning maximizes the similarity between positive pairs and minimizes the similarity between negative pairs, which meets our needs. Given a mini-batch containing K samples $\{x_{pc}^{(j)}\}_{j=1}^{K}$, we randomly add occlusion to $\{x_{pc}^{(j)}\}_{j=1}^{K}$ to obtain

$\{x_{po}^{(j)}\}_{j=1}^K$. Then both $x_{po}^{(j)}$ and $x_{pc}^{(j)}$ are fed into extractor F to extract features $h_{po}^{(j)} = F(x_{po}^{(j)})$ and $h_{pc}^{(j)} = F(x_{pc}^{(j)})$. We use projection head R to obtain facial representations $z_{po}^{(j)} = R(h_{po}^{(j)})$ and $z_{pc}^{(j)} = R(h_{pc}^{(j)})$. Because $x_{po}^{(j)}$ is transformed from $x_{pc}^{(j)}$, we consider $(z_{po}^{(j)}, z_{pc}^{(j)})$ as the positive pair, $(z_{po}^{(j)}, z)_{z \in \{z_{po}^i, z_{pc}^i\}_{i=1}^K / \{z_{po}^j, z_{pc}^j\}}$ and $(z_{pc}^{(j)}, z)_{z \in \{z_{po}^i, z_{pc}^i\}_{i=1}^K / \{z_{po}^j, z_{pc}^j\}}$ as the negative pair. In our case, the positive pair has a higher similarity than the negative pair. We adopt cosine similarity to measure the similarity.

The similarity loss is

$$\mathcal{L}_{SS} = \frac{1}{2K} \sum_{j=1}^K (\mathcal{L}_{spo}^j + \mathcal{L}_{spc}^j) \tag{1}$$

where $\mathcal{L}_{spo}^j$ and $\mathcal{L}_{spc}^j$ are the similarity loss of $z_{po}^{(j)}$ and $z_{pc}^{(j)}$, respectively. The specific forms of $\mathcal{L}_{spo}^j$ and $\mathcal{L}_{spc}^j$ are

$$
\begin{aligned}
\mathcal{L}_{spo}^j &= -\log \frac{\exp\left(sim\left(z_{po}^{(j)}, z_{pc}^{(j)}\right)/\tau\right)}{\sum\limits_{z \neq z_{po}^{(j)}} \exp\left(sim\left(z_{po}^{(j)}, z\right)/\tau\right)} \\[2mm]
\mathcal{L}_{spc}^j &= -\log \frac{\exp\left(sim\left(z_{po}^{(j)}, z_{pc}^{(j)}\right)/\tau\right)}{\sum\limits_{z \neq z_{pc}^{(j)}} \exp\left(sim\left(z_{pc}^{(j)}, z\right)/\tau\right)}
\end{aligned}
\tag{2}
$$

where τ is the temperature parameter and $sim(u, v) = u^\top v / \|u\| \|v\|$ denotes the cosine similarity between two vectors u and v. By minimizing $\mathcal{L}_{spo}^j$ and $\mathcal{L}_{spc}^j$, the similarity of positive pairs in the numerator is increased and the similarity of negative pairs in the denominator is decreased.

Occlusion Loss. In facial images, the occluded area typically contains less or even no information about the facial expression. In the pretext task, the occluded position of synthesized occluded facial images is easy to obtain. When generating a synthesized occluded facial image, we can obtain a binary mask $\mathbf{M}$ about the occlusion position, where 1 represents occlusion and 0 represents no occlusion. We expect that the network can contain the occluded position information; hence, we add a mask recognition network U_o to obtain $\hat{\mathbf{M}}$ to predict binary mask $\mathbf{M}$. Because $\mathbf{M}$ is a binary mask, we use the cross-entropy loss to optimize the result. The occlusion loss is

$$
\begin{aligned}
\mathcal{L}_{mask} = -\frac{1}{H \times W} \sum_{j,k} (&\mathbf{M}[j,k] \log \hat{\mathbf{M}}[j,k] \\
&+ (1 - \mathbf{M}[j,k]) \log(1 - \hat{\mathbf{M}}[j,k]))
\end{aligned}
\tag{3}
$$

where $\mathbf{M}[j, k]$ denotes whether the point (j, k) is occluded. $\hat{\mathbf{M}}[j, k]$ represents the probability of predicting whether the point (j, k) belongs to occlusion. The point (j, k) represents the point in the kth row and j^{th} column.

The overall pretext task loss is defined as

$$\mathcal{L}_{pre} = \mathcal{L}_{SS} + \lambda \mathcal{L}_{mask} \tag{4}$$

where λ is a hyperparameter that balances the trade-off between the two losses.

The overall loss considers both the occluded position and facial features. We optimize the model by minimizing the overall pretext task loss.

3.3 Downstream Task

We set occluded expression recognition as the downstream task. In the pretext task, we obtain extractor F for occluded facial images. In the downstream task, we obtain feature extractor F obtained in the previous step and add a classifier C after it. Then we use the training set to fine-tune extractor F and classifier C. $\hat{y} = C(F(x))$ performs facial expression recognition for facial image x, where $x \in \mathcal{D}_{fine\_train}$. We use cross-entropy loss to measure the difference between it and the truth label. The classification loss is

$$\mathcal{L}_{cla} = \mathcal{L}_{CE}\left(\hat{y}, y\right) \tag{5}$$

3.4 Optimization

Among the losses, we only use $\mathcal{L}_{SS}$ and $\mathcal{L}_{mask}$ in the pretext task, and $\mathcal{L}_{cla}$ in the downstream task. We first use $\mathcal{L}_{SS}$ and $\mathcal{L}_{mask}$ to obtain extractor F, and then use $\mathcal{L}_{cla}$ to fine-tune extractor F and classifier C. We use Adam [7] to update the parameters.

4 Experiment

4.1 Experimental Conditions

For the pretext task, we chose a large-scale face recognition dataset VGGFace2 [1] as the pre-training set. Following the experimental conditions in Pan *et al.*'s study [16] and Li *et al.*'s study [11], we conducted within-database experiments on synthesized occluded databases, that is, the Real-world Affective Faces Database (RAF-DB) [9], AffectNet database [15], and extended Cohn-Kanade database (CK+) [13]. When testing our method on the Facial Expression Dataset with Real Occlusions (FED-RO) [10], we merged AffectNet and RAF-DB as the training set. We also tested our methods on the original test databases, that is, RAF-DB and AffectNet. Following the experimental conditions in Wang *et al.*'s study [18], we tested our model on Occlusion-AffectNet [18] and Occlusion-RAF-DB [18]. The details are as follows:

VGGFace2 includes 3.31M images from 9,131 subjects. Each subject has an average of 362.6 images. The database is downloaded from Google Image Search and has large variations in ethnicity, age, and pose. In our experiment, we used the VGGFace2 database as our pre-training set.

RAF-DB contains approximately 30K real-world images annotated with basic or compound expressions using 40 annotators. In our experiment, we only used images with seven basic expressions (i.e., neutral, happiness, sadness, surprise, fear, disgust and anger); hence, we used 12,271 images as the training set and 3,068 images as the test set.

AffectNet was collected from the internet by querying expression-related keywords. It contains more than 1M facial images, of which approximately 450K images were annotated by 12 human experts. In our experiment, we used facial images with seven basic expressions, which included approximately 280K images as the training set and 3,500 as the test set.

CK+ consists of 593 sequences from 123 subjects. The image sequence begins with the onset frame and ends with the apex frame. We collected onset and apex frames as neutral and target expressions. In our experiment on CK+, we collected 636 facial images and adopted 10-fold subject-independent cross-validation.

Occlusion-AffectNet and Occlusion-RAF-DB were selected from the validation set of AffectNet and test set of the RAF-DB by Wang et al. [18]. In these two test sets, we used the same experimental conditions as those in the Wang et al.' study. Occlusion-AffectNet includes 683 realistic occluded facial images with eight basic expressions (i.e., neutral, happiness, sadness, surprise, fear, disgust, anger and contempt). Because Occlusion-AffectNet contains eight basic expressions, we used images with eight basic expressions from AffectNet, which included approximately 287K images as the training set and Occlusion-AffectNet as the test set. Occlusion-RAF-DB contains 735 realistic occluded facial images with seven basic expressions. In our experiment, we used images from RAF-DB with seven basic expressions, which included 12,271 images as the training set and Occlusion-RAF-DB as the test set.

Fig. 2. Examples of realistic occluded facial images in FED-RO and the synthesized occluded facial images in AffectNet. The first and second rows are realistic occluded facial images, and the third and fourth rows are synthesized occluded facial images.

FED-RO is the first facial expression dataset to present real occlusions in the wild, and was collected by Li *et al.* [11]. FED-RO contains 400 images, which are labeled with seven basic expressions. Because FED-RO is small, we only used it for cross-database evaluation.

To mimic real-world scenarios, we artificially synthesized occluded facial images by adding occluding objects at random locations in all databases except the three real occlusion facial expression databases, that is, FED-RO, Occlusion-AffectNet, and Occlusion-RAF-DB. We used a variety of occlusion types to synthesize occluded facial images. The position of occlusion in each facial image was random. To compare our method with Pan *et al.*'s method [16], we used the same type of occlusion as that in Pan *et al.*'s study: food, hands and drinks. In Fig. 2, we show some examples of realistic occluded facial images and synthesized occluded facial images in FED-RO and AffectNet. Because the Occlusion-AffectNet database on Wang *et al.*'s work and the AffectNet databases on Pan *et al.*'s work use different facial expression numbers, we use C7 to denote the seven classification task and C8 to denote the eight classification task. Affect-Net(C7) and Occlusion-AffectNet(C8) represent seven and eight facial expression recognition tasks separately.

We conducted ablation experiments to verify the effect of the pretext task and the two different loss functions in the pretext task, that is, occlusion loss and similarity loss on AffectNet, Occlusion-AffectNet, RAF-DB, Occlusion-RAF-DB and FED-RO. First, we directly used ResNet-34 without pre-training as the baseline, which we refer to as the non-pretext task. Second, we used $\mathcal{L}_{SS}$ or $\mathcal{L}_{mask}$ as the pretext task, denoted by $\mathcal{L}_{SS}$ or $\mathcal{L}_{mask}$. Finally, we used $\mathcal{L}_{SS}$ and $\mathcal{L}_{mask}$ as the pretext task, denoted by $\mathcal{L}_{SS} + \mathcal{L}_{mask}$.

The implementation of the proposed method is based on the PyTorch framework. Because the AffectNet database is imbalanced, we resampled the data during training. We exploited ResNet-34 [5] to extract features. We built two fully connected layers as the small neural network projection head R to obtain the dimensional representation; two convolutional layers and an upsampling layer as the mask recognition network U_o and two fully connected layers as the classifier C. We add the mask recognition network U_o after the Conv1 layer of the ResNet-34. In our experiments, we resized the facial images in CK+ to 48×48 pixels and other images to 224×224 pixels. When conducting experiments on CK+, we used five types of occlusion, that is, 8×8 occlusion, 16×16 occlusion, 24×24 occlusion, eye occlusion and mouth occlusion to illustrate the robustness of the model to occlusions of different sizes. The batch size was 64. The hyperparameter λ was 0.2. The temperature parameter τ was 2. The learning rate of the network was $1e^{-4}$. We determined the hyperparameter in the loss function using grid search.

4.2   Experimental Results and Analysis

Analysis of Facial Expression Recognition Without Occlusions. The experimental results of facial expression recognition without occlusions are shown in Table 1. We trained the method on AffectNet(C7) and RAF-DB and tested it on the original validation set of AffectNet(C7) and the original test set of RAF-DB, separately. Table 1 yields the following observations.

Table 1. Experimental results of facial expression recognition without occlusions on the AffectNet(C7) and the RAF-DB databases. (C7 represents seven classifications.)

Methods	AffectNet(C7)	RAF-DB
PG-CNN [11]	55.33	83.27
gACNN [10]	58.78	85.07
Non-pretext	57.09	82.53
$\mathcal{L}_{SS}$	59.66	84.09
$\mathcal{L}_{mask}$	58.40	83.18
$\mathcal{L}_{SS} + \mathcal{L}_{mask}$	**60.20**	**85.95**

First, using $\mathcal{L}_{SS}$ or $\mathcal{L}_{mask}$ led to an improvement compared with the baseline non-pretext task. Specifically, the accuracies of $\mathcal{L}_{SS}$ and $\mathcal{L}_{mask}$ were 2.57% and 1.31% higher than that of the non-pretext task on AffectNet(C7). The experimental results on RAF-DB databases demonstrated a similar trend. Guidance regarding both the feature and occluded position helped the extractor F to learn more robust feature representations.

Table 2. Experimental results of facial expression recognition under synthesized occlusions on RAF-DB, AffectNet(C7) and CK+. (R8, R16 and R24 denote the sizes of the occlusions: 8×8, 16×16 and 24×24 respectively. The AffectNet(C7) represents seven classifications.)

Methods	RAF-DB	AffectNet(C7)	CK+				
			R8	R16	R24	eye occluded	mouth occluded
RGBT [20]	72.56	49.21	92.00	82.00	62.50	88.00	30.30
WLS-RF [3]	74.66	51.74	92.20	86.40	74.80	87.90	72.70
PG-CNN [11]	78.05	52.47	96.58	95.70	92.86	96.50	93.92
gACNN [10]	80.54	54.84	96.58	95.97	94.82	96.57	93.88
Pan et al.'s work [16]	81.97	56.42	97.80	**96.86**	94.03	96.86	93.55
Xia et al.'s work [19]	82.74	57.46	98.01	96.22	**95.91**	97.17	95.44
Non-pretext	81.45	54.06	96.91	94.65	94.18	95.12	94.33
$\mathcal{L}_{SS}$	83.18	57.09	97.64	96.07	94.97	96.23	95.60
$\mathcal{L}_{mask}$	82.53	55.03	97.17	95.28	94.65	95.75	95.12
$\mathcal{L}_{SS} + \mathcal{L}_{mask}$	**84.06**	**58.40**	**98.27**	96.70	95.59	**97.33**	**96.07**

Second, similarity loss was more effective than occlusion loss, which were 1.26% and 0.91% higher than using only occlusion loss on AffectNet(C7) and RAF-DB. This may be because the test image was non-occluded; hence occlusion loss had little effect.

Table 3. Experimental results of facial expression recognition under realistic occlusions on FED-RO, Occlusion-AffectNet(C8) and Occlusion-RAF-DB. (C8 represents eight classifications.)

Methods	FED-RO	Occlusion-AffectNet(C8)	Occlusion-RAF-DB
PG-CNN [11]	64.25	–	–
gACNN [10]	66.50	–	–
Pan *et al.*'s work [16]	69.75	–	–
Xia *et al.*'s work [19]	**70.50**	–	–
RAN [18]	67.98	58.50	**82.72**
Non-pretext	68.25	55.93	79.73
$\mathcal{L}_{SS}$	69.25	58.86	81.09
$\mathcal{L}_{mask}$	69.00	57.25	80.41
$\mathcal{L}_{SS} + \mathcal{L}_{mask}$	70.00	**59.30**	82.45

Table 4. Experimental results of λ sensitivity analysis on Occlusion-AffectNet(C8)

λ	Acc(%)
0.02	58.51
0.2	59.30
2	58.42
20	57.54

Third, our method achieved the best performance using similarity loss and occlusion loss together. Specifically, the accuracies of our method were 3.11%, 0.54% and 1.80% higher than those of the non-pretext task, $\mathcal{L}_{SS}$ and $\mathcal{L}_{mask}$ on AffectNet(C7), and 3.42%, 1.86% and 2.77% higher on RAF-DB, respectively. The pretext task of using both similarity loss and occlusion loss helped the downstream task to learn a more robust facial representation and make better predictions.

Analysis of Facial Expression Recognition with Synthesized Occlusions. The experimental results of facial expression recognition with synthesized occlusions are shown in Table 2. We trained our method on AffectNet(C7) and RAF-DB, and tested it on the synthesized occluded AffectNet(C7) and RAF-DB test set. The table yields the following observations:

First, our method achieved the best performance using both occlusion loss and similarity loss. Specifically, the accuracies of our method were 2.61%, 0.88% and 1.53% higher than that of non-pretext, $\mathcal{L}_{SS}$ and $\mathcal{L}_{mask}$ on RAF-DB, and 4.34%, 1.31% and 3.37% higher on AffectNet(C7), respectively . Such observations are consistent with experiments without occlusion.

Second, the experiment on CK+ obtained good results. The results demonstrated that our method was robust to occlusions of different sizes. As the size

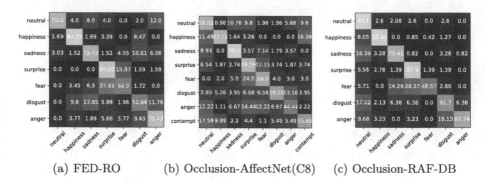

(a) FED-RO   (b) Occlusion-AffectNet(C8)   (c) Occlusion-RAF-DB

Fig. 3. Confusion matrix on three realistic occlusion databases. (Row indices represent the ground truth labels, whereas column indices represent the predictions.)

of the occlusion increased, the classification accuracy decreased. This demonstrated that the larger the size of the occlusion, the more useful information about the facial expression was occluded, and the more difficult it was to recognize facial expressions. The result of mouth occluded is lower than eye occluded. This demonstrated that the mouth contains more expression information than the eyes. The experimental results for synthesized occluded facial expression recognition were similar to those of non-occluded facial expression recognition. This demonstrated that our method was helpful in both non-occluded facial expression recognition and occluded facial expression recognition.

Analysis of Facial Expression Recognition with Realistic Occlusions.
The experimental results of facial expression recognition with realistic occlusions are shown in Table 3. We trained the method on AffectNet(C8) and RAF-DB and tested it on Occlusion-AffectNet(C8) and Occlusion-RAF-DB separately. We merged AffectNet(C7) and RAF-DB to obtain the training set and used FED-RO as the test set for the cross-database experiment. Table 3 yields the following observations.

First, the performance using both occlusion loss and similarity loss as the pretext task obtained better classification accuracy. For example, the accuracies of our method were 3.37%, 0.44% and 2.05% higher than those of the non-pretext task, $\mathcal{L}_{SS}$, and $\mathcal{L}_{mask}$ on Occlusion-AffectNet(C8); 1.75%, 0.75% and 1.00% higher on FED-RO; and 2.72%, 1.36% and 2.04% higher on Occlusion-RAF-DB, respectively. In addition, we also make a sensitivity analysis on λ on Occlusion-AffectNet(C8), and the results are shown in Table 4. The results show that the lambda we set can well balance the trade-off between the two losses.

Second, we investigated the per expression category classification performance on FED-RO, Occlusion-AffectNet(C8) and Occlusion-RAF-DB. The confusion matrices based on our method are shown in Fig. 3. These matrices show that our method achieved high accuracy in the happiness category. Many images of fear were mistakenly classified as surprise. The results demonstrated that it was difficult to distinguish between fear and surprise. The correct rate of disgust was relatively low on the three test sets. This may be because the numbers of facial images showing disgust in AffectNet and the RAF-DB were relatively small.

Third, we obtained good results for realistic occlusions, synthesized occlusions, and the original test sets, which indecates the effectiveness of our method. The pretext task helped facial expression recognition. To visualize the perception ability of the mask recognition network U_o, we also created a result map of the generated mask, which is shown in Fig. 4. Our network found the location of the synthesized occlusions. Our network also detected part of the real occluded positions.

Fig. 4. Examples of the learned occluded masks on AffectNet and FED-RO. The first two rows are the synthesized occluded images and corresponding learned occluded areas. The last two rows are realistic occluded images and corresponding learned occluded areas.

4.3   Comparison with Related Work

To illustrate the superiority of our method, we compared it with state-of-the-art methods, that is, RGBT [20], WLS-RF [3], PG-CNN [11] and gACNN [10] on realistic occluded databases, synthesized occluded databases, and the original (non-occluded) test set. Because the Occlusion-AffectNet(C8) and the Occlusion-RAF-DB databases are new collected by Wang *et al.* [18], on these databases, we only compare our method to RAN [18].

Our method obtained better results on most datasets. Table 1 shows that our method obtained better performance on RAF-DB and AffectNet(C7) for non-occluded facial images. Specifically, the accuracies of our method were 4.87% and 1.42% higher than those of PG-CNN and gACNN on AffectNet(C7), and 2.68% and 0.88% higher on RAF-DB, respectively. We also compared our method on synthesized occluded databases. As shown in Table 2, our method achieved better accuracy than PG-CNN, gACNN, Pan *et al.*'s method, and Xia *et al.*'s method by 6.01%, 3.52%, 2.09% and 1.32% on RAF-DB, and 5.93%, 3.56%, 1.98% and 0.94% on AffectNet(C7), respectively. Our method also achieved superior performance under four types of occlusion: eye occluded, mouth occluded, 8×8 size of occlusion and 24×24 size of occlusion on CK+.

RGBT uses hand crafted features to help facial expression recognition, which lacks generalization. WLS-RF does not use the guidance of non-occluded facial images and the entire framework is not trained end to end. PG-CNN uses an attention network to extract local features from facial regions of the convolutional feature maps. gACNN introduces the global gated unit to complement the global information of facial images, which extends PGCNN. Although these two methods use the attention mechanism to pay attention to non-occluded regions, these methods typically require facial landmarks to locate sub-regions. Pan *et al.*'s method uses non-occluded images to guide the occluded expression classifier. However, it trains two distinct networks for non-occluded and occluded images, so it cannot predict occluded and non-occluded facial emotion in a single network. Xia *et al.*'s method divides the dataset into three parts based on the difficulty of the database, and then the network is learned in three stages. The end of each stage may not be optimal and requires human control. Our method obtained competitive results and recognized occluded and non-occluded facial expressions end to end, and effectively used the occluded and non-occluded facial image information.

Finally, we compared the generalization ability of our method with that of related methods for realistic facial images. The experimental results with realistic occlusions in Table 3 demonstrate that our method outperformed most related methods. Specifically, our method achieved better accuracy than PG-CNN, gACNN and Pan *et al.*'s method by 5.75%, 3.50% and 0.25% on FED-RO, respectively. Our method also achieved 2.02% and 0.80% higher accuracy than RAN on FED-RO and Occlusion-AffectNet(C8), respectively. These results demonstrate that our method also obtained good results on realistic facial images.

5   Conclusion

In this study, we proposed an occluded expression recognition method through self-supervised learning. We designed pretext tasks related to occluded facial images. We adopted similarity loss to make the representation of facial image and those of its variations with synthesized occlusions similar, and used occlusion loss to optimize the occlusion detection. Our method achieved better results than most state-of-the-art methods on both occluded and non-occluded facial images, which demonstrates its superiority.

Acknowledgements. This work was supported by National Natural Science Foundation of China 92048203 and project from Anhui Science and Technology Agency 202104h04020011.

References

1. Cao, Q., Shen, L., Xie, W., Parkhi, O.M., Zisserman, A.: VggFace2: a dataset for recognising faces across pose and age. In: 2018 13th IEEE International Conference on Automatic Face & Gesture Recognition (FG 2018), pp. 67–74. IEEE (2018)
2. Cornejo, J.Y.R., Pedrini, H.: Recognition of occluded facial expressions based on centrist features. In: 2016 IEEE International Conference on Acoustics, Speech and Signal Processing (ICASSP), pp. 1298–1302. IEEE (2016)
3. Dapogny, A., Bailly, K., Dubuisson, S.: Confidence-weighted local expression predictions for occlusion handling in expression recognition and action unit detection. Int. J. Comput. Vis. **126**(2), 255–271 (2018)
4. Georgescu, M.I., Ionescu, R.T., Popescu, M.: Local learning with deep and hand-crafted features for facial expression recognition. IEEE Access **7**, 64827–64836 (2019)
5. He, K., Zhang, X., Ren, S., Sun, J.: Deep residual learning for image recognition. In: Proceedings of the IEEE Conference on Computer Vision and Pattern Recognition, pp. 770–778 (2016)
6. Jing, L., Tian, Y.: Self-supervised visual feature learning with deep neural networks: a survey. IEEE Trans. Pattern Anal. Mach. Intell. (2020)
7. Kingma, D.P., Ba, J.: Adam: a method for stochastic optimization. In: 3rd International Conference on Learning Representations, ICLR 2015, San Diego, CA, USA, May 7–9, 2015, Conference Track Proceedings (2015)
8. Li, S., Deng, W.: Deep facial expression recognition: a survey. IEEE Trans. Affect. Comput. (2020)
9. Li, S., Deng, W., Du, J.: Reliable crowdsourcing and deep locality-preserving learning for expression recognition in the wild. In: Proceedings of the IEEE Conference on Computer Vision and Pattern Recognition, pp. 2852–2861 (2017)
10. Li, Y., Zeng, J., Shan, S., Chen, X.: Occlusion aware facial expression recognition using CNN with attention mechanism. IEEE Trans. Image Process. **28**(5), 2439–2450 (2018)
11. Li, Y., Zeng, J., Shan, S., Chen, X.: Patch-gated CNN for occlusion-aware facial expression recognition. In: 2018 24th International Conference on Pattern Recognition (ICPR), pp. 2209–2214. IEEE (2018)
12. Lu, Y., Wang, S., Zhao, W., Zhao, Y.: WGAN-based robust occluded facial expression recognition. IEEE Access **7**, 93594–93610 (2019)
13. Lucey, P., Cohn, J.F., Kanade, T., Saragih, J., Ambadar, Z., Matthews, I.: The extended cohn-kanade dataset (ck+): a complete dataset for action unit and emotion-specified expression. In: 2010 IEEE Computer Society Conference on Computer Vision and Pattern Recognition-Workshops, pp. 94–101. IEEE (2010)
14. Mao, X., Li, Q., Xie, H., Lau, R.Y., Wang, Z., Paul Smolley, S.: Least squares generative adversarial networks. In: Proceedings of the IEEE International Conference on Computer Vision, pp. 2794–2802 (2017)
15. Mollahosseini, A., Hasani, B., Mahoor, M.H.: AffectNet: a database for facial expression, valence, and arousal computing in the wild. IEEE Trans. Affect. Comput. **10**(1), 18–31 (2017)
16. Pan, B., Wang, S., Xia, B.: Occluded facial expression recognition enhanced through privileged information. In: Proceedings of the 27th ACM International Conference on Multimedia, pp. 566–573 (2019)
17. Ranzato, M., Susskind, J., Mnih, V., Hinton, G.: On deep generative models with applications to recognition. In: CVPR 2011, pp. 2857–2864. IEEE (2011)

18. Wang, K., Peng, X., Yang, J., Meng, D., Qiao, Y.: Region attention networks for pose and occlusion robust facial expression recognition. IEEE Trans. Image Process. **29**, 4057–4069 (2020)
19. Xia, B., Wang, S.: Occluded facial expression recognition with step-wise assistance from unpaired non-occluded images. In: Proceedings of the 28th ACM International Conference on Multimedia, pp. 2927–2935 (2020)
20. Zhang, L., Tjondronegoro, D., Chandran, V.: Random Gabor based templates for facial expression recognition in images with facial occlusion. Neurocomputing 145(Dec. 5), 451–464 (2014)
21. Zhao, R., Liu, T., Xiao, J., Lun, D.P.K., Lam, K.M.: Deep multi-task learning for facial expression recognition and synthesis based on selective feature sharing. Int. Conf. Pattern Recogn. (2020)

Heterogeneous Avatar Synthesis Based on Disentanglement of Topology and Rendering

Nan Gao⬡, Zhi Zeng⬡, GuiXuan Zhang^(✉)⬡, and ShuWu Zhang⬡

Institute of Automation Chinese Academy of Sciences, Beijing, China
{nan.gao,zhi.zeng,guixuan.zhang,shuwu.zhang}@ia.ac.cn

Abstract. There are obviously structural and color discrepancies among different heterogeneous domains. In this paper, we explore the challenging heterogeneous avatar synthesis (HAS) task considering topology and rendering transfer. HAS transfers the topology as well as rendering styles of the referenced face to the source face, to produce high-fidelity heterogeneous avatars. Specifically, first, we utilize a Rendering Transfer Network (RT-Net) to render the grayscale source face based on the color palette of the referenced face. The grayscale features and color style are injected into RT-Net based on adaptive feature modulation. Second, we apply a Topology Transfer Network (TT-Net) to conduct heterogeneous facial topology transfer, where the image content of RT-Net is transferred based on AdaIN controlled by heterogeneous identity embedding. Comprehensive experimental results show that the disentanglement of rendering and topology is beneficial to the HAS task, and our HAS-Net has comparable performance compared with other state-of-the-art methods.

Keywords: Image synthesis · Style transfer · Disentanglement representation learning

1 Introduction

Avatar means another stylized identity in the heterogeneous domain. As for face images, there are various topology patterns for facial components, such as 3D cartoon, 2D anime, sketch, nir, real-world or other domains. Style transfer methods [11,13,19,20,26] change the textural style and preserve the content of the source image, guided by the referenced image. These methods are universal to different heterogeneous domains but ignore the facial topology transfer. To adapt to the target face distribution, some GAN-based methods [17,24,28] are proposed to address two-domain face translations. However, these methods have obvious topology and color distortions.

The high-fidelity HAS task is also a style transfer task, which is supposed to possess high-fidelity facial topology and global color consistencies with the reference-domain face, while preserving as much as other content and attributes

ⓒ The Author(s), under exclusive license to Springer Nature Switzerland AG 2023
L. Wang et al. (Eds.): ACCV 2022, LNCS 13844, pp. 137–152, 2023.
https://doi.org/10.1007/978-3-031-26316-3_9

Fig. 1. HASNet realizes high-fidelity heterogeneous avatar synthesis based on topology and rendering transfer, where HASNet handles various heterogeneous domains.

of the source face, e.g., facial pose, expression, background and hair. Heterogeneous faces have diverse facial component topology patterns. Moreover, since the different domains might have a large color distribution discrepancy, it poses us a non-trivial challenge to implement HAS task without considering rendering consistency. Therefore, to produce more vivid heterogeneous avatars, both rendering and topology styles deserve to be manipulated. As for color rendering, ColorThief is used to extract prominent color palette values. As for topology adaptation, the pretrained identity extractor is capable of capturing discriminative facial component shapes, e.g., Arcface extracts the prominent structural feature of a face. In our paper, we conduct rendering transfer based on the color palette from ColorThief[1], and implement topology transfer based on the manifold of Arcface [6]. This disentanglement framework realizes high-fidelity heterogeneous avatar synthesis. As shown in Fig. 1, the results of HASNet have a good performance on style transfer of global color and facial topology, while preserving other contents and attributes of the source faces.

Our contributions are three folds as follows.

– We propose a two-stage framework to explore the challenging and meaningful heterogeneous avatar synthesis task. We propose an effective rendering transfer framework. By considering adaptive modulation of color and spa-

---

[1] http://lokeshdhakar.com/projects/color-thief/.

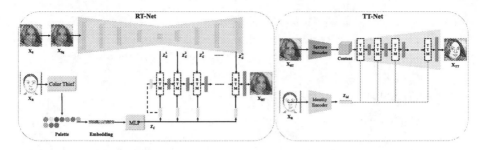

Fig. 2. Two-stage framework of HASNet. RT-Net helps HASNnet implement higher-fidelity heterogeneous rendering. TT-Net refines the facial topology style of the heterogeneous avatars.

tial content, RT-Net achieves topology-aware and high-quality colorization in various heterogeneous domains.

- Based on the first stage, we propose TT-Net to realize high-fidelity and controllable facial topology transfer across various heterogeneous domains.
- We observe HASNet variants and conduct comparative experiments with state-of-the-art universal and GAN-based style transfer methods. Experimental results demonstrate a good performance of HASNet both in visual and quantitative manners.

2 Related Work

2.1 Universal Style Transfer

There are some universal style transfer methods [11,13,19,26]. [11] explores the domain-aware characteristics from the texture and topology features of referenced images to handle both artistic and photo-realistic style transfer. AdaIN [13] makes the variance and mean values of the source features aligned with those of the reference features, which achieves real-time universal style transfer. WCT [19] enables universal style transfer based on disentangled whitening and coloring operations. SANet [26] integrates the style patterns according to the semantic spatial attentional map of the source content.

2.2 GAN-Based Methods

StyleGANs [14–16] synthesize high-resolution images based on noise disentanglement and style modulation. UGATIT [17] modulates the generator applying the discriminative domain-aware style embedding and proposes AdaILN. NICE-GAN [5] reuses the discriminator to implement unsupervised image-to-image translation. [7] proposes an unsupervised face rotation method considering disentangling the facial shape and identity. Toonify [28] trains the generator and discriminator for 3D cartoon and real-world domain, respectively. Then the base and transferred models are used to create the interpolated translation model. FS-Ada [24] proposes a transfer learning schedule to impose the adaption to diverse target domains from the source model, based on limited data.

2.3 Cross-Domain Image Synthesis

As for VIS-NIR translation, [8,9] demonstrate that dual heterogeneous face mapping is beneficial to recognizing heterogeneous faces. [33] utilizes the inter-supervision and intra-supervision style transfer for Sketch-VIS task. [32] designs a memorized Grayscale-VIS colorization model with AdaIN modulation. As for heterogeneous datasets for research, CASIA NIR-VIS 2.0 [18], Oulu-CASIA NIR-VIS [4] and BUAA-VisNir Face [12] databases are widely used in the NIR-VIS task. CUHK Face Sketch FERET (CUFSF) [34], IIIT-D Viewed Sketch database [2] and XM2VTS database [22] are established in Sketch-VIS research. [17] proposes the selfie2anime dataset in the anime domain. IIIT-CFW [23] provides some faces with exaggerated drawings. Tufts dataset [25] presents subjects captured in different imaging domains, e.g. RGB, NIR and Thermal.

3   Approach

Source   Ref   RT-Net   w/o RT-Net   w/ RT-Net

Fig. 3. RT-Net helps HASNnet implement higher-fidelity heterogeneous rendering.

We design a two-stage framework to synthesize heterogeneous avatars, as shown in Fig. 2. In stage I, given a source face X_S, we obtain the grayscale image X_{Sg} by means of the transformation from RGB to Lab space. We apply ColorThief to extract the dominant color palette of the reference face X_R. Based on the spatial features of X_{Sg} and color style z_C, we modulate RT-Net to obtain the rendering transfer face X_{RT}. In stage II, we conduct topology transfer based on X_{RT} to synthesize X_{TT} whose facial topology distribution is consistent with X_R. In this way, HASNet generates high-fidelity heterogeneous avatars.

3.1   RT-Net

As shown in Fig. 3, the color consistency between the facial area and the background is higher for w/ RT-Net than w/o RT-Net, which indicates the importance of RT-Net for HAS task. More examples are shown in Fig. 5.

In RT-Net, X_{Sg} is mapped to the multi-scale spatial feature maps $z_G = \{z_G^1, z_G^2, ..., z_G^n\}$ via an Unet [29]. A multi-FC mapping network MLP is leveraged to convert the extracted color palette embedding $C_{X_R} \in \mathbb{R}^{20 \times 3}$ to color style $z_C \in \mathbb{R}^{256}$, which is used to modulate multi-scale decoder of RT-Net. We integrate these two embeddings adopting the Rendering Transfer Module (RTM) in an adaptive alignment manner.

Specifically, the input feature $h^i \in \mathbb{R}^{C_h^i \times H^i \times W^i}$ of RTM_i is aligned with guidance of two groups of learning parameters from z_G^i and z_C, respectively. It is formulated as:

$$h_I^i = \frac{h^i - \mu_I^i}{\sigma_I^i}, \tag{1}$$

$$\{G, C\}_I^i = \gamma_{\{G,C\}_I}^i \odot h_I^i + \beta_{\{G,C\}_I}^i, \tag{2}$$

where μ_I^i and σ_I^i, i.e., the mean and standard deviation of h^i, are used as instance-aware normalization parameters. Let $z_G^i \in \mathbb{R}^{C_G^i \times H_G^i \times W_G^i}$ and $z_C \in \mathbb{R}^{C_C \times 1}$ be the grayscale and color embedding, respectively. Furthermore, affine transform parameters $\gamma_{G_I}^i$ and $\beta_{G_I}^i \in \mathbb{R}^{C_h^i \times H^i \times W^i}$ are obtained from z_G^i using a convolutional layer, similar with SPADE [27]. Meanwhile, $\gamma_{C_I}^i$ and $\beta_{C_I}^i \in \mathbb{R}^{C_h^i \times H^i \times W^i}$ are mapped from z_C using a full connection transform, inspired by AdaIN [13]. To adaptively integrate the spatial and color modulation, the attentional maps $M_{G_I}^i$ and $M_{C_I}^i$ from h_I^i arc generated after passing a convolution layer followed by a sigmoid operation. The instance denormalization result H_I^i is denoted as

$$H_I^i = G_I^i \odot M_{G_I}^i + C_I^i \odot M_{C_I}^i, \tag{3}$$

where $\odot$ represents the element-wise product, and $M_{G_I}^i + M_{C_I}^i = 1$.

As for the objective functions, in the training stage of RT-Net, we use Huber loss where $\triangle_{X_{RT}} = |X_{RT} - X_S|$, to constrain the colored image reconstruction via

$$\mathcal{L}_{rec} = \begin{cases} \frac{1}{2}\triangle_{X_{RT}}^2 & \triangle_{X_{RT}} \leq \delta \\ \delta \cdot [\triangle_{X_{RT}} - \frac{1}{2}\delta] & otherwise \end{cases}. \tag{4}$$

Moreover, we optimize a discriminator for the generated X_{RT} with conditions of X_{Sg} and C_{X_S}. The total loss of RT-Net is as follows

$$\mathcal{L}_{RT-Net} = \mathcal{L}_{adv}(X_{Sg}, C_{X_S}, X_{RT}) + \lambda_{rec}\mathcal{L}_{rec}. \tag{5}$$

In the test time, the color style is extracted from the heterogeneous reference face X_R, to lessen the color gap between the source and reference domains.

3.2   TT-Net

After stage I, the global rendering style has been transferred to the reference heterogeneous domain. Stage II designs TT-Net to further transfer the facial topology style to produce vivid HAS results.

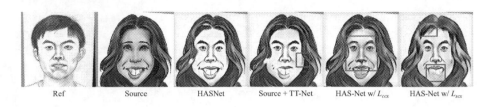

| Ref | Source | HASNet | Source + TT-Net | HAS-Net w/ L_{ccx} | HAS-Net w/ L_{scx} |

Fig. 4. $\mathcal{L}_{SCX}$ helps HASNnet implement higher-fidelity topology transfer.

In the multi-level topology transfer modulation (TTM) module, we inject z_{id} to each feature level of TT-Net by implementing layer-aware AdaIN operations, where we use StyledConv block of StyleGAN [16]. It is denoted as:

$$AdaIN(Con^i, \gamma_{id}^i, \beta_{id}^i) = \gamma_{id}^i \odot \frac{Con^i - \mu^i}{\sigma^i} + \beta_{id}^i, \qquad (6)$$

where $Con^i \in \mathbb{R}^{C_{Con}^i \times H^i \times W^i}$ is the spatial information of TTM_i. σ^i and μ^i are the standard deviation and mean values of Con^i. γ_{id}^i and $\beta_{id}^i \in \mathbb{R}^{C_{Con}^i \times H^i \times W^i}$ based on z_{id} are sent to address the instance denormalization. The initial content feature Con^0 is extracted by VGG [30] for X_{RT}.

As for the objective functions, we use the reconstruction loss as the pixel-level supersision between X_{TT}^{ij} and X_S^i, when the source face X_S^i and reference face X_R^j are same. It is denoted as

$$\mathcal{L}_{rec} = \begin{cases} \frac{1}{2} \left\| X_{TT}^{ij} - X_S^i \right\|_2^2 & if \ i = j \\ 0 & otherwise \end{cases} . \qquad (7)$$

We use identity consistency loss on X_{TT}^{ij} for HAS face synthesis as follows. Specifically, we measure the distance between the Arcface manifolds of X_{TT}^{ij} and X_R^j via

$$\mathcal{L}_{id} = 1 - \langle z_{id}(X_{TT}^{ij}), z_{id}(X_R^j) \rangle, \qquad (8)$$

where $\langle \cdot, \cdot \rangle$ means cosine similarity, z_{id} is the pretrained Arcface [6].

Furthermore, TT-Net utilizes the contextual similarity loss [21] for X_{TT}^{ij} and X_R^j to improve the contextual distribution perception. It is denoted as

$$\mathcal{L}_{SCX} = -\log(CX(F_{vgg}^l(X_{TT}^{ij}), F_{vgg}^l(X_R^j))), \qquad (9)$$

where l means $relu_{3\_2}$ and $relu_{4\_2}$ layers of the pretrained VGG19 model [30]. At the same time, we introduce a content contextual similarity loss between X_{TT}^{ij} and X_S^i for content preservation of the source face.

$$\mathcal{L}_{CCX} = -\log(CX(F_{vgg}^l(X_{TT}^{ij}), F_{vgg}^l(X_S^i))). \qquad (10)$$

As shown in Fig. 4, only with $\mathcal{L}_{CCX}$ fails to synthesize the topology of the sketch domain, while only with $\mathcal{L}_{SCX}$ transfers excessive reference styles ignoring

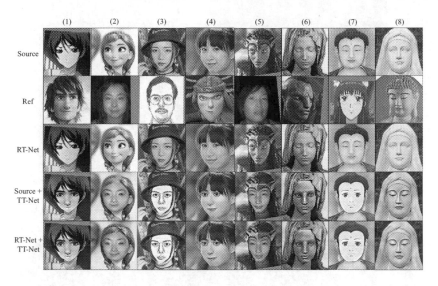

Fig. 5. Stage I and stage II of HASNet. RT-Net helps HASNet implement higher-fidelity heterogeneous rendering.

the spatial content preservation of the source face. More ablation analyses are shown in Fig. 9. Therefore, TT-Net combines both $\mathcal{L}_{CCX}$ and $\mathcal{L}_{SCX}$.

The total loss of TT-Net is as follows

$$\mathcal{L}_{TT-Net} = \mathcal{L}_{adv} + \lambda_{id}\mathcal{L}_{id} + \lambda_{rec}\mathcal{L}_{rec} + \\ \lambda_{CCX}\mathcal{L}_{CCX} + \lambda_{SCX}\mathcal{L}_{SCX}. \tag{11}$$

4 Experiment

4.1 Dataset

We collect three kinds of heterogeneous domains to conduct HAS task as follows:

- Lighting condition variants. We select daytime portraits with variable attributes and identities from FFHQ [15]. We collect some night portraits taken in the evening or at night. And we randomly select some samples in CASIA NIR-VIS 2.0 dataset [18].
- Art drawing variants. 3D cartoons, anime images, sketches, exaggerated drawings and oil paintings are collected from the Toonify dataset [28], selfie2 anime dataset [17], CUHK Face Sketch FERET (CUFSF) [34], IIIT-CFW [23] and MetFaces[2], respectively. Moreover, there are some sculptures of famous people (only for research). And we collect some role faces of Beijing Opera. Note that we obtain the super-resolution counterparts using [31] for the low-quality face samples.

---

[2] https://github.com/postite/metfaces-dataset.

Fig. 6. Ablation study of HAS-Net concerning the dimension of initial content feature in TT-Net.

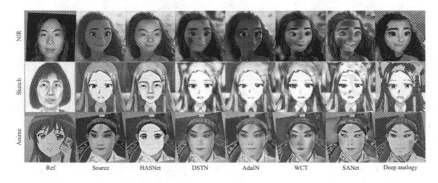

Fig. 7. Qualitative comparison with other state-of-the-art universal style transfer methods in NIR, sketch and anime style domains.

- Life dimension variants. We collect some Buddha statues images (only for research) from the Internet. We collect the faces from the movie *Avatar*. And we select some drawings of ancient people in Chinese culture.

Note that we take all kinds of heterogeneous faces, i.e. about 4000 images, as a whole training set in Stage I, where the true color palette of the source face is used to modulate RT-Net. While TT-Net takes random different heterogeneous-domain faces as the source and reference faces, respectively. There are 13 kinds of heterogeneous domains considered in the single generator of TT-Net, which is different from other GAN-based methods [17, 24, 28] that focus on certain two-domain translations.

Fig. 8. Additional comparison results of HASNet and other methods. HASNet realizes high-fidelity HAS based on topology and rendering transfer, where the identity and attributes are more controllable with better background preservation.

4.2   Implementation Details

The source and target faces of our dataset are aligned and cropped based on 5 facial landmarks [31]. We reshape all samples to 512×512 resolution. We set $Con^0 \in \mathbb{R}^{512 \times 32 \times 32}$, $z_{id} \in \mathbb{R}^{512 \times 1}$, $z_{Palette} \in \mathbb{R}^{20 \times 3}$, and $z_C \in \mathbb{R}^{256 \times 1}$. There are 7 RTM and 10 TTM modules. In Eq. 5, $\lambda_{rec} = 10$. In Eq. 11, $\lambda_{adv} = \lambda_{SCX} = 1$, $\lambda_{rec} = 100$, $\lambda_{id} = 10$, and $\lambda_{CCX} = 0.5$. We set batchsize to 4. RT-Net and TT-Net are trained separately. We adapt the architecture of our discriminator networks from [1] and [15] for RT-Net and TT-Net, respectively.

In the architecture of StyleGAN [15], the coarse-resolution (4×4–32×32) layers (1–7) are used to control shape modulation. Moreover, we conduct a toy experiment and find that the dimensions of the initial content feature of X_{RT} have an important impact on the TT-Net. As shown in Fig. 6, Con_{16} has more background distortion, which is not enough to conduct the high-fidelity HAS task that considers background topology preservation. Furthermore, Con_{64} provides more constraints to the source content, so that it fails to transfer the shape style of eyes in the Avatar style domain. Our HASNet uses the initial Con^0 with 32×32 resolution and shows visually satisfying HAS behavior.

4.3 Qualitative Evaluation

As shown in Fig. 7, DSTN [11] synthesizes distorted and messy faces with the wrong global color style in NIR and sketch domains. AdaIN [13] achieves a good textural transfer in the global view but is not competent to transferring shape styles of eyes, nose and mouth. WCT [19] also has obvious color distortion, especially in the NIR domain. SANet [26] has some artifacts on the face area and background. Deep analogy [20] is another universal style transfer method that conducts patch matching based on vgg features, which is time-consuming and easy to cause matching errors. Furthermore, although the local component is more vivid than [11,13,19,26], the eye size is still similar to that of the source faces. HASNet realizes sufficient topology and rendering transfer. RT-Net bridges the color gap between the source and reference images. And TT-Net modifies the source facial topology to the reference component distribution.

As for the GAN-based method, we compare HASNet with Toonify [28], FS-Ada [24] and U-GAT-IT [17], as shown in Fig. 8. Toonify finds an optimized latent code of the source, and then feeds this code to the reference domain generator. Its results have obvious background distortions (rows 7, 9), and bad heterogeneous rendering effects (rows 1–6). And the nose of avatars are not identifiable enough compared with HASNet. FS-Ada has easily detectable artifacts concerning topology and color, especially in anime domains. U-GAT-IT has an unstable cross-domain translation performance, e.g., the topological and color artifacts on the facial area (rows 2&7), and does not respect the facial occlusions (row 2). HASNet has higher topology and rendering consistencies with the reference faces, while better preserving the background content of the source images. From the above qualitative evaluations, we find a user study is deserved to be surveyed, considering that human eyes are highly sensitive to the topology and color appearance. More details are introduced in Sect. 4.4.

4.4 User Study

User study plays an important role in the quality evaluation of the HAS task. First, we briefly explain the HAS task, and invite 10 users to carefully observe the source, reference and HAS faces. Each type of face has 30 samples where NIR, sketch and anime domains have 10 samples respectively. These observers need to give scores recorded as 1–10, from four aspects: (a) transferred topology perception, (b) content preservation of the source face, (c) color transfer of the reference face, and (d) overall preference, where the higher HAS quality is reflected by higher scores. Finally, we collect 300 score tables, where each table contains 52 human decisions about different methods and indexes. The average scores are displayed in Table 1.

Specifically, transferred topology perception measures the identification degree and topological completeness of the HAS results referenced to the target

Table 1. User study results of HASNet variants, as well as the universal and GAN-based style transfer methods.

Methods	Topology↑	Content↑	Color↑	Preference↑
DSTN [11]	1.2	1.9	7.8	2.7
AdaIN [13]	0.9	4.4	8.3	3.2
WCT [19]	1.8	3.7	8.4	3.3
SANet [26]	1.9	1.7	7.9	2.5
Deep analogy [26]	2.2	4.6	8.1	4.5
Toonify [28]	8.0	4.5	3.5	7.1
FS-Ada [24]	6.5	4.9	7.6	7.9
U-GAT-IT [17]	7.2	5.5	5.8	8.4
HASNet	**8.9**	**9.1**	**8.5**	**9.0**
w/o RT-Net	8.8	8.9	5.1	8.4
w/ $\mathcal{L}_{CCX}$	7.3	8.9	4.7	4.7
w/ $\mathcal{L}_{SCX}$	8.5	7.1	7.0	3.3
w/o $\mathcal{L}_{CX}$	6.9	6.5	3.1	2.0

style images. Content preservation of the source face focuses on the face pose, expression, hair and other background content. These contents are supposed to be faithful to the source image to some extent. HAS task mainly transfers the global color and facial component topology. The color index represents the color obedience guided by the color style of the reference faces, as well as the diffusion degree of the colorization. The preference score means the overall preference degree for users. HASNet has the best scores on topology, content and color aspects. Moreover, HASNet has better perception scores than Toonify and U-GAT-IT. We show more cross-domain HAS results in Fig. 10, which indicates that our approach synthesizes controllable heterogeneous avatars.

4.5   Ablation Study

As shown in Fig. 9, we show some diverse HAS results of variants of HASNet. Specifically, RT-Net successfully transfers the color style to the source faces based on the reference images. As shown in row 5, HASNet has higher consistency between the facial area and the background (cols 4&5). While direct TT-Net based on the original source image will result in a large color discrepancy between the results and the reference faces, which hinders high-fidelity heterogeneous avatar synthesis. w/ $\mathcal{L}_{CCX}$ maintains more spatial grayscale information of the faces of RT-Net (col 4), which stays far from the topology style of the referenced sketch face. Moreover, this variant produces low-fidelity cartoon eyes (cols 6–8). w/ $\mathcal{L}_{SCX}$ transfers excessive reference style, e.g., hair (cols 1–3), and carries uncontrollable artifacts (cols 5–8). w/o $\mathcal{L}_{CX}$ means HASNet without $\mathcal{L}_{SCX}$ and $\mathcal{L}_{CCX}$, which results in more color artifacts.

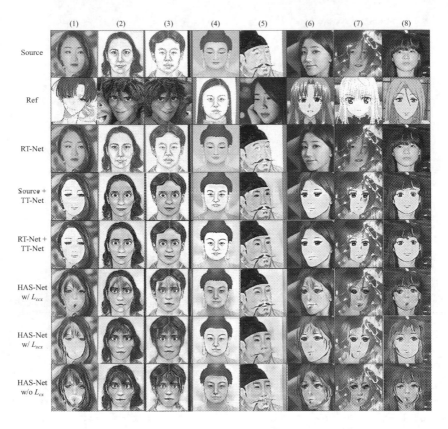

Fig. 9. Ablation study of HAS-Net. We show the results of RT-Net, Source+TT-Net, RT-Net+TT-Net (HASNet), only w/ L_{CCX}, only w/ L_{SCX} and w/o L_{CX}.

4.6 Quantitative Evaluation

Quantitative results are only the evaluation reference for performance. It is more scientific to consider both the objective score and user study. If one kind of heterogeneous domain is the reference domain, we select 10 images from each of other 12 kinds of domains as the source dataset in the test stage of HAS task. We employ two evaluation metrics considering the image fidelity (FID [10], KID [3]). They are used in [17] to measure the cross-domain synthesis quality. As for the evaluation of topology transfer, we measure the identity distance of HAS results and the reference face based on Arcface. Moreover, we calculate the color distance of the first 20 prominent color values between the HAS result and the reference face based on ColorThief. As shown in Table 2, we compare HASNet with good-performance style transfer approaches including fine-tuned DSTN [11], AdaIN [13], WCT [19] and SANet [26] in three distinctive domains, i.e., NIR, sketch and Anime. Note that Deep analogy [20] needs to consume lots of inference time, so we only show the visual comparisons in Fig. 7.

Table 2. Quantitative evaluation on CASIA NIR-VIS 2.0 [18], CUHK [34] and selfie2anime [17] datasets. Our model has better ID scores and comparable color scores. As for FID, due to the specificity of the HAS problem, we need to preserve the background, which has an impact on FID because FID score is calculated based on the whole image. In our opinion, a controllable HAS task is supposed to consider both the vivid avatar synthesis and the topology preservation of the background. Otherwise, the essential heterogeneous scene information of the source will be seriously lost (Fig. 8).

Domains	Methods	ID × 10↓	Color × 10↓	FID ↓	KID × 100 ↓	Overall↓
NIR	DSTN [11]	9.3	0.3	185.48	**14.6**	209.68
	AdaIN [13]	9.3	0.4	212.52	19.01	241.23
	WCT [19]	9.1	**0.3**	210.72	18.54	238.66
	SANet [26]	9.2	0.5	199.25	18.17	227.12
	HASNet	4.5	0.6	195.7	18.11	218.91
	w/o RT-Net	4.4	1.2	206.89	19.41	231.9
	w/ $\mathcal{L}_{CCX}$	3.3	1.4	206.95	19.13	230.78
	w/ $\mathcal{L}_{SCX}$	2.8	0.5	**177.26**	16.28	**196.84**
	w/o $\mathcal{L}_{CX}$	**2.6**	1.2	210.72	21.11	235.63
CUHK	DSTN [11]	8.7	0.4	**162.74**	**12.3**	**184.14**
	AdaIN [13]	9	0.5	187.42	16.93	213.85
	WCT [19]	8.8	0.5	185.07	16.22	210.59
	SANet [26]	8.6	**0.4**	166.71	15.2	190.91
	HASNet	5	0.7	166.73	15	187.43
	w/o RT-Net	4.9	1.3	177.37	16.14	199.71
	w/ $\mathcal{L}_{CCX}$	3.8	1.2	177.81	15.8	198.61
	w/ $\mathcal{L}_{SCX}$	3.1	0.6	173.09	17.07	193.86
	w/o $\mathcal{L}_{CX}$	**2.8**	1.3	186.43	18.1	208.63
Anime	DSTN [11]	7.9	0.4	138.5	7.45	154.25
	AdaIN [13]	7.6	0.6	115.29	**5.56**	129.05
	WCT [19]	7.4	**0.3**	124.65	7.72	140.07
	SANet [26]	7.6	0.7	113.47	6.09	127.86
	HASNet	4.6	0.7	112.3	7.36	124.96
	w/o RT-Net	4.7	1.2	111.92	7.24	125.06
	w/ $\mathcal{L}_{CCX}$	2.5	1.2	155.46	12.11	171.27
	w/ $\mathcal{L}_{SCX}$	2.1	0.5	**101.8**	6.19	**110.59**
	w/o $\mathcal{L}_{CX}$	**1.8**	1.1	128.75	9.90	141.55

Concretely, [11,13,19,26] have a slightly better color consistency score than HASNet. However, as for DSTN [11] in the NIR domain, the results have purple color, which has a worse perception in human eyes. This demonstrates the necessity of the user study. These state-of-the-art universal style transfer methods have worse identity consistency scores than HASNet. DSTN has better FID

Table 3. Quantitative evaluation with GAN-based methods on CASIA NIR-VIS 2.0 [18], CUHK [34] and selfie2anime [17] datasets. However, a better FID score does not prove better performance in the HAS task. U-GAT-IT has better FID scores, but with visual distortions of color, shape, and background (Fig. 8).

Domains	Methods	ID × 10↓	Color × 10↓	FID ↓	KID × 100 ↓
NIR	Toonify [28]	8.3	0.9	**109.73**	**7.36**
	FS-Ada [24]	8.4	1.2	191.53	19.5
	U-GAT-IT [17]	7.7	0.8	118.94	9.9
	HASNet	**4.3**	**0.7**	195.46	17.94
CUHK	Toonify [28]	7.3	1.7	105.06	8.47
	FS-Ada [24]	7.3	0.3	117.46	7.82
	U-GAT-IT [17]	7.6	**0.3**	**89.84**	**6.77**
	HASNet	**4.8**	0.8	160.61	13.64
Anime	Toonify [28]	5.7	1.3	173.3	13.65
	FS-Ada [24]	5.2	1.6	192.43	17.61
	U-GAT-IT [17]	4.7	1.3	**98.48**	**3.79**
	HASNet	**4.5**	**0.9**	125.21	8.67

Fig. 10. More HAS results across diverse heterogeneous domains.

and KID scores than that of HASNet in NIR and sketch domains, but with messy visual imaging, as shown in Fig. 7. w/o RT-Net has a color score drop, as well as obvious performance degradation of FID and KID in NIR and sketch domains. w/ $\mathcal{L}_{CCX}$ and w/o $\mathcal{L}_{CX}$ cause many color artifacts (Fig. 9) because of no constraint of $\mathcal{L}_{SCX}$, which get worse FID and KID scores. However, only $\mathcal{L}_{SCX}$ has lower FID and KID scores, e.g., in the anime domain, but with severe topology degradations (Fig. 9).

We further evaluate the performance compared with GAN-based methods, as shown in Table 3. There are 100 real-world faces from FFHQ as the source images. Toonify and U-GAT-IT have good FID and KID scores, and our HASNet has better identity and color consistency scores.

5   Conclusion

We explore the challenging heterogeneous avatar synthesis (HAS) task considering topology and rendering transfer. RT-Net and TT-Net alleviate the color and structural discrepancies between the HAS results and the reference faces, while preserving other source contents and attributes. Comprehensive experimental results in various heterogeneous domains show that the disentanglement of rendering and topology is beneficial to the HAS task. HASNet produces controllable and high-fidelity heterogeneous avatars.

Acknowledgments. This work is supported by the National Key R&D Program of China (2019YFB1406202).

References

1. Bahng, H., Yoo, S., Cho, W., et al.: Coloring with words: guiding image colorization through text-based palette generation. In: Proceedings of ECCV, pp. 431–447 (2018)
2. Bhatt, H.S., Bharadwaj, S., Singh, R., et al.: Memetically optimized MCWLD for matching sketches with digital face images. IEEE TIFS **7**(5), 1522–1535 (2012)
3. Bińkowski, M., Sutherland, D.J., Arbel, M., Gretton, A.: Demystifying mmd GANs. In: ICLR (2018)
4. Chen, J., Yi, D., Yang, J., et al.: Learning mappings for face synthesis from near infrared to visual light images. In: 2009 IEEE Conference on CVPR, pp. 156–163. IEEE (2009)
5. Chen, R., Huang, W., Huang, B., et al.: Reusing discriminators for encoding: towards unsupervised image-to-image translation. In: Proceedings of CVPR, pp. 8168–8177 (2020)
6. Deng, J., Guo, J., Xue, N., et al.: Arcface: additive angular margin loss for deep face recognition. In: Proceedings of CVPR, pp. 4690–4699 (2019)
7. Duan, B., Fu, C., Li, Y., et al.: Cross-spectral face hallucination via disentangling independent factors. In: Proceedings of CVPR, pp. 7930–7938 (2020)
8. Fu, C., Wu, X., Hu, Y., et al.: Dual variational generation for low shot heterogeneous face recognition. In: NIPS (2019)
9. Fu, C., Wu, X., Hu, Y., et al.: DVG-face: dual variational generation for heterogeneous face recognition. IEEE Trans. PAMI **44**(6), 2938–2952 (2021)
10. Heusel, M., Ramsauer, H., Unterthiner, T., et al.: GANs trained by a two time-scale update rule converge to a local nash equilibrium. In: NIPS (2017)
11. Hong, K., Jeon, S., Yang, H., et al.: Domain-aware universal style transfer. In: Proceedings of ICCV, pp. 14609–14617 (2021)
12. Huang, D., Sun, J., Wang, Y.: The BUAA-VisNir face database instructions. School Computer Science and Engineering, Beihang University, Beijing, China, Technical report, IRIP-TR-12-FR-001, vol. 3 (2012)
13. Huang, X., Belongie, S.: Arbitrary style transfer in real-time with adaptive instance normalization. In: Proceedings of ICCV, pp. 1501–1510 (2017)

14. Karras, T., Aittala, M., Laine, S., et al.: Alias-free generative adversarial networks. In: NIPS (2021)
15. Karras, T., Laine, S., Aila, T.: A style-based generator architecture for generative adversarial networks. In: Proceedings of CVPR, pp. 4401–4410 (2019)
16. Karras, T., Laine, S., Aittala, M., et al.: Analyzing and improving the image quality of StyleGAN. In: Proceedings of CVPR, pp. 8110–8119 (2020)
17. Kim, J., Kim, M., Kang, H., et al.: U-GAT-IT: Unsupervised generative attentional networks with adaptive layer-instance normalization for image-to-image translation. In: ICLR (2020)
18. Li, S., Yi, D., Lei, Z., et al.: The CASIA NIR-VIS 2.0 face database. In: Proceedings of the IEEE Conference on CVPR Workshops, pp. 348–353 (2013)
19. Li, Y., Fang, C., Yang, et al.: Universal style transfer via feature transforms. In: NIPS, pp. 386–396 (2017)
20. Liao, J., Yao, Y., Yuan, L., et al.: Visual attribute transfer through deep image analogy. In: SIGGRAPH (2017)
21. Mechrez, R., Talmi, I., Zelnik-Manor, L.: The contextual loss for image transformation with non-aligned data. In: Proceedings of ECCV, pp. 768–783 (2018)
22. Messer, K., Matas, J., Kittler, J., et al.: XM2VTSDB: the extended M2VTS database. In: Second International Conference on Audio and Video-Based Biometric Person Authentication, vol. 964, pp. 965–966. Citeseer (1999)
23. Mishra, A., Rai, S.N., Mishra, A., Jawahar, C.V.: IIIT-CFW: a benchmark database of cartoon faces in the wild. In: Hua, G., Jégou, H. (eds.) ECCV 2016. LNCS, vol. 9913, pp. 35–47. Springer, Cham (2016). https://doi.org/10.1007/978-3-319-46604-0_3
24. Ojha, U., Li, Y., Lu, J., et al.: Few-shot image generation via cross-domain correspondence. In: Proceedings of CVPR, pp. 10743–10752 (2021)
25. Panetta, K., Wan, Q., Agaian, S., et al.: A comprehensive database for benchmarking imaging systems. IEEE Trans. PAMI $42(3)$, 509–520 (2018)
26. Park, D.Y., Lee, K.H.: Arbitrary style transfer with style-attentional networks. In: Proceedings of CVPR, pp. 5880–5888 (2019)
27. Park, T., Liu, M.Y., Wang, T.C., et al.: Semantic image synthesis with spatially-adaptive normalization. In: Proceedings of CVPR, pp. 2337–2346 (2019)
28. Pinkney, J.N., Adler, D.: Resolution dependent gan interpolation for controllable image synthesis between domains. arXiv preprint arXiv:2010.05334 (2020)
29. Ronneberger, O., Fischer, P., Brox, T.: U-Net: convolutional networks for biomedical image segmentation. In: Navab, N., Hornegger, J., Wells, W.M., Frangi, A.F. (eds.) MICCAI 2015. LNCS, vol. 9351, pp. 234–241. Springer, Cham (2015). https://doi.org/10.1007/978-3-319-24574-4_28
30. Simonyan, K., Zisserman, A.: Very deep convolutional networks for large-scale image recognition. In: 3rd ICLR (2015)
31. Wang, X., Li, Y., Zhang, H., et al.: Towards real-world blind face restoration with generative facial prior. In: CVPR (2021)
32. Yoo, S., Bahng, H., Chung, S., et al.: Coloring with limited data: few-shot colorization via memory augmented networks. In: Proceedings of CVPR, pp. 11283–11292 (2019)
33. Zhang, M., Wang, R., Gao, X., et al.: Dual-transfer face sketch-photo synthesis. IEEE Trans. Image Process. $28(2)$, 642–657 (2018)
34. Zhang, W., Wang, X., Tang, X.: Coupled information-theoretic encoding for face photo-sketch recognition. In: CVPR 2011, pp. 513–520. IEEE (2011)

Pose and Action

Focal and Global Spatial-Temporal Transformer for Skeleton-Based Action Recognition

Zhimin Gao[1]📷, Peitao Wang[1], Pei Lv[1], Xiaoheng Jiang[1], Qidong Liu[1], Pichao Wang[2], Mingliang Xu[1(✉)], and Wanqing Li[3]

[1] Zhengzhou University, Zhengzhou, China
{iegaozhimin,ielvpei,jiangxiaoheng,ieqdliu,iexumingliang}@zzu.edu.cn
[2] DAMO Academy, Alibaba Group (U.S.) Inc., Bellevue, USA
[3] AMRL, University of Wollongong, Wollongong, Australia
wanqing@uow.edu.au

Abstract. Despite great progress achieved by transformer in various vision tasks, it is still underexplored for skeleton-based action recognition with only a few attempts. Besides, these methods directly calculate the pair-wise global self-attention equally for all the joints in both the spatial and temporal dimensions, undervaluing the effect of discriminative local joints and the short-range temporal dynamics. In this work, we propose a novel **F**ocal and **G**lobal **S**patial-**T**emporal Trans**former** network (FG-STFormer), that is equipped with two key components: (1) FG-SFormer: focal joints and global parts coupling spatial transformer. It forces the network to focus on modelling correlations for both the learned discriminative spatial joints and human body parts respectively. The selective focal joints eliminate the negative effect of non-informative ones during accumulating the correlations. Meanwhile, the interactions between the focal joints and body parts are incorporated to enhance the spatial dependencies via mutual cross-attention. (2) FG-TFormer: focal and global temporal transformer. Dilated temporal convolution is integrated into the global self-attention mechanism to explicitly capture the local temporal motion patterns of joints or body parts, which is found to be vital important to make temporal transformer work. Extensive experimental results on three benchmarks, namely NTU-60, NTU-120 and NW-UCLA, show our FG-STFormer surpasses all existing transformer-based methods, and compares favourably with state-of-the-art GCN-based methods.

Keywords: Action recognition · Skeleton · Spatial-temporal transformer · Focal joints · Motion patterns

1 Introduction

Human action recognition has long been a crucial and active research field in video understanding since it has a broad range of applications, such as human-computer interaction, intelligent video surveillance and robotics [4,34,44].

L. Wang et al. (Eds.): ACCV 2022, LNCS 13844, pp. 155–171, 2023.
https://doi.org/10.1007/978-3-031-26316-3_10

In recent years, skeleton-based action recognition has gained increasing attention with advent of cost-effective depth cameras like Microsoft Kinect [52] and advanced pose estimation techniques [2], which make skeleton data more accurate and accessible. By representing the action as a sequence of joint coordinates of human body, the highly abstracted skeleton data is compact and robust to illumination, human appearance changes and background noises.

Effectively modelling the spatial-temporal correlations and dynamics of joints is crucial for recognizing actions from skeleton sequences. The dominant solutions to it in recent years are the graph convolutional networks (GCNs) [46], as they can model the irregular topology of the human skeleton. Via designing advanced graph topology or traversal rules, the recognition performance is greatly improved by GCN-based methods [30,40]. Meanwhile, the recent success of Transformer [41] has gained significant interest and performance boost in various computer vision tasks [3,9,29,32]. For skeleton-based action recognition, one would expect that the self-attention mechanism in transformer shall naturally capture effective correlations of joints in both spatial and temporal dimensions for action categorization, without enforcing the articulating constrains of human body like GCN. However, there are only a few transformer-based attempts [33,38,51], and they devise hybrid model of GCN and transformer [33] or multi-task learning framework [51]. How to utilize self-attention to learn effective spatial-temporal relations of joints and representative motion features is still a thorny problem. Moreover, most of these Transformer based methods directly calculate the global one-to-one relations of joints for spatial and temporal dimensions respectively. Such strategy undervalues the spatial interactions of discriminative local joints and short-term temporal dynamics for identifying crucial action-related patterns. On the one hand, since not all joints are informative for recognizing actions [16,27], these methods suffer from the influence of irrelevant or noisy joints by accumulating the correlations with them via attention mechanism, which could harm the recognition. On the other hand, with the fact that the vanilla transformer lacks of inductive bias [29] to capture the locality of temporal structural data, it is difficult for these methods to directly model effective temporal relations of joints globally over long input sequence.

To tackle these issues, we propose a novel end-to-end **F**ocal and **G**lobal **S**patial-**T**emporal Trans**former** network, dubbed as FG-STFormer, to effectively capture relations of the crucial local joints and the global contextual information in both spatial and temporal dimensions for skeleton-based action recognition. It is composed of two components: FG-SFormer and FG-TFormer. Intuitively, each action can be distinguished by the co-movement of: (1) some critical local joints, (2) global body parts, and (or) (3) joint-part interaction. For example, as shown in Fig. 1, actions such as *taking a selfie* and *kicking* mainly involve important joints of hands, head and feet, as well as related body parts of arms and legs, while the actions like *sit down* primarily require understanding of body parts cooperation and dynamics. Based on the above observations, at the late stage of the network, we adaptively sample the informative spatial local joints (focal joints) for each action, and force the network to focus on modelling the correlations among them via multi-head self-attention without involving non-informative joints. Meanwhile, in order to compensate for the missing global co-movement and spatial structure information, we incorporate the dependencies among human body parts using self-attention. Furthermore, interactions between the body parts and the focal joints are explicitly modelled via mutual cross-attention to enhance their spatial collaboration. All of these are achieved by the proposed FG-SFormer.

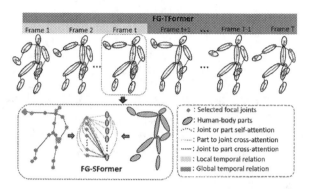

Fig. 1. The proposed FG-SFormer (bottom) learns correlations for both adaptively selected focal joints and body parts, as well as the joint-part interactions via cross-attention. FG-TFormer (top) models the explicit local temporal relations of joints or parts, as well as the global temporal dynamics.

The FG-TFormer is designed to model the temporal dynamics of joints or body parts. It is found that straightforwardly using the vanilla temporal transformer leads to ineffective temporal relations and poor recognition performance. We found one of the key culprits lying in the absence of local bias, making it challenging for transformer to focus on effective temporal motion patterns in the long input. Taking these factors into consideration, we integrate the dilated temporal convolutions into multi-head self-attention mechanism to explicitly encode the short-term temporal motions of a joint or part from their neighbors respectively, which equips transformer with local inductive bias. The short-range feature representations of all the frames are further fused by the global self-attention weights to embrace the global contextual motion information into the representations. Thus, the designed strategy enables transformer to learn both important local and effective global temporal relations of the joints and human body parts in a unified structure, which is validated critical to make temporal transformer work.

To summarize, the contributions of this work lie in four aspects:

1. We propose a novel FG-STFormer network for skeleton-based action recognition, that can effectively capture the discriminative correlations of focal joints as well as the global contextual motion information in both the spatial and temporal dimensions.

2. We design a focal joints and global parts coupling spatial transformer, namely FG-SFormer, to model the correlations of adaptively selected focal joints and that of human body parts. The joint-part mutual cross-attention is integrated to enhance the spatial collaboration.
3. We introduce a FG-TFormer to explicitly capture both the short and long range temporal dependencies of the joints and body parts effectively.
4. The extensive experimental results on three datasets highlight the effectiveness of our method, that surpasses all existing transformer-based methods.

2   Related Work

Skeleton-Based Action Recognition. With great progress achieved in skeleton-based action recognition, existing works can be broadly divided into three groups, i.e., RNNs, CNNs, and GCNs based methods. RNNs concatenate the coordinates of all joints in one frame and treat the sequence as time series [10,19,24,49,53]. Some works design specialized network structure, like trees [26,42] to make RNN aware of spatial information. CNN based methods transform one skeleton sequence to a pseudo-image via hand-crafted manners [11,13,18,21,22,28,45], and then use popular networks to learn spatial and temporal dynamics in it.

The appearance of GCN based methods, like ST-GCN [46], enables more natural spatial topology representation of skeleton joints by organizing them as a non-Euclidean graph. The spatial correlation is modelled for bone-connected joints. As the fixed graph topology (or adjacency matrix) is not flexible to model the dependencies among spatially disconnected joints, many subsequent methods focus on designing high-order or multi-scale adjacency matrix [12,15,20,23,30], and dynamically adjusted graph topology [5,23,37,48,50]. Nevertheless, these manually devised joint traversal rules limit the flexibility to learn more effective spatial-temporal dynamics of joints for action recognition.

Transformer Based Methods. Several recent works extend Transformer [41] to spatial and temporal dimensions of skeleton-based action recognition. Among them, DSTA [38] is the first to use self-attention to learn joint relations, whereas in practice spatial transformer interleaved with temporal convolution is employed for some typical datasets. ST-TR [33] adopts a hybrid architecture of GCN and transformer in a two-stream network, with each stream replacing the GCN or temporal convolution with spatial or temporal self-attention. STST [51] introduces a transformer network that the spatial and temporal dimensions are parallelly separated. Besides, the network is trained together with multi-task self-supervised learning tasks.

3   Proposed Method

In this section, we first briefly review the basic spatial and temporal Transformer blocks (referred to as Basic-SFormer and Basic-TFormer blocks respectively) used by most existing skeleton-based action recognition methods [33,38], which is also the basics of our network. Then the proposed Focal and Global Spatial-Temporal Transformer (FG-STFormer) is introduced in detail.

3.1   Basic Spatial-Temporal Transformer on Skeleton Data

Vanilla Transformer (V-Former) Block. The vanilla transformer [41] block consists of two important modules: multi-head self-attention (MSA) and point-wise feed-forward network (FFN). Let an input composed of N elements and C-dimensional features be $X \in \mathbb{R}^{N \times C}$. For a MSA having H heads, X is first linearly projected to a set of queries Q, keys K and values V. Then, the scaled dot-product attention of head h is calculated as:

$$\text{Attention}(Q^h, K^h, V^h) = \text{softmax}(\frac{Q^h K^{h^T}}{\sqrt{d}}) V^h = A^h V^h, \tag{1}$$

where Q^h, K^h, $V^h \in \mathbb{R}^{N \times d}$ with $d = C/H$ being the feature dimension of one head. $A^h \in \mathbb{R}^{N \times N}$ is the attention map.

MSA concatenates the output of all the heads and feeds into FFN module, that generally consists of a number of linear layers to transform the features.

Basic Spatial Transformer (Basic-SFormer) Block. For a skeleton sequence of T frames and N joints, let the input of C-dimension be $X = \{X_t \in \mathbb{R}^{N \times C}\}_{t=1}^{T}$. The Basic-SFormer block extends the V-Former block [41] to spatial dimension. It computes the inter-joint correlations for each frame X_t via Eq. (1) and generates an attention map $A_t^h \in \mathbb{R}^{N \times N}$, with each element $(A_t^h)_{ij}$ representing the spatial correlation score between joints i and j. Then, the features of each joint are updated as the weighted sum of values of all the joints. For the entire skeleton sequence, T spatial attention maps are produced.

Basic Temporal Transformer (Basic-TFormer) Block. By extending the V-Former to the temporal dimension, one Basic-TFormer learns global-range dynamics of a joint along the entire sequence. It rearranges the input as $X = \{X_n \in \mathbb{R}^{T \times C}\}_{n=1}^{N}$ to tackle temporal dimension. With Eq. (1), one of the N attention map $A_n^h \in \mathbb{R}^{T \times T}$ is computed for the nth joint. Each row in it stands for the dependencies of this joint across all the frames.

3.2   Focal and Global Spatial-Temporal Transformer Overview

The overview of the proposed FG-STFormer network is depicted in Fig. 2. It consists of two stages, in which our two primary components are FG-SFormer block and FG-TFormer block. The former is designed for the network late stage to model both the correlations of the sampled focal joints and the co-movement of human body parts globally in spatial dimension, as well as the interactions between the focal joints and body parts. The latter is devised to learn important local relations explicitly and global motion dynamics in temporal dimension, and is used in both stages. Therefore, the two stages are assigned specific responsibilities. That is, stage 1 aims to learn correlations for all joint pairs as generally done, so as to provide effective representations for stage 2 to mine reliable focal joints and part embeddings. Stage 2 targets at modelling both the discriminative

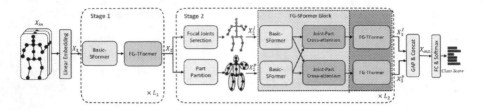

Fig. 2. Architecture of the proposed Focal and Global Spatial-Temporal Transformer (FG-STFormer). L_1 and L_2 are the number of layers in Stage 1 and Stage 2 respectively.

relations among focal joints and the global movement information of body parts. These two stages cooperate with each other to make the network learn discriminative and comprehensive spatial-temporal motion patterns for recognition.

Specifically, given a raw skeleton sequence $X_{\mathrm{in}} \in \mathbb{R}^{N \times T \times C_0}$, a linear layer is first applied to project it from the 2D or 3D joint coordinates of C_0 to a higher dimension C_1, generating feature $X_1 \in \mathbb{R}^{N \times T \times C_1}$. Then, X_1 goes through the two successive stages of FG-STFormer. Stage 1 sequentially stacks L_1 layers with each consisting of a Basic-SFormer block and an our FG-TFormer block.

At the end of stage 1, we obtain the high-level feature representations $X_2 \in \mathbb{R}^{N \times T \times C_2}$. It is then passed into stage 2, where the network is split into two branches. One branch adaptively selects K focal joints for each frame of the sequence and discards the remaining non-informative ones, producing features $X_2^J \in \mathbb{R}^{K \times T \times C_2}$. Meanwhile, the other branch partitions the joints into P global-level human body parts and generates feature tensor $X_2^P \in \mathbb{R}^{P \times T \times C_2}$. X_2^J and X_2^P are then passed through L_2 layers that interleave FG-SFormer and FG-TFormer blocks. In particular, one FG-SFormer block consists of a Basic-SFormer sub-block and a joint-part cross-attention sub-block to sufficiently model the spatial interaction information of actions. Stage 2 then produces output features X_3^J and X_3^P from the two branches respectively. They are applied global average pooling (GAP), and then concatenated along feature channels producing features $X_{\mathrm{out}} \in \mathbb{R}^{1 \times 1 \times C_{\mathrm{out}}}$. With which, FG-STFormer finally performs classification using two fully connected layers and a Softmax classier.

3.3   Focal and Global Spatial Transformer (FG-SFormer)

The proposed FG-SFormer block designed for network stage 2 learns critical and comprehensive spatial structure and motion patterns based on facts in two aspects. For one aspect, there is often a subset of key joints that play a vital role in action categorization [16,27], while the other joints are irrelevant or even noisy for action analysis. Especially, for transformer-based methods, the features of one joint could be influenced by those non-informative ones when integrating features of all the joints. Therefore, it is beneficial to identify the focal joints and concentrate on them at the deep layers of the network after the shallow layers have sufficiently learned the relationships among all the joints.

For the other aspect, it is not enough to just focus on the movement of focal joints. The movement of human body parts carry crucial global contextual motion information for recognizing an action [10,14]. Meanwhile, the interactions between joints and parts convey rich kinematic information, that could be exploited to fully mine action-related patterns.

Therefore, we propose to learn relations for adaptively identified focal joints and for human body parts, as well as their interactions in spatial dimension. Three modules to achieve this are designed: (i) Focal joints selection; (ii) Global-level part partition encoding; and (iii) Joint-part cross-attention.

Focal Joints Selection. In the joint branch of stage 2, we design a ranking based strategy to adaptively sample the focal joints subset for each frame in an action sequence with the input $X_2 \in \mathbb{R}^{N \times T \times C_2}$, and discard the non-informative ones. To achieve this, we leverage a trainable projection vector $W_p \in \mathbb{R}^{C_2 \times 1}$ and sigmoid function to predict the informativeness scores $S \in \mathbb{R}^{N \times T}$ for all the joints in individual frame as:

$$S = \text{sigmoid}(X_2 W_p / \|W_p\|), \tag{2}$$

Each element S_{ij} represents the informativeness score of ith joint in j^{th} frame. The larger the score is, the more informative the joint is. We sort the scores of all the joints for each frame and take the features corresponding to the top K joints having the largest scores to form the features $X_2^J \in \mathbb{R}^{K \times T \times C_2}$ of the focal joints subset as:

$$
\begin{aligned}
\text{idx} &= \text{sort}(S, K), \\
X_2^J &= X_2(\text{idx}, :, :),
\end{aligned}
\tag{3}
$$

where idx is the indices of the selected joints with largest scores.

X_2^J is then fed into the Basic-SFormer sub-block introduced in Sect. 3.1 to calculate the correlations only for those focal joints and update their feature embeddings. The Basic-SFormer block/sub-block used in both stages 1 and 2 is depicted in Fig. 3 (a). It uses the sine and cosine position encoding [41] to encode the joint type information. In the MSA module with H heads, the spatial attention map A_t is calculated for each frame. As in [38], we add a global regularization attention map A_g shared by all the sequences. The FFN module consists of a linear layer followed by applying activation function of Leaky ReLU [31].

Global-Level Part Partition Encoding. We explicitly model the correlations between global-level body parts in the other branch of stage 2 in our FG-STFormer. The joints are partitioned into P parts based on the physical skeleton structure and human prior. To obtain feature embeddings of the P parts with X_2, we concatenate the features of joints belonging to the same body part and then transform them into one part-level feature embedding via a linear layer shared by all parts. This generates the part embedding $X_2^P \in \mathbb{R}^{P \times T \times C_2}$, which is then passed into the Basic-SFormer sub-block shown in Fig. 3 (a) to model the one-to-one part relations and update features correspondingly.

Joint-Part Cross-Attention. To enable information diffusion across the focal joints and body parts to model their co-movement, we devise a joint-part cross-attention sub-block, termed as JP-CA. It uses multi-head cross-attention to

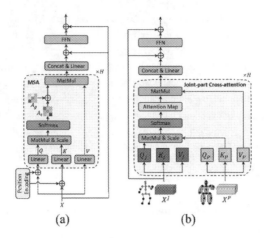

Fig. 3. (a) The Basic-SFormer block/sub-block used in Stages 1 and 2. (b) The Joint-part cross-attention (JP-CA) sub-block used in FG-SFormer block.

interact and diffuse features of the two branches. Here, we present JP-CA from the part branch to the focal joint branch as an example, as shown in Fig. 3 (b). For notational convenience, we omit the subscripts of X_2^J and X_2^P. Let Q_J, K_J and V_J be the queries, keys and values mapped from the joint-branch features X^J, and Q_P, K_P and V_P be those from the part-branch features X^P respectively. The part-to-joint cross-attention takes the Q_J as queries, and K_P and V_P as keys and values, and is calculated as:

$$\text{Attention}(Q_J, K_P, V_P) = \text{softmax}(\frac{Q_J K_P{}^T}{\sqrt{d}})V_P = A_{jp}V_P, \tag{4}$$

where d is the feature dimension of one head.

The attention map $A_{jp} \in \mathbb{R}^{K \times P}$ models the joint-part correlations and is used to aggregate part features for each focal joint. Other operations in this sub-block is same as those in the adopted Basic-SFormer sub-block. Notably, JP-CA is adaptive to actions, which is flexible to capture distinct collaborative patterns for input actions. Analogously, the cross-attention from the joint-branch to part-branch can be defined in similar operations.

3.4   Focal and Global Temporal Transformer (FG-TFormer)

Though temporal transformer has been applied in skeleton-based action recognition in existing works [1,33,51], it is rarely effectively deployed solely with spatial transformer in a single-stream architecture or in pure transformer-based models, largely because: (i) it is difficult for the self-attention to directly model effective temporal relations globally for distinguishing actions over the long input sequence; and (ii) the lack of inductive biases of transformer.

To address these issues, we propose to assist transformer in focusing on both the important local and the global temporal relations of joints explicitly, and design the component of focal and global temporal self-attention (FG-TSA), as depicted in Fig. 4. It utilizes the dilated temporal convolution to generate the values V in MSA, that works in two aspects: (i) explicitly learning the short-term temporal motion representation of a joint from its neighboring frames; and (ii) introducing beneficial local inductive biases to transformer. Meanwhile, the attention map generated by MSA models the global temporal correlations. Therefore, the resulting fused joint representations integrate both local temporal relation and global contextual information. The same effect is also achieved for the body part representations when FG-TFormer is applied to the part-branch.

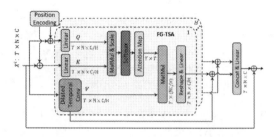

Fig. 4. The pipeline of the proposed FG-TFormer block.

Specifically, let the input feature tensor of the FG-TFormer block in layer l ($l = 1, 2, ..., L$) be $X^l \in \mathbb{R}^{N \times T \times C}$, where L is the total number of layers. First, the absolute position encoding is used to encode the temporal order information. And then, for the queries and keys Q_h^l, $K_h^l \in \mathbb{R}^{N \times T \times \frac{C}{H}}$ in head h among the H ones, they are generated via the usual linear projection in FG-TSA. Whereas for the V_h^l, different from existing works, we utilize dilated temporal convolution with kernel size $k_t \times 1$ and dilation rate d_t to obtain it, denoted as $V_h^l = \text{TCN}_{\text{dilate}}(X^l)_h \in \mathbb{R}^{N \times T \times \frac{C}{H}}$. Then, the global self-attention map is calculated via Eq. (1) and utilized to fuse the local feature representation V_h^l. Hence, each joint representation is injected with its global contextual dynamics. The features are then reshaped and linearly transformed with $W_h \in \mathbb{R}^{\frac{C}{H} \times \frac{C}{H}}$. This is followed by concatenating the output features of all the heads and conducting linear transform with $W_O \in \mathbb{R}^{C \times C}$ using activation function of Leaky ReLU. The output of the FG-TFormer block X^{l+1} is obtained by adding the shortcut from the input X^l. The whole process is formulated as:

$$X^{l+1} = \text{Concat}[\text{head}(X^l)_1, ..., \text{head}(X^l)_H]W_O + X^l,$$
$$\text{head}(X^l)_h = [\text{Attention}(Q_h^l, K_h^l)\text{TCN}_{\text{dilate}}(X^l)_h]W_h + \text{TCN}_{\text{dilate}}(X^l)_h. \tag{5}$$

Besides, we halve the temporal resolution of a sequence during generating V with convolution of stride 2 when the feature channels are doubled for a FG-TFormer block. This hierarchical structure reduces the computational cost.

4   Experiments

4.1   Dataset

NTU-RGB+D 60 (NTU-60)   [35] contains 56,880 sequences in 60 classes. It is collected from 40 subjects and provides the 3D locations of 25 human body joints. It recommends two benchmarks for evaluation: (1) Cross-subject (X-Sub): training data is from 20 subjects and test data from the other 20 subjects. (2) Cross-view (X-View): sequences captured by camera views 2 and 3 are taken as training data, and those captured by camera view 1 as testing data.

NTU-RGB+D 120 (NTU-120)   [25] has 120 classes and 113,945 samples captured from 32 camera setups and 106 subjects. It recommends two benchmarks: (1) Cross-subject (X-Sub): training data is from 53 subjects, and test data from the other 53 subjects. (2) Cross-setup (X-Set): samples with even setup IDs are used for training data, and samples with odd setup IDs for test.

Northwestern-UCLA (NW-UCLA)   [43] is captured by three Kinect cameras from three viewpoins. It contains 1,494 sequences in 10 action categories, with each performed by 10 actors. The same evaluation protocol in [26] is used: training data from the first two cameras and test data from the other camera.

4.2   Implementation Details

Our FG-STFormer model consists of 8 layers in two stages. Stage 1 contains $L_1 = 6$ layers and stage 2 consists of $L_2 = 2$ layers. The channel dimensions of each layer are 64, 64, 128, 128, 256, 256, 256 and 256. The number of frames is halved at the third and fifth layers. The number of spatial attention heads for the Basic-SFormer and FG-SFormer blocks is set to be 3. Each FG-TFormer block uses 2 attention heads, which adopt temporal kernel size of $k_t = 7$, and dilation rates of $d_t = 1$ and $d_t = 2$ respectively. The numbers of focal joints and body parts in the two branches of stage 2 are $K = 15$ and $P = 10$ respectively.

All experiments are conducted on one RTX 3090 GPU with PyTorch framework. We use SGD with Nesterov momentum of 0.9 and weight decay of 0.0005 to train our model for 80 epochs. Warm up strategy is used for the first 5 epochs. The initial learning rate is set to 0.01 and decays by a factor of 10 at the 50th and 70th epochs. For NTU-60 and NTU-120, the batch size is 32. The sequences are sampled to 128 frames, and we use the data pre-processing method in [5]. For NW-UCLA, the batch size is 32, we use the same data pre-processing in [43].

4.3   Ablation Study

In this section, the effectiveness of individual component of FG-STFormer is evaluated under X-Sub protocol of NTU-60 dataset, using only the joint stream.

Effectiveness of FG-SFormer Block.   To examine the effectiveness of the proposed FG-SFormer, we evaluate the important components in it, i.e., focal joints selection, part branch and joint-part cross-attention (JP-CA). We employ the Basic-SFormer as baseline, which calculates the correlations for all the joints at every layer of the network without using part branch and JP-CA. For temporal modelling, our FG-TFormer is used. We gradually replace the baseline by adding our designs one-by-one. The experimental results are shown in Table 1.

Table 1. Ablation study of different components in FG-SFormer block.

Methods	Focal Joints Selection	Part Branch	JP-CA	Acc (%)
Basic-SFormer	–	–	–	87.8
A	√	–	–	88.3
B	√	√	–	89.1
C	√	√	√	**89.5**

As seen, model A selects the focal joints at stage 2 of the network and improves the performance of Basic-SFormer by 0.5%. This indicates that it is beneficial to identify discriminative joints. Then, model B introduces the part branch to network stage 2. This provides performance improvement of 0.8% and reflects the spatial relations of intra-parts carry helpful global motion patterns. Finally, by adding the JP-CA into model C, the accuracy is further increased by 0.4%. This implies that the interactions between body parts and the selected focal joints are helpful for distinguishing actions.

Impact of Number of Focal Joints. To explore the effect of selecting different number of focal joints, we test the models using different K in FG-SFormer blocks at stage 2. Note that $K = 25$ means all the joints are used. As shown in Table 2, the accuracy gradually improves as K increases from 3 to 15, and then decreases when it further increases. This implies that the redundant or noisy joints indeed harm the recognition performance. In addition, too small number of focal joints are not enough to accurately identify the actions.

Table 2. Comparison of classification accuracy using different number of focal joints.

K	3	6	9	12	15	18	21	25
Acc (%)	88.6	88.9	89.0	89.2	**89.5**	89.2	89.3	89.1

Effectiveness of FG-TFormer Block. To evaluate the efficacy of FG-TFormer block, we build up experiments based on the complete network by modifying the FG-TFormer block only. The model using Basic-TFormer is taken as the baseline, which solely adopts the global MSA in temporal dimension. According to the results shown in Table 3, without the TCN_{dilate} in MSA, the Basic-TFormer performs significantly worse than our FG-TFormer with a large margin of -6.7%. Besides, by replacing Basic-TFormer with TCN_{dilate}, the performance is greatly improved by 6.3%. Finally, our FG-TFormer further achieves improvement of 0.4% by integrating TCN_{dilate} into self-attention mechanism.

Table 3. Ablation study for components in FG-TFormer block.

Methods	Global MSA	TCN$_{dilate}$	Acc (%)
Basic-TFormer	$\checkmark$	$\times$	82.8
FG-TFormer	$\times$	$\checkmark$	89.1
	$\checkmark$	$\checkmark$	**89.5**

Configuration Exploration. We explore different network configurations for stages 1 and 2 in our FG-STFormer by adjusting the number of layers L_1 and L_2. The total number of layers is fixed as 8. The results are shown in Table 4. Comparing models A, B and C, we can find that higher performance is obtained with more than 4 layers used in stage 1, and the best performance is achieved by $L_1 = 6$ and $L_2 = 2$. The accuracy drops down when stage 2 is assigned less layers in model D. These observations indicate that it is necessary for stage 1 to sufficiently learn the relations among all the joints, otherwise the performance could be harmed by focusing on unreliable focal joints and part collaborations.

Table 4. Comparison of different network configurations of our FG-STFormer.

Methods	Stage 1	Stage 2	Acc (%)	Methods	Stage 1	Stage 2	Acc (%)
	L_1	L_2			L_1	L_2	
A	4	4	88.8	C	6	2	**89.5**
B	5	3	89.0	D	7	1	89.0

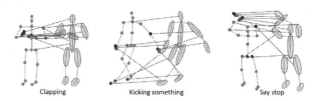

Fig. 5. The selected focal joints and learned joint-part interactions of actions. (Color figure online)

4.4 Visualization and Analysis

To validate what the focal joints are concentrated on at stage 2, we visualize the sampled 13 focal joints having largest scores for three actions in Fig. 5. These focal joints are depicted as coloured dots in the left skeleton of each action. The darker the dot is, the higher the informativeness score is of the joint. We can see that the actions *Clapping, Kicking something* and *Say stop* mainly select hands, shoulders, elbows and feet as the focal joints. Besides, Fig. 5 illustrates the learned attention weights from parts to focal joints of these actions. Attentions with large values are shown as green lines. As seen, the actions *Clapping* and *Say stop* mainly build interactions between focal joints and upper limbs, while action *Kicking something* interacts between focal joints and the whole body parts. These results verify that the spatial relations between the key joints and the global contextual movement information are captured by our FG-SFormer.

Moreover, we compare the performance of the Basic-TFormer with our FG-TFormer on action classes that the former has low accuracy. As shown in Fig. 6, our network improves the performance of those exhibited classes, which mainly involve the subtle and fine-grained motions of hands, feet and head. This concludes that our FG-TFormer can capture those subtle interaction patterns via explicitly embedding the neighboring relations into it.

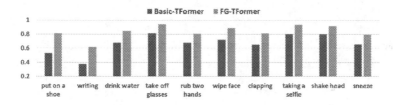

Fig. 6. Accuracy comparison between the Basic-TFormer and our FG-TFormer blocks.

Table 5. Comparison to state-of-the-arts on NW-UCLA dataset.

Methods	Year	NW-UCLA Top-1 (%)
HBRNN-L [10]	2015	78.5
Ensemble TS-LSTM [19]	2017	89.2
AGC-LSTM [39]	2019	93.3
Shift-GCN [8]	2020	94.6
DC-GCN+ADG [7]	2020	95.3
CTR-GCN [5]	2021	96.5
FG-STFormer (ours)	2022	**97.0**

4.5 Comparison with the State-of-the-Arts

We compare our FG-STFormer with existing state-of-the-art (SOTA) methods on three datasets: NW-UCLA, NTU-60 and NTU-120. Following the previous works [30,38,51], we fuse results of four modalities, i.e., joint, bone, joint motion, and bone motion. The results are shown in Tables 5 and 6. As seen, our method outperforms all existing transformer-based methods under nearly all evaluation benchmarks on NTU-60 and NTU-120, including the latest method STST [51] which uses not only the parallel spatial and temporal transformers but also multiple self-supervised learning tasks, and ST-TR [33] which adopts hybrid architecture of spatial-temporal transformer and GCN. Our method surpasses DSTA [38] by 2.4% and 1.6% on the two evaluation protocols of NTU-120.

Table 6. Performance comparisons against the SOTA methods on NTU- 60 and 120.

Methods	Year	NTU-60		NTU-120	
		X-Sub (%)	X-View (%)	X-Sub (%)	X-Set (%)
GCN-based methods					
ST-GCN [46]	2018	81.5	88.3	70.7	73.2
2s-AGCN [37]	2019	88.5	95.1	82.9	84.9
DGNN [36]	2019	89.9	96.1	–	–
Shift-GCN [8]	2020	90.7	96.5	85.9	87.6
Dynamic GCN [48]	2020	91.5	96.0	87.3	88.6
MS-G3D [30]	2020	91.5	96.2	86.9	88.4
MST-GCN [6]	2021	91.5	96.6	87.5	88.8
CTR-GCN [5]	2021	92.4	96.8	88.9	**90.6**
STF [17]	2022	92.5	**96.9**	88.9	90.0
Transformer-based methods					
DSTA [38]	2020	91.5	96.4	86.6	89.0
ST-TR [33]	2021	89.9	96.1	82.7	84.7
UNIK [47]	2021	86.8	94.4	80.8	86.5
STST [51]	2021	91.9	96.8	–	–
FG-STFormer (ours)	2022	**92.6**	96.7	**89.0**	**90.6**

Moreover, compared to GCN-based methods, the performance of our FG-STFormer is also at the top. It compares favourably with current state-of-the-art STF [17] and CTR-GCN [5] on NTU-60 and NTU-120, and even outperforms the latter on NW-UCLA by 0.5%, verifying the effectiveness of FG-STFormer.

5   Conclusion

In this work, we present a novel focal and global spatial-temporal transformer network (FG-STFormer) for skeleton-based action recognition. In spatial dimension, it learns intra- and inter- correlations for adaptively sampled focal joints and global body parts, which captures the discriminative and comprehensive spatial dependencies. In temporal dimension, it explicitly learns both the local and global temporal relations, enabling the network to capture rich motion patterns effectively. On three datasets, the proposed FG-STFormer achieves the state-of-the-art performance, demonstrating the effectiveness of our method.

Acknowledgement. The work of Zhimin Gao and Peitao Wang was supported in part by the National Natural Science Foundation of China (NSFC) under Grant 61906173. The work of Xiaoheng Jiang, Qidong Liu, and Mingliang Xu was supported by NSFC under U21B2037, 62276238 and 62036010 respectively.

References

1. Bai, R., et al.: GCST: graph convolutional skeleton transformer for action recognition. arXiv preprint arXiv:2109.02860 (2021)
2. Cao, Z., Hidalgo, G., Simon, T., Wei, S.E., Sheikh, Y.: OpenPose: realtime multi-person 2D pose estimation using part affinity fields. IEEE Trans. PAMI **43**(1), 172–186 (2019)
3. Carion, N., Massa, F., Synnaeve, G., Usunier, N., Kirillov, A., Zagoruyko, S.: End-to-end object detection with transformers. In: Vedaldi, A., Bischof, H., Brox, T., Frahm, J.-M. (eds.) ECCV 2020. LNCS, vol. 12346, pp. 213–229. Springer, Cham (2020). https://doi.org/10.1007/978-3-030-58452-8_13
4. Carreira, J., Zisserman, A.: Quo vadis, action recognition? A new model and the kinetics dataset. In: Proceedings of CVPR, pp. 6299–6308 (2017)
5. Chen, Y., Zhang, Z., Yuan, C., Li, B., Deng, Y., Hu, W.: Channel-wise topology refinement graph convolution for skeleton-based action recognition. In: Proceedings of ICCV, pp. 13359–13368 (2021)
6. Chen, Z., Li, S., Yang, B., Li, Q., Liu, H.: Multi-scale spatial temporal graph convolutional network for skeleton-based action recognition. In: Proceedings of AAAI, vol. 35, pp. 1113–1122 (2021)
7. Cheng, K., Zhang, Y., Cao, C., Shi, L., Cheng, J., Lu, H.: Decoupling GCN with DropGraph module for skeleton-based action recognition. In: Vedaldi, A., Bischof, H., Brox, T., Frahm, J.-M. (eds.) ECCV 2020. LNCS, vol. 12369, pp. 536–553. Springer, Cham (2020). https://doi.org/10.1007/978-3-030-58586-0_32
8. Cheng, K., Zhang, Y., He, X., Chen, W., Cheng, J., Lu, H.: Skeleton-based action recognition with shift graph convolutional network. In: Proceedings of CVPR, pp. 183–192 (2020)
9. Dosovitskiy, A., et al.: An image is worth 16 × 16 words: transformers for image recognition at scale. In: ICLR (2020)
10. Du, Y., Wang, W., Wang, L.: Hierarchical recurrent neural network for skeleton based action recognition. In: CVPR, pp. 1110–1118 (2015)
11. Duan, H., Zhao, Y., Chen, K., Shao, D., Lin, D., Dai, B.: Revisiting skeleton-based action recognition. arXiv preprint arXiv:2104.13586 (2021)
12. Gao, X., Hu, W., Tang, J., Liu, J., Guo, Z.: Optimized skeleton-based action recognition via sparsified graph regression. In: Proceedings of ACM MM, pp. 601–610 (2019)
13. Hou, Y., Li, Z., Wang, P., Li, W.: Skeleton optical spectra-based action recognition using convolutional neural networks. IEEE Trans. CSVT **28**(3), 807–811 (2018)
14. Huang, L., Huang, Y., Ouyang, W., Wang, L.: Part-level graph convolutional network for skeleton-based action recognition. In: Proceedings of AAAI, vol. 34, pp. 11045–11052 (2020)
15. Huang, Z., Shen, X., Tian, X., Li, H., Huang, J., Hua, X.S.: Spatio-temporal inception graph convolutional networks for skeleton-based action recognition. In: Proceedings of ACM MM, pp. 2122–2130 (2020)

16. Jiang, M., Kong, J., Bebis, G., Huo, H.: Informative joints based human action recognition using skeleton contexts. Signal Process. Image Commun. **33**, 29–40 (2015)

17. Ke, L., Peng, K.C., Lyu, S.: Towards to-at spatio-temporal focus for skeleton-based action recognition. In: Proceedings of AAAI (2022)

18. Ke, Q., Bennamoun, M., An, S., Sohel, F., Boussaid, F.: A new representation of skeleton sequences for 3D action recognition. In: Proceedings of CVPR, pp. 3288–3297 (2017)

19. Lee, I., Kim, D., Kang, S., Lee, S.: Ensemble deep learning for skeleton-based action recognition using temporal sliding LSTM networks. In: Proceedings of ICCV, pp. 1012–1020 (2017)

20. Li, B., Li, X., Zhang, Z., Wu, F.: Spatio-temporal graph routing for skeleton-based action recognition. In: Proceedings of AAAI, vol. 33, pp. 8561–8568 (2019)

21. Li, C., Xie, C., Zhang, B., Han, J., Zhen, X., Chen, J.: Memory attention networks for skeleton-based action recognition. IEEE Trans. NNLS **33**(9), 4800–4814 (2021)

22. Li, C., Zhong, Q., Xie, D., Pu, S.: Skeleton-based action recognition with convolutional neural networks. In: ICMEW, pp. 597–600. IEEE (2017)

23. Li, M., Chen, S., Chen, X., Zhang, Y., Wang, Y., Tian, Q.: Actional-structural graph convolutional networks for skeleton-based action recognition. In: Proceedings of CVPR, pp. 3595–3603 (2019)

24. Li, S., Li, W., Cook, C., Zhu, C., Gao, Y.: Independently recurrent neural network (IndRNN): building a longer and deeper RNN. In: Proceedings of CVPR, pp. 5457–5466 (2018)

25. Liu, J., Shahroudy, A., Perez, M., Wang, G., Duan, L.Y., Kot, A.C.: NTU RGB+D 120: a large-scale benchmark for 3D human activity understanding. IEEE Trans. PAMI **42**(10), 2684–2701 (2019)

26. Liu, J., Shahroudy, A., Xu, D., Wang, G.: Spatio-temporal LSTM with trust gates for 3D human action recognition. In: Leibe, B., Matas, J., Sebe, N., Welling, M. (eds.) ECCV 2016. LNCS, vol. 9907, pp. 816–833. Springer, Cham (2016). https://doi.org/10.1007/978-3-319-46487-9_50

27. Liu, J., Wang, G., Hu, P., Duan, L.Y., Kot, A.C.: Global context-aware attention LSTM networks for 3D action recognition. In: Proceedings of CVPR (2017)

28. Liu, M., Liu, H., Chen, C.: Enhanced skeleton visualization for view invariant human action recognition. Pattern Recogn. **68**, 346–362 (2017)

29. Liu, Z., et al.: Swin transformer: hierarchical vision transformer using shifted windows. In: Proceedings of ICCV, pp. 10012–10022 (2021)

30. Liu, Z., Zhang, H., Chen, Z., Wang, Z., Ouyang, W.: Disentangling and unifying graph convolutions for skeleton-based action recognition. In: Proceedings of CVPR, pp. 143–152 (2020)

31. Maas, A.L., Hannun, A.Y., Ng, A.Y., et al.: Rectifier nonlinearities improve neural network acoustic models. In: Proceedings of ICML, vol. 30, p. 3 (2013)

32. Neimark, D., Bar, O., Zohar, M., Asselmann, D.: Video transformer network. In: Proceedings of ICCV, pp. 3163–3172 (2021)

33. Plizzari, C., Cannici, M., Matteucci, M.: Skeleton-based action recognition via spatial and temporal transformer networks. Comput. Vis. Image Underst. **208**, 103219 (2021)

34. Poppe, R.: A survey on vision-based human action recognition. Image Vis. Comput. **28**(6), 976–990 (2010)

35. Shahroudy, A., Liu, J., Ng, T.T., Wang, G.: NTU RGB+D: a large scale dataset for 3D human activity analysis. In: Proceedings of CVPR, pp. 1010–1019 (2016)

36. Shi, L., Zhang, Y., Cheng, J., Lu, H.: Skeleton-based action recognition with directed graph neural networks. In: Proceedings of CVPR, pp. 7912–7921 (2019)

37. Shi, L., Zhang, Y., Cheng, J., Lu, H.: Two-stream adaptive graph convolutional networks for skeleton-based action recognition. In: Proceedings of CVPR, pp. 12026–12035 (2019)

38. Shi, L., Zhang, Y., Cheng, J., Lu, H.: Decoupled spatial-temporal attention network for skeleton-based action-gesture recognition. In: Proceedings of ACCV (2020)

39. Si, C., Chen, W., Wang, W., Wang, L., Tan, T.: An attention enhanced graph convolutional LSTM network for skeleton-based action recognition. In: Proceedings of CVPR, pp. 1227–1236 (2019)

40. Song, Y.F., Zhang, Z., Shan, C., Wang, L.: Constructing stronger and faster baselines for skeleton-based action recognition. IEEE Trans. PAMI **45**(2), 1474–1488 (2022)

41. Vaswani, A., et al.: Attention is all you need. In: Advances in Neural Information Processing Systems, pp. 5998–6008 (2017)

42. Wang, H., Wang, L.: Modeling temporal dynamics and spatial configurations of actions using two-stream recurrent neural networks. In: Proceedings of CVPR, pp. 499–508 (2017)

43. Wang, J., Nie, X., Xia, Y., Wu, Y., Zhu, S.C.: Cross-view action modeling, learning and recognition. In: Proceedings of CVPR, pp. 2649–2656 (2014)

44. Wang, P., Li, W., Ogunbona, P., Wan, J., Escalera, S.: RGB-D-based human motion recognition with deep learning: a survey. Comput. Vis. Image Underst. **171**, 118–139 (2018)

45. Wang, P., Li, Z., Hou, Y., Li, W.: Action recognition based on joint trajectory maps using convolutional neural networks. In: Proceedings of ACM MM, pp. 102–106 (2016)

46. Yan, S., Xiong, Y., Lin, D.: Spatial temporal graph convolutional networks for skeleton-based action recognition. In: Proceedings of AAAI (2018)

47. Yang, D., Wang, Y., Dantcheva, A., Garattoni, L., Francesca, G., Bremond, F.: Unik: a unified framework for real-world skeleton-based action recognition. In: Proceedings of BMVC (2021)

48. Ye, F., Pu, S., Zhong, Q., Li, C., Xie, D., Tang, H.: Dynamic GCN: context-enriched topology learning for skeleton-based action recognition. In: Proceedings of ACM MM, pp. 55–63 (2020)

49. Zhang, P., Lan, C., Xing, J., Zeng, W., Xue, J., Zheng, N.: View adaptive recurrent neural networks for high performance human action recognition from skeleton data. In: Proceedings of ICCV, pp. 2117–2126 (2017)

50. Zhang, P., Lan, C., Zeng, W., Xing, J., Xue, J., Zheng, N.: Semantics-guided neural networks for efficient skeleton-based human action recognition. In: Proceedings of CVPR, pp. 1112–1121 (2020)

51. Zhang, Y., Wu, B., Li, W., Duan, L., Gan, C.: STST: spatial-temporal specialized transformer for skeleton-based action recognition. In: Proceedings of ACM MM, pp. 3229–3237 (2021)

52. Zhang, Z.: Microsoft Kinect sensor and its effect. IEEE Multimedia **19**(2), 4–10 (2012)

53. Zhu, W., et al.: Co-occurrence feature learning for skeleton based action recognition using regularized deep LSTM networks. In: Proceedings of AAAI, vol. 30 (2016)

Spatial-Temporal Adaptive Graph Convolutional Network for Skeleton-Based Action Recognition

Rui Hang and MinXian Li[✉]

Nanjing University of Science and Technology, Nanjing, China
{hangrui,minxianli}@njust.edu.cn

Abstract. Skeleton-based action recognition approaches usually construct the skeleton sequence as spatial-temporal graphs and perform graph convolution on these graphs to extract discriminative features. However, due to the fixed topology shared among different poses and the lack of direct long-range temporal dependencies, it is not trivial to learn the robust spatial-temporal feature. Therefore, we present a spatial-temporal adaptive graph convolutional network (STA-GCN) to learn adaptive spatial and temporal topologies and effectively aggregate features for skeleton-based action recognition. The proposed network is composed of spatial adaptive graph convolution (SA-GC) and temporal adaptive graph convolution (TA-GC) with an adaptive topology encoder. The SA-GC can extract the spatial feature for each pose with the spatial adaptive topology, while the TA-GC can learn the temporal feature by modeling the direct long-range temporal dependencies adaptively. On three large-scale skeleton action recognition datasets: NTU RGB+D 60, NTU RGB+D 120, and Kinetics Skeleton, the STA-GCN outperforms the existing state-of-the-art methods. The code is available at https://github.com/hang-rui/STA-GCN.

Keywords: Action recognition · Adaptive topology · Graph convolution

1 Introduction

Action recognition is an essential task in human-centered computing and computer vision, which plays an increasingly crucial role in video surveillance, human-computer interaction, video analysis, and other applications [1,26,38]. In recent years, skeleton-based human action recognition has attracted much attention due to the development of depth sensors [46] and pose estimation algorithms [2,33]. Conventional deep learning methods adopt re-current neural networks (RNN) [8,20,43] and convolutional neural networks (CNN) [14–16] to analyze the skeleton sequence by representing it as vector sequence or pseudo-image. However, the skeleton sequence is naturally structured as a spatial-temporal

Supplementary Information The online version contains supplementary material available at https://doi.org/10.1007/978-3-031-26316-3_11.

L. Wang et al. (Eds.): ACCV 2022, LNCS 13844, pp. 172–188, 2023.
https://doi.org/10.1007/978-3-031-26316-3_11

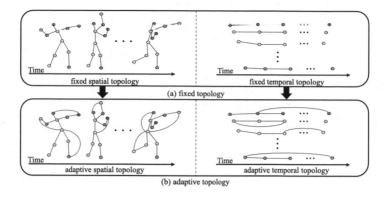

Fig. 1. Illustration of (a) the fixed spatial and temporal topology and (b) the adaptive spatial and temporal topology. Different color points indicate the different human joints, and the lines correspond to the spatial and temporal direct correlation between joints. Best viewed in color.

graph. For this reason, Yan *et al.* [40] firstly proposed the spatial-temporal graph convolutional network (ST-GCN) to model the motion patterns of action on a skeleton spatial-temporal graph. After that, a series of graph convolutional networks (GCN) methods based on spatial-temporal graphs have been proposed for skeleton-based action recognition [7, 22, 29].

However, there are still two disadvantages to the feature extraction operations on the spatial-temporal graph: (1) In the stage of learning spatial features, the fixed spatial topology is shared among all poses, which may not be optimal for the action with large changes in pose. For action such as "throw", before and after the throw, the human pose presents two forms of backward-leaning and forward-leaning, which represents different semantics. Using the fixed spatial topology may mistakenly enhance irrelevant connections or weaken critical ones, failing to accurately represent the spatial dependencies. This fact suggests that the spatial topology should be adaptive to each pose in the skeleton sequence. (2) In the stage of learning temporal features, existing methods apply temporal convolution with a fixed small kernel to extract the short-range temporal feature. It leads to the weak capacity to model temporal long-range joint dependencies vital for action recognition.

To learn the robust feature representation in the spatial and temporal dimensions, we propose a spatial-temporal Adaptive Graph Convolutional Network (STA-GCN) in this work. The proposed network is composed of two key modules: Spatial Adaptive Graph Convolution (SA-GC) and Temporal Adaptive Graph Convolution (TA-GC). Both SA-GC and TA-GC have a critical embedded component: Topology Adaptive Encoder (TAE). The SA-GC module is designed to extract spatial features by modeling the spatial adaptive joint dependencies. The TA-GC module is designed to learn temporal features by capturing the direct long-range joint dependencies in the temporal dimension. Combined with SA-GC and TA-GC modules, the proposed model can learn the discriminative features both in the spatial and temporal dimensions.

The TAE component is proposed to learn spatial adaptive topology and temporal adaptive topology. The existing GCN methods use fixed spatial and temporal topology (Fig. 1(a)). This fixed spatial topology forces each skeleton frame to adopt the same spatial topology, while the fixed temporal topology forces all trajectories to use the same temporal topology. We argue that this fixed topology is insufficient to represent the joint dependencies per pose or per trajectory. Therefore, we propose the TAE to solve this problem by learning the spatial adaptive topology and the temporal adaptive topology (Fig. 1(b)). The spatial adaptive topology can generate the pose-specific dependencies for each frame in the skeleton sequence to learn discriminative spatial features. The temporal adaptive topology can model the direct long-range dependencies between any two joints in the trajectory graph to extract robust temporal features.

The main contributions of this work are summarized as follows:

- A spatial adaptive graph convolution (SA-GC) module is proposed to extract the spatial feature for each pose with the spatial adaptive topology.
- A temporal adaptive graph convolution (TA-GC) module is proposed to learn the temporal feature by modeling the direct long-range temporal dependencies.
- A topology adaptive encoder (TAE) embedded into graph convolution is proposed to generate the adaptive spatial topology and the adaptive temporal topology.
- We propose a Spatial-Temporal Adaptive Graph Convolutional Network (STA-GCN) composed of SA-GC and TA-GC, which outperforms state-of-the-art approaches on three large-scale skeleton action recognition datasets: NTU RGB+D [27], NTU RGB+D 120 [19], and Kinetics-skeleton [13].

2   Related Work

2.1   Skeleton-Based Action Recognition

With the development of deep learning based video understanding technology, a series of video-based methods were proposed for action recognition. Specifically, 2D CNNs [18,36] efficiently recognize actions by modeling the relationships in the temporal dimension, while 3D CNNs [3,34] capture motion information in a unified network through a simple extension from the spatial domain to the spatial-temporal domain. Recently skeleton-based methods [7,8,14–16,20,22,24,29,39, 40,42,43] have been developed extensively since skeleton data are more computationally efficient and exhibit stronger robustness. Skeleton data can eliminate the influences of variations of illumination, camera viewpoints, background changes, and clothing variance in real-world videos [21,35,37]. Therefore, we adopt the skeleton-based action recognition approach in this paper.

2.2   Graph Convolution Networks in Action Recognition

The type of skeleton-based action recognition approaches is divided into three categories: RNN-based, CNN-based, and GCN-based. RNN represents the skeleton sequence as a vector sequence [8, 20, 43], while CNN represents it as a pseudo-image [14–16]. However, both RNN and CNN methods ignore the information of skeleton topology among joints. GCN-based methods represent the skeleton sequence as a spatial-temporal graph and extract features in spatial and temporal dimensions by graph convolution and temporal convolution respectively. Yan [40] *et al.* firstly introduce graph convolution and temporal convolution into the skeleton-based action recognition to model the spatial configurations and temporal dynamics simultaneously. The following works improved the model by performing multi-hop methods to extract multi-scale features [12, 17], applying additional mechanisms to adaptively capture the relations between distant joints [17, 41, 44], and adding additional edges between adjacent frames to extract features based on the extended graph [9, 22, 23]. However, all these methods only apply Graph Convolution Network (GCN) in the spatial dimension. In the temporal dimension, Temporal Convolution Network (TCN) is used to learn the temporal feature. It is not powerful due to the use of a fixed small temporal convolution kernel. In this work, we apply graph convolution both in spatial and temporal dimensions. Temporal graph convolution can effectively learn the temporal feature by modeling the direct long-range temporal dependencies.

2.3 Topology Adaptive-Based Methods

The topology-based adaptive methods [17, 29, 41, 45] aim to generate the appropriate spatial topology based on the input data. This spatial topology indicates whether and how important a connection exists between any two joints in the graph. Existing methods focus only on the spatial dimension, and they mainly use parametric adjacency matrices to learn a spatial topology optimized for all data [29] or generate a specific spatial topology for each action sample based on the input data [17, 41, 45]. These methods alleviate the limitation caused by the fixed pre-defined spatial topology of GCN methods. However, existing methods force all frames to share the same spatial topology, which is unreasonable because each pose represents a different semantic and cannot use the same spatial topology to extract features. Moreover, these approaches do not consider topology in the temporal dimension, which hinders the long-range dependencies modeling in the temporal dimension. To address these problems, we propose the Topology Adaptive Encoder (TAE) to learn spatial adaptive topology for each frame in the spatial graph and temporal adaptive topology for each trajectory in the temporal graph. The spatial adaptive topology can efficiently help spatial graph convolution extract the features of different poses in action. On the other hand, the temporal adaptive topology is used in temporal graph convolution to extract long-range dependencies.

3   Methodology

3.1   Preliminaries

Notations. A human skeleton with N joints is presented as an undirected spatial graph $\mathcal{G} = (\mathcal{V}, \mathcal{E})$, where $\mathcal{V} = \{v_i | i = 1, \ldots, N\}$ is the set of N vertices

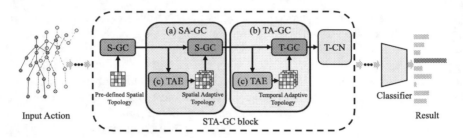

Fig. 2. The overview of the proposed Spatial-Temporal Adaptive Graph Convolutional Network (STA-GCN). (a) The spatial adaptive graph convolution (SA-GC) module is adopted to learn the spatial feature for each pose with the spatial adaptive topology. (b) The temporal adaptive graph convolution (TA-GC) module is subsequently adopted to learn the temporal feature for each trajectory with the temporal adaptive topology. (c) The Topology Adaptive Encoder (TAE) is used both in SA-GC and TA-GC modules to learn the spatial adaptive topology and the temporal adaptive topology.

representing joints and $\mathcal{E}$ is the edges set representing bones. The topology of $\mathcal{G}$ is formulated as an adjacency matrix $\mathbf{A} \in \mathbb{R}^{N \times N}$ and its element $a_{ij} \in (0,1)$ denoting weather an edge exists between vertex v_i and v_j.

Graph Convolution. According to Yan *et al.* [40], the spatial dependencies of the joints in each frame can be conveniently encoded with Graph Convolution. The operation of Graph Convolution is formulated as:

$$\mathbf{X}^{(l+1)} = \sigma \left(\sum_{k=0}^{K} \widetilde{\mathbf{A}}_{(k)} \mathbf{X}^{(l)} \mathbf{W}_{(k)}^{(l)} \right) \tag{1}$$

where $\mathbf{X}$ is the feature of each layer, $\sigma(\cdot)$ is an activation function. K denotes the pre-defined maximum graphic distance. $\widetilde{\mathbf{A}}_{(k)} = \mathbf{\Lambda}_{(k)}^{-\frac{1}{2}} (\mathbf{A}_{(k)} + \mathbf{I}_{(k)}) \mathbf{\Lambda}_{(k)}^{-\frac{1}{2}}$ is the k-th order normalized adjacency matrix, where $\mathbf{A} + \mathbf{I}$ is the skeleton graph with added self-loops to keep identity features, $\mathbf{\Lambda}$ is the diagonal degree matrix of $(\mathbf{A} + \mathbf{I})$. $\mathbf{W}_{(k)}$ is learnable parameters to implement the convolution operation.

3.2   Overview

To learn discriminative features on the spatial and temporal graph, we propose a Spatial-Temporal Adaptive Graph Convolutional Network (STA-GCN). An overview of our proposed method is illustrated in Fig. 2. After extracting local skeleton features using a pre-defined spatial topology (i.e. human skeleton structure), we adopt Spatial Adaptive Graph Convolution (SA-GC) and Temporal Adaptive Graph Convolution (TA-GC) to learn discriminative features. The SA-GC is proposed to extract the spatial feature for each pose with the spatial adaptive topology. The TA-GC is proposed to learn the temporal feature by modeling the direct long-range temporal dependencies. Both SA-GC and

TA-GC have an embedded Topology Adaptive Encoder (TAE), which generates a unique and appropriate topology for each pose and each trajectory. On the top of SA-GC and TA-GC, we apply a temporal convolutional convolution to aggregate the features in the temporal dimension. Based on the discriminative features extracted by several STA-GC blocks, we use a fully connected layer with softmax activation function to obtain the final class. The method will be discussed in detail in subsequent sections.

3.3   Spatial and Temporal Graph Convolution

Most existing works treat the human skeleton sequence as a spatial-temporal graph where features are extracted through spatial graph convolution and temporal convolution. However, the temporal convolution is not powerful to learn the temporal feature due to the use of a fixed small temporal convolution kernel. Therefore, in this section, we introduce a more robust feature extraction operator on the spatial and temporal graph. Let us first consider a graph convolution in spatial-temporal graph $\mathcal{G}^{st} = (\mathcal{V}^{st}, \mathcal{E}^{st})$ where $\mathcal{V}^{st} = \{v_{nt} | n = 1, \ldots, N, t = 1, \ldots, T, \}$ is the set of all nodes across T frames in the skeleton sequence and $\mathcal{E}^{st}$ is the spatial-temporal edge set.

We further deconstruct a spatial-temporal graph into T spatial graphs across time and N temporal graphs across joints. The spatial graphs are represented as $\mathcal{G}^s = (\mathcal{V}^{st}, \mathcal{E}^s)$ where $\mathcal{E}^s$ is the spatial edge set and is formulated as a spatial adjacency matrix $\mathbf{A}^s \in \mathbb{R}^{T \times N \times N}$. Note that the spatial adjacency matrix is degraded to vanilla form $\mathbf{A} \in \mathbb{R}^{N \times N}$ when all spatial graphs have the same spatial correlations. Similarly, the temporal graphs are represented as $\mathcal{G}^t = (\mathcal{V}^{st}, \mathcal{E}^t)$ and the temporal adjacency matrix $\mathbf{A}^t \in \mathbb{R}^{N \times T \times T}$ can be formulated. After the graph decomposition, we develop two graph convolutions: spatial graph convolution (S-GC) and temporal graph convolution (T-GC). S-GC and T-GC are respectively formulated as:

$$f_{out}(v_{nt}) = \sum_{p=1}^{N} a^s_{(pt)(nt)} f_{in}(v_{pt}) \mathbf{w}(v_{pt}) \tag{2}$$

$$f_{out}(v_{nt}) = \sum_{q=1}^{T} a^t_{(nq)(nt)} f_{in}(v_{nq}) \mathbf{w}(v_{nq}) \tag{3}$$

where $a^s_{(pt)(nt)}$ and $a^t_{(nq)(nt)}$ are elements of $\mathbf{A}^s$ and $\mathbf{A}^t$, respectively. Features on the spatial-temporal graph can be extracted by employing S-GC and T-GC. The whole process of feature extraction is formulated as:

$$f_{out}(v_{nt}) = \sum_{q=1}^{T} a^t_{(nq)(nt)} (\sum_{p=1}^{N} a^s_{(pq)(nq)} f_{in}(v_{pq}) \mathbf{w}_1(v_{pq})) \mathbf{w}_2(v_{nq}) \tag{4}$$

3.4   Topology Adaptive Encoder

Previous works force each pose in the skeleton sequence to share a fixed spatial topology. However, different poses represent different semantics, and using the same topology will incorrectly extract spatial features for different poses, which leads to weak performance in recognizing action with large changes in pose. Moreover, existing methods do not apply graph convolution in the temporal dimension. If graph convolution is to be performed in the temporal dimension, a suitable temporal topology is needed to represent the relationships between joints in the temporal graph.

To generate a more detailed topology, we propose a unified topology generation module, Topology Adaptive Encoder (TAE), which applies an unshared topology generation strategy. Specifically, the TAE module generates the spatial topology for each pose in spatial graph convolution and the temporal topology for each trajectory in temporal graph convolution. Technically, the TAE applies self-attention operations to extract correlations between joints in the embedding space. The function is formulated as:

$$A = \ell_2\text{\_norm}(XW_\phi W_\psi^T X^T) \tag{5}$$

where X is the input feature, we first utilize two linear transformation functions ϕ and ψ to embed input feature into the embedding space, W_ϕ and W_ψ are the parameters of the embedding functions ϕ and ψ, respectively. Then, the two embedded feature matrices are multiplied to obtain a topology matrix A. Finally, the ℓ_2 normalization is applied to each row of the adjacency matrix, which eases the optimization and with the help of ℓ_2 normalization, the normalization of node degree is unnecessary.

3.5   Spatial and Temporal Adaptive Graph Convolution

We integrate the proposed TAE into S-GC and T-GC to obtain a pair of adaptive graph convolution operations on the spatial-temporal graph: Spatial Adaptive Graph Convolution (SA-GC) (Fig. 3 (a)) and Temporal Adaptive Graph Convolution (TA-GC) (Fig. 3 (b)). Since SA-GC on the spatial graph and TA-GC on the temporal graph are equivalent operations, for simplicity, we only introduce SA-GC in the rest part. Its counterpart TA-GC can be deduced naturally.

Specifically, our SA-GC contains three parts: (1) feature transformation with function $\mathcal{T}^s(\cdot)$, (2) topology adaptive Encoding with function $\mathcal{M}^s(\cdot)$, (3) feature aggregation with function $\mathcal{A}^s(\cdot)$. Given the input feature $X \in \mathbb{R}^{T \times N \times C}$, the output $Y \in \mathbb{R}^{T \times N \times C'}$ of SA-GC is formulated as:

$$Y = \mathcal{A}^s(\mathcal{T}^s(X), \mathcal{M}^s(X)) \tag{6}$$

Feature Transformation. The goal of feature transformation is to transform input features into high-level representations using function $\mathcal{T}^s(\cdot)$. Here we use the simple linear transformation function which is formulated as:

$$F^s = \mathcal{T}^s(X) = XW^s \tag{7}$$

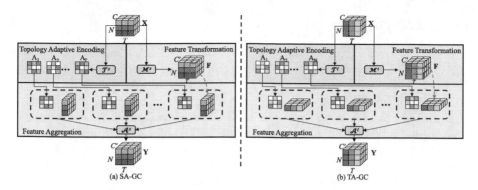

Fig. 3. (a) The Spatial Adaptive Graph Convolution (SA-GC) module. (b)The Temporal Adaptive Graph Convolution (TA-GC) module.

where $F^s \in \mathbb{R}^{T \times N \times C'}$ is the transformed high-level features and $W^s \in \mathbb{R}^{C \times C'}$ is the weight matrix.

Topology Adaptive Encoding. The topology adaptive encoding part in SA-GC is to use the proposed TAE module to generate a topology optimized for each skeleton individually. The function is formulated as:

$$A^s = \mathcal{M}^s(X) = \ell_2\text{-norm}(XW_\phi W_\psi^T X^T) \tag{8}$$

where X is the input feature, W_ϕ and W_ψ are the parameters of the embedding functions ϕ and ψ to embed the features. Then, the two feature maps are reshaped to matrix $M_\phi \in \mathbb{R}^{T \times N \times C}$ and $M_\psi \in \mathbb{R}^{T \times C \times N}$. By multiplying these two matrices to obtain a topology matrix $A^s \in \mathbb{R}^{T \times N \times N}$, whose element represents the correlation between joints on a specific frame. Finally, the ℓ_2 normalization is applied to each row of the adjacency matrix.

Feature Aggregation. As indicated in Fig. 3, given the temporal-wise spatial adaptive adjacency matrix A^s from input samples, we aggregate high-level features F^s with temporal-wise feature aggregation function $\mathcal{A}^s$. The function is formulated as:

$$Y = \mathcal{A}^s(A^s, F^s) = [A_1^s F_1^s \|_t A_2^s F_2^s \|_t \cdots \|_t A_T^s F_T^s] \tag{9}$$

where $\|_t$ is concatenation operation along the temporal dimension. $A_t^s \in \mathbb{R}^{N \times N}$ and $F_t^s \in \mathbb{R}^{N \times C}$ are respectively from t-th frame of A^s and F^s. During the whole process, the topology is optimized for each skeleton individually. Therefore, the proposed SA-GC can effectively distinguish actions, especially those with large changes in pose.

3.6   Model Details

The entire network consists of five STA-GC blocks, and the number of output channels of five blocks are 64-64-64-128-256. We apply a data BatchNorm layer at the start to normalize the input data. The temporal dimension is halved at the 4-th and 5-th blocks by strided temporal convolution. The temporal convolutional network used (TCN) in the STA-GC block is designed as multi-scale temporal convolutions following [22]. The main difference is that we reduce the number of channels and fuse six branches with a point-wise convolution. We also add extra residual connections to facilitate training. Due to the operations in SA-GC and TA-GC are equivalent, we only introduce a detailed implementation of SA-GC. We first utilize two linear transformation functions to transform input features into two neatly compact representations. Then, the two embedded features are multiplied to obtain spatial adaptive topology. We further apply ℓ_2 normalization to normalize the adjacency matrix, and the resulting adjacency matrix is used to apply the graph convolution.

4   Experiments

4.1   Datasets

NTU RGB+D 60 [27] is currently the most widely used indoor skeleton-based action recognition dataset, which contains $56,880$ skeleton action sequences with 60 action classes performed by 40 volunteers and captured by three Microsoft Kinect v2 cameras from different views concurrently. Each sample contains one action with two subjects at most, and each skeleton is composed of 25 joints. The dataset is separated into two benchmarks: (1) Cross-subject (X-Sub): $40,320$ samples performed by 20 subjects are separated into the training set, and the other $16,560$ samples performed by different 20 subjects belong to the test set. (2) Cross-view (X-View): the training set contains $37,920$ samples from camera views 2 and 3, and the test set contains $18,960$ samples from camera view 1.

NTU RGB+D 120 [19] is the largest indoor skeleton-based action recognition dataset, which extends NTU RGB+D 60 with additional 57,367 skeleton sequences over 60 extra action classes, totalling contains $114,480$ skeleton action sequences in 120 action classes performed by 106 volunteers, and has 32 different camera setups, each setup representing a specific location and background. Similarly, the dataset is separated into two benchmarks: (1) Cross-subject (X-Sub): $63,026$ samples performed by 53 subjects are separated into the training set, and the other $50,922$ samples performed by different 53 subjects belong to the test set. (2) Cross-setup (X-Set): the training set contains $54,471$ samples with even setup IDs, and the test set contains $59,477$ samples with odd setup IDs.

Kinetics Skeleton. Kinetics 400 [13] is a large-scale human action dataset that contains $300,000$ video clips of 400 classes collected from the Internet. After applying Openpose [2] pose-estimation algorithm on Kinetics 400, the Kinetics

Skeleton dataset obtain 240, 436 training and 19, 796 evaluation skeleton clips, where each skeleton graph contains 18 body joints, along with their 2D coordinates and confidence score.

4.2 Implementation Details

All experiments are implemented on two RTX 3090 GPUs with the PyTorch deep learning framework. The stochastic gradient descent (SGD) with the momentum of 0.9 and the weight decay of 0.0001 is used for optimization. The model is trained for 70 epochs in total. The initial learning rate is set to 0.1 and decays with a cosine schedule after the 10-th epoch. Moreover, a warm-up strategy [10] was applied over the first 10 epochs, gradually increasing the learning rate from 0 to the initial value in order to make the training procedure more stable. The batch size is set to 32. Input data are preprocessed following [32], and cross-entropy loss is employed.

4.3 Ablation Studies

We analyze the individual components and their configurations in the final architecture. The performance is reported as Top-1 and Top-5 classification accuracy on the Cross-Subject benchmark of NTU RGB+D 60 using only the joint data.

Table 1. Comparison of the accuracy when gradually adding STA-GC and only adding SA-GC or TA-GC on the X-Sub of NTU RGB+D 60.

Methods	Params	Top-1 (%)	Top-5 (%)
Baseline	1.44	88.1	98.2
+ 1 STA-GC	1.30	88.7	98.0
+ 2 STA-GC	1.36	89.1	98.4
+ 3 STA-GC	1.38	89.3	98.4
STA-GCN with SA-GC only	1.21	88.7	98.0
STA-GCN with TA-GC only	1.21	89.0	98.3
STA-GCN	1.40	**89.5**	**98.4**

Effectiveness of STA-GC. To verify the effectiveness of the proposed STA-GC block, we build up the model incrementally with its individual modules. We employ ST-GCN [40] as the baseline for controlled experiments. For a fair comparison, we add residual connections in ST-GCN and replace its temporal modeling module with temporal convolution described in Sect. 3.6. The experimental results are shown in Table 1. We first gradually add STA-GC into the baseline. For a fair comparison, we halved the original ten stages in the baseline to five after adding STA-GC to control for parameters, and this also alleviates the over-smoothing problem caused by adding the new graph convolution.

We observe that accuracies increase steadily, and the accuracy is substantially improved after each graph convolution has been added with the STA-GC module, which validates the effectiveness of STA-GC. Then we validate the effects of the SA-GC and the TA-GC respectively by adding either of them into the baseline. We observed performance raise of 1.4% and 0.4% respectively, indicating that our proposed SA-GC and TA-GC can effectively extract features in spatial and temporal dimension respectively. Moreover, the SA-GC and TA-GC are complementary and their combination can promote each other to achieve better performance for effective motion feature learning.

Table 2. Comparison of the accuracy when STA-GCN applies the fixed topology or the adaptive topology on the X-Sub of NTU RGB+D 60.

Spatial topology	Temporal topology	Top-1 (%)	Top-5 (%)
Fixed	Fixed	86.6	97.7
Adaptive	Fixed	88.0	97.8
Fixed	Adaptive	87.7	98.0
Adaptive	Adaptive	**89.5**	**98.4**

Effectiveness of TAE. To verify the effectiveness of our proposed TAE, we keep the backbone of the STA-GCN and apply the adaptive topologies generated by TAE or the fixed parameterized topologies in SA-GC and TA-GC respectively. As shown in Table 2, the models only using TAE in the spatial dimension or temporal dimension outperform the model using only the fixed topologies, and the model applying TAE both in spatial and temporal dimensions achieves the best results. It demonstrates that TAE can effectively generate appropriate spatial topology and temporal topology for robust feature learning.

Table 3. Comparison of the accuracy when STA-GCN applies different topology adaptive methods on the X-Sub of NTU RGB+D 60.

Methods	Top-1 (%)	Top-5 (%)
2s-AGCN [29]	88.9	97.9
Dynamic-GCN [41]	80.0	95.4
TAE	**89.5**	**98.4**

Comparison with Other Topology Adaptive Methods. To validate the effectiveness of our TAE, we also compare the performance of different topology adaptive methods in Table 3. Specifically, we keep the backbone of the STA-GCN and only replace the topology adaptive method in graph convolution for a fair comparison. From Table 3, we observe that TAE outperforms other topology adaptive methods from 2s-AGCN and Dynamic-GCN, proving that TAE is effective in generating adaptive typologies.

Table 4. Comparison of the accuracy when STA-GCN applies shared or unshared topology generation strategy on the X-Sub of NTU RGB+D 60.

Spatial topology	Temporal topology	Top-1 (%)	Top-5 (%)
Shared	Shared	88.6	98.1
Unshared	Shared	89.0	98.3
Shared	Unshared	88.8	98.2
Unshared	Unshared	**89.5**	**98.4**

Shared Topology vs Unshared Topology. We also verify the topology unshared strategy of TAE. Specifically, we keep the backbone of the STA-GCN and compare the model's performance when TAE uses shared or unshared topology generation strategies. The topology-unshared strategy means that TAE generates a specific topology for each skeleton or trajectory in each action sample, while the topology-shared strategy represents that TAE generates the same topology for all poses or trajectories in each action sample. As shown in Table 4, the topology-unshared strategy achieve better performance than the topology-shared strategy, indicating the importance of generating a specific topology for each skeleton or trajectory.

Table 5. Comparisons of the Top-1 accuracy (%) on actions with large changes in pose on the X-Sub of NTU RGB+D 60.

Methods	Actions				
	throw	stand up	hopping	pick up	falling down
ST-GCN [40]	88.3	96.8	97.1	92.8	97.1
STA-GCN	**91.0**	**98.5**	**98.6**	**95.2**	**99.6**

Performance for Recognizing Actions with Large Changes in Pose. To further verify that the proposed STA-GCN can recognize actions with large pose changes more effectively, we compare with ST-GCN on several actions. For a fair comparison, we added additional residual connections and applied the same temporal convolution module to the ST-GCN. The results are shown in Table 5, our method outperforms ST-GCN in Top1 accuracy on five poses (throw, stand up, hopping, pick up, and falling down), demonstrating that our approach can effectively identify actions with large changes in pose.

Table 6. Comparisons of the Top-1 accuracy with the state-of-the-art methods on the NTU RGB+D 60 and NTU RGB+D 120 datasets.

Methods	NTU RGB+D 60		NTU RGB+D 120	
	X-Sub (%)	X-View (%)	X-Sub (%)	X-Set (%)
ST-GCN [40]	81.5	88.3	–	–
AS-GCN [17]	86.8	94.2	–	–
2s-AGCN [29]	88.5	95.1	–	–
AGC-LSTM [30]	89.2	95.0	–	–
DGNN [28]	89.9	96.1	–	–
PL GCN [11]	89.2	95.0	–	–
NAS-GCN [25]	89.4	95.7	–	–
SGN [44]	89.0	94.5	79.2	81.5
Shift-GCN [7]	90.7	96.5	85.9	87.6
MS-G3D [22]	91.5	96.2	86.9	88.4
DC-GCN+ADG [6]	90.8	96.6	86.5	88.1
PA-ResGCN-B19 [31]	90.9	96.0	87.3	88.3
Dynamic-GCN [41]	91.5	96.0	87.3	88.6
MST-GCN [5]	91.5	96.6	87.5	88.8
EfficientGCN-B4 [32]	92.1	96.1	88.7	88.9
CTR-GCN [4]	92.4	96.8	88.9	90.6
STA-GCN(J)	89.5	95.6	85.0	86.2
STA-GCN(B)	90.2	95.4	85.5	87.1
STA-GCN(JM)	88.3	94.3	82.9	84.0
STA-GCN(BM)	88.6	94.1	83.1	84.9
2s-STA-GCN(J, B)	91.6	96.2	88.1	89.6
3s-STA-GCN(J, B, JM)	92.7	96.9	89.2	90.6
4s-STA-GCN(J, B, JM, BM)	**92.8**	**97.0**	**89.4**	**90.8**

4.4   Comparisons with SOTA Methods

For fair comparisons, we follow the same multi-stream fusion strategy as [4,7,41]. Specifically, we use four modality streams, *i.e.*, joint stream (J), bone stream (B), joint motion stream (JM), and bone motion stream (BM). A simple score-level fusion strategy is adopted to obtain the fused score. The 1-stream model uses the individual stream of four modalities as input data. The 2-stream model fuses the joint and bone stream. The 3-stream model fuses the joint, bone, and joint motion stream. The 4-stream model fuses all four modality streams.

We compare our models with the state-of-the-art methods on NTU RGB+D 60, NTU RGB+D 120, and Kinetics Skeleton in Tables 6 and 7 respectively. On NTU RGB+D 60 and NTU RGB+D 120, our STA-GCN using three modality streams outperforms the previous state-of-the-art methods using four modality streams (*i.e.*, CTR-GCN [4] and MST-GCN [5]). Our final 4s-STA-GCN achieves new state-of-the-art performance. On Kinetics Skeleton, our model with four streams fusion outperforms current state-of-the-art MST-GCN [5] by 2.1% and 2.7% on Top-1 and Top-5 accuracy respectively. All these experimental results demonstrate the superiority of the STA-GCN.

Table 7. Comparisons of the Top-1 and Top-5 accuracy with the state-of-the-art methods on the Kinetics dataset.

Methods	Kinetics skeleton	
	Top-1 (%)	Top-5 (%)
ST-GCN [40]	30.7	52.8
AS-GCN [17]	34.8	56.5
2s-AGCN [29]	36.1	58.7
DGNN [28]	36.9	59.6
NAS-GCN [25]	37.1	60.1
MS-G3D [22]	38.0	60.9
MST-GCN [5]	38.1	60.8
STA-GCN(J)	36.0	58.5
STA-GCN(B)	34.9	57.3
STA-GCN(JM)	33.1	56.4
STA-GCN(BM)	33.6	56.6
2s-STA-GCN(J, B)	38.5	61.5
3s-STA-GCN(J, B, JM)	40.0	63.0
4s-STA-GCN(J, B, JM, BM)	**40.2**	**63.5**

5   Conclusion

In this work, we present a Spatial-Temporal Adaptive Graph Convolutional Network (STA-GCN) to capture robust motion patterns for skeleton action recognition. The STA-GCN is composed of spatial adaptive graph convolution (SA-GC) and temporal adaptive graph convolution (TA-GC). The SA-GC module is designed to extract spatial features by modeling the spatial adaptive joint dependencies. The TA-GC module is designed to learn temporal features by capturing the direct long-range joint dependencies in the temporal dimension. Both SA-GC and TA-GC have a critical embedded component: Topology Adaptive Encoder (TAE). The TAE is adopted to generate spatial adaptive topology and temporal adaptive topology for learning the discriminative features. Extensive experimental results demonstrate the effectiveness of the proposed modules. On three large-scale skeleton action recognition datasets, the proposed STA-GCN achieves the state-of-the-art performance.

Acknowledgements. This work is supported by National Natural Science Foundation of China (Project No. 62076132) and Natural Science Foundation of Jiangsu (Project No. BK20211194).

References

1. Aggarwal, J.K., Ryoo, M.S.: Human activity analysis: a review. ACM Comput. Surv. (CSUR) **43**(3), 1–43 (2011)
2. Cao, Z., Simon, T., Wei, S.E., Sheikh, Y.: Realtime multi-person 2D pose estimation using part affinity fields. In: Proceedings of the IEEE Conference on Computer Vision and Pattern Recognition, pp. 7291–7299 (2017)
3. Carreira, J., Zisserman, A.: Quo vadis, action recognition? A new model and the kinetics dataset. In: Proceedings of the IEEE Conference on Computer Vision and Pattern Recognition, pp. 6299–6308 (2017)
4. Chen, Y., Zhang, Z., Yuan, C., Li, B., Deng, Y., Hu, W.: Channel-wise topology refinement graph convolution for skeleton-based action recognition. In: Proceedings of the IEEE/CVF International Conference on Computer Vision, pp. 13359–13368 (2021)
5. Chen, Z., Li, S., Yang, B., Li, Q., Liu, H.: Multi-scale spatial temporal graph convolutional network for skeleton-based action recognition. In: Proceedings of the AAAI Conference on Artificial Intelligence, vol. 35, pp. 1113–1122 (2021)
6. Cheng, K., Zhang, Y., Cao, C., Shi, L., Cheng, J., Lu, H.: Decoupling GCN with DropGraph module for skeleton-based action recognition. In: Vedaldi, A., Bischof, H., Brox, T., Frahm, J.-M. (eds.) ECCV 2020. LNCS, vol. 12369, pp. 536–553. Springer, Cham (2020). https://doi.org/10.1007/978-3-030-58586-0_32
7. Cheng, K., Zhang, Y., He, X., Chen, W., Cheng, J., Lu, H.: Skeleton-based action recognition with shift graph convolutional network. In: Proceedings of the IEEE/CVF Conference on Computer Vision and Pattern Recognition, pp. 183–192 (2020)
8. Du, Y., Wang, W., Wang, L.: Hierarchical recurrent neural network for skeleton based action recognition. In: Proceedings of the IEEE Conference on Computer Vision and Pattern Recognition, pp. 1110–1118 (2015)
9. Gao, X., Hu, W., Tang, J., Liu, J., Guo, Z.: Optimized skeleton-based action recognition via sparsified graph regression. In: Proceedings of the 27th ACM International Conference on Multimedia, pp. 601–610 (2019)
10. He, K., Zhang, X., Ren, S., Sun, J.: Deep residual learning for image recognition. In: Proceedings of the IEEE Conference on Computer Vision and Pattern Recognition, pp. 770–778 (2016)
11. Huang, L., Huang, Y., Ouyang, W., Wang, L.: Part-level graph convolutional network for skeleton-based action recognition. In: Proceedings of the AAAI Conference on Artificial Intelligence, vol. 34, pp. 11045–11052 (2020)
12. Huang, Z., Shen, X., Tian, X., Li, H., Huang, J., Hua, X.S.: Spatio-temporal inception graph convolutional networks for skeleton-based action recognition. In: Proceedings of the 28th ACM International Conference on Multimedia, pp. 2122–2130 (2020)
13. Kay, W., et al.: The kinetics human action video dataset. arXiv preprint arXiv:1705.06950 (2017)
14. Ke, Q., Bennamoun, M., An, S., Sohel, F., Boussaid, F.: A new representation of skeleton sequences for 3d action recognition. In: Proceedings of the IEEE Conference on Computer Vision and Pattern Recognition, pp. 3288–3297 (2017)
15. Kim, T.S., Reiter, A.: Interpretable 3D human action analysis with temporal convolutional networks. In: 2017 IEEE Conference on Computer Vision and Pattern Recognition Workshops (CVPRW), pp. 1623–1631. IEEE (2017)

16. Li, C., Zhong, Q., Xie, D., Pu, S.: Skeleton-based action recognition with convolutional neural networks. In: 2017 IEEE International Conference on Multimedia & Expo Workshops (ICMEW), pp. 597–600. IEEE (2017)

17. Li, M., Chen, S., Chen, X., Zhang, Y., Wang, Y., Tian, Q.: Actional-structural graph convolutional networks for skeleton-based action recognition. In: Proceedings of the IEEE/CVF Conference on Computer Vision and Pattern Recognition, pp. 3595–3603 (2019)

18. Lin, J., Gan, C., Han, S.: TSM: temporal shift module for efficient video understanding. In: Proceedings of the IEEE/CVF International Conference on Computer Vision, pp. 7083–7093 (2019)

19. Liu, J., Shahroudy, A., Perez, M., Wang, G., Duan, L.Y., Kot, A.C.: NTU RGB+ D 120: a large-scale benchmark for 3D human activity understanding. IEEE Trans. Pattern Anal. Mach. Intell. **42**(10), 2684–2701 (2019)

20. Liu, J., Wang, G., Hu, P., Duan, L.Y., Kot, A.C.: Global context-aware attention LSTM networks for 3D action recognition. In: Proceedings of the IEEE Conference on Computer Vision and Pattern Recognition, pp. 1647–1656 (2017)

21. Liu, M., Liu, H., Chen, C.: Enhanced skeleton visualization for view invariant human action recognition. Pattern Recogn. **68**, 346–362 (2017)

22. Liu, Z., Zhang, H., Chen, Z., Wang, Z., Ouyang, W.: Disentangling and unifying graph convolutions for skeleton-based action recognition. In: Proceedings of the IEEE/CVF Conference on Computer Vision and Pattern Recognition, pp. 143–152 (2020)

23. Obinata, Y., Yamamoto, T.: Temporal extension module for skeleton-based action recognition. In: 2020 25th International Conference on Pattern Recognition (ICPR), pp. 534–540. IEEE (2021)

24. Ofli, F., Chaudhry, R., Kurillo, G., Vidal, R., Bajcsy, R.: Sequence of the most informative joints (SMIJ): a new representation for human skeletal action recognition. J. Visual Commun. Image Representation **25**(1), 24–38 (2014)

25. Peng, W., Hong, X., Chen, H., Zhao, G.: Learning graph convolutional network for skeleton-based human action recognition by neural searching. In: Proceedings of the AAAI Conference on Artificial Intelligence, vol. 34, pp. 2669–2676 (2020)

26. Poppe, R.: A survey on vision-based human action recognition. Image Vision Comput. **28**(6), 976–990 (2010)

27. Shahroudy, A., Liu, J., Ng, T.T., Wang, G.: NTU RGB+D: a large scale dataset for 3D human activity analysis. In: Proceedings of the IEEE Conference on Computer Vision and Pattern Recognition, pp. 1010–1019 (2016)

28. Shi, L., Zhang, Y., Cheng, J., Lu, H.: Skeleton-based action recognition with directed graph neural networks. In: Proceedings of the IEEE/CVF Conference on Computer Vision and Pattern Recognition, pp. 7912–7921 (2019)

29. Shi, L., Zhang, Y., Cheng, J., Lu, H.: Two-stream adaptive graph convolutional networks for skeleton-based action recognition. In: Proceedings of the IEEE/CVF Conference on Computer Vision and Pattern Recognition, pp. 12026–12035 (2019)

30. Si, C., Chen, W., Wang, W., Wang, L., Tan, T.: An attention enhanced graph convolutional LSTM network for skeleton-based action recognition. In: Proceedings of the IEEE/CVF Conference on Computer Vision and Pattern Recognition, pp. 1227–1236 (2019)

31. Song, Y.F., Zhang, Z., Shan, C., Wang, L.: Stronger, faster and more explainable: a graph convolutional baseline for skeleton-based action recognition. In: Proceedings of the 28th ACM International Conference on Multimedia, pp. 1625–1633 (2020)

32. Song, Y.F., Zhang, Z., Shan, C., Wang, L.: Constructing stronger and faster baselines for skeleton-based action recognition. IEEE Trans. Pattern Anal. Mach. Intell. **45**(2), 1474–1488 (2022)
33. Sun, K., Xiao, B., Liu, D., Wang, J.: Deep high-resolution representation learning for human pose estimation. In: Proceedings of the IEEE/CVF Conference on Computer Vision and Pattern Recognition, pp. 5693–5703 (2019)
34. Tran, D., Bourdev, L., Fergus, R., Torresani, L., Paluri, M.: Learning spatiotemporal features with 3D convolutional networks. In: Proceedings of the IEEE International Conference on Computer Vision, pp. 4489–4497 (2015)
35. Vemulapalli, R., Arrate, F., Chellappa, R.: Human action recognition by representing 3D skeletons as points in a lie group. In: Proceedings of the IEEE Conference on Computer Vision and Pattern Recognition, pp. 588–595 (2014)
36. Wang, L., et al.: Temporal segment networks for action recognition in videos. IEEE Trans. Pattern Anal. Mach. Intell. **41**(11), 2740–2755 (2018)
37. Wang, P., Li, W., Ogunbona, P., Wan, J., Escalera, S.: RGB-D-based human motion recognition with deep learning: a survey. Comput. Vis. Image Underst. **171**, 118–139 (2018)
38. Weinland, D., Ronfard, R., Boyer, E.: A survey of vision-based methods for action representation, segmentation and recognition. Comput. Vis. Image Underst. **115**(2), 224–241 (2011)
39. Xia, L., Chen, C.C., Aggarwal, J.K.: View invariant human action recognition using histograms of 3d joints. In: 2012 IEEE Computer Society Conference on Computer Vision and Pattern Recognition Workshops, pp. 20–27. IEEE (2012)
40. Yan, S., Xiong, Y., Lin, D.: Spatial temporal graph convolutional networks for skeleton-based action recognition. In: Thirty-Second AAAI Conference on Artificial Intelligence (2018)
41. Ye, F., Pu, S., Zhong, Q., Li, C., Xie, D., Tang, H.: Dynamic GCN: context-enriched topology learning for skeleton-based action recognition. In: Proceedings of the 28th ACM International Conference on Multimedia, pp. 55–63 (2020)
42. Zanfir, M., Leordeanu, M., Sminchisescu, C.: The moving pose: an efficient 3D kinematics descriptor for low-latency action recognition and detection. In: Proceedings of the IEEE International Conference on Computer Vision, pp. 2752–2759 (2013)
43. Zhang, P., Lan, C., Xing, J., Zeng, W., Xue, J., Zheng, N.: View adaptive recurrent neural networks for high performance human action recognition from skeleton data. In: Proceedings of the IEEE International Conference on Computer Vision, pp. 2117–2126 (2017)
44. Zhang, P., Lan, C., Zeng, W., Xing, J., Xue, J., Zheng, N.: Semantics-guided neural networks for efficient skeleton-based human action recognition. In: Proceedings of the IEEE/CVF Conference on Computer Vision and Pattern Recognition, pp. 1112–1121 (2020)
45. Zhang, X., Xu, C., Tao, D.: Context aware graph convolution for skeleton-based action recognition. In: Proceedings of the IEEE/CVF Conference on Computer Vision and Pattern Recognition, pp. 14333–14342 (2020)
46. Zhang, Z.: Microsoft kinect sensor and its effect. IEEE Multimedia **19**(2), 4–10 (2012)

3D Pose Based Feedback for Physical Exercises

Ziyi Zhao[1,2], Sena Kiciroglu[1(✉)], Hugues Vinzant[3], Yuan Cheng[4], Isinsu Katircioglu[5],
Mathieu Salzmann[1,6], and Pascal Fua[1]

[1] CVLab, EPFL, Lausanne, Switzerland
sena.kiciroglu@epfl.ch
[2] Tongji University, Lausanne, Switzerland
zhaozi1@tongji.edu.cn
[3] a-gO, Corps-Nuds, France
[4] KTH, Stockholm, Sweden
[5] SDSC, Lausanne, Switzerland
[6] Clearspace, Lausanne, Switzerland

Unsupervised self-rehabilitation exercises and physical training can cause serious injuries if performed incorrectly. We introduce a learning-based framework that identifies the mistakes made by a user and proposes corrective measures for easier and safer individual training. Our framework does not rely on hard-coded, heuristic rules. Instead, it learns them from data, which facilitates its adaptation to specific user needs. To this end, we use a Graph Convolutional Network (GCN) architecture acting on the user's pose sequence to model the relationship between the body joints trajectories. To evaluate our approach, we introduce a dataset with 3 different physical exercises. Our approach yields 90.9% mistake identification accuracy and successfully corrects 94.2% of the mistakes. Our code and dataset are available at https://github.com/Jacoo-Zhao/3D-Pose-Based-Feedback-For-Physical-Exercises.

1 Introduction

Being able to perform exercises without requiring the supervision of a physical trainer is a convenience many people enjoy, especially after the COVID-19 pandemic. However, the lack of effective supervision and feedback can end up doing more harm than good, which may include causing serious injuries. There is therefore a growing need for computer-aided exercise feedback strategies.

A few recent works have addressed this problem [1–6]. However, they focus only on identifying whether an exercise is performed correctly or not [1,6], or they rely on hard-coded rules based on joint angles that cannot easily be extended to new exercises [2–4]. In this work, we therefore leverage recent advances in the fields of pose estimation [7–9], action recognition [10,11] and motion prediction [12–14] to design a framework that provides automated and personalized feedback to supervise physical exercises.

Specifically, we developed a method that not only points out mistakes but also offers suggestions on how to fix them without relying on hard-coded, heuristic rules to define

H. Vinzant, Y. Cheng and I. Katircioglu—Former members of CVLab, EPFL.

Supplementary Information The online version contains supplementary material available at https://doi.org/10.1007/978-3-031-26316-3_12.

L. Wang et al. (Eds.): ACCV 2022, LNCS 13844, pp. 189–205, 2023.
https://doi.org/10.1007/978-3-031-26316-3_12

what a successful exercise sequence should be. Instead, it learns from data. To this end, we use a two-branch deep network. One branch is an action classifier that tells users what kind of errors they are making. The other proposes corrective measures. They both rely on Graph Convolutional Networks (GCNs) that can learn to exploit the relationships between the trajectories of individual joints. Figure 1 depicts the kind of output our network produces.

Fig. 1. Example results from our framework depicting frames from the a) squat, b) lunge, and c) plank classes. The red poses correspond to the exercises performed incorrectly while the green poses correspond to our corrections. Note that although we display a single pose from each mistake type, our framework operates on entire sequences.

To showcase our framework's performance, we recorded a physical exercise dataset with 3D poses and instruction label annotations. Our dataset features 3 types of exercises; squats, lunges and planks. Each exercise type is performed correctly and with mistakes following specific instructions by 4 different subjects. Our approach achieves 90.9% mistake recognition accuracy on a test set. Furthermore, we use the classification branch of our framework to evaluate the performance of the correction branch, considering the correction to be successful if the corrected motion is classified as "correct". Under this metric, our approach successfully corrects 94.2% of users' mistakes. We will make our code and dataset publicly available upon acceptance.

2   Related Work

Our work is at the intersection of several sub-fields of computer vision: (i) We draw inspiration from GCN based **human motion prediction** architectures; (ii) we identify the users' mistakes in an **action recognition** fashion; and (iii) we address the task of **physical exercise analysis**. We therefore discuss these three topics below.

2.1 Human Motion Prediction

Human motion prediction is a complex task due to the inherent uncertainty in forecasting into the future. In recent years, many deep learning methods based on recurrent neural networks (RNNs) [14–20], variational auto encoders (VAEs) [21–24], transformers [12], and graph convolutional networks (GCNs) [13, 25–27] have been proposed. We focus our discussion on GCN based ones, as we also exploit the graph-like connections of human joints with GCNs in our approach.

In [25], Mao *et al.* proposed to exploit a GCN to model the relationships across joint trajectories by representing them with Discrete Cosine Transform (DCT) coefficients. The approach was in [13] by integrating an attention module, and in [27, 28] by using cross-subject attention for multi-person motion prediction. In [26], the input coefficients to the GCN were extracted via an inception module instead of the DCT. Our motion correction branch is inspired by [25], but instead of forecasting future motion, we predict correctly performed exercises.

2.2 Action Recognition

Although there is a vast literature on image-based action recognition, here we focus on its skeleton-based counterpart, as our approach also processes 3D poses. Early deep learning based approaches to skeleton-based action recognition mostly relied on RNNs [29–33]. Li *et al.* [10] used convolutional neural networks (CNNs) to extract features hierarchically by first finding local point-level features and gradually extracting global spatial and temporal features. Zhang *et al.* [34] designed CNN and RNN networks that are robust to viewpoint changes.

Recently, [11, 35, 36] employed GCNs for action recognition. Specifically, Tang *et al.* [35] designed a reinforcement learning scheme to select the most informative frames and feed them to a GCN. Li *et al.* [36] developed a GCN framework that not only models human joint connections, but also learns to infer "actional-links", which are joint dependencies learned from the data.

Zhang *et al.* [11] designed a two-module network, consisting of a first GCN-based module that extracts joint-level information and a second frame-level module capturing temporal information via convolutional layers and spatial and temporal max-pooling. Our classification branch borrows ideas from Mao *et al.*'s [25] and Zhang *et al.*'s [11] architectures. It is composed of graph convolutional blocks as proposed by Mao *et al.* [25] combined with the frame-level module architecture proposed by Zhang *et al.* [11].

2.3 Physical Exercise Analysis

Physical exercise analysis aims to prevent injuries that may arise when a person performs motions incorrectly. In its simplest form, such an analysis amounts to detecting whether the subject performs the exercise correctly or not. This was achieved several works [1, 5, 6] by exploiting 2D poses extracted from the input images. In particular, Dittakavi *et al.* [5] detected which joints need to be fixed by finding the overall joint angle distribution of the dataset and detecting poses in which a joint angle is an anomaly.

This framework operates on single frames, as opposed to our method which operates on entire sequences. In [37], Zell *et al.* represented the human body as a mass-spring model and analyzed the extension torque on certain joints, allowing them to classify whether a motion is performed correctly or not. While useful, such classification-based approaches offer limited information to the user, as they do not provide them with any feedback about the specific type of mistakes they made. Moreover, most of existing works operate on 2D pose inputs [1–3,5,6]. Similar to [4], we also design our framework to work with 3D poses enabling us to be robust to ambiguities found in 2D poses.

While a few works took some steps toward giving feedback [2–4], this was achieved in a hard-coded fashion, by thresholding angles between some of the body joints. As such, this approach relies on manually defining such thresholds, and thus does not easily extend to new exercises. Furthermore, it does not provide the user personalized corrective measures in a visual manner, by demonstrating the correct version of their performance. We address this by following a data driven approach able to automatically learn the different "correct" forms of an exercise, and that can easily extend to different types of exercises and mistakes. To the best of our knowledge, our framework is the first to both identify mistakes and suggest personalized corrections to the user.

3   Methodology

Before we introduce our framework in detail, let us formally define the tasks of motion classification and correction. Motion classification seeks to predict the action class c of a sequence of 3D poses from $t = 1$ to $t = N$, denoted as $\mathbf{P}_{1:N}$. We can write this as

$$c = F_{\text{class}}(\mathbf{P}_{1:N}) ,$$

where F_{class} is the classification function.

We define motion correction as the task of finding the "correct" version of a sequence, which can be written as

$$\hat{\mathbf{P}}_{1:N} = F_{\text{corr}}(\mathbf{P}_{1:N}) ,$$

where F_{corr} is the correction function and $\hat{\mathbf{P}}_{1:N}$ is the corrected sequence. Ideally, the corrected sequence should be of class "correct". We can use the classification function to verify that this is the case, i.e.,

$$c_{\text{correct}} = F_{\text{class}}(\hat{\mathbf{P}}_{1:N}) ,$$

where c_{correct} is the label corresponding to a correctly performed exercise.

Given these definitions, we now describe the framework we designed to address these tasks and discuss our training and implementation details.

3.1   Exercise Analysis Framework

Our framework for providing exercise feedback relies on GCNs and consists of two branches: One that predicts whether the input motion is correct or incorrect, specifying the mistake being made in the latter case, and one that outputs a corrected 3D pose sequence, providing a detailed feedack to the user. We refer to these two branches as the "classifier" and "corrector" models, respectively.

Inspired by Mao *et al.* [25], we use the DCT coefficients of joint trajectories, rather than the 3D joint positions, as input to our model. This allows us to easily process sequences of different lengths. The corrector model outputs DCT coefficient residuals, which are then summed with the input coefficients and undergo an inverse-DCT transform to be converted back to a series of 3D poses.

To reduce the time and space complexity of training the classifier and the corrector separately and to improve the accuracy of the model, we combine the classification and correction branches into a single end-to-end trainable model. Figure 2 depicts our overall framework. It takes the DCT coefficients of each joint trajectory as input. The first layers are shared by the two models, and the framework then splits into the classification and correction branches.

Furthermore, we feed the predicted action labels coming from the classification branch to the correction branch. We depict this in Fig. 2 as the "Feedback Module". Specifically, we first find the label with the maximum score predicted by the classification branch, convert this label into a one-hot encoding, and feed it to a fully-connected layer. The resulting tensor is concatenated to the output of the first graph convolutional blocks (GCB) of the correction branch. This process allows us to explicitly provide label information to the correction module, enabling us to further improve the accuracy of the corrected motion.

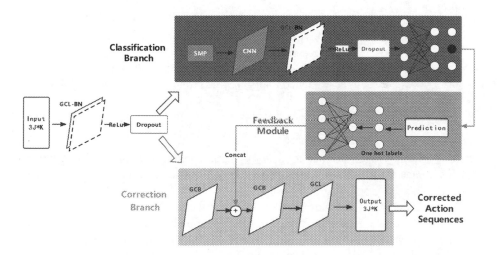

Fig. 2. Our framework consists of a classification and a correction branch. They share several graph convolutional layers are then split such that the classification branch identifies the type of mistakes made by the user and the correction branch outputs a corrected pose sequence. The result of the classification branch is fed to the correction branch via a feedback module.

Implementation and Training Details. We primarily use GCB similar to those presented in [25] in our network architecture, depicted in Fig. 3. These modules allow us to learn the connectivity between different joint trajectories. Each graph convolutional layer is set to have 256 hidden features. Additionally, our classification branch borrows ideas from Zhang *et al.*'s [11] action recognition model. It is a combination of GCB modules and the frame-level module architecture of [11] consisting of convolutional layers and spatial max-pooling layers.

We train our network in a supervised manner by using pairs of incorrectly performed and correctly performed actions. However, it is not straightforward to find these pairs of motions. The motion sequences are often of different lengths, and we face the task of matching incorrectly performed actions to the closest correctly performed action from the same actor. To do so, we make use of Dynamic Time Warping (DTW) [38], which enables us to find the minimal alignment cost between two time series of different lengths, using dynamic programming. We compute the DTW loss between each incorrect and correct action pair candidate and select the pair with the smallest loss value.

We use the following loss functions to train our model.

- E_{corr}: The loss of the correction branch, which aims to minimize the soft-DTW [39] loss between the corrected output sequence and the closest correct motion sequence, determined as described previously. The soft-DTW loss is a differentiable version of the DTW loss, implemented by replacing the minimum operation by a soft minimum.
- E_{smooth}: The smoothness loss on the output of the correction branch, to ensure the produced motion is smooth and realistic. It penalizes the velocities of the output motion by imposing an L2 loss on them.
- E_{class}: The loss of the classification branch, which aims to minimize the cross entropy loss between the predicted logits and the ground-truth instruction label.

We combine these losses into

$$E_{loss} = w_{corr}E_{corr} + w_{class}E_{class} + w_{smooth}E_{smooth}, \tag{1}$$

where E_{loss} is the overall loss and w_{corr}, w_{class}, w_{smooth} are the weights of the correction, classification, and smoothness losses, respectively. For our experiments we set $w_{corr} = 1$, $w_{class} = 1$, and $w_{smooth} = 1e - 3$.

During training, we use curriculum learning in the feedback module: Initially the ground-truth instruction labels are given to the correction branch. We then use a scheduled sampling strategy similar to [40], where the probability of using the ground-truth labels instead of the predicted ones decreases from 1 to 0 linearly as the epochs increase. In other words, the ground-truth labels are progressively substituted with the labels predicted by the classification branch, until only the predicted labels are used. During inference, only the predicted labels are given to the correction branch.

We use Adam [41] as our optimizer. The learning rate is initially set to 0.01 and decays according to the equation $lr = 0.01 \cdot 0.9^{i/s}$, where lr is the learning rate, i is the epoch and s is the decay step, which is set to 5. To increase robustness and avoid overfitting, we also use drop-out layers with probability 0.5. We use a batch size of 32 and train for 50 epochs.

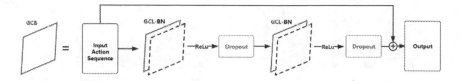

Fig. 3. Graph Convolutional Block (GCB) consisting of graph convolutional layers, batch normalization layers, ReLUs and drop-outs.

4 EC3D Dataset

Existing sports datasets such as Yoga-82 [42], FineGym [43], FSD-1O [44], and Diving48 [45] often include correct performances of exercises but do not include incorrect sequences. They are also not annotated with 3D poses. Therefore to evaluate our approach, we recorded and processed a dataset of physical exercises performed both correctly and incorrectly, and named the "EC3D" (Exercise Correction in **3D**) dataset.

Specifically, this dataset contains 3 types of actions, each with 4 subjects who repeatedly performed a particular correct or incorrect motion as instructed. We show the number of sequences per action and the instructions for each subject in Table 1. The dataset contains a total of 132 squat, 127 lunge, and 103 plank action sequences, split across 11 instruction labels.

The videos were captured by 4 GoPro cameras placed in a ring around the subject, using a frame rate of 30 fps and a 1920 × 1080 image resolution. Figure 4 depicts example images taken from the dataset with their corresponding 2D and 3D skeleton representation. The cameras' intrinsics were obtained by recording a chessboard pattern and using standard calibration methods implemented in OpenCV [46].

Table 1. The EC3D dataset with the number of sequences per instruction of each subject, the total number of sequences per instruction and the total number of sequences per action. We reserve Subjects 1, 2, and 3 for training and 4 for testing.

Exercise	Instruction label	Subject 1	Subject 2	Subject 3	Subject 4	Total (per instruction)	Total (per action)
Squats	Correct	10	10	11	10	41	132
	Feet too wide	5	8	5	5	23	
	Knees inward	6	7	5	5	23	
	Not low enough	5	7	5	4	21	
	Front bent	5	6	6	7	24	
Lunges	Correct	12	11	11	12	46	127
	Not low enough	10	10	10	10	40	
	Knee passes toe	10	10	11	10	41	
Planks	Correct	7	8	11	7	33	103
	Arched back	5	5	11	9	30	
	Hunch back	10	10	11	9	40	

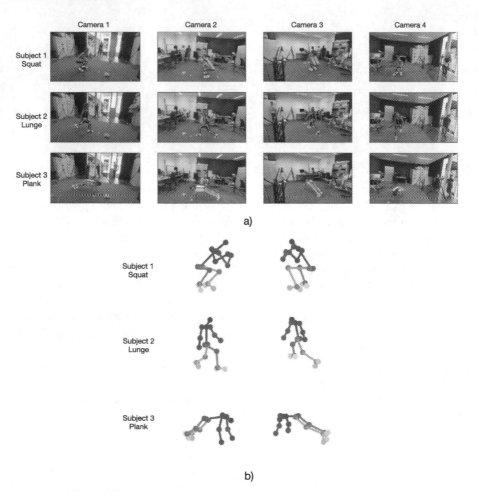

Fig. 4. Examples images from the EC3D dataset, depicting images from the SQUAT, lunge, and plank classes with their corresponding 3D pose visualizations. a) Images for each exercise type from the dataset from each camera viewpoint, with the 2D poses overlayed. b) The corresponding 3D poses, visualized from two different viewpoints.

We annotated the 3D poses in an automated manner, whereas the action and instruction labels were annotated manually. Specifically, the 3D pose annotation was performed as follows: First, the 2D joint positions were extracted from the images captured by each camera using OpenPose [47], an off-the-shelf 2D pose estimation network. We then used bundle adjustment to determine the cameras' extrinsics. For the bundle adjustment algorithm to converge quickly and successfully, additional annotations were made on static landmarks in 5 frames. Since the cameras were static during recording, for each camera, we averaged the extrinsics optimized for each of these frames. Afterwards, these values were kept constant, and we triangulated the 2D poses to compute the 3D poses.

During the triangulation process, we detected whenever any joint had a high reprojection error to catch mistakes in the 2D pose estimates. Such 2D pose annotations were discarded to prevent mistakes in the 3D pose optimization. The obtained 3D pose values were afterwards smoothed using a Hamming filter to avoid jittery motion. Finally, we manually went through the extracted 3D pose sequences in order to ensure that there are no mistakes and that they are consistent with the desired motion.

To make the resulting 3D poses uniform, we further normalized, centred and rotated them. As the different heights and body sizes of the different subjects cause differences in skeletal lengths, a random benchmark was selected to normalize the skeletal lengths while maintaining the connections between joints. Furthermore, we centered all poses on their hip joint and rotated them so that the spine was perpendicular to the ground and all movements performed in the same direction.

5   Evaluation

5.1   Dataset and Metrics

We use the EC3D dataset to evaluate our model performance both quantitatively and qualitatively. We use subjects 1, 2, and 3 for training and subject 4 for evaluation.

We use top-1 classification accuracy to evaluate the results of the instruction classification task, as used by other action classification works [10,11]. For the motion correction task, we make use of the action classifier branch: If the corrected motion is classified as "correct" by our classification branch, we count the correction as successful. We report the percentage of successfully corrected motions as the evaluation metric for this task.

5.2   Quantitative Results

We achieve an average mistake recognition accuracy of 90.9% when classifying sequences in EC3D, as shown by the detailed results for each specific exercise instruction in Table 2. In the same table, we also show that 94.2% of the corrected results are classified as "correct" by our classification model. The high classification accuracy and correction success show that our framework is indeed capable of analyzing physical exercises and giving useful feedback.

As no existing works have proposed detailed correction strategies, we compare our framework to a simple correction baseline consisting of retrieving the closest "correct" sequence from the training data. The closest sequence is determined as the sequence with the lowest DTW loss value to the input sequence. In Table 3, we provide the DTW values between the incorrectly performed input and the corrected output. The DTW loss acts as an evaluation of the accuracy of joint positions, as it is an L2 loss on the time aligned sequences. For this metric, the lower, the better, i.e., the output motion should be as close as possible to the original one while being corrected as necessary. Our framework yields a high success rate of correction together with a lower DTW loss than the baseline, thus supporting our claims. Note that we do not evaluate the baseline's correction success percentage because it retrieves the same sequences that were used to train the network, to which the classification branch might have already overfit.

Table 2. Results of our classification and correction branches on the EC3D dataset. We achieve 90.9% recognition accuracy on average and successfully correct 94.2% of the mistakes.

Exercise	Mistake label	Classification accuracy (%)	Correction success (%)
Squats	Correct	90.0	100
	Feet too wide	100	100
	Knees inward	100	100
	Not low enough	100	100
	Front bent	57.1	85.7
Lunges	Correct	66.7	100
	Not low enough	100	60.0
	Knee passes toe	100	90.0
Planks	Correct	85.7	100
	Arched back	100	100
	Hunch back	100	100
Average		90.9	94.2

5.3   Qualitative Results

In Fig. 5, we provide qualitative results corresponding to all the incorrect motion examples from each action category. Note that the incorrect motions are successfully corrected, yet still close to the original sequence. This makes it possible for the user to easily recognize their own motion and mistakes.

5.4   Ablation Studies

We have tried various versions of our framework and recorded our results in Table 4. In this section, we present the different experiments, also depicted in Fig. 6, and the discussions around these experiments.

Separated Models. We first analyze the results of separated classification networks. According to Table 4, our separated classification branch architecture is denoted as "separated classification." We have also evaluated a simpler, fully GCN based separated action classifier branch, denoted as "separated classification (simple)". We show that the results of the classification branch degrade slightly when separated from the correction branch. This indicates that the classification branch also sees a minor benefit from being part of a combined model. The simpler classification network performs worse than our architecture inspired by [11], showing that the pooling module improves the classification accuracy.

Afterwards, we analyze the results of a separated correction network, denoted as "separated correction". Here the difference is quite profound; we see that separating the correction model from the classification model degrades correction success significantly. We note that 50 epochs was not enough for the separated corrector framework to converge, therefore we trained it for a total of 150 epochs.

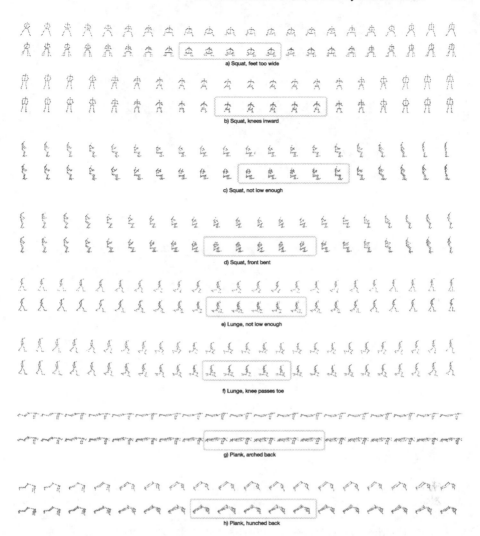

Fig. 5. Qualitative results from our framework depicting incorrect input motions and corrected output motions from categories a–d) squats, e,f) lunges, g, h) planks. We present the incorrect input sequences (red) in the top row. The corrected sequences (green) overlaid on top of the incorrect input sequences (red) are presented in the bottom row. The most significant corrections are highlighted with a yellow bounding box. We find that our proposals are successful in correcting the incorrect sequences. This figure is best viewed in color and zoomed in on a screen.

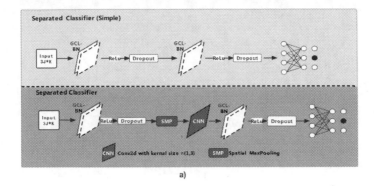

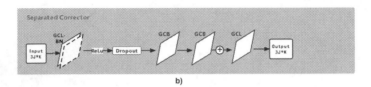

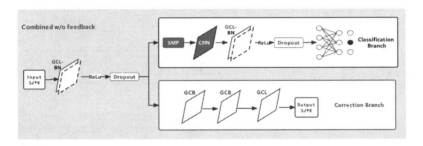

Fig. 6. Ablation study frameworks. We depict the different architectures we evaluated for the ablation studies: a) Separated classifier (simple) and separated classifier. b) Separated corrector. c) Combined model without feedback. This model does not include a "Feedback Module", the classification branch's results are not explicitly fed to the correction branch.

Table 3. DTW results of the correction branch. We compare our framework to a simple baseline retrieving the best matching "correct" sequence from the training dataset depending on the classification label. We report the DTW loss between the input and the output sequences (lower is better). Our framework successfully corrects the subject's mistakes, while not changing the input so drastically that the subject would not be able to recognize their own performance.

Exercise	Mistake label	Retrieval baseline	Our framework
Squats	Correct	1.28	**0.56**
	Feet too wide	4.23	**1.46**
	Knees inward	1.61	**0.66**
	Not low enough	1.83	**0.61**
	Front bent	4.74	**2.53**
Lunges	Correct	1.94	**1.82**
	Not low enough	1.86	**1.31**
	Knee passes toe	2.27	**1.48**
Planks	Correct	2.41	**1.79**
	Arched back	12.20	**1.53**
	Hunch back	4.10	**1.09**
Average		3.49	**1.35**

Combined Models. We train our framework without the feedback module ("combined w/o feedback"), and without the smoothness loss ("combined w/o smoothness"). We find that these perform worse than our model with the feedback module and with the smoothness loss in terms of correction success. This shows that using the classification results as feedback as well as the smoothness loss for training is useful for more successful corrections. We notice that our framework trained without smoothness loss has higher classification accuracy, despite having a lower correction success. We believe this is due to the fact that the smoothness loss acts as a regularizer on the framework, therefore causing slight performance losses to the classification branch. However, the results of the correction success are significantly higher with the smoothness loss. We also evaluate our trained model by passing random incorrect instruction labels to the correction branch instead of the labels predicted by the classification branch ("combined with random incorrect feedback"). The correction success drops significantly, showing that the classification results are indeed very useful for the correction branch.

Further qualitative and quantitative results are presented in the supplementary material.

5.5   Limitations and Future Work

While our current results are quite impressive, there is still room for improvement in terms of performance. In particular, our framework struggles in correcting specific types of motions, such as not low enough lunges. We plan to explore different additions to our framework, such as an attention module, to better correct these types of mistakes.

Table 4. Results of the ablation studies with several variations of our framework. We report the classification accuracy (%) and the correction success rate (%), where higher is better for both metrics. Our framework benefits greatly from combining the two tasks in a end-to-end learning fashion, from using a feedback module, and from using a pooling layer in the classification branch. The smoothness loss causes slight degradation in classification accuracy but is greatly beneficial for the correction success.

		Classification accuracy (%)	Correction success (%)
Separated	Separated classification (simple)	88.6	–
	Separated classification	89.8	–
	Separated correction	–	83.5
Combined	Combined w/o feedback	82.3	85.3
	Combined with random incorrect feedback	90.9	87.3
	Combined w/o smoothness	**93.4**	87.5
	Ours	90.9	**94.2**

Our future work will consist of expanding the dataset to include more action sequences and more types of mistakes, performed by a larger set of subjects. We believe that this will allow us to further improve our framework and add components that address the shortcomings we will discover using such a dataset.

6   Conclusion

We have presented a 3D pose based feedback framework for physical exercises. We have designed this framework to output feedback in two branches; a classification branch to identify a potential mistake and a correction branch to output a corrected sequence. Through ablation studies, we have validated our network architectural choices and presented detailed experimental results, making a strong case for the soundness of our framework design. We have also introduced a dataset of physical exercises, on which we have achieved 90.9% classification accuracy and 94.2% correction success.

Acknowledgements. This work was supported in part by the Swiss National Science Foundation.

References

1. Chen, S., Yang, R.R.: Pose Trainer: correcting exercise posture using pose estimation, arXiv Preprint (2020)
2. Yang, L., Li, Y., Zeng, D., Wang, D.: Human exercise posture analysis based on pose estimation. In: 2021 IEEE 5th Advanced Information Technology, Electronic and Automation Control Conference (IAEAC) (2021)
3. Kanase, R., Kumavat, A., Sinalkar, R., Somani, S.: Pose Estimation and Correcting Exercise Posture. ITM Web Conf. **40**, 03031 (2021)

4. Fieraru, M., Zanfir, M., Pirlea, S.C., Olaru, V., Sminchisescu, C.: AIFit: automatic 3D human-interpretable feedback models for fitness training. In: Conference on Computer Vision and Pattern Recognition (2021)
5. Dittakavi, B., et al.: Pose Tutor: an explainable system for pose correction in the wild. In: Proceedings of the IEEE/CVF Conference on Computer Vision and Pattern Recognition (CVPR) Workshops (2022)
6. Rangari, T., Kumar, S., Roy, P., Dogra, D., Kim, B.G.: Video based exercise recognition and correct pose detection. Multim. Tools Appl. **81** (2022)
7. Kocabas, M., Athanasiou, N., Black, M.J.: VIBE: video inference for human body pose and shape estimation. In: Conference on Computer Vision and Pattern Recognition, pp. 5252–5262 (2020)
8. Zheng, C., Zhu, S., Mendieta, M., Yang, T., Chen, C., Ding, Z.: 3D human pose estimation with spatial and temporal transformers. In: International Conference on Computer Vision, pp. 11656–11665 (2021)
9. Gong, K., et al.: PoseTriplet: co-Evolving 3D human pose estimation, Imitation, and Hallucination Under Self-Supervision. In: Conference on Computer Vision and Pattern Recognition (2022)
10. Li, Y., Yuan, L., Vasconcelos, N.: Co-Occurrence feature learning from skeleton data for action recognition and detection with hierarchical aggregation. In: International Joint Conference on Artificial Intelligence (2018)
11. Zhang, P., Lan, C., Zeng, W., Xing, J., Xue, J., Zheng, N.: Semantics-guided neural networks for efficient skeleton-based human action recognition. In: Conference on Computer Vision and Pattern Recognition (2020)
12. Aksan, E., Kaufmann, M., Cao, P., Hilliges, O.: A Spatio-temporal transformer for 3D human motion prediction. In: International Conference on 3D Vision (3DV) (2021)
13. Mao, W., Liu, M., Salzmann, M.: History repeats itself: human motion prediction via motion attention. In: Vedaldi, A., Bischof, H., Brox, T., Frahm, J.-M. (eds.) ECCV 2020. LNCS, vol. 12359, pp. 474–489. Springer, Cham (2020). https://doi.org/10.1007/978-3-030-58568-6_28
14. Kiciroglu, S., Wang, W., Salzmann, M., Fua, P.: Long term motion prediction using keyposes. In: International Conference on 3D Vision (3DV) (2022)
15. Fragkiadaki, K., Levine, S., Felsen, P., Malik, J.: Recurrent network models for human dynamics. In: International Conference on Computer Vision (2015)
16. Jain, A., Zamir, A., ADN Saxena, S.S.A.: Structural-RNN: deep learning on spatio-temporal graphs. In: Conference on Computer Vision and Pattern Recognition (2016)
17. Ghosh, P., Song, J., Aksan, E., Hilliges, O.: Learning human motion models for long-term predictions. In: International Conference on 3D Vision (2017)
18. Martinez, J., Black, M., Romero, J.: On human motion prediction using recurrent neural networks. In: Conference on Computer Vision and Pattern Recognition (2017)
19. Wang, B., Adeli, E., Chiu, H.K., Huang, D.A., Niebles, J.C.: Imitation learning for human pose prediction. In: International Conference on Computer Vision, pp. 7123–7132 (2019)
20. Barsoum, E., Kender, J., Liu, Z.: HP-GAN: probabilistic 3D human motion prediction via GAN. In: Conference on Computer Vision and Pattern Recognition (2017)
21. Butepage, J., Black, M., Kragic, D., Kjellstrom, H.: Deep representation learning for human motion prediction and classification. In: Conference on Computer Vision and Pattern Recognition (2017)
22. Bütepage, J., Kjellström, H., Kragic, D.: Anticipating many futures: online human motion prediction and generation for human-robot interaction. In: International Conference on Robotics and Automation (2018)

23. Bütepage, J., Kjellström, H., Kragic, D.: Predicting the what and how - a probabilistic semi-supervised approach to multi-task human activity modeling. In: Conference on Computer Vision and Pattern Recognition, pp. 2923–2926 (2019)
24. Aliakbarian, S., Saleh, F.S., Salzmann, M., Petersson, L., Gould, S.: A stochastic conditioning scheme for diverse human motion prediction. In: Conference on Computer Vision and Pattern Recognition (2020)
25. Mao, W., Liu, M., Salzmann, M., Li, H.: Learning trajectory dependencies for human motion prediction. In: International Conference on Computer Vision (2019)
26. Lebailly, T., Kiciroglu, S., Salzmann, M., Fua, P., Wang, W.: Motion prediction using temporal inception module. In: Asian Conference on Computer Vision (2020)
27. Katircioglu, I., Georgantas, C., Salzmann, M., Fua, P.: Dyadic human motion prediction, arXiv Preprint (2022)
28. Guo, W., Bie, X., Alameda-Pineda, X., Moreno-Noguer, F.: Multi-person extreme motion prediction. In: Proceedings of the IEEE/CVF Conference on Computer Vision and Pattern Recognition (CVPR) (2022)
29. Du, Y., Wang, W., Wang, L.: Hierarchical recurrent neural network for skeleton based action recognition. In: Conference on Computer Vision and Pattern Recognition (2015)
30. Shahroudy, A., Liu, J., Ng, T., Wang, G.: NTU RGB+D: a large scale dataset for 3D human activity analysis. In: Conference on Computer Vision and Pattern Recognition, pp. 1010–1019 (2016)
31. Liu, J., Shahroudy, A., Xu, D., Wang, G.: Spatio-temporal LSTM with trust gates for 3d human action recognition. In: Leibe, B., Matas, J., Sebe, N., Welling, M. (eds.) ECCV 2016. LNCS, vol. 9907, pp. 816–833. Springer, Cham (2016). https://doi.org/10.1007/978-3-319-46487-9_50
32. Song, S., Lan, C., Xing, J., Zeng, W., Liu, J.: An end-to-end spatio-temporal attention model for human action recognition from skeleton data. In: AAAI Conference on Artificial Intelligence (2017)
33. Liu, J., Wang, G., Hu, P., Duan, L., Kot, A.: Global Context-aware attention LSTM networks for 3D action recognition. In: Conference on Computer Vision and Pattern Recognition (2017)
34. Zhang, P., Lan, C., Xing, J., Zeng, W., Xue, J., Zheng, N.: View adaptive neural networks for high performance skeleton-based human action recognition. . Trans. Pattern, Anal. Mach. Intell. **41** 1963–1978 (2019)
35. Tang, Y., Tian, Y., Lu, J., Li, P., Zhou, J.: Deep progressive reinforcement learning for skeleton-based action recognition. In: Conference on Computer Vision and Pattern Recognition (2018)
36. Li, M., Chen, S., Chen, X., Zhang, Y., Wang, Y., Tian, Q.: Actional-structural graph convolutional networks for skeleton-based action recognition. In: Conference on Computer Vision and Pattern Recognition (2019)
37. Zell, P., Wandt, B., Rosenhahn, B.: Joint 3D human motion capture and physical analysis from monocular videos. In: 2017 IEEE Conference on Computer Vision and Pattern Recognition Workshops (CVPRW) (2017)
38. Sakoe, H., Chiba, S.: Dynamic programming algorithm optimization for spoken word recognition. In: IEEE Trans. Acoust. Speech Sig. Proc. **26**, 43–49 (1978)
39. Cuturi, M., Blondel, M.: Soft-DTW: a differentiable loss function for time-series. In: International Conference on Machine Learning (2017)
40. Bengio, S., Vinyals, O., Jaitly, N., Shazeer, N.: Scheduled sampling for sequence prediction with recurrent neural networks. In: Advances in Neural Information Processing Systems (2015)
41. Kingma, D.P., Ba, J.: Adam: A Method for stochastic optimisation. In: International Conference on Learning Representations (2015)

42. Verma, M., Kumawat, S., Nakashima, Y., Raman, S.: Yoga-82: a new dataset for fine-grained classification of human poses. In: The IEEE/CVF Conference on Computer Vision and Pattern Recognition (CVPR) Workshops (2020)
43. Shao, D., Zhao, Y., Dai, B., Lin, D.: FineGym: a Hierarchical Video Dataset for Fine-grained Action Understanding. In: IEEE Conference on Computer Vision and Pattern Recognition (CVPR) (2020)
44. Shenglan, L., et al.: FSD-10: a fine-grained classification dataset for figure skating. Neurocomputing **413**, 360–367 (2020)
45. Li, Y., Li, Y., Vasconcelos, N.: RESOUND: towards action recognition without representation bias. In: Ferrari, V., Hebert, M., Sminchisescu, C., Weiss, Y. (eds.) ECCV 2018. LNCS, vol. 11210, pp. 520–535. Springer, Cham (2018). https://doi.org/10.1007/978-3-030-01231-1_32
46. (Open Source Computer Vision Library). http://opencv.org
47. Cao, Z., Simon, T., Wei, S., Sheikh, Y.: Realtime multi-person 2D pose estimation using part affinity fields. In: Conference on Computer Vision and Pattern Recognition, pp. 1302–1310 (2017)

Generating Multiple Hypotheses for 3D Human Mesh and Pose Using Conditional Generative Adversarial Nets

Xu Zheng, Yali Zheng$^{(\boxtimes)}$ (iD), and Shubing Yang

School of Automation Engineering, University of Electronic and Scientific Technology of China, Chengdu, China
zhengyl@uestc.edu.cn

Abstract. Despite recent successes in 3D human mesh/pose recovery, the human mesh/pose reconstruction ambiguity is a challenging problem that can not be avoided as lighting, occlusion or self-occlusion in scenes happens. We argue that there could be multiple 3D human meshes corresponding a single image from a view point, because we really do not know what happens in extreme lighting or behind occlusion/self occlusion. In this paper, we address the problem using Conditional Generative Adversarial Nets (CGANs) to generate multiple hypotheses for 3D human mesh and pose from a single image under the condition of 2D joints and relative depth of adjacent joints. The initial estimation of 2D human skeletons, relative depth and features is taken as input of CGANs to train the generator and discriminator in the first stage. Then the generator of CGANs is used to generate multiple human meshes via different conditions which are consistent with human silhouette and 2D joint points in the second stage. Selecting and clustering are utilized to eliminate abnormal and redundant human meshes. The number of hypothesis is not unified for each single image, and it is dependent on 2D pose ambiguity. Unlike the existing end-to-end 3D human mesh recovery methods, our approach consists of three task-specific deep networks trained separately to mitigate the training burden in terms of time and datasets. Our approach has been evaluated not only on the datasets of laboratory and real scenes but also on Internet images qualitatively and quantitatively, and experimental results demonstrate the effectiveness of our approach.

Keywords: Human mesh · CGAN · Multiple hypotheses

1 Introduction

Recovering 3D human mesh and pose from images is a fundamental problem in the field of computer vision. It is challenging to reconstruct 3D human mesh accurately and robustly from a single image, which has drawn a lot of attention from researchers. Essentially this is an ill-posed problem whether using traditional 3D geometric methods or using deep learning methods to restore. The number of geometric constraints is less than the number of unknown variables, which causes the uncertainty of 3D human mesh and pose reconstruction. Further, as extreme lighting, occlusion or self-occlusion in real scenes happens, it exacerbates the uncertainty [10,51]. Most of the existing methods are only able to recover one plausible human mesh from a single image, however, there

L. Wang et al. (Eds.): ACCV 2022, LNCS 13844, pp. 206–222, 2023.
https://doi.org/10.1007/978-3-031-26316-3_13

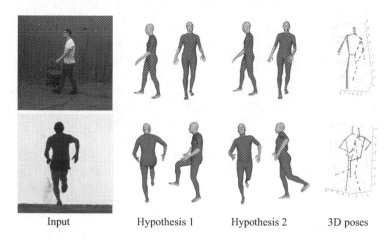

| Input | Hypothesis 1 | Hypothesis 2 | 3D poses |

Fig. 1. Our approach takes a single image as input, and recovers multiple diverse hypotheses of 3D human mesh and pose. Here two hypotheses for human mesh are shown in the picture in two different views. (Best view in color)

exists multiple possible reconstruction for a number of images, see the first column in Fig. 1, that is one image leads to multiple reconstructions.

Generative Adversarial Nets are the popular frameworks to train generative model in an alternative manner, however, they suffer from the *mode collapse* easily, and they are hard to train. So Mirza and Osindero proposed CGANs, and conditioning could be class labels, some part of data or even data from different modality [29]. CGANs have many benefits. They can generate models under the control of condition, and speed the convergence of model training. This property of CGANs is quite suitable to solve the problem of multiple hypotheses reconstruction for human mesh and pose. In this paper, we try to generate multiple hypotheses for reasonable 3D human mesh via CGANs which are consistent with human silhouette and 2D joint points detected from images, see Fig. 1.

Unlike the existing end-to-end methods for 3D human mesh reconstruction, we propose a two-stage approach to generate multiple hypotheses of 3D human mesh under the CGANs framework. In the first stage, human silhouette, 2D joints and features are detected automatically, and CGANs are trained to generate human mesh bases in an adversarial manner. Since 2D joints and relative depth may be estimated incorrectly due to the extreme light or occlusion/self-occlusion, so we expand the conditions to control the generation of 3D human meshes and poses. So in the second stage, limited hypotheses for 3D human mesh and pose are produced by varying the conditions of CGANs due to being tolerant to detection ambiguity, then clustered in a cascade pipeline. Here, the conditions include 2D joints and the relative depth of adjacent joints. One of the benefits of our system is that the background influence is removed since only human silhouette, 2D joints and relative depth of 2D joints are taken as input for CGANs, it is easily generalized to different scenarios. Our contributions include:

1) We propose a two-stage weakly-supervised approach to address the problem of multiple hypotheses for human mesh and pose by CGANs, which only takes human silhouette as input under the constraints of 2D joints and relative depth without any 3D supervision.

2) We generate multiple 3D human body meshes by importing limited and discrete hypotheses with varying conditions of CGANs and clustering, instead of sampling from a probability distribution. In our approach, the number of conditions is limited and discrete, since the hypotheses space is limited and discrete.

3) We evaluate throughout the experiments that our system is suitable for multiple hypotheses generation for 3D human mesh and pose not only from image datasets of the laboratory and real scenarios, but also from Internet images without networks retraining.

2   Related Work

3D Mesh Reconstruction. The reconstruction methods of 3D human body mesh from a single image have made great progress in recent years along with the development of deep learning technique. These methods can be roughly divided into two categories: one is the human body shape reconstruction method based on parametric models [2, 13, 25, 37], the other is the human body shape reconstruction methods based on non-parameters [9, 19, 41, 43, 51]. The former methods try to encode a human mesh into a parametric model. The advantage of this type of methods is that it is easy to reconstruct the complete shape of human body from a single image, even if occlusion happens in the image. The disadvantage is that parametric models suffer from the ability of limited detail representation. A human shape is only represented as a linear combination of 10 shape bases, so details of the human shape may be recovered insufficiently. The model-free methods describe the details better, but require to learn the large amount of variables.

Model-Based Mesh Reconstruction. A number of human mesh models have been proposed in the field, such as SCAPE [2], Skinned Multi-Person Linear model (SMPL) [25], ADAM [13] and so on [31, 37, 52]. The most popular SMPL model [25] has 82-dimensional parameters, including 10 shape base parameters and 24×3 joint rotation parameters, and it has been extended into SMPL-H [37], SMPLify [3] and SMPLify-X [31]. SMPLify, a multi-stage optimization method, was proposed in [3], which used DeepCut [36] to estimate 2D pose first, and then optimized reconstruct results. HMR [14] used an end-to-end CNN network to regress SMPL parameters from RGB images with discriminator constraints. TexturePose [32] took texture information to enhance CNN network ability. SPIN [18] combined a learning-based algorithm and optimization algorithm into an iterative framework to obtain SMPL parameters, and achieved state-of-the-art results among the methods of model-based human mesh recovering. SMPLify-X [31] extended the SMPL-X model, which had a human mesh with hands and face, and took 2D human pose to optimize with prior constraints and penalty of mesh collision. MTC [48] optimized ADAM [13] by 3D pose direction vector and 2D pose estimating by CNN-based network.

Model-Free Mesh Reconstruction. The model-free shape recovery methods consider a human body consisting of voxel volumes, and learn the body surface directly. BodyNet [43] presented an end-to-end CNN based network to estimate volumetric representation of human with 2D pose, depth and 3D pose. Kolotouros et al. proposed that taking CNN and graph neural network was significantly easy to regress 3D location of human mesh [19]. Gabeur et al. estimated the "visible" and "hidden" depth maps, and combined into a full-body 3D point cloud [9]. Zhu et al. [51] proposed a hierarchical method to capture the detailed human body. Tan et al. presented a self-supervised method to relax the dependance on ground truth data from videos [41].

3D Pose Estimation. While the problem of 2D human pose estimation has been well solved, 3D human pose estimation is still challenging due to 3D reconstruction ambiguity caused by the variation of viewpoint, human body and clothing. Deep learning technique is a popular way to estimate 3D pose from a single image, and these approaches achieve good results. Pavlakos et al. [34] applied the stack hourglass network to estimate every voxel likelihood for each joint. Mehta et al. [28] argued that the algorithm generalizability constrained by available 3D pose datasets, and proposed a benchmark "MPI-INF-3DHP" (MPII-3D for short in this paper) which covered outdoor and indoor scenes. Zhou et al. explored the problem of 3D pose estimation in the wild, and proposed a weakly-supervised transfer learning framework due to the lack of training data [50]. Kacabas et al. proposed EpipolarPose to estimate 3D human poses and camera matrix with the constraints of epipolar geometry [17]. Jahangiri and Yuille addressed the problem to estimate multiple diverse 3D poses in [10], and started initial 3D pose to generate multiple hypotheses according to prior sampling, while Li and Lee [21] generated multiple hypotheses for 3D human pose based on a mixture density network. As transformer technique arises, it is regarded as a powerful backbone in vision field. Zheng et al. [49] and Li et al. [22] proposed to estimate human pose via transformer, and achieve big approvement. However, transformer-based methods always have a big model, which are resource-consuming and have high requirements on GPU, while our approach has less training variables in weakly-supervised way.

3 The Proposed Approach

In this section, we introduce our proposed weakly-supervised approach in detail. We give the work flow firstly, then explain how to estimate human mesh and pose using CGANs, and how to generate the multiple hypotheses by CGANs by selecting and clustering. Finally we give the implementation details of each parts.

3.1 Overview

We present a cascade framework to recover multiple hypotheses for 3D human shapes that are consistent with the 2D human silhouette and 2D joints heat map extracted from images. As shown in Fig. 2, there are two stages in our pipeline, consisting of three parts $\mathcal{N}_1, \mathcal{N}_2, \mathcal{N}_3$:

$\mathcal{N}_1$: an encoder-decoder structure, takes the original image as input to regress 30 2D joints heat map $\mathcal{F}_J$ and human silhouette S, and learns features $\mathcal{F}$ (Note that 30 joints

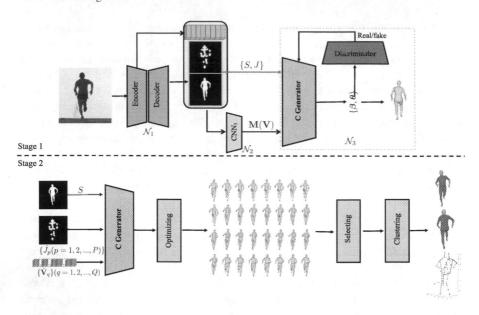

Fig. 2. Schematic diagram of the proposed approach. There are two stages in our pipeline, consisting of three parts: the encoder-decoder $\mathcal{N}_1$ to regress 2D human silhouette and 2D joint heat map, the CNN $\mathcal{N}_2$ to estimate a relative depth vector, the CGANs $\mathcal{N}_3$ to generate human mesh. Firstly 2D human joints, human silhouette learnt by $\mathcal{N}_1$, and the relative depth vector learnt by $\mathcal{N}_2$ are taken as the inputs to train the generator and discriminator of CGANs $\mathcal{N}_3$. In the second stage, multiple hypotheses for human mesh are generated by rationally expanding the relative depth vectors $\{\tilde{\mathbf{V}}_q\}$ and 2D joints sets $\{\tilde{J}_p\}$. Then abnormal 3D human meshes are removed by selecting steps, and the close human meshes are clustered to output the final mesh and pose hypotheses.

including 24 points of SMPL model, 5 points for one nose, two eyes, two ears, and 1 point on head).

$\mathcal{N}_2$: a CNN, takes 2D human silhouette S, joint heat map $\mathcal{F}_J$ and features $\mathcal{F}$ learnt by $\mathcal{N}_1$ as input to estimate a relative depth matrix $\mathbf{M}$. $\mathbf{V}$ is a binary vector form of $\mathbf{M}$, which only has the relative depth between adjacent joints.

$\mathcal{N}_3$: the CGANs, consisting of a conditional generator and a discriminator. The generator takes feature maps of joints, heat map $\mathcal{F}_J$, human silhouette S and the relative depth vector $\mathbf{V}$ to regress the parameters β and θ of SMPL model to generate human mesh. The discriminator distinguishes meshes from β and θ of SMPL model as real or fake.

In the first stage, 2D human joint heat map and human silhouette learnt by $\mathcal{N}_1$ and the relative depth vector learnt by $\mathcal{N}_2$ are taken as inputs to train the generator and discriminator of CGANs $\mathcal{N}_3$. In the second stage, multiple hypotheses for human meshes are generated reasonably by expanding the relative depth vector and the 2D skeleton set. And the optimization is following to make human meshes better. Then abnormal 3D human meshes are rejected by selecting step after discarding human meshes with crossing through parts, and the close human meshes are clustered to output the final mesh hypotheses after detail refining. The whole pipeline is shown in Fig. 2.

Since 2D joints and silhouette for humans are well studied, and a number of 2D skeleton and human segment datasets are published by different institutions, so it is not difficult to learn $\mathcal{N}_1$ and $\mathcal{N}_2$. The implementation details of $\mathcal{N}_1$ and $\mathcal{N}_2$ will be described in Sect. 3.5. Once $\mathcal{N}_1$ is trained to learn feature map of 2D joints $\mathcal{F}_J$, the silhouette image S, and a 2048D feature vector $\mathcal{F}$, the location set of joints denoted by J is derived by integrating over the corresponding feature map of joints $\mathcal{F}_J$ [39]. $\mathcal{F}_J$, S and $\mathcal{F}$ are taken, as input, to train $\mathcal{N}_2$ to estimate the matrix $\mathbf{M}$ of relative depths. Each element between $[0, 1]$ in $\mathbf{M}_{ij}$, can be thought of as a probability of ith joint farther than jth joint in the camera coordinate [33]. Let $\mathbf{M}'_{ij}$ be a sparse binary matrix, in which $0/1$ represents that the ith joint farther/closer than the jth joint in the camera coordinate. In the following subsections, the key parts are introduced, including human mesh estimation via CGANs, multiple hypotheses generation, and human mesh selecting and clustering.

3.2 Human Mesh Estimation Using CGANs

In our CGANs, human mesh is encoded by SMPL parametric model, and the model parameters – β (shape) and θ (pose) are learnt to represent human mesh from human silhouette S constrained by J and $\mathbf{M}'$. The discriminator distinguishes human meshes from β and θ as real or fake. The modified objective function for CGANs is minimized as follows,

$$\min_{\mathcal{G}} \max_{\mathcal{D}} L(\mathcal{D}, \mathcal{G}) = E[\log \mathcal{D}(\beta, \theta)] + E[\log(1 - \mathcal{D}(\mathcal{G}(S|J, \mathbf{M}')))] \qquad (1)$$

In the formula, the generator $\mathcal{G}$ is conditioned by 2D joints J and the relative depth $\mathbf{M}'$, while $\mathcal{D}$ is a classifier learnt by taking samples from (β, θ) generated by $\mathcal{G}$ as negative samples, and samples from Mosh dataset as positive ones. The reason that we do not take condition constraint on discriminator $\mathcal{D}$ is because we found it is sufficient to take (β, θ) for classifying real or fake one. The generator is built on ResNet-18, and the discriminator is just a multi-layer perceptron net. Please refer the details in Sect. 3.5. Then the generator $\mathcal{G}$ and the discriminator $\mathcal{D}$ are trained under the GANs framework in an alternative manner. As the discriminator becomes stronger, the generator gets stronger. The generator is trained under the control of J and $\mathbf{M}'$ in weak supervision of 2D joints J and the human silhouette S, and the generator enables us to generate multiple hypotheses by varying the conditions.

Further, the loss for the conditional generator is minimized over β, θ as follows:

$$L_{\mathcal{G}}(\beta, \theta) = L_{2D} + L_{dep} + L_{reg} \qquad (2)$$

The $L_{\mathcal{G}}$ consists of three terms, 2D constraint loss L_{2D}, the relative depth constraint loss L_{dep}, and the regularization term L_{reg}. They have to be balanced in the training.

Each term is defined as follows,

$$L_{2D} = ||\mathfrak{J}(\mathcal{M}(\beta, \theta)) - J_{2D}||_2 + ||\mathfrak{S}(\mathcal{M}(\beta, \theta)) - S||_2$$

$$L_{dep} = \sum_{l=1}^{|J_{2D}|} \sum_{J_k \in \mathbb{N}(J_l)} max((J_l - J_k) \cdot (2 \cdot \mathbf{M}'_{l,k} - 1), 0)$$

$$L_{reg} = \gamma_\beta ||\beta||_2 + \gamma_\theta ||\theta||_2 \tag{3}$$

In which, $\mathcal{M}$ is an operation of recovering a mesh from (β, θ), $\mathfrak{J}$ is an operation to project 3D joints of mesh into 2D, and $\mathfrak{S}$ is an operation to render a silhouette binary image from the generated mesh [16]. $\mathbb{N}$ denotes the neighbourhood set of a joint, since only the adjacent joints are considered in our approach. $\mathbf{M}'_{l,k}$ in L_{dep} estimated in the first stage, which is 1 or 0, would constrain the joint J_l is closer or farther than the joint J_k. And γ_β and γ_θ are the balance factors in L_{reg}. Once CGANs are trained in the first stage, the generator is exploited to produce multiple possible human meshes and poses in the second stage.

3.3   Condition Expansion for Multiple Hypotheses

Now we explain how to generate multiple hypotheses for human mesh and pose from a single image. We only care about the relative depth between adjacent joints, and transform matrix $\mathbf{M}'$ into a vector form $\mathbf{V}$. Each element in $\mathbf{V}$ represents the relative depth relationship between to adjacent joints, and a limited space is generated with the relative depth vector $\mathbf{V}_q$ ($q = 1, 2, ..., Q$) and 2D joints sets J_p ($p = 1, 2, ..., P, P <= 4$) to help multiple human mesh and pose recovery. Since the relative depth estimation and 2D joints detection are not always correct, so it should be tolerant reasonably. As has been observed, each element in $\mathbf{M}_{ij}$ describes an estimate confidence of ith joint farther than jth joint in the camera coordinates. When $\mathbf{M}_{ij}$ is close to 1, which means the ith joint is more likely to be closer than the jth one, and when $\mathbf{M}_{ij}$ is close to 0, which means the ith joint is more likely to be farther than the jth one. When $\mathbf{M}_{ij}$ is near around 0.5, it is ambiguous to determine which one is closer. Let δ be an ambiguity capacity factor, then we have

$$\tilde{\mathbf{M}}_{ij} = \begin{cases} 0, & \mathbf{M}_{ij} \in [0, 0.5 - \delta), \\ 0 \ or \ 1, & \mathbf{M}_{ij} \in [0.5 - \delta, 0.5 + \delta], \\ 1, & \mathbf{M}_{ij} \in (0.5 + \delta, 1]. \end{cases} \tag{4}$$

$\tilde{\mathbf{M}}$ is also transformed into a vector form $\mathbf{V}$. Assume that $\tilde{\mathbf{M}}$ has Q possible values when $\mathbf{M}_{ij}(\in [0.5 - \delta, 0.5 + \delta])$, then it leads to $\tilde{\mathbf{V}}_q, q = 1, 2, ..., Q$. Note that we only care about the relative depth between the adjacent joints in the vector. In this way, a single $\mathbf{V}$ in the first stage is expanded into a set space $\{\tilde{\mathbf{V}}_1, \tilde{\mathbf{V}}_2, ..., \tilde{\mathbf{V}}_Q\}$, which makes the relative depth be tolerant to a certain estimation error.

Further, the initial estimation of 2D pose J shows much ambiguous as lighting, occlusion or self-occlusion as well. The human mesh may be confused with the body of the right and left parts due to the fact that a human body is a symmetrical object.

There are three extra cases: 1) the right leg is confused with the left leg; 2) the right arm is confused with the left arm; 3) the right body is confused with the left body. So the initial estimation of 2D joints is expanded into a space of 4 cases, that is $\{\tilde{J}_1, ..., \tilde{J}_P\}$ and $P = 4$, 2D joints of right leg is flipped with ones of left leg, the right arm is flipped with joints on the left arm. However, we know if a human to be recovered is facing forward or unknown by the detection from the relative depth between eyes and ears. If the human faces forward confidently, then one of four cases is ignored, that is $\{\tilde{J}_1, ..., \tilde{J}_P\}$ and $P = 3$. The expansions of relative depth and 2D skeleton are taken as the reconstruction condition for the conditional generator to produce multiple hypotheses for human mesh, which enable our approach be tolerant to the estimation error and be resistant to the ambiguity of body structure.

All generated human meshes are optimized, and the objective function is the same as Eq. 2, minimized over β and θ by gradient descend as follows,

$$\min_{\beta, \theta} \quad L_{\mathcal{G}} \tag{5}$$

The optimized β and θ are used to generate more accurate human mesh.

3.4 Human Mesh Selecting and Clustering

By varying $\{\tilde{\mathbf{V}}_1, ..., \tilde{\mathbf{V}}_Q\}$ and $\{\tilde{J}_1, ..., \tilde{J}_P\}$, QP 3D human meshes (each one has 6890 $3D$ points) are generated in the previous subsection. Obviously, besides of the positive effect of withstanding generation ambiguity, the expansions may cause some redundancy, and generate abnormal meshes.

The abnormal meshes include crossing through parts, abnormal posture and so on. Crossing through parts would be rejected partially by body collision detection [31]. Also the rest meshes have to be classified by classifier in [4], and abnormal posture would be rejected and removed from the human mesh set. Assume that N human body meshes are passed through the selection as normal, and they are clustered using mean-shift, each category has N_k human meshes, $k = 1, 2, ..., K$. K is the number of classes from mean-shift clustering. Obviously, the method of mean-shift is not necessary to cluster samples into a fixed number of class. Take the one with the least projection error of 2D silhouette and 2D human joints to represent each class, the objective function is as follows:

$$\min_{n_k \in [1, ..., N_k]} ||\mathfrak{J}(\mathcal{M}_{n_k}) - J||_2 + \gamma ||\mathfrak{S}(\mathcal{M}_{n_k}) - S||_2 \tag{6}$$

where γ is a balance number, and need to be set in advance. Once K human meshes are clustered, they output the 3D human poses by determining 3D joints from human mesh.

3.5 Implementation Details

Four different parts $\mathcal{N}_1, \mathcal{N}_2, \mathcal{N}_3$ are trained separately. The $\mathcal{N}_1$ takes the pre-trained ResNet-50 as the encoder, and 9 layers of deconvolution as the decoder to regress the 2D silhouettes/segments S and feature maps of 2D human joints $\mathcal{F}_J$. ResNet-18 is used in the second CNN $\mathcal{N}_2$ to predict the relative depth of adjacent human joints $\mathbf{V}$ and the distance from the camera T. The conditional generator, a modified ResNet, is a stack of

ResNet-18 and four fully connected layers, and end up with a regressor like in [18]. The discriminator, which is a multi-layer perceptron based network with 1024 neurons. The generator and discriminator in $\mathcal{N}_3$ are trained alternatively by minimizing the objective Eq. 2. In order to overcome the challenge of unbalanced laboratory and real scene datasets, the similar number of images are sampled from the laboratory dataset and the real scene dataset in each epoch as [6]. In the optimization step, each optimizing only run ten times. δ is set as 0.2 through all experiments. The learning rate for conditional generator is $1e-4$, and learning rate for discriminator is $2e-4$. All networks are trained with GPU of RTX TITAN.

4   Experimental Results

4.1   Datasets

A variety of datasets are assembled to train different deep networks in our experiments. All 2D pose human datasets are all made to be consistent with our 30 points human skeleton, the occlusion ones will be ignored. LSP [11] and LSPET [12] are 2D human pose datasets of 14 joint points, while MPII [1] is a 2D human pose dataset of 16 joint points. The UP [20] and COCO [23] dataset consist of 2D human pose and 2D segmentation. MTC [45] and MPII-3D [28] are laboratory multi-view 3D pose datasets, while the former has a large range of viewing angle, and the latter has more than 1.3M images in total. The Mosh [24] is a synthetic 3D human mesh dataset, while SURREAL [44] renders the original human meshes in Mosh dataset into real scenes. 3DPW [26] is a 3D pose dataset captured from real scenes. 2D pose datasets are utilized to train 2D human skeleton detection. Because our method is a weakly-supervised algorithm, we only use 2D pose and silhouette to weakly supervise the training process of $\mathcal{N}_3$.

LSP, LSPET, MPII, COCO, MPII-3D, MTC, SURREAL, 3DPW datasets are used to train $\mathcal{N}_1$. MPII-3D, MTC, SURREAL, 3DPW datasets are used to train $\mathcal{N}_2$. The datasets including LSP, LSPET, MPII-3D, SURREAL, 3DPW are used for $\mathcal{N}_3$. Although MPII-3D, MTC, SURREAL, 3DPW are both 2D and 3D pose datasets, we do not use 3D pose but 2D pose only through all experiments.

4.2   Qualitative Results

We not only reconstruct multiple human meshes from a single image, but also recover multiple 3D human poses. Figure 3 shows more qualitative results on challenging images from the laboratory dataset (MPII-3D), real scene datasets (LSP, LSPET, MPII), and Fig. 4 shows more results on Internet images. The first column shows the original input images, the second column shows an example of 2D projection of 3D human mesh and detected 2D pose. Four hypotheses of human meshes in two different view generated from our approach are shown from the third to tenth column. The last two columns show all 3D human poses in the same coordinates to compare the difference between poses in two different views. The dash line represents the right part of a human body, the solid line represents the left part of a human body.

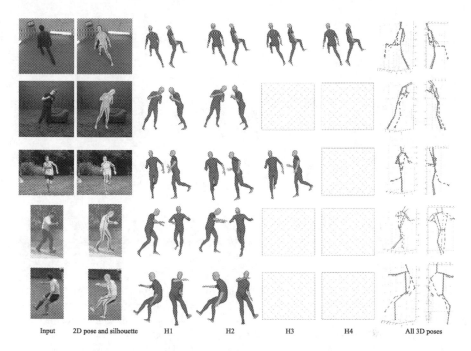

Input   2D pose and silhouette   H1   H2   H3   H4   All 3D poses

Fig. 3. Results from images of laboratory and real scenes dataset. The first column shows the original input images, the second column shows an example of 2D projection of 3D human mesh and 2D pose. The hypotheses for human mesh and 3D pose generated from our approach are shown from the third to tenth column in two different view. The last two columns show all 3D human poses in the same coordinates to compare the difference between poses in two different views. The first image has four hypotheses generation of human mesh, the third image has three hypotheses generation, while the second, fourth and fifth have two hypotheses generation.

4.3   Statistic Analysis of Multiple Hypotheses

We also do statistic analysis on LSP, MPII-3D and 3DPW datasets, and count the number of generation hypotheses for all images in these datasets. Statistic analysis is reported in Fig. 5. Most of images have 2 human mesh hypotheses generated by our method on all datasets. More than 85% images have $1-3$ hypotheses, and few images (about 2% or less) have more than 5 hypotheses. In order to show the statistic figure better, the number of images in MPII-3D and 3DPW is reduced to one tenth just for better viewing. We investigate that these images with more than 5 hypotheses are often caused by the low-quality generation for human mesh results from the generator, which have extreme large 2D projection error.

4.4   Quantitative Comparison

Our approach is evaluated quantitatively with respect to errors of human mesh reprojection and 3D joints location on three datasets, which are popular ways in the field. Since our method has generated multiple human meshes and 3D poses, the best one

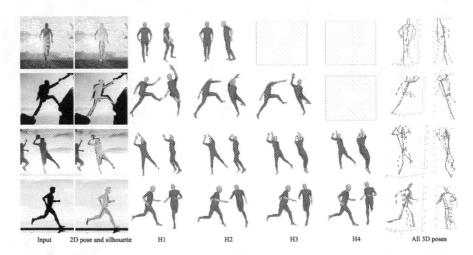

Input   2D pose and silhouette   H1   H2   H3   H4   All 3D poses

Fig. 4. Results from Internet images. The first column shows the original input images, the second column shows an example of 2D projection of 3D human mesh and 2D pose. The hypotheses for human mesh and 3D pose generated from our approach are shown from the third to tenth column in two different view. The last two columns show all 3D human poses in the same coordinates to compare the difference between poses in two different views. The first image has two hypotheses generation of human mesh, the second image has three hypotheses generation, while the third and fourth images have four hypotheses generation. The dash line represents the right part of a human body, the solid line represents the left part of a human body.

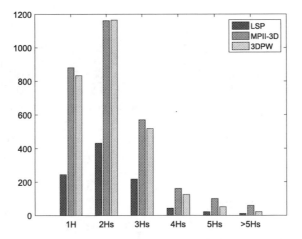

Fig. 5. Statistic analysis of hypotheses generation after selecting and clustering by our approach on three datasets. The number is the number of images with different generation hypotheses in each dataset.

is selected to compare with the ground truth. The test dataset of LSP has 972 images, MPII-3D has 24111 test images, while 3DPW has 26549 test images totally, we use all test images to do a quantitative evaluation.

We compare with SMPLify and its variations [3,20,35], HMR [14] and SPIN [18] in terms of the silhouette Intersection over Union (sil-IoU) on LSP test dataset, which measures the matching rate of the projected silhouette of the predicted 3D human mesh and image human segment, and report the results in Table 1. Our method shows best results in both accuracy and F1 measurement on foreground-background and part segmentation on LSP dataset, and increases about 2% compared with SPIN in terms of accuracy. We also compare with PoseNet3D [42], HMR [14], HMR-Video [15], CMR [19], HM-LGD [38], SPIN [18], PARE [5], ROMP [47], ProHMR [7,30,40], with respect to 3D joint error on 3DPW in Table 2. And we report the comparison results on MPII-3D datasets in Table 3 with PoseNet3D [42], HMR [14], SPIN [18], Vnect [8], DenseRaC [27,46]. 3D joint error is measured by the least mean per joint position error (MPJPE) between the generated 3D poses and the ground truth before and after rigid alignment. Because our method is a weakly-supervised algorithm, we do not use 3D human pose in training process, and all data are unpaired through our pipeline.

Table 1. Evaluation on foreground-background and six-part segmentation on LSP test set.

Methods	FB segments		Part Segments	
	Acc	F1	Acc	F1
SMPLify [3]	91.89	0.88	87.71	0.64
SMPLify oracle [20]	92.17	0.88	88.82	0.67
SMPLify+anchor [35]	92.17	0.88	88.24	0.64
HMR [14]	91.67	0.87	87.12	0.60
SPIN [18]	91.83	0.87	89.41	0.68
Our-*Stage1*	92.01	0.87	89.16	0.67
Ours-*es1*	93.83	0.90	91.00	0.72
Ours-*es2*	93.75	0.90	90.78	0.71
Ours	**93.94**	**0.91**	**91.92**	**0.72**

Ablative Analysis. We examine that if we only have the generator to produce human meshes in the first stage. And the key point of our method is that multiple hypotheses J_p and $\mathbf{V}_q$ are used to guide the generator to create multiple human meshes and poses, which help to improve the human mesh accuracy. Here we try to study how multiple 2D joints J_p and multiple relative depth $\mathbf{V}_q$ are helpful to generate human meshes. 2D human joints, the relative depth and 2D silhouette are all needed as the inputs for the generator, so we fix one when examining the other. They are denoted by two experimental setups as follows,

Experimental setup 1 (es1): When the effect of multiple hypotheses of $\mathbf{V}$ is tested, J is fixed to the initial estimation of J', which is the estimation from the encoder-decoder network $\mathcal{N}_1$.

Table 2. Evaluation on 3DPW dataset. The numbers are MPJPE after rigid alignment.

Supervision	Methods	3D data	PAMPJPE
Strong	HMR [14]	paired, with 3D	81.3
	HM-LGD [38]	paired, with 3D	55.9
	SPIN-static fits [18]	with $\{\beta, \theta\}$	66.3
	SPIN in the loop [18]	with $\{\beta, \theta\}$	59.2
	Doersch et al. [7]	with 3D	74.7
	HMR-Video [15]	with 3D	72.6
	CMR [19]	with 3D	70.2
	Sun et al. [40]	with 3D	66.3
	PARE [5]	with 3D and $\{\beta, \theta\}$	57.1
	ROMP [47]	with 3D and $\{\beta, \theta\}$	56.8
	ProHMR [30]	with 3D and $\{\beta, \theta\}$	59.8
Weak	PoseNet3D [42]	No	63.2
	Our-*stage1*	No	64.35
	Ours-*es1*	No	61.48
	Ours-*es2*	No	64.73
	Ours	No	59.78

Experimental setup 2 (es2): When the effect of multiple hypotheses of J is tested, $\mathbf{V}$ is fixed to $\mathbf{V}'$, which is the original estimation from $\mathcal{N}_2$.

The quantitative results are reported in Table 1, Table 2 and Table 3 on different datasets as well. It can be seen that the space expansion of 2D skeleton and relative depth are helpful for generating more accurate human segments and 3D joints.

4.5   Failure Case

Figure 6 shows two failure examples. Most of failure cases happen for these special images. The human in these two images makes some unusual pose, which is less learnt by CGANs from the dataset, and even it is difficult for 2D human joints detection. Then the generator is not able to generate an reasonable initial human mesh. So training better networks for 2D joint detection and CGANs with more various data is still a fundamental task.

Table 3. Evaluation on MPII-3D dataset. The numbers are 3D Percentage of Correct Keypoints (PCK) and mean per joint position error (MPJPE) before and after rigid alignment.

Supervision	Methods	3D data	Absolute		Rigid alignment	
			PCK	MPJPE	PCK	PAMPJPE
Strong	HMR [14]	paired,with 3D,$\{\beta,\theta\}$	72.9	124.2	86.3	89.8
	SPIN	paired, with $\{\beta,\theta\}$	76.4	105.2	92.5	67.5
	DenseRaC [46]	with 3D	–	–	89.0	83.5
	Mehta et al. [27]	with 3D	75.7	117.6	–	–
Weak	HMR [14]	unpaired, with 3D,$\{\beta,\theta\}$	59.6	169.5	77.1	113.2
	Vnect [8]	No (but video)	76.6	124.7	83.9	98
	PoseNet3D [42]	No	–	–	81.9	102.4
	Our-*stage1*	No	64.68	148.51	84.45	92.58
	Ours-*es1*	No	68.68	140.96	87.24	87.32
	Ours-*es2*	No	58.84	163.2	79.72	104.14
	Ours	No	70.81	133.91	88.58	84.88

Fig. 6. Failure cases in our experiments.

5   Conclusion

We propose a two-stage weakly-supervised pipeline to generate multiple hypotheses via CGANs for 3D human mesh and pose in this paper. The CGANs are trained to generate a single human mesh by taking 2D silhouette as input under the control of 2D joints and relative depth without any 3D supervision. With a reasonable assumption, the conditions of inputs are expanded into a bigger discrete space for generating multiple hypotheses. Then the generated abnormal meshes are rejected by collision detection and classifier, redundant human meshes are clustered. The benefit of our system is that the background influence is removed, and it is easy to generalize to Internet images. The main limitation is that, it generates multiple hypotheses for human mesh and pose, but the details of meshes are still not sufficient, and some unusual meshes and poses are still hard to reconstruct. This is our future work to discover more elaborate human meshes.

Acknowledgement. This work is supported by the National Natural Science Foundation of China (NSFC No. 61971106). Authors would like to thank all reviewers for their valuable and meticulous comments.

References

1. Andriluka, M., Pishchulin, L., Gehler, P., Schiele, B.: 2D human pose estimation: new benchmark and state of the art analysis. In: IEEE Conference on Computer Vision and Pattern Recognition (CVPR) (2014)
2. Anguelov, D., Srinivasan, P., Koller, D., Thrun, S., Rodgers, J., Davis, J.: Scape: shape completion and animation of people. In: ACM SIGGRAPH 2005 Papers (2005)
3. Bogo, F., Kanazawa, A., Lassner, C., Gehler, P., Romero, J., Black, M.J.: Keep It SMPL: automatic estimation of 3d human pose and shape from a single image. In: Leibe, B., Matas, J., Sebe, N., Welling, M. (eds.) ECCV 2016. LNCS, vol. 9909, pp. 561–578. Springer, Cham (2016). https://doi.org/10.1007/978-3-319-46454-1_34
4. Bouritsas, G., Bokhnyak, S., Ploumpis, S., Bronstein, M., Zafeiriou, S.: Neural 3D morphable models: spiral convolutional networks for 3d shape representation learning and generation. In: International Conference on Computer Vision (ICCV) (2019)
5. Bouritsas, G., Bokhnyak, S., Ploumpis, S., Bronstein, M., Zafeiriou, S.: Pare: Part attention regressor for 3d human body estimation. In: International Conference on Computer Vision (ICCV) (2021)
6. Cao, Z., Hidalgo, G., Simon, T., Wei, S.E., Sheikh, Y.: OpenPose: realtime multi-person 2D pose estimation using Part Affinity Fields. IEEE Trans. Pattern Anal. Mach. Intell. **43**(1), 172–186 (2021)
7. Doersch, C., Zisserman, A.: Sim2real transfer learning for 3d human pose estimation: motion to the rescue. In: Advances in Neural Information Processing Systems (NIPS) (2019)
8. Dushyant, M., et al.: VNect: real-time 3d human pose estimation with a single RGB camera. ACM Trans. Graphics **36**(4), 33–51 (2017)
9. Gabeur, V., Franco, J.S., Martin, X., Schmid, C., Rogez, G.: Moulding humans: Nonparametric 3D human shape estimation from single images. In: IEEE International Conference on Computer Vision (ICCV) (2019)
10. Jahangiri, E., Yuille, A.L.: Generating multiple diverse hypotheses for human 3D pose consistent with 2D joint detections. In: IEEE International Conference on Computer Vision (ICCV) (2017)
11. Johnson, S., Everingham, M.: Clustered pose and nonlinear appearance models for human pose estimation. In: The British Machine Vision Conference (BMVC) (2010)
12. Johnson, S., Everingham, M.: Learning effective human pose estimation from inaccurate annotation. In: IEEE Conference on Computer Vision and Pattern Recognition (CVPR) (2011)
13. Joo, H., Simon, T., Sheikh, Y.: Total capture: a 3D deformation model for tracking faces, hands, and bodies. In: IEEE Conference on Computer Vision and Pattern Recognition (CVPR) (2018)
14. Kanazawa, A., Black, M.J., Jacobs, D.W., Malik, J.: End-to-end recovery of human shape and pose. In: IEEE Conference on Computer Vision and Pattern Recognition (CVPR) (2018)
15. Kanazawa, A., Zhang, J.Y., Felsen, P., Malik, J.: Learning 3D human dynamics from video. In: IEEE Conference on Computer Vision and Pattern Recognition (CVPR) (2019)
16. Kato, H., Ushiku, Y., Harada, T.: Neural 3D mesh renderer. In: IEEE Conference on Computer Vision and Pattern Recognition (CVPR) (2018)

17. Kocabas, M., Karagoz, S., Akbas, E.: Self-supervised learning of 3D human pose using multi-view geometry. In: IEEE Conference on Computer Vision and Pattern Recognition (CVPR) (2019)
18. Kolotouros, N., Pavlakos, G., Black, M.J., Daniilidis, K.: Learning to reconstruct 3D human pose and shape via model-fitting in the loop. In: IEEE Conference on Computer Vision and Pattern Recognition (CVPR) (2019)
19. Kolotouros, N., Pavlakos, G., Daniilidis, K.: Convolutional mesh regression for single-image human shape reconstruction. In: IEEE Conference on Computer Vision and Pattern Recognition (CVPR) (2019)
20. Lassner, C., Romero, J., Kiefel, M., Bogo, F., Black, M.J., Gehler, P.V.: Unite the people: Closing the loop between 3D and 2D human representations. In: IEEE Conference on Computer Vision and Pattern Recognition (CVPR) (2017)
21. Li, C., Lee, G.H.: Generating multiple hypotheses for 3D human pose estimation with mixture density network. In: IEEE Conference on Computer Vision and Pattern Recognition (CVPR) (2019)
22. Li, W., Liu, H., Tang, H., Wang, P., Gool, L.V.: Mhformer: multi-hypothesis transformer for 3D human pose estimation. In: IEEE Conference on Computer Vision and Pattern Recognition (CVPR) (2022)
23. Lin, T.V., et al.: Microsoft COCO: common objects in context. In: Fleet, D., Pajdla, T., Schiele, B., Tuytelaars, T. (eds.) ECCV 2014. LNCS, vol. 8693, pp. 740–755. Springer, Cham (2014). https://doi.org/10.1007/978-3-319-10602-1_48
24. Loper, M., Mahmood, N., Black, M.J.: Mosh: motion and shape capture from sparse markers. ACM Trans. Graphics **33**(6), 1–13 (2014)
25. Loper, M., Mahmood, N., Romero, J., Pons-Moll, G., Black, M.J.: SMPL: a skinned multi-person linear model. ACM Trans. Graphics **34**(6), 1–16 (2015)
26. von Marcard, T., Henschel, R., Black, M.J., Rosenhahn, B., Pons-Moll, G.: Recovering accurate 3d human pose in the wild using imus and a moving camera. In: Ferrari, V., Hebert, M., Sminchisescu, C., Weiss, Y. (eds.) ECCV 2018. LNCS, vol. 11214, pp. 614–631. Springer, Cham (2018). https://doi.org/10.1007/978-3-030-01249-6_37
27. Mehta, D., et al.: Monocular 3D human pose estimation in the wild using improved CNN supervision. In: International Conference on 3D vision (3DV) (2017)
28. Mehta, D., et al.: Monocular 3D human pose estimation in the wild using improved CNN supervision. In: International Conference on 3D Vision (3DV) (2017)
29. Mirza, M., S., O.: Conditional generative adversarial nets (2014). https://arxiv.org/abs/1411.1784
30. Nikos Kolotouros, Georgios Pavlakos, D.J., Daniilidis, K.: Probabilistic modeling for human mesh recovery. In: International Conference on Computer Vision (ICCV) (2021)
31. Pavlakos, G., et al.: Expressive body capture: 3D hands, face, and body from a single image. In: IEEE Conference on Computer Vision and Pattern Recognition (CVPR) (2019)
32. Pavlakos, G., Kolotouros, N., Daniilidis, K.: Texturepose: supervising human mesh estimation with texture consistency. In: IEEE International Conference on Computer Vision (ICCV) (2019)
33. Pavlakos, G., Zhou, X., Daniilidis, K.: Ordinal depth supervision for 3d human pose estimation. In: IEEE Conference on Computer Vision and Pattern Recognition (CVPR) (2018)
34. Pavlakos, G., Zhou, X., Derpanis, K.G., Daniilidis, K.: Coarse-to-fine volumetric prediction for single-image 3D human pose. In: IEEE Conference on Computer Vision and Pattern Recognition (CVPR) (2017)
35. Pavlakos, G., Zhu, L., Zhou, X., Daniilidis, K.: Learning to estimate 3d human pose and shape from a single color image. In: IEEE Conference on Computer Vision and Pattern Recognition (CVPR) (2018)

36. Pishchulin, L., et al.: Deepcut: Joint subset partition and labeling for multi person pose esti-mation. In: IEEE Conference on Computer Vision and Pattern Recognition (CVPR) (2016)
37. Romero, J., Tzionas, D., Black, M.J.: Embodied hands: modeling and capturing hands and bodies together. ACM Trans. Graph. **36**(6), 245 (2017)
38. Song, J., Chen, X., Hilliges, O.: Human body model fitting by learned gradient descent. In: Vedaldi, A., Bischof, H., Brox, T., Frahm, J.-M. (eds.) ECCV 2020. LNCS, vol. 12365, pp. 744–760. Springer, Cham (2020). https://doi.org/10.1007/978-3-030-58565-5_44
39. Sun, X., Xiao, B., Wei, F., Liang, S., Wei, Y.: Integral human pose regression. In: Ferrari, V., Hebert, M., Sminchisescu, C., Weiss, Y. (eds.) ECCV 2018. LNCS, vol. 11210, pp. 536–553. Springer, Cham (2018). https://doi.org/10.1007/978-3-030-01231-1_33
40. Sun, Y., Ye, Y., Liu, W., Gao, W., Fu, Y., Mei, T.: Human mesh recovery from monocular images via a skeleton disentangled representation. In: International Conference on Computer Vision (ICCV) (2019)
41. Tan, F., Zhu, H., Cui, Z., Zhu, S., Pollefeys, M., Tan, P.: Self-supervised human depth estima-tion from monocular videos. In: IEEE Conference on Computer Vision and Pattern Recog-nition (CVPR) (2020)
42. Tripathi, S., Ranade1, S., Tyagi, A., Agrawal, A.: Posenet 3D Learning temporally consis-tent 3D human pose via knowledge distillation. In: International Conference on 3D Vision (IC3DV) (2020)
43. Varol, G., et al.: Bodynet: Volumetric inference of 3d human body shapes. In: European Conference on Computer Vision (ECCV) (2018)
44. Varol, G., Romero, J., Martin, X., Mahmood, N., Black, M.J., Laptev, I., Schmid, C.: Learn-ing from synthetic humans. In: IEEE Conference on Computer Vision and Pattern Recogni-tion (CVPR) (2017)
45. Xiang, D., Joo, H., Sheikh, Y.: Monocular total capture: Posing face, body, and hands in the wild. In: IEEE Conference on Computer Vision and Pattern Recognition (CVPR) (2019)
46. Xu, Y., Zhu, S.C., Tung, T.: Denserac: Joint 3d pose and shape estimation by dense render-and-compare. In: IEEE International Conference on Computer Vision (ICCV) (2019)
47. Yu Sun, Qian Bao, W.L.Y.F.M.J.B., Mei, T.: Monocular, one-stage, regression of multiple 3d people. In: International Conference on Computer Vision (ICCV) (2021)
48. Zanfir, A., Marinoiu, E., Sminchisescu, C.: Monocular 3D pose and shape estimation of multiple people in natural scenes-the importance of multiple scene constraints. In: IEEE Conference on Computer Vision and Pattern Recognition (CVPR) (2018)
49. Zheng, C., Zhu, S., Mendieta, M., Yang, T., Chen, C., Ding, Z.: 3D human pose estimation with spatial and temporal transformers. In: International Conference on Computer Vision (ICCV) (2021)
50. Zhou, X., Huang, Q., Sun, X., Xue, X., Wei, Y.: Towards 3D human pose estimation in the wild: a weakly-supervised approach. In: IEEE International Conference on Computer Vision (ICCV) (2017)
51. Zhu, H., Zuo, X., Wang, S., Cao, X., Yang, R.: Detailed human shape estimation from a single image by hierarchical mesh deformation. In: IEEE Conference on Computer Vision and Pattern Recognition (CVPR) (2019)
52. Zuffi, S., Black, M.J.: The stitched puppet: A graphical model of 3d human shape and pose. In: IEEE Conference on Computer Vision and Pattern Recognition (CVPR) (2015)

SCOAD: Single-Frame Click Supervision for Online Action Detection

Na Ye[1], Xing Zhang[1], Dawei Yan[1], Wei Dong[1]($\boxtimes$),
and Qingsen Yan[2]($\boxtimes$)

[1] Xi'an University of Architecture and Technology, Xi'an, China
dongwei156@outlook.com
[2] Northwestern Polytechnical University, Xi'an, China
qingsenyan@nwpu.edu.cn

Abstract. Online action detection based on supervised learning requires heavy manual annotation, which is difficult to obtain and may be impractical in real applications. Weakly supervised online action detection (WOAD) can effectively mitigate the problem of substantial labeling costs by using video-level labels. In this paper, we revisit WOAD and propose a weakly supervised online action detection using click-level labels for training, named Single-frame Click Supervision for Online Action Detection (SCOAD). Comparatively, click-level labels can effectively improve prediction accuracy by carrying a small amount of temporal information without massively increase the difficulty and cost of annotation. Specifically, SCOAD includes two joint training modules, *i.e.*, Action Instance Miner (AIM) and Online Action Detector (OAD). To provide more guidance for training network as accuracy as possible, AIM mines pseudo-action instances under the supervision of click labels. Meanwhile, we generate video similarity instances offline by the similarity between video frames and use it to perform finer granularity filtering of error instances generated by AIM. OAD is trained jointly with AIM for online action detection by the pseudo frame-level labels converted from the filtered pseudo-action instances. We conduct extensive experiments on two benchmark datasets to demonstrate that SCOAD can effectively mine and utilize the small amount of temporal information in click-level labels. Code is available at https://github.com/zstarN70/SCOAD.git.

Keywords: Online action detection · Weakly supervised learning

1 Introduction

Online action detection aims to report the presence of action instances in an untrimmed streaming video until the end. Unlike offline Temporal Action Local-

This work is supported by the Fundamental Research Funds for the Central Universities (No. D5000220444) and the Natural Science Basic Research Program of Shaanxi (2021JLM-16) and the Yulin Science and Technology Plan Project (CXY-2020-063).

Supplementary Information The online version contains supplementary material available at https://doi.org/10.1007/978-3-031-26316-3_14.

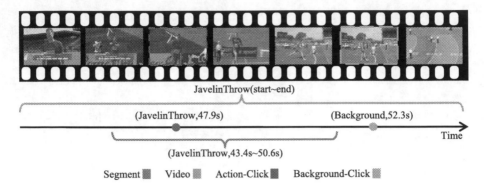

Fig. 1. Illustration of different annotation methods. (a) Segment-level annotation needs to label the action type and the precise time boundary. (b) Video-level annotation only needs to give the type of action present in the whole video. (c) Action-click labels need to give the timestamp and category corresponding to the frame where the action occurred. Similarly, Background-click happens in the background area.

ization (TAL), online action detection can only access historical frames that have been observed instead of all frames. This feature also makes it widely used in scenarios with high real-time requirements such as autonomous driving [13], anomaly detection [14] and video surveillance [21]. Therefore, this also has higher requirements for the accuracy of the algorithm and inference speed.

Current fully-supervised online action detection algorithms rely on extensive manual segment-level annotations, which are expensive to annotate. Recently, Gao *et al.* [8] considered that with the help of existing text retrieval technology, many video-level labels are relatively easy to obtain on the Internet. Therefore, WOAD was proposed to use only video-level labels, including two joint training modules, *i.e.*, Temporal Proposal Generator (TPG) and Online Action Recognizer (OAR). TPG generates pseudo temporal proposals, and OAR performs online action detection with pseudo frame-level labels generated from the pseudo temporal proposals. Although video-level labels can provide category information, almost without providing effective temporal information. Thus, WOAD tends to obtain the blurred temporal boundaries of temporal proposals generated by TPG. The current methods are limited by this problem and have poor performance for online action detection, but fewer attempts have been made to solve it. As shown in Fig. 1, click-level annotation denotes the labels of the action category and the corresponding timestamp that occurred at a random point on the video. Existing works [18,32] show that for a one-minute video, the times required to annotate video-level, click-level, and frame-level are 45 s, 50 s, and 300 s, respectively. It demonstrates that click-level annotation without increase significantly the cost of annotation compared with video-level annotation, but click-level annotation provides more information and has capacity to improve the prediction accuracy.

With the supervision of click labels, we propose an elaborate framework, termed SCOAD, to perform online action detection through two jointly trained modules, *i.e.*, Action Instance Miner (AIM) and Online Action Detector (OAD). In our method, AIM mines the pseudo-action instances and devotes itself to understanding the video from three perspectives: the entire video, the action area, and the background area. On the one hand, AIM uses the top-k strategy to mine potential action frames on the entire video, while aggregating them by increasing the response of action-click frames. On the other hand, reducing the response of the background area can forces the network to distinguish foreground and background. These operations in AIM provides more clear boundaries of the pseudo-action instances. Furthermore, to effectively eliminate noise in the generated action instances and ensure maximum expression of network. We obtain video similarity instances based on the assumptions that actions are continuous, actions and backgrounds are separated, and actions are affine to each other. We compute the IoU value between video similarity instances and the pseudo-action instances of AIM, and reduce the noise present in the pseudo-action instance by threshold filtering. The filtered pseudo-action instances will be converted to pseudo-frame labels for OAD learning. The OAD performs online action detection under the supervision of these pseudo-frame labels and uses GRU [4] as a prediction cell for online action detection. Compared with LSTM [10], GRU has a smaller number of parameters and shorter inference time, which is suitable for scenarios with high real-time performance.

Therefore, AIM and OAD are jointly trained for online action detection under the supervision of click-label. We only use the OAD during inference, so there is no increase in inference time. We test the efficacy of Thumos14 and ActivityNet1.2 and achieve state-of-the-art performance in weakly supervised online action detection.

The contributions of this paper are as follows:

- We initially explored the application of click annotations in online action detection. In our method, we propose the SCOAD consisting of two joint training modules, which generates pseudo-action instances by AIM and performs online action detection by OAD.
- Our algorithm is more flexible. When a video is manually annotated, it can be freely switched between fully-supervision and weakly-supervision.
- Extensive experiments on two benchmarks. Compared with existing weakly-supervised methods using video-level labels, we effectively improve prediction accuracy without a significant increase in annotation cost.

2   Related Work

Online Action Detection. Online action detection is pretty popular among various computer vision tasks [12,16,19,23,26–30,33]. Given an untrimmed streaming video, action instances and their classes are reported through historical and current frames that have been observed. Geest *et al.* [9] first described this problem as online action detection. Gao *et al.* proposed RED [6] to predict future sequences by taking multiple historical sequences as input. Similarly, TRN [25] improves action recognition at the current time by predicting future actions. IDU [5] considered that the input sequence may contain background and irrelevant actions, and used the traditional GRU cell [4] to judge whether to accumulate input information according to the correlation between the input sequence and the current information. Recently, Wang *et al.* noticed the existence of non-parallelism and gradient vanishing in traditional Recurrent Neural Network (RNN), that OadTR [24] was proposed to model long-term temporal dependencies based on Transformers. Focusing on category-level modeling, Yang *et al.* proposed Colar [31], an advisory paradigm mechanism. Besides, Gao *et al.* concerned that it is equally important to accurately identify the start time of an action, and proposed StartNet [7]. However, the above methods all rely on a large number of manual annotations for training, while our work uses click annotations to jointly identify the beginning of an action instance and continue to the end.

Weakly Supervised Online Action Detection. Comparatively, few weakly supervised online action detection are available. As a pioneering work, Gao *et al.* first proposed WOAD [8] in online action detection using video-level annotation. In this paper, Temporal Proposal Generator (TPG) and Online Action Recognizer (OAR) are jointly trained. The former mines action instances through video-level annotations to generate pseudo-annotations for the latter. Moreover, OAR performs pooling operations in the temporal dimension for action start prediction. Although the annotations used by this method are easy to obtain, only using video-level annotations still suffers from the problem of blurred temporal boundaries.

Click-Level Supervision. Weakly supervised temporal action detection has been extensive research [15,20,22,34,35]. Click-level supervision is an intermediate weakly supervised learning paradigm between fully supervised and weakly supervised, as Bearman *et al.* [1] first utilized point supervision for image semantic segmentation. At the video level, individual frames can also be viewed as points on the graph. SF-Net [18] is pioneering work in weakly supervised temporal action localization, exploits click-frame to mine pseudo action and background frames through supervised classification, which is further used to train a classifier. BackTAL [32] models location information and feature information through a score separation module and an affinity module, respectively.

3   Method

3.1   Overview

Given an untrimmed streaming video $\mathbf{V}_i$, the online action detection through historical and observed current frame reports of the probability of occurrence

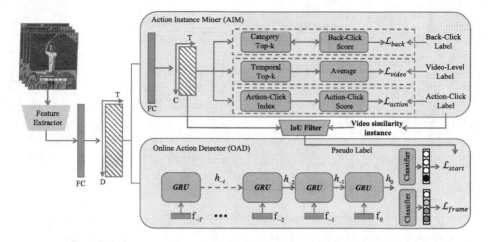

Fig. 2. The framework of click-supervised online action detection. Given a video, AIM generates pseudo-action instances under the supervision of click-level and video-level labels. After the pseudo-action instances filters itself for noise through the IoU filter, will be converted into pseudo-frame labels for OAD learning. The OAD performs online action detection under the supervision of these pseudo-frame labels.

and its category $\mathbf{y}_i = [y_0, y_1, \cdots y_c]$ for the action instance, where $y_c \in \{0, 1\}$ represents whether current frame feature $\mathbf{f}_t$ belongs to the cth category.

As shown in Fig. 2, our method includes two joint training modules, *i.e.*, Action Instance Miner (AIM) and Online Action Detector (OAD). During training, AIM generates pseudo-action instances under the supervision of click labels. Then we use the IoU filter to remove noise (e.g., error in estimated results) in the pseudo-action instances, these refined labels will be converted into pseudo-frame labels for OAD learning. During inference, network is restricted from using future information and only OAD is used for online action detection.

3.2   Action Instance Miner

In this work, we attempt to mine action instances with clearer boundaries by three constraints in the Action Instance Miner (AIM). We will introduce the three loss: video-level Multiple Instance Learning [20] loss $\mathcal{L}_{\text{video}}$, action frame classification loss $\mathcal{L}_{\text{action}}$ and background score separation loss [32] $\mathcal{L}_{\text{back}}$.

Give a video $\mathbf{V}_i$, AIM takes the feature sequence $\mathbf{F}_i = [\mathbf{f}_{-T}, \ldots, \mathbf{f}_0]$ of this $\mathbf{V}_i$ as input and outputs the corresponding class score $\mathbf{s} = \{s_t\}_{t=-T}^{0}$. Its corresponding action-click labels can be expressed as $\mathbf{A}_i = \{a_t\}_{t=-T}^{0}$, in which only the frames with click annotations will be set to the corresponding c^{th}, and the rest are 0, indicating that it is uncertain whether the frame belongs to action or background. Similarly, the background labels can be denoted as $\mathbf{B}_i = \{b_t\}_{t=-T}^{0}$, $b_t \in \{0, 1\}$ indicates whether it belongs to the background.

Video-Level Loss. $\mathcal{L}_{\text{video}}$ We calculate the video-level classification scores $\mathbf{s}_i^c$ in the temporal axis using top-k strategy for the cth category of ith video:

$$s_i^c = \frac{1}{k} \max_{\substack{\mathcal{M} \subset \mathbf{s}[c,:] \\ |\mathcal{M}|=k}} \sum_{l=1}^{k} \mathcal{M}_l, \tag{1}$$

where $\mathcal{M}_l$ indicates the lth element in the set $\mathcal{M}$. Finally, $\mathcal{L}_{\text{video}}$ is the cross-entropy between the predicted $\hat{\mathbf{s}}_i^c$ and video-level labels:

$$\mathcal{L}_{\text{video}} = -\frac{1}{K} \sum_{i}^{K} \mathbf{y}_i^c \log \hat{\mathbf{s}}_i^c, \tag{2}$$

where $\hat{\mathbf{s}}_i^c$ indicates the video-level classification scores after softmax normalization, $\mathbf{y}_i^c$ indicates the video includes action labels.

Action-Level Loss. $\mathcal{L}_{\text{action}}$ Let us assume there are N action-click frames in $\mathbf{V}_i$, the cross-entropy loss for N action-click frames:

$$\mathcal{L}_{\text{action}} = -\frac{1}{N} \sum_{j}^{N} a_j \log \hat{\mathbf{s}}_j, \tag{3}$$

where $\hat{\mathbf{s}}_j$ indicates the action frame scores after softmax normalization.

Background-Level Loss. $\mathcal{L}_{\text{back}}$ Although $\mathcal{L}_{\text{video}}$ and $\mathcal{L}_{\text{action}}$ can employ the top-k selected positions to move closer to the region of the clicked label in the early training stage, mature models will select similar top-k positions later in training. As pointed out in BackTAL [32] study, within these regions, the model will confidently show high responses for action frames and discreetly low responses for background frames. Therefore, $\mathcal{L}_{\text{back}}$ encourages the model to classify the responses of background and action frames distinctly.

Specifically, given a video that contains M background frames, we calculate the mean pseudo-action frame score p_{act} using the top-k strategy and the mean score p_{bg} of M background frames:

$$p_{\text{act}} = \frac{1}{k} \sum_{\forall b_t=0} s_t^c, \quad p_{\text{bg}} = \frac{1}{M} \sum_{\forall b_t=1} s_t^c. \tag{4}$$

Then, we guide $\hat{p}_{\text{act}}$ to be one while $\hat{p}_{\text{bg}}$ to be zero as follows:

$$\mathcal{L}_{\text{back}} = -\log \hat{p}_{\text{act}} - \log\left(1 - \hat{p}_{\text{bg}}\right), \tag{5}$$

where $\hat{p}_{\text{bg}}$ and $\hat{p}_{\text{act}}$ indicate the mean pseudo-action frame score and mean score of M background frames after softmax malization. Finally, we combine $\mathcal{L}_{\text{action}}$, $\mathcal{L}_{\text{video}}$ and $\mathcal{L}_{\text{back}}$ to form $\mathcal{L}_{\text{AIM}}$:

$$\mathcal{L}_{\text{AIM}} = \mathcal{L}_{\text{action}} + \mathcal{L}_{\text{video}} + \mathcal{L}_{\text{back}}. \tag{6}$$

Algorithm 1 Similarity Instances Mining.

Input:
    Video feature sequence: $\mathbf{V} = \{\mathbf{f}_t\}_{t=-T}^{0}$, $\mathbf{f_t} \in \mathbb{R}^{1 \times N}$
    Action-click label: $\mathbf{A} = \{c\}_{t=-T}^{0}$

Output:
    Video similarity instances: $\mathbf{S} = \{y\}_{t=-T}^{0}$

1: **function** GENERATE($\mathbf{S}$)
2:    **for** i *where* $\mathbf{A} > 0$ **do**
3:       $k \leftarrow j \leftarrow i$
4:       $s \leftarrow \cos(\frac{\mathbf{V} \cdot \mathbf{f}_i}{\|\mathbf{V}\|\|\mathbf{f}_i\|})$
5:       $\tau = \text{mean}(s[s > 0])$
6:       $s[s < \tau] \leftarrow 0$
7:       $s[s > \tau] \leftarrow 1$
8:       **while** $s[k] \neq 0$ **do**
9:          $k \leftarrow k - 1$
10:      **end while**
11:      **while** $s[j] \neq 0$ **do**
12:         $j \leftarrow j - 1$
13:      **end while**
14:      $\mathbf{S}[k...j][c^{th}] \leftarrow s[k...j]$
15:    **end for**
16: **return** $\mathbf{S}$;

3.3 Pseudo Labels Generation

During the early stage of training, we use the action-click frame to calculate the similarity score with the whole video frame and generate similarity instances. The detailed process of the algorithm is summarized in Algorithm 1. During training, AIM obtains pseudo-action instances through a two-stage threshold strategy. First, categories of video-level small confidence scores are filtered using thresholds. Naturally, short instances that cannot constitute an action are filtered using threshold. Finally get the pseudo-action instances $\mathbf{I} = \{y\}_{t=-T}^{0}$, to calculate its IoU value with the video similarity instances $\theta = \text{IoU}(\mathbf{I}, \mathbf{S})$. When θ is greater than the set IoU threshold, the pseudo-action instances are converted into a pseudo-frame label for OAD learning, otherwise, the video similarity instances is converted.

3.4 Online Action Detector

Online Action Detector (OAD) takes a series of continuous feature sequence $\mathbf{F} = [\mathbf{f}_{-T}, \dots, \mathbf{f}_0]$ as input. It outputs the corresponding action category score y_t and the probability of whether belong action start, which T is the sequence length.

In this work, OAD uses GRU as the prediction cell. The GRU updates its hidden layer h_t at each time step as:

$$h_t = \text{GRU}(h_{t-1}, \mathbf{f}_t). \tag{7}$$

Next, the fully connected layer is used to classify h_t at the current time t to obtain a_t and s_t, where a_t and s_t represent the action category score and probability of whether it belongs action start, respectively.

At the end of each training epoch, OAD obtain pseudo-action-frame labels $\mathbf{y}_{ja}^p$ and pseudo-action-start label y_{js}^p for training video from the action instances generated by AIM, where $j = \{1, 2, .., \widetilde{T}\}$ indicates the index of a frame in the training video and $\widetilde{T}$ is the total number of frames, and $y \in \{0, 1\}$ indicates the action non-start or start. Following previous work [8], we calculate the cross-entropy loss between $\mathbf{y}_{ja}^p$ and the action category score a_t as frame loss $\mathcal{L}_{\text{frame}}$:

$$\mathcal{L}_{\text{frame}} = -\frac{1}{\widetilde{T}} \sum_{j=1}^{\widetilde{T}} \sum_{c=0}^{C} \mathbf{y}_{ja}^p \log a_{jc}. \tag{8}$$

At the same time, we utilize focal loss [17] between y_{js}^p and predicted probability whether belong action start s_t as start loss $\mathcal{L}_{\text{start}}$:

$$\mathcal{L}_{\text{start}} = -\frac{1}{\widetilde{T}} \sum_{j=1}^{\widetilde{T}} \sum_{m=0}^{1} y_{js}^p (1 - s_{jm})^\gamma \log s_{jm}, \tag{9}$$

where γ is a hyper parameter. Finally, we combine $\mathcal{L}_{\text{frame}}$ and $\mathcal{L}_{\text{action}}$ as $\mathcal{L}_{\text{OAD}}$:

$$\mathcal{L}_{\text{OAD}} = \mathcal{L}_{\text{frame}} + \mathcal{L}_{\text{start}}. \tag{10}$$

3.5   Trianing and Inference

Training. In the early stages of training, we utilize $\mathcal{L}_{\text{AIM}}$ to optimize AIM generator pseudo-action instances. After action instances is first generated, we jointly train AIM and OAD through $\mathcal{L}_{\text{total}}$:

$$L_{\text{total}} = L_{\text{OAD}} + \lambda L_{\text{AIM}} \tag{11}$$

As shown in Fig. 2, pseudo-action instances are continuously generated by AIM. To reduce computation, we update the pseudo-action instances after every training epoch and take *Iter* iterations as an epoch. Although we have not use the co-activity similarity loss mentioned in WOAD, that is to ensure that videos of the same category of action appear in each batch, we still split the dataset in each batch in the same way for a fair comparison.

Inference. During the inference phase, only the OAD is required for online action detection tasks. At the each time step t, OAD outputs a_t and s_t, where a_t can be used directly as the action frame prediction score. Following previous works [7,8], we obtain action start score $\widetilde{s}_t$, where $\widetilde{s}_{t(1:c)} = a_{t(1:c)} * s_{t1}$ and $\widetilde{s}_{t0} = a_{t0} * s_{t0}$ indicates c^{th} action start score and background score respectively.

Table 1. The respective performances of our method and several existing methods on THUMOS14 under different label formats are compared.

Methods	Feature	Supervison	pAP@Time threshold (Seconds)										mAP
			1.0	2.0	3.0	4.0	5.0	6.0	7.0	8.0	9.0	10.0	
W-TALC [20]ECCV18	I3D	Video-level	16.2	26.0	31.3	34.6	36.2	37.6	38.6	39.3	39.9	40.3	48.0
WOAD [8]CVPR21	I3D	Video-level	21.9	32.9	40.5	44.4	48.1	49.8	50.8	51.7	52.4	53.1	54.4
SCOAD	I3D	Video-level+Click-level	**24.4**	**39.2**	**44.8**	**49.0**	**50.7**	**51.6**	**52.4**	**53.0**	**53.6**	**54.0**	**61.9**
StartNet [7]ICCV19	I3D	Frame-Level	21.9	33.5	39.6	42.5	46.2	46.6	47.7	48.3	48.6	49.0	–
TRN [25]ICCV19	I3D	Frame-Level	–										51.0
WOAD [8]CVPR21	I3D	Frame-Level	28.0	40.6	45.7	48.0	50.1	51.0	51.9	52.4	53.0	53.1	67.1
SCOAD	I3D	Frame-Level	**30.6**	**42.3**	**48.2**	**51.9**	**54.5**	**55.4**	**56.0**	**56.5**	**56.9**	**57.0**	**69.9**

Table 2. mAP is compared on THUMOS14 with strongly supervised and weakly supervised methods. (+x%Frame) means that x% of the videos have frame-level (strong) annotations, while the others keep their original annotations.

Methods	mAP@ Supervision (+x%Frame-Level)				
	+0%	+10%	+30%	+50%	+100%
TRN [25]ICCV19	–				51.0
WOAD [8]CVPR21	54.4	55.0	59.3	62.6	67.1
SCOAD	**61.9**	**63.7**	**65.2**	**66.8**	**69.9**

4 Experiments

Datasets. We conduct experiments on two widely used benchmarks, THU-MOS14 [11] and ActivityNet1.2 [2]. THUMOS14 includes more than 254 h of 20 sports category videos collected from YouTube. Following previous works [5,6,8,24,31], we trained the model on the validation set (200 videos) and evaluate it on the test set (212 videos). ActivatyNet1.2 contains 9682 videos of complex human activities in 100 categories. We train on the training set (4819 videos) and evaluate on the validation set (2383 videos). The two datasets face different challenges: THUMOS14 mainly stems from the dramatic change in the duration of action instances. ActivityNet1.2 is for numerous action categories, massive intra-class changes, etc.

Evaluation Metrics. Following previous works [5,6,8,9,24,25], we report per-frame mean average prevision (mAP) and point-based average precision (pAP) to measure the performance of action category and action start, where mAP calculates the precision and recall for the sorting results classification scores of all frames, and then calculates the average precision of interpolation to obtain the average of the category score (AP) as the mAP. Similar to the bounding box-based AP in the object detection task, the pAP measures the accuracy of the prediction of the action start by the temporal discrepancy. We follow WOAD [8] to report pAP at these ten thresholds of [1.0–10.0] seconds.

Table 3. The respective performances of our method and several existing methods on ActivityNet1.2 under different label formats are compared.

Methods	Feature	Supervison	pAP@Time Threshold(Seconds)										mAP
			1.0	2.0	3.0	4.0	5.0	6.0	7.0	8.0	9.0	10.0	
W-TALC [20]ECCV18	I3D	Video-level	5.2	8.5	10.7	12.8	14.5	15.9	17.1	18.1	19.1	20.1	53.8
WOAD [8]CVPR21	I3D	Video-level	7.9	11.6	14.3	16.4	18.8	20.3	22.2	23.4	24.7	25.3	66.7
SCOAD	I3D	Video-level+Click-level	**9.2**	**12.9**	**15.9**	**18.8**	**21.1**	**22.3**	**23.7**	**24.7**	**25.6**	**26.2**	**68.7**
StartNet [7]ICCV19	I3D	Frame-Level	7.5	11.5	14.1	16.5	18.4	19.7	20.9	21.8	22.9	23.6	–
TRN [25]ICCV19	I3D	Frame-Level	–										69.1
WOAD [8]CVPR21	I3D	Frame-Level	8.7	13.6	17	19.7	21.6	23	24.7	25.8	26.8	27.7	70.7
SCOAD	I3D	Frame-Level	**12.4**	**17.4**	**21.2**	**24.3**	**26.9**	**29**	**30.8**	**32.3**	**33.6**	**34.5**	**72.2**

Table 4. mAP is compared on ActivityNet1.2 with strongly supervised and weakly supervised methods. (+x%Frame) means that x% of the videos have frame-level (strong) annotations, while the others keep their original annotations.

Methods	mAP@ Supervision(+x%Frame)				
	+0%	+30%	+50%	+70%	+100%
TRN [25]ICCV19	–				69.1
WOAD [8]CVPR21	66.7	66.9	68.5	69.3	70.7
SCOAD	**68.7**	**70.2**	**71.1**	**71.6**	**72.2**

Baseline. Since the framework of our method comes from the recent work WOAD [8], we used it as the baseline for click supervision for experiments. In our method, the LSTM cells in WOAD are replaced by GRU cells, and only the video-level Multiple Instance Learning (MIL) [20] loss is retained, and experiments are conducted to verify the effectiveness of our method.

Implementation Details. We conducted experiments using two-stream (RGB and optical flow) features extracted from the I3D network [3] pre-trained on the Kinetics-400 [3] dataset on THUMOS14 and ActivityNet1.2. Video frames are extracted with a frame rate of 25 fps and chunk size is 16.

Our method is implemented on PyTorch and optimized by the Adam algorithm. We benchmark our model on an NVIDIA RTX 3090 GPU. We set batch size as 10, learning rate as $3e-4$ and weight decay as $1e-4$. The update proposal generation with $Iter = 100$ as an epoch on THUMOS14 and $Iter = 500$ for ActivityNet1.2 as an epoch. For OAD, we follow WOAD [8] to set hyper parameter $\gamma = 2$ in Eq. 9. The h_t dimension of the hidden layer is set to 4096, and the length of training sequence for GRU is 64. Since only few action starts exist in the video, we use all positive frames and randomly sample 3 times the number of negative frames to calculate the start loss in each training process. In addition, the click labels used in the results reported on THUMOS14 in the paper

Table 5. Compare parameters and inference time of our method with strong and weak supervision, respectively. The reported times do not include the processing time of feature extraction.

Methods	Supervision	Param	Infer time
TRN [25]ICCV19	Frame	314M	2.60 ms
StartNet [7]ICCV19	Frame	118M	0.56 ms
WOAD [8]CVPR21	Video	110M	0.40 ms
SCOAD	Video+Click	80M	0.32 ms

Table 6. Ablation study on the efficacy of each component of the AIM on the THU-MOS14 dataset.

Baseline	$\mathcal{L}_{action}$	$\mathcal{L}_{back}$	IoU Filter	mAP	pAP@ 1.0
✓				49.6	17.9
✓		✓		53.7	20.7
✓	✓			56.2	22.8
✓	✓	✓		59.3	23.5
✓	✓	✓	✓	61.9	24.4

come from the human click annotations provided by [18,32]. ActivityNet1.2 is the click label that we randomly generate using ground truth.

4.1 Comparison Experiments

Quantitative Comparisons. We compare with recent state-of-the-art methods for weakly supervised online action detection and consistently obtain significant performance on THUMOS14 and ActivityNet1.2. As shown in Table 1, Our model achieves state-of-the-art performance and improves mAP on the THUMOS14 dataset from 54.4% to 61.9% compared to the WOAD. It can be seen that the more accurate click annotations can bring very intuitive performance improvements, which give a accurate clue of time information for action detection. In Table 3, compared with WOAD, the mAP of our model is improved 2.0% with click labels on ActivityNey1.2, which shows the effectiveness of the proposed method. In addition, the proposed method also can be used for action detection with frame labels. As shown in Table 3, our method outperforms WOAD with 1.5% improvement on ActivityNet1.2.

For action start prediction, our model comprehensively surpasses WOAD at every threshold in Tables 1 and 3. Especially when the threshold is 1.0 s, our method pAP improves by 2.5% and 1.3% compared to WOAD on THUMOS14 and ActivityNet1.2. This shows that the classification accuracy of our method is significant. Meanwhile, compared with WOAD, our method outperforms its strongly supervised methods on both click label and frame label on THUMOS14, ActivityNet1.2. This phenomenon shows that choosing a suitable classifier is critical.

Table 7. Ablation study on the efficacy of each component of the OAD on the THU-MOS14 dataset.

Methods	Supervision	mAP	pAP@ 1.0
SCOAD LSTM	Video-Level+Click-Level	61.6	23.9
SCOAD Temp.pool	Video-Level+Click-Level	61.3	20.7
SCOAD	Video-Level+Click-Level	61.9	24.4
SCOAD LSTM	Frame-Level	69.6	30.4
SCOAD Temp.pool	Frame-Level	69.4	28.6
SCOAD	Frame-Level	69.9	30.6

Table 8. Randomly generated click labels on the ground truth of THUMOS14 using different random seeds.

Method	Supervision	Seed	mAP	pAP@ 1.0
SCOAD	Video-Level+Click-Level	1	63.1	23.9
	Video-Level+Click-Level	10	63.1	22.6
	Video-Level+Click-Level	100	60.6	21.3
	Video-Level+Click-Level	1000	61.5	22.7

Effectiveness and Efficiency. To further illustrate the flexibility and effectiveness of our method, we also evaluate the performance using mixed annotations, as shown in Tables 2 and 4. This shows that our method can improve performance by improving the annotation accuracy. Compared to WOAD, our method without using frame-level labels approaches its performance of using 50% frame-level labels on THUMOS14 and ActivityNet1.2.

At the same time, we compare the number of parameters and computation of the model with the baseline in Table 5, and our Param and Inference time are both lower than the baseline methods and strongly supervised methods. For a fair comparison, we use the same NVIDIA Tesla V100 GPU as WOAD to calculate the average inference times on the entire THUMOS14. Benefiting from the GRU cell with fewer parameters and faster convergence, our model is the fastest, 0.08 ms faster than WOAD, and has only 80M parameters.

4.2   Ablation Experiments

Although we use a similar structure to WOAD, our performance outperforms WOAD. A specific reason is that our label information is more robust than WOAD. We will explore deeper reasons below through ablation experiments.

AIM of Each Component. In Table 6, the influence of each loss in AIM on action instances generation is studied. When using the background constraint on the top-k score, we can see that both mAP and pAP grow, 4.1% and 2.8% respectively. When adding action frame constraints, mAP and pAP are improved by 6.6% and 4.9%, respectively.

Therefore, it can be inferred that although the top-k strategy can simply constrain the action classes, but a real action position is difficult to give effectively. However, although there are only a few click table labels, it can constrain the source of the top-k region. With the further restriction of back-click labels, the boundaries of actions are clearer. After that, action instances with higher confidence are further selected under the filtering of video similarity proposals. We visualize this process in Fig. 3 and it will be described in detail in Sect. 4.4.

OAD of Each Component. Compared with WOAD, we remove the max-pooling operation in the temporal dimension, directly use h_t for action start prediction in Eq. 7, and use GRU cell instead of LSTM cell. We conducted detailed experiments on the two improvements on THUMOS14, and the results are shown in Table 7. We experimented the results with different labels separately. When using click-level labels and LSTM network for prediction, mAP and pAP decreased by 0.3% and 0.5%, respectively. For action start prediction, temporal pooling will increase the convergence difficulty of the network, shown as a joint reduction of 0.6% and 3.7% in mAP and pAP. As expected, the above phenomenon also occurs when frame-level labels are used, thus validating our inferences.

4.3   Random Influence of Clicking Labels

Since a huge variability in clicks from individual persons, elements of randomness are inevitable in our method. We devote to verifying that the performance improvement from click supervision is robust. Similar to previous work [18,32], we randomly generate several sets of click labels with different random seeds on the ground truth of THUMOS14. The experiment results are shown in Table 8. Each action area contains at least one action-click label, and each video contains at least one background-click label. It can be seen that the click labels generated by different random seeds bring about 2.5% and 2.6% fluctuations of mAP and pAP, respectively. But it's enough to show that click label is influential for prediction. However, how to eliminate the random factor to ensure the network converges to the same position as much as possible is still worth studying.

4.4   Qualitative Results

Figure 3 provides a quantitative analysis of our action instances generation process. As shown in Fig. 3(a), although the top-k strategy can mine action regions, it has the blurred time boundaries problem. The constraint of back-level loss can significantly respond to the background area. But it also makes the response of the action area less confident. Action-level loss constraints do not seem to fully address this problem, as shown in Fig. 3(b). When the background loss and action loss are jointly employed, the predicted scores were significantly expressed in Fig. 3(c). The predicted scores showed more confident responses at action regions and background boundaries. Finally, after filtering through the IoU threshold, it is converted into a pseudo-label in Fig. 3(d).

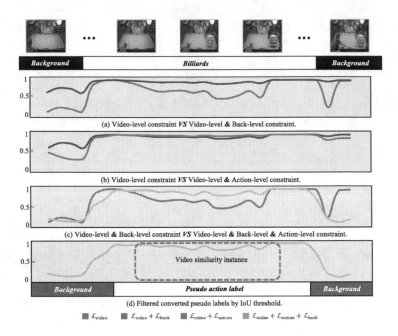

Fig. 3. Visualize the prediction scores of the AIM module under different loss constraints.

5   Conclusion

This paper proposes a method for online action detection using click labels. Extensive experiments have demonstrated that the using click labels does not significantly increase the cost of manual annotation, but it can effectively improve the accuracy of prediction. The proposed method consists of pseudo label generation and action prediction. AIM finds potential action regions through a top-k strategy for video label and foreground label, then uses background labels to reduce the response of background regions. Thanks to these operations, the proposed method can effectively avoid the blurry boundaries. To refine the estimated results, we generates the final pseudo-labels under the filtering of similarity instances. Remarkably, exploiting the offline generated video similarity instances for action clicks brings enormous performance gains to our model. However, this video similarity proposal has prior knowledge, but it also leaves a promising direction for future research.

References

1. Bearman, A., Russakovsky, O., Ferrari, V., Fei-Fei, L.: What's the point: semantic segmentation with point supervision. In: Leibe, B., Matas, J., Sebe, N., Welling, M. (eds.) ECCV 2016. LNCS, vol. 9911, pp. 549–565. Springer, Cham (2016). https://doi.org/10.1007/978-3-319-46478-7_34

2. Caba Heilbron, F., Escorcia, V., Ghanem, B., Carlos Niebles, J.: ActivityNet: a large-scale video benchmark for human activity understanding. In: CVPR, pp. 961–970 (2015)
3. Carreira, J., Zisserman, A.: Quo vadis, action recognition? A new model and the kinetics dataset. In: CVPR, pp. 6299–6308 (2017)
4. Cho, K., et al.: Learning phrase representations using RNN encoder-decoder for statistical machine translation. In: EMNLP, pp. 1724–1734, October 2014
5. Eun, H., Moon, J., Park, J., Jung, C., Kim, C.: Learning to discriminate information for online action detection. In: CVPR, pp. 809–818 (2020)
6. Gao, J., Yang, Z., Nevatia, R.: RED: reinforced encoder-decoder networks for action anticipation. In: BMVC (2017)
7. Gao, M., Xu, M., Davis, L.S., Socher, R., Xiong, C.: StartNet: online detection of action start in untrimmed videos. In: ICCV, pp. 5542–5551 (2019)
8. Gao, M., Zhou, Y., Xu, R., Socher, R., Xiong, C.: WOAD: weakly supervised online action detection in untrimmed videos. In: CVPR, pp. 1915–1923 (2021)
9. De Geest, R., Gavves, E., Ghodrati, A., Li, Z., Snoek, C., Tuytelaars, T.: Online action detection. In: Leibe, B., Matas, J., Sebe, N., Welling, M. (eds.) ECCV 2016. LNCS, vol. 9909, pp. 269–284. Springer, Cham (2016). https://doi.org/10.1007/978-3-319-46454-1_17
10. Hochreiter, S., Schmidhuber, J.: Long short-term memory. In: Neural Computation, pp. 1735–1780, November 1997
11. Jiang, Y.G., et al.: THUMOS challenge: action recognition with a large number of classes (2014). http://crcv.ucf.edu/THUMOS14/
12. Kim, H.-U., Koh, Y.J., Kim, C.-S.: Global and local enhancement networks for paired and unpaired image enhancement. In: Vedaldi, A., Bischof, H., Brox, T., Frahm, J.-M. (eds.) ECCV 2020. LNCS, vol. 12370, pp. 339–354. Springer, Cham (2020). https://doi.org/10.1007/978-3-030-58595-2_21
13. Kim, J., Misu, T., Chen, Y.T., Tawari, A., Canny, J.: Grounding human-to-vehicle advice for self-driving vehicles. In: CVPR, pp. 10591–10599 (2019)
14. Ko, K.E., Sim, K.B.: Deep convolutional framework for abnormal behavior detection in a smart surveillance system. Eng. Appl. Artif. Intell. **67**, 226–234 (2018)
15. Lee, P., Byun, H.: Learning action completeness from points for weakly-supervised temporal action localization. In: ICCV, pp. 13648–13657 (2021)
16. Li, J., Han, K., Wang, P., Liu, Y., Yuan, X.: Anisotropic convolutional networks for 3D semantic scene completion. In: CVPR, pp. 3351–3359 (2020)
17. Lin, T.Y., Goyal, P., Girshick, R., He, K., Dollár, P.: Focal loss for dense object detection. In: ICCV, pp. 2980–2988 (2017)
18. Ma, F., et al.: SF-Net: single-frame supervision for temporal action localization. In: Vedaldi, A., Bischof, H., Brox, T., Frahm, J.-M. (eds.) ECCV 2020. LNCS, vol. 12349, pp. 420–437. Springer, Cham (2020). https://doi.org/10.1007/978-3-030-58548-8_25
19. Moran, S., Marza, P., McDonagh, S., Parisot, S., Slabaugh, G.: DeepLPF: deep local parametric filters for image enhancement. In: CVPR, pp. 12826–12835 (2020)
20. Paul, S., Roy, S., Roy-Chowdhury, A.K.: W-TALC: weakly-supervised temporal activity localization and classification. In: Ferrari, V., Hebert, M., Sminchisescu, C., Weiss, Y. (eds.) ECCV 2018. LNCS, vol. 11208, pp. 588–607. Springer, Cham (2018). https://doi.org/10.1007/978-3-030-01225-0_35
21. Shu, T., Xie, D., Rothrock, B., Todorovic, S., Chun Zhu, S.: Joint inference of groups, events and human roles in aerial videos. In: CVPR, pp. 4576–4584 (2015)
22. Wang, L., Xiong, Y., Lin, D., Van Gool, L.: UntrimmedNets for weakly supervised action recognition and detection. In: CVPR, pp. 4325–4334 (2017)

23. Wang, P., Liu, L., Shen, C., Shen, H.T.: Order-aware convolutional pooling for video based action recognition. Pattern Recogn. **91**, 357–365 (2019)
24. Wang, X., et al.: OadTR: online action detection with transformers. In: ICCV, pp. 7565–7575 (2021)
25. Xu, M., Gao, M., Chen, Y.T., Davis, L.S., Crandall, D.J.: Temporal recurrent networks for online action detection. In: ICCV, pp. 5532–5541 (2019)
26. Yan, Q., Gong, D., Liu, Y., van den Hengel, A., Shi, J.Q.: Learning Bayesian sparse networks with full experience replay for continual learning. In: Proceedings of the IEEE/CVF Conference on Computer Vision and Pattern Recognition, pp. 109–118 (2022)
27. Yan, Q., et al.: High dynamic range imaging via gradient-aware context aggregation network. Pattern Recogn. **122**, 108342 (2022)
28. Yan, Q., et al.: Attention-guided network for ghost-free high dynamic range imaging. In: Proceedings of the IEEE/CVF Conference on Computer Vision and Pattern Recognition, pp. 1751–1760 (2019)
29. Yan, Q., Gong, D., Zhang, Y.: Two-stream convolutional networks for blind image quality assessment. IEEE Trans. Image Process. **28**(5), 2200–2211 (2018)
30. Yan, Q., et al.: Deep HDR imaging via a non-local network. IEEE Trans. Image Process. **29**, 4308–4322 (2020)
31. Yang, L., Han, J., Zhang, D.: Colar: effective and efficient online action detection by consulting exemplars. In: CVPR (2022)
32. Yang, L., et al.: Background-click supervision for temporal action localization. IEEE Trans. Pattern Anal. Mach. Intell. **44**(12), 9814–9829 (2021)
33. Yu, L., Yang, Y., Huang, Z., Wang, P., Song, J., Shen, H.T.: Web video event recognition by semantic analysis from ubiquitous documents. IEEE Trans. Image Process. **25**(12), 5689–5701 (2016)
34. Yuan, Y., Lyu, Y., Shen, X., Tsang, I., Yeung, D.Y.: Marginalized average attentional network for weakly-supervised learning. In: ICLR (2019)
35. Zhang, C., Cao, M., Yang, D., Chen, J., Zou, Y.: Cola: weakly-supervised temporal action localization with snippet contrastive learning. In: CVPR, pp. 16010–16019 (2021)

Neural Puppeteer: Keypoint-Based Neural Rendering of Dynamic Shapes

Simon Giebenhain, Urs Waldmann$^{(\boxtimes)}$ ⓘ, Ole Johannsen ⓘ,
and Bastian Goldluecke ⓘ

University of Konstanz, Konstanz, Germany
urs.waldmann@uni-konstanz.de

Abstract. We introduce Neural Puppeteer, an efficient neural rendering pipeline for articulated shapes. By inverse rendering, we can predict 3D keypoints from multi-view 2D silhouettes alone, without requiring texture information. Furthermore, we can easily predict 3D keypoints of the same class of shapes with one and the same trained model and generalize more easily from training with synthetic data which we demonstrate by successfully applying zero-shot synthetic to real-world experiments. We demonstrate the flexibility of our method by fitting models to synthetic videos of different animals and a human, and achieve quantitative results which outperform our baselines. Our method uses 3D keypoints in conjunction with individual local feature vectors and a global latent code to allow for an efficient representation of time-varying and articulated shapes such as humans and animals. In contrast to previous work, we do not perform reconstruction in the 3D domain, but project the 3D features into 2D cameras and perform reconstruction of 2D RGB-D images from these projected features, which is significantly faster than volumetric rendering. Our synthetic dataset will be publicly available, to further develop the evolving field of animal pose and shape reconstruction.

1 Introduction

Neural scene representations became an emerging trend in computer vision during the last couple of years. They allow to represent scenes through neural networks which operate on 3D space, allowing for tasks like novel view synthesis of static content, generalization over object and scene classes, body, hand and face modelling and relighting and material editing. For a detailed overview please refer to [53,56]. While most such methods rely on time and memory intensive volumetric rendering, [50] proposes a method of rendering with a single network evaluation per ray. In this paper we propose a single-evaluation rendering

---

S. Giebenhain and U. Waldmann–Authors contributed equally.

We acknowledge funding by the Deutsche Forschungsgemeinschaft (DFG, German Research Foundation) under Germany's Excellence Strategy – EXC 2117 – 422037984, and the SFB Transregio 161 "Quantitative Methods for Visual Computing", Project B5.

Supplementary Information The online version contains supplementary material available at https://doi.org/10.1007/978-3-031-26316-3_15.

L. Wang et al. (Eds.): ACCV 2022, LNCS 13844, pp. 239–256, 2023.
https://doi.org/10.1007/978-3-031-26316-3_15

(a) Input View   (b) Given Mask   (c) Reconstruction   (d) Novel view

Fig. 1. Zero-Shot Synthetic to Real Experiment: Giraffe. Reconstruction of a giraffe in different poses. From left to right: input image, segmentation mask used as input for keypoint estimation, estimated keypoints and shape, rendering from different perspective (cf. supplementary for more real-world examples).

formulation based on keypoints in order to represent dynamically deforming objects, like humans and animals. This allows us to render not only novel views of a known shape but also new and unseen poses by adjusting the positions of 3D keypoints. We apply this flexible approach to prevalent tasks in computer vision: the representation and reconstruction of pose for humans and animals. More specifically, we propose a silhouette based 3D keypoint detector that is based on inverse neural rendering. We decide to factor out appearance variation for keypoint detection, in order to ease the bridging of domain gaps and handle animals with few data available. Despite only relying on silhouettes, our proposed approach shows promising results when compared to the state-of-the-art 3D keypoint detector [17]. Furthermore, our approach is capable of zero-shot synthetic to real generalization, see Fig. 1.

Contributions. Our contribution is a flexible keypoint based neural scene representation and neural rendering framework called Neural Puppeteer (NePu).

We demonstrate that NePu provides valuable gradients as a differential forward map in an inverse rendering approach for 3D keypoint detection. Since we formulate the inverse rendering exclusively on 2D silhouettes, the resulting 3D keypoint detector is inherently robust with respect to transformations or domain shifts. Note that common 3D keypoint estimators require a huge amount of training samples with different texture in order to predict keypoints of the same class of shapes. Another advantage of being independent of texture is that it is easier to generalize from training with synthetic data. This can be particularly useful in cases where it is highly challenging to obtain a sufficient amount of real-world annotations, such as for wild animals (cf. Fig. 1). For animal shapes, we outperform a state-of-the-art 3D multi-view keypoint estimator in terms of Mean Per Joint Position Error (MPJPE) [17].

Unlike common practice, we shift rendering from the 3D to 2D domain, requiring only a single neural network evaluation per ray. In this sense, our approach can be interpreted as a locally conditioned version of Light Field Networks (LFNs) [50]. Our formulation is capable of learning the inter-pose variations of a single instance (constant shape and appearance) under constant lighting conditions, similar to [52]. In contrast, LFNs learn a prior over inter-object variations. We retain the rendering efficiency of LFNs and are capable of rendering color, depth and occupancy simultaneously at 20 ms per 256^2 image. This is significantly faster than NeRF-like approaches such as [52], which typically achieve less than 1 fps. Due to our fast renderer, fitting a target pose by inverse rendering can be done at ~1 fps using 8 cameras. Furthermore, we show that our keypoint-based local conditioning significantly improves the neural rendering of articulated objects, with visibly more details and quantitative improvements in PSNR and MAE over LFNs.

Code and data sets [12] to reproduce the results in the paper are publicly available at https://urs-waldmann.github.io/NePu/. We hope to inspire further work on animal pose and shape reconstruction, where our synthetic dataset can serve as a controlled environment for evaluation purposes and to experiment with novel ideas.

2   Related Work

2.1   3D Keypoint Estimation

Human keypoint estimation is a vast field with many applications [2,6,55,57]. For further reading, we refer the reader to [9,19,25,54]. The current state-of-the-art methods for 3D human keypoint estimation when trained on a single data set, i.e. the famous Human3.6M data set [16], are [14,17,45] with an average MPJPE of 18.7 mm, 20.8 mm and 26.9 mm respectively. At the time of writing, there was no code available for [45]. That is why we choose LToHP [17] as a baseline to quantitatively compare our model to. With the huge success of human keypoint estimation, the prediction of 3D keypoints for animals became a sub-branch of its own [3,13,20,21,34]. We notice that all these 3D frameworks for animals exploit 2D keypoints with standard 3D reconstruction techniques. That is why we also choose LToHP [17] as our baseline for animals, since it uses a learnable approach for the 3D reconstruction part.

Please keep in mind that our pose estimation only relies on multi-view silhouettes, while the above methods require RGB images. Using silhouettes gives more robustness to changes in texture and lighting. The only other work we are aware of that extracts keypoints from silhouettes is [5]. While [5] extracts 2D keypoints from silhouettes for quadrupeds using a template mesh and a stacked hourglass network [35], we are able to predict 3D coordinates for arbitrary shapes.

2.2   Morphable Models

The seminal morphable models for animals and humans are SMAL [65] and SMPL [26] respectively. An extension of SMPL, called SMPL-X [41], includes hands [46] and face [24]. These models have been used to estimate the 3D pose and shape from a single image [4,47,62], from multiple unconstrained images in the wild [49,64] or in an unsupervised manner from a sparse set of landmarks [29]. Because creating these models is tedious, the authors of [63] present an unsupervised disentanglement of pose and 3D mesh shape. Neural Body [43] uses implicit neural representations where different sparse multi-view frames of a video share the same set of latent codes anchored to the vertices of the SMPL model. In this way the SMPL model provides a geometric prior for their model with which they can reconstruct 3D geometry and appearance and synthesize novel views.

While these representations are comparatively easy to handle, the need of a parametric template mesh limits them to a pre-defined class of shapes. In particular, models for quadrupeds can fit a large variety of relatively similar animals like cats and horses, but run into problems if uncommon shapes are present [65]. For example, the fitting of elephants or giraffes pose significant problems due to their additional features (trunk) or different shapes (long neck). Similar problems arise in case of humans and clothing, e.g. a free flowing coat.

On the contrary, [23,58] exploit implicit representations and reduce manual intervention completely. They disentangled dynamic objects into a signed distance field defined in canonical space and a latent pose code represented by the flow field from a canonical pose to a given shaped pose of the same identity.

2.3 Neural Fields

Neural fields are an emergent research area that has become a popular representation for neural rendering [32,33,50,53,60], 3D reconstruction and scene representation [31,38,51], geometry aware generative modelling [7,36,48] and many more. For a detailed overview, we refer the reader to [56].

While these representations often rely on a global latent vector to represent the information of interest [31,38,50,51], the importance of locally conditioning the neural field has been demonstrated in [8,11,33,36,44]. By relying on local information drawn from geometrically aligned latent vectors, these methods often obtain higher quality reconstructions and generalize better due to their translation equivariance.

Neural fields have an especially strong impact on neural rendering with the formulation of radiance-based integration over rays introduced in NeRF [32]. While NeRF and many follow-ups [33,60] achieve high-quality renderings of single scenes, [59] match their performance with relying on neural networks, implying that NeRF's core competence are meaningful gradients for optimization. [52] combines the well known NeRF pipeline [32] with a keypoint based skeleton. This allows them to reconstruct articulate 3D representation of humans by representing the pose through the 3D positions of the keypoints. However, [52] optimizes 3D coordinates in the camera system obtained from fitting the SMPL model [26] while we predict 3D world coordinates from multi-view silhouettes.

In contrast to the volumetric rendering of NeRF typically requiring hundreds of evaluations of the neural network, LFN [50] propose an alternative by instantly predicting a pixels color given the corresponding ray's origin and direction. This approach results in a much faster rendering compared to NeRF-like approaches. In this work, we embrace the idea of single-evaluation rendering [50], and employ ideas from [11] for an efficient representation and architecture. Our approach of rendering is thus much faster than [52], allowing also for faster solutions to the inverse rendering problem of estimating keypoint locations from images. Since [50] is the only other single-evaluation-rendering method we are aware of, we quantitatively compare our method in Table 1 in terms of color and depth.

3   Neural Puppeteer

We will describe Neural Puppeteer in three parts. First, we discuss the encoding of the pose and latent codes, as can be seen on the upper half of Fig. 2. Second, we describe our keypoint-based neural rendering, as is depicted in the lower half of the figure. Third and last, we discuss how our pipeline can be inverted to perform keypoint estimation by fitting the pose generated from 3D keypoints to input silhouette data.

3.1   3D Keypoint Encoding

Given 3D keypoint coordinates $\mathbf{x} \in \mathbb{R}^{K \times 3}$ of a subject with K keypoints, we aim to learn an encoder network

$$\text{enc} : \mathbb{R}^{K \times 3} \to \mathbb{R}^{d_z}, \ \mathbf{x} \mapsto \mathbf{z} \tag{1}$$

that encodes a pose $\mathbf{x}$ as a d_z-dimensional global representation $\mathbf{z}$, as well as to learn a decoder network

$$\text{dec} : \mathbb{R}^{d_z} \to \mathbb{R}^{K \times 3} \times \mathbb{R}^{K \times d_f}, \ \mathbf{z} \mapsto (\hat{\mathbf{x}}, \mathbf{f}) \tag{2}$$

that reconstructs the pose $\hat{\mathbf{x}}$ and obtains local features $\mathbf{f}_k \in \mathbb{R}^{d_f}$ for each keypoint. Subsequently, we use the representation $(\mathbf{z}, \mathbf{f})$ to locally condition our neural rendering on the pose $\mathbf{x}$, as explained in Sect. 3.2. Please note that we do not require a skeleton model, i.e. connectivity between key points, since the pose is only interpreted as a point cloud.

We build our encoder upon the vector self-attention

$$\text{VSA} : \mathbb{R}^{K \times 3} \times \mathbb{R}^{K \times d_f} \to \mathbb{R}^{K \times d_f}, \ (\mathbf{x}, \mathbf{f}) \mapsto \mathbf{f}' \tag{3}$$

introduced in [61] as a means of a neural geometric point cloud operator. Consequently, our encoder which consists of L layers produces features

$$\mathbf{f}^{(l+1)} = \text{ET}_l(\text{BN}_l(\mathbf{f}^{(l)} + \text{VSA}_l(\mathbf{x}, \mathbf{f}^{(l)}))), l \in \{0, \cdots, L-1\}, \tag{4}$$

where BN denotes a BatchNorm layer [15] and ET denotes element-wise transformations containing a 2-layer MLP, residual connection and another BatchNorm. Initial features $\mathbf{f}^{(0)} \in \mathbb{R}^{K \times d_f}$ are learned as free parameters. The final global representation $\mathbf{z}$ is obtained via dimension-wise global maxpooling,

$$\mathbf{z} = \max_{k=1,\ldots,K} \mathbf{f}_k^{(L)}. \tag{5}$$

We decode the latent vector $\mathbf{z}$ using two 3-layer MLPs. Keypoints are reconstructed using $\hat{\mathbf{x}} = \text{MLP}_{pos}(\mathbf{z})$ and additionally features $\tilde{\mathbf{f}} = \text{MLP}_{\text{feats}}(\mathbf{z})$ holding information for the subsequent rendering are extracted. Finally, these features are refined to $\mathbf{f}$ using 3 further VSA layers, which completes the decoder $\text{dec}(\mathbf{z}) = (\hat{\mathbf{x}}, \mathbf{f})$. For more architectural details we refer to the supplementary.

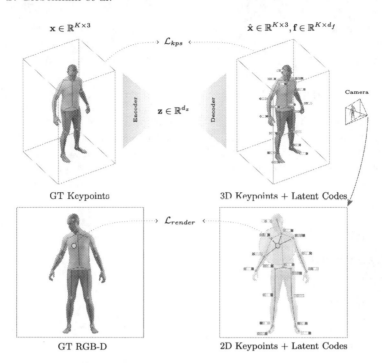

Fig. 2. Neural Puppeteer: Our method takes 3D keypoints (top left) and learns individual latent codes for each keypoint (top right). We project them into arbitrary camera views and perform neural rendering to reconstruct 2D RGB-D images (bottom right). The pipeline can be used to perform pose estimation by inverse rendering, optimizing the 3D keypoint positions by performing gradient descent in the latent code $\mathbf{z} \in \mathbb{R}^{d_z}$. For rendering, we use only the closest keypoints, illustrated by the yellow circle around the point in question (yellow dot). Connections between keypoints are shown just for visualization and not used by the network. (Color figure online)

3.2 Keypoint-Based Neural Rendering

Contrary to many recent neural rendering approaches, we do not rely on costly NeRF-style volumetric rendering [32]. Instead we adopt an approach similar to LFN [50], that predicts a pixel color with a single neural network evaluation.

While LFN uses a global latent vector and the ray's origin and orientation as network inputs, we directly operate in pixel coordinates. Perspective information is incorporated by projecting keypoints $\mathbf{x}$ and corresponding local features $\mathbf{f}$ into pixel coordinates. More specifically, given camera extrinsics $\mathbf{E}$ and intrinsics $\mathbf{K}$ and the resulting projection as $\pi_{\mathbf{E},\mathbf{K}}$ we obtain 2D keypoints

$$\mathbf{x}_{2D} = \pi_{\mathbf{E},\mathbf{K}}(\mathbf{x}). \tag{6}$$

Additionally, we append the depth values $\mathbf{d}$, such that the keypoint's positions are $\mathbf{x}_{2D}^* = (\mathbf{x}_{2D}, \mathbf{d}) \in \mathbb{R}^{K \times 3}$. Using this positional information we further refine the features in pixel coordinates using $L = 3$ layers of VSA. Specifically, we define $\mathbf{f}_{2D}^{(0)} = \mathbf{f}$ and

$$\mathbf{f}_{2D}^{(l+1)} = \text{VSA}_l(\mathbf{x}_{2D}^*, \mathbf{f}_{2D}^{(l)}), l \in \{0, 1, 2\}. \tag{7}$$

The resulting refined features $\mathbf{f}_{2D}^{(L)}$ and coordinates $\mathbf{x}_{2D}^*$ are the basis for our single-evaluation rendering, as described next.

Given a pixel coordinate $\mathbf{q} \in \mathbb{R}^2$, we follow [11] to interpolate meaningful information from nearby features $\mathbf{f}_{2D}^{(L)}$ and the global representation $\mathbf{z}$. To be more specific, the relevant information is

$$\mathbf{y} = \text{VCA}((\mathbf{q}, 0), \mathbf{z}, \mathbf{x}_{2D}^*, \mathbf{f}_{2D}^{(L)}) \in \mathbb{R}^{d_f}, \tag{8}$$

where VCA denotes the vector cross-attention from [11] and we set the depth for $\mathbf{q}$ to zero.

Finally, the predictions $\hat{\mathbf{c}}$ for color, $\hat{\mathbf{d}}$ for depth and $\hat{\mathbf{o}}$ for 2D occupancy values are predicted using three feed-forward network heads

$$\text{FFN}_{\text{col}} : \mathbb{R}^{d_f} \to [0,1]^3, \text{FFN}_{\text{dep}} : \mathbb{R}^{d_f} \to [0,1], \text{ and FFN}_{\text{occ}} : \mathbb{R}^{d_f} \to [0,1], \tag{9}$$

respectively, using the same architecture as in [44]. For convenience we define the color rendering function

$$\mathcal{C}_{\mathbf{E},\mathbf{K}} : \mathbb{R}^{K \times 3} \times \mathbb{R}^{K \times d_f} \times \mathbb{R}^{d_z} \to [0,1]^{H \times W \times 3}, \tag{10}$$

which renders an image seen with extrinsics $\mathbf{E}$ and intrinsics $\mathbf{K}$ conditioned on keypoints $\mathbf{x}$, encoded features $\mathbf{f}$ and global representation $\mathbf{z}$, by executing FFN_{col} for all pixels $\mathbf{q}$. For the depth and silhouette modalities we similarly define $\mathcal{D}_{\mathbf{E},\mathbf{K}}$ and $\mathcal{S}_{\mathbf{E},\mathbf{K}}$, respectively. Note, that our silhouettes contain probabilities for a pixel lying on the object.

3.3   Training

We consider a dataset consisting of M poses $\mathbf{x}_m \in \mathbb{R}^{K \times 3}$ captured by C cameras with extrinsics $\mathbf{E}_c$ and intrinsics $\mathbf{K}_c$. For each view c and pose m we have 2D observations

$$\mathbf{I}_{m,c}, \ \mathbf{D}_{m,c}, \ \mathbf{S}_{m,c}, \ m \in \{1, \dots, M\}, \ c \in \{1, \dots, C\}, \tag{11}$$

corresponding to color, depth and silhouette, respectively.

All model parameters are trained jointly to minimize the composite loss

$$\mathcal{L} = \lambda_{\text{pos}}\mathcal{L}_{\text{pos}} + \lambda_{\text{col}}\mathcal{L}_{\text{col}} + \lambda_{\text{dep}}\mathcal{L}_{\text{dep}} + \lambda_{\text{sil}}\mathcal{L}_{\text{sil}} + \lambda_{\text{reg}}\|\mathbf{z}\|_2, \tag{12}$$

where the different positive numbers λ are hyperparameters to balance the influence of the different losses. The keypoint reconstruction loss

$$\mathcal{L}_{\text{pos}} = \sum_{m=1}^{M} \|\mathbf{x}_m - \hat{\mathbf{x}}_m\|_2 \tag{13}$$

minimizes the mean Euclidean distance between the ground truth and reconstructed keypoint positions. The color rendering loss

$$\mathcal{L}_{\text{col}} = \sum_{m=1}^{M} \sum_{c=1}^{C} \|\mathcal{C}_{\mathbf{E}_c, \mathbf{K}_c}(\mathbf{x}_m, \mathbf{f}_m, \mathbf{z}_m) - \mathbf{I}_{m,c}\|_2^2 \qquad (14)$$

is the squared pixel-wise difference over all color channels, the depth loss $\mathcal{L}_{\text{dep}}$ is given by a structurally identical formula. Finally, the silhouette loss

$$\mathcal{L}_{\text{sil}} = \sum_{m-1}^{M} \sum_{c-1}^{C} \text{BCE}\left(\mathcal{S}_{\mathbf{E}_c, \mathbf{K}_c}(\mathbf{x}_m, \mathbf{f}_m, \mathbf{z}_m), \mathbf{S}_{m,c}\right) \qquad (15)$$

measures the binary cross entropy $\text{BCE}(\hat{o}, o) = -[o \cdot \log(\hat{o}) + (1 - o) \cdot \log(1 - \hat{o})]$ over all pixels. Hence the silhouette renderer is trained to classify pixels into inside and outside points, similar to [31].

3.4   Pose Reconstruction and Tracking

While the proposed model learns a prior over poses along with their appearance and geometry and thus can be used to render from 3D keypoints, we can also infer 3D keypoints by solving an inverse problem, using NePu as a differentiable forward map from keypoints to images. We are especially interested in silhouette-based inverse rendering, in order to obtain robustness against transformations that leave silhouettes unchanged.

Given observed silhouettes $\mathbf{S}_c$, extrinsics $\mathbf{E}_c$ and intrinsics $\mathbf{K}_c$ for cameras $c \in \{1, \ldots, C\}$, we optimize for

$$\hat{\mathbf{z}} = \underset{\mathbf{z}}{\text{argmin}} \sum_{c=1,\ldots,C} \text{BCE}\left(\mathcal{S}_{\mathbf{E}_c, \mathbf{K}_c}(\text{dec}(\mathbf{z}), \mathbf{z}), \mathbf{S}_c\right) + \lambda \|\mathbf{z}\|_2. \qquad (16)$$

Following [41], we use the PyTorch [39] implementation of the limited-memory BFGS (L-BFGS) [37] to solve the optimization problem. Once $\hat{\mathbf{z}}$ has been obtained, we recover the corresponding pose as $\hat{\mathbf{x}} = \text{MLP}_{\text{pos}}(\hat{\mathbf{z}})$.

Since the above mentioned optimization problem does not assume any prior knowledge about the pose, the choice of an initial value is critical. For initial values too far from the ground truth pose, we observe an unstable optimization behavior that frequently diverges or gets stuck in local minima. To overcome this obstacle, we run the optimization with I different starting conditions $\mathbf{z}^{(1)}, \ldots, \mathbf{z}^{(I)}$, which we obtain by clustering the 2-dimensional t-SNE [28] embedding of the global latent vectors over the training set using affinity propagation [10]. We obtain different solutions $\hat{\mathbf{z}}_i$ from minimization of Eq. (16), and as the final optimum choose the one with best IoU to the target silhouettes,

$$\hat{\mathbf{z}} = \underset{\hat{\mathbf{z}}_i}{\text{argmax}} \sum_{c=1,\ldots,C} \text{IoU}\left(\mathcal{S}_{\mathbf{E}_c, \mathbf{K}_c}(\text{dec}(\hat{\mathbf{z}}_i), \hat{\mathbf{z}}_i), \mathbf{S}_c\right). \qquad (17)$$

While such a procedure carries a significant overhead, the workload is amortized in a tracking scenario. When provided with a sequence of silhouettes

Table 1. *Quantitative results of the rendering on the test sets.* Comparison of the color PSNR [db] between the reconstructed RGB images and mean absolut error (MAE) [mm] for the reconstructed depth by LFN* and NePu. The local conditioning significantly improves reconstruction accuracy. See text for a discussion of the results.

	Color PSNR [dB]			Depth MAE [mm]		
	LFN*	NePu	Δ	LFN*	NePu	Δ
Sy. Cow	14.95	**19.17**	+4.22	43.4	**22.3**	−21.1
Sy. Giraffe	15.99	**21.23**	+5.24	92.1	**35.4**	−56.7
Sy. Pigeon	21.47	**28.03**	+6.56	6.9	**2.5**	−4.4
Sy. Human	20.23	**27.49**	+7.26	77.8	**20.9**	−56.9
Average	18.16	**23.98**	+5.82	55.1	**20.3**	−34.8

$\mathbf{S}_{c,1}, \ldots, \mathbf{S}_{c,T}, c \in \{1, \ldots, C\}$ of T frames, we use the above method to determine $\hat{\mathbf{z}}_1$. For subsequent frames we initialize $\hat{\mathbf{z}}_{t+1} = \hat{\mathbf{z}}_t$ and fine-tune by minimizing Eq. (16) using a few steps of gradient descent. Our unoptimized implementation of the tracking approach runs at roughly 1 s per frame using 8 cameras.

4   Experiments

In this section, we evaluate different aspects of our method. To evaluate pose estimation on previously unseen data, we compare our method to the state-of-the-art multi-view keypoint detector [17], which we train for each individual subjects using the same dataset as for NePu.

The fundamental claims of our methodology are shown by rendering novel views and poses, quantitatively evaluating color and depth estimates and comparing against a version of our framework that utilizes the LFN [50] formulation for neural rendering. For visual evidence that our method produces temporally consistent views and additional qualitative results we refer to videos and experiments in our supplemental material. In addition we compare NePu to AniN-eRF [42] on Human3.6M [16].

4.1   Datasets

Our method can be trained for any kind of shape data that can be described by keypoints or skeletons. Connectivity between the keypoints neither needs to be known, nor do we make use of it in any form. We perform a thorough evaluation of our method on multiple datasets, for two types of shapes where a keypoint description often plays a major role: humans and animals.

For the human data we use the SMPL-X Blender add-on [41]. Here, we evaluate both on individual poses and captured motion sequences. The poses were obtained through the AGORA dataset [40], the animations through the

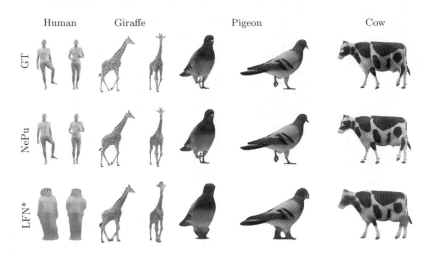

Fig. 3. Novel view synthesis for novel poses. Our method is capable of generating realistic renderings of the captured subject. In this figure, we show novel poses from new perspectives. See text for a discussion of the results.

AMASS dataset, and originally captured by ACCAD and BMLmovi [1,22,30]. We describe the human pose by 33 keypoints, see additional material for details.

For the training of animal data we use hand-crafted animated 3D models from different sources. To capture a variety of different animal shapes, we render datasets with a cow, a giraffe, and a pigeon. In particular the giraffe is a challenging animal for template mesh based methods, as the neck is much longer compared to most other quadrupeds. For each animal we created animations which include idle movement (e.g. looking around), walking, trotting, running, cleaning, and eating and render between 910 and 1900 timesteps. We use between 19 and 26 keypoints to describe the poses.

The keypoints for all shapes were created by tracking individual vertices. As keypoints in the interior of the shape are typically preferred, we average the position of two vertices on opposing sides of the body part. All datasets were rendered using Blender (www.blender.org) and include multi-view data of 24 cameras placed on three rings at different heights around the object. For each view and time step we generate ground truth RGB-D data as well as silhouettes of the main object. The camera parameters, and 3D and 2D keypoints for each view and timestep are also included.

4.2   Implementation Details

We implement all our models in PyTorch [39] and train using the AdamW optimizer [27] with a weight decay of 0.005 and a batch size of 64 for a total of 2000 epochs. We set the initial learning rate to $5e^{-4}$, which is decayed with a factor of 0.2 every 500 epochs. We weight the training loss in Eq. (12) with $\lambda_{pos} = 2$,

Fig. 4. 3D point cloud reconstruction. The 2D depth estimations from our method can be used to perform 3D reconstruction of the captured shape. The raw 3D point clouds generated by projecting the estimates from all cameras into the common reference frame already yield a good representation. The outliers originate from depth estimates at occlusion boundaries as the L_2 loss encourages smoothed depth maps and could be easily removed by filtering the point cloud.

$\lambda_{col} = \lambda_{dep} = 1$, $\lambda_{sil} = 3$ and $\lambda_{reg} = 1/16$. Other hyperparameters and architectural details are presented in the supplementary.

Instead of rendering complete images during training to compute $\mathcal{L}_{col}$, $\mathcal{L}_{dep}$ and $\mathcal{L}_{sil}$, we only render a randomly sampled subset of pixels. Both for color and depth we sample uniformly from all pixels in the ground truth mask. Hence the color and depth rendering is unconstrained outside of the silhouette. To compute $\mathcal{L}_{sil}$ we sample areas near the boundary of the silhouette more thoroughly, similar to [8].

4.3   Baselines

We use different baselines: LToHP [17] for our pose estimation and tracking approach and [50] for our keypoint-based locally conditioned rendering results. In addition we compare NePu to AniNeRF [42] on Human3.6M [16].

LToHP. LToHP [17] presents two solutions for multi-view 3D human pose estimation; an algebraic and volumetric one. For details see our supplementary and [17]. We use their implementation from [18] with their provided configuration file for hyperparameters. In order to obtain the quantitative results in Table 2, we individually fine-tune [17] on every animal, using the same data that we trained NePu on. In animal pose estimation it is common practice to fine-tune a network that was pretrained, as in state of the art animal pose estimators like DeepLabCut [34], due to a lack of animal pose data. For all models, including our models, we select the epoch with minimum validation error for test.

LFNs. For the comparison to [50], we integrate their rendering formulation in our framework, resulting in two differences. First, their rendering operates in Plücker coordinates instead of pixel values. Second, and more importantly, we do not use local features **f** for conditioning in this baseline, but global conditioning via concatenation using **z**. In the following, we denote this model by LFN*.

AniNeRF. We trained NePu using the same training regime as AniNeRF [42]: We only trained on a single subject, using every $5th$ of the first 1300 frames of

Table 2. *Quantitative results for 3D keypoint estimation on the test sets.* Comparison to LToHP [17]. Values are given in MPJPE [mm] and its median of all samples [mm]. Delta shows difference to [17]. alg., vol. and vol.^{gt} indicate that [17] is trained with its algebraic model, volumetric model with root joint from alg. results and from ground truth respectively.

	MPJPE [mm]			Median [mm]		
	LToHP [17]	NePu	Δ	LToHP [17]	NePu	Δ
Sy. Cow	124 (vol.)	**15**	-109	110 (vol.)	**9**	-101
Sy. Giraffe	190 (vol.^{gt})	**67**	-123	154 (vol.^{gt})	**31**	-123
Sy. Pigeon	10.3 (alg.)	**1.3**	-9.0	6.2 (alg.)	**1.1**	-5.1
Sy. Human	**28** (alg.)	46	$+18$	**22** (alg.)	28	$+6$
Average	88	**32**	-56	73	**17**	-56

sequence "Posing" of subject "S9" for training and every $5th$ of the following 665 frames for testing. Like AniNeRF, we also only used cameras 0–2 for training and camera 3 for testing.

4.4   3D Keypoint and Pose Estimation

Quantitative results for the 3D keypoint estimation are shown in Table 2. We report the MPJPE in mm and its median [mm] over all test samples. For LToHP [17] we evaluate both the algebraic and volumetric model and report the better result. In addition we report the average of MPJPE and median over all objects. We achieve a better average MPJPE and median (32 mm and 17 mm respectively) over all objects than LToHP (88 mm and 73 mm respectively). Note, however, that [17] achieves better results for humans only. We hypothesize two reasons for that. First, the human-specific pre-training of LToHP transfers well to the human data we evaluate on. Secondly, the extremities of the human body (especially arms and hands) vanish more often in silhouettes than for the animals we worked with. Example qualitative results can be found in Fig. 5.

4.5   Keypoint-Based Neural Rendering

Quantitative results for the keypoint-based neural rendering part are shown in Table 1. For color comparison we report the PSNR [dB] over all test samples, while for depth comparison we report the MAE [mm]. We achieve a better average PSNR and MAE (23.98 dB and 20.3 mm respectively) over all objects than LFN* (18.16 dB and 55.1 mm respectively).

Qualitative results for color rendering and depth reconstruction can be found in Fig. 3 and Sec. 3.2 of our supplementary respectively. Comparing our method to the implementation without local conditioning shows the importance of the local conditioning, the renderings are much more detailed, with more precise

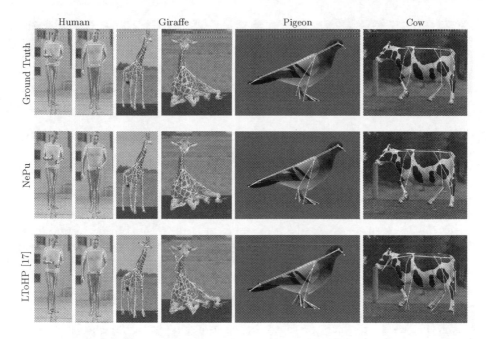

Fig. 5. Results for pose estimation. We show projected keypoints of the ground truth, NePu and LToHP [17] in the first, second and third row, respectively, for a human, a giraffe, a pigeon and a cow. Color coded dots indicate the keypoint location and its 3D error to the ground truth. Red dots indicate a high, green dots a low 3D error. The connections between keypoints are just for visualization and not used by the networks. (Color figure online)

boundaries. Figure 4 shows projections of multiple depth maps from different viewing directions as 3D point clouds. The individual views align up nicely, yielding good input for further processing and analysis. The results of the novel view and pose comparison with AniNeRF [42] on Human3.6M [16] are shown in Fig. 6. Even though our rendering formulation is fundamentally different and much less developed, our results look promising, but cannot meet the quality of AniNeRF. While AniNeRF leverages the SMPL parameters to restrict the problem to computing blend weight fields, our method has to solve a more complex problem. In principle our rendering could also be formulated in canonical space and leverage SMPL to model deformations. In addition, part segmentation maps, as well as, the relation of view direction and body orientation could further help to reduce ambiguities in the 2D rendering.

Compared to A-NeRF [52] and AniNeRF [42] the fundamental differences in the rendering pipeline result in a significant speed increase. Both render at 0.25–1 fps and 1 fps at $512^2 px$, respectively, while we render at 50 fps at $256^2 px$. In contrast to both methods we do not make use of optimization techniques that constrain the rendering to a bounding box of the 3D keypoints. Employing this would result in a total speed increase of 50–$200\times$ at $512^2 px$.

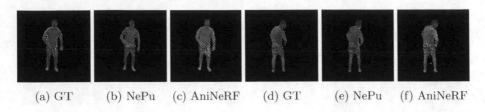

(a) GT   (b) NePu   (c) AniNeRF   (d) GT   (e) NePu   (f) AniNeRF

Fig. 6. Novel view and pose comparison with AniNeRF. Novel view (a–c) and novel view + novel pose (d–f) rendering on Human3.6M dataset.

5   Limitations and Future Work

In the future we plan to extend our model to account for instance specific shape and color variations, by incorporating the respective information in additional latent spaces. Moreover, our 3D keypoint encoder architecture (cf. Sect. 3.1) can for example be further improved for humans in a similar fashion to [52]. Such an approach would intrinsically be rotation equivariant and better capture the piece-wise rigidity of skeletons. Finally, while 2D rendering is much faster, it also means that we do not have guaranteed consistency between the generated views of the scene, in the sense that they are not necessarily exact renderings of the same 3D shape. This is an inherent limitation and cannot easily be circumvented, but our quantitative results indicate that it does not seem to impact quality of the rendering by much.

6   Conclusions

In this paper we present a neural rendering framework called Neural Puppeteer that projects keypoints and features into a 2D view for local conditioning. We demonstrate that NePu can detect 3D keypoints with an inverse rendering approach that takes only 2D silhouettes as input. In contrast to common 3D keypoint estimators, this is by design robust with respect to change in texture, lighting or domain shifts (e.g. synthetic vs. real-world data), provided that silhouettes can be detected. Due to our single-evaluation neural rendering, inverse rendering for downstream tasks becomes feasible. For animal shapes, we outperform a state-of-the-art 3D multi-view keypoint estimator in terms of MPJPE [17], despite only relying on silhouettes.

In addition, we render color, depth and occupancy simultaneously at 20 ms per 256^2 image, significantly faster than NeRF-like approaches, which typically achieve less than 1 fps. The proposed keypoint-based local conditioning significantly improves neural rendering of articulated objects quantitatively and qualitatively, compared to a globally conditioned baseline.

References

1. Advanced Computing Center for the Arts and Design: ACCAD MoCap Dataset. https://accad.osu.edu/research/motion-lab/mocap-system-and-data
2. Artacho, B., Savakis, A.: UniPose+: a unified framework for 2D and 3D human pose estimation in images and videos. IEEE TPAMI **44**(12), 9641–9653 (2021)
3. Bala, P.C., Eisenreich, B.R., Yoo, S.B.M., Hayden, B.Y., Park, H.S., Zimmermann, J.: Automated markerless pose estimation in freely moving macaques with Open-MonkeyStudio. Nat. Commun. **11**, 4560 (2020)
4. Biggs, B., Boyne, O., Charles, J., Fitzgibbon, A., Cipolla, R.: Who left the dogs out? 3D animal reconstruction with expectation maximization in the loop. In: Vedaldi, A., Bischof, H., Brox, T., Frahm, J.-M. (eds.) ECCV 2020. LNCS, vol. 12356, pp. 195–211. Springer, Cham (2020). https://doi.org/10.1007/978-3-030-58621-8_12
5. Biggs, B., Roddick, T., Fitzgibbon, A., Cipolla, R.: Creatures great and SMAL: recovering the shape and motion of animals from video. In: Jawahar, C.V., Li, H., Mori, G., Schindler, K. (eds.) ACCV 2018. LNCS, vol. 11365, pp. 3–19. Springer, Cham (2019). https://doi.org/10.1007/978-3-030-20873-8_1
6. Cao, Z., Simon, T., Wei, S.E., Sheikh, Y.: Realtime multi-person 2D pose estimation using part affinity fields. In: CVPR (2017)
7. Chan, E., Monteiro, M., Kellnhofer, P., Wu, J., Wetzstein, G.: pi-GAN: periodic implicit generative adversarial networks for 3D-aware image synthesis. arXiv (2020)
8. Chibane, J., Alldieck, T., Pons-Moll, G.: Implicit functions in feature space for 3D shape reconstruction and completion. In: CVPR (2020)
9. Cormier, M., Clepe, A., Specker, A., Beyerer, J.: Where are we with human pose estimation in real-world surveillance? In: Proceedings of the IEEE/CVF Winter Conference on Applications of Computer Vision (WACV) Workshops, pp. 591–601 (2022)
10. Frey, B.J., Dueck, D.: Clustering by passing messages between data points. Science **315**, 972–976 (2007)
11. Giebenhain, S., Goldlücke, B.: AIR-Nets: an attention-based framework for locally conditioned implicit representations. In: 2021 International Conference on 3D Vision (3DV), pp. 1054–1064 (2021)
12. Giebenhain, S., Waldmann, U., Johannsen, O., Goldlücke, B.: Neural puppeteer: keypoint-based neural rendering of dynamic shapes (datset), October 2022. https://doi.org/10.5281/zenodo.7149178
13. Günel, S., Rhodin, H., Morales, D., Campagnolo, J., Ramdya, P., Fua, P.: DeepFly3D, a deep learning-based approach for 3D limb and appendage tracking in tethered, adult *Drosophila*. Elife **8**, e48571 (2019)
14. He, Y., Yan, R., Fragkiadaki, K., Yu, S.I.: Epipolar transformers. In: CVPR (2020)
15. Ioffe, S., Szegedy, C.: Batch normalization: accelerating deep network training by reducing internal covariate shift. In: Proceedings of the 32nd International Conference on Machine Learning (ICML). Proceedings of Machine Learning Research, vol. 37 (2015)
16. Ionescu, C., Papava, D., Olaru, V., Sminchisescu, C.: Human3.6M: large scale datasets and predictive methods for 3D human sensing in natural environments. IEEE TPAMI **36**(7), 1325–1339 (2014)
17. Iskakov, K., Burkov, E., Lempitsky, V., Malkov, Y.: Learnable triangulation of human pose. In: ICCV (2019)

18. Iskakov, K., Burkov, E., Lempitsky, V., Malkov, Y.: Learnable triangulation of human pose (2019). https://github.com/karfly/learnable-triangulation-pytorch

19. Ji, X., Fang, Q., Dong, J., Shuai, Q., Jiang, W., Zhou, X.: A survey on monocular 3D human pose estimation. Virtual Reality Intell. Hardw. **2**(6), 471–500 (2020)

20. Joska, D., et al.: AcinoSet: a 3D pose estimation dataset and baseline models for cheetahs in the wild. In: 2021 IEEE International Conference on Robotics and Automation (ICRA), pp. 13901–13908 (2021). https://doi.org/10.1109/ICRA48506.2021.9561338

21. Karashchuk, P., et al.: Anipose: a toolkit for robust markerless 3D pose estimation. Cell Rep. **36**(13), 109730 (2021)

22. BioMotionLab: BMLmovi Motion Capture Database. https://www.biomotionlab.ca//

23. Lei, J., Daniilidis, K.: CaDeX: learning canonical deformation coordinate space for dynamic surface representation via neural homeomorphism. In: CVPR, pp. 6624–6634, June 2022

24. Li, T., Bolkart, T., Black, M.J., Li, H., Romero, J.: Learning a model of facial shape and expression from 4D scans. ACM Trans. Graph. **36**(6), 194:1–194:17 (2017). Two first authors contributed equally

25. Liu, Z., Zhu, J., Bu, J., Chen, C.: A survey of human pose estimation: the body parts parsing based methods. J. Vis. Commun. Image Represent. **32**, 10–19 (2015)

26. Loper, M., Mahmood, N., Romero, J., Pons-Moll, G., Black, M.J.: SMPL: a skinned multi-person linear model. ACM Trans. Graph. **34**(6), 1–16 (2015). https://doi.org/10.1145/2816795.2818013

27. Loshchilov, I., Hutter, F.: Decoupled weight decay regularization. arXiv preprint arXiv:1711.05101 (2017)

28. van der Maaten, L., Hinton, G.: Visualizing data using t-SNE. J. Mach. Learn. Res. **9**(86) (2008). https://jmlr.org/papers/v9/vandermaaten08a.html

29. Madadi, M., Bertiche, H., Escalera, S.: Deep unsupervised 3D human body reconstruction from a sparse set of landmarks. Int. J. Comput. Vis. **129**(8), 2499–2512 (2021). https://doi.org/10.1007/s11263-021-01488-2

30. Mahmood, N., Ghorbani, N., Troje, N.F., Pons-Moll, G., Black, M.J.: AMASS: archive of motion capture as surface shapes. In: ICCV, pp. 5441–5450, October 2019. https://doi.org/10.1109/ICCV.2019.00554

31. Mescheder, L., Oechsle, M., Niemeyer, M., Nowozin, S., Geiger, A.: Occupancy networks: learning 3D reconstruction in function space. In: CVPR (2019)

32. Mildenhall, B., Srinivasan, P.P., Tancik, M., Barron, J.T., Ramamoorthi, R., Ng, R.: NeRF: representing scenes as neural radiance fields for view synthesis. In: Vedaldi, A., Bischof, H., Brox, T., Frahm, J.-M. (eds.) ECCV 2020. LNCS, vol. 12346, pp. 405–421. Springer, Cham (2020). https://doi.org/10.1007/978-3-030-58452-8_24

33. Müller, T., Evans, A., Schied, C., Keller, A.: Instant neural graphics primitives with a multiresolution hash encoding. ACM Trans. Graph. **41**(4), 102:1–102:15 (2022). https://doi.org/10.1145/3528223.3530127

34. Nath, T., Mathis, A., Chen, A.C., Patel, A., Bethge, M., Mathis, M.W.: Using DeepLabCut for 3D markerless pose estimation across species and behaviors. Nat. Protoc. **14**, 2152–2176 (2019)

35. Newell, A., Yang, K., Deng, J.: Stacked hourglass networks for human pose estimation. In: Leibe, B., Matas, J., Sebe, N., Welling, M. (eds.) ECCV 2016. LNCS, vol. 9912, pp. 483–499. Springer, Cham (2016). https://doi.org/10.1007/978-3-319-46484-8_29

36. Niemeyer, M., Geiger, A.: GIRAFFE: representing scenes as compositional generative neural feature fields. In: CVPR (2021)
37. Nocedal, J., Wright, S.J.: Numerical Optimization, 2nd edn. Springer, New York (2006). https://doi.org/10.1007/978-0-387-40065-5
38. Park, J.J., Florence, P., Straub, J., Newcombe, R., Lovegrove, S.: DeepSDF: learning continuous signed distance functions for shape representation. In: CVPR (2019)
39. Paszke, A., et al.: PyTorch: an imperative style, high-performance deep learning library. In: NeurIPS (2019)
40. Patel, P., Huang, C.H.P., Tesch, J., Hoffmann, D.T., Tripathi, S., Black, M.J.: AGORA: avatars in geography optimized for regression analysis. In: CVPR, June 2021
41. Pavlakos, G., et al.: Expressive body capture: 3D hands, face, and body from a single image. In: CVPR (2019)
42. Peng, S., et al.: Animatable neural radiance fields for modeling dynamic human bodies. In: ICCV, pp. 14314–14323, October 2021
43. Peng, S., et al.: Neural body: implicit neural representations with structured latent codes for novel view synthesis of dynamic humans. In: CVPR (2021)
44. Peng, S., Niemeyer, M., Mescheder, L., Pollefeys, M., Geiger, A.: Convolutional occupancy networks. In: Vedaldi, A., Bischof, H., Brox, T., Frahm, J.-M. (eds.) ECCV 2020. LNCS, vol. 12348, pp. 523–540. Springer, Cham (2020). https://doi.org/10.1007/978-3-030-58580-8_31
45. Reddy, N.D., Guigues, L., Pishchulin, L., Eledath, J., Narasimhan, S.G.: TesseTrack: end-to-end learnable multi-person articulated 3D pose tracking. In: CVPR, pp. 15190–15200 (2021)
46. Romero, J., Tzionas, D., Black, M.J.: Embodied hands: modeling and capturing hands and bodies together. ACM Trans. Graph. (Proc. SIGGRAPH Asia) **36**(6), 245:1–245:17 (2017). https://doi.org/10.1145/3130800.3130883
47. Rüegg, N., Zuffi, S., Schindler, K., Black, M.J.: BARC: learning to regress 3D dog shape from images by exploiting breed information. In: Proceedings of the IEEE/CVF Conference on Computer Vision and Pattern Recognition (CVPR), pp. 3876–3884, June 2022
48. Schwarz, K., Liao, Y., Niemeyer, M., Geiger, A.: GRAF: generative radiance fields for 3D-aware image synthesis. In: Advances in Neural Information Processing Systems (NeurIPS) (2020)
49. Sengupta, A., Budvytis, I., Cipolla, R.: Probabilistic 3D human shape and pose estimation from multiple unconstrained images in the wild. In: CVPR, pp. 16094–16104, June 2021
50. Sitzmann, V., Rezchikov, S., Freeman, W.T., Tenenbaum, J.B., Durand, F.: Light field networks: neural scene representations with single-evaluation rendering. In: Advances in Neural Information Processing Systems (2021). https://openreview.net/forum?id=q0h6av9Vi8
51. Sitzmann, V., Zollhoefer, M., Wetzstein, G.: Scene representation networks: continuous 3D-structure-aware neural scene representations. In: Advances in Neural Information Processing Systems (NeurIPS), vol. 32. Curran Associates, Inc. (2019). https://proceedings.neurips.cc/paper/2019/file/b5dc4e5d9b495d0196f61d45b26ef33e-Paper.pdf
52. Su, S.Y., Yu, F., Zollhöfer, M., Rhodin, H.: A-NeRF: articulated neural radiance fields for learning human shape, appearance, and pose. In: NeurIPS (2021)
53. Tewari, A., et al.: Advances in neural rendering. arXiv preprint arXiv:2111.05849 (2021)

54. Toshpulatov, M., Lee, W., Lee, S., Roudsari, A.H.: Human pose, hand and mesh estimation using deep learning: a survey. J. Supercomput. **78**, 7616–7654 (2022). https://doi.org/10.1007/s11227-021-04184-7

55. Tu, H., Wang, C., Zeng, W.: VoxelPose: towards multi-camera 3D human pose estimation in wild environment. In: Vedaldi, A., Bischof, H., Brox, T., Frahm, J.-M. (eds.) ECCV 2020. LNCS, vol. 12346, pp. 197–212. Springer, Cham (2020). https://doi.org/10.1007/978-3-030-58452-8_12

56. Xie, Y., et al.: Neural fields in visual computing and beyond. https://arxiv.org/abs/2111.11426 (2021). https://arxiv.org/abs/2111.11426

57. Yang, S., Quan, Z., Nie, M., Yang, W.: Transpose: keypoint localization via transformer. In: ICCV, pp. 11802–11812 (2021)

58. Yenamandra, T., et al.: i3DMM: deep implicit 3D morphable model of human heads. In: CVPR, pp. 12803–12813 (2021)

59. Yu, A., Fridovich-Keil, S., Tancik, M., Chen, Q., Recht, B., Kanazawa, A.: Plenoxels: radiance fields without neural networks (2021)

60. Yu, A., Ye, V., Tancik, M., Kanazawa, A.: pixelNeRF: neural radiance fields from one or few images. In: CVPR (2021)

61. Zhao, H., Jiang, L., Jia, J., Torr, P., Koltun, V.: Point transformer. In: International Conference on Computer Vision (ICCV) (2021)

62. Zheng, Z., Yu, T., Wei, Y., Dai, Q., Liu, Y.: DeepHuman: 3D human reconstruction from a single image. In: Proceedings of the IEEE/CVF International Conference on Computer Vision (ICCV), October 2019

63. Zhou, K., Bhatnagar, B.L., Pons-Moll, G.: Unsupervised shape and pose disentanglement for 3D meshes. In: Vedaldi, A., Bischof, H., Brox, T., Frahm, J.-M. (eds.) ECCV 2020. LNCS, vol. 12367, pp. 341–357. Springer, Cham (2020). https://doi.org/10.1007/978-3-030-58542-6_21

64. Zuffi, S., Kanazawa, A., Black, M.J.: Lions and tigers and bears: capturing non-rigid, 3D, articulated shape from images. In: CVPR (2018)

65. Zuffi, S., Kanazawa, A., Jacobs, D.W., Black, M.J.: 3D menagerie: modeling the 3D shape and pose of animals. In: CVPR (2017)

Decanus to Legatus: Synthetic Training for 2D-3D Human Pose Lifting

Yue Zhu$^{(\boxtimes)}$ and David Picard

LIGM, Ecole des Ponts, Univ Gustave Eiffel, CNRS, Marne-la-Vallée, France
{yue.zhu,david.picard}@enpc.fr
https://github.com/Zhuyue0324/Decanus-to-Legatus

Abstract. 3D human pose estimation is a challenging task because of the difficulty to acquire ground-truth data outside of controlled environments. A number of further issues have been hindering progress in building a universal and robust model for this task, including domain gaps between different datasets, unseen actions between train and test datasets, various hardware settings and high cost of annotation, etc. In this paper, we propose an algorithm to generate infinite 3D synthetic human poses (Legatus) from a 3D pose distribution based on 10 initial handcrafted 3D poses (Decanus) during the training of a 2D to 3D human pose lifter neural network. Our results show that we can achieve 3D pose estimation performance comparable to methods using real data from specialized datasets but in a zero-shot setup, showing the generalization potential of our framework.

Keywords: 3D Human pose · Synthetic training · Zero-shot

1 Introduction

3D Human pose estimation from single images [1] is a challenging and yet very important topic in computer vision because of its numerous applications from pedestrian movement prediction to sports analysis. Given an RGB image, the system predicts the 3D positions of the key body joints of human(s) in the image. Recent works on deep learning methods have shown very promising results on this topic [6, 21, 26, 48–50]. Current existing discriminative 3D human pose estimation methods, in which the neural network directly outputs the positions, can be put into two categories: One stage methods which directly estimate the 3D poses inside the world or camera space [29, 34], or two stage methods which first estimate 2D human poses in the camera space, then lift 2D estimated skeletons to 3D [18].

However, all these approaches require massive amount of supervision data to train the neural network. Contrarily to 2D annotations, obtaining the 3D annotations for training and evaluating these methods is usually limited to controlled

This work was granted access to the HPC resources of IDRIS under the allocation 2021-AD011012640 made by GENCI, and was supported and funded by Ergonova.

Supplementary Information The online version contains supplementary material available at https://doi.org/10.1007/978-3-031-26316-3_16.

L. Wang et al. (Eds.): ACCV 2022, LNCS 13844, pp. 257–274, 2023.
https://doi.org/10.1007/978-3-031-26316-3_16

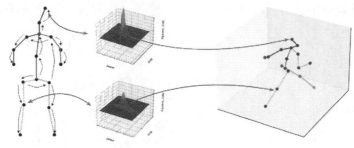

Markov tree & associated joint distributions   Synthetic skeleton sampling

Fig. 1. The main idea of our synthetic generation method: use a hierarchic probabilistic tree and its per joint distribution to generate realistic synthetic 3D human poses.

environments for technical reasons (Motion capture systems, camera calibration, etc.). This brings a weakness in generalization to in-the-wild images, where there can be more unseen scenarios with different kinds of human appearances, backgrounds and camera parameters.

In comparison, obtaining 2D annotations is much easier, and there are much more diverse existing 2D datasets in the wild [3,22,51]. This makes 2D to 3D pose lifting very appealing since they can benefit from the more diverse 2D data at least for their 2D detection part. Since the lifting part does not require the input image but only the 2D keypoints, we infer that it can be trained without any real ground-truth 3D information. Training 3D lifting without using explicit 3D ground-truth has previously been realized by using multiple views and cross-view consistency to ensure correct 3D reconstructions [45]. However, multiple views can be cumbersome to acquire and are also limited to controlled environments.

In order to tackle this problem, we propose an algorithm which generates infinite synthetic 3D human skeletons on the fly during the training of the lifter from just a few initial handcrafted poses. This generator provides enough data to train a lifter to invert 2D projections of these generated skeletons back to 3D, and can also be used to generate multiple views for cross-view consistency. We introduce a Markov chain with a tree structure (Markov tree) type of model, following a hierarchical parent-child joint order which allows us to generate skeletons with a distribution that we evolve through time so as to increase the complexity of the generated poses (see Fig. 1). We evaluate our approach on the two benchmark datasets Human3.6M and MPI-INF-3DHP and achieve zero-shot results that are competitive with that of weakly supervised methods. To summarize, our contributions are:

- A 3D human pose generation algorithm following a probabilistic hierarchical architecture and a set of distributions, which uses zero real 3D pose data.
- A Markov tree model of distributions that evolve through time, allowing generation of unseen human poses.
- A semi-automatic way to handcraft few 3D poses to seed initial distribution.
- Zero-shot results that are competitive with methods using real data.

2   Related Work

Monocular 3D Human Pose Estimation. In recent years, monocular 3D human pose estimation has been widely explored in the community. The models can be mainly categorized into generative models [2,4,7,24,33,39,47] which fit 3D parametric models to the image, and discriminative models which directly learn 3D positions from image [1,38]. Generative models try to fit the shape of the entire body and as such are great for augmented reality or animation purpose [35]. However, they tend to be less precise than discriminative models. On the other hand, a difficulty that the discriminative models have is that depth information is hard to infer from a single image when it is not explicitly modeled, and thus additional bias must be learned using 3D supervision [25,26], multiview spatial consistency [13,45,48] or temporal consistency [1,9,23]. Discriminative models can also be categorized into one stage models which predict directly 3D poses from images [14,25,29,34] and two stage methods which first learn a 2D pose estimator, then lift the obtained 2D poses to 3D [18,28,45,48,49,52]. Lifting 2D pose to 3D is somewhat of an ill-posed problem because of depth ambiguity ambiguity. But the larger quantity and diversity of 2D datasets [3,22,51], as well as the already achieved much better performance in 2D human pose estimation provide a strong argument for focusing on lifting 2D human poses to 3D.

Weak Supervision Methods. Since obtaining precise 3D annotations of human poses are hard due to technical reasons and are mostly limited to controlled environments, many research proposals tackled this problem by designing weak supervision methods to avoid using 3D annotations. For example, Iqbal et al. [18] apply a rigid-aligned multiview consistency 3D loss between multiple 3D poses estimated from different 2D views of the same 3D sample. Mitra et al. [30] learn 3D pose in a canonical form and ensure same predicted poses from different views. Fang et al. [13] propose a virtual mirror so that the estimated 3D poses, after being symmetrically projected into the other side of the mirror, should also look correctly, thus simulating another way of 'multiview' consistency. Finally, Wandt et al. [45] learn lifted 3D poses in a canonical form as well as a camera position so that every 3D pose lifted from a different view of a same 3D sample should still have 2D reprojection consistencies. For us, in addition to 3D supervision obtained from our synthetical generation, we also use multiview consistency to improve our training performance.

Synthetic Human Pose Training. Since the early days of the Kinect, synthetic training has been a popular option for estimating 3D human body pose [40]. The most common strategy is to perform data augmentation in order to increase the size and diversity of real datasets [16]. Others like Sminchisescu *et al.* [43] render synthetically generated poses on natural indoor and outdoor image backgrounds. Okada *et al.* [32] generate synthetic human poses in a subspace constructed by PCA using the walking sequences extracted from the CMU Mocap dataset [19]. Du *et al.* [12] create a synthetic height-map dataset to train a dual-stream convolutional network for 2D joints localization. Ghezelghieh *et al.* [15] utilize

3D graphic software and the CMU Mocap dataset to synthesize humans with different 3D poses and viewpoints. Pumarola *et al.* [36] created 3DPeople, a large-scale synthetic dataset of photo-realistic images with a large variety of subjects, activities and human outfits. Both [11] and [25] use pressure maps as input to estimate 3D human pose with synthetic data. In this paper, we are only interested in generating realistic 3D poses as a set of keypoints so as to train a 2D to 3D lifting neural network. As such, we do not need to render visually realistic humans with meshes, textures and colors for this much simpler task.

Human Pose Prior. Since the human body is highly constrained, it can be leveraged as an inductive bias in pose estimation. Bregler *et al.* [8] use kinematic-chain human pose model that follow the skeletal structure, extended by Sigal *et al.* [42] with interpenetration constraints. Chow *et al.* [10] introduced Chow-Liu tree, the maximum spanning tree of all-pairwise-mutual-information tree to model pairs of joints that exhibit a high flow of information. Lehrmann *et al.* [20] use a Chow-Liu tree that maximize an entropy function depending on nearest neighbor distances and learn local conditional distributions from data based on this tree structure. Sidenblahn *et al.* [41] use cylinders and spheres to model human body. Akhter *et al.* [2] learn joint-angle limits prior under local coordinate systems of 3 human body parts as torso, head, and upper-legs. We use a variant of kinematic model because the 3D limb lengths are fixed no matter the view, which can facilitate the generation process of synthetic skeleton.

Cross Dataset Generalization. Due to the diversity of human appearances and view points, cross-dataset generalization has recently been the center of attention of several works. Wang *et al.* [46] learn to predict camera views so as to auto-adjust to different datasets. Li *et al.* [21] and Gong *et al.* [16] perform data augmentation to cover the possible unseen poses in test dataset. Rapczyński *et al.* [37] discuss several methods including normalisation, viewpoint estimation, etc., for improving cross-dataset generalization. In our method, since we use purely synthetic data, we are always in a cross-dataset generalization setup.

3   Proposed Method

The goal of our method is to create a simple synthetic human pose generation model allowing us to train on pure synthetic data without any real 3D human pose data information during the whole training procedure.

3.1   Synthetic Human Pose Generation Model

Local Spherical Coordinate System. Without loss of generalization, we use Human3.6M skeleton layout shown in Fig. 2 (a) throughout the paper. To simplify human pose generation, we set the pelvis joint (joint 0) as root joint and the origin of the global Cartesian coordinate system from which a tree structure is applied to generate joints one by one. We suppose that the position of one joint

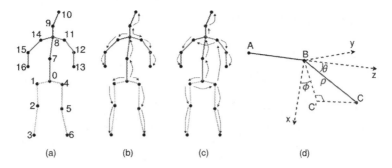

Fig. 2. (a) The 17-joint model of Human3.6M that we use (b) The **parent-child joint relation** graph. With parent joint's coordinate as origin of local spherical coordinate system, it generates child joint's position. (c) The **parent-child ρ, θ , ϕ relation** graph. With parent joint's ρ, θ , ϕ information, it samples child joint's ρ, θ , ϕ. (d) An example of how child joint is generated with sampled ρ, θ , ϕ from relationship in (c) under the local spherical coordinate system with it's parent joint in (b) as origin.

depends on the position of the joint which is directly connected to it but closer (in geodesic meaning) to the root joint. We call this kinematic chain **parent-child joint relations**, as shown in Fig. 2 (b). With this relationship, we propose to generate the child joint in a local spherical coordinate system (ρ, θ, ϕ) centered on its parent joint (see Fig. 2 (d)). The ρ, θ , ϕ values are sampled with respect to a conditional distribution $P(x_{child}|x_{parent})$. This produces a Markov chain indexed by a tree structure which we denote as a Markov Tree.

Our motivation to use a local spherical coordinate system for joint generation is that each human body branch has a fixed length ρ no matter the movement. Also, since the supination and the pronation of the branches are not encoded in skeleton representation, the new joint position can be parameterized with polar angle θ and azimuthal angle ϕ. Furthermore, by using an axis system depending on 'grandparent-parent' branch instead of global coordinate system, the possible angle interval of θ and ϕ achieved by human is more limited than in a global coordinate system. Finally, our local spherical coordinate system is entirely bijective with global coordinate system.

Hierarchic Probabilistic Skeleton Sampling Model. Generating a human pose in our local spherical coordinate system is equivalent to generating a set of (ρ, θ, ϕ). We thus propose to sample these values from a distribution that approximate that of real human poses. To retain plausible poses, we limit the range of (ρ, θ, ϕ) for each joint based on what is on average biologically achievable.

Since body joints follow a tree-like structure, it is unlikely that sampling each joint independently of the others leads to realistic poses. Instead, we propose to model the distribution of the joints by a Markov chain index by a tree following the skeleton, where probability of sampling a tuple (ρ, θ, ϕ) for a joint depends on the values sampled for its parent. More formally, denoting a child joint c and its parent $p(c)$ following the tree structure, we have:

$$(\rho_c, \theta_c, \phi_c) \sim P((\rho, \theta, \phi) | (\rho_{p(c)}, \theta_{p(c)}, \phi_{p(c)})) \tag{1}$$

Please note that the tree structure used for accounting the dependencies between joints as shown on Fig. 2 (c) is slightly different than the kinematic one. We found in practice that it is better to condition the position of one shoulder on the position of the same side hip, and to condition symmetrical shoulder/hip on their already generated counterpart rather than on their common parent. Intuitively, this seems to better encode global consistency.

To facilitate modeling distribution $P((\rho, \theta, \phi) | (\rho_{p(c)}, \theta_{p(c)}, \phi_{p(c)}))$, we make further assumption that all 3 components only depend on their parent counter parts. More formally:

$$\rho_c \sim P(\rho | \rho_{p(c)}), \quad \theta_c \sim P(\theta | \theta_{p(c)}), \quad \phi_c \sim P(\phi | \phi_{p(c)}) \tag{2}$$

This allows us to model each distribution with a simple non-parametric model consisting of a simple 2D histogram representing the probability of sampling, e.g., ρ_c knowing the value of $\rho_{p(c)}$. In practice, we use 50 bins histograms for each value, totalling to $3 \times 16 = 48$ 2D histograms of size 50×50. When there is no ambiguity, we use the same notation $P(\cdot|\cdot)$ for the histogram and the probability.

3.2   Pseudo-realistic 3D Human Pose Sampling

The next step is to estimate a distribution that can approximate the real 3D pose distribution, and from which our model can sample, so that the generated poses look like real human actions. Under the constraint of zero-shot 3D real data, we choose to make breakthrough by looking at limited amount of 2D real poses and 'manually' lift them into 3D to make our distribution. However, it is impossible for us to tell the exact depths of keypoints from an image with our eye, and it is also a huge amount of work to do if we check a lot of images one by one. Instead, we choose a 3-step procedure to get our handcrafted 3D pose:

High-Variance 2D Poses. We randomly sample 1000 sets of 10 2D-human poses from the target dataset (e.g., Human3.6M). We then compute the total variance for each set and pick the sets with largest variance as our candidates. This ensure our initial pose set has high diversity.

Semi-automatic 2D to 3D Seed Pose Lifting. Next, we use a semi-automatic way to lift samples in each seed set to 3D. The idea is as follows: from an image for which we already know the 2D distances between connected joints, and if we can estimate the 3D length of each branch who connects the joints as well as the proportion λ_{prop} between the 2D length in the image (in pixel) and the 3D length (in centimeter), we can estimate the relative depth between connected joints using Pythagorean theorems under the assumption that the camera produces an almost orthogonal projection. The ambiguity about the sign of these depths, which decide if one joint is in front of or in the back of its parent joint, can easily be manually annotated.

To estimate the 3D length, we define a set of fixed value representing branch lengths ($||c-p(c)||_2, \forall c$ except the root joint) of the human body based on biological data. Since we later calculate under a proportionality assumption between 3D and 2D, we only need it to roughly represent the proportionality between different human bone length. We also manually annotate $sign_c$ for each keypoint c, denoting if it is relatively further or closer to the camera compared to its parent joint $p(c)$. Finally the 2D-3D size proportion λ_{prop} is calculated under the assumption that the 3 joints around the head (head top, nose and neck) form a triangle of known ratio which is independent of rotation and view, visually shown in Fig. 3. This is reasonable since there are no largely moving articulated part in this triplet. We choose $AB = 1$ the unit length and we suppose the proportion between AB, BC and CA is fixed ($BC = \alpha AB, AC = \beta AB$). Noting $d_B = B'B - A'A$ and $d_C = C'C - A'A$, for the 2D skeleton we know $A'B', B'C'$ and $A'C'$, then we have 3 unknown variables d_B, d_C, and $\lambda_{prop} = \frac{A'B'(pixels)}{A'B'(meters)}$ and 3 equations:

$$d_B^2 = AB^2 - (\frac{A'B'}{\lambda_{prop}})^2, \quad d_C^2 = (\beta AB)^2 - (\frac{A'C'}{\lambda_{prop}})^2,$$

$$(d_B - d_C)^2 = (\alpha AB)^2 - (\frac{B'C'}{\lambda_{prop}})^2 \tag{3}$$

Then we can solve λ_{prop}. In practice, we set $\alpha = 1$ and $\beta = 5/3$.

After obtaining these depths, we apply Pythagorean theorem to get the final depth value of all joints with the kinematic order. Examples of semi-automatic lifted 3D poses are shown on Fig. 4. Since there are only a few keypoints to label as *in front of* or *behind* their parent joint, the labeling process is very easy and takes about 3 min per image only.

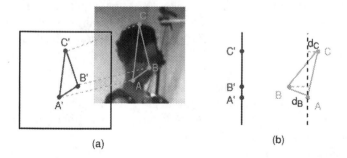

(a)   (b)

Fig. 3. (a) 3D poses (red A,B and C, unit in centimeters) of 3 joints of the head projected onto 2D camera plan (blue A', B' and C', unit in pixels). (b) same but right side view after 90° rotation. (Color figure online)

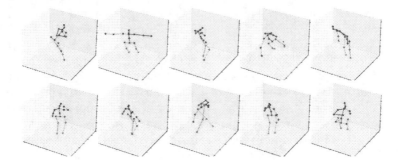

Fig. 4. A example of a set of 10 semi-automatic lifted 3D poses. This set of seeds is also the one which produce our best score on Human3.6M dataset. These 10 lifted samples have a 79.42mm MPJPE error compare to the groundtruth.

Distribution Diffusion. We then transform 3D poses into the local spherical coordinate system and used each seed set as initial distribution to populate the histograms. Since the sampling of a new skeleton follows the Markov tree structure and different limbs have a weak correlation between them in our model, it is possible to sample skeletons that look like combinations of the original 10 samples within the seed set.

However, these initial samplings are by no mean complete, and we run the risk of overfitting the lifter network to these poses only. To alleviate this problem, we introduce a diffusion process among each 2D histogram such that the probability of adjacent parameters is raised over time. More formally:

$$P(x_c|x_{p(c)})_{t+1} = P(x_c|x_{p(c)})_t + \alpha_{x_c}\Delta P(x_c|x_{p(c)})_t, \ x \in \{\rho, \theta, \phi\} \qquad (4)$$

where Δ is the Laplacien operator and α_{x_c} is the diffusion coefficient. This idea is derived from the heat diffusion equation in thermodynamics, in which bins with a higher probability diffuse to their neighbours (Laplacian operator), making the generation process more and more likely to generate samples out of initial bin.

The main reason behind our diffusion process is that of curriculum learning [5]. At first, the diversity of sampled skeletons is low and the neural network is able to quickly learn how to lift these poses. At later stage, the diffusion process allows the sampling process to generate more diverse skeletons that are progressive extensions of the initial pose angles, avoiding overfitting the original poses. We show in Fig. 5 an example of evolution of the histogram and increase of generation variety through diffusion.

3.3 Training with Synthetic Data

The training setup of 2D-3D lifter network l_w is shown on Fig. 6 and consists of 3 main components: (1) Sampling a batch of skeletons at each step; (2) sampling different virtual cameras to project the generated skeletons into 2D; and finally

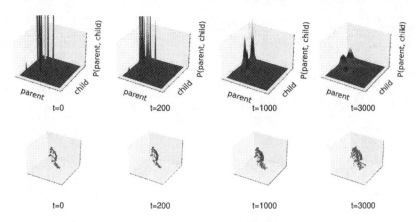

Fig. 5. First row is an example of the distribution histogram of a joint after 0, 200, 1000 and 3000 steps of diffusion. Second row shows an example of slightly increased generation variety when sampling from a single bin and generating 10 samples each time after 0, 200, 1000 and 3000 steps of diffusion.

Algorithm 1 Sampling algorithm

Require: True distribution P_t, empirical distribution P_e;
  $bins \leftarrow$ where $P_t > 0$ and $P_e \leq P_t$
  $b \sim \mathcal{U}(bins)$
  return Random sample from b

---

(3) the different losses used to optimize l_w. In practice, l_w is a simple 8-layer MLP with 1 in-layer, 3 basic residual blocks of width 1024, and 1 out-layer, adapted from [45].

When sampling a new batch of skeleton using our generator, we have to keep in mind that the distribution of the generator varies through time because of the diffusion process introduced in Eq. 4. To avoid over-sampling or under-sampling bins with low density, we propose to track the amount of skeletons that have been generated in each bin and adjust the sampling strategy accordingly. More formally, let us denote P_t the *true distribution* obtained by Eq. 4, and P_e the *empirical distribution* obtained by tracking the generation process. The corrected sampling algorithm is shown in Algorithm 1 and basically selects uniformly a plausible bin ($P_t > 0$) that has not been over-sampled ($P_e \leq P_t$). The whole generation process simply loops over all joints using the Markov tree and is shown on Algorithm 2.

At initialization, we sample 5000 real 2D poses, compute the proportion of nearest neighbour within each pose seed, and use it to initialize the histogram to give more importance to more frequent poses.

Regarding the projection of the batch into 2D, we propose to sample a set of batch-wise rotation matrices $R_{1,...,N}$, mostly rotating around the vertical axis, to simulate different viewpoints. Then, the rotated 3D skeletons are just simply: $X_{3D,i} = R_i X_{3D,0}, \quad i \in \{1, \ldots, N\}$, with $X_{3D,0}$ being the original skeleton in

---

Algorithm 2 Pose generation algorithm

Require: True distribution P_t, empirical distribution P_e, Markov tree structure T,
sampling algorithm S

  $X \leftarrow 0_{(J,3)}$ ▷ $3 = \rho, \theta, \phi$

  for $i \in \rho, \theta, \phi$ **do**   ▷ root joint

    $X[0, i] \leftarrow S(\ P_t(X_0), P_e(X_0))$

  end for

  for (p,c) in T **do**   ▷ parent-child relations in T

    for $i \in \rho, \theta, \phi$ **do**

      $X[c, i] \leftarrow S(P_t(X_{(c,i)}|X_{(p,i)}), P_e(X_{(c,i)}|X_{(p,i)}))$

      Update $P_e(X_{(c,i)}|X_{(p,i)})$

    end for

  end for

  return X in Cartesian coordinates

---

global Cartesian coordinates. To simulate the cameras, we follow [45] and use a scaleless orthogonal projection:

$$X_{2D,i} = \frac{WX_{3D,i}}{\|WX_{3D,i}\|_F}, \quad W = \begin{pmatrix} 1 & 0 & 0 \\ 0 & 1 & 0 \end{pmatrix}, \tag{5}$$

where W is the orthogonal projection matrix and $\|\cdot\|_F$ is the Frobenius norm. Normalizing by the Frobenius norm allows us to be independent of the global scale of $X_{2D,i}$ while retaining the relative scale of each bone with respect to each other. In practice, we found that uniformly sampling random rotation matrices at each batch renders the training much more difficult. Instead, we sample view with a small noise around the identity matrix and let the noise increase as the training goes on to generate more complex views at later stages.

Finally, to train the network, we leverage several losses. First, since we have the 3D ground-truth associated with each generated skeleton:

$$\mathcal{L}_{3D} = \frac{1}{N} \sum_{i=1..N} \left\| \frac{\hat{X}_{3D,i}}{\|\hat{X}_{3D,i}\|_F} - \frac{X_{3D,i}}{\|X_{3D,i}\|_F} \right\|_1, \tag{6}$$

with $\hat{X}_{3D,i} = l_w(X_{2D,i})$ being the output of the lifter l_w, and $\|\cdot\|_1$ the ℓ_1 norm. 3D skeletons are normalized before being compared because the input of the lifter is scaleless and as such it would make no sense to expect the lifter to recover the global scale of X_{3D}. Then, we use the multiple views generated thanks to R_i to enforce a multiview consistency loss. Calling $\hat{X}_{2D,i,j} = WR_j R_i^{-1}\hat{X}_{3D,i}$ the projection of the lifted skeleton from view i into view j, we optimize the cross-view projection error:

$$\mathcal{L}_{2D} = \frac{1}{N^2} \sum_{i=1}^{N} \sum_{j=1}^{N} \left\| \frac{\hat{X}_{2D,i,j}}{\|\hat{X}_{2D,i,j}\|_F} - \frac{X_{2D,j}}{\|X_{2D,j}\|_F} \right\|_1 \tag{7}$$

The global synthetic training loss we use is the following combination:

$$\mathcal{L} = \mathcal{L}_{2D} + \lambda_{3D}\mathcal{L}_{3D} \tag{8}$$

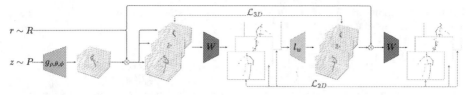

$g_{\rho,\theta,\phi}$: synthetic pose generator W: scaleless 2D projection l_w: 3D pose lifter $\mathcal{L}_*$: losses

Fig. 6. Our whole training process with synthetic data. Our generator g generates a 3D human pose following given distributions P of ρ, θ and ϕ. It will be applied with multiple different random generated r to project into different camera view. Projector W will projects them into scaleless 2D coordinates and they are the network inputs. The output estimated 3D poses will be applied with scaleless 3D supervision loss $\mathcal{L}_{3D}$, and also cross-view scaleless 2D reprojection loss $\mathcal{L}_{2D}$, which rotate estimated 3D pose from one view to another with known r and apply 2D supervision after projection W.

4   Experiments

4.1   Datasets

We use two widely used dataset Human3.6M [17] and MPI-INF-3DHP [29] to quantitatively evaluate our method.

We only use our generated synthetic samples for training and evaluate on S9 and S11 of Human3.6M and TS1-TS6 on MPI-INF-3DHP with their common protocols. In order to compare the quality of our generated skeletons with real 2D data, We also use the COCO [22] and MPII [3] datasets to check the generalizability of our method with qualitative evaluation.

4.2   Evaluation Metrics

For the quantitative evaluation on both Human3.6M and MPI-INF-3DHP we use MPJPE, i.e. the mean euclidean distance between the reconstructed and ground-truth 3D pose coordinates after the root joint is aligned ($P1$ evaluation protocol of Human3.6M dataset). Since we train the network with a scaleless loss, we follow [45] and scale the output 3D pose's Forbenius norm into the ground-truth 3D pose's Forbenius norm in order to compute the MPJPE. We also report PCK, i.e. the percentage of keypoints with the distance between predicted 3D pose and ground-truth 3D pose is less or equal to half of the head's length.

4.3   Implementation Details

We use a batch-size of 32 and we train for 10 epochs on a single 16G GPU using Adam optimizer and a learning rate of 10^{-4}. We set the number of views $N = 4$ and the total number of synthetic 2D input samples for each epoch is the same as the number of H36M training samples to make a fair comparison. The distribution diffusion coefficient α_{x_c} is a joint-wise loss dependent value, set to

Table 1. Comparison of our results with the state-of-the-arts under the common protocol 1 on Human 3.6M and MPI-INF-3DHP. The value before and after ± symbol are mean and standard deviation values.

		Weak supervision			Synthetic training				Ours	
		[18]	[30]	[45]	[21]	[15]	[12]	[44]	10 sets	Best run
H36M	MPJPE↓	67.4	120.95	65.9	106.8	$\geq$ 78.13	126.47	111.6	95.4 ± 13.5	60.8
3DHP	MPJPE↓	109.3	-	104.0	-	-	-	-	148.4 ± 7.6	132.8
	PCK↑	79.5	-	77.0	-	-	-	-	57.7 ± 2.3	61.9

$10^{-5} \times 10^{|\delta\mathcal{L}|/(10 \times N)}$ where $\delta\mathcal{L}$ is the joint-wise difference between loss of the last batch and the current batch, and the rotation R are sampled with a noise that increases in $\frac{1}{2 \times \#batch}$ after each step, with $\#batch$ the number of elapsed batches in the current epoch. For the loss, $\lambda_{3D} = 0.1$ is set empirically. To account for the variation due to the selection of the 2D pose using total variance, we keep the 10 sets with highest variance and show averaged results. Our method trains on about 100k generated samples per hour on a V100 GPU, whereas inference time for lifting is negligible.

4.4 Comparison with the State-of-the Art

We compare our results with the state-of-the-art methods with synthetic supervision for training in Table 1. We present several weak supervision methods which also do not use real 3D annotations, and instead use other sort of real data supervision whereas we do not. We can see that our method outperforms these synthetic training methods and achieves the performance on par with weakly supervised methods on H36M, while never using a real example for training.

We show qualitative results on the COCO dataset on Fig. 7. Since the COCO layout is different from that of H36M, we use a linear interpolation of existing joints to localize the missing joints. We can see that our model still achieves good qualitative performances on zero shot lifting of human poses in the wild (first 2 rows). Failed predictions (last row) tend to bend the legs backward even when the human is standing still, which may be a bias of the generator.

5 Ablation Studies

5.1 Synthetic Poses Realism

We want to see how similar our synthetic skeletons are to real skeletons. Qualitatively we compare our distribution after diffusion with the distribution of the whole Human3.6M and MPI-INF-3DHP datasets, for some of the joints as shown in Fig. 8. We can see that, even though there are many poses in MPI-INF-3DHP have never appear in Human3.6M, the distributions of angles θ and ϕ of these two real datasets have very similar shapes, which means our local spherical coordinate system successfully models the invariance of the biological achievable

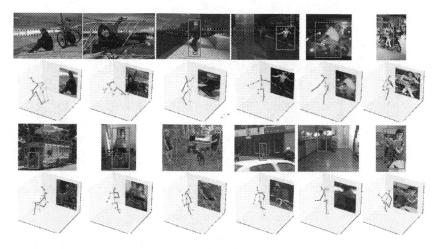

Fig. 7. Example of zero shot lifting in the wild on images from the COCO dataset. The first row are visually correct prediction, while the last row presents 'failure' cases, mostly due to right leg learnt a bias of leaning backward.

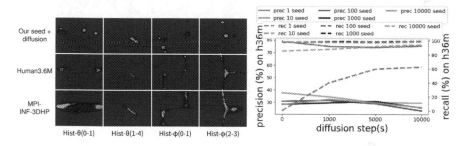

Fig. 8. Left: Examples of distributions of angle θ and ϕ from same parent-child pairs computed on Human3.6M, MPI-INF-3DHP, and our diffusion process. **Right**: Precision and recall evaluated with 5k generated samples and 5k real 2D samples from h36m.

human pose angles and their frequencies which are independent of camera view point. Our seeds+diffuse strategy produces a Gaussian mixture which succeed in covering big parts of real dataset's distribution.

Quantitatively we apply a precision/recall test, as is common practice with GANs [31]. We sample 5000 real and 5000 synthetic poses and project them to 2D plane using the scaleless projection in 5 and the Euclidean distance. Precision (resp. Recall) is defined as percentage of synthetic samples (resp. real samples) inside the union of the balls centered on each real sample (resp. synthetic sample) and with a radius of the distance to its 10-th nearest real sample neighbor (resp. synthetic sample neighbor). In our case, we already know that most synthetic skeleton generated by our Markov tree are biologically possible thanks to the limits in the generation intervals. As such, we are more interested in a very high recall so as to not miss the diversity of real skeletons. All our seed sets have

Table 2. Results on the 24-keypoint SMPL model, compared to the state-of-the-art

Method	Labeled training data	MPJPE↓
CLIFF (ECCV22)	H36m + 3DHP + COCO + MPII + 3DPW	**52.8**
DynaBOA (TPAMI22)	H36m + 3DPW	65.5
Ours	24 samples from 3DPW	61.09 ± 2.16

more than 70% recall and highest one achieves 91.8% recall. The precision, on the other hand, is around 40%, with 47.1% as the highest, which is still good considering we only start with 10 manually lifted initial poses for each seed.

5.2 Effect of Diffusion

We want to see why diffusion process is essential to our method. We take respectively 1, 10, 100, 1000 and 10000 samples of 3D poses on Human3.6M dataset as initial seed to make distribution graphs, and apply our 2D precision recall test after diffusion process. The result is shown in Fig. 8. We can see that diffusion generally increase recall value at the cost of precision value. The distribution using 1 samples as seed is much worse with the others in recall which means it can only cover around 60% of samples from real dataset even with diffusion process, while the distribution using 100 samples or more are close in performances. The diffusion process can reduce the gap between the distribution using 10 samples as seeds and those using 100 or more samples, which is important to us considering we want to avoid handcrafting a lot of initial poses.

5.3 Layout Adaptation

We show that our synthetic generation and training method also work on a different keypoint layout by applying the whole process on a newly defined hierarchic Markov tree based on 24 keypoints of SMPL model [24] and evaluating on 3DPW dataset [27]. We use 24 samples from its training set (one frame from each video) using our 2D variance based criterion for the seeds. Since our training method is scaleless, we rescale the predicted 3D poses by the average Forbenius norm of the 24 samples in the seed. The average MPJPE of 10 different seeds is shown in Table 2. This validates the generalization capability of our method.

6   Conclusion

We present an algorithm which allows to generate synthetic 3D human skeletons on the fly during the training, following a Markov-tree type distribution which evolve through out time to create unseen poses. We propose a scaleless multiview training process based on purely synthetic data generated from a few handcrafted poses. We evaluate our approach on the two benchmark datasets Human3.6M and MPI-INF-3DHP and achieve promising results in a zero shot setup.

References

1. Agarwal, A., Triggs, B.: Recovering 3d human pose from monocular images. IEEE Trans. Pattern Anal. Mach. Intell. **28**, 44–58 (2006)
2. Akhter, I., Black, M.J.: Pose-conditioned joint angle limits for 3d human pose reconstruction. In: Proceedings of the IEEE Conference on Computer Vision and Pattern Recognition (CVPR) (2015)
3. Andriluka, M., Pishchulin, L., Gehler, P., Schiele, B.: 2d human pose estimation: new benchmark and state of the art analysis. In: Proceedings of the IEEE/CVF Conference on Computer Vision and Pattern Recognition (CVPR) (2014)
4. Anguelov, D., Srinivasan, P., Koller, D., Thrun, S., Rodgers, J., Davis, J.: Scape: shape completion and animation of people. In: ACM Transactions on Graph (2005)
5. Bengio, Y., Louradour, J., Collobert, R., Weston, J.: Curriculum learning. In: Proceedings of the 26th Annual International Conference on Machine Learning (2009)
6. Bhatnagar, B.L., Sminchisescu, C., Theobalt, C., Pons-Moll, G.: Combining implicit function learning and parametric models for 3D human reconstruction. In: Vedaldi, A., Bischof, H., Brox, T., Frahm, J.-M. (eds.) ECCV 2020. LNCS, vol. 12347, pp. 311–329. Springer, Cham (2020). https://doi.org/10.1007/978-3-030-58536-5_19
7. Bogo, F., Kanazawa, A., Lassner, C., Gehler, P., Romero, J., Black, M.J.: Keep It SMPL: automatic estimation of 3D human pose and shape from a single image. In: Leibe, B., Matas, J., Sebe, N., Welling, M. (eds.) ECCV 2016. LNCS, vol. 9909, pp. 561–578. Springer, Cham (2016). https://doi.org/10.1007/978-3-319-46454-1_34
8. Bregler, C., Malik, J.: Tracking people with twists and exponential maps. In: Proceedings of 1998 IEEE Computer Society Conference on Computer Vision and Pattern Recognition (Cat. No. 98CB36231) (1998)
9. Choi, H., Moon, G., Chang, J.Y., Lee, K.M.: Beyond static features for temporally consistent 3d human pose and shape from a video. In: Proceedings of the IEEE/CVF Conference on Computer Vision and Pattern Recognition (CVPR) (2021)
10. Chow, C., Liu, C.: Approximating discrete probability distributions with dependence trees. IEEE Trans. Inf. Theory. **14**, 462–467 (1968)
11. Clever, H.M., Erickson, Z., Kapusta, A., Turk, G., Liu, K., Kemp, C.C.: Bodies at rest: 3d human pose and shape estimation from a pressure image using synthetic data. In: IEEE/CVF Conference on Computer Vision and Pattern Recognition (CVPR) (2020)
12. Du, Y., et al.: Marker-less 3D human motion capture with monocular image sequence and height-maps. In: Leibe, B., Matas, J., Sebe, N., Welling, M. (eds.) ECCV 2016. LNCS, vol. 9908, pp. 20–36. Springer, Cham (2016). https://doi.org/10.1007/978-3-319-46493-0_2
13. Fang, Q., Shuai, Q., Dong, J., Bao, H., Zhou, X.: Reconstructing 3d human pose by watching humans in the mirror. In: Proceedings of the IEEE/CVF Conference on Computer Vision and Pattern Recognition (CVPR) (2021)
14. Gärtner, E., Pirinen, A., Sminchisescu, C.: Deep reinforcement learning for active human pose estimation. In: AAAI (2020)
15. Ghezelghieh, M.F., Kasturi, R., Sarkar, S.: Learning camera viewpoint using CNN to improve 3d body pose estimation. In: 3D Vision (2016)
16. Gong, K., Zhang, J., Feng, J.: PoseAug: a differentiable pose augmentation framework for 3d human pose estimation. In: Proceedings of the IEEE/CVF Conference on Computer Vision and Pattern Recognition (CVPR) (2021)

17. Ionescu, C., Papava, D., Olaru, V., Sminchisescu, C.: Human3.6 m: large scale datasets and predictive methods for 3d human sensing in natural environments. IEEE Trans. Pattern. Anal. Mach. Intell. **36**, 1325–1339 (2014)

18. Iqbal, U., Molchanov, P., Kautz, J.: Weakly-supervised 3d human pose learning via multi-view images in the wild. In: IEEE/CVF Conference on Computer Vision and Pattern Recognition (CVPR) (2020)

19. Lab, C.G.: Motion capture database (2001). http://mocap.cs.cmu.edu

20. Lehrmann, A.M., Gehler, P.V., Nowozin, S.: A non-parametric Bayesian network prior of human pose. In: 2013 IEEE International Conference on Computer Vision (2013)

21. Li, S., et al.: Cascaded deep monocular 3d human pose estimation with evolutionary training data. In: 2020 IEEE/CVF Conference on Computer Vision and Pattern Recognition (CVPR) (2020)

22. Lin, T.Y., Maire, M., Belongie, S.J., Hays, J., Perona, P., Ramanan, D., Dollár, P., Zitnick, C.L.: Microsoft coco: Common objects in context. In: ECCV (2014)

23. Liu, R., Shen, J., Wang, H., Chen, C., Cheung, S.c., Asari, V.: Attention mechanism exploits temporal contexts: Real-time 3d human pose reconstruction. In: IEEE/CVF Conference on Computer Vision and Pattern Recognition (CVPR) (2020)

24. Loper, M., Mahmood, N., Romero, J., Pons-Moll, G., Black, M.J.: SMPL: A skinned multi-person linear model. ACM Trans. Graphics (Proc. SIGGRAPH Asia) (2015)

25. Luo, Y., Li, Y., Foshey, M., Shou, W., Sharma, P., Palacios, T., Torralba, A., Matusik, W.: Intelligent carpet: Inferring 3d human pose from tactile signals. In: Proceedings of the IEEE/CVF Conference on Computer Vision and Pattern Recognition (CVPR) (2021)

26. Ma, X., Su, J., Wang, C., Ci, H., Wang, Y.: Context modeling in 3d human pose estimation: A unified perspective. In: Proceedings of the IEEE/CVF Conference on Computer Vision and Pattern Recognition (CVPR) (2021)

27. von Marcard, T., Henschel, R., Black, M.J., Rosenhahn, B., Pons-Moll, G.: Recovering accurate 3D human pose in the wild using IMUs and a moving camera. In: Ferrari, V., Hebert, M., Sminchisescu, C., Weiss, Y. (eds.) ECCV 2018. LNCS, vol. 11214, pp. 614–631. Springer, Cham (2018). https://doi.org/10.1007/978-3-030-01249-6_37

28. Martinez, J., Hossain, R., Romero, J., Little, J.J.: A simple yet effective baseline for 3d human pose estimation. In: Proceedings IEEE International Conference on Computer Vision (ICCV) (2017)

29. Mehta, D., et al.: Monocular 3d human pose estimation in the wild using improved CNN supervision. In: 2017 Fifth International Conference on 3D Vision (3DV) (2017)

30. Mitra, R., Gundavarapu, N.B., Sharma, A., Jain, A.: Multiview-consistent semi-supervised learning for 3d human pose estimation. In: IEEE/CVF Conference on Computer Vision and Pattern Recognition (CVPR) (2020)

31. Naeem, M.F., Oh, S.J., Uh, Y., Choi, Y., Yoo, J.: Reliable fidelity and diversity metrics for generative models. In: International Conference on Machine Learning (2020)

32. Okada, R., Soatto, S.: Relevant feature selection for human pose estimation and localization in cluttered images. In: Forsyth, D., Torr, P., Zisserman, A. (eds.) ECCV 2008. LNCS, vol. 5303, pp. 434–445. Springer, Heidelberg (2008). https://doi.org/10.1007/978-3-540-88688-4_32

33. Pavlakos, G., et al.: Expressive body capture: 3d hands, face, and body from a single image. In: CVPR (2019)
34. Pavlakos, G., Zhou, X., Derpanis, K.G., Daniilidis, K.: Coarse-to-fine volumetric prediction for single-image 3d human pose. In: CVPR (2017)
35. Petrovich, M., Black, M.J., Varol, G.: Action-conditioned 3D human motion synthesis with transformer VAE. In: ICCV (2021)
36. Pumarola, A., Sanchez, J., Choi, G., Sanfeliu, A., Moreno-Noguer, F.: 3DPeople: modeling the geometry of dressed humans. In: International Conference in Computer Vision (ICCV) (2019)
37. Rapczyński, M., Werner, P., Handrich, S., Al-Hamadi, A.: A baseline for cross-database 3d human pose estimation. Sensors. **31**, 3769 (2021)
38. Rhodin, H., et al.: Learning monocular 3d human pose estimation from multi-view images. In: Proceedings/CVPR, IEEE Computer Society Conference on Computer Vision and Pattern Recognition. IEEE Computer Society Conference on Computer Vision and Pattern Recognition (2018)
39. Schmidtke, L., Vlontzos, A., Ellershaw, S., Lukens, A., Arichi, T., Kainz, B.: Unsupervised human pose estimation through transforming shape templates. In: Proceedings of the IEEE/CVF Conference on Computer Vision and Pattern Recognition (CVPR) (2021)
40. Shotton, J., et al.: Real-time human pose recognition in parts from single depth images. In: CVPR 2011 (2011)
41. Sidenbladh, H., Black, M.J., Fleet, D.J.: Stochastic tracking of 3D human figures using 2D image motion. In: Vernon, D. (ed.) ECCV 2000. LNCS, vol. 1843, pp. 702–718. Springer, Heidelberg (2000). https://doi.org/10.1007/3-540-45053-X_45
42. Sigal, L., Isard, M., Haussecker, H., Black, M.J.: Loose-limbed people: estimating 3D human pose and motion using non-parametric belief propagation. Int. J. Comput. Vision. **98**, 15–48 (2011)
43. Sminchisescu, C., Kanaujia, A., Metaxas, D.: Learning joint top-down and bottom-up processes for 3d visual inference. In: 2006 IEEE Computer Society Conference on Computer Vision and Pattern Recognition (CVPR 2006) (2006)
44. Varol, G., et al.: Learning from synthetic humans. In: CVPR (2017)
45. Wandt, B., Rudolph, M., Zell, P., Rhodin, H., Rosenhahn, B.: CanonPose: self-supervised monocular 3d human pose estimation in the wild. In: Proceedings of the IEEE/CVF Conference on Computer Vision and Pattern Recognition (CVPR) (2021)
46. Wang, Z., Shin, D., Fowlkes, C.C.: Predicting camera viewpoint improves cross-dataset generalization for 3d human pose estimation. CoRR (2020)
47. Xu, H., Bazavan, E.G., Zanfir, A., Freeman, W.T., Sukthankar, R., Sminchisescu, C.: Ghum & Ghuml: generative 3d human shape and articulated pose models. In: 2020 IEEE/CVF Conference on Computer Vision and Pattern Recognition (CVPR) (2020)
48. Xu, J., Yu, Z., Ni, B., Yang, J., Yang, X., Zhang, W.: Deep kinematics analysis for monocular 3d human pose estimation. In: IEEE/CVF Conference on Computer Vision and Pattern Recognition (CVPR) (2020)
49. Xu, T., Takano, W.: Graph stacked hourglass networks for 3d human pose estimation. In: Proceedings of the IEEE/CVF Conference on Computer Vision and Pattern Recognition (CVPR) (2021)

50. Zanfir, A., Bazavan, E.G., Xu, H., Freeman, W.T., Sukthankar, R., Sminchisescu, C.: Weakly supervised 3D human pose and shape reconstruction with normalizing flows. In: Vedaldi, A., Bischof, H., Brox, T., Frahm, J.-M. (eds.) ECCV 2020. LNCS, vol. 12351, pp. 465–481. Springer, Cham (2020). https://doi.org/10.1007/978-3-030-58539-6_28
51. Zhang, S.H., et al.: Pose2seg: detection free human instance segmentation. In: Proceedings of the IEEE/CVF Conference on Computer Vision and Pattern Recognition (CVPR) (2019)
52. Zhou, X., Zhu, M., Leonardos, S., Derpanis, K.G., Daniilidis, K.: Sparseness meets deepness: 3d human pose estimation from monocular video. In: 2016 IEEE Conference on Computer Vision and Pattern Recognition (CVPR) (2016)

Social Aware Multi-modal Pedestrian Crossing Behavior Prediction

Xiaolin Zhai[1,2], Zhengxi Hu[1,2], Dingye Yang[1,2], Lei Zhou[1,2], and Jingtai Liu[1,2]($\boxtimes$)

[1] Institute of Robotics and Automatic Information System, College of Artificial Intelligence, Nankai University, Tianjin, China
{2120210410,hzx,1711502}@mail.nankai.edu.cn, liujt@nankai.edu.cn
[2] Tianjin Key Laboratory of Intelligent Robotics, Nankai University, Tianjin, China

Abstract. With the development of self-driving vehicles, pedestrian behavior prediction plays a vital role in constructing a safe human-robot interactive environment. Previous methods ignored the inherent uncertainty of pedestrian future actions and the temporal correlations of spatial interactions. To solve the aforementioned problems, we propose a novel social aware multi-modal pedestrian crossing behavior prediction network. In this research field, our network innovatively explores the multimodality nature of pedestrian future action prediction and forecasts diverse and plausible futures. Also, to model the social aware context in both the spatial and temporal domain, we construct a spatial-temporal heterogeneous graph, bridging the spatial-temporal gap between the scene and the pedestrian. Experiments show that our model achieves state-of-the-art performance on pedestrian action detection and prediction task. The code is available at https://github.com/zxll0106/Pedestrian_Crossing_Behavior_Prediction.

Keywords: Pedestrian crossing behavior prediction · Video understanding

1 Introduction

Predicting pedestrian behaviors plays a critical role in the human-robot interactive scene. Pedestrians in the urban traffic scenario can extract useful information from the surroundings to infer others' motion patterns and make reasonable decisions. We hope that autonomous systems can acquire the capability to mimic human perception to predict pedestrian behaviors, which is important for creating a safer environment for both robots and pedestrians.

This work is supported in part by National Key Research and Development Project under Grant 2019YFB1310604, in part by National Natural Science Foundation of China under Grant 62173189.

Supplementary Information The online version contains supplementary material available at https://doi.org/10.1007/978-3-031-26316-3_17.

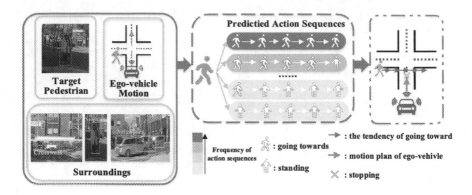

Fig. 1. Our model captures useful information not only from the target pedestrian but also from the social aware context. Interactions with the surrounding traffic objects and the ego-vehicle motion are utilized to enhance the dynamical representation of the target pedestrian. Our model also considers the uncertainty of pedestrian future actions and performs diverse predictions. We choose the most frequently appearing action sequence as the final prediction result. The ego-vehicle will make reactive decisions based on the predicted action sequence of the target pedestrian. Only *standing* and *going towards* are involved in this figure and other action labels are not involved.

Many works [10,13,14,22] explored pedestrian crossing behaviors and achieved significant improvement in the pedestrian behavior understanding. SF-GRU [13] designed the stacked RNNs to gradually fuse multiple inputs to estimate pedestrian intention. MMH-PAP [14] modeled the temporal dynamics of different inputs and designed an attention module to calculate the weights of each input to predict binary crossing action. Yao *et al.* [22] proposed an intention estimation and action detection network. A soft-attention module is designed to capture spatial interaction between different traffic objects. However, they not only neglect the multi-modal nature of pedestrian future actions, but also ignore the temporal continuity of relations between traffic objects.

Different from previous works, we take account into the inherent uncertainty of pedestrian behaviors which is a critical cue for inferring future actions, as shown in Fig. 1. Under the same history states, this uncertainty can cause diverse and plausible futures. For example, when the pedestrian comes to the crossroad, he may cross directly or wait for the red traffic light. Deterministic models are not suitable for capturing a one-to-many mapping and producing probabilistic inference. As a result, we propose Multi-Modal Conditional Generative Module to learn multiple modes of pedestrian future actions. In this module, we introduce multiple latent variables to model the multimodality of pedestrian future and perform diverse action predictions.

Spatial-temporal interactions between traffic objects play a vital role in refining the scene information and enhancing the pedestrian representation. Pedestrians observe the current and past states of other traffic objects to perceive the surrounding environment and make reasonable decisions. To enable the robot with the ability of perception, we propose Social Aware Encoder to extract the pedestrian-specific contextual information. Spatial relations between traffic objects and the temporal continuity of these relations should also be fully considered. In Social Aware Encoder, we construct a spatial-temporal heterogeneous graph to model spatial-temporal interactions between the target pedestrian and heterogeneous traffic objects.

The key contributions of our work are threefold:

- We highlight a new direction in pedestrian behavior understanding, namely the multi-modal pedestrian action prediction. Multi-Modal Conditional Generative Module is proposed to capture the multimodality of pedestrian future actions.
- Our proposed Social Aware Encoder can jointly model spatial-temporal contextual interactions and augment the target pedestrian representation.
- To aggregate the above modules, we propose a social aware multi-modal pedestrian behavior prediction network. Experiment results on the PIE dataset and JAAD dataset show that our network achieves state-of-the-art performance on the action detection and action prediction task. On the intention estimation task, our model improve by 1.1%–5.7% based on different metrics.

2   Related Work

2.1   Pedestrian Intention Estimation

Pedestrian intention estimation plays a critical role in helping the autonomous system perform safer decisions and construct a safe urban traffic environment. Previous works [4,8,20,24] took the destination of the trajectory as pedestrian intentions and lacked a deeper semantic interpretation of pedestrian intentions. PIE [10] extracted pedestrian intention from RGB images, captured temporal dynamics of intention features, and utilized intentions to guide pedestrian motion prediction. ST-DenseNet [17] proposed a real-time network that incorporates pedestrian detection, tracking, and intention estimation. They utilized YOLOv3 [15] to detect pedestrians, used SORT [21] algorithm to track them, and designed a spatial-temporal DenseNet [5] to estimate their intentions. SF-GRU [13] collected visual inputs from pedestrians and their surrounding scenes. Then they gradually concatenated inputs and fused them into the stacked RNNs. Kotseruba et al. [6] analyzed the influence of the pedestrian self and the environment on intentions. They evaluate the impact of the gaze, location, orientation, and interaction of pedestrians. And locations of designated crosswalks and curbs in the environment are also utilized to estimate pedestrian intention. They combined the influences of these factors and used a logistic regression classifier to infer pedestrian intentions. FuSSI-Net [9] utilized the pose estimation network to obtain human joint coordinates and designed different strategies to fuse human joint features and visual features. Liu et al. [7] estimated intention from the pedestrian-centric and location-centric perspectives. They constructed a pedestrian-centric graph based on pedestrian position relation and appearance relation. Then the pedestrian-centric graph is modified to the location-centric graph to predict whether there are pedestrians crossing in front of the ego-vehicle.

Considering that intentions are forward-looking, intentions will eventually be reflected in actions. They ignored the interaction between intentions and actions. Our network produces the multi-modal future action sequences under the direction of intentions which fully account for relations between pedestrian intentions and actions.

2.2   Pedestrian Behavior Prediction

In the human-robot interactive scene, uncertainties of pedestrian behaviors are an important challenge for autonomous systems. Pedestrian behavior understanding can provide a reasonable inference of pedestrian behaviors which is beneficial for robot navigation.

MMH-PAP [14] utilized LSTM to integrate temporal dynamics of visual features and position features and applied an attention mechanism to generate weighted representations of each input. Graph-SIM [23] proposed a pedestrian action prediction network based on graph neural networks. Given their locations and speeds, they clustered road users and assigned importance weights to relations with the target pedestrian. CIR-Net [2] paid particular attention to the relation between pedestrian actions and trajectories. To couple actions and trajectories, they aligned changes in trajectories with changes in actions. The multi-task network proposed by [22] can estimate pedestrian intention and detect crossing action simultaneously. They utilized the soft attention network [3] to model spatial relations between the target pedestrian and other road components. Previous works ignored the inherent multimodality of pedestrian future actions and their deterministic models are not suitable for modeling a distribution over diverse futures. We take consideration into the uncertainty of pedestrian futures and design Multi-Modal Conditional Generative Module to produce diverse and plausible future action sequences.

3   Method

Our goal is to produce diverse and plausible action predictions of the target pedestrian P. At time t, we summarize P's current state s_t and history states $s_{t-1}, ..., s_{t-H}$ for H history time steps as input $X_t = s_{(t-H:t)}$. There is also additional surrounding information S_t of P, including the position and type of other traffic objects, and the ego-vehicle's motion to which autonomous systems have access. Given X_t and S_t, we aim to predict the target pedestrian's actions $Y_t = \widehat{A}_{(t+1:t+F)}$ for the future F time steps, which is referred to $P(Y_t|X_t, S_t)$.

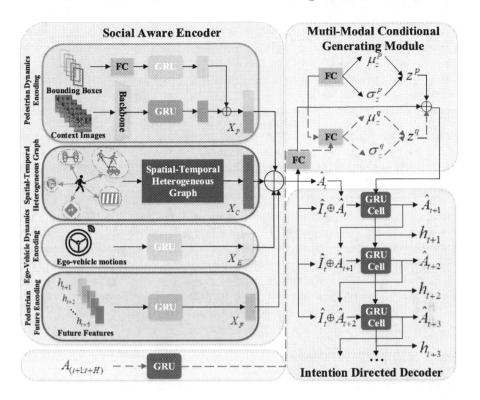

Fig. 2. Overview of our social-aware multi-modal pedestrian behavior prediction network. Our proposed network contains 3 parts, namely Social Aware Encoder, Multi-Modal Conditional Generative Module, and Intention Directed Decoder. Social Aware Encoder is composed of four modules: Pedestrian Dynamics Encoding Module (green), Spatial-Temporal Heterogeneous Graph Module (blue), Ego-Vehicle Dynamics Encoding Module (yellow), and Pedestrian Future Fusion Module (brown). $\oplus$ denotes the concatenation operation. (Color figure online)

3.1 Social Aware Encoder

Considering the target pedestrian and the surrounding environment, we proposed Social Aware Encoder which is composed of four modules in parallel, including Pedestrian Dynamics Encoding, Spatial-Temporal Heterogeneous Graph, Ego-vehicle Dynamics Encoding, and Pedestrian Future Fusion, as shown in Fig. 2. We explore the temporal dynamics of the target pedestrian P to enhance the individual representation X_P in the Pedestrian Dynamics Encoding (PDE). Simultaneously, Spatial-Temporal Heterogeneous Graph (STHG) models pairwise relation between P with heterogeneous traffic objects to obtain context encoded feature X_C. Meanwhile, Ego-Vehicle Dynamics Encoding (EVDE) considers decision making of autonomous systems and models the temporal dependency X_E of the ego-vehicle's motion. We aggregate future features of P to capture rich latent information X_F from the future in the Pedestrian Future Fusion (PFF).

We concatenate X_P, X_C, X_E and X_F and apply a fully-connected layer to obtain the social-aware feature X_S. Finally, an action classifier and an intent classifier are applied on X_S to estimate the current intention $\widehat{I_t}$ and the current action $\widehat{A_t}$, respectively.

Pedestrian Dynamics Encoding (PDE). This module models temporal dynamics of the current and history states of the target pedestrian. We note that the current state lacks temporal context, so we merge the information of the history states in the temporal domain. The states $s_{(t-H:t)}$ contain position features $pos_{(t-H:t)}$ and visual features $v_{(t-H:t)}$. Position features are the embedded bounding box of the target pedestrian through a fully-connected layer. Visual features are captured by pretrained CNN backbone network [18] on the cropped image patch which includes the pedestrian and the surrounding environment. These two features have different semantic attributes, so we model their temporal dynamics separately. We input these feature sequences into the corresponding GRU network to enrich them with temporal dynamical evolution clues. Outputs of two GRU networks are concatenated to obtain the Pedestrian Dynamical Feature X_P which aggregates temporal dynamics from multiple inputs.

Spatial-Temporal Heterogeneous Graph (STHG). We propose the Spatial-Temporal Heterogeneous Graph to capture other traffic objects' influence on the target pedestrian. There are various types of traffic objects, so their inherent heterogeneity can not be ignored. For example, traffic objects of the same type have the same semantic attributes and relations with the target person. Hence, we regard traffic objects in the traffic scenario as a heterogeneous multi-instance system. Taking account into the inherent heterogeneity of traffic objects, all traffic objects of the same type are aggregated in the local graph to capture intra-type interaction. Then we model the high-order relation between the target pedestrian and traffic objects of the same type in the global graph.

We construct a local graph $\mathcal{G}^c = (\mathcal{V}^c, \mathcal{E}^c)$ for each type $c \in C$, where the type set C includes 5 types, namely traffic neighbors (pedestrians, cyclists, vehicles, *etc.*), traffic signs, traffic lights, stations, and crosswalks. $\mathcal{V}^c = \{v_1, v_2, ..., v_{N_c}\}$ contains traffic objects of type c at the same time step. $v_i = \{x_{tl}, y_{tl}, x_{br}, y_{br}, c\}$ represents the bounding box coordinates and the type of v_i. In the edge set $\mathcal{E}^c$, we allow traffic objects belonging to the same type to connect with each other. We propose an information propagation mechanism on the local graph $\mathcal{G}^c$ to aggregate contextual information from the neighbor objects:

$$v_c^{(l)} = Avgpool\left(\phi_c(v_1^{(l)}), \ldots, \phi_c(v_{N_c}^{(l)})\right) \tag{1}$$

$$v_i^{(l+1)} = concat\left(\phi_c(v_i^{(l)}), v_c^{(l)}\right) \tag{2}$$

where $v_i^{(l)}$ is the feature of object i in the l-th layer of the local graph, $v_c^{(l)}$ is the spatial interaction feature of the type c, and $v_i^{(0)}$ is the initial feature v_i.

ϕ_c is the fully connected layer and embeds the object feature of type c into the high-dimensional spaces. Average pooling is utilized to propagate spatial context information from neighbor objects. We stack multiple layers of the local graph $\mathcal{G}^c$ to fuse features of different objects in the different layers.

In the global graph, we integrate the interaction between the target pedestrian and traffic objects of type c into the global feature X_c. We note that the attention mechanism can assign an adaptive weight to each object and understand underlying relations better. Hence, we utilize the attention mechanism to calculate traffic objects' weights of type c based on relations with target pedestrian:

$$A_i = \frac{\exp(\theta_p(v_P)^T \theta_q(v_i)/\sqrt{d_\theta})}{\sum_{j=1}^{N_c} \exp(\theta_p(v_P)^T \theta_q(v_j)/\sqrt{d_\theta})} \tag{3}$$

$$X_c = \sum_{i=1}^{N_c} A_i \cdot v_i \tag{4}$$

where $v_i \in \mathcal{V}^c$, v_P is the target pedestrian feature, and θ_p and θ_q embed them into the d_θ-dimension space.

Previous works considered only spatial interaction [22]. Different from them, we capture the spatial-temporal dynamics integrally to build the bridge between the spatial and temporal context. We utilize GRU to aggregate contextual information of X_c in the temporal domain to obtain the final context encoded feature.

$$X_C = concat\left(\{GRU_c(X_c), c \in C\}\right) \tag{5}$$

Finally, we obtain the Context Encoded Feature X_C which aggregates spatial-temporal dependency of relations between the target pedestrian and heterogeneous traffic objects.

Ego-Vehicle Dynamics Encoding (EVDE). Interactions between pedestrians and the ego-vehicle contain important latent information. For example, when observing the crossing action of the pedestrian in front of the ego-vehicle, the controller will slow down or make way accordingly. Similarly, the pedestrian observes the motion tendency of the ego-vehicle and then may wait for the ego-vehicle to pass. As a result, we take into account ego-vehicle future motion plans and infer the pedestrian intention and action from them. We consider that the attribute of ego-vehicle motion is different from other road users, so this module utilizes a separate bi-directional GRU to encode the ego-vehicle future motions $M_{E(t+1:t+F)}$, including speed, acceleration, yaw rate, and yaw acceleration. We refer to the output of the GRU as Ego-vehicle Dynamical Feature X_E which integrates the bi-directional long-term dependency of the future motion plan.

Pedestrian Future Fusion (PFF). Considering that future actions can reflect pedestrian intentions, we collect future features of the target pedestrian, namely hidden states h of GRU cells in the Intention-Directed Decoder for the future F time steps. Future features for the future F time steps are denoted as $H_F = h_{(t+1:t+F)}$. We apply a bi-directional GRU to model the temporal dynamics of the target pedestrian's future features. In a bi-directional GRU, information flows in forward and backward two directions, so bi-directional information can be integrated. We obtain the last hidden state of the GRU and refer to it as Pedestrian Future Features X_F.

3.2   Multi-modal Conditional Generating Module

There are multiple uncertainties in action prediction since the prediction is inherently probabilistic. For example, when the pedestrian faces the crosswalk, the pedestrian may cross the road, wait for a red light or wait for other vehicles to pass. Hence, under the same situation, there will be diverse future scenarios where each one contains a reasonable explanation. A deterministic function that projects one input to one output may not have the adequate capacity to represent the diverse latent space. To learn a one-to-many mapping, we adopt a deep conditional generative model, conditional variational auto-encoder [19].

This module contains the prior network $P_\nu(z|X_t)$, and the recognition network $Q_\phi(z|X_t, Y_t)$. ϕ, ν refer to the weights of corresponding networks. The latent variables z play a critical role in modeling the inherent multimodality of the pedestrian future. In the training stage, we utilize a bi-directional GRU to encode the ground truth future actions $A_{(t+1:t+F)}$ obtaining Y_t. $Q_\phi(z|X_t, Y_t)$ takes X_t and Y_t as inputs to predict the mean μ_z^q and covariance σ_z^q and then sample the latent variable z_q from $N(\mu_z^q, \sigma_z^q)$. The goal of Q_ϕ is to learn a distribution from observations and ground truth to the latent variable z_q. With no prior knowledge of ground truth future actions, $P_\nu(z|X_t)$ predicts the mean μ_z^p and covariance σ_z^p based on observations. Similarly, the latent variable z^p is sampled from the distribution $N(\mu_z^p, \sigma_z^p)$. We optimize Kullback-Leibler divergence between $N(\mu_z^q, \sigma_z^q)$ and $N(\mu_z^p, \sigma_z^p)$ to make z_p produced by prior network learn the distribution modeled by recognition network. During the training stage, the latent variables z are sampled from the recognition network and then concatenated with X_S from the encoder. In the testing stage, we sample z from the prior network because we can not get access to ground truth future actions.

3.3   Intention-Directed Decoder

The intention is prospective and action can be described as following along line to attain a certain intention. Under the guidance of intentions, generation network $P_\theta(Y_t|X_t, \widehat{I}_t, z)$ is utilized to model the diversity of future actions. Multiple latent variables z are used to generate future actions with multiple modes. And pedestrian intention $\widehat{I}_t$ provides a direction for the development of future actions. Considering that future actions can be regarded as a temporal sequence, we adopt GRU cells to construct the Intention-Directed Decoder. The input of the decoder is the concatenation of estimated intention $\widehat{I}_t$ and the predicted action at the last prediction time step. Under the guidance of pedestrian intention, the decoder iteratively predicts future actions. And we collect the hidden states $h_{(t+1:t+F)}$ of the GRU cell and pass them to Social Aware Encoder to extract important information in the future.

3.4   Multi-task Loss

The loss function of our model contains Kullback-Leibler divergence (KLD) between Q_ϕ and P_ν, binary cross entropy loss $\mathcal{L}_I$ for intention estimation at time step t, cross entropy loss $\mathcal{L}_A$ for action detection at time step t and action prediction for future time steps from $t+1$ to $t+F$:

$$\mathcal{L} = \sum_{t=1}^{T}(\lambda_1 KLD(Q_\phi(z|X,Y), P_\nu(z|X)) + \lambda_2\mathcal{L}_I(\widehat{I}_t, I_t)$$
$$+\lambda_3\mathcal{L}_A(\widehat{A}_t, A_t) + \lambda_4\mathcal{L}_A(\widehat{A}_{(t+1:t+F)}, A_{(t+1:t+F)})) \qquad (6)$$

where T is the total sample length, $\widehat{I}$ and $\widehat{A}$ are the predicted value of intention and action, and I and A are the ground truth of intention and action. $\lambda_1, \lambda_2, \lambda_3$, and λ_4 are utilized to balance multiple tasks.

4   Experiments

We conduct experiments on the PIE and JAAD datasets under the original data setting and the 'time to event' setting to verify the effectiveness of our pedestrian behavior prediction model. Firstly, we introduce experiment settings, including datasets, implementation details, data sampling strategies, and evaluation metrics. Secondly, we perform quantitative experiments to compare our model with the state-of-the-art methods and then demonstrate the effectiveness of our proposed modules through the ablation study. Finally, we present qualitative experiments to visualize the efficiency of our proposed modules.

4.1   Experiment Settings

Datasets. There are two publicly available naturalistic pedestrian behavior datasets, namely Pedestrian Intention Estimation (PIE) [10] and Joint Attention in Autonomous Dataset (JAAD) [11,12].

PIE contains 6 h of driving videos under the first-person perspective which are shot by a monocular dashboard camera. The dataset provides annotations of 1842 pedestrians and traffic objects appearing at the same time. Pedestrian annotations contain the bounding box coordinates, intentions, and actions. Traffic objects consist of cyclists, vehicles, signs, traffic lights, crosswalks, and stations. Annotations of traffic objects contain the bounding box and the type. In addition, the ego-vehicle motion is also captured by the onboard diagnostics sensor. We follow [22] to split 1842 pedestrians into 880 for training, 243 for validation, and 719 for testing.

JAAD provides 300 video clips of 686 pedestrians which range from 5 to 15 s. The dataset also collects the video from the first-person perspective. There are annotations of pedestrians, including the bounding box, intention, and action labels. In addition, the dataset provides contextual tags which contain the traffic light state and the sign type. The ego-vehicle motion state is also accessible. We follow [22] to split 686 pedestrians into 188 for training, 32 for validation, and 126 for testing.

Implementation Details. The image patches in the input contain the pedestrian and the surrounding environment. We follow [10] to expand the bounding box to twice its original size and crop the image based on the enlarged bounding box. The VGG-16 [18] pretrained on ImageNet [16] is used as the backbone network to extract the feature map of the image patch. In Multi-Modal Conditional Generative Module, we sample $K = 20$ latent variables z in the prior network and recognition network. Inspired by [22], we expand 2 action labels (*walking* and *standing*) to 7 semantic action labels (*standing, waiting, going towards the crossing point, crossing, crossed and standing, crossed and waiting,* and *other walking*). And the pedestrian intention consists of two binary labels, namely *crossing* and *not-crossing*. At each frame, we detect the intention and action of the current frame and predict actions for the future $F = 5$ frames. To balance the loss of multiple tasks, we set $\lambda_1, \lambda_2, \lambda_3$, and λ_4 as 1. To train our model, we set the batch size as 64, learning rate as 10^{-5}, and adopt RMSprop optimizer with $\alpha = 0.9, \epsilon = 10^{-7}$. In the testing stage, we consider the inconsistent length of trajectories, so we input one trajectory at each testing time. We take the future action sequence which appears most in K predicted sequences as the final prediction result. We collect action prediction results at each time step of the trajectory to calculate mAP of action prediction.

Data Sampling Strategy. We adopt two data sampling strategies that previous works widely used, and we list the results under the two strategies: (1) the PIE strategy: all of the original data was utilized in the PIE [10]. In the training stage, we sample the trajectories which are truncated to the length T to make batch training feasible, where we set T as 15 and 30. (2) the time to event (TTE) strategy: During training, we follow [13,14] to sample trajectories with 1–2 s before the crossing event.

Evaluation Metrics. Pedestrian actions contain multiple labels, so we utilize mAP to evaluate the results of action prediction and detection. Pedestrian intention estimation is a binary classification problem, so we adopt accuracy, F1 score, precision, and area under curve (AUC) as evaluation metrics.

Table 1. Results of pedestrian action detection and prediction are compared with state-of-the-art methods on the PIE dataset using the PIE sampling strategy. '*' indicates the result produced by the re-implemented model. 'Det' represents pedestrian action detection and 'Pred' represents pedestrian action prediction.

Method	T = 15		T = 30	
	mAP(Det)	mAP(Pred)	mAP(Det)	mAP(Pred)
Yao *et al.* [22]	0.23*	0.22*	0.24	0.23*
Ours	**0.29**	**0.25**	**0.29**	**0.26**

Table 2. Results of pedestrian intention estimation are compared with state-of-the-art methods. 'PIE(PIE)', 'PIE(TTE)', 'JAAD(TTE)' in the first row indicate tested on the PIE dataset using the PIE sampling strategy, on the PIE dataset using the time to event (TTE) sampling strategy, on the JAAD dataset using the TTE sampling strategy, respectively.

Method	PIE(PIE)				PIE(TTE)				JAAD(TTE)			
	Acc	F1	Prec	AUC	Acc	F1	Prec	AUC	Acc	F1	Prec	AUC
I3D [1]	–	–	–	–	0.63	0.42	0.37	0.58	0.79	0.49	0.42	0.71
PIE [10]	0.79	0.87	0.86	0.73	–	–	–	–	–	–	–	–
SF-GRU [13]	–	–	–	–	0.87	0.78	0.74	0.85	0.83	0.59	0.50	0.79
MMH-PAP [14]	–	–	–	–	**0.89**	0.81	0.77	0.88	0.84	0.62	0.54	0.8
Yao *et al.* [22]	0.82	0.88	**0.94**	0.83	0.84	0.90	**0.96**	0.88	0.87	0.70	0.66	**0.92**
Ours	**0.85**	**0.91**	0.92	**0.87**	0.85	**0.91**	0.93	**0.89**	**0.88**	**0.74**	**0.67**	0.91

4.2   Performance Evaluation and Analysis

Quantitative Results on Pedestrian Action Detection and Prediction.
To verify the effectiveness of our model on the pedestrian action detection and prediction tasks, we compare our model with state-of-art models on the PIE dataset using the PIE sampling strategy in Table 1. Our model surpasses the previous works by a good margin on pedestrian action detection and prediction under $T = 15$ and $T = 30$ sampling length. Compared with [22], our model captures the spatial-temporal interaction between the target pedestrian with other traffic objects. The uncertainty of action prediction is also modeled in Conditional Multi-Modal Generating Model to sample multi-modal future action sequences. It is clear that our model incorporates the social aware context and the multimodality of pedestrian futures, which significantly improves the effectiveness of the pedestrian detection and prediction task.

Quantitative Results on Pedestrian Intention Estimation. In Table 2, we also conduct pedestrian intention estimation experiments on the PIE and JAAD datasets adopting different sampling strategies. On the PIE dataset under the PIE sampling strategy, our model surpasses the state-of-the-art method [22] and achieves the best 0.85 accuracy, 0.91 F1 score, and 0.87 AUC. F1 score

Table 3. Ablation study of Spatial-Temporal Heterogeneous Graph module on the PIE dataset using the PIE sampling strategy and sampling length $T = 30$. The third row shows the result without traffic objects, namely removing the STHG module in our model. Row 4–8 show the effect of using one type of traffic object alone. The last row is the result with all traffic object types.

Traffic objects					Action		Intention			
Traffic neighbor	Crosswalk	Traffic light	Traffic sign	Station	mAP(Det)	mAP(Pred)	Acc	F1	Prec	AUC
–	–	–	–	–	0.24	0.17	0.78	0.87	0.86	0.72
✓	–	–	–	–	0.25	0.21	0.80	0.88	0.88	0.75
–	✓	–	–	–	0.26	0.21	0.83	0.89	0.89	0.82
–	–	✓	–	–	0.27	0.22	0.84	0.90	0.90	0.82
–	–	–	✓	–	0.25	0.21	0.81	0.88	0.88	0.76
–	–	–	–	✓	0.26	0.21	0.81	0.88	0.88	0.77
✓	✓	✓	✓	✓	**0.29**	**0.26**	**0.85**	**0.91**	**0.92**	**0.87**

Table 4. Ablation study of the local graph in Saptial-Temporal Heterogeneous Graph on the PIE dataset using the PIE sampling strategy and sampling length $T = 30$.

Layers of local graphs	mAP(Det)	mAP(Pred)	Acc	F1	Prec	AUC
$L = 1$	0.27	0.23	0.84	0.90	0.91	0.86
$L = 2$	0.28	0.24	**0.85**	**0.91**	0.91	0.87
$L = 3$	**0.29**	**0.26**	**0.85**	**0.91**	**0.92**	**0.87**
$L = 4$	0.28	0.24	**0.85**	0.90	**0.92**	0.86

and AUC metrics are also increased by our model on the PIE dataset using the TTE strategy. On the JAAD dataset with the TTE sampling strategy, our model outperforms previous methods by 1.1% to 5.7% on the multiple metrics. Experiment results on intention estimation demonstrate the effectiveness of our multi-task model.

Ablation Study on Spatial-Temporal Heterogeneous Graph. To explore the impact of traffic objects belonging to different types, we conduct the ablation study on the STHG module. Results in Table 3 show that adding any type of traffic object enhances the effectiveness of the STHG module. Among them, crosswalks and traffic lights both make a significant performance boost. We consider that the crosswalk determines where pedestrians will cross, and the traffic light determines when pedestrians can cross the road. Results also demonstrate that our model can mimic human perception to observe the surrounding environment and extract useful information. In addition, we investigate the impact of different layers of the local graph in the STHG module in Table 4. As the number of layers increases, the effectiveness on multiple metrics will improve. And the performance achieves the best when the number of layers is three.

Table 5. Ablation study on Multi-Modal Conditional Generative Module (MMCGM) on the PIE dataset using PIE sampling strategy and sampling length $T - 30$.

MMCGM	Action		Intention			
	mAP(Det)	mAP(Pred)	Acc	F1	Prec	AUC
$K = 1$	0.28	0.24	**0.85**	**0.91**	0.91	**0.87**
$K = 20$	**0.29**	**0.26**	**0.85**	**0.91**	**0.92**	**0.87**

Ablation Study on Multi-modal Conditional Generating Module. As shown in Table 5, we conduct the ablation study on Multi-Modal Conditional Generating Module (MMCGM). The performance of the deterministic form ($K = 1$) on the action prediction is degraded. We consider that the deterministic form may not fully account for the inherent stochasticity of the future. We set the number of samples K as 20 to evaluate the effectiveness of the multi-modal action prediction. The result suggests that the multi-modal prediction makes a performance improvement from 0.24 mAP to 0.26 mAP. The introduction of multiple latent variables plays a critical role in approximating the one-to-many distribution to model the diversity of future actions.

4.3 Qualitative Example

Figure 3 shows the visualization of the STHG module and the predicted multi-modal action sequences. In the frame $t - 2 \sim t + 2$, the target pedestrian facing the crosswalk intends to cross the road. Observing green traffic lights for vehicles, he is waiting for crossing. At the time step t, STHG module pays attention to the traffic lights and the crosswalk. In addition, traffic neighbors in the urban scenario also deserve our attention. Traffic neighbor 1 and 2 waiting for crossing are highlighted, since the pedestrian usually observes surrounding neighbor pedestrians to infer the current scene information. STHG module also notes that there are many passing vehicles, which suggests that pedestrians should wait for vehicles to pass. At the bottom of the Fig. 3, we present multi-modal predictions of future action sequences. At the time step $t - 2$, our model (K = 20) outperforms the deterministic form (K = 1). It is clear that our multi-modal model considers the uncertainty of pedestrians and conducts probabilistic predictions. In the probabilistic outputs, we take the future action sequence which appears most as the final prediction result. The first action sequence appears 7 times and is highly likely to happen. We consider that the multi-modal model which can produce probabilistic inference is more preferable than the deterministic form in some application scenarios. At the time step t, both the multi-modal and deterministic model perform well, since the two models can capture the useful spatial-temporal interactions and infer the correct future actions from them.

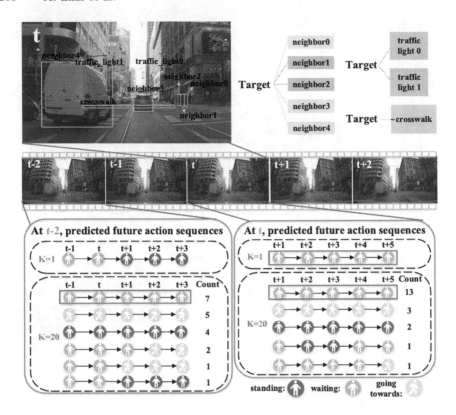

Fig. 3. At the top of the figure, we present the visualization of attention matrices of Spatial-Temporal Heterogeneous Graph Module at the t frame. The bottom of the figure is the visualization of the deterministic output ($K = 1$) and the probabilistic output ($K = 20$). Red bounding boxes indicate the correctly predicted action sequence. 'Count' represents the frequency the action sequence appeared. (Color figure online)

5   Conclusion

In this work, we propose a social aware multi-modal pedestrian behavior prediction network. In the Social Aware Encoder, we capture the spatial-temporal interaction between the target pedestrian and heterogeneous traffic objects. Multi-Modal Conditional Generative Module is designed to model the inherent uncertainties of future action sequences. Experiments demonstrate that our model outperforms previous methods on multiple tasks on the PIE and JAAD datasets using multiple metrics. The comprehensive ablation study and visualization also verify the effectiveness of our proposed modules.

References

1. Carreira, J., Zisserman, A.: Quo vadis, action recognition? A new model and the kinetics dataset. In: Proceedings of the IEEE Conference on Computer Vision and Pattern Recognition, pp. 6299–6308 (2017)
2. Chen, B., Li, D., He, Y.: Simultaneous prediction of pedestrian trajectory and actions based on context information iterative reasoning. In: 2021 IEEE/RSJ International Conference on Intelligent Robots and Systems (IROS), pp. 1007–1014. IEEE (2021)
3. Chen, L., et al.: SCA-CNN: spatial and channel-wise attention in convolutional networks for image captioning. In: Proceedings of the IEEE Conference on Computer Vision and Pattern Recognition, pp. 5659–5667 (2017)
4. Gu, J., Sun, C., Zhao, H.: DenseTNT: end-to-end trajectory prediction from dense goal sets. In: Proceedings of the IEEE/CVF International Conference on Computer Vision, pp. 15303–15312 (2021)
5. Huang, G., Liu, Z., van der Maaten, L., Weinberger, K.Q.: Densely connected convolutional networks. In: 2017 IEEE Conference on Computer Vision and Pattern Recognition, CVPR 2017, Honolulu, HI, USA, 21–26 July 2017, pp. 2261–2269. IEEE Computer Society (2017). https://doi.org/10.1109/CVPR.2017.243
6. Kotseruba, I., Rasouli, A., Tsotsos, J.K.: Do they want to cross? Understanding pedestrian intention for behavior prediction. In: 2020 IEEE Intelligent Vehicles Symposium (IV), pp. 1688–1693. IEEE (2020)
7. Liu, B., et al.: Spatiotemporal relationship reasoning for pedestrian intent prediction. IEEE Robot. Autom. Lett. 5(2), 3485–3492 (2020)
8. Mangalam, K., et al.: It is not the journey but the destination: endpoint conditioned trajectory prediction. In: Vedaldi, A., Bischof, H., Brox, T., Frahm, J.-M. (eds.) ECCV 2020. LNCS, vol. 12347, pp. 759–776. Springer, Cham (2020). https://doi.org/10.1007/978-3-030-58536-5_45
9. Piccoli, F., et al.: FuSSi-Net: fusion of spatio-temporal skeletons for intention prediction network. In: 2020 54th Asilomar Conference on Signals, Systems, and Computers, pp. 68–72. IEEE (2020)
10. Rasouli, A., Kotseruba, I., Kunic, T., Tsotsos, J.K.: PIE: a large-scale dataset and models for pedestrian intention estimation and trajectory prediction. In: Proceedings of the IEEE/CVF International Conference on Computer Vision, pp. 6262–6271 (2019)
11. Rasouli, A., Kotseruba, I., Tsotsos, J.K.: Are they going to cross? A benchmark dataset and baseline for pedestrian crosswalk behavior. In: ICCVW, pp. 206–213 (2017)
12. Rasouli, A., Kotseruba, I., Tsotsos, J.K.: It's not all about size: on the role of data properties in pedestrian detection. In: Leal-Taixé, L., Roth, S. (eds.) ECCV 2018. LNCS, vol. 11129, pp. 210–225. Springer, Cham (2019). https://doi.org/10.1007/978-3-030-11009-3_12
13. Rasouli, A., Kotseruba, I., Tsotsos, J.K.: Pedestrian action anticipation using contextual feature fusion in stacked RNNs. arXiv preprint arXiv:2005.06582 (2020)
14. Rasouli, A., Yau, T., Rohani, M., Luo, J.: Multi-modal hybrid architecture for pedestrian action prediction. In: 2022 IEEE Intelligent Vehicles Symposium (IV), pp. 91–97. IEEE (2022)
15. Redmon, J., Farhadi, A.: YOLOv3: an incremental improvement. CoRR abs/1804.02767 (2018). http://arxiv.org/abs/1804.02767

16. Russakovsky, O., et al.: ImageNet large scale visual recognition challenge. Int. J. Comput. Vis. **115**(3), 211–252 (2015). https://doi.org/10.1007/s11263-015-0816-y
17. Saleh, K., Hossny, M., Nahavandi, S.: Real-time intent prediction of pedestrians for autonomous ground vehicles via spatio-temporal DenseNet. In: 2019 International Conference on Robotics and Automation (ICRA), pp. 9704–9710. IEEE (2019)
18. Simonyan, K., Zisserman, A.: Very deep convolutional networks for large-scale image recognition. arXiv preprint arXiv:1409.1556 (2014)
19. Sohn, K., Lee, H., Yan, X.: Learning structured output representation using deep conditional generative models. In: Advances in Neural Information Processing Systems, vol. 28 (2015)
20. Wang, C., Wang, Y., Xu, M., Crandall, D.: Stepwise goal-driven networks for trajectory prediction. IEEE Robot. Autom. Lett. **7**(2), 2716–2723 (2022)
21. Wojke, N., Bewley, A., Paulus, D.: Simple online and realtime tracking with a deep association metric. In: 2017 IEEE International Conference on Image Processing, ICIP 2017, Beijing, China, 17–20 September 2017, pp. 3645–3649. IEEE (2017). https://doi.org/10.1109/ICIP.2017.8296962
22. Yao, Y., Atkins, E., Roberson, M.J., Vasudevan, R., Du, X.: Coupling intent and action for pedestrian crossing behavior prediction. arXiv preprint arXiv:2105.04133 (2021)
23. Yau, T., Malekmohammadi, S., Rasouli, A., Lakner, P., Rohani, M., Luo, J.: Graph-SIM: a graph-based spatiotemporal interaction modelling for pedestrian action prediction. In: 2021 IEEE International Conference on Robotics and Automation (ICRA), pp. 8580–8586. IEEE (2021)
24. Zhao, H., et al.: TNT: target-driven trajectory prediction. arXiv preprint arXiv:2008.08294 (2020)

Action Representing by Constrained Conditional Mutual Information

Haoyuan Gao[1], Yifaan Zhang[1,3(✉)], Linhui Sun[1,2], and Jian Cheng[1,3]

[1] Institute of Automation, Chinese Academy of Sciences, Beijing, China
{gaohaoyuan2019,sunlinhui2018}@ia.ac.cn,
{yfzhang,jcheng}@nlpr.ia.ac.cn
[2] School of Artificial Intelligence, University of Chinese Academy of Sciences, Beijing, China
[3] AIRIA, Guangzhou, China

Abstract. Contrastive learning achieves a remarkable performance for representation learning by constructing the InfoNCE loss function. It enables learned representations to describe the invariance in data transformation without labels. Contrastive learning also been employed in self-supervised learning of action recognition. However, this kind of method fails to introduce assumptions according to human knowledge about the prior distribution of representations in the training process. For solving this problem, this paper proposes a self-supervised learning framework, which can achieve different self-supervised learning methods by choosing different assumptions about the prior distribution of representations, while still learning the description of invariance in data transformation as contrastive learning. This framework minimizes the CCMI (Constrained Conditional Mutual Information) loss function, which represents the conditional mutual information between input augmented samples of the same sample and the output representations of the encoder while the prior distribution of representations is constrained. By theoretical analysis of the framework, it is proved that traditional contrastive learning by InfoNCE is a special case without human knowledge constraint of this framework. The Gaussian Mixture Model on Unit Hyper-sphere is chosen as the representation prior distribution to achieve the self-supervised method called CoMInG. Compared with the existing methods, the performance of the learned representation by this method in the downstream task of action recognition is significantly improved.

1 Introduction

Action recognition is widely used in video surveillance, human-computer interaction, and video understanding. The methods of action recognition include RGB image-based, depth image-based and skeleton-based methods. Recently, skeleton-based methods have attracted increasing attention due to their lower computation consumption and higher robustness against viewpoint variations and noisy backgrounds than the methods using RGB images and depth images [22]. Various skeleton-based works were proposed and achieved significant performance in recent years [6] [29] [32]. However, most of them utilize the supervised learning paradigm to learn action representations, which require massive labeled samples for training. This leads to some problems, including the high cost of labeling and the risk of mislabeling due to the high inter-class similarity of the actions. In addition, massive valuable information for learning implied

© The Author(s), under exclusive license to Springer Nature Switzerland AG 2023
L. Wang et al. (Eds.): ACCV 2022, LNCS 13844, pp. 291–306, 2023.
https://doi.org/10.1007/978-3-031-26316-3_18

in unlabeled data is not utilized in the training. Therefore, self-supervised learning of action representation attracted increasing attention.

Self-supervised learning is a type of unsupervised representation learning that creates learning target without human annotation to train the encoder network to obtain effective representation for downstream tasks. Only a few works focus on self-supervised learning of action recognition [21] [25] [33]. The best performer [21] of them used contrastive learning for action representing, constructing the InfoNCE loss function to enable learned representations to describe the invariance in data transformation. However, such contrastive learning methods fail to introduce assumptions according to human knowledge about the prior distribution of representations in the training process.

For solving this problem, this paper proposes a self-supervised learning framework, which can achieve different self-supervised learning methods by choosing different assumptions about the prior distribution of representations, while still learning the description of invariance in data transformation as contrastive learning. This framework minimizes the CCMI (Constrained Conditional Mutual Information) loss function, which represents the conditional mutual information between input augmented samples of the same sample and the output representations of the encoder while the prior distribution of representations is constrained. By theoretical analysis of the framework, it is proved that traditional contrastive learning by InfoNCE is a special case without human knowledge constraint of this framework. In this paper, the Gaussian Mixture Model on Unit Hyper-sphere is chosen as the representation prior distribution to introduce a self-supervised method named CoMInG (**Co**nstrained Conditional **M**utual **In**formation Minimizing with **G**aussian Mixture Model on Unit Hyper-sphere as Representation Prior), and employ it for action representing of skeleton-based action recognition.

Our contributions can be summarized as follows.

- This paper proposes a self-supervised learning framework by minimizing CCMI, which can achieve different self-supervised learning methods by choosing different assumptions about the prior distribution of representations, while still learning the description of invariance in data transformation as contrastive learning.
- By theoretical analysis of the framework, it is proved that traditional contrastive learning by InfoNCE is a special case without human knowledge constraint of this framework. This conclusion enhances the theoretical credibility of the proposed framework.
- This paper proposes a self-supervised learning method CoMInG, by assuming the prior of representations as a Gaussian Mixture Model on Unit Hyper-sphere, and comprehensively evaluates the effectiveness of CoMInG on three public datasets: NTU RGB+D 60, NTU RGB+D 120 and SBU datasets. Under the linear evaluation protocol, the proposed CoMInG achieves the best performance than existing self-supervised learning methods.

The rest of this paper is organized as follows: Sect. 2 introduces the previous works related to the proposed work. Section 3 introduces the details of CCMI loss and CoMInG method. Section 4 proves the relationship between InfoNCE and CCMI. Section 5 compares CoMInG with existing baselines. Ablation studies are also provided in Sect. 6. The conclusion of the proposed work is shown in Sect. 6.

2 Related Work

2.1 Self-Supervised Learning

Self-supervised learning aims to learn an effective representation from unlabeled data by using a self-supervised proxy task to train an encoder network. The learned representation can be transferred and beneficial to the downstream tasks [12].

Self-supervised learning methods include generative-based methods, contrastive-based methods, and clustering-based methods [18]. Generative-based methods employ an encoder-decoder structure or generative adversarial network to learn the representation. For example, [14] proposed a method to use an autoencoder as a generator with a discriminator for the automatic colorization of images. For the clustering-based methods, [2] presented DeepCluster, which iteratively groups the features with a standard clustering algorithm, K-means, and uses the subsequent assignments as supervision to update the weights of the encoder network. [1] proposed a method obtained by maximizing the information between labels and input data indices, using a fast variant of the Sinkhorn-Knopp algorithm. [3] proposed a framework both using the contrastive-based and clustering-based method, SwAV, which employs the Sinkhorn-Knopp algorithm to cluster the data and uses the cluster codeID of the other augmentations fromof the same image to guide the representations. PCL combined MoCo with an off-line k-means clustering process to propose the ProtoNCE loss [15]. Most of the state-of-the-art self-supervised learning methods are contrastive-based methods. Contrastive multiview coding (CMC) enforced the different views of the same image close to each other [26]. Momentum contrastive (MoCo) improved contrastive learning by introducing a momentum encoder and a queue-based memory bank [10]. Chen et al. proposed SimCLR, which adds a projector network behind the encoder and redesigns a stronger augmentation strategy for contrastive learning [4]. [8] introduced BYOL, which relies on two neural networks, referred to as online network and target network. It trains the online network to predict the target network representation of the same image under a different augmented view and updates the target network with a slow-moving average of the online network. Barlow Twins maximized the similarity between the correlation matrix of the output representations and the identity matrix, avoiding the complete collapse that is likely to appear in contrastive learning [31].

2.2 Action Recognition

Traditional skeleton-based methods recognized the pattern of action by designing hand-crafted descriptors, such as the method proposed by Oreifej et al. which learns features by a modified histogram of oriented gradients (HOG) algorithm [20]. Due to the significant development of deep learning, numerous methods for supervised skeleton-based action recognition by the deep neural network were proposed, such as Directed Graph

Neural Network (DGNN) proposed by Shi et al. [24], DeCoupling Graph Convolutional Networks (DC-GCN) proposed by Cheng et al. [5] and Spatial-temporal Graph Convolutional Networks (STGCN) proposed by Yan et al. [29].

Only a few self-supervised methods were proposed in recent years. [33] proposed a method by both using an encoder-decoder structure and a generative adversarial structure to reconstruct a masked input sequence. [25] forced the encoder to learn the action representation by using an autoencoder to re-generate the skeleton sequence, and additionally proposed a decoder-weakening mechanism by fixing the decoder weights or decoder states. [16] proposed to integrate multiple self-supervised tasks, that are motion prediction, jigsaw puzzle recognition, and contrastive learning, to learn more general representations. [21] maximized the similarity between augmented instances of the same input skeleton sequence by a queue-based memory bank and momentum encoder. [28] proposed a self-supervised framework, which not only reconstructs sequence by an autoencoder but also regards the K-means clustering results of action representations as pseudo labels to train the encoder. Contrastive learning by InfoNCE has best performance in self-supervised action recoginition [21], but it cannot introduce assumptions according to human knowledge about the prior distribution of representations in the training process.

3   Methodology

3.1   Traditional Contrastive Learning by Minimizing InfoNCE

InfoNCE is a widely used loss function in self-supervised learning, which enables learned representations to describe the invariance in data transformation, that is:

$$L_{NCE} = -\frac{1}{I} \sum_{i=1}^{I} \log \frac{\exp\left(sim(z^{(i,1)}, z^{(i,2)})\right)}{\sum_{j=1}^{I} \exp\left(sim(z^{(i,1)}, z^{(j,2)})\right)} \tag{1}$$

I is the size of the dataset. $z^{(i,1)}$ and $z^{(i,2)}$ are respectively the representations of two randomly augmented version of ith sample extracted by the encoder network [4]. $sim(a, b)$ measures the similarity between two variables. The self-supervised methods using InfoNCE are called contrastive learning methods. Although contrastive learning achieves a remarkable performance for representation learning in multiple applications, it fails to introduce assumptions according to human knowledge about the prior distribution of representations in the training process. Therefore, this paper proposes the CCMI loss to solve this problem, and we will prove InfoNCE is a special case of CCMI in Sect. 4.

3.2   A Learning Framework by Minimizing CCMI

This subsection proposes a simple framework for self-supervised learning by minimizing loss function called CCMI (Constrained Conditional Mutual Information). This framework can introduce various different self-supervised learning methods by different choices of representation prior according to human knowledge, while still learning the description of invariance in data transformation as contrastive learning.

Let us consider one dataset $X = \{x^i\}_{i=1}^{I}$ consisting of I i.i.d. samples of some continuous variable x, which is a skeleton sequence in our task. We assume that each sample is generated by a random process including an unobserved continuous random variable z. In addition, a variable v denotes the augmented sample for the sample x after some random data augmentation. Specifically, the generating process of x and v can be divided into three steps: (1) a value z^i is sampled from distribution $p_{\theta^*}(z)$; (2) a value x^i is sampled from distribution $p_{\theta^*}(x|z)$; (3) a value v^i is sampled from distribution $p_{\theta^*}(v|x,z)$. We assume that the $p_{\theta^*}(z)$, $p_{\theta^*}(x|z)$ and $p_{\theta^*}(v|x,z)$ come from parametric families of distributions $p_\theta(z)$, $p_\theta(x|z)$ and $p_\theta(v|x,z)$. Based on the used data augmentation strategy, we can reasonably assume that for any x^i and x^j ($i \neq j$), $v^i = v^j$ will never happen if v^i and v^j are respectively sampled from $p_{\theta^*}(v|x^i)$ and $p_{\theta^*}(v|x^j)$. Therefore, if $g(v)$ is a function that can find the only sample x corresponding to the augmented sample v, the conditional distribution $p_{\theta^*}(x|v)$ is $p_{\theta^*}(x|v) = 1$ if $x = g(v)$ $else$ 0. Then we can obtain that $p_{\theta^*}(z|v) = \int p_{\theta^*}(x|v)p_{\theta^*}(z|x,v)dx = p_{\theta^*}(z|x,v)$.

Because we wish that the information of the representation vector z is only affected by the semantic information in the sample x that does not change with the data augmentation, the learning target is that when x is known, v and z have **conditional independence**, that is, $p_\theta(v|x,z) = p_\theta(v|x)$. Therefore, the learning target we set is:

$$\theta^* = \arg\min_\theta KL(p_\theta(v|x,z)\|p_\theta(v|x))$$

$$= \arg\min_\theta \mathop{E}_{p_\theta(z,x,v)} [\log \frac{p_\theta(v|x,z)}{p_\theta(v|x)}]$$

$$= \arg\min_\theta \mathop{E}_{p_\theta(z,x,v)} [\log \frac{p_\theta(v|x,z)p_\theta(z,x)}{p_\theta(v|x)p_\theta(z,x)}]$$

$$= \arg\min_\theta \mathop{E}_{p_\theta(z,x,v)} [\log \frac{p_\theta(z|x,v)p_\theta(v|x)p_\theta(x)}{p_\theta(v|x)p_\theta(z|x)p(x)}]$$

$$= \arg\min_\theta \mathop{E}_{p_\theta(z,x,v)} [\log \frac{p_\theta(z|v)p_\theta(v|x)}{p_\theta(v|x)p_\theta(z|x)}]$$

$$= \arg\min_\theta \int p_\theta(x)[\iint p_\theta(z,v|x) \log \frac{p_\theta(z,v|x)}{p_\theta(v|x)p_\theta(z|x)} dzdv]dx$$

$$= \arg\min_\theta I(Z;V|X) \tag{2}$$

where $I(Z;V|X)$ is the conditional mutual information between z and v, while x is known. When $I(Z;V|X)$ reaches its minimum value 0, $p_\theta(v|x,z) = p_\theta(v|x)$.

However, only minimizing $I(Z,V|X)$ can easily achieve the trivial solution, that is, whatever the value of v and x are, $p_\theta(z|v,x)$ is the same. A simple way to solve this problem is to inject our human knowledge into the hypothesis of prior distribution $p_\theta(z)$. We can choose a distribution $q_\phi(z)$ based on the human knowledge, and then set the optimization problem as:

$$\theta^*, \phi^* = \arg\min_{\theta,\phi} I(Z;V|X) \quad s.t. \ KL(p_\theta(z)\|q_\phi(z)) = 0 \tag{3}$$

Thanks to the constraint $KL(p_\theta(z)\|q_\phi(z)) = 0$, the trivial solution is avoided. For transform the optimization problem to a loss function for encoder training, we need to reform it. The Monte-Carlo Estimate [9] shows that:

$$\underset{p(x)}{E}[f(x)] = \int p(x)f(x)dx \approx \frac{1}{N}\sum_{i=1}^{N}f(x^i) \tag{4}$$

where, x^i is sampled from $p(x)$.

Based on the Lagrangian multiplier method and Monte-Carlo Estimate, the above optimization problem is equal to:

$$\theta^*, \phi^*, \lambda^* = \underset{\theta,\phi,\lambda}{\arg\min}\, I(Z;V|X) + \lambda \underset{p_\theta(x,v|z)}{E}[KL(p_\theta(z)\|q_\phi(z))]$$

$$= \underset{\theta,\phi,\lambda}{\arg\min} \iiint p_\theta(x)p_\theta(v|x)p_\theta(z|v)[\log\frac{p_\theta(z|v)}{p_\theta(z|x)} + \lambda\log\frac{p_\theta(z)}{q_\phi(z)}]dzdvdx$$

$$= \underset{\theta,\phi,\lambda}{\arg\min} \iiint p_\theta(x)p_\theta(v|x)p_\theta(z|v)[\log p_\theta(z|v) - \log p_\theta(z|x)$$

$$+ \lambda\log p_\theta(z) - \lambda\log q_\phi(z)]$$

$$= \underset{\theta,\phi,\lambda}{\arg\min} \iiint p_\theta(x)p_\theta(v|x)p_\theta(z|v)[\log p_\theta(z|v) - \log\left(\int p_\theta(v|x)p_\theta(z|v)dv\right)$$

$$+ \lambda\log\left(\iint p_\theta(x)p_\theta(v|x)p_\theta(z|v)dvdx\right) - \lambda\log q_\phi(z)]dzdvdx$$

$$= \underset{\theta,\phi,\lambda}{\arg\min} \frac{1}{IML}\sum_{i=1}^{I}\sum_{m=1}^{M}\sum_{l=1}^{L}[-\log\sum_{n=1}^{N}p_\theta(z^{(i,m,l)}|v^{(i,n)})$$

$$+ \lambda\log\sum_{j=1}^{I}\sum_{n=1}^{N}p_\theta(z^{(i,m,l)}|v^{(j,n)}) - \lambda\log q_\phi(z^{(i,m,l)}) + \frac{1}{L}H(p_\theta(z|v^{(i,m)})) - \log N]$$

$$\tag{5}$$

This loss function is called as CCMI (Constrained Conditional Mutual Information), where the $x^{(i)}$, $v^{(i,m)}$ and $z^{(i,m,l)}$ are sampled from $p_\theta(x)$, $p_\theta(v|x^{(i)})$ and $p_\theta(z|v^{(i,m)})$ respectively. I denotes the size of the dataset. Both M and N denote the number of times that sample v. L denotes the number of times that sample z. $H(p_\theta(z|v))$ denotes the entropy of $p_\theta(z|v)$. The illustration of the framework by minimizing CCMI loss is shown in Fig. 1.

3.3 A Self-Supervised Method: CoMInG

In this subsection, we choose the Gaussian Mixture Model on Unit Hyper-sphere as the representation prior of CCMI to introduce a novel self-supervised method named CoM-InG (**Co**nstrained Conditional **Mu**tual **I**nformation **Mi**nimizing with **G**aussian Mixture Model on Unit Hyper-sphere as Representation Prior).

For the training of encoder, the distribution $p_\theta(z|v)$ and $q_\phi(z)$ need to be determined. Following the paper of Variational Auto-Encoder [13], the $p_\theta(z|v)$ is set as:

$$p_\theta(z|v) = N(f_\eta(v), I) \tag{6}$$

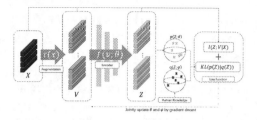

Fig. 1. Illustration of the proposed method.

where the f_η denotes our encoder network with the parameter η. The $N(\mu, S)$ denotes a multivariate Gaussian distribution with μ as the mean vector and S as the covariance matrix. In addition, we set the covariance matrix of the Gaussian distribution as identity matrix I. The reason of this setting is, in CCMI loss, the fourth part, the entropy of $p_\theta(z|v)$, needs to be minimized, which means the variance of Gaussian distribution needs to be minimized. For this target, we can primarily set the variance of Gaussian distribution as a small constant and we don't need to optimize this part in the training process.

The $q_\phi(z)$ is determined as a GMM (Gaussian Mixture Model) on Unit Hyper-sphere, due to the category information included in representation. We assume that each category match one Gaussian distribution in the GMM. We regard z as a sampling from one of the categorics c, that is:

$$q_\phi(z) = \sum_{c \in C} q(c)q(z|c) = \frac{1}{K}\sum_{k=1}^{K} N(w^k, I)$$

$$= \frac{1}{K}\sum_{k=1}^{K}\frac{1}{2\pi^{\frac{D}{2}}}\exp\left(-\frac{(z-w^k)^{\mathrm{T}}(z-w^k)}{2}\right) \tag{7}$$

We assume that for any k, $q(c^k) = 1/K$. I denotes the identity matrix, and D denotes the dimension of z. $W = \{w^k\}_{k=1}^{K}$ are K vectors that denote the mean vectors of the K Gaussian distributions. These mean vectors need to be optimized, which are same as the parameter η of the encoder network.

However, previous works prove that for contrastive learning, employing cosine similarity is better than using Euclidean distance in experiments [4], so we replace Euclidean distance in $q_\phi(z)$ with cosine similarity, that is:

$$q_\phi(z) = \sum_{c \in C} q(c)q(z|c) = \frac{1}{K}\sum_{k=1}^{K}\frac{1}{2\pi^{\frac{D}{2}}}\exp\left(\frac{sim(z, w^k)}{2}\right) \tag{8}$$

where $sim(z, w^k) = \frac{z^{\mathrm{T}}w^k}{\|z\|\|w^k\|}$.

However, due to the replacement, $\int q(z|c)dc = 1$ is no longer available. Therefore, $q(z|c)$ need to be normalized, that is:

$$q_\phi(z) = \sum_{c \in C} q(c)q(z|c) = \frac{1}{K}\sum_{k=1}^{K}\frac{1}{2\pi^{\frac{D}{2}}S}\exp\left(\frac{sim(z, w^k)}{2}\right) \tag{9}$$

where $S = \int \frac{1}{2\pi^{\frac{D}{2}}} \exp\left(\frac{sim(z,w^k)}{2}\right)dz$. This makes $\int q(z|c)dc = 1$ available again. Here Gaussian Mixture Model is transformed to the Gaussian Mixture Model on Unit Hyper-sphere. Other distributions can be chosen as $q_\phi(z)$ based on human knowledge, such as uniform distribution, exponential distribution, etc. According to the conclusion in f-gan [19], variational inference can be utilized to introduce all kinds of distribution as the target of prior.

According to the choice of $p_\theta(z|v)$ and $q_\phi(z)$, the loss function is:

$$
\begin{aligned}
L_{CCMI} &= \frac{1}{IML}\sum_{i=1}^{I}\sum_{m=1}^{M}\sum_{l=1}^{L}[-\log\sum_{n=1}^{N}p_\theta(z^{(i,m,l)}|v^{(i,n)}) \\
&+ \lambda\log\sum_{j=1}^{I}\sum_{n=1}^{N}p_\theta(z^{(i,m,l)}|v^{(j,n)}) \\
&- \lambda\log q_\phi(z^{(i,m,l)}) + \frac{1}{L}H(p_\theta(z|v^{(i,m)})) - \log N] \\
&= \frac{1}{IML}\sum_{i=1}^{I}\sum_{m=1}^{M}\sum_{l=1}^{L}[-\log\sum_{n=1}^{N}\exp\left(sim(z^{(i,m,l)}, f_\eta(v^{(i,n)}))\right) \\
&+ \lambda\log\sum_{j=1}^{I}\sum_{n=1}^{N}\exp\left(sim(z^{(i,m,l)}, f_\eta(v^{(j,n)}))\right) \\
&- \lambda\log\sum_{k=1}^{K}\frac{1}{S}\exp\left(sim(z^{(i,m,l)}, w^{(k)})\right) + constant]
\end{aligned}
\tag{10}
$$

where D is the dimension of representations.

3.4   Details of Training Process

Following previous contrastive learning works [4] [10] and for fair comparison, in our experiments, we set $M = N = 1$. According to the result in the paper of Variational Auto-Encoder [13], we set $L = 1$. In addition, according to the ablation study, we set $\lambda = 1$. Based on this setting, when the constant term is omitted, the CCMI loss function becomes:

$$
\begin{aligned}
L_{CCMI} &= \frac{1}{I}\sum_{i=1}^{I}[-sim(z^{(i,1)}, z^{(i,2)}) + \log\sum_{j=1}^{I}\exp\left(sim(z^{(i,1)}, z^{(j,2)})\right) \\
&- \log\sum_{k=1}^{K}\exp\left(sim(z^{(i,1)}, w^{(k)})\right)]
\end{aligned}
\tag{11}
$$

where $z^{(i,1)}$ is the representation vector of the augmented sample obtained by x after the first random data augmentation through the encoder, and $z^{(i,2)}$ is the representation vector of the augmented sample obtained by x after the second random data augmentation. Then the first and second parts of CCMI can be optimized directly by gradient descent, but the third part of CCMI loss cannot. Thus we minimize the third part by EM algorithm.

The optimization problem of the third part of CCMI loss is:

$$\eta^*, \phi^* = \arg\min_{\eta,\phi} \log q_\phi(z) = \arg\min_{\eta,\phi} \log q_\phi(f_\eta(v)) \tag{12}$$

It is hard to optimize this function directly, so we use a surrogate function to higher-bound it:

$$\log q_\phi(f_\eta(v)) = \log \sum_{c\in C} q_\phi(f_\eta(v), c) = \log \sum_{c\in C} q_\phi(c|f_\eta(v)) \frac{q_\phi(f_\eta(v), c)}{q_\phi(c|f_\eta(v))}$$

$$\leq \sum_{c\in C} q_\phi(c|f_\eta(v)) \log \frac{q_\phi(f_\eta(v), c)}{q_\phi(c|f_\eta(v))} \tag{13}$$

Then the E-step and M-step can be obtained:

- **E-Step**

$$q_{\phi^{(t)}}(c|f_{\eta^{(t)}}(v)) = q_{\phi^{(t)}}(c, f_{\eta^{(t)}}(v))/q_{\phi^{(t)}}(f_{\eta^{(t)}}(v)) \tag{14}$$

- **M-Step**

$$\eta^{(t+1)}, \phi^{(t+1)} = \arg\min_{\eta,\phi} \sum_{c\in C} q_{\phi^{(t)}}(c|f_{\eta^{(t)}}(v)) \log\left(q_\phi(f_\eta(v)|c)\right) \tag{15}$$

The pseudo-code of complete training process is shown in Algorithm 1. The parameter τ denotes a temperature parameter, which is widely used in previous contrastive learning methods to adjust the loss function [4]. The E-step is the calculation of γ^{ik} in the training process, and the M-step is the gradient descent in the training process.

4   Relationship Between CCMI Loss and InfoNCE Loss

InfoNCE is a widely used loss function in previous self-supervised methods such as CPC, SimCLR and MoCo. This section proves that InfoNCE is a special case of CCMI.

In CoMInG, we set $p_\theta(z|v)$ as Gaussian distribution and set $\lambda = 1$, $M = N = 1$ and $L = 1$ for CCMI. If we change the choice of $q_\phi(z)$, and set that $q_\phi(z) = p_\theta(z|v)$, which means we have no human knowledge about prior distribution and set no limitation for it, then the CCMI loss function is:

$$L_{CCMI} = \frac{1}{IML} \sum_{i=1}^{I} \sum_{m=1}^{M} \sum_{l=1}^{L} [-\log \sum_{n=1}^{N} P_\theta(z^{(i,m,l)}|v^{(i,n)})$$

$$+ \lambda \log \sum_{j=1}^{I} \sum_{n=1}^{N} P_\theta(z^{(i,m,l)}|v^{(j,n)}) - \lambda \log q_\phi(z^{(i,m,l)})$$

$$+ \frac{1}{L} H(p_\theta(z|v^{(i,m)})) - \log N]$$

Algorithm 1. CoMInG

Input: initialized encoder parameters η, initialized mean vectors for Gaussian Mixture Model $W = \{w^k\}_{k=1}^K$, batch size B, learning rate α, temperature parameter τ, augmentation strategy T, similarity measurement $sim(a,b)$

1: **for** all minibatch in one epoch **do**
2:     Sample augmentation t^1 and t^2 from T
3:     **for** sampled minibatch $\{x^i\}_{i=1}^B$ **do**
4:         **for** sampled minibatch $\{x^j\}_{j=1}^B$ **do**
5:             $v^{(i,1)} = t^1(x^{(i)}), v^{(i,2)} = t^2(x^{(i)}), v^{(j,2)} = t^2(x^{(j)})$
6:             $z^{(i,1)} = f_\eta(v^{(i,1)}), z^{(i,2)} = f_\eta(v^{(i,2)}), z^{(j,2)} = f_\eta(v^{(j,2)})$
7:             **for** $k = 1 : K$ **do**
8:                 $\gamma^{ik} = \frac{\exp\left(sim(z^{(i,1)}, w^{(k)})\right)}{\sum_{l=1}^K \exp\left(sim(z^{(i,2)}, w^{(l)})\right)}$
9:             **end for**
10:        **end for**
11:    **end for**
12:    $L_{ccmi} = -\frac{1}{B}[\sum_{i=1}^B sim(z^{(i,1)}, z^{(i,2)})/\tau$
13:        $+ \log \sum_{j=1}^B \exp\left(sim(z^{(i,1)}, z^{(j,2)})/\tau\right)]$
14:        $- \log \sum_{k=1}^K \gamma^{ik} \exp\left(sim(z^{(i,1)}, w^{(k)})/\tau\right)]$
15:    $W = W - \alpha \frac{\partial L_{ccmi}}{\partial W}, \eta = \eta - \alpha \frac{\partial L_{ccmi}}{\partial \eta}$
16: **end for**
17: **return** encoder parameterss η

$$
= \frac{1}{IML} \sum_{i=1}^I \sum_{m=1}^M \sum_{l=1}^L [-\log \sum_{n=1}^N P_\theta(z^{(i,m,l)}|v^{(i,n)})
$$

$$
+ \log \sum_{j=1}^I \sum_{n=1}^N P_\theta(z^{(i,m,l)}|v^{(j,n)}) - \frac{1}{L} \int p_\theta(z|v^{(i,m)}) \log p_\theta(z|v^{(i,m)}) dz
$$

$$
+ \frac{1}{L} H(p_\theta(z|v^{(i,m)})) - \log N]
$$

$$
= \frac{1}{IML} \sum_{i=1}^I \sum_{m=1}^M \sum_{l=1}^L [-\log \sum_{n=1}^N P_\theta(z^{(i,m,l)}|v^{(i,n)})
$$

$$
+ \log \sum_{j=1}^I \sum_{n=1}^N P_\theta(z^{(i,m,l)}|v^{(j,n)}) + constant]
$$

$$
= \frac{1}{IML} \sum_{i=1}^I \sum_{m=1}^M \sum_{l=1}^L [-\log \sum_{n=1}^N \exp\left(sim(z^{(i,m,l)}, f_\eta(v^{(i,n)}))\right)
$$

$$
+ \log \sum_{j=1}^I \sum_{n=1}^N \exp\left(sim(z^{(i,m,l)}, f_\eta(v^{(j,n)}))\right) + constant]
$$

$$
= -\frac{1}{I} \sum_{i=1}^I \log \frac{\exp\left(sim(z^{(i,1)}, z^{(i,2)})\right)}{\sum_{j=1}^I \exp\left(sim(z^{(i,1)}, z^{(j,2)})\right)} + constant \tag{16}
$$

This loss function equals to InfoNCE, so that InfoNCE is a special case of CCMI when we set $\lambda = 1$, $M = N = 1$, $L = 1$, $p_\theta(z|v) = N(f_\eta(v), I)$ and $q_\phi(z) = p_\theta(z|v)$. This conclusion reveals that the essence of minimizing InfoNCE is to minimize the constrained conditional mutual information without injecting human knowledge to constrain the representation prior distribution.

5   Experiments

5.1   The Setting of Experiments

For evaluation, we conduct our experiments on commonly used three datasets: NTU60 dataset [23] (56578 samples, 60 categories), NTU120 dataset [17] (113945 samples, 120 categories) and SBU dataset [30] (282 samples, 8 categories).

In the experiments, the sequence length is set to 150, 150, and 40 for NTU60, NTU120, and SBU, respectively. The coordinate of the middle spine joint is subtracted by the coordinates of all joints for normalizing the skeleton sequences. Our encoder network adopts the LSTM with 512 hidden units.

In self-supervised pre-training, the encoder is trained by CoMInG. The batch size is 32, 32, and 128 for NTU60, NTU120, and SBU respectively. The network is trained by the SGD optimizer. The weight decay and momentum are set to 1e-4 and 0.9, respectively. We run the pre-training process for 60 epochs and the learning rate is multiplied by 0.1 per 30 epochs with 0.01 as the initialization. According to the results of ablation studies, we set the temperature τ as 0.06. The number of Gaussian of representation prior to 120, 150, and 30 for NTU60, NTU120, and SBU datasets respectively. The random data augmentations used in pre-training are 'Axis2Zero' and 'Shear'. For each joint in each sample, 'Axis2Zero' randomly chooses one of the axes of the 3D coordinates of the joint and changes it to zero, and the 'Shear' augmentation replaces each joint in a fixed direction [21].

After self-supervised pre-training, we use the linear evaluation to test our method. Specifically, the pre-trained encoder network by self-supervised learning is attached to a linear classifier, and we train the linear classifier for 90 epochs using skeleton sequences and labels in the training set while the parameter of the encoder is frozen. No augmentation is adopted in the training of the linear classifier. Then the Top-1 accuracy on the testing set is used to evaluate the effectiveness of the representations. The optimizer for training is stochastic gradient descent with a Nesterov momentum set as 0.9. The initialization of the learning rate is 1, and the learning rate decays at 15, 35, 60, and 75 epochs by 0.5.

5.2   Comparison with Existing Methods

Table 1, Table 2 and Table 3 compare the results of our approach with supervised approaches (using RNN as the backbone) and previous self-supervised approaches on SBU, NTU60 and NTU120 datasets in the linear evaluation setting respectively. The compared self-supervised approaches include all state-of-the-art approaches in this domain. "⋆" represents that we use the code shared by authors of original papers to obtain

Table 1. Comparison with supervised, and self-supervised methods on SBU dataset. Bold numbers refer to the best performers.

ID	Method	Fold					
		1	2	3	4	5	Avg
	Supervised						
1	RNN	40.0	42.3	26.8	27.8	35.4	34.5
2	GRU	40.0	40.4	28.6	33.3	40.0	36.5
3	LSTM	49.1	53.2	37.5	42.0	53.8	47.1
	Self-supervised						
4	*P&C FW [25]	16.4	15.4	21.4	27.8	15.6	19.3
5	*PCRP [28]	16.4	36.5	21.4	24.1	29.7	25.6
6	ASCAL [21]	52.7	46.2	41.1	31.5	41.5	42.6
7	⋆InfoNCE	61.8	53.8	46.4	50.0	53.1	53.0
8	Ours	**65.5**	**63.5**	**51.8**	**61.1**	**53.1**	**59.0**

these results, because in the original papers authors don't show the performance of the methods on these datasets. "⋆" represents that this method is coded by us, using the setting of our method. The results in Table 1 show that our method achieves significant improvement over previous self-supervised approaches and supervised baselines on all testing folds on SBU datasets. The results in Table 2 and Table 3 show that our method outperforms all self-supervised approaches and supervised approaches with RNN backbone on both cross-view and cross-subject settings of NTU60 and both cross-set and cross-subject settings of NTU120. It is worth noting that our method outperforms the traditional contrastive learning with InfoNCE on both SBU (ID=7) and NTU60 (ID=10) datasets.

5.3   Ablations

Hyperparameters. Table 4 shows the performance with different λ on three datasets. λ measures the weight of the constraint of $p_\theta(z)$ in the loss function. The training process will pay more attention on the $KL(p_\theta(z)\|q_\phi(z))$ and less attention on $I(Z;V|X)$ when λ becomes larger. The results show that it performs best when $\lambda = 1$, CoMInG achieves the best result. In Table 5, the results show that when τ set as 0.06, the best performance is obtained.

In Table 6, we show the performance with different batch size on three datasets. The best batch size of NTU60 and SBU is 32 and 128 respectively. An interesting phenomenon is, for specific training epochs (here is 60), for large-scale dataset NTU60, smaller batch size has a significant advantage over the larger ones, and for small-scale datasets SBU, larger batch size has a significant advantage over the smaller ones. This phenomenon was also shown in previous paper [21].

The Setting of Prior Distribution. In CoMInG, the GMM (Gaussian Mixture Model) on unit hyper-sphere is chosen as $q_\phi(z)$. Here we try different numbers of Gaussian

Table 2. Comparison with supervised, and self-supervised methods on NTU60 dataset. Bold numbers refer to the best performers.

ID	Method	CView Acc(%)	CSub Acc(%)
	Supervised		
1	Lie Group [27]	52.8	50.1
2	HBRNN [7]	64.0	59.0
3	Deep RNN [17]	64.1	56.3
	Self-Supervised		
4	*PCL [28]	53.7	\
5	LongT GAN [33]	48.1	39.1
6	P&C FW [25]	44.3	50.8
7	MS2L [16]	\	52.6
8	PCRP [28]	63.5	53.9
9	ASCAL [21]	63.6	58.0
10	⋆InfoNCE	61.0	56.9
11	Ours	**69.4**	**59.8**

Table 3. Comparison with supervised, and self-supervised methods on NTU120 dataset. Bold numbers refer to the best performers.

ID	Method	CSet Acc(%)	CSub Acc(%)
	Supervised		
1	Soft RNN [11]	44.9	36.3
2	PA LSTM [23]	26.3	25.5
	Self-Supervised		
3	P&C FW [25]	42.7	41.7
4	PCRP [28]	45.1	41.7
5	ASCAL [21]	49.2	48.3
6	Ours	**50.7**	**49.4**

Table 4. Performances for using different λ on two datasets for training 60 epochs. For SBU and NTU60, the results are for fold1 and cross-view setting respectively.

Dataset	λ			
Acc(%)	1	2	3	4
NTU60	**69.4**	65.6	57.1	42.8

Table 5. Performances for using different τ on two datasets for training 60 epochs. For SBU and NTU60, the results are for fold1 and cross-view setting respectively.

Dataset	τ			
Acc(%)	0.03	0.06	0.1	0.3
SBU	60.0	**65.5**	61.8	56.4
NTU60	67.3	**69.4**	63.6	55.8

Table 6. Performances for using different batch size on two datasets for training 60 epochs. For SBU and NTU60, the results are for fold1 and cross-view setting respectively.

Dataset	batch size			
Acc(%)	32	64	128	256
SBU	23.6	56.4	**65.5**	58.2
NTU60	**69.4**	67.2	65.1	62.8

Table 7. Performances for using different number of Gaussian on three datasets for training 60 epochs. For SBU, NTU60 and NTU120, the results are for fold1, cross-view setting and cross-set setting respectively.

SBU	Num of Gauss				
Acc(%)	10	20	30	40	50
Top1	61.8	61.8	**65.5**	60.0	60.0
NTU60	Num of Gauss				
Acc(%)	30	60	90	120	150
Top1	65.4	67.4	68.2	**69.4**	67.9
NTU120	Num of Gauss				
Acc(%)	90	120	150	180	210
Top1	48.8	50.2	**50.7**	49.8	49.6

for our mixture model as the prior distribution. The choices of best performances are shown in Table 7.

6 Conclusion

We propose a framework for self-supervised learning by minimizing the constrained conditional mutual information between input augmented samples of the same sample and the output representations of the encode, which can achieve different self-supervised learning methods by choosing different assumptions about the prior distribution of representations, while still learning the description of invariance in data transformation as contrastive learning. Theoretical analysis shows that contrastive learning by InfoNCE is a special case of the proposed framework without human knowledge constraint. Based on this framework, we introduce a self-supervised method by choosing the Gaussian Mixture Model on Unit Hyper-sphere as the prior distribution of representations, and employ it for unsupervised action representing of skeleton-based action recognition. Experimental results of the proposed method show significant improvement on various commonly used datasets for action recognition.

Acknowledgements. This work was supported in part by NSFC 62273347, the National Key Research and Development Program of China (2020AAA0103402), Jiangsu Leading Technology Basic Research Project (BK20192004), and NSFC 61876182.

References

1. Asano, Y.M., Rupprecht, C., Vedaldi, A.: Self-labelling via simultaneous clustering and representation learning. arXiv preprint arXiv:1911.05371 (2019)
2. Caron, M., Bojanowski, P., Joulin, A., Douze, M.: Deep clustering for unsupervised learning of visual features. In: Proceedings of the European Conference on Computer Vision (ECCV), pp. 132–149 (2018)
3. Caron, M., Misra, I., Mairal, J., Goyal, P., Bojanowski, P., Joulin, A.: Unsupervised learning of visual features by contrasting cluster assignments. arXiv preprint arXiv:2006.09882 (2020)
4. Chen, T., Kornblith, S., Norouzi, M., Hinton, G.: A simple framework for contrastive learning of visual representations. In: International Conference on Machine Learning, pp. 1597–1607. PMLR (2020)
5. Cheng, K., Zhang, Y., Cao, C., Shi, L., Cheng, J., Lu, H.: Decoupling GCN with DropGraph module for skeleton-based action recognition. In: Vedaldi, A., Bischof, H., Brox, T., Frahm, J.-M. (eds.) ECCV 2020. LNCS, vol. 12369, pp. 536–553. Springer, Cham (2020). https://doi.org/10.1007/978-3-030-58586-0_32
6. Cheng, K., Zhang, Y., He, X., Chen, W., Cheng, J., Lu, H.: Skeleton-based action recognition with shift graph convolutional network. In: Proceedings of the IEEE/CVF Conference on Computer Vision and Pattern Recognition, pp. 183–192 (2020)
7. Du, Y., Wang, W., Wang, L.: Hierarchical recurrent neural network for skeleton based action recognition. In: Proceedings of the IEEE Conference on Computer Vision and Pattern Recognition, pp. 1110–1118 (2015)
8. Grill, J.B., et al.: Bootstrap your own latent: a new approach to self-supervised learning. arXiv preprint arXiv:2006.07733 (2020)

9. Hammersley, J., Morton, K.: A new monte Carlo technique: antithetic variates. In: Mathematical proceedings of the Cambridge philosophical society, vol. 52, pp. 449–475. Cambridge University Press (1956)

10. He, K., Fan, H., Wu, Y., Xie, S., Girshick, R.: Momentum contrast for unsupervised visual representation learning. In: Proceedings of the IEEE/CVF Conference on Computer Vision and Pattern Recognition, pp. 9729–9738 (2020)

11. Hu, J.F., Zheng, W.S., Ma, L., Wang, G., Lai, J., Zhang, J.: Early action prediction by soft regression. IEEE Trans. Pattern Anal. Mach. Intell. **41**(11), 2568–2583 (2018)

12. Jing, L., Tian, Y.: Self-supervised visual feature learning with deep neural networks: a survey. IEEE Trans. Pattern Anal. Mach. Intell. (2020)

13. Kingma, D.P., Welling, M.: Auto-encoding variational bayes. arXiv preprint arXiv:1312.6114 (2013)

14. Larsson, G., Maire, M., Shakhnarovich, G.: Colorization as a proxy task for visual understanding. In: Proceedings of the IEEE Conference on Computer Vision and Pattern Recognition, pp. 6874–6883 (2017)

15. Li, J., Zhou, P., Xiong, C., Hoi, S.C.: Prototypical contrastive learning of unsupervised representations. arXiv preprint arXiv:2005.04966 (2020)

16. Lin, L., Song, S., Yang, W., Liu, J.: MS2L: Multi-task self-supervised learning for skeleton based action recognition. In: Proceedings of the 28th ACM International Conference on Multimedia, pp. 2490–2498 (2020)

17. Liu, J., Shahroudy, A., Perez, M., Wang, G., Duan, L.Y., Kot, A.C.: NTU RGB+ D 120: a large-scale benchmark for 3D human activity understanding. IEEE Trans. Pattern Anal. Mach. Intell. **42**(10), 2684–2701 (2019)

18. Liu, X., et al.: Self-supervised learning: generative or contrastive. IEEE Trans. Knowl. Data Eng. (2021)

19. Nowozin, S., Cseke, B., Tomioka, R.: F-GAN: training generative neural samplers using variational divergence minimization. In: Proceedings of the 30th International Conference on Neural Information Processing Systems, pp. 271–279 (2016)

20. Ohn-Bar, E., Trivedi, M.: Joint angles similarities and HOG2 for action recognition. In: Proceedings of the IEEE Conference on Computer Vision and Pattern Recognition Workshops, pp. 465–470 (2013)

21. Rao, H., Xu, S., Hu, X., Cheng, J., Hu, B.: Augmented skeleton based contrastive action learning with momentum LSTM for unsupervised action recognition. Inf. Sci. **569**, 90–109 (2021)

22. Ren, B., Liu, M., Ding, R., Liu, H.: A survey on 3D skeleton-based action recognition using learning method. arXiv preprint arXiv:2002.05907 (2020)

23. Shahroudy, A., Liu, J., Ng, T.T., Wang, G.: NTU RGB+ D: A large scale dataset for 3D human activity analysis. In: Proceedings of the IEEE Conference on Computer Vision and Pattern Recognition, pp. 1010–1019 (2016)

24. Shi, L., Zhang, Y., Cheng, J., Lu, H.: Skeleton-based action recognition with directed graph neural networks. In: Proceedings of the IEEE/CVF Conference on Computer Vision and Pattern Recognition, pp. 7912–7921 (2019)

25. Su, K., Liu, X., Shlizerman, E.: Predict & Cluster: unsupervised skeleton based action recognition. In: Proceedings of the IEEE/CVF Conference on Computer Vision and Pattern Recognition, pp. 9631–9640 (2020)

26. Tian, Y., Krishnan, D., Isola, P.: Contrastive multiview coding. In: Vedaldi, A., Bischof, H., Brox, T., Frahm, J.-M. (eds.) ECCV 2020. LNCS, vol. 12356, pp. 776–794. Springer, Cham (2020). https://doi.org/10.1007/978-3-030-58621-8_45

27. Vemulapalli, R., Arrate, F., Chellappa, R.: Human action recognition by representing 3D skeletons as points in a lie group. In: Proceedings of the IEEE Conference on Computer Vision and Pattern Recognition, pp. 588–595 (2014)

28. Xu, S., Rao, H., Hu, X., Hu, B.: Prototypical contrast and reverse prediction: unsupervised skeleton based action recognition. arXiv preprint arXiv:2011.07236 (2020)

29. Yan, S., Xiong, Y., Lin, D.: Spatial temporal graph convolutional networks for skeleton-based action recognition. In: Thirty-second AAAI Conference on Artificial Intelligence (2018)

30. Yun, K., Honorio, J., Chattopadhyay, D., Berg, T.L., Samaras, D.: Two-person interaction detection using body-pose features and multiple instance learning. In: 2012 IEEE Computer Society Conference on Computer Vision and Pattern Recognition Workshops, pp. 28–35. IEEE (2012)

31. Zbontar, J., Jing, L., Misra, I., LeCun, Y., Deny, S.: Barlow twins: self-supervised learning via redundancy reduction. In: International Conference on Machine Learning, pp. 12310–12320. PMLR (2021)

32. Zhang, P., Lan, C., Xing, J., Zeng, W., Xue, J., Zheng, N.: View adaptive recurrent neural networks for high performance human action recognition from skeleton data. In: Proceedings of the IEEE International Conference on Computer Vision, pp. 2117–2126 (2017)

33. Zheng, N., Wen, J., Liu, R., Long, L., Dai, J., Gong, Z.: Unsupervised representation learning with long-term dynamics for skeleton based action recognition. In: Proceedings of the AAAI Conference on Artificial Intelligence, vol. 32 (2018)

Temporal-Viewpoint Transportation Plan for Skeletal Few-Shot Action Recognition

Lei Wang[1,2] 🆔 and Piotr Koniusz[1,2(✉)] 🆔

[1] Australian National University, Canberra, Australia
[2] Data61/CSIRO, Sydney, Australia
{lei.wang,piotr.koniusz}@data61.csiro.au

Abstract. We propose a Few-shot Learning pipeline for 3D skeleton-based action recognition by Joint tEmporal and cAmera viewpoiNt alIgnmEnt (JEANIE). To factor out misalignment between query and support sequences of 3D body joints, we propose an advanced variant of Dynamic Time Warping which jointly models each smooth path between the query and support frames to achieve simultaneously the best alignment in the temporal and simulated camera viewpoint spaces for end-to-end learning under the limited few-shot training data. Sequences are encoded with a temporal block encoder based on Simple Spectral Graph Convolution, a lightweight linear Graph Neural Network backbone. We also include a setting with a transformer. Finally, we propose a similarity-based loss which encourages the alignment of sequences of the same class while preventing the alignment of unrelated sequences. We show state-of-the-art results on NTU-60, NTU-120, Kinetics-skeleton and UWA3D Multiview Activity II.

1 Introduction

Action recognition is arguably among key topics in computer vision due to applications in video surveillance [63,65], human-computer interaction, sports analysis, virtual reality and robotics. Many pipelines [7,18,19,30,59,64] perform action classification given the large amount of labeled training data. However, manually collecting and labeling videos for 3D skeleton sequences is laborious, and such pipelines need to be retrained or fine-tuned for new class concepts. Popular action recognition networks include two-stream neural networks [18,19,71] and 3D convolutional networks (3D CNNs) [7,59], which aggregate frame-wise and temporal block representations, respectively. However, such networks indeed must be trained on large-scale datasets such as Kinetics [7,31,66,68] under a fixed set of training class concepts.

Thus, there exists a growing interest in devising effective Few-shot Learning (FSL) for action recognition, termed Few-shot Action Recognition (FSAR), that rapidly adapts to novel classes given a few training samples [5,14,23,47,67,73,79]. However, FSAR for videos is scarce due to the volumetric nature of videos and large intra-class variations.

Supplementary Information The online version contains supplementary material available at https://doi.org/10.1007/978-3-031-26316-3_19.

FSL for image recognition has been widely studied [3,17,20,33,34,46] including contemporary CNN-based FSL methods [21,29,54,57,61,76], which use meta-learning, prototype-based learning and feature representation learning. Just in 2020–2022, many FSL methods [5,13,15,16,22,24,32,35,37,41,42,52,58,70,78,86] have been dedicated to image classification or detection [75,77,82–84]. Noteworthy mentioning is the incremental learning paradigm that can also tackle novel classes [51]. In this paper, we aim at advancing few-shot recognition of articulated set of connected 3D body joints.

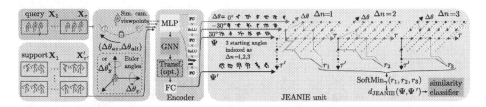

Fig. 1. Our 3D skeleton-based FSAR with JEANIE. Frames from a query sequence and a support sequence are split into short-term temporal blocks $\mathbf{X}_1, ..., \mathbf{X}_\tau$ and $\mathbf{X}'_1, ..., \mathbf{X}'_{\tau'}$ of length M given stride S. Subsequently, we generate (i) multiple rotations by $(\Delta\theta_x, \Delta\theta_y)$ of each query skeleton by either Euler angles (baseline approach) or (ii) simulated camera views (gray cameras) by camera shifts $(\Delta\theta_{az}, \Delta\theta_{alt})$ w.r.t.the assumed average camera location (black camera). We pass all skeletons via Encoding Network (with an optional transformer) to obtain feature tensors $\mathbf{\Psi}$ and $\mathbf{\Psi}'$, which are directed to JEANIE. We note that the temporal-viewpoint alignment takes place in 4D space (we show a 3D case with three views: $-30°, 0°, 30°$). Temporally-wise, JEANIE starts from the same $t = (1, 1)$ and finishes at $t = (\tau, \tau')$ (as in DTW). Viewpoint-wise, JEANIE starts from every possible camera shift $\Delta\theta \in \{-30°, 0°, 30°\}$ (we do not know the true correct pose) and finishes at one of possible camera shifts. At each step, the path may move by no more than $(\pm\Delta\theta_{az}, \pm\Delta\theta_{alt})$ to prevent erroneous alignments. Finally, SoftMin picks up the smallest distance.

With an exception of very recent models [38,39,44,45,48,67], FSAR approaches that learn from skeleton-based 3D body joints are scarce. The above situation prevails despite action recognition from articulated sets of connected body joints, expressed as 3D coordinates, does offer a number of advantages over videos such as (i) the lack of the background clutter, (ii) the volume of data being several orders of magnitude smaller, and (iii) the 3D geometric manipulations of sequences being relatively friendly.

Thus, we propose a FSAR approach that learns on skeleton-based 3D body joints via Joint tEmporal and cAmera viewpoiNt alIgnmEnt (JEANIE). As FSL is based on learning similarity between support-query pairs, to achieve good matching of queries with support sequences representing the same action class, we propose to simultaneously model the optimal (i) temporal and (ii) viewpoint alignments. To this end, we build on soft-DTW [11], a differentiable variant of Dynamic Time Warping (DTW) [10]. Unlike soft-DTW, we exploit the projective camera geometry. We assume that the best smooth path in DTW should simultaneously provide the best temporal and viewpoint alignment, as sequences that are being matched might have been captured under different camera viewpoints or subjects might have followed different trajectories.

To obtain skeletons under several viewpoints, we rotate skeletons (zero-centered by hip) by Euler angles [1] w.r.t.x, y and z axes, or generate skeleton locations given simulated camera positions, according to the algebra of stereo projections [2].

We note that view-adaptive models for action recognition do exist. View Adaptive Recurrent Neural Networks [80,81] is a classification model equipped with a view-adaptive subnetwork that contains the rotation and translation switches within its RNN backbone, and the main LSTM-based network. Temporal Segment Network [62] models long-range temporal structures with a new segment-based sampling and aggregation module. However, such pipelines require a large number of training samples with varying viewpoints and temporal shifts to learn a robust model. Their limitations become evident when a network trained under a fixed set of action classes has to be adapted to samples of novel classes. Our JEANIE does not suffer from such a limitation.

Our pipeline consists of an MLP which takes neighboring frames to form a temporal block. Firstly, we sample desired Euler rotations or simulated camera viewpoints, generate multiple skeleton views, and pass them to the MLP to get block-wise feature maps, next forwarded to a Graph Neural Network (GNN), *e.g.*, GCN [27], Fisher-Bures GCN [56], SGC [72], APPNP [28] or S^2GC [85,87], followed by an optional transformer [12], and an FC layer to obtain graph-based representations passed to JEANIE.

JEANIE builds on Reproducing Kernel Hilbert Spaces (RKHS) [53] which scale gracefully to FSAR problems which, by their setting, learn to match pairs of sequences rather than predict class labels. JEANIE builds on Optimal Transport [60] by using a transportation plan for temporal and viewpoint alignment in skeletal action recognition.

Below are our contributions:

i. We propose a Few-shot Action Recognition approach for learning on skeleton-based articulated 3D body joints via JEANIE, which performs the joint alignment of temporal blocks and simulated viewpoint indexes of skeletons between support-query sequences to select the smoothest path without abrupt jumps in matching temporal locations and view indexes. Warping jointly temporal locations and simulated viewpoint indexes helps meta-learning with limited samples of novel classes.

ii. To simulate different viewpoints of 3D skeleton sequences, we consider rotating them (1) by Euler angles within a specified range along x and y axes, or (2) towards the simulated camera locations based on the algebra of stereo projection.

iii. We investigate several different GNN backbones (including transformer), as well as the optimal temporal size and stride for temporal blocks encoded by a simple 3-layer MLP unit before forwarding them to GNN.

iv. We propose a simple similarity-based loss encouraging the alignment of within-class sequences and preventing the alignment of between-class sequences.

We achieve the state of the art on large-scale NTU-60 [50], NTU-120 [39], Kinetics-skeleton [74] and UWA3D Multiview Activity II [49]. As far as we can tell, the simultaneous alignment in the joint temporal-viewpoint space for FSAR is a novel proposition.

2   Related Works

Below, we describe 3D skeleton-based action recognition, FSAR approaches and GNNs.

Action Recognition (3D Skeletons). 3D skeleton-based action recognition pipelines often use GCNs [27], *e.g.*, spatio-temporal GCN [74], an a-links inference model [36], shift-graph model [9] and multi-scale aggregation node [40]. However, such models rely on large-scale datasets, and cannot be easily adapted to novel class concepts.

FSAR (Videos). Approaches [23,47,73] use a generative model, graph matching on 3D coordinates and dilated networks, respectively. Approach [88] uses a compound memory network. ProtoGAN [14] generates action prototypes. Model [79] uses permutation-invariant attention and second-order aggregation of temporal video blocks, whereas approach [5] proposes a modified temporal alignment for query-support pairs via DTW.

FSAR (3D Skeletons). Few FSAR models use 3D skeletons [38,39,44,45]. Global Con-text-Aware Attention LSTM [38] selectively focuses on informative joints. Action-Part Semantic Relevance-aware (APSR) model [39] uses the semantic relevance between each body part and action class at the distributed word embedding level. Signal Level Deep Metric Learning (DML) [45] and Skeleton-DML [44] one-shot FSL approaches encode signals into images, extract features using CNN and apply multi-similarity miner losses. In contrast, we use temporal blocks of 3D body joints of skeletons encoded by GNNs under multiple viewpoints of skeletons to simultaneously perform temporal and viewpoint-wise alignment of query-support in the meta-learning regime.

Graph Neural Networks. GNNs are popular in the skeleton-based action recognition. We build on GNNs in this paper due to their excellent ability to represent graph-structured data such as interconnected body joints. GCN [27] applies graph convolution in the spectral domain, and enjoys the depth-efficiency when stacking multiple layers due to non-linearities. However, depth-efficiency costs speed due to backpropagation through consecutive layers. In contrast, a very recent family of so-called spectral filters do not require depth-efficiency but apply filters based on heat diffusion to the graph Laplacian. As a result, they are fast linear models as learnable weights act on filtered node representations. SGC [72], APPNP [28] and S^2GC [85] are three methods from this family which we investigate for the backbone.

Multi-view Action Recognition. Multi-modal sensors enable multi-view action recognition [64,80]. A Generative Multi-View Action Recognition framework [69] integrates complementary information from RGB and depth sensors by View Correlation Discovery Network. Some works exploit multiple views of the subject [39,50,69,81] to overcome the viewpoint variations for action recognition on large training datasets. In contrast, our JEANIE learns to perform jointly the temporal and simulated viewpoint alignment in an end-to-end meta-learning setting. This is a novel paradigm based on similarity learning of support-query pairs rather than learning class concepts.

3   Approach

To learn similarity/dissimilarity between pairs of sequences of 3D body joints representing query and support samples from episodes, our goal is to find a smooth joint viewpoint-temporal alignment of query and support and minimize or maximize the matching distance d_{JEANIE} (end-to-end setting) for same or different support-query labels, respectively. Figure 2 (top) shows that sometimes matching of query and support may be as easy as rotating one trajectory onto another, in order to achieve viewpoint invariance. A viewpoint invariant distance [25] can be defined as:

$$d_{\text{inv}}(\boldsymbol{\Psi},\boldsymbol{\Psi}') = \inf_{\gamma,\gamma' \in T} d\big(\gamma(\boldsymbol{\Psi}),\gamma'(\boldsymbol{\Psi}')\big), \tag{1}$$

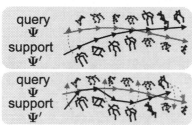
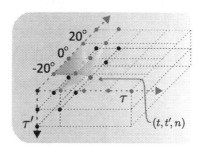

Fig. 2. (*top*) In viewpoint-invariant learning, the distance between query features $\boldsymbol{\Psi}$ and support features $\boldsymbol{\Psi}'$ has to be computed. The blue arrow indicates that trajectories of both actions need alignment. (*bottom*) In real life, subject's 3D body joints deviate from one ideal trajectory, and so advanced viewpoint alignment strategy is needed.

Fig. 3. JEANIE (1-max shift). We loop over all points. At (t,t',n) (green point) we add its base distance to the minimum of accumulated distances at $(t,t'-1,n-1)$, $(t,t'-1,n)$, $(t,t'-1,n+1)$ (orange plane), $(t-1,t'-1,n-1)$, $(t-1,t'-1,n)$, $(t-1,t'-1,n+1)$ (red plane) and $(t-1,t',n-1)$, $(t-1,t',n)$, $(t-1,t',n+1)$ (blue plane).

where T is a set of transformations required to achieve a viewpoint invariance, $d(\cdot,\cdot)$ is some base distance, *e.g.*, the Euclidean distance, and $\boldsymbol{\Psi}$ and $\boldsymbol{\Psi}'$ are features describing query and support pair of sequences. Typically, T may include 3D rotations to rotate one trajectory onto the other. However, such a global viewpoint alignment of two sequences is suboptimal. Trajectories are unlikely to be straight 2D lines in the 3D space. Figure 2 (bottom) shows that 3D body joints locally follow complicated non-linear paths.

Thus, we propose JEANIE that aligns and warps query/support sequences based on the feature similarity. One can think of JEANIE as performing Eq. (1) with T containing camera viewpoint rotations, and the base distance $d(\cdot,\cdot)$ being a joint temporal-viewpoint variant of soft-DTW to account for local temporal-viewpoint variations of 3D body joint trajectories. JEANIE unit in Fig. 1 realizes such a strategy (SoftMin operation is equivalent of Eq. (1)). While such an idea sounds simple, it is effective, it has not been done before. Figure 3 (discussed later in the text) shows one step of the temporal-viewpoint computations of JEANIE.

We present a necessary background on Euler angles and the algebra of stereo projection, GNNs and the formulation of soft-DTW in Appendix Sect. A. Below, we detail our pipeline shown in Fig. 1, explain the proposed JEANIE and our loss function.

Notations. $\mathcal{I}_K$ stands for the index set $\{1,2,...,K\}$. Concatenation of α_i is denoted by $[\alpha_i]_{i \in \mathcal{I}_I}$, whereas $\mathbf{X}_{:,i}$ means we extract/access column i of matrix $\boldsymbol{D}$. Calligraphic mathcal fonts denote tensors (*e.g.*, $\mathcal{D}$), capitalized bold symbols are matrices (*e.g.*, $\boldsymbol{D}$), lowercase bold symbols are vectors (*e.g.*, $\boldsymbol{\psi}$), and regular fonts denote scalars.

Encoding Network (EN). We start by generating $K \times K'$ Euler rotations or $K \times K'$ simulated camera views (moved gradually from the estimated camera location) of query skeletons. Our EN contains a simple 3-layer MLP unit (FC, ReLU, FC, ReLU, Dropout, FC), GNN, optional Transformer [12] and FC. The MLP unit takes M neighboring frames, each with J 3D skeleton body joints, forming one temporal block. In total, depending on stride S, we obtain some τ temporal blocks which capture the short temporal dependency, whereas the long temporal dependency is modeled with our JEANIE. Each temporal block is encoded by the MLP into a $d \times J$ dimensional feature map. Subsequently, query feature maps of size $K \times K' \times \tau$ and support feature maps of size τ' are forwarded to a GNN, optional Transformer (similar to ViT [12], instead of using image patches, we feed each body joint encoded by GNN into the transformer), and an FC layer, which returns $\boldsymbol{\Psi} \in \mathbb{R}^{d' \times K \times K' \times \tau}$ query feature maps and $\boldsymbol{\Psi}' \in \mathbb{R}^{d' \times \tau'}$ support feature maps. Feature maps are passed to JEANIE and the similarity classifier.

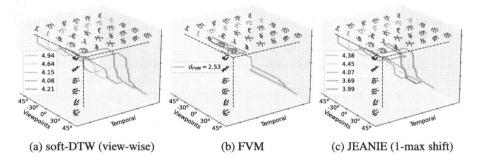

(a) soft-DTW (view-wise)   (b) FVM   (c) JEANIE (1-max shift)

Fig. 4. A comparison of paths in 3D for soft-DTW, Free Viewpoint Matching (FVM) and our JEANIE. For a given support skeleton sequence (green color), we choose viewing angles between $-45°$ and $45°$ for the camera viewpoint simulation. The support skeleton sequence is shown in black color. (a) soft-DTW finds each individual alignment per viewpoint fixed throughout alignment: $d_{\text{shortest}} = 4.08$. (b) FVM is a greedy matching algorithm that in each time step seeks the best alignment pose from all viewpoints which leads to unrealistic zigzag path (person cannot jump from front to back view suddenly): $d_{\text{FVM}} = 2.53$. (c) Our JEANIE (1-max shift) is able to find smooth joint viewpoint-temporal alignment between support and query sequences. We show each optimal path for each possible starting position: $d_{\text{JEANIE}} = 3.69$. While $d_{\text{FVM}} = 2.53$ for FVM is overoptimistic, $d_{\text{shortest}} = 4.08$ for fixed-view matching is too pessimistic, whereas JEANIE strikes the right matching balance with $d_{\text{JEANIE}} = 3.69$.

Let support maps $\boldsymbol{\Psi}'$ be $[f(\boldsymbol{X}'_1; \mathcal{F}), ..., f(\boldsymbol{X}'_{\tau'}; \mathcal{F})] \in \mathbb{R}^{d' \times \tau'}$ and query maps $\boldsymbol{\Psi}$ be $[f(\boldsymbol{X}_1; \mathcal{F}), ..., f(\boldsymbol{X}_\tau; \mathcal{F})] \in \mathbb{R}^{d' \times K \times K' \times \tau}$, for query and support frames per block $\boldsymbol{X}, \boldsymbol{X}' \in \mathbb{R}^{3 \times J \times M}$. Moreover, we define $f(\mathbf{X}; \mathcal{F}) = \text{FC}(\text{Transf}(\text{GNN}(\text{MLP}(\mathbf{X}; \mathcal{F}_{MLP}); \mathcal{F}_{GNN}); \mathcal{F}_{Transf}); \mathcal{F}_{FC})$, $\mathcal{F} \equiv [\mathcal{F}_{MLP}, \mathcal{F}_{GNN}, \mathcal{F}_{Transf}, \mathcal{F}_{FC}]$ is the set of parameters of EN (note optional Transformer [12]). As GNN, we try GCN [27], SGC [72], APPNP [28] or S^{2}GC [85].

JEANIE. Matching query-support pairs requires temporal alignment due to potential offset in locations of discriminative parts of actions, and due to potentially different dynamics/speed of actions taking place. The same concerns the direction of the dominant action trajectory w.r.t.the camera. Thus, JEANIE, our advanced soft-DTW, has the transportation plan $\mathcal{A}' \equiv \mathcal{A}_{\tau,\tau',K,K'}$, where apart from temporal block counts τ and τ', for query sequences, we have possible η_{az} left and η_{az} right steps from the initial camera azimuth, and η_{alt} up and η_{alt} down steps from the initial camera altitude. Thus, $K = 2\eta_{az}+1$, $K' = 2\eta_{alt}+1$. For the variant with Euler angles, we simply have $\mathcal{A}'' \equiv \mathcal{A}_{\tau,\tau',K,K'}$ where $K = 2\eta_x+1$, $K' = 2\eta_y+1$ instead. Then, JEANIE is given as:

$$d_{\text{JEANIE}}(\boldsymbol{\Psi},\boldsymbol{\Psi}') = \operatorname*{SoftMin}_{\mathbf{A}\in\mathcal{A}'}{}_{\gamma}\langle \mathbf{A}, \mathcal{D}(\boldsymbol{\Psi},\boldsymbol{\Psi}')\rangle, \qquad (2)$$

where $\mathcal{D} \in \mathbb{R}_+^{K \times K' \times \tau \times \tau'} \equiv [d_{\text{base}}(\boldsymbol{\psi}_{m,k,k'}, \boldsymbol{\psi}'_n)]_{\substack{(m,n)\in\mathcal{I}_\tau\times\mathcal{I}_{\tau'} \\ (k,k')\in\mathcal{I}_K\times\mathcal{I}_{K'}}}$,and $\mathcal{D}$ contains distances.

Figure 3 shows one step of JEANIE (1-max shift). Suppose the given viewing angle set is $\{-40°, -20°, 0°, 20°, 40°\}$. For 1-max shift, we loop over (t, t', n). At location (t, t', n), we extract the base distance and add it together with the minimum of aggregated distances at the shown 9 predecessor points. We store that total distance at (t, t', n), and we move to the next point. Note that for viewpoint index n, we look up $(n-1, n, n+1)$. Extension to the ι-max shift is straightforward.

Algorithm 1 illustrates JEANIE. For brevity, let us tackle the camera viewpoint alignment in a single space, e.g., for some shifting steps $-\eta, ..., \eta$, each with size $\Delta\theta_{az}$. The maximum viewpoint change from block to block is ι-max shift (smoothness). As we have no way to know the initial optimal camera shift, we initialize all possible origins of shifts in accumulator $r_{n,1,1} = d_{\text{base}}(\boldsymbol{\psi}_{n,1},\boldsymbol{\psi}'_1)$ for all $n \in \{-\eta, ..., \eta\}$. Subsequently, a phase related to soft-DTW (temporal-viewpoint alignment) takes place. Finally, we choose the path with the smallest distance over all possible viewpoint ends by selecting a soft-minimum over $[r_{n,\tau,\tau'}]_{n\in\{-\eta,...,\eta\}}$. Notice that accumulator $\mathcal{R} \in \mathbb{R}^{(2\iota+1)\times\tau\times\tau'}$. Moreover, whenever either index $n-i$, $t-j$ or $t'-k$ in $r_{n-i,t-j,t'-k}$ (see algorithm) is out of bounds, we define $r_{n-i,t-j,t'-k} = \infty$.

JEANIE. Matching query-support pairs requires temporal alignment due to potential offset in locations of discriminative parts of actions, and due to potentially different dynamics/speed of actions taking place. The same concerns the direction of the dominant action trajectory w.r.t.the camera. Thus, JEANIE, our advanced soft-DTW, has the transportation plan $\mathcal{A}' \equiv \mathcal{A}_{\tau,\tau',K,K'}$, where apart from temporal block counts τ and τ', for query sequences, we have possible η_{az} left and η_{az} right steps from the initial camera azimuth, and η_{alt} up and η_{alt} down steps from the initial camera altitude. Thus, $K = 2\eta_{az}+1$, $K' = 2\eta_{alt}+1$. For the variant with Euler angles, we simply have $\mathcal{A}'' \equiv \mathcal{A}_{\tau,\tau',K,K'}$ where $K = 2\eta_x+1$, $K' = 2\eta_y+1$ instead. Then, JEANIE is given as:

$$d_{\text{JEANIE}}(\boldsymbol{\Psi},\boldsymbol{\Psi}') = \operatorname*{SoftMin}_{\mathbf{A}\in\mathcal{A}'}{}_{\gamma}\langle \mathbf{A}, \mathcal{D}(\boldsymbol{\Psi},\boldsymbol{\Psi}')\rangle, \qquad (3)$$

where $\mathcal{D} \in \mathbb{R}_+^{K \times K' \times \tau \times \tau'} \equiv [d_{\text{base}}(\boldsymbol{\psi}_{m,k,k'}, \boldsymbol{\psi}'_n)]_{\substack{(m,n)\in\mathcal{I}_\tau\times\mathcal{I}_{\tau'} \\ (k,k')\in\mathcal{I}_K\times\mathcal{I}_{K'}}}$,and $\mathcal{D}$ contains distances.

Figure 3 shows one step of JEANIE (1-max shift). Suppose the given viewing angle set is $\{-40°, -20°, 0°, 20°, 40°\}$. For 1-max shift, we loop over (t, t', n). At location (t, t', n), we extract the base distance and add it together with the minimum of aggregated distances at the shown 9 predecessor points. We store that total distance at (t, t', n), and we move to the next point. Note that for viewpoint index n, we look up $(n-1, n, n+1)$. Extension to the ι-max shift is straightforward.

Algorithm 1 illustrates JEANIE. For brevity, let us tackle the camera viewpoint alignment in a single space, e.g., for some shifting steps $-\eta, ..., \eta$, each with size $\Delta\theta_{az}$. The maximum viewpoint change from block to block is ι-max shift (smoothness). As we have no way to know the initial optimal camera shift, we initialize all possible origins of shifts in accumulator $r_{n,1,1} = d_{\text{base}}(\psi_{n,1}, \psi'_1)$ for all $n \in \{-\eta, ..., \eta\}$. Subsequently, a phase related to soft-DTW (temporal-viewpoint alignment) takes place. Finally, we choose the path with the smallest distance over all possible viewpoint ends by selecting a soft-minimum over $[r_{n,\tau,\tau'}]_{n \in \{-\eta,...,\eta\}}$. Notice that accumulator $\mathcal{R} \in \mathbb{R}^{(2\iota+1) \times \tau \times \tau'}$. Moreover, whenever either index $n-i$, $t-j$ or $t'-k$ in $r_{n-i,t-j,t'-k}$ (see algorithm) is out of bounds, we define $r_{n-i,t-j,t'-k} = \infty$.

FVM. To ascertain whether JEANIE is better than performing separately the temporal and simulated viewpoint alignments, we introduce a baseline called the Free Viewpoint Matching (FVM). FVM, for every step of DTW, seeks the best local viewpoint alignment, thus realizing non-smooth temporal-viewpoint path in contrast to JEANIE. To this end, we apply DTW in Eq. (3) with the base distance replaced by:

$$d_{\text{FVM}(\psi_t, \psi'_{t'})} = \underset{m,n,m',n' \in \{-\eta,...,\eta\}}{\text{SoftMin}_{\bar{\gamma}}} d_{\text{base}}(\psi_{m,n,t}, \psi'_{m',n',t'}), \qquad (4)$$

where $\Psi \in \mathbb{R}^{d' \times K \times K' \times \tau}$ and $\Psi' \in \mathbb{R}^{d' \times K \times K' \times \tau'}$ are query and support feature maps. We abuse the notation by writing $d_{\text{FVM}(\psi_t, \psi'_{t'})}$ as we minimize over viewpoint indexes in Eq. (4). We compute the distance matrix $D \in \mathbb{R}_+^{\tau \times \tau'} \equiv [d_{\text{FVM}}(\psi_t, \psi'_{t'})]_{(t,t') \in \mathcal{I}_\tau \times \mathcal{I}_{\tau'}}$.

Figure 4 shows the comparison between soft-DTW (view-wise), FVM and our JEANIE. FVM is a greedy matching method which leads to complex zigzag path in 3D space (assuming the camera viewpoint single space in $\psi_{n,t}$ and no viewpoint in $\psi'_{t'}$). Although FVM is able to find the smallest distance path compared to soft-DTW and JEANIE, it suffers from several issues (i) It is unreasonable for poses in a given sequence to match under sudden jumps in viewpoints. (ii) Suppose the two sequences are from two different classes, FVM still yields the smallest distance (decreased interclass variance).

Loss Function. For the N-way Z-shot problem, we have one query feature map and $N \times Z$ support feature maps per episode. We form a mini-batch containing B episodes. Thus, we have query feature maps $\{\Psi_b\}_{b \in \mathcal{I}_B}$ and support feature maps $\{\Psi'_{b,n,z}\}_{b \in \mathcal{I}_B, n \in \mathcal{I}_N, z \in \mathcal{I}_Z}$. Moreover, Ψ_b and $\Psi'_{b,1,:}$ share the same class, one of N classes drawn per episode, forming the subset $C^\ddagger \equiv \{c_1, ..., c_N\} \subset \mathcal{I}_C \equiv \mathcal{C}$. To be precise, labels $y(\Psi_b) = y(\Psi'_{b,1,z}), \forall b \in \mathcal{I}_B, z \in \mathcal{I}_Z$ while $y(\Psi_b) \neq y(\Psi'_{b,n,z}), \forall b \in \mathcal{I}_B, n \in \mathcal{I}_N \setminus \{1\}, z \in \mathcal{I}_Z$. In most cases, $y(\Psi_b) \neq y(\Psi_{b'})$ if $b \neq b'$ and $b, b' \in \mathcal{I}_B$. Selection of $C^\ddagger$ per episode is random. For the N-way Z-shot protocol, we minimize:

Algorithm 1. Joint tEmporal and cAmera viewpoiNt alIgnmEnt (JEANIE).

Input (forward pass): $\boldsymbol{\Psi}, \boldsymbol{\Psi}', \gamma > 0, d_{\text{base}}(\cdot, \cdot)$, ι-max shift.

1: $r_{:,:,:} = \infty, r_{n,1,1} = d_{\text{base}}(\boldsymbol{\psi}_{n,1}, \boldsymbol{\psi}'_1), \forall n \in \{-\eta, ..., \eta\}$

2: $\Pi \equiv \{-\iota, ..., 0, ..., \iota\} \times \{(0,1), (1,0), (1,1)\}$

3: **for** $t \in \mathcal{I}_\tau$:

4:      **for** $t' \in \mathcal{I}_{\tau'}$:

5:          **if** $t \neq 1$ or $t' \neq 1$:

6:              **for** $n \in \{-\eta, ..., \eta\}$:

7:                  $r_{n,t,t'} = d_{\text{base}}(\boldsymbol{\psi}_{n,t}, \boldsymbol{\psi}'_{t'}) + \text{SoftMin}_\gamma \left([r_{n-i,t-j,t'-k}]_{(i,j,k) \in \Pi} \right)$

Output: $\text{SoftMin}_\gamma \left([r_{n,\tau,\tau'}]_{n \in \{-\eta, ..., \eta\}} \right)$

$$l(\boldsymbol{d}^+, \boldsymbol{d}^-) = \left(\mu(\boldsymbol{d}^+) - \{\mu(\text{TopMin}_\beta(\boldsymbol{d}^+))\} \right)^2 \tag{5}$$
$$+ \left(\mu(\boldsymbol{d}^-) - \{\mu(\text{TopMax}_{\text{NZ}\beta}(\boldsymbol{d}^-))\} \right)^2,$$

where $\boldsymbol{d}^+ = [d_{\text{JEANIE}}(\boldsymbol{\Psi}_b, \boldsymbol{\Psi}'_{b,1,z})]_{\substack{b \in \mathcal{I}_B \\ z \in \mathcal{I}_Z}}$ and $\boldsymbol{d}^- = [d_{\text{JEANIE}}(\boldsymbol{\Psi}_b, \boldsymbol{\Psi}'_{b,n,z})]_{\substack{b \in \mathcal{I}_B, \\ n \in \mathcal{I}_N \setminus \{1\}, z \in \mathcal{I}_Z}},$

$$(6)$$

where $\boldsymbol{d}^+$ is a set of within-class distances for the mini-batch of size B given N-way Z-shot learning protocol. By analogy, $\boldsymbol{d}^-$ is a set of between-class distances. Function $\mu(\cdot)$ is simply the mean over coefficients of the input vector, $\{\cdot\}$ detaches the graph during the backpropagation step, whereas $\text{TopMin}_\beta(\cdot)$ and $\text{TopMax}_{\text{NZ}\beta}(\cdot)$ return β smallest and $NZ\beta$ largest coefficients from the input vectors, respectively. Thus, Eq. (5) promotes the within-class similarity while Eq. (6) reduces the between-class similarity. Integer $\beta \geq 0$ controls the focus on difficult examples, e.g., $\beta = 1$ encourages all within-class distances in Eq. (5) to be close to the positive target $\mu(\text{TopMin}_\beta(\cdot))$, the smallest observed within-class distance in the mini-batch. If $\beta > 1$, this means we relax our positive target. By analogy, if $\beta = 1$, we encourage all between-class distances in Eq. (6) to approach the negative target $\mu(\text{TopMax}_{\text{NZ}\beta}(\cdot))$, the average over the largest NZ between-class distances. If $\beta > 1$, the negative target is relaxed.

4 Experiments

We provide network configurations and training details in Appendix Section H. Below, we describe the datasets and evaluation protocols on which we validate our JEANIE.

Datasets. Appendix Section B. and Table 9. contain details of datasets described below.

i. *UWA3D Multiview Activity II* [49] contains 30 actions performed by 9 people in a cluttered environment. In this dataset, the Kinect camera was moved to different positions to capture the actions from 4 different views: front view (V_1), left view (V_2), right view (V_3), and top view (V_4).

ii. *NTU RGB+D (NTU-60)* [50] contains 56,880 video sequences and over 4 million frames. This dataset has variable sequence lengths and high intra-class variations.

iii. *NTU RGB+D 120 (NTU-120)* [39], an extension of NTU-60, contains 120 action classes (daily/health-related), and 114,480 RGB+D video samples captured with 106 distinct human subjects from 155 different camera viewpoints.

iv. *Kinetics* [26] is a large-scale collection of 650,000 video clips that cover 400/600/700 human action classes. It includes human-object interactions such as *playing instruments*, as well as human-human interactions such as *shaking hands* and *hugging*. As the Kinetics-400 dataset provides only the raw videos, we follow approach [74] and use the estimated joint locations in the pixel coordinate system as the input to our pipeline. To obtain the joint locations, we first resize all videos to the resolution of 340×256, and convert the frame rate to 30 FPS. Then we use the publicly available *OpenPose* [6] toolbox to estimate the location of 18 joints on every frame of the clips. As OpenPose produces the 2D body joint coordinates and Kinetics-400 does not offer multiview or depth data, we use a network of Martinez *et al.*[43] pre-trained on Human3.6M [8], combined with the 2D OpenPose output to estimate 3D coordinates from 2D coordinates. The 2D OpenPose and the latter network give us (x, y) and z coordinates, respectively.

Evaluation Protocols. For the UWA3D Multiview Activity II, we use standard multi-view classification protocol [49,63,64], but we apply it to one-shot learning as the view combinations for training and testing sets are disjoint. For NTU-120, we follow the standard one-shot protocol [39]. Based on this protocol, we create a similar one-shot protocol for NTU-60, with 50/10 action classes used for training/testing respectively. To evaluate the effectiveness of the proposed method on viewpoint alignment, we also create two new protocols on NTU-120, for which we group the whole dataset based on (i) horizontal camera views into left, center and right views, (ii) vertical camera views into top, center and bottom views. We conduct two sets of experiments on such disjoint view-wise splits: (i) using 100 action classes for training, and testing on the same 100 action classes (ii) training on 100 action classes but testing on the rest unseen 20 classes. Appendix Section G details new/additional eval. protocols on NTU-60/NTU-120.

Stereo Projections. For simulating different camera viewpoints, we estimate the funda-mental matrix F (Eq. 7 in Appendix), which relies on camera parameters. Thus, we use the Camera Calibrator from MATLAB to estimate intrinsic, extrinsic and lens distortion parameters. For a given skeleton dataset, we compute the range of spatial coordinates x and y, respectively. We then split them into 3 equally-sized groups to form roughly left, center, right views and other 3 groups for bottom, center, top views. We choose $\sim$15 frame images from each corresponding group, upload them to the Camera Calibrator, and export camera parameters. We then compute the average distance/depth and height per group to estimate the camera position. On NTU-60 and NTU-120, we simply group the whole dataset into 3 cameras, which are left, center and right views, as provided in [39], and then we compute the average distance per camera view based on the height and distance settings given in the table in [39].

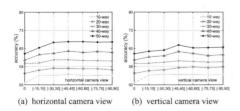

(a) horizontal camera view   (b) vertical camera view

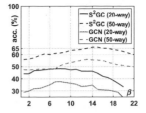

Fig. 5. The impact of viewing angles on NTU-6

Fig. 6. The impact of β in loss function on NTU-60 with S^2GC and GCN.

Table 1. Experimental results on NTU-60 (left) and NTU-120 (right) for different camera viewpoint simulations. Below the dashed line are ablated few variants of JEANIE.

	NTU-60					NTU-120				
# Training classes	10	20	30	40	50	20	40	60	80	100
Euler simple $(K+K')$	54.3	56.2	60.4	64.0	68.1	30.7	36.8	39.5	44.3	46.9
Euler $(K \times K')$	**60.8**	67.4	67.5	70.3	**75.0**	32.9	39.2	43.5	48.4	50.2
CamVPC $(K \times K')$	59.7	**68.7**	**68.4**	70.4	73.2	**33.1**	**40.8**	**43.7**	**48.4**	**51.4**
V(Euler)	54.0	56.0	60.2	63.8	67.8	30.6	36.7	39.2	44.0	47.0
2V(Euler simple)	54.3	56.2	60.4	64.0	68.1	30.7	36.8	39.5	44.3	46.9
2V(Euler)	60.8	67.4	67.5	70.3	75.0	32.9	39.2	43.5	48.4	50.2
2V(CamVPC)	59.7	68.7	68.4	70.4	73.2	33.1	40.8	43.7	48.4	51.4
2V(CamVPC+crossval.)	63.4	72.4	73.5	73.2	78.1	37.2	43.0	49.2	50.0	55.2
2V(CamVPC+crossval.)+Transf.	**65.0**	**75.2**	**76.7**	**78.9**	**80.0**	**38.5**	**44.1**	**50.3**	**51.2**	**57.0**

4.1 Ablation Study

We start our experiments by investigating the GNN backbones (Appendix Section C.1), camera viewpoint simulation and their hyper-parameters (Appendix Section C.3, C.4, C.5).

Camera Viewpoint Simulations. We choose $15°$ as the step size for the viewpoints simulation. The ranges of camera azimuth/altitude are in $[-90°, 90°]$. Where stated, we perform a grid search on camera azimuth/altitude with Hyperopt. Below, we explore the choice of the angle ranges for both horizontal and vertical views. Figure 5a and 5b (evaluations on the NTU-60 dataset) show that the angle range $[-45°, 45°]$ performs the best, and widening the range in both views does not increase the performance any further. Table 1 (top) shows results for the chosen range $[-45°, 45°]$ of camera viewpoint simulations. Euler simple $(K+K')$ denotes a simple concatenation of features from both horizontal and vertical views, whereas Euler/CamVPC$(K \times K')$ represents the grid search of all possible views. It shows that Euler angles for the viewpoint augmentation outperform Euler simple, and CamVPC (viewpoints of query sequences are generated by the stereo projection geometry) outperforms Euler angles in almost all the experiments on NTU-60 and NTU-120. This proves the effectiveness of using the stereo projection geometry for the viewpoint augmentation. More baseline experiments with/without viewpoint alignment are in Appendix Sec. C.2.

Table 2. Experimental results on NTU-60 (left) and NTU-120 (right) for ι-max shift. ι-max shift is the max. Viewpoint shift from block to block in JEANIE.

	NTU-60					NTU-120				
	10	20	30	40	50	20	40	60	80	100
$\iota=1$	60.8	70.7	72.5	72.9	75.2	36.3	42.5	48.7	**50.0**	54.8
$\iota=2$	**63.8**	**72.9**	**74.0**	**73.4**	**78.1**	**37.2**	**43.0**	**49.2**	**50.0**	**55.2**
$\iota=3$	55.2	58.9	65.7	67.1	72.5	36.7	**43.0**	48.5	49.0	54.9
$\iota=4$	54.5	57.8	63.5	65.2	70.4	36.5	42.9	48.3	48.9	54.3

Table 3. The impact of the number of frames M in temporal block under stride step S on results (NTU-60). $S=pM$, where $1-p$ describes the temporal block overlap percentage. Higher p means fewer overlap frames between temporal blocks.

	$S=M$		$S=0.8M$		$S=0.6M$		$S=0.4M$		$S=0.2M$	
M	50-class	20-class	50-class	20-class	50-class	20-class	50-class	20-class	50-class	20-class
5	69.0	55.7	71.8	57.2	69.2	59.6	73.0	60.8	71.2	61.2
6	69.4	54.0	65.4	54.1	67.8	58.0	72.0	57.8	**73.0**	**63.0**
8	67.0	52.7	67.0	52.5	**73.8**	**61.8**	67.8	60.3	68.4	59.4
10	62.2	44.5	63.6	50.9	65.2	48.4	62.4	57.0	70.4	56.7
15	62.0	43.5	62.6	48.9	64.7	47.9	62.4	57.2	68.3	56.7
30	55.6	42.8	57.2	44.8	59.2	43.9	58.8	55.3	60.2	53.8
45	50.0	39.8	50.5	40.6	52.3	39.9	53.0	42.1	54.0	45.2

Evaluation of β. Figure 6 shows that if $\beta=8$ and 14, our loss function performs the best on 20- and 50-class protocol, respectively, on NTU-60 for the S^2GC and GCN backbone. Moreover, β is not affected by backbone.

The ι-max shift. Table 2 shows the evaluations of ι for the maximum shift. We notice that $\iota=2$ yields the best results for all the experimental settings on both NTU-60 and NTU-120. Increasing ι does not help improve the performance.

Block Size and Strides. Table 3 shows evaluations of block size M and stride S, and indicates that the best performance (both 50- and 20-class) is achieved for smaller block size (frame count in the block) and smaller stride. Longer temporal blocks decrease the performance due to the temporal information not reaching the temporal alignment step. Our block encoder encodes each temporal block for learning the local temporal motions, and aggregate these block features finally to form the global temporal motion cues. Smaller stride helps capture more local motion patterns. Considering the computational cost and the performance, we choose $M=8$ and $S=0.6M$.

Euler vs CamVPC. Table 1 (bottom) shows that using the viewpoint alignment simultaneously in two dimensions, x and y for Euler angles, or azimuth and altitude the stereo projection geometry (*CamVPC*), improves the performance by 5–8% compared to (*Euler simple*), a variant where the best viewpoint alignment path was chosen from the best alignment path along x and the best alignment path along y. Euler simple is better than Euler with y rotations only ((*V*) includes rotations along y while (*2V*) includes rotations along two axes). Using HyperOpt [4] to search for the best angle range in which we perform the viewpoint alignment (*CamVPC+crossval.*) improves results. Enabling the viewpoint alignment for support sequences yields extra improvement. With Transformer (*2V+Transf.*), JEANIE boosts results by $\sim 2\%$.

Table 4. Results on NTU-60 (S^2GC backbone). Models use temporal alignment by soft-DTW or JEANIE (joint temporal-viewpoint alignment) except if indicated otherwise.

# Training Classes	10	20	30	40	50
Each frame to frontal view	52.9	53.3	54.6	54.2	58.3
Each block to frontal view	53.9	56.1	60.1	63.8	68.0
Traj. aligned baseline (video-level)	36.1	40.3	44.5	48.0	50.2
Traj. aligned baseline (block-level)	52.9	55.8	59.4	63.6	66.7
Matching Nets [61]	46.1	48.6	53.3	56.2	58.8
Matching Nets [61]+2V	47.2	50.7	55.4	57.7	60.2
Prototypical Net [54]	47.2	51.1	54.3	58.9	63.0
Prototypical Net [54]+2V	49.8	53.1	56.7	60.9	64.3
TAP [55]	54.2	57.3	61.7	64.7	68.3
S^2GC (no soft-DTW)	50.8	54.7	58.8	60.2	62.8
soft-DTW	53.7	56.2	60.0	63.9	67.8
(no soft-DTW)+Transf.	56.0	64.2	67.3	70.2	72.9
soft-DTW+Transf.	57.3	66.1	68.8	72.3	74.0
JEANIE+Transf.	**65.0**	**75.2**	**76.7**	**78.9**	**80.0**

4.2 Comparisons with the State-of-the-Art Methods

One-Shot Action Recognition (NTU-60). Table 4 shows that aligning query and support trajectories by the angle of torso 3D joint, denoted (*Traj. aligned baseline*) is not very powerful, as alluded to in Fig. 2 (top). Aligning piece-wise parts (blocks) is better than aligning entire trajectories. In fact, aligning individual frames by torso to the frontal view (*Each frame to frontal view*) and aligning block average of torso direction to the frontal view (*Each block to frontal view*)) were marginally better. We note these baselines use soft-DTW. We show more comparisons in Appendix Sec. E. Our JEANIE with Transformer (*JEANIE+Transf.*) outperforms soft-DTW with Transformer (*soft-DTW+Transf.*) by 7.46% on average.

One-Shot Action Recognition (NTU-120). Table 5 shows that JEANIE outperforms recent SL-DML and Skeleton-DML by 6.1% and 2.8% respectively (100 training classes). For comparisons, we extended the view adaptive neural networks [81] by combining them with Prototypical Net [54]. VA-RNN+VA-CNN [81] uses 0.47M+24M parameters with random rotation augmentations while JEANIE uses 0.25–0.5M params. Their *rotation+translation* keys are not proven to perform smooth optimal alignment as JEANIE. In contrast, d_{JEANIE} performs jointly a smooth viewpoint-temporal alignment via a principled transportation plan (≥ 3 dim. space) by design. Their use Euler angles which are a worse option than the camera projection of JEANIE. We notice that ProtoNet+VA backbones is 12% worse than our JEANIE. Even if we split skeletons into blocks to let soft-DTW perform temporal alignment of prototypes and query, JEANIE is still 4–6% better. JEANIE outperforms FVM by 2–4%. This shows that seeking jointly the best temporal-viewpoint alignment is more valuable than considering viewpoint alignment as a local task (free range alignment per each step of soft-DTW).

Table 5. Experimental results on NTU-120 (S^2GC backbone). Methods use temporal alignment by soft-DTW or JEANIE (joint temporal-viewpoint alignment) except VA [80,81] and other cited works. For VA*, we used soft-DTW on temporal blocks while VA generated temporal blocks.

# Training Classes	20	40	60	80	100
APSR [39]	29.1	34.8	39.2	42.8	45.3
SL-DML [45]	36.7	42.4	49.0	46.4	50.9
Skeleton-DML [44]	28.6	37.5	48.6	48.0	54.2
Prototypical Net+VA-RNN(aug.) [80]	25.3	28.6	32.5	35.2	38.0
Prototypical Net+VA-CNN(aug.) [81]	29.7	33.0	39.3	41.5	42.8
Prototypical Net+VA-fusion(aug.) [81]	29.8	33.2	39.5	41.7	43.0
Prototypical Net+VA*-fusion(aug.) [81]	33.3	38.7	45.2	46.3	49.8
TAP [55]	31.2	37.7	40.9	44.5	47.3
S^2GC(no soft-DTW)	30.0	35.9	39.2	43.6	46.4
soft-DTW	30.3	37.2	39.7	44.0	46.8
(no soft-DTW)+Transf.	31.2	37.5	42.3	47.0	50.1
soft-DTW+Transf.	31.6	38.0	43.2	47.8	51.3
FVM+Transf.	34.5	41.9	44.2	48.7	52.0
JEANIE+Transf.	**38.5**	**44.1**	**50.3**	**51.2**	**57.0**

Table 6. Experiments on 2D and 3D Kinetics-skeleton. Note that we have no results on JEANIE or FVM for 2D coordinates (aligning viewpoints is an operation in 3D).

	S^2GC (no soft-DTW)	soft-DTW	*FVM*	JEANIE	JEANIE +Transf.
2D skel.	32.8	34.7	-	-	-
3D skel.	35.9	39.6	44.1	**50.3**	**52.5**

JEANIE on the Kinetics-Skeleton. We evaluate our proposed model on both 2D and 3D Kinetics-skeleton. We split the whole dataset into 200 actions for training, and the rest half for testing. As we are unable to estimate the camera location, we simply use Euler angles for the camera viewpoint simulation. Table 6 shows that using 3D skeletons outperforms the use of 2D skeletons by 3–4%, and JEANIE outperforms the baseline (temporal alignment only) and Free Viewpoint Matching (FVM, for every step of DTW, seeks the best local viewpoint alignment, thus realizing non-smooth temporal-viewpoint path in contrast to JEANIE) by around 5% and 6%, respectively. With the transformer, JEANIE further boosts results by 2%.

Few-Shot Multiview Classification. Table 7 (UWA3D Multiview Activity II) shows that adding temporal alignment to SGC, APPNP and S^2GC improves the performance, and the big performance gain is obtained by further adding the viewpoint alignment. As this dataset is challenging in recognizing the actions from a novel view point, our proposed method performs consistently well on all different combinations of training/testing viewpoint variants. This is predictable as our method aligns both temporal and camera viewpoints which allows a robust classification. JEANIE outperforms FVM by 4.2%, and outperforms the baseline (with temporal alignment only) by 7% on average.

Table 7. Experiments on the UWA3D Multiview Activity II.

Training view	V_1 & V_2		V_1 & V_3		V_1 & V_4		V_2 & V_3		V_2 & V_4		V_3 & V_4		Mean
Testing view	V_3	V_4	V_2	V_4	V_2	V_3	V_1	V_4	V_1	V_3	V_1	V_2	
GCN	36.4	26.2	20.6	30.2	33.7	22.4	43.1	26.6	16.9	12.8	26.3	36.5	27.6
SGC	40.9	60.3	44.1	52.6	48.5	38.7	50.6	52.8	52.8	37.2	57.8	49.6	48.8
+soft-DTW	43.9	60.8	48.1	54.6	52.6	45.7	54.0	58.2	56.7	40.2	60.2	51.1	52.2
+JEANIE	47.0	62.8	50.4	57.8	53.6	47.0	57.9	62.3	57.0	44.8	61.7	52.3	54.6
APPNP	42.9	61.9	47.8	58.7	53.8	44.0	52.3	60.3	55.1	38.2	58.3	47.9	51.8
+soft-DTW	44.3	63.2	50.7	62.3	53.9	45.0	56.9	62.8	56.4	39.3	60.1	51.9	53.9
+JEANIE	46.8	64.6	51.3	65.1	54.7	46.4	58.2	65.1	58.8	43.9	60.3	52.5	55.6
S^2GC	45.5	64.4	46.8	61.6	49.5	43.2	57.3	61.2	51.0	42.9	57.0	49.2	52.5
+soft-DTW	48.2	67.2	51.2	67.0	53.2	46.8	62.4	66.2	57.8	45.0	62.2	53.0	56.7
+FVM	50.7	68.8	56.3	69.2	55.8	47.1	63.7	68.8	62.5	51.4	63.8	55.7	59.5
+JEANIE	**55.3**	**70.2**	**61.4**	**72.5**	**60.9**	**50.8**	**66.4**	**73.9**	**68.8**	**57.2**	**66.7**	**60.2**	**63.7**

Table 8. Results on NTU-120 (multiview classification). Baseline is soft-DTW + S^2GC.

Training view	bott.	bott.	bott.& cent.	left	left	left & cent.
Testing view	Cent.	Top	Top	Cent.	Right	Right
100/same 100 (baseline)	74.2	73.8	75.0	58.3	57.2	68.9
100/same 100 (*FVM*)	79.9	78.2	80.0	65.9	63.9	75.0
100/same 100 (*JEANIE*)	**81.5**	**79.2**	**83.9**	**67.7**	**66.9**	**79.2**
100/novel 20 (baseline)	58.2	58.2	61.3	51.3	47.2	53.7
100/novel 20 (*FVM*)	66.0	65.3	68.2	58.8	53.9	60.1
100/novel 20 (*JEANIE*)	**67.8**	**65.8**	**70.8**	**59.5**	**55.0**	**62.7**

Table 8 (NTU-120) shows that adding more camera viewpoints to the training process helps the multiview classification. Using bottom and center views for training and top view for testing, or using left and center views for training and the right view for testing yields 4% gain ('*same 100*' means the same train/test classes but different views). Testing on 20 novel classes ('*novel 20*' never used in training) yields 62.7% and 70.8% for multiview classification in horizontal and vertical camera viewpoints, respectively.

5 Conclusions

We have proposed a Few-shot Action Recognition (FSAR) approach for learning on 3D skeletons via JEANIE. We have demonstrated that the joint alignment of temporal blocks and simulated viewpoints of skeletons between support-query sequences is efficient in the meta-learning setting where the alignment has to be performed on new action classes under the low number of samples. Our experiments have shown that using the stereo camera geometry is more efficient than simply generating multiple views by Euler angles in the meta-learning regime. Most importantly, we have introduced a novel FSAR approach that learns on articulated 3D body joints.

Acknowledgements. We thank Dr. Jun Liu (SUTD) for discussions on FSAR for 3D skeletons, and CSIRO's Machine Learning and Artificial Intelligence Future Science Platform (MLAI FSP).

References

1. Euler angles. Wikipedia. https://en.wikipedia.org/wiki/Euler_angles. Accessed 08 Mar 2022
2. Lecture 12: Camera projection. On-line. https://www.cse.psu.edu/~rtc12/CSE486/lecture12.pdf. Accessed: 08 Mar 2022
3. Bart, E., Ullman, S.: Cross-generalization: Learning novel classes from a single example by feature replacement. In: CVPR, pp. 672–679 (2005)
4. Bergstra, J., Komer, B., Eliasmith, C., Yamins, D., Cox, D.D.: Hyperopt: a python library for model selection and hyperparameter optimization. Comput. Sci. Discov. **8**(1), 014008 (2015)
5. Cao, K., Ji, J., Cao, Z., Chang, C.Y., Niebles, J.C.: Few-shot video classification via temporal alignment. In: CVPR (2020)
6. Cao, Z., Simon, T., Wei, S.E., Sheikh, Y.: Realtime multi-person 2d pose estimation using part affinity fields. In: CVPR (2017)
7. Carreira, J., Zisserman, A.: Quo vadis, action recognition? A new model and the kinetics dataset. In: CVPR (2017)
8. Catalin, I., Dragos, P., Vlad, O., Cristian, S.: Human3.6m: large scale datasets and predictive methods for 3D human sensing in natural environments. IEEE TPAMI (2014)
9. Cheng, K., Zhang, Y., He, X., Chen, W., Cheng, J., Lu, H.: Skeleton-based action recognition with shift graph convolutional network. In: CVPR (2020)
10. Cuturi, M.: Fast global alignment kernels. In: ICML (2011)
11. Cuturi, M., Blondel, M.: Soft-DTW: a differentiable loss function for time-series. In: ICML (2017)
12. Dosovitskiy, A., et al.: An image is worth 16x16 words: transformers for image recognition at scale. In: ICLR (2020)
13. Dvornik, N., Schmid, C., Mairal, J.: Selecting relevant features from a multi-domain representation for few-shot classification. In: Vedaldi, A., Bischof, H., Brox, T., Frahm, J.-M. (eds.) ECCV 2020. LNCS, vol. 12355, pp. 769–786. Springer, Cham (2020). https://doi.org/10.1007/978-3-030-58607-2_45
14. Dwivedi, S.K., Gupta, V., Mitra, R., Ahmed, S., Jain, A.: Protogan: towards few shot learning for action recognition. arXiv (2019)
15. Elsken, T., Staffler, B., Metzen, J.H., Hutter, F.: Meta-learning of neural architectures for few-shot learning. In: CVPR (2020)
16. Fei, N., Guan, J., Lu, Z., Gao, Y.: Few-shot zero-shot learning: Knowledge transfer with less supervision. In: ACCV (2020)
17. Fei-Fei, L., Fergus, R., Perona, P.: One-shot learning of object categories. IEEE Trans. Pattern Anal. Mach. Intell. **28**(4), 594–611 (2006)
18. Feichtenhofer, C., Pinz, A., Wildes, R.P.: Spatiotemporal multiplier networks for video action recognition. In: CVPR (2017)
19. Feichtenhofer, C., Pinz, A., Zisserman, A.: Convolutional two-stream network fusion for video action recognition. In: CVPR (2016)
20. Fink, M.: Object classification from a single example utilizing class relevance metrics. In: NeurIPS, pp. 449–456 (2005)
21. Finn, C., Abbeel, P., Levine, S.: Model-agnostic meta-learning for fast adaptation of deep networks. In: Precup, D., Teh, Y.W. (eds.) ICML, vol. 70, pp. 1126–1135. PMLR (2017)
22. Guan, J., Zhang, M., Lu, Z.: Large-scale cross-domain few-shot learning. In: ACCV (2020)

23. Guo, M., Chou, E., Huang, D.-A., Song, S., Yeung, S., Fei-Fei, L.: Neural graph matching networks for fewshot 3d action recognition. In: Ferrari, V., Hebert, M., Sminchisescu, C., Weiss, Y. (eds.) ECCV 2018. LNCS, vol. 11205, pp. 673–689. Springer, Cham (2018). https://doi.org/10.1007/978-3-030-01246-5_40

24. Guo, Y., Codella, N.C., Karlinsky, L., Codella, J.V., Smith, J.R., Saenko, K., Rosing, T., Feris, R.: A broader study of cross-domain few-shot learning. In: Vedaldi, A., Bischof, H., Brox, T., Frahm, J.-M. (eds.) ECCV 2020. LNCS, vol. 12372, pp. 124–141. Springer, Cham (2020). https://doi.org/10.1007/978-3-030-58583-9_8

25. Haasdonk, B., Burkhardt, H.: Invariant kernel functions for pattern analysis and machine learning. Mach. Learn. **68**(1), 35–61 (2007)

26. Kay, W., et al.: The kinetics human action video dataset. arXiv (2017)

27. Kipf, T.N., Welling, M.: Semi-supervised classification with graph convolutional networks. In: ICLR (2017)

28. Klicpera, J., Bojchevski, A., Gunnemann, S.: Predict then propagate: graph neural networks meet personalized pagerank. In: ICLR (2019)

29. Koch, G., Zemel, R., Salakhutdinov, R.: Siamese neural networks for one-shot image recognition. In: ICML Deep Learning Workshop, vol. 2 (2015)

30. Koniusz, P., Wang, L., Cherian, A.: Tensor representations for action recognition. IEEE TPAMI (2020)

31. Koniusz, P., Wang, L., Sun, K.: High-order tensor pooling with attention for action recognition. arXiv (2021)

32. Koniusz, P., Zhang, H.: Power normalizations in fine-grained image, few-shot image and graph classification. IEEE Trans. Pattern Anal. Mach. Intell. **44**(2), 591–609 (2022)

33. Lake, B.M., Salakhutdinov, R., Gross, J., Tenenbaum, J.B.: One shot learning of simple visual concepts. CogSci (2011)

34. Li, F.F., VanRullen, R., Koch, C., Perona, P.: Rapid natural scene categorization in the near absence of attention. Proc. Natl. Acad. Sci. **99**(14), 9596–9601 (2002)

35. Li, K., Zhang, Y., Li, K., Fu, Y.: Adversarial feature hallucination networks for few-shot learning. In: CVPR (2020)

36. Li, M., Chen, S., Chen, X., Zhang, Y., Wang, Y., Tian, Q.: Actional-structural graph convolutional networks for skeleton-based action recognition. In: CVPR (2019)

37. Lichtenstein, M., Sattigeri, P., Feris, R., Giryes, R., Karlinsky, L.: TAFSSL: task-adaptive feature sub-space learning for few-shot classification. In: Vedaldi, A., Bischof, H., Brox, T., Frahm, J.-M. (eds.) ECCV 2020. LNCS, vol. 12352, pp. 522–539. Springer, Cham (2020). https://doi.org/10.1007/978-3-030-58571-6_31

38. Liu, J., Wang, G., Hu, P., Duan, L., Kot, A.C.: Global context-aware attention LSTM networks for 3d action recognition. In: CVPR, pp. 3671–3680 (2017)

39. Liu, J., Shahroudy, A., Perez, M., Wang, G., Duan, L.Y., Kot, A.C.: NTU RGB+D 120: a large-scale benchmark for 3D human activity understanding. IEEE Trans. Pattern Anal. Mach. Intell. (2019)

40. Liu, Z., Zhang, H., Chen, Z., Wang, Z., Ouyang, W.: Disentangling and unifying graph convolutions for skeleton-based action recognition. In: CVPR (2020)

41. Lu, C., Koniusz, P.: Few-shot keypoint detection with uncertainty learning for unseen species. In: CVPR (2022)

42. Luo, Q., Wang, L., Lv, J., Xiang, S., Pan, C.: Few-shot learning via feature hallucination with variational inference. In: WACV (2021)

43. Martinez, J., Hossain, R., Romero, J., Little, J.J.: A simple yet effective baseline for 3d human pose estimation. In: ICCV. pp. 2659–2668 (2017)

44. Memmesheimer, R., Häring, S., Theisen, N., Paulus, D.: Skeleton-DML: deep metric learning for skeleton-based one-shot action recognition. arXiv (2021)

45. Memmesheimer, R., Theisen, N., Paulus, D.: Signal level deep metric learning for multi-modal one-shot action recognition. arXiv (2020)
46. Miller, E.G., Matsakis, N.E., Viola, P.A.: Learning from one example through shared densities on transforms. In: CVPR, vol. 1, pp. 464–471 (2000)
47. Mishra, A., Verma, V.K., Reddy, M.S.K., Arulkumar, S., Rai, P., Mittal, A.: A generative approach to zero-shot and few-shot action recognition. In: WACV, pp. 372–380 (2018)
48. Qin, Z., et al.: Fusing higher-order features in graph neural networks for skeleton-based action recognition. IEEE Trans. Neural Netw. Learn. Syst. (99), 1–15 (2022)
49. Rahmani, H., Mahmood, A., Huynh, D.Q., Mian, A.: Histogram of Oriented Principal Components for Cross-View Action Recognition. IEEE Trans. Pattern Anal. Mach. Intell. **38**, 2430–2443 (2016)
50. Shahroudy, A., Liu, J., Ng, T.T., Wang, G.: NTU RGB+D: a large scale dataset for 3D human activity analysis. In: CVPR (2016)
51. Simon, C., Koniusz, P., Harandi, M.: On learning the geodesic path for incremental learning. In: CVPR, pp. 1591–1600 (2021)
52. Simon, C., Koniusz, P., Nock, R., Harandi, M.: On Modulating the gradient for meta-learning. In: Vedaldi, A., Bischof, H., Brox, T., Frahm, J.-M. (eds.) ECCV 2020. LNCS, vol. 12353, pp. 556–572. Springer, Cham (2020). https://doi.org/10.1007/978-3-030-58598-3_33
53. Smola, A.J., Kondor, R.: Kernels and regularization on graphs. In: Schölkopf, B., Warmuth, M.K. (eds.) COLT-Kernel 2003. LNCS (LNAI), vol. 2777, pp. 144–158. Springer, Heidelberg (2003). https://doi.org/10.1007/978-3-540-45167-9_12
54. Snell, J., Swersky, K., Zemel, R.S.: Prototypical networks for few-shot learning. In: Guyon, I., et al.: (eds.) NeurIPS, pp. 4077–4087 (2017)
55. Su, B., Wen, J.R.: Temporal alignment prediction for supervised representation learning and few-shot sequence classification. In: ICLR (2022)
56. Sun, K., Koniusz, P., Wang, Z.: Fisher-Bures adversary graph convolutional networks. In: Conference on Uncertainty in Artificial Intelligence, Israel, vol. 115, pp. 465–475 (2019)
57. Sung, F., Yang, Y., Zhang, L., Xiang, T., Torr, P.H.S., Hospedales, T.M.: Learning to compare: Relation network for few-shot learning. In: CVPR, pp. 1199–1208 (2018)
58. Tang, L., Wertheimer, D., Hariharan, B.: Revisiting pose-normalization for fine-grained few-shot recognition. In: CVPR (2020)
59. Tran, D., Bourdev, L., Fergus, R., Torresani, L., Paluri, M.: Learning spatiotemporal features with 3d convolutional networks. In: ICCV (2015)
60. Villani, C.: Optimal Transport Old and New. Springer, Heidelberg (2009). https://doi.org/10.1007/978-3-540-71050-9
61. Vinyals, O., Blundell, C., Lillicrap, T., Kavukcuoglu, K., Wierstra, D.: Matching networks for one shot learning. In: Lee, D.D., Sugiyama, M., von Luxburg, U., Guyon, I., Garnett, R. (eds.) NeurIPS, pp. 3630–3638 (2016)
62. Wang, L., Xiong, Y., Wang, Z., Qiao, Y., Lin, D., Tang, X., Van Gool, L.: Temporal segment networks for action recognition in videos. IEEE Trans. Pattern. Anal. Mach. Intell. **41**(11), 2740–2755 (2019)
63. Wang, L.: Analysis and evaluation of kinect-based action recognition algorithms. Master's thesis, School of the Computer Science and Software Engineering, The University of Western Australia (2017)
64. Wang, L., Huynh, D.Q., Koniusz, P.: A comparative review of recent kinect-based action recognition algorithms. IEEE Trans. Image Process. **29**, 15–28 (2020)
65. Wang, L., Huynh, D.Q., Mansour, M.R.: Loss switching fusion with similarity search for video classification. In: ICIP (2019)
66. Wang, L., Koniusz, P.: Self-supervising action recognition by statistical moment and subspace descriptors. In: ACM-MM, pp. 4324–4333 (2021)

67. Wang, L., Koniusz, P.: Uncertainty-DTW for time series and sequences. In: Avidan, S., Brostow, G., Cisse, M., Farinella, G.M., Hassner, T. (eds) Computer Vision–ECCV 2022. ECCV 2022. LNCS, vol. 13681. Springer, Cham (2022). https://doi.org/10.1007/978-3-031-19803-8_11

68. Wang, L., Koniusz, P., Huynh, D.Q.: Hallucinating IDT descriptors and I3D optical flow features for action recognition with CNNs. In: ICCV (2019)

69. Wang, L., Ding, Z., Tao, Z., Liu, Y., Fu, Y.: Generative multi-view human action recognition. In: ICCV (2019)

70. Wang, S., Yue, J., Liu, J., Tian, Q., Wang, M.: Large-scale few-shot learning via multi-modal knowledge discovery. In: Vedaldi, A., Bischof, H., Brox, T., Frahm, J.-M. (eds.) ECCV 2020. LNCS, vol. 12355, pp. 718–734. Springer, Cham (2020). https://doi.org/10.1007/978-3-030-58607-2_42

71. Wang, Y., Long, M., Wang, J., Yu, P.S.: Spatiotemporal pyramid network for video action recognition. In: CVPR (2017)

72. Wu, F., Zhang, T., de Souza Jr., A.H., Fifty, C., Yu, T., Weinberger, K.Q.: Simplifying graph convolutional networks. In: ICML (2019)

73. Xu, B., Ye, H., Zheng, Y., Wang, H., Luwang, T., Jiang, Y.G.: Dense dilated network for few shot action recognition. In: ACM ICMR, pp. 379–387 (2018)

74. Yan, S., Xiong, Y., Lin, D.: Spatial Temporal Graph Convolutional Networks for Skeleton-Based Action Recognition. In: AAAI (2018)

75. Yu, X., Zhuang, Z., Koniusz, P., Li, H.: 6DoF object pose estimation via differentiable proxy voting regularizer. In: BMVC. BMVA Press (2020)

76. Zhang, H., Koniusz, P.: Power normalizing second-order similarity network for few-shot learning. In: WACV, pp. 1185–1193 (2019)

77. Zhang, H., Koniusz, P., Jian, S., Li, H., Torr, P.H.S.: Rethinking class relations: absolute-relative supervised and unsupervised few-shot learning. In: CVPR, pp. 9432–9441 (June 2021)

78. Zhang, H., Li, H., Koniusz, P.: Multi-level second-order few-shot learning. IEEE Trans. Multim. (99), 1 (2022)

79. Zhang, H., Zhang, L., Qi, X., Li, H., Torr, P.H.S., Koniusz, P.: Few-shot action recognition with permutation-invariant attention. In: Vedaldi, A., Bischof, H., Brox, T., Frahm, J.-M. (eds.) ECCV 2020. LNCS, vol. 12350, pp. 525–542. Springer, Cham (2020). https://doi.org/10.1007/978-3-030-58558-7_31

80. Zhang, P., Lan, C., Xing, J., Zeng, W., Xue, J., Zheng, N.: View adaptive recurrent neural networks for high performance human action recognition from skeleton data. In: ICCV (2017)

81. Zhang, P., Lan, C., Xing, J., Zeng, W., Xue, J., Zheng, N.: View adaptive neural networks for high performance skeleton-based human action recognition. IEEE Trans. Pattern Aanal. Mach. Intell. **41**(8), 1963–1978 (2019)

82. Zhang, S., Luo, D., Wang, L., Koniusz, P.: Few-shot object detection by second-order pooling. In: Ishikawa, H., Liu, C.-L., Pajdla, T., Shi, J. (eds.) ACCV 2020. LNCS, vol. 12625, pp. 369–387. Springer, Cham (2021). https://doi.org/10.1007/978-3-030-69538-5_23

83. Zhang, S., Murray, N., Wang, L., Koniusz, P.: Time-rEversed diffusioN tEnsor transformer: a new TENET of few-shot object detection. In: Avidan, S., Brostow, G., Cisse, M., Farinella, G.M., Hassner, T. (eds.) Computer Vision–ECCV 2022. ECCV 2022. LNCS, vol. 13680. Springer, Cham (2022). https://doi.org/10.1007/978-3-031-20044-1_18

84. Zhang, S., Wang, L., Murray, N., Koniusz, P.: Kernelized few-shot object detection with efficient integral aggregation. In: CVPR, pp. 19207–19216 (June 2022)

85. Zhu, H., Koniusz, P.: Simple spectral graph convolution. In: ICLR (2021)

86. Zhu, H., Koniusz, P.: EASE: unsupervised discriminant subspace learning for transductive few-shot learning. In: CVPR (2022)
87. Zhu, H., Sun, K., Koniusz, P.: Contrastive laplacian eigenmaps. In: NeurIPS, pp. 5682–5695 (2021)
88. Zhu, L., Yang, Y.: Compound memory networks for few-shot video classification. In: Ferrari, V., Hebert, M., Sminchisescu, C., Weiss, Y. (eds.) ECCV 2018. LNCS, vol. 11211, pp. 782–797. Springer, Cham (2018). https://doi.org/10.1007/978-3-030-01234-2_46

Video Analysis and Event Recognition

Spatial Temporal Network for Image and Skeleton Based Group Activity Recognition

Xiaolin Zhai[1,2]![ORCID], Zhengxi Hu[1,2]![ORCID], Dingye Yang[1,2]![ORCID], Lei Zhou[1,2]![ORCID], and Jingtai Liu[1,2]([✉])![ORCID]

[1] Institute of Robotics and Automatic Information System,
College of Artificial Intelligence, Nankai University, Tianjin, China
{2120210410,hzx,1711502}@mail.nankai.edu.cn, liujt@nankai.edu.cn
[2] Tianjin Key Laboratory of Intelligent Robotics, Nankai University, Tianjin, China

Abstract. Group activity recognition aims to infer group activity in multi-person scenes. Previous methods usually model inter-person relations and integrate individuals' features into group representations. However, they neglect intra-person relations contained in the human skeleton. Individual representations can also be inferred by analyzing the evolution of human skeletons. In this paper, we utilize RGB images and human skeletons as the inputs which contain complementary information. Considering different semantic attributes of the two inputs, we design two diverse branches, respectively. For RGB images, we propose Scene Encoded Transformer, Spatial Transformer, and Temporal Transformer to explore inter-person spatial and temporal relations. For skeleton inputs, we capture the intra-person spatial and temporal dynamics by designing Spatial and Temporal GCN. Our main contributions are: i) we propose a spatial-temporal network with two branches for group activity recognition utilizing RGB images and human skeletons. Experiments show that our model achieves 97.1% MCA and 96.1% MPCA on the Collective Activity dataset and 94.0% MCA and 94.4% MPCA on the Volleyball dataset. ii) we extend the two datasets by introducing human skeleton annotations, namely human joint coordinates and confidence, which can also be used in the action recognition task. The code is available at https://github.com/zxll0106/Image_and_Skeleton_Based_Group_Activity_Recognition.

Keywords: Group activity recognition · Video analysis · Scene understanding

1 Introduction

Group activity recognition [6] is widely used in social behavior understanding, service robots, and autonomous driving cars, therefore playing a vital role in

This work is supported in part by National Key Research and Development Project under Grant 2019YFB1310604, in part by National Natural Science Foundation of China under Grant 62173189.

L. Wang et al. (Eds.): ACCV 2022, LNCS 13844, pp. 329–346, 2023.
https://doi.org/10.1007/978-3-031-26316-3_20

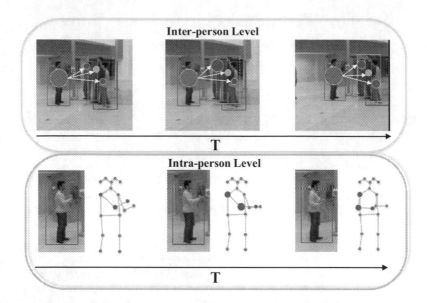

Fig. 1. Intuitive examples of a *talking* video clip in the Collective Activity dataset. On the inter-person level, we capture the spatial relation in the group at the same frame, and temporal dynamics of interaction between consecutive frames. On the intra-person level, the human skeleton reveals the more detailed evolution of the individual action, so we model the complex and diverse spatial-temporal individual features.

video analysis and scene understanding. The goal of group activity recognition is to understand what they are doing in the multi-person scene. In the previous methods, great efforts have been made to verify the effectiveness of modeling inter-person relations. However, intra-person relations in the human skeleton convey fine-grained actions information, but receive less attention in the group activity recognition task.

To model the relationship between individuals, graph neural networks are applied to represent inter-person social relations. Nodes in the graph are individuals and the edges are formed by the relationship between them. Recent attention in group activity recognition has also focused on Transformer [28] which was proposed in the field of natural language processing. Transformer in group activity recognition [11,19,37] is utilized to model the inter-person relation and combine individuals' features. In addition, some methods [2,11,19] utilized optical flow frames which contain different motion features from RGB images. Previous works show impressive improvement, which suggests the effectiveness of modeling inter-person relations. However, they neglect that intra-person information of the human skeleton contains fine-grained motion dynamics.

Human skeletons convey information complementary to RGB images and optical flow frames. Human bodies can be viewed as a whole composed of a trunk and limbs, while human actions can be regarded as motions of joints and bones. And skeletons can be represented by 2D position sequences of human joints. With the development of pose estimation networks, we can obtain more accurate human joint coordinates. From reliable skeleton data, we can model intra-person spatial-temporal dynamics and extract an effective representation.

Accordingly, we propose a spatial-temporal network with two branches to capture the intra-person and inter-person relations. For RGB image inputs, we propose Scene Encoded Transformer to model the interaction between individuals and the surrounding scene, and extract informative scene features. As shown in Fig. 1, individuals also take consideration into the reaction of others to determine their behaviors, besides interacting with the scene. We adopt Spatial and Temporal Transformer to infer spatial and temporal inter-person relations, respectively. Moreover, for skeleton inputs, Spatial and Temporal GCN are designed to model intra-person spatial and temporal relations in the human skeleton. In Spatial GCN, spatial information is propagated from one node to another along intra-person connections. Temporal GCN passes temporal dynamics between consecutive frames of the same node. As shown in Fig. 1, exploiting intra-person spatial-temporal relations is important to modeling individuals' features.

The contributions of this work can be summarized as:

(1) We propose a spatial-temporal network with two branches for group activity recognition utilizing RGB images and human skeletons. For RGB image inputs, we propose Scene Encoded Transformer, Spatial Transformer, and Temporal Transformer to model the inter-person relation. For skeleton inputs, we propose Spatial and Temporal GCN to capture intra-person spatial and temporal relations which lack further exploration in the group activity recognition task. Experiment results show that our proposed model achieves 97.1% MCA and 96.1% MPCA on the Collective Activity dataset and 94.0% MCA and 94.4% MPCA on the Volleyball dataset.
(2) We extend the two datasets by introducing human skeleton annotations, namely human joint coordinates and confidences. Extended datasets can be not only used in group activity recognition but also used in skeleton-based action recognition.

2   Related Work

2.1   Group Activity Recognition

Machine learning plays an important role in addressing the issue of group activity recognition. Initially, many approaches designed hand-crafted features and applied probabilistic graphical models [1,4,5,7,12,17]. With the development of deep neural networks, many approaches [3,8,15,16,24] used CNNs to extract the feature map of the video clip. RNNs were also used to infer temporal dynamics of individual actions and group activity [3,8,15,16,24]. [16] utilized the person LSTM layer and the group LSTM layer to extract individual and group features, respectively. [3] designed a unified framework where RNN can reason the probability of individual actions. The semantic graph extracted from text labels and images was applied in stagNet [24] and structural-RNN was used to capture temporal dynamics.

Recent development in graph neural networks has improved the ability of modeling relation between individuals [9,14,23,30,33,38]. ARG [30] constructed actor relation graphs to capture appearance and position relations between actors. [9] utilized I3D as the backbone network to extract spatial-temporal features of a video clip, used self-attention to integrate individuals' features, and used graph attention to model relations between individuals. [23] used the mean-field conditional random fields to infer temporal relations and spatial relations. The social adaptive module designed in [33] has the same structure in the spatial and temporal domain and can infer key instances under the weakly supervised setting. In PRL [14], individuals' relation was represented on the semantic relation graph. Dynamic Inference Network was proposed in [38] to construct person-specific spatial and temporal graphs by designing Dynamic Relation (DR) module and Dynamic Walk (DW) module.

Meanwhile, researchers in group activity recognition [11,19,27,37,40] have shown an increased interest in Transformer [28] which was proposed in the field of natural language processing. Actor-Transformer [11] used RGB frames, optical flow frames, and pose features as input and used Transformer to model group representation. GroupFormer [19] designed a Transformer encoder to capture spatial and temporal features and adopted a Transformer decoder in a cross manner to capture spatial-temporal interactive features. [37] enhanced individuals' representations with the global contextual information and aggregated the relation between individuals using Spatial-Temporal Bi-linear Pooling module.

Previous works paid more attention to exploring the inter-person relation and confirmed its effectiveness in inferring group features. However, they neglected that the human skeleton which contains intra-person relations also conveys different fine-grained information. Different from them, we capture spatial and temporal relations both on intra-person and inter-person levels in parallel by designing two branches.

2.2   Modeling of Interaction

Modeling interaction relationships is an important component in the multi-instance problem, such as action recognition [18,21,34], pedestrian intent estimation [35,36], and trajectory prediction [10,39]. To model the interaction between joint nodes, [21] proposed the MS-G3D network which can disentangle the node representations in different spatial-temporal neighborhoods. Actional links were utilized in [18] to propagate actional information between different joint nodes and structural links were used to expand the respective fields. To exploit the interaction between the target individual and other traffic users, [35] proposed Attentive Relation Network which adopted soft attention to assign the weight of multiple traffic users. AgentFormer [39] utilized individual-aware attention to capture individual-to-itself and individual-to-others relations. Previous works have considered only one of the intra-person and inter-person relations.

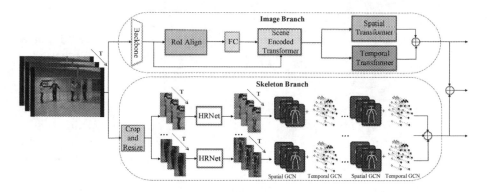

Fig. 2. Overview of the proposed network, which contains the Image Branch and the Skeleton Branch. The Image Branch is proposed to model the inter-person interaction and the Skeleton Branch is designed to extract the intra-person relation.

Different from them, we capture the interaction on the intra-person and inter-peron levels. Considering different semantic attributes of the two levels, we propose the two-branch model to refine different-grained spatial-temporal dynamics.

3 Method

In this section, we outline the pipeline of our method. The overall framework of our method is illustrated in Fig. 2. Our method consists of two branches, namely the Image Branch and the Skeleton Branch. The Image Branch takes RGB frames as input, captures the inter-person interaction, and obtains the spatial-temporal contextual individual and group features. The Skeleton branch takes human skeletons as input, adopts Spatial and Temporal GCN which are suitable for modeling intra-person relations, and obtains complementary individual and group representations. We concatenate the outputs of two branches and pass them to classifiers getting individuals' actions and group activity.

3.1 Image Branch

First, we take T-frame images as the input and obtain individual features from Image Feature Extractor. Then, Scene Encoded Transformer incorporates contextual information in the scene to enhance the individual feature. Considering the continuity and the interactivity of the individual action and group activity, Spatial and Temporal Transformer mine spatial and temporal relations of enhanced individual features. Finally, we concatenate the outputs of Spatial and Temporal Transformer and then apply a pooling layer to get the group feature of the Image Branch.

Image Feature Extractor. The input of the Image Branch is T-frame images that are centered on the labeled frame. We utilize the backbone network to extract feature maps $X \in \mathbb{R}^{T \times D \times H \times W}$ of the input frames. Given bounding boxes of N individuals, we apply RoIAlign [13] to extract individuals' features from feature maps. Then a fully-connected layer is adopted to get d-dimensional individuals' features $X_I \in \mathbb{R}^{T \times N \times d}$.

Scene Encoded Transformer. Individual actions are influenced by multiple social and environmental elements, in particular the interaction with the scene. Exploring the impact of the scene from a global view lacks further exploration in previous works [11,15,30,38]. To extract relational features of the surrounding scene, we propose Scene Encoded Transformer which captures useful information from the surrounding scene to enhance the individual features. Inspired by Transformer, we note that the attention mechanism plays a critical role in encoding the relative important elements effectively. Taking account into the unique importance of each scene element, we utilize the attention mechanism in Scene Encoded Transformer to calculate the weight based on relations between individuals and the scene. We consider that the scene feature X is composed of $H \times W$ scene elements, and each element contains the visual feature and the position feature. Different scene elements contribute to the target individual unequally, so the attention mechanism is adopted to assign an adaptive weight to each element. To align individuals' features X_I and feature maps X, we project them to d_E dimensional subspaces, obtaining $X_I' \in \mathbb{R}^{T \times N \times d_E}$ and $X' \in \mathbb{R}^{T \times HW \times d_E}$. We view X_I' as the query Q, and view X' as the key K the value V. Then we utilize the similarity between the query Q and the key K to assign the adaptive weight of the value V and compute the weighted sum of value V.

It is difficult to capture order information of inputs because there is no recurrence and no convolution in Transformer [28]. As a result, we encode their position features as follows:

$$
\begin{aligned}
PE_{(pos,2k)} &= \sin(\frac{pos}{10000^{2k/L}}) \\
PE_{(pos,2k+1)} &= \cos(\frac{pos}{1000^{2k/L}})
\end{aligned}
\tag{1}
$$

For the ith individuals' features, we encode center coordinates (x_i, y_i) of his bounding box.

$$
pos = \begin{cases} x_i, & k \in \{0, 1, \ldots, L/4 - 1\} \\ y_i, & k \in \{L/4, L/4 + 1, \ldots, L/2 - 1\} \end{cases}
\tag{2}
$$

For the scene feature map, we encode each scene element's coordinate (x, y) of the feature map, where $x \in [0, H]$, $y \in [0, W]$.

$$
pos = \begin{cases} x, & k \in \{0, 1, \ldots, L/4 - 1\} \\ y, & k \in \{L/4, d/4 + 1, \ldots, L/2 - 1\} \end{cases}
\tag{3}
$$

Finally, we obtain the scene encoded individual features $X_E \in \mathbb{R}^{T \times N \times d_E}$ which integrate the scene contextual information to individual features.

Spatial Transformer. In order to propagate information between different individuals in the scene, we propose Spatial Transformer to regard individuals as nodes of the graph and aggregate information from surrounding individuals. Our Spatial Transformer captures the target individual's interaction with surrounding individuals and assigns the relative importance weight to them. Then inter-person spatial relations can be integrated into individuals' features. Spatial Transformer views the temporal dimension of scene encoded individuals' features X_E as the batch dimension. We adopt different projection functions to map X_E to Q_S, K_S and V_S, calculate the attention weights of neighbor individuals, and obtain the feature X_S which incorporate spatial information. Positional encoding follows Scene Encoded Transformer.

Temporal Transformer. Considering the temporal correlations of individual actions and group activities, we propose Temporal Transformer to encode the evolution of individual features in the temporal domain. Some previous works [9,19] utilized I3D to integrate the temporal feature of the input images, but they ignored individual-level temporal dynamics. Compared to the pooling operation in the temporal domain [30], the attention mechanism in Temporal Transformer can calculate the adaptive importance of individuals in the consecutive frames. Temporal Transformer views the spatial dimension of scene encoded individuals' features X_E as the batch dimension, so we swap the temporal and spatial dimension of X_E reshaping them to $X'_E \in \mathbb{R}^{N \times T \times d_E}$. The self-attention mechanism takes X'_E as inputs and captures contextual temporal information to enrich individuals' features.

Positional encoding in Temporal Transformer uses time order as the position information of different frames.

$$pos = t, t \in [0, T] \tag{4}$$

3.2   Skeleton Branch

We obtain human skeleton data from RGB images using Skeleton Data Extractor. Spatial and Temporal GCN are utilized to model intra-person spatial-temporal features. These features are pooled to form the group feature of the Skeleton Branch.

Skeleton Data Extractor. The skeleton data contains intra-person interaction information which is different from inter-person interaction modeled in the Image Branch. It can be represented by 2D coordinates sequences of human joints. We apply HRNet [26] pretrained on the COCO dataset as a pose estimation network. Given the bounding box of each individual, we crop the RGB image and resize it to 192×256. HRNet receives cropped images and then predicts human joints coordinates with confidence which are utilized as the input of Spatial and Temporal GCN.

Graph Convolutional Network. The input of graph convolutional network (GCN) is graph-structured data which is different from common image-structured data. Layer-wise propagation rules of GCN have a simple structure:

$$H^{(l+1)} = \sigma(\tilde{D}^{-\frac{1}{2}}\tilde{A}\tilde{D}^{-\frac{1}{2}}H^{(l)}W^{(l)}) \tag{5}$$

where $\tilde{A} = A + I$, A is the adjacent matrix of the input graph, I is the identity matrix, $\tilde{D}_{ii} = \sum_j \tilde{A}_{ij}$ is the degree matrix of $\tilde{A}$, $W^{(l)}$ is the learned weight matrix, $H^{(l)}$ is feature vectors of nodes in the lth layer, and $\sigma(\cdot)$ is the activation function $ReLU(\cdot) = \max(0, \cdot)$.

Spatial GCN. Spatial relations of human joints can be represented as a graph, so we utilize Spatial GCN to integrate intra-person spatial relations. We construct the spatial graph $G_t = (V_t, E_t)$ on human joints in the tth frame. $V_t = \{v_{ti}|v_{ti}, i = 1, \ldots, N^{joint}\}$ represents joint nodes, and the attribute of v_{ti} is the coordinates and estimate confidence. The edge set $E_t = \{e_{ij}|i, j = 1, \ldots, N^{joint}\}$ represents the relation between the joint node v_{ti} and v_{tj}. Constructing relation matrix A^S is the key component in GCN. We note that human joints can only move within certain limits and interact with the surrounding joints. Hence, we only consider interaction between the node v_{ti} and its neighborhood set $N(v_{ti}) = \{v_{tj}|d(v_{ti}, v_{tj}) \leq 1, j \neq i\}$. Noting that the motion of joints is driven by the muscles, the movements of muscles are split into two types, namely shortening and lengthening. Muscles shorten pulling the weight towards the body center and muscles lengthen keeping the weight away from the body center. So we divide nodes in the neighborhood set $N(v_{ti})$ into two categories: the proximal set and the distal set.

$$\begin{aligned} \text{Proximal} &= \{v_{tj}|d(v_{tj}, v_{center}) < d(v_{ti}, v_{center}), v_{tj} \in N(v_{ti})\} \\ \text{Distal} &= \{v_{tj}|d(v_{tj}, v_{center}) > d(v_{ti}, v_{center}), v_{tj} \in N(v_{ti})\} \end{aligned} \tag{6}$$

where we regard the neck node as the center v_{center} of the body. The value of A^S_{ij} depends on which set the neighbor node v_{tj} belongs to. v_{ti} in the tth frame are stacked to form the $H^{(0)}$.

Temporal GCN. Temporal GCN captures the temporal dynamics of the same joint node in different frames. We construct the temporal graph $G_i = (V_i, E_i)$ on the ith joint node in the different frames, where $V_i = \{v_{ti}|t = 1, 2, \ldots, T\}$, $E_i = \{e_{tt'}|t, t' = 1, 2, \ldots, T, t' \neq t\}$. We only consider M frames around the tth frame, namely

$$A^T_{tt'} = \begin{cases} 1, & t - \frac{M-1}{2} \leqslant t' \leqslant t + \frac{M-1}{2} \\ 0, & else \end{cases} \tag{7}$$

v_{ti} on the ith joint node are stacked to form $H^{(0)}$.

3.3 Training Loss

To train an end-to-end model, we calculate the loss using the cross-entropy loss:

$$\mathcal{L} = \mathcal{L}_1(y^G, \widehat{y}^G) + \lambda \mathcal{L}_2(y^I, \widehat{y}^I) \tag{8}$$

where $\mathcal{L}_1$ and $\mathcal{L}_2$ are the cross-entropy loss, y^G and $\widehat{y}^G$ are the ground truth and the prediction value of the group activity, y^I and $\widehat{y}^I$ are the ground truth and the prediction value of the individual action, and λ is a hyper-parameter balancing the two loss.

4 Experiments

In this section, we conduct experiments on the Collective Activity dataset and the Volleyball dataset. First, we introduce the two datasets which are widely used in group activity recognition. Second, we provide the implementation details of our proposed model. Third, we compare our proposed method with state-of-the-art methods. Finally, we present the ablation study and visualization of our model to verify the effectiveness of our proposed modules.

4.1 Datasets

Collective Activity Dataset. The Collective Activity dataset [6] consists of 44 clips composed of frames that range from 194 to 1814. The middle frame of every ten frames contains the bounding box coordinates annotations and individuals' action labels (*NA, waiting, talking, queuing, crossing*, and *walking*). Actions of most individuals in the same scene determine the group activity (*waiting, talking, queuing, crossing*, and *walking*). We follow [29,31,32] to merge the label *crossing* with *walking* to the label *moving*. The test set is composed of 1/3 of the video clips and the training set is composed of the rest of the video clips following [24].

Volleyball Dataset. The Volleyball dataset [15] consists of multiple volleyball clips whose length is 41 frames. The middle frame of each clip contains bounding box coordinates, individual action labels, and group activity labels. Individual action labels contain 9 actions: *setting, digging, falling, jumping, blocking, moving, spiking, waiting*, and *standing*. Group activity labels contain 8 activities, namely *right set, right pass, right spike, right winpoint, left set, left pass, left spike*, and *left winpoint*. We split the dataset into the training set composed of 3493 clips and the testing set composed of 1337 clips.

4.2 Implementation Details

Referring to the previous methods, we resize the images in the Volleyball dataset to $H \times W = 720 \times 1280$ and images in the Collective Activity dataset to $H \times W = 480 \times 720$. We select $T = 10$ frames in the clips, which contain 5 frames before

Table 1. Comparison with state-of-the-art models on the Collective Activity dataset(CAD) and Volleyball dataset(VD) using Multi-class Classification Accuracy(MCA) and Mean Per Class Accuracy(MPCA) metrics.

Method	MCA-CAD	MPCA-CAD	MCA-VD	MPCA-VD
HDTM [16]	81.5	-	81.9	-
SBGAR [20]	86.1	-	66.9	-
StagNet [24]	89.1	-	89.3	-
CRM [2]	85.8	94.2	93.0	-
ARG [30]	91.0	-	92.6	-
M. Ehsanpour et al [9]	89.4	-	93.1	-
PRL [14]	-	93.8	91.4	91.8
Actor-Transformer [11]	91.0	-	93.5	-
DIN [38]	-	95.9	93.6	93.8
Ours-Image	93.7	93.1	92.9	93.5
Ours-Skeleton	95.0	92.8	83.9	83.0
Ours-Image+Skeleton	**97.1**	**96.1**	**94.0**	**94.4**

the middle frames and 4 frames after the middle frames. In the Image Branch, we adopt pretrained VGG16 [25] as the backbone and apply RoIAlign with the crop size $K \times K = 5 \times 5$ on the feature map. After the fully connected layer, the dimension of individuals' features d is 1024. We set scene encoded individual feature dimension $d_E = 1024$. For simplicity, the number of layers in Scene Encoded/Spatial/Temporal Transformer is set as 1. In the Skeleton Branch, we adopt HRNet model *pose_hrnet_w32* which is pretrained on the COCO dataset. Given the bounding boxes of individuals, we crop images and resize them to $H \times W = 256 \times 192$. For the training loss, we set $\lambda = 1$ to balance two tasks. In addition, we set the dropout rate as 0.3 to reduce overfitting.

For training on the Volleyball dataset, we adopt Adam optimizer with $\beta_1 = 0.9$, $\beta_2 = 0.999$, $\epsilon = 10^{-8}$. We finetune the backbone network with the learning rate 10^{-5} in 200 epochs and then train the whole model in 30 epochs with the learning rate ranging from 10^{-4} to 10^{-5}. For training on the Collective Activity dataset, we adopt Adam optimizer with the same hyper-parameters. We finetune the backbone network with the same learning rate in 100 epochs and then train the whole model in 30 epochs with the fixed learning rate 10^{-5}. After training the Image Branch and Skeleton Branch separately, we freeze the parameters in the two branches, concatenate the group features extracted by the two branches, and train the classifier layer to obtain the result of the fusion.

4.3   Comparison with the State-of-the-Arts

Collective Activity Dataset. To verify the effectiveness of our model, we compare our model with the state-of-the-art models on the Collective Activity

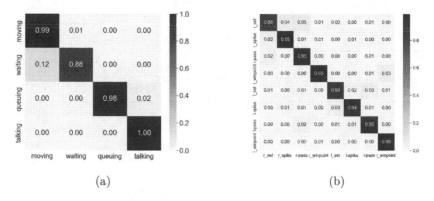

(a)   (b)

Fig. 3. (a) The confusion matrix of our proposed model on Collective Activity dataset. (b) The confusion matrix of our proposed model on Volleyball dataset.

dataset. The result measured by the Multi-class Classification Accuracy(MCA) and Mean Per Class Accuracy(MPCA) metrics is shown in Table 1. We list the results of the Image Branch, the Skeleton Branch, and the fusion of the two branches. Our Image Branch model and Skeleton Branch model achieves 93.7% MCA and 95.0% MCA respectively, and outperform other methods. Intra-person spatial-temporal dynamics extracted in the Skeleton Branch play a pivotal role in understanding individual and group behaviors. And inter-person relation captured in the Image Branch also models important contextual interaction. The fusion of two branches surpasses all methods by 6.1% MCA and 0.2% MPCA, which demonstrates that the two branches play a complementary role.

To show the effectiveness of our model, we draw the confusion matrix on the Collective Activity dataset shown in Fig. 3(a). It indicates that our model can distinguish well on most classes.

Volleyball Dataset. We further evaluate the effectiveness of our model on the Volleyball dataset and list the results compared with the state-of-the-art models in Table 1. MCA and MPCA metrics are also used to evaluate the results on the Volleyball dataset. Our Image Branch model achieves 92.9% MCA and 93.5% MPCA. It shows that context encoded features and spatial-temporal relations between individuals are pivotal for inferring group features. Our skeleton branch model performs not well on the Volleyball dataset, as we think there are overlapping instances in some images, which will lead to a decline of recognition accuracy. The fusion of two branches achieves 94.0% MCA and 94.4% MPCA and outperforms previous methods, which also demonstrates that inter-person and intra-person relations are complementary.

As shown in Fig. 3(b), the confusion matrix of the Volleyball dataset shows that the accuracy of our model achieves 90% in most classes. Our model rarely confuses the left and right sides because our model integrates the whole scene feature with individuals' features and captures the relation between individuals.

Table 2. Ablation study of the Image Branch on the Volleyball dataset. The base model contains the backbone network, utilizes RoIAlign to extract individuals' features, and applies a classifier layer to obtain the group activity.

Scene encoded transformer	Spatial transformer	Temporal transformer	MCA	MPCA
			88.0	88.6
✓			92.5	93.1
✓	✓		92.2	92.6
✓		✓	92.5	92.8
✓	✓	✓	**92.9**	**93.5**

Table 3. Ablation study of the skeleton branch on the collective activity dataset. **SGCN** denotes Spatial GCN module. **TGCN** denotes Temporal GCN module. **STGCN** denotes a layer composed of Spatial and Temporal GCN.

Components	SGCN	SGCN+TGCN	Layers of STCGN					
	-	-	L=2	L=4	L=6	L=8	L=10	L=12
MCA/MPCA	53.7/42.2	69.9/50.6	78.0/67.9	86.7/77.0	87.2/78.7	88.2/79.0	**95.0/92.8**	93.7/90.0

4.4 Ablation Studies

To evaluate the effectiveness of our proposed modules, we perform ablation study using MCA and MPCA metrics.

Scene Encoded Transformer. To study the performance of Scene Encoded Transformer, we append this module to the base model. As shown in Table 2, Scene Encoded Transformer improves 4.5% MCA and 4.5% MPCA. Considering that individuals always interact with the scene, the scene contains latent information related to individuals' actions and group activity. Hence, utilizing relations between individuals and the scene can enhance the group representation.

Spatial and Temporal Transformer. In Table 2, Spatial and Temporal Transformer are appended after Scene Encoded Transformer, respectively. Using Spatial or Temporal Transformer alone impacts negatively on the effectiveness of our model, as we consider that spatial and temporal dependency are not considered simultaneously. After the concatenation of Spatial Transformer and Temporal Transformer, our model achieves 92.9% MCA and 93.5% MPCA on the Volleyball dataset, which indicates that fusing the complex spatial-temporal interaction can improve the effectiveness of our model.

Spatial and Temporal GCN. We conduct the ablation study on the Skeleton Branch to verify the effectiveness of Spatial and Temporal GCN. As can be seen in Table 3, adding Temporal GCN after Spatial GCN improves the performance from 42.2% MPCA to 50.6% MPCA. It demonstrates that integrating spatial

Fig. 4. Visualization results of the attention matrix in Scene Encoded Transformer on the Volleyball dataset.

and temporal intra-person relations is fundamental to modeling individuals' representations. In addition, we gradually increase the number of layers, and observe that the performance achieves the best result when the number of layers is set to 10. Stacked layers of Spatial and Temporal GCN can jointly exploit intra-person spatial-temporal relations and integrate them into individuals' features.

4.5 Visualization

Scene Encoded Transformer. We visualize the attention map of Scene Encoded Transformer on the Volleyball dataset in Fig. 4. Scene Encoded Transformer emphasizes the location of the volleyball, spectator seats, and referees on the volleyball court. We observe that Scene Encoded Transformer pays more attention to the volleyball at the moment of individual spiking the volleyball. It is clear that key scene features can be captured by Scene Encoded Transformer.

Spatial and Temporal Transformer. Visualization results of the *left spike* activity on the Volleyball dataset are shown in Fig. 5. We observe that individual 4 is spiking the volleyball in frame 0. And Spatial Transformer highlights individual 4 and relation with others. This shows that Spatial Transformer pays attention to the key instances and utilizes relations with each other to infer the group activity. In frame 9, individual 6 who is blocking the ball is highlighted by Spatial Transformer. Interaction between individual 6 and individual 4 can also confirm that the group activity is *left spike*. Additionally, Temporal Transformer takes into account the inter-frame influence.

Spatial and Temporal GCN. As shown in Fig. 6, we project group features extracted by Spatial and Temporal GCN on the Collective Activity dataset to 2D dimensions using t-SNE [22]. Spatial and Temporal GCN aggregate intra-person spatial and temporal dynamics to the group representations and cluster the representations well. As the layers get deeper, group features are more discriminative. These visualization results verify the effectiveness of our Spatial and Temporal GCN in the Skeleton Branch.

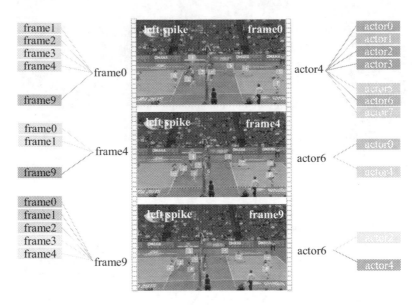

Fig. 5. Visualization results of the attention matrix in Spatial and Temporal Transformer on the Volleyball dataset.

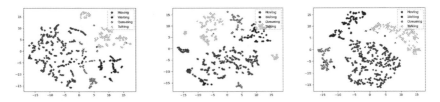

Fig. 6. From left to right, they are t-SNE visualization of 2 layers, 6 layers and 10 layers of Spatial and Temporal GCN on the Collective Activity Dataset.

5   Conclusion

This paper proposes a spatial-temporal network with two branches taking RGB images and human skeletons as input. For RGB image inputs, Scene Encoded Transformer is proposed to incorporate scene features into individuals' features. Spatial and Temporal Transformer are designed to extract spatial and temporal information, respectively. Furthermore, we utilize innovative human skeletons as input and capture spatial and temporal dynamics by utilizing Spatial and Temporal GCN. Experiments demonstrate our proposed model achieves outstanding performance while it can capture spatial and temporal dependencies on the intra-person and inter-person levels.

References

1. Amer, M.R., Lei, P., Todorovic, S.: HiRF: hierarchical random field for collective activity recognition in videos. In: Fleet, D., Pajdla, T., Schiele, B., Tuytelaars, T. (eds.) ECCV 2014. LNCS, vol. 8694, pp. 572–585. Springer, Cham (2014). https://doi.org/10.1007/978-3-319-10599-4_37

2. Azar, S.M., Atigh, M.G., Nickabadi, A., Alahi, A.: Convolutional relational machine for group activity recognition. In: IEEE Conference on Computer Vision and Pattern Recognition, CVPR 2019, Long Beach, CA, USA, June 16–20, 2019, pp. 7892–7901. Computer Vision Foundation / IEEE (2019). https://doi.org/10.1109/CVPR.2019.00808,https://openaccess.thecvf.com/content_CVPR_2019/html/zar_Convolutional_Relational_Machine_for_Group_Activity_Recognition_CVPR_2019_paper.html

3. Bagautdinov, T.M., Alahi, A., Fleuret, F., Fua, P., Savarese, S.: Social scene understanding: End-to-end multi-person action localization and collective activity recognition. In: 2017 IEEE Conference on Computer Vision and Pattern Recognition, CVPR 2017, Honolulu, HI, USA, 21–26 July 2017. pp. 3425–3434. IEEE Computer Society (2017). https://doi.org/10.1109/CVPR.2017.365,https://doi.org/10.1109/CVPR.2017.365

4. Choi, W., Savarese, S.: A Unified framework for multi-target tracking and collective activity recognition. In: Fitzgibbon, A., Lazebnik, S., Perona, P., Sato, Y., Schmid, C. (eds.) ECCV 2012. LNCS, vol. 7575, pp. 215–230. Springer, Heidelberg (2012). https://doi.org/10.1007/978-3-642-33765-9_16

5. Choi, W., Savarese, S.: Understanding collective activitiesof people from videos. IEEE Trans. Pattern Anal. Mach. Intell. **36**(6), 1242–1257 (2014) 10.1109/TPAMI.2013.220, https://doi.org/10.1109/TPAMI.2013.220

6. Choi, W., Shahid, K., Savarese, S.: What are they doing? : collective activity classification using spatio-temporal relationship among people. In: 12th IEEE International Conference on Computer Vision Workshops, ICCV Workshops 2009, Kyoto, Japan, September 27 - October 4, 2009, pp. 1282–1289. IEEE Computer Society (2009). https://doi.org/10.1109/ICCVW.2009.5457461,https://doi.org/10.1109/ICCVW.2009.5457461

7. Choi, W., Shahid, K., Savarese, S.: Learning context for collective activity recognition. In: The 24th IEEE Conference on Computer Vision and Pattern Recognition, CVPR 2011, Colorado Springs, CO, USA, 20–25 June 2011. pp. 3273–3280. IEEE Computer Society (2011). https://doi.org/10.1109/CVPR.2011.5995707,https://doi.org/10.1109/CVPR.2011.5995707

8. Deng, Z., Vahdat, A., Hu, H., Mori, G.: Structure inference machines: Recurrent neural networks for analyzing relations in group activity recognition. In: 2016 IEEE Conference on Computer Vision and Pattern Recognition, CVPR 2016, Las Vegas, NV, USA, 27–30 June 2016, pp. 4772–4781. IEEE Computer Society (2016). https://doi.org/10.1109/CVPR.2016.516,https://doi.org/10.1109/CVPR.2016.516

9. Ehsanpour, M., Abedin, A., Saleh, F., Shi, J., Reid, I., Rezatofighi, H.: Joint learning of social groups, individuals action and sub-group activities in videos. In: Vedaldi, A., Bischof, H., Brox, T., Frahm, J.-M. (eds.) ECCV 2020. LNCS, vol. 12354, pp. 177–195. Springer, Cham (2020). https://doi.org/10.1007/978-3-030-58545-7_11

10. Gao, J., Sun, C., Zhao, H., Shen, Y., Anguelov, D., Li, C., Schmid, C.: Vectornet: encoding HD maps and agent dynamics from vectorized representation. In: 2020 IEEE/CVF Conference on Computer Vision and Pattern Recognition, CVPR 2020, Seattle, WA, USA, June 13–19, 2020. pp. 11522–11530. Computer Vision Foundation / IEEE (2020). https://doi.org/10.1109/CVPR42600.2020.01154,https://openaccess.thecvf.com/content_CVPR_2020/html/Gao_VectorNet_Encoding_HD_Maps_and_Agent_Dynamics_From_Vectorized_Representation_CVPR_2020_paper.html

11. Gavrilyuk, K., Sanford, R., Javan, M., Snoek, C.G.M.: Actor-transformers for group activity recognition. In: 2020 IEEE/CVF Conference on Computer Vision and Pattern Recognition, CVPR 2020, Seattle, WA, USA, June 13–19, 2020. pp. 836–845. Computer Vision Foundation / IEEE (2020). https://doi.org/10.1109/CVPR42600.2020.00092,https://openaccess.thecvf.com/content_CVPR_2020/html/Gavrilyuk_Actor-Transformers_for_Group_Activity_Recognition_CVPR_2020_paper.html

12. Hajimirsadeghi, H., Yan, W., Vahdat, A., Mori, G.: Visual recognition by counting instances: a multi-instance cardinality potential kernel. In: IEEE Conference on Computer Vision and Pattern Recognition, CVPR 2015, Boston, MA, USA, 7–12 June 2015, pp. 2596–2605. IEEE Computer Society (2015). https://doi.org/10.1109/CVPR.2015.7298875,https://doi.org/10.1109/CVPR.2015.7298875

13. He, K., Gkioxari, G., Dollár, P., Girshick, R.: Mask r-CNN. In: Proceedings of the IEEE International Conference on Computer Vision, pp. 2961–2969 (2017)

14. Hu, G., Cui, B., He, Y., Yu, S.: Progressive relation learning for group activity recognition. In: 2020 IEEE/CVF Conference on Computer Vision and Pattern Recognition, CVPR 2020, Seattle, WA, USA, June 13–19, 2020. pp. 977–986. Computer Vision Foundation / IEEE (2020). https://doi.org/10.1109/CVPR42600.2020.00106,https://openaccess.thecvf.com/content_CVPR_2020/html/Hu_Progressive_Relation_Learning_for_Group_Activity_Recognition_CVPR_2020_paper.html

15. Ibrahim, M.S., Mori, G.: Hierarchical relational networks for group activity recognition and retrieval. In: Ferrari, V., Hebert, M., Sminchisescu, C., Weiss, Y. (eds.) ECCV 2018. LNCS, vol. 11207, pp. 742–758. Springer, Cham (2018). https://doi.org/10.1007/978-3-030-01219-9_44

16. Ibrahim, M.S., Muralidharan, S., Deng, Z., Vahdat, A., Mori, G.: A hierarchical deep temporal model for group activity recognition. In: 2016 IEEE Conference on Computer Vision and Pattern Recognition, CVPR 2016, Las Vegas, NV, USA, 27–30 June 2016, pp. 1971–1980. IEEE Computer Society (2016). https://doi.org/10.1109/CVPR.2016.217,https://doi.org/10.1109/CVPR.2016.217

17. Lan, T., Sigal, L., Mori, G.: Social roles in hierarchical models for human activity recognition. In: 2012 IEEE Conference on Computer Vision and Pattern Recognition, Providence, RI, USA, 16–21 June 2012, pp. 1354–1361. IEEE Computer Society (2012). https://doi.org/10.1109/CVPR.2012.6247821,https://doi.org/10.1109/CVPR.2012.6247821

18. Li, M., Chen, S., Chen, X., Zhang, Y., Wang, Y., Tian, Q.: Actional-structural graph convolutional networks for skeleton-based action recognition. In: IEEE Conference on Computer Vision and Pattern Recognition, CVPR 2019, Long Beach, CA, USA, 16–20 June 2019, pp. 3595–3603. Computer Vision Foundation/IEEE (2019). https://doi.org/10.1109/CVPR.2019.00371,https://openaccess.thecvf.com/content_CVPR_2019/html/Li_Actional-Structural_Graph_Convolutional_Networks_for_Skeleton-Based_Action_Recognition_CVPR_2019_paper.html

19. Li, S., Cao, Q., Liu, L., Yang, K., Liu, S., Hou, J., Yi, S.: Groupformer: group activity recognition with clustered spatial-temporal transformer. In: Proceedings of the IEEE/CVF International Conference on Computer Vision, pp. 13668–13677 (2021)
20. Li, X., Chuah, M.C.: SBGAR: semantics based group activity recognition. In: IEEE International Conference on Computer Vision, ICCV 2017, Venice, Italy, 22–29 October 2017, pp. 2895–2904. IEEE Computer Society (2017). https://doi.org/10.1109/ICCV.2017.313, https://doi.org/10.1109/ICCV.2017.313
21. Liu, Z., Zhang, H., Chen, Z., Wang, Z., Ouyang, W.: Disentangling and unifying graph convolutions for skeleton-based action recognition. In: Proceedings of the IEEE/CVF Conference on Computer Vision and Pattern Recognition, pp. 143–152 (2020)
22. Van der Maaten, L., Hinton, G.: Visualizing data using t-SNE. J. Mach. Learn. Res. **9**(11) (2008)
23. Pramono, R.R.A., Chen, Y.T., Fang, W.H.: Empowering relational network by self-attention augmented conditional random fields for group activity recognition. In: Vedaldi, A., Bischof, H., Brox, T., Frahm, J.-M. (eds.) ECCV 2020. LNCS, vol. 12346, pp. 71–90. Springer, Cham (2020). https://doi.org/10.1007/978-3-030-58452-8_5
24. Qi, M., Qin, J., Li, A., Wang, Y., Luo, J., Van Gool, L.: stagNet: an attentive semantic rnn for group activity recognition. In: Ferrari, V., Hebert, M., Sminchisescu, C., Weiss, Y. (eds.) ECCV 2018. LNCS, vol. 11214, pp. 104–120. Springer, Cham (2018). https://doi.org/10.1007/978-3-030-01249-6_7
25. Simonyan, K., Zisserman, A.: Very deep convolutional networks for large-scale image recognition. arXiv preprint arXiv:1409.1556 (2014)
26. Sun, K., Xiao, B., Liu, D., Wang, J.: Deep high-resolution representation learning for human pose estimation. In: IEEE Conference on Computer Vision and Pattern Recognition, CVPR 2019, Long Beach, CA, USA, 16–22 June 2019, pp. 5693–5703. Computer Vision Foundation / IEEE (2019). https://doi.org/10.1109/CVPR.2019.00584, https://openaccess.thecvf.com/content_CVPR_2019/html/Sun_Deep_High-Resolution_Representation_Learning_for_Human_Pose_Estimation_CVPR_2019_paper.html
27. Tamura, M., Vishwakarma, R., Vennelakanti, R.: Hunting group clues with transformers for social group activity recognition. CoRR abs/2207.05254 (2022). 10.48550/arXiv. 2207.05254, https://doi.org/10.48550/arXiv.2207.05254
28. Vaswani, A., et al.: Attention is all you need. In: Advances in neural information processing systems, pp. 5998–6008 (2017)
29. Wang, M., Ni, B., Yang, X.: Recurrent modeling of interaction context for collective activity recognition. In: 2017 IEEE Conference on Computer Vision and Pattern Recognition, CVPR 2017, Honolulu, HI, USA, July 21–26, 2017, pp. 7408–7416. IEEE Computer Society (2017). https://doi.org/10.1109/CVPR.2017.783, https://doi.org/10.1109/CVPR.2017.783
30. Wu, J., Wang, L., Wang, L., Guo, J., Wu, G.: Learning actor relation graphs for group activity recognition. In: IEEE Conference on Computer Vision and Pattern Recognition, CVPR 2019, Long Beach, CA, USA, 16–20 June 2019, pp. 9964–9974. Computer Vision Foundation / IEEE (2019). https://doi.org/10.1109/CVPR.2019.01020, https://openaccess.thecvf.com/content_CVPR_2019/html/Wu_Learning_Actor_Relation_Graphs_for_Group_Activity_Recognition_CVPR_2019_paper.html

31. Yan, R., Tang, J., Shu, X., Li, Z., Tian, Q.: Participation-contributed temporal dynamic model for group activity recognition. In: Boll, S., et al. (eds.) 2018 ACM Multimedia Conference on Multimedia Conference, MM 2018, Seoul, Republic of Korea, 22–26 October 2018. pp. 1292–1300. ACM (2018). https://doi.org/10.1145/3240508.3240572,https://doi.org/10.1145/3240508.3240572

32. Yan, R., Xie, L., Tang, J., Shu, X., Tian, Q.: HIGCIN: hierarchical graph-based cross inference network for group activity recognition. IEEE Trans. Pattern Anal. Mach. Intell. Early Access, 1–1 (2020). https://doi.org/10.1109/TPAMI.2020.3034233

33. Yan, R., Xie, L., Tang, J., Shu, X., Tian, Q.: Social adaptive module for weakly-supervised group activity recognition. In: Vedaldi, A., Bischof, H., Brox, T., Frahm, J.-M. (eds.) ECCV 2020. LNCS, vol. 12353, pp. 208–224. Springer, Cham (2020). https://doi.org/10.1007/978-3-030-58598-3_13

34. Yan, S., Xiong, Y., Lin, D.: Spatial temporal graph convolutional networks for skeleton-based action recognition. In: McIlraith, S.A., Weinberger, K.Q. (eds.) Proceedings of the Thirty-Second AAAI Conference on Artificial Intelligence, (AAAI-2018), the 30th Innovative Applications of Artificial Intelligence (IAAI-18), and the 8th AAAI Symposium on Educational Advances in Artificial Intelligence (EAAI-18), New Orleans, Louisiana, USA, 2–7 February 2018, pp. 7444–7452. AAAI Press (2018). https://www.aaai.org/ocs/index.php/AAAI/AAAI18/paper/view/17135

35. Yao, Y., Atkins, E., Roberson, M.J., Vasudevan, R., Du, X.: Coupling intent and action for pedestrian crossing behavior prediction. arXiv preprint arXiv:2105.04133 (2021)

36. Yau, T., Malekmohammadi, S., Rasouli, A., Lakner, P., Rohani, M., Luo, J.: Graphsim: a graph-based spatiotemporal interaction modelling for pedestrian action prediction. In: 2021 IEEE International Conference on Robotics and Automation (ICRA), pp. 8580–8586. IEEE (2021)

37. Yuan, H., Ni, D.: Learning visual context for group activity recognition. In: Thirty-Fifth AAAI Conference on Artificial Intelligence, AAAI 2021, Thirty-Third Conference on Innovative Applications of Artificial Intelligence, IAAI 2021, The Eleventh Symposium on Educational Advances in Artificial Intelligence, EAAI 2021, Virtual Event, 2–9 February 2021, pp. 3261–3269. AAAI Press (2021). https://ojs.aaai.org/index.php/AAAI/article/view/16437

38. Yuan, H., Ni, D., Wang, M.: Spatio-temporal dynamic inference network for group activity recognition. In: Proceedings of the IEEE/CVF International Conference on Computer Vision, pp. 7476–7485 (2021)

39. Yuan, Y., Weng, X., Ou, Y., Kitani, K.: Agentformer: agent-aware transformers for socio-temporal multi-agent forecasting. In: 2021 IEEE/CVF International Conference on Computer Vision, ICCV 2021, Montreal, QC, Canada, 10–17 October 2021, pp. 9793–9803. IEEE (2021). https://doi.org/10.1109/ICCV48922.2021.00967,https://doi.org/10.1109/ICCV48922.2021.00967

40. Zhou, H., et al.: Composer: compositional reasoning of group activity in videos with keypoint-only modality. In: Proceedings of the 17th European Conference on Computer Vision (ECCV 2022) (2022). https://doi.org/10.1007/978-3-031-19833-5_15

Learning Using Privileged Information for Zero-Shot Action Recognition

Zhiyi Gao[1], Yonghong Hou[1(✉)], Wanqing Li[2], Zihui Guo[1], and Bin Yu[1]

[1] School of Electrical and Information Engineering, Tianjin University, Tianjin, China
{zhiyigao,houroy,gzihui,yubin_1449508506}@tju.edu.cn
[2] Advanced Multimedia Research Laboratory, University of Wollongong, Wollongong, Australia
wanqing@uow.edu.au

Abstract. Zero-Shot Action Recognition (ZSAR) aims to recognize video actions that have never been seen during training. Most existing methods assume a shared semantic space between seen and unseen actions and intend to directly learn a mapping from a visual space to the semantic space. This approach has been challenged by the semantic gap between the visual space and semantic space. This paper presents a novel method that uses object semantics as privileged information to narrow the semantic gap and, hence, effectively, assist the learning. In particular, a simple hallucination network is proposed to implicitly extract object semantics during testing without explicitly extracting objects and a cross-attention module is developed to augment visual feature with the object semantics. Experiments on the Olympic Sports, HMDB51 and UCF101 datasets have shown that the proposed method outperforms the state-of-the-art methods by a large margin.

Keywords: Action recognition · Zero-shot learning · Privileged information · Cross-attention

1 Introduction

Research on action recognition from videos has made rapid progress in the past years with ceiling performance being reached on some datasets [1–5]. However, with the growing number of action categories, traditional supervised approach suffers from scalability problem. Moreover, annotating sufficient examples for the ever-growing new categories in real-world is cost-expensive and time-consuming. Therefore, extending a well-trained model to new/unseen classes, known as Zero-Shot Action Recognition (ZSAR), is gaining increasing interest recently.

In general, ZSAR assumes that both seen and unseen classes of actions share a common semantic space, e.g. a high dimensional vector space, and is achieved by learning a mapping from visual feature to the semantic space. Once the mapping is learned, an unseen sample is classified by the nearest neighbor search in the semantic space. There are a number of ways to define the semantic space.

L. Wang et al. (Eds.): ACCV 2022, LNCS 13844, pp. 347–362, 2023.
https://doi.org/10.1007/978-3-031-26316-3_21

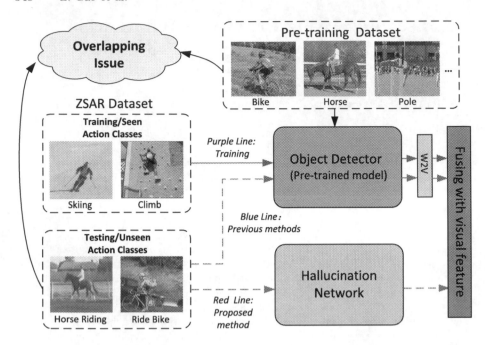

Fig. 1. Proposed method vs existing methods. An object detector pretrained on ImageNet is used to extract objects during training (purple line). When testing, the unseen actions are high-related to ImageNet classes. The previous methods (blue line) still require an object detector. Instead, the proposed method (red line) uses a hallucination network to extract object related information. (Color figure online)

Typical early works [6–8] define the semantic space using hand-crafted attributes. Recently, the semantic space is often defined as the word embedding space of action names, i.e. word2vec, using a pretrained language model [9]. As for the mapping, direct learning from seen actions has been studied in [10]. However, visual observation and action names or hand-crafted attributes are two different modalities. The semantic gap between the two modalities has been challenging the effectiveness and robustness of learning such a mapping from visual feature to semantic representation.

To mitigate the semantic gap, several approaches have been reported recently. One approach is to define a semantic space using example images [11] e.g. considering the visual feature of the images as the semantic space. The second approach is to amend the visual feature with semantic information. For instance, in [12], a pretrained object detector is employed to recognize objects from action videos, and embedding of object names is treated as the visual representation without considering spatio-temporal information of the actions. In [13], objects are also extracted and the embedding of their names is concatenated to the feature representing poses to form the visual feature. Since such visual feature carries some amount of semantic information, the semantic gap is expected to be narrowed.

It has to be pointed out that both [12] and [13] require an object detector to extract objects and word-embed their names during testing. The detector is pretrained on a large dataset such as the ImageNet [14], this practice has raised a question of validity of their methods being truly ZSAR because the large-scale dataset that is used to train the object detector likely contains images high-related to unseen action classes (see Fig. 1). For instance, images representing actions "Ride Bike" and "Horse Riding" are found in ImageNet [14] and these actions are likely to be considered as unseen actions in the random training-test split of ZSAR evaluation.

Nevertheless, using information of objects relevant to actions is a promising strategy. It not only narrows the semantic gap via word-embedding of the extracted object names, but also provides information in addition to the spatio-temporal visual information of the actions. As a true ZSAR, it is desirable that the object detector pretrained on a dataset should not be employed during testing for the sake of computation and avoiding potential licenses fee in commercial applications. This paper presents such a method based on the paradigm of learning using privileged information (LUPI) [15]. Specifically, objects are annotated or extracted offline from seen actions and their names are word-embedded into a vector in the visual space as privileged information (PI) in training. Unlike the methods in [12,13], our method *does not need the object detector during testing phase*, instead it uses a hallucination network to mimic the extraction of related semantic information. The output of the hallucination network is fused with the visual feature by a cross-attention module to narrow the semantic gap and assist the mapping from visual feature to the semantic space. Experiments on the widely used Olympic Sports [16], HMDB51 [17] and UCF101 [18] datasets show that the proposed method achieves the state-of-the-art results and outperforms the existing methods by a large margin.

The main contributions of this paper are as follows:

- A novel ZSAR method that implicitly leverages object information to narrow the semantic gaps and assists learning of a mapping from visual feature to a semantic space.
- A hallucination network with the privileged information of objects related to actions to "imitate" the object semantics during testing.
- A cross-attention module to fuse visual feature with object semantics or the feature from the hallucination network.
- Extensive evaluation on the widely used Olympic Sports, HMDB51 and UCF101 benchmarks achieves the state-of-the-art results.

2 Related Work

2.1 Zero-Shot Action Recognition

In general, ZSAR is achieved by defining a common semantic space for seen and unseen actions and learning a mapping between the visual space and semantic space. In relation to the semantic space, early works define it based on hand-crafted attributes [6–8]. However, the definition and annotation of hand-crafted attributes are subjective and labour-intensive. Recently, word-embedding of action names [10, 19, 20] is favorably adopted. In addition, textual descriptions of human actions or visual features extracted from still images are studied as a semantic space [11]. In [21, 22], semantically meaningful poses extracted from associated textual instructions of actions are also defined as a semantic space. Chen and Huang [23] use the textural description, called Elaborative Description (ED), of actions and embed them to define a semantic space. As for the visual space, it is generally defined through hand-crafted features [24] (e.g. the Dense Trajectory Features (DTF) [25] and the Improved Dense Trajectories (IDT) [26]) or deep features [27] extracted with the pre-trained 3D CNN-based Networks (e.g. Convolutional 3D Network (C3D) [28], Inflated 3D Network (I3D) [2]). After constructing semantic space and visual space, most existing works directly learn a mapping from the visual space to semantic space [10, 29]. However, this approach has been challenged by the semantic gap between the visual space and semantic space. Recently, some works attempt to amend the visual feature with object semantics in order to narrow the semantic gap between the visual space and semantic space. For example, Jain et al. [12] extract objects from action videos by a pretrained object detector and embed object names to form a visual representation. In [13], the object information and pose information are simultaneously extracted to form visual feature. However, both of them utilize the pretrained object detector at test time, which is questionable as illustrated in the introduction. In contrast, the proposed method in this paper does not use any object detector during testing.

2.2   Learning Using Privileged Information

Vapnik and Vashist [15] introduce the paradigm of learning using privileged information (LUPI). It assumes that there are additional data available during training, referred to as privileged information (PI), but not available in testing. A number of approaches have been proposed for action recognition by using privileged information. Niu et al. [30] utilize textual features extracted from the contextual descriptions of web images and videos as privileged information to train a robust action detector. Motiian et al. [31] explore several types of privileged information such as motion information or 3D skeletons for improving model learning. Crasto et al. [32] regard optical flow as privileged information, along with RGB for training, but only RGB is used in test to avoid flow computation. Similarly, Garcia et al. [33] consider that depth data is not going to be always available when a model is deployed in real scenarios. Therefore, depth data is used as privileged information along with RGB data during training, while employing only RGB data at test time. Different from the above methods, the proposed method uses object semantics from seen actions as privileged information to assist learning visual feature as well as a mapping between the visual feature and semantic space. In addition, a hallucination network is proposed to implicitly extract semantics during testing instead of relying on a pretrained object detector.

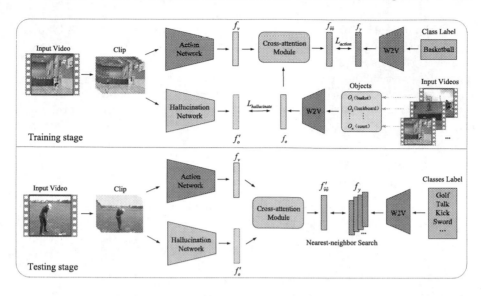

Fig. 2. The illustration of the proposed framework. It takes visual feature and the corresponding object semantics as input during training. A hallucination network is simultaneously trained to implicitly extract object related semantics and replace the object detector at test time. A cross-attention module is designed to fuse visual feature with object semantics or the feature from hallucination network. At test time, the nearest neighbor search is used for prediction in the semantic space. Gray blocks represent modules that are fixed during training. Colors (green, red, purple) blocks indicate modules trained. (Color figure online)

3   Method

3.1   Overview

Let the training set $D_s = \{(v_s, y_s)|v_s \in \mathcal{V}_s, y_s \in \mathcal{Y}_s\}$ consist of videos v_s in video space $\mathcal{V}_s$ with labels from seen action classes $\mathcal{Y}_s$. Similarly, $D_u = \{(v_u, y_u)|v_u \in \mathcal{V}_u, y_u \in \mathcal{Y}_u\}$ is the test set, which consists of videos v_u with labels from unseen action classes $\mathcal{Y}_u$. Here, seen and unseen action classes are disjoint, i.e. $\mathcal{Y}_s \cap \mathcal{Y}_u = \emptyset$. The task of ZSAR is to train a model only on D_s, and predict the class labels of the unseen videos v_u. The proposed architecture is illustrated in Fig. 2, where the embedding space f_y of action names is used as the shared semantic space, visual feature f_v extracted by the action network is used as visual space. In addition, the object semantic f_o for each known action class is utilized as privileged information to assist learning of a mapping from visual feature to the semantic space. The object semantic f_o is the word embedding of object o which may be obtained offline using an object detector. Specifically, in the training stage, video features f_v and object semantics f_o are fed to a cross-attention module and fused into a joint representation $f_{\hat{v}\hat{o}}$ being matched with the semantic representations f_y in the semantic space. Meanwhile, a hallucination network is trained to "imitate" the object semantics f_o. At test time, the trained hallucination network is used to implicitly extract object related semantic f'_o, which is fused with the visual feature f_v by the cross-attention module to obtain the joint representation $f'_{\hat{v}\hat{o}}$ as well. Finally, nearest-neighbor search method is used in the semantic space to recognize unseen action class. In the rest of the section, the five key components of the proposed framework including video and semantic embedding, object embedding, hallucination network, cross-attention module, and training and testing of the framework will be described in detail.

3.2 Video and Semantic Embedding

Video Embedding. Given a video clip $v \in \mathbb{R}^{T \times H \times W}$ of T frames with a height and width of H and W respectively, the action network consists of a pretrained R(2+1)D [10] network followed by a fully connected layer for extracting a video spatio-temporal feature $f_v \in \mathbb{R}^{300}$.

Semantic Embedding. Following the majority of works [19], word2vec is applied to embed action names as semantic representation. Specifically, the skip-gram language model [9] trained on 100 billion words from Google News articles is used to map each word into a 300 dimensional semantic space. For the class name containing multiple words, the vectors of all words are averaged to obtain the semantic embedding. That is, a class name consisting of n words $y = \{y_1, \cdots, y_n\}$ can be embedded as $f_y = \frac{1}{n} \sum_{i=1}^{n} \text{W2V}(y_i) \in \mathbb{R}^{300}$.

3.3 Object Embedding as PI

Objects are extracted by a pretrained object detector from all video clips with t frames uniformly sampled, and the top k objects with the highest probabilities are reserved. In order to associate the extracted objects with action classes, we calculate the frequency of the objects extracted from all clips of the same action class and preserve the top m objects $o = \{o_1, \cdots, o_m\}$. Then the objects are word-embedded as the privileged information of individual action class. Similarly, the object semantic of each action class can be presented by $f_o = \frac{1}{m} \sum_{i=1}^{m} \text{W2V}(o_i) \in \mathbb{R}^{300}$.

3.4 Hallucination Network

Since the ImageNet dataset from which the object detector is pretrained contains highly related categories with unseen action classes, we argue that the object detector should not be employed during testing phase for truly ZSAR. To replace the object detector, we introduce a hallucination network, which is trained to "imitate" the object semantics. Thus at test time, it can extract the object related semantics f'_o to fuse with visual features. The hallucination network consists of a fixed pretrained R(2+1)D [10] network and four fully connected layers.

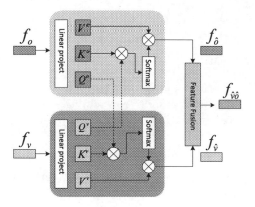

Fig. 3. Illustration of the proposed cross-attention module. It consists of a mutual-attention layer and feature fusion operation. Visual features and object semantics or features from hallucination network are fed to the attention layer to exchange the information. Then the features are fused together to obtain the joint representations.

3.5   Cross-Attention Module

We propose a cross-attention module to fuse visual feature f_v with object semantics f_o (training stage) or the feature f'_o (testing stage) from the hallucination network using mutual-attentional mechanism. As shown in Fig. 3, the proposed module contains two parts, mutual-attention layer and feature fusion operation. Take the training phase as an example, mathematically, visual feature f_v and object semantic f_o are first mapped to a series vectors query (Q^v/Q^o) and key (K^v/K^o) of dimension d_k, and value (V^v/V^o) of dimension d_v using unshared learnable linear projections. Then the mutual-attention is applied to retrieve the information from vectors (key K^v and value V^v) of visual feature related to query vector Q^o of object semantics and vice the verse. The dot-product of the Q^o vector and K^v vector goes through a softmax function to obtain the attention weights, and have a weighted sum of V^v. The process can be expressed as

$$f_{\hat{v}} = \text{Softmax}\left(\frac{Q^o(K^v)^T}{\sqrt{d_k}}\right)V^v \tag{1}$$

Similarly, we can calculate the attention weights on each object semantic with respect to visual feature to obtain $f_{\hat{o}}$.

$$f_{\hat{o}} = \text{Softmax}\left(\frac{Q^v(K^o)^T}{\sqrt{d_k}}\right)V^o \tag{2}$$

Such a mutual-attention scheme takes cross-modal interactions into consideration. It calculates the visual feature attention in semantic modality and object semantic attention in visual modality, and produces corresponding cross-modal features. After mutual-attention layer, features from both modalities are fused via simply feature addition operation.

$$f_{\hat{v}\hat{o}} = f_{\hat{v}} + f_{\hat{o}} \tag{3}$$

3.6 Training and Testing

As illustrated in Fig. 2, in the training stage, the visual feature f_v and object semantic f_o are fused to obtain a joint representation $f_{\hat{v}\hat{o}}$ in the semantic space to match their equivalent semantic representation f_y. The loss function $\mathcal{L}_{action}$ is defined as

$$\mathcal{L}_{action} = \|f_y - f_{\hat{v}\hat{o}}\|_2^2 \tag{4}$$

In addition, a regression loss between the object semantics f_o and feature f'_o from the hallucination network is defined

$$\mathcal{L}_{hallucinate} = \|f_o - f'_o\|_2^2 \tag{5}$$

The full model is end-to-end trained, and the overall loss of the proposed model is defined as

$$\mathcal{L} = \mathcal{L}_{action} + \mathcal{L}_{hallucinate} \tag{6}$$

At test time, a test video goes through both the action network and the hallucination network to produce the visual feature f_v and object related semantics f'_o. After the cross-attention module, a joint representation $f'_{\hat{v}\hat{o}}$ is obtained. Finally, nearest-neighbor search is used in the semantic space to predict the label $\mathcal{P}$ of unseen action class as

$$\mathcal{P} = \operatorname*{argmin}_{y \in D_u} \cos(f'_{\hat{v}\hat{o}}, f_y) \tag{7}$$

where $\cos(\cdot)$ is the cosine distance.

4 Experiments

4.1 Setup

Datasets and Splits. The proposed method is extensively evaluated on the Olympic Sports [16], HMDB51 [17] and UCF101 [18] datasets. The Olympic Sports dataset consists of 783 videos, which are divided into 16 sports actions. The HMDB51 dataset contains 6766 realistic videos distributed in 51 actions. The UCF101 dataset has 101 action classes with a total of 13,320 videos, which are collected from YouTube. To compare the proposed method with the state-of-the-art methods, we follow the same 50/50 data splits as used in [29], i.e., videos of 50% categories are used for training and the rest 50% categories are considered as the unseen for testing. Specifically, experiments are conducted on the cases of 8/8, 26/25 and 51/50 splits for Olympic Sports, HMDB51 and UCF101, respectively. For each case, 30 independent splits for each dataset [34] are randomly generated and the average accuracy and standard deviation are reported for experimental evaluation.

ZSAR Settings. There are two ZSAR settings: inductive setting and transductive setting. The former assumes that only the labeled videos from the seen categories are available during training while the latter can use the unlabeled data of the unseen categories for model training. Specifically, in this work, we focus on inductive ZSAR [10,13,23,35] and do not discuss the transductive approach [7,29].

4.2 Implementation Details

In the experiments, we use the pretrained R(2+1)D network [10] followed by a fully connected layer to extract spatio-temporal features. The object detector is the BiT image model [36] pretrained on ImageNet21k [14]. When extracting objects, frame t of each video clip is set to 8 and top k of all action clips is set to 20, top m of each action class is set to 5. Both the action class and extracted objects are embedded using skip-gram word2vec model [9], which is pretrained on the Google News dataset. The word2vec model generates a 300-dimensional word vector representation for each word. For the input videos, we select a clip of 16 frames from each video, uniformly sampled in time, and each frame is cropped to 112 pixels × 112 pixels. The experiments are conducted on two TitanX GPUs. The networks are trained for 10 epochs on HMDB51 and UCF101 datasets, 20 epochs on the Olympic Sports dataset with a batch size of 16.

We use Adam optimizer [37] to train all networks. The learning rate is initially set to 1e-4 and is decayed by a factor of 0.5 every 5 epochs. The average Top-1 and Top-5 accuracy (%) ± standard deviation are reported.

Table 1. ZSAR performances on the three existing benchmarks compared with state-of-the-art methods. VE and SE represent Visual Embedding and Semantic Embeding. FV represents Fisher Vectors, obj represents objects. A represents the human annotated attribute vectors, W represents the word2vec embedding, and ED represents elaborative description. The average Top-1 accuracy (%) ± standard deviation is reported.

Methods	VE	SE	Olympic sports	HMDB51	UCF101
O2A [12]	Obj	W	N/A	15.6	30.3
MTE [38]	FV	W	44.3 ± 8.1	19.7 ± 1.6	15.8 ± 1.3
ASR [11]	C3D	W	N/A	21.8 ± 0.9	24.4 ± 1.0
GMM [34]	C3D	W	N/A	19.3 ± 2.1	17.3 ± 1.1
UAR [39]	FV	W	N/A	24.4 ± 1.6	17.5 ± 1.6
CEWGAN [40]	I3D	W	50.5 ± 6.9	30.2 ± 2.7	26.9 ± 2.8
TS-GCN [41]	GCN	W	56.5 ± 6.6	23.2 ± 3.0	34.2 ± 3.1
BD-GAN [35]	C3D	A	49.80 ± 10.72	23.86 ± 2.95	18.73 ± 3.73
E2E [10]	R(2+1)D	W	N/A	32.7	48
VDARN [13]	GCN	W	57.6 ± 3.4	21.6 ± 2.8	26.4 ±2.8
ER [23]	TSM	ED	60.2 ± 8.9	35.3 ± 4.6	51.8 ± 2.9
Ours	R(2+1)D	W	**61.9 ± 7.8**	**38.8 ± 4.6**	**52.6 ± 2.4**

4.3 Comparison with the State-of-the-Art

The comparison results are shown in Table 1. Overall, the proposed method performs best against state-of-the-art methods on three widely used datasets [16–18]. When using the same word embedding methods, our approach outperforms recent VDARN [13], and TS-GCN [41] methods on three datasets. Compared with other attributes-based [35] and ED-based [23] methods which are not scalable and labour-intensive, the proposed method uses the word embedding of action names as semantic space and still obtains better performance than them. Compared with E2E [10] method which also uses the R(2+1)D network and directly maps from visual features to the semantic space, our method improves by 6.1 and 4.6% points on the HMDB51 and UCF101 datasets respectively, indicating that amending the visual feature with semantic information can effectively narrow the semantic gap. Note that the O2A [12] and the VDARN [13] methods also use object information as part of the visual feature, but they use object detector during testing as well. Our method uses a hallucination network to extract object related information during testing and still performs better than them. In addition, the proposed method not only outperforms the very recent method ER [23] in classification accuracy, but also has a smaller standard deviation, which indicates that our method has a relatively stable performance under different training and testing data partitions.

Table 2. Ablation study of the proposed object semantics and hallucination network on HMDB51 and UCF101 datasets. "Baseline" refers to directly mapping visual features to semantic space; "OS (Object Semantics)" refers to extracting object semantics through an object detector to fuse with visual features during training, but the detector is not used in the test stage; "HN (Hallucination Network)" refers to implicitly extracting object semantics to fuse with visual features at test time. The average Top-1 and Top-5 accuracy (%) ± standard deviation are reported.

Baseline	OS	HN	HMDB51		UCF101	
			Top-1	Top-5	Top-1	Top-5
✓	×	×	32.7 ± 3.0	56.3 ± 3.4	47.9 ± 2.2	72.4 ± 3.4
✓	✓	×	36.7 ± 4.3	65.6 ± 5.0	49.8 ± 2.5	76.9 ± 2.7
✓	✓	✓	$\mathbf{38.8 \pm 4.6}$	$\mathbf{66.9 \pm 3.7}$	$\mathbf{52.6 \pm 2.4}$	$\mathbf{77.1 \pm 2.9}$

4.4 Ablation Studies

Impact of Object Semantics and Hallucination Network. To verify the effectiveness of the proposed object semantics and hallucination network, ablation studies are conducted on HMDB51 and UCF101 datasets. As shown in the second row of Table 2, leveraging object semantics to augment visual feature only at the training stage can significantly improve classification accuracy on the two datasets. The hallucination network contributes further improvement by 4.9/2.1/2.8% points for Top-1 accuracy and 2.1/1.3/0.2% points for

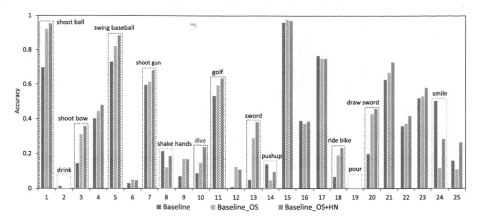

Fig. 4. Accuracy comparison of each unseen action class on the HMDB51 dataset between the proposed method and baseline. Here, OS represents Object Semantics, HN represents Hallucination Network. The experimental settings of OS and HN are the same as Table 2.

Top-5 accuracy on the three datasets respectively (see the last row of Table 2). Figure 4 further shows the accuracy comparison of each unseen action. It can be seen that the recognition accuracy of our method is significantly higher than the baseline in the red-marked categories, such as "shooting ball", "shooting bow", "sword", "ride bike" etc. These actions involve interaction with objects thus the hallucination network extracts the semantics of the related object to augment the visual feature and improve the recognition. In addition, some actions such as "dive" (marked in blue) occur in a specific scene, and information on the scene is extracted by the hallucination network to augment the visual information. However, for some actions such as "smile" and "shake hands" (marked in yellow), these actions do not involve any interaction with the surrounding environment. In this case, the information extracted by the hallucination network may interfere with the visual feature. The interference can be minimized by the choice of the objects in the training (See below Sect. 4.4 Impact of Object Number).

Impact of Fusion Strategy. In order to verify that the proposed cross-attention module is better than other fusion strategies, we perform another ablation experiment on the HMDB51 and UCF101 datasets, the results are shown in Table 3. Compared with multiplication, concatenation and addition fusing schemes, the proposed fusion strategy achieves the best performance. The reason is that other fusion strategies neglect the modality discrepancy. The proposed cross-attention block can effectively mitigate such an issue, in which the visual features and object semantics can exchange information through the

Table 3. Ablation study of the proposed cross-attention module with different fusion manners on HMDB51 and UCF101 datasets. The average Top-1 and Top-5 accuracy (%) ± standard deviation are reported.

Methods	HMDB51		UCF101	
	Top-1	Top-5	Top-1	Top-5
Multiplication	34.1 ± 4.6	56.6 ± 3.6	49.4 ± 3.4	73.2 ± 3.6
Concatenation	35.3 ± 4.7	58.6 ± 5.4	51.8±2.8	75.4±3.4
Addition	37.1 ± 4.3	64.5 ± 4.4	51.9 ± 2.5	76.9 ± 3.5
Cross-attention	**38.8 ± 4.6**	**66.9 ± 3.7**	**52.6 ± 2.4**	**77.1 ± 2.9**

mutual-attention layer to reduce the feature variations. Then the cross-modal features are fused to produce the conjoint feature representation. The results also show that the proposed cross-attention module has the smallest standard deviation on the UCF101 dataset.

Impact of Object Number. Figure 5 shows the impact of the object number used for each action class. It can be seen that the performance first increases as the number of objects increases and then decreases when the number of objects is over five for the HMDB51 dataset. This is probably due to that each action class is related to only a limited number of objects, and irrelevant objects will become noise interference and degrade performance. However, the performance degradation is insignificant (only about 0.3% points) even if the object number is increased to nine.

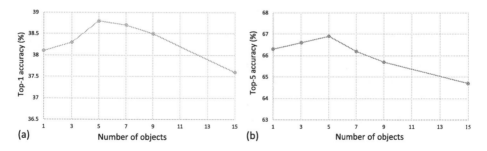

Fig. 5. Evaluate the proposed method with different number of objects on HMDB51 dataset. The average Top-1 (a) and Top-5 (b) accuracy (%) are illustrated.

4.5   Qualitative Results

We further analyze the effectiveness of the proposed method via the qualitative visualizations of visual feature distribution in the semantic space [42], as shown in Fig. 6. For the sake of visualizations, we randomly sample 8 unseen

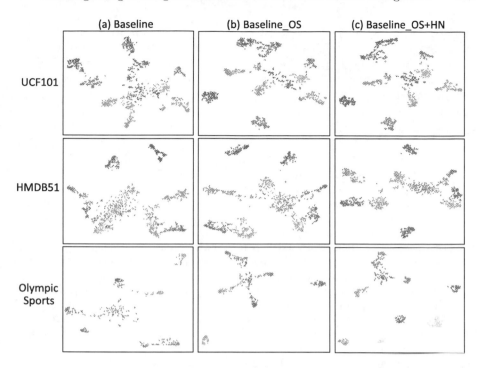

Fig. 6. The t-SNE visualization of the visual feature distribution of unseen classes of the three datasets. Different classes are shown as dots in different colors. Here, OS represents Object Semantics, HN represents Hallucination Network. The experimental settings of OS and HN are the same as Sect. 4.4.

classes from the Olympic Sports dataset and 10 unseen classes from HMDB51 and UCF101 datasets. As can be seen from Fig. 6(a) and Fig. 6(b), with the aid of the object detector during training, the baseline network yield improved feature representations. In Fig. 6(c), we further use a hallucination network to mimic the object semantics to fuse with visual features at test time, which can obtain the tightest intra-class clusters and separable inter-class clusters in the semantic space. The visualization results prove that the proposed method can effectively use object semantics to amend visual representation, thus alleviating the semantic gap.

5   Conclusions

In this paper, we propose a novel method for ZSAR. It is built upon the paradigm of learning using privileged information (LUPI). Specifically, the object semantics are used as privileged information to narrow the semantic gaps and assist learning of a mapping from visual feature to a semantic space. A hallucination network is trained to implicitly extract object related semantics at test time instead of using a pretrained object detector. Moreover, a cross-attention module is designed to fuse visual feature with object semantics or the feature from hallucination network. Empirical experiments on the Olympic Sports, HMDB51 and UCF101 benchmarks have shown that the proposed method achieves the state-of-the-art results.

References

1. Feichtenhofer, C., Pinz, A., Zisserman, A.: Convolutional two-stream network fusion for video action recognition. In: Proceedings of the IEEE Conference on Computer Vision and Pattern Recognition, pp. 1933–1941 (2016)
2. Carreira, J., Zisserman, A.: Quo vadis, action recognition? a new model and the kinetics dataset. In: Proceedings of the IEEE Conference on Computer Vision and Pattern Recognition, pp. 6299–6308 (2017)
3. Feichtenhofer, C., Pinz, A., Wildes, R.P.: Spatiotemporal multiplier networks for video action recognition. In: Proceedings of the IEEE Conference on Computer Vision and Pattern Recognition, pp. 4768–4777 (2017)
4. Wang, Y., Long, M., Wang, J., Yu, P.S.: Spatiotemporal pyramid network for video action recognition. In: Proceedings of the IEEE Conference on Computer Vision and Pattern Recognition, pp. 529–1538 (2017) 1
5. Zhu, Y., et al.: A comprehensive study of deep video action recognition. arXiv preprint arXiv:2012.06567 (2020)
6. Liu, J., Kuipers, B., Savarese, S.: Recognizing human actions by attributes. In: Proceedings of the IEEE Conference on Computer Vision and Pattern Recognition, pp. 3337–3344 (2011)
7. Fu, Y., Hospedales, T.M., Xiang, T., Fu, Z., Gong, S.: Transductive multi-view embedding for zero-shot recognition and annotation. In: Fleet, D., Pajdla, T., Schiele, B., Tuytelaars, T. (eds.) ECCV 2014. LNCS, vol. 8690, pp. 584–599. Springer, Cham (2014). https://doi.org/10.1007/978-3-319-10605-2_38
8. Wang, Q., Chen, K.: Zero-shot visual recognition via bidirectional latent embedding. Int. J. Comput. Vis. **124**, 356–383 (2017)
9. Mikolov, T., Chen, K., Corrado, G., Dean, J.: Efficient estimation of word representations in vector space. arXiv preprint arXiv:1301.3781 (2013)
10. Brattoli, B., Tighe, J., Zhdanov, F., Perona, P., Chalupka, K.: Rethinking zero-shot video classification: End-to-end training for realistic applications. In: Proceedings of the IEEE Conference on Computer Vision and Pattern Recognition, pp. 4613–4623 (2020)
11. Wang, Q., Chen, K.: Alternative semantic representations for zero-shot human action recognition. In: Ceci, M., Hollmén, J., Todorovski, L., Vens, C., Džeroski, S. (eds.) ECML PKDD 2017. LNCS (LNAI), vol. 10534, pp. 87–102. Springer, Cham (2017). https://doi.org/10.1007/978-3-319-71249-9_6
12. Jain, M., Van Gemert, J.C., Mensink, T., Snoek, C.G.: Objects2action: classifying and localizing actions without any video example. In: Proceedings of the IEEE International Conference on Computer Vision, pp. 4588–4596 (2015)
13. Su, Y., Xing, M., An, S., Peng, W., Feng, Z.: Vdarn: video disentangling attentive relation network for few-shot and zero-shot action recognition. Ad Hoc Netw. **113**, 102380 (2021)
14. Deng, J., Dong, W., Socher, R., Li, L.J., Li, K., Fei-Fei, L.: Imagenet: a large-scale hierarchical image database. In: Proceedings of the IEEE Conference on Computer Vision and Pattern Recognition, pp. 248–255. IEEE (2009)

15. Vapnik, V., Vashist, A.: A new learning paradigm: learning using privileged information. Neural Netw. **22**, 544–557 (2009)
16. Niebles, J.C., Chen, C.W., Fei-Fei, L.: Modeling temporal structure of decomposable motion segments for activity classification. In: Proceedings of the European Conference on Computer Vision, Springer (2010) 392–405
17. Kuehne, H., Jhuang, H., Garrote, E., Poggio, T., Serre, T.: Hmdb: a large video database for human motion recognition. In: Proceedings of the IEEE International Conference on Computer Vision, pp. 2556–2563. IEEE (2011)
18. Soomro, K., Zamir, A.R., Shah, M.: Ucf101: a dataset of 101 human actions classes from videos in the wild. arXiv preprint arXiv:1212.0402 (2012)
19. Xu, X., Hospedales, T., Gong, S.: Semantic embedding space for zero-shot action recognition. In: 2015 IEEE International Conference on Image Processing, pp. 63–67. IEEE (2015)
20. Bishay, M., Zoumpourlis, G., Patras, I.: Tarn: Temporal attentive relation network for few-shot and zero-shot action recognition. arXiv preprint arXiv:1907.09021 (2019)
21. Zhou, L., Li, W., Ogunbona, P., Zhang, Z.: Semantic action recognition by learning a pose lexicon. Pattern Recogn. **72**, 548–562 (2017)
22. Zhou, L., Li, W., Ogunbona, P., Zhang, Z.: Jointly learning visual poses and pose lexicon for semantic action recognition. IEEE Trans. Circuits Syst. Video Technol. **30**, 457–467 (2019)
23. Chen, S., Huang, D.: Elaborative rehearsal for zero-shot action recognition. In: Proceedings of the IEEE International Conference on Computer Vision, pp. 13638–13647 (2021)
24. Lampert, C.H., Nickisch, H., Harmeling, S.: Learning to detect unseen object classes by between-class attribute transfer. In: Proceedings of the IEEE Conference on Computer Vision and Pattern Recognition, pp. 951–958 (2009)
25. Wang, H., Schmid, C.: Action recognition with improved trajectories. In: Proceedings of the IEEE International Conference on Computer Vision, pp. 3551–3558 (2013)
26. Wang, H., Oneata, D., Verbeek, J., Schmid, C.: A robust and efficient video representation for action recognition. Int. J. Comput. Vis. **119**, 219–238 (2016)
27. Tran, D., Wang, H., Torresani, L., Ray, J., LeCun, Y., Paluri, M.: A closer look at spatiotemporal convolutions for action recognition. In: Proceedings of the IEEE Conference on Computer Vision and Pattern Recognition, pp. 6450–6459 (2018)
28. Tran, D., Bourdev, L., Fergus, R., Torresani, L., Paluri, M.: Learning spatiotemporal features with 3d convolutional networks. In: Proceedings of the IEEE International Conference on Computer Vision, pp. 4489–4497 (2015)
29. Xu, X., Hospedales, T., Gong, S.: Transductive zero-shot action recognition by word-vector embedding. International Journal of Computer Vision **123**, 309–333 (2017)
30. Niu, L., Li, W., Xu, D.: Visual recognition by learning from web data: a weakly supervised domain generalization approach. In: Proceedings of the IEEE Conference on Computer Vision and Pattern Recognition, pp. 2774–2783 (2015)
31. Motiian, S., Piccirilli, M., Adjeroh, D.A., Doretto, G.: Information bottleneck learning using privileged information for visual recognition. In: Proceedings of the IEEE Conference on Computer Vision and Pattern Recognition, pp. 1496–1505 (2016)
32. Crasto, N., Weinzaepfel, P., Alahari, K., Schmid, C.: Mars: motion-augmented RGB stream for action recognition. In: Proceedings of the IEEE Conference on Computer Vision and Pattern Recognition, pp. 7882–7891 (2019)

362   Z. Gao et al.

33. Garcia, N.C., Morerio, P., Murino, V.: Learning with privileged information via adversarial discriminative modality distillation. IEEE Trans. Pattern Anal. Mach. Intell. **42**, 2581–2593 (2020)
34. Mishra, A., Verma, V.K., Reddy, M.S.K., Arulkumar, S., Rai, P., Mittal, A.: A generative approach to zero-shot and few-shot action recognition. In: Proceedings of the IEEE Winter Conference on Applications of Computer Vision, pp. 372–380 (2018)
35. Mishra, A., Pandey, A., Murthy, H.A.: Zero-shot learning for action recognition using synthesized features. Neurocomputing **390**, 117–130 (2020)
36. Kolesnikov, A., et al.: Big transfer (BiT): general visual representation learning. In: Vedaldi, A., Bischof, H., Brox, T., Frahm, J.-M. (eds.) ECCV 2020. LNCS, vol. 12350, pp. 491–507. Springer, Cham (2020). https://doi.org/10.1007/978-3-030-58558-7_29
37. Kingma, D.P., Ba, J.: Adam: a method for stochastic optimization. arXiv preprint arXiv:1412.6980 (2014)
38. Xu, X., Hospedales, T.M., Gong, S.: Multi-task zero-shot action recognition with prioritised data augmentation. In: Leibe, B., Matas, J., Sebe, N., Welling, M. (eds.) ECCV 2016. LNCS, vol. 9906, pp. 343–359. Springer, Cham (2016). https://doi.org/10.1007/978-3-319-46475-6_22
39. Zhu, Y., Long, Y., Guan, Y., Newsam, S., Shao, L.: Towards universal representation for unseen action recognition. In: Proceedings of the IEEE Conference on Computer Vision and Pattern Recognition, pp. 9436–9445 (2018)
40. Mandal, D., et al.: Out-of-distribution detection for generalized zero-shot action recognition. In: Proceedings of the IEEE Conference on Computer Vision and Pattern Recognition, pp. 9985–9993 (2019)
41. Gao, J., Zhang, T., Xu, C.: I know the relationships: zero-shot action recognition via two-stream graph convolutional networks and knowledge graphs. Proc. AAAI Conf. Artif. Intell. **33**, 8303–8311 (2019)
42. Van der Maaten, L., Hinton, G.: Visualizing data using t-sne. J. Mach. Learn. Res. **9**, 2579–2605 (2008)

MGTR: End-to-End Mutual Gaze Detection with Transformer

Hang Guo, Zhengxi Hu, and Jingtai Liu[✉]

Nankai University, Tianjin, China
{1911610,hzx}@mail.nankai.edu.cn, liujt@nankai.edu.cn

Abstract. People's looking at each other or mutual gaze is ubiquitous in our daily interactions, and detecting mutual gaze is of great significance for understanding human social scenes. Current mutual gaze detection methods focus on two-stage methods, whose inference speed is limited by the two-stage pipeline and the performance in the second stage is affected by the first one. In this paper, we propose a novel one-stage mutual gaze detection framework called Mutual Gaze TRansformer or MGTR to perform mutual gaze detection in an end-to-end manner. By designing mutual gaze instance triples, MGTR can detect each human head bounding box and simultaneously infer mutual gaze relationship based on global image information, which streamlines the whole process with simplicity. Experimental results on two mutual gaze datasets show that our method is able to accelerate mutual gaze detection process without losing performance. Ablation study shows that different components of MGTR can capture different levels of semantic information in images. Code is available at https://github.com/Gmbition/MGTR.

Keywords: End-to-End mutual gaze detection · One-stage method · Mutual gaze instance match

1 Introduction

Containing rich information, the gaze plays an important role in reflecting the attention, intention and emotion of one person [1,2]. Among all kinds of gaze, mutual gaze is indispensable in building the bridge between two minds [3,17]. From mutual gaze, one can infer the willingness to interact and the strength of the relationship. Moreover, mutual gaze can also be used for people connection analysis in social scene interpretation and is an important clue for Human-Robot-Interaction. For these reasons, it is very promising to achieve automatic mutual gaze detection.

This work is supported in part by National Key Research and Development Project under Grant 2019YFB1310604, in part by National Natural Science Foundation of China under Grant 62173189.

Supplementary Information The online version contains supplementary material available at https://doi.org/10.1007/978-3-031-26316-3_22.

L. Wang et al. (Eds.): ACCV 2022, LNCS 13844, pp. 363–378, 2023.
https://doi.org/10.1007/978-3-031-26316-3_22

The target for end-to-end mutual gaze detection is to detect all the human heads in the scene and then recognize whether any two people are looking at each other. Previous studies [6,19,20] have got favorable results by dividing the whole process into two stages: detect all human heads in the scene and then regard the recognition of mutual gaze as a binary classification problem. Specifically, Marin *et al.* proposed a video based method [19] (Fig. 1(a)) that first detects all human heads through a pretrained head detector and then enumerates all head-crop-pairs and put them into a classification network to identify whether two people are looking at each other. The image based mutual gaze detection work in Doosti *et al.* [6] (Fig. 1(b)) uses pseudo 3D gaze to boost mutual gaze detection and also adapts a two-stage strategy.

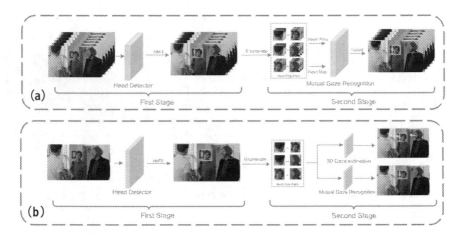

Fig. 1. An illustration of two-stage methods. The method in (a) uses a pretrained head detector and performs mutual gaze recognition by exploiting pose and position information of enumerated head-crop-pairs. The method in (b) also uses a head detector and utilizes pseudo 3D gaze to boost mutual gaze detection. It can be seen that methods in (a) and (b) both perform head detection in the first stage and need to enumerate all head-crop-pairs in one image which slows down the inference process.

Although these two-stage methods have achieved promising results, their designs have some shortcomings. Firstly, the classifier in the second stage makes inferences based on the local information of the head crop instead of the image global information. For example, body posture is also an important clue for judging mutual gaze. Moreover, the performance of the classification results in the second stage depends on the localization accuracy of the first stage. Furthermore, when there are many people in the scene, the computational cost will also increase due to the need to enumerate all detected heads which will slow down the inference process.

To overcome these limitations, inspired by the attention mechanism in Transformer Network [25], we propose a *one-stage* mutual gaze detection model called Mutual Gaze TRansformer or MGTR (Fig. 2) which can detect all human heads in the scene and simultaneously identify whether there is a mutual gaze based

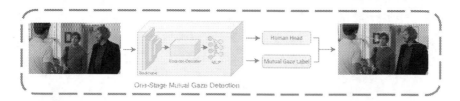

Fig. 2. An illustration of proposed one-stage method. Utilizing the designed mutual gaze instance triples, our proposed one-stage method can detect all human heads and the corresponding mutual gaze labels in parallel, which accelerates the pipeline of mutual gaze detection.

on the global image information. By designing mutual gaze instance triples, we improve the mutual gaze detection process from serial to parallel, which greatly accelerates the inference speed without losing performance.

Our proposed MGTR consists of four following modules: A Backbone for feature extraction, Transformer Encoder, Transformer Decoder, and a Fully Connected Neural Network for mutual gaze instance prediction. First, a convolutional neural network is used to extract image features, then a one-dimensional vector is generated by flattening the feature map and we combine it with the positional encoding [4, 22] to get the input of Encoder. After that, the output of Encoder combined with learnable mutual gaze queries are passed through Decoder to model connections between different people, and finally, the encoded mutual gaze queries are passed through a fully connected neural network to output the mutual gaze instance as the result of our model.

Overall, our main contributions are as follows:

- We build a *one-stage* model called MGTR which combines the human head detection and the mutual gaze recognition. To the best of our knowledge, this is the first work that integrates mutual gaze detection task into a one stage method.
- Modeling the location information of people and the relationship between them using global information instead of head image crops.
- Our model outperforms the state-of-the-art method on end-to-end mutual gaze detection task. Moreover, MGTR can perform faster mutual gaze detection.

2   Related Work

In this section, we first review methods for gaze estimation which encompass a variety of gaze types (Sect. 2.1). Then go down to the literature of mutual gaze detection (Sect. 2.2). At last one-stage detection methods (Sect. 2.3) will be reviewed.

2.1   Gaze Estimation in Social Scenarios

Eye gaze can convey rich information and is closely related to the attention, intention, and emotion of a person, even people from different cultures may share a similar meaning of eye gaze [10]. Recently, in the computer vision community, there are also a lot of research focusing on social scenario gaze estimation and yielding promising results. For example, Lian et al. [13] proposed a solution for gaze point prediction of the target persons. Fan et al. [7] proposed a spatial-temporal modeling method to detect people's looking at the same target simultaneously. Zhuang et al. proposed MUGGLE [28], an approach that is suitable for massive people's shared-gazing. In order to detect whether two people in the video are looking at each other, Marin et al. [19,20] proposed a method based on the spatial and temporal information to solve this problem. To understand different types of eye gaze in a group of people, Fan et al. [8] proposed a method to detect multiple types of human gaze, such as single gaze, shared gaze, etc. In this work, we focus on image based one-stage mutual gaze detection.

2.2   Mutual Gaze Detection

Mutual gaze is one of the most common types of social scenarios gaze communication and there are also methods trying to achieve automatic mutual gaze detection. These methods are all two-stage methods consisting of a human head detector in the first stage and a mutual gaze classifier in the second stage. Specifically, Marin et al. proposed viode-based LAEO-Net [20] and get a promising result in mutual gaze detection by considering both temporal and spatial information. After that, they further modified LAEO-Net to get LAEO-Net++ [19], which achieved better performance. Moreover, the image-based approach proposed by Doosti et al. [6] takes advantage of multi-task learning by using pseudo 3D gaze to boost mutual gaze detection. However, these methods suffer from the lack of global image information and slow inference speed due to the sequential two-stage architecture.

2.3   One-Stage Detection Method

Recently, in the field of computer vision, many research designs have followed a one-stage idea to speed up the processing pipeline. For example, in the field of object detection, the favorable results of SSD [16], YOLO [23], RetinaNet [14], DETR [5] and other methods have demonstrated the advantages of one-stage detection methods. Compared with two-stage detection methods, one-stage methods can perform the detection and classification tasks by using only one network and the pipeline is generally simpler, faster, and more computationally efficient so that it is easier to adopt for real-world applications.

3   Method

The task of one-stage mutual gaze detection is to give an image and then detect all mutual maze instances in the image in an end-to-end way. In this section,

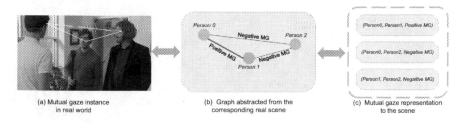

(a) Mutual gaze instance
in real world

(b) Graph abstracted from the
corresponding real scene

(c) Mutual gaze representation
to the scene

Fig. 3. Representation of Mutual Gaze Instances in the scene. In the social scene, each individual is uniquely numbered and the enumerated head pairs are order-independent. *MG* means Mutual Gaze.

we will first describe the Representation of Mutual Gaze Instances (Sect. 3.1) followed by the detailed Model Architecture (Sect. 3.2), after that we will introduce the Strategy for Mutual Gaze Instances Match (Sect. 3.3) and at last the Loss Function Setting (Sect. 3.4).

3.1   Representation of Mutual Gaze Instances

We define a mutual gaze instance as a triple, namely (Person1, Person2, Mutual Gaze Label), where Person1 and Person2 contain the bounding box coordinates and the class confidence of a person, and Mutual Gaze Label is one when Person1 and Person2 are looking at each other otherwise zero. It is noteworthy that under the mutual gaze detection task, it seems useless to predict the class of each box, however, this setting can be used as a detection threshold when we conduct the test phase in which we need the person class confidence of each box. A detailed example of representing a real scene with a mutual gaze instance is given in Fig. 3. Additionally, a mutual gaze instance is unordered, that is to say, the relationship between i-th Person and j-th Person only needs to be recorded once as (Person$i$, Person$j$, Mutual Gaze label between i and j).

3.2   Model Architecture

Our proposed MGTR mainly consists of four parts: a Backbone, an Encoder module, a Decoder module, and MLP. Figure 4 shows an overview of MGTR architecture.

Backbone. A convolutional neural network is used to extract features from an input image of original size $[H, W, 3]$. After the convolutional neural network, we get a feature map of size $[C, H, W]$. We then reduce the channel dimension of the feature map from C to d by a 1×1 convolution kernel resulting the new feature map of size $[d, H, W]$. Since Transformer Encoder requires a sequence as input data, we compress the last two dimensions of the new feature map to obtain a flatten feature called *input embedding* of size$[d, HW]$.

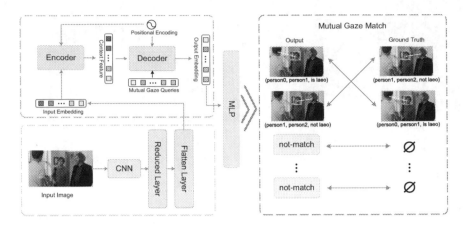

Fig. 4. An overview of MGTR architecture. It consists of four components: a Backbone to convert input image into an one-dimensional input embedding, an Encoder followed by a Decoder to get the encoded Mutual Gaze Queries, and MLP to predict mutual gaze instances. *laeo* means Looking At Each Other. See Sect. 3.2 for more details.

Encoder. The Encoder layer in MGTR is the same as the standard Transformer Encoder layer, including a multi-head self-attention layer and a feed forward network(FFN). Due to the permutation invariance of the Transformer, we add the *positional encoding* to the input embedding to obtain the Query and Key of Encoder and only use the input embedding as the Value. For the convenience of description, we represent the output of the Encoder as the *context feature*.

Decoder. The Decoder layer in MGTR is also the same as the standard Transformer Decoder layer which contains two multi-head attention layers and a feed-forward network. We refer to the N learnable positional embeddings as *mutual gaze queries*. In the multi-head self-attention layer, the Query, Key, and Value all come from either the mutual gaze queries or the sum of the previous decoder layer's output and mutual gaze queries. As for the encoder-decoder cross attention layer, the Value comes from the context feature generated from the Encoder, the Key is the sum of context feature and positional encoding and the Query is the sum of the multi-head self-attention layer's output and mutual gaze queries. We denote the output of the Decoder as *output embedding*.

The self-attention mechanism in the Encoder and Decoder can help us model the positions of different people in the image and the relationship between them. The N output embeddings encoded from N mutual gaze queries are then converted into a mutual gaze instance by the subsequent MLP so that we get N final mutual gaze instance results and we will discuss this part next.

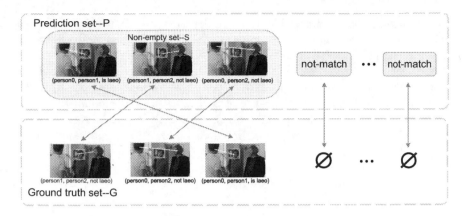

Fig. 5. An example explaining the mutual gaze instances match. By padding the set **G** with ∅, we make **P** and **G** the same size, which can be transformed into a bipartite graph matching problem. *laeo* means Looking At Each Other

MLP for Mutual Gaze Prediction. After passing the N mutual gaze queries through the Decoder, we get N output embedding, which contains information about the position of the head bounding boxes and the relationship between different people in the image. Then we pass the output embedding into the MLP to predict the mutual gaze instance. Specifically, three one-layer fully-connected neural networks are used to predict the confidence score of Person1, Person2, and Mutual Gaze Label respectively, and two three-layer fully-connected neural networks are used to predict the head bounding boxes of Person1 and Person2. For the human confidence score prediction branch, there are three classes: whether it is a person and *not-match* (the meaning of not-match will be described later). For the mutual gaze confidence prediction branch, there are also three categories that indicate whether Person1 and Person2 are looking at each other and not-match. We then apply Softmax to the results of all confidence prediction branches to obtain normalized confidence. For the branch of head bounding boxes regression, the output dimension of the MLP is four, which represent the normalized center coordinates, width, and height of the bounding boxes respectively.

3.3   Strategy for Mutual Gaze Instances Match

After passing the output embedding through MLP, we get N predicted mutual gaze instances. However, the number of ground truth mutual gaze instances is not necessarily N (often less than N). So it requires some predicted mutual gaze instances representing empty which we denoted as *no-match*, indicating that these mutual gaze instances do not match any ground truth. To be precise, we denote the set of ground truth instances as **G**, and the size of **G** is M, the set of predicted instances is **P** and the size of **P** is N, then a satisfactory model should output a set which contains $N - M$ instances representing *not-match*.

After solving the problem that the number of ground truth instances and predicted instances are not always equal by designing the *not-match* class, the next key problem is how to match the predicted instances with the ground truth instances. Specifically, we denote the set of elements predicted to be non-empty in **P** as **S**, then the problem we have to solve is how to build a map σ from **S** to **G**. It is worth mentioning that the size of **S** is not necessarily equal to the size of **G**. We solve this matching problem in another way: by padding the ground truth set **G** with $\varnothing$, we make **G** and **P** equal in size. So the above matching problem is transformed into a one-to-one bipartite matching problem between **P** and **G**. In this work, we use the Hungarian algorithm [12] to solve this problem. A more concrete example can be seen in Fig. 5.

3.4   Loss Function Setting

Assume the mapping from the predicted set **P** to the ground-truth set **G** is denoted as $\sigma(i)$, which means the i-th element in the set **P** will be mapped to the $\sigma(i)$-th element in **G**. We design the matching cost of the Hungarian algorithm as follows.

$$\mathcal{L}_{match} = \sum_{i=1}^{N}[\beta_1 \mathcal{L}_{class}(c_i, c_{\sigma(i)}) + \beta_2 \mathcal{L}_{box}(b_i, b_{\sigma(i)})] \tag{1}$$

where $\mathcal{L}_{class}(c_i, c_{\sigma(i)})$ represents the cost between the i-th mutual gaze instance from **P** and the $\sigma(i)$-th ground truth from **G** in human class confidence and mutual gaze confidence, and we call it the *class loss function*. $\mathcal{L}_{box}(b_i, b_{\sigma(i)})$ represents the cost between the i-th mutual gaze instance from **P** and the $\sigma(i)$-th from **G** in head bounding boxes regression, and we call it the *head bounding box regression loss function*. β_1 and β_2 are hyperparameters used to measure the weight between these two types of losses. The specific forms of these two losses are discussed below.

For the class loss function, we denote the value of mutual gaze confidence as p_i^{gaze} and use $p_i^{h_1}$ and $p_i^{h_2}$ to represent the human class confidence. Under this representation, the class loss function is defined as follow.

$$\mathcal{L}_{class}(c_i, c_{\sigma(i)}) = \alpha_1 p_i^{h_1} + \alpha_2 p_i^{h_2} + \alpha_3 p_i^{gaze} \tag{2}$$

For the head bounding box regression loss function, the definition is as follow.

$$\mathcal{L}_{box}(b_i, b_{\sigma(i)}) = \gamma_1 \ell_1(b_i, b_{\sigma(i)}) + \gamma_2 \text{GIoU}(b_i, b_{\sigma(i)}) \tag{3}$$

where $\ell_1(\cdot)$ is the L_1 loss. We also added the GIoU loss [24] into the head bounding box regression loss function, and we will confirm the importance of GIoU loss in the later Ablation Study.

By defining the above costs from the i-th element in **P** to the $\sigma(i)$-th element in **G**, we can solve the following optimal bipartite match problem based on the Hungarian algorithm.

$$\sigma^* = \underset{\sigma}{\text{argmin}}\, \mathcal{L}_{match} \tag{4}$$

Fig. 6. Some examples of UCO-LAEO and AVA-LAEO datasets. The images in the first row come from the UCO-LAEO dataset which contains both head bounding boxes and mutual gaze labels. The images in the second row are from the AVA-LAEO dataset which only contains mutual gaze labels and we approximate the ground-truth head bounding box coordinates by using a pre-trained head detector [16].

After the optimal bipartite matching problem is solved by the Hungarian algorithm, we can next calculate the loss function for training. In this work, the definition of loss function is almost the same as in Eq. 1, the difference is that in $\mathcal{L}_{class}$ we use cross-entropy loss instead of negative predicted probability and there are also some subtle differences in the settings of hyperparameters as well (more details can be seen in Sect. 4.4).

Different from previous mutual gaze detection methods which first train a head detector, and then freeze the head detector parameters and train the mutual gaze classifier. The method in our work optimizes the class loss function and the head bounding box regression loss function simultaneously during the model training process.

4 Experiments

4.1 Datasets

UCO-LAEO. [20] This dataset consists of detailed mutual gaze instance annotations with human head bounding boxes and mutual gaze labels. The original dataset has been manually divided into positive and negative instances for sample balance, which means there may be some instances in one image that are not annotated as negative instances. In this work, since we need to detect all the instances in one image, so we only use positive samples from original annotations and treat all remaining instances as negative samples.

AVA-LAEO. [20] This dataset has a broad coverage that reaching 50,797 video frames generated from *Atomic Visual Actions* dataset(AVA v2.2) [21]. Since there are no manual head bounding box annotations in the original AVA v2.2 dataset, we first use a pre-trained head detector [16] to generate all head bounding box coordinates in each frame as head bounding box ground truth before our training starts. Similarly, we also only use positive instances from the original dataset annotations and regard all the remaining as negative.

Some examples of UCO-LAEO and AVA-LAEO datasets are shown in Fig. 6.

4.2   Evaluation Metric

In this work, we use the *mean average precision*(mAP) as a measure of how well our model performs on the two datasets. It is worth mentioning that although our task is similar to a binary classification task, we do not use the AP of a single class as an evaluation metric. This is because our task is to simultaneously predict accurate human head positions as well as mutual gaze labels. So we need to consider both classes of AP to include the evaluation of head position localization accuracy. For example, if it is known that the model predicts the mutual gaze label in the current instance to be positive but in fact it is a wrong prediction, it does not necessarily mean that the ground-truth mutual gaze label is negative because it is also possible that the model does not match the head bounding box with ground truth when predicting. So, an instance predicts correctly if and only if it locates the head boxes of the two people in the correct position and predicts correctly whether the two are looking at each other. This criterion requires both the AP in two classes should be high.

4.3   Current State-of-the-Art Approach

The current state-of-the-art approach for video-based mutual gaze detection task is LAEO-Net++ proposed by Marin *et al.* [19]. The image-based state-of-the-art method is proposed by Doosti *et al.* [6] and we call it *Pseudo 3D Gaze* in this paper.

The two aforementioned works are different from ours. Specifically, both works above only detect mutual gaze without detecting human head position, while MGTR detects above both whose task is more difficult. Moreover, LAEO-Net++ is a video-based method which uses ground-truth human head box as model input and utilizes temporal connection among neighbor video frames resulting in the input containing more prior information, while ours is an image-based method whose input is only one single image. Since these differences, we modify them accordingly for a fair comparison. Specifically, as for LAEO-Net++, since we focus on image-based mutual gaze detection in this work, so we plan to use its image-based version introduced in [6]. However, we cannot get the modified LAEO-Net++ for comparison since the code in [6] is not open source. So we directly use the performance numbers of image-based *LAEO-Net* reported in [6]. As for Pseudo 3D Gaze which focuses on the performance of the second stage, we add a head detector in front of the original model for end-to-end mutual gaze detection, a detailed description of this model can be seen in Sect. 4.4.

4.4   Implementation Details

Data Augmentation. We normalize the input image by using the mean and std from ImageNet [11]. To improve the robustness of the model, we randomly apply horizontal flipping, adjusting brightness and contrast, random cropping, and random resize (to enable the model to detect instances at multiple scales).

Hyperparameters Settings. In order to balance the class cost and head bounding box regression cost, we set $\beta_1 = 1.2$, $\beta_2 = 1.0$ in the Hungarian algorithm matching process and $\beta_1 = \beta_2 = 1.0$ in the training loss function. At the same time, we set $\alpha_1 = \alpha_2 = 1.0$, $\alpha_3 = 2.0$ both in the Hungarian algorithm and loss function to make our model focus more on judging the existence of mutual gaze. In the head bounding box regression loss function, we set $\gamma_1 = 5.0$, $\gamma_2 = 2.0$.

Training Settings. We used Resnet50 [9] with frozen batchnorm layer as Backbone for MGTR, the number of Encoder and Decoder layers are both set to 6, the same as in [29]. Both Backbone and Encoder-Decoder use the pretrained parameters from COCO [15] pretrained DETR [5]. The number of mutual gaze queries is set to 100. In the training phase, we choose the batch size to be 8, and we use AdamW [18] as an optimizer, with a constant learning rate of 1e-4 in Encoder-Decoder and 1e-5 in Backbone, we train the model until the performance on the test set no longer improves.

Description for Pseudo 3D Gaze Using Head Detector. Since the Pseudo 3D Gaze method in [6] uses the ground truth head bounding boxes, for a fair comparison, we set this baseline that consists of two parts: a head detector and a mutual gaze classifier. We use the head detector proposed in [26] for head box detection. The classifier of this baseline adopts the same network as the one proposed by Doosti *et al.* and also uses the pseudo 3D gaze to boost the training process. During training, we only train the mutual gaze classifier with randomly initialized weights by using the ground truth head bounding boxes. During testing, we first detect each head bounding boxes through the pretrained detector and then pass the paired detected head crops through the mutual gaze classifier to get the result for each mutual gaze instance.

4.5   Comparison with State-of-the-Art Method

Table 1 shows the quantitative results compared with the state-of-the-art method in terms of FPS and mAP on UCO-LAEO and AVA-LAEO datasets.

For UCO-LAEO dataset, MGTR processing is more than 150 times faster than the two-stage approach in extreme social scenes with more than four people and nearly 17 times faster in scenes averaged across the test set. Moreover, MGTR achieves 64.8% mAP, an 18.1% increase over the Pseudo 3D Gaze using detected head bounding boxes. Meanwhile, our method is also comparable with image-based SoTA methods who use ground truth head bounding boxes, with only a difference of 0.3%. The good performance of MGTR on mAP shows that MGTR can handle the imbalance of positive and negative samples in the training set well.

Table 1. Comparison with State-of-the-Art Method. FPS_{ext} refers to the number of images processed per second in the extreme social scene with more than four people, and FPS_{all} refers to the number of pictures processed per second in scenes averaged across the whole test set. The FPS is evaluated using NVIDIA 3090TI GPU. Since we have no access to the code of image based LAEO-Net in [6], we do not evaluate the FPS of image based LAEO-Net. w/ GT and w/t GT respectively indicate whether to use ground truth head bounding boxes as model input. The mAP represents positive class's AP in two-stage methods and two classes's average AP in one-stage methods. *Number reported from [6].

Method		End-to-end training	FPS_{ext}		FPS_{all}		mAP			
			UCO	AVA	UCO	AVA	UCO-LAEO		AVA-LAEO	
							w/ GT	w/t GT	w/ GT	w/t GT
$Two-stage^*$	Iamge based LAEO-Net	✗	-	-	-	-	55.9	-	70.2	-
	Pseudo 3D Gaze	✗	0.51	0.93	0.49	0.89	**65.1**	46.7	**72.2**	52.3
$One-stage$	MGTR (ours)	✓	**78.06**	**14.56**	**9.18**	**10.81**	-	64.8	-	**66.2**

As for AVA-LAEO dataset, MGTR processes each image more than 15 times faster than the two-stage method in more than four people social scenes and more than 12 times in the average scene across the test set. Besides, MGTR gets a 66.2% mAP score, a 13.9% increase compared with the baseline using detected head bounding boxes.

4.6   Ablation Study

In this part, we design some ablation methods to study how data augmentation, different components of MGTR and loss function setting will affect the performance. We choose the UCO-LAEO dataset and use MGTR with Resnet50 backbone as a base model. Ablation baselines are as follows: (1) No multi-scale and random cropping in data augmentation (2) Using Resnet101 as a backbone, we design this part to study how the complexity of the model will affect performance (3) No GIoU loss, we only use BCE Loss and L_1 Loss as loss function. (4) Use DIoU loss [27] instead of original GIoU loss and keep other parts consistent with base model. We design this part to explore the effect of different IoU losses. (5) Use CIoU loss [27] to replace GIoU loss and keep other parts unchanged.

The results of the ablation study are provided in Table 2. It can be seen that data augmentation including multi-scale resize and random cropping are important for training, without which resulting in an 11.4% drop in mAP. At the same time, the use of Resnet101 as the backbone leads to a decrease on performance, so the more complex the model is not always the better. As can be seen from the fourth-to-last row of Table 2, when the GIoU loss is removed from the loss function, the performance of MGTR on mAP drops by 9.0%, which indicates that the GIoU loss is indispensable for MGTR to accurately locate each person's head bounding box. When using DIoU loss or CIoU loss to

Table 2. Ablation Study of MGTR. mAP refers to the average AP in the two classes, AP_{rare} refers to the AP of the minority category (usually the positive mutual gaze label), and AP_{normal} refers to the majority category, and Recall refers to the average Recall over two classes.

Model setting	#param	mAP	AP_{rare}	AP_{normal}	Recall
NoDataAugmentation	41.4M	53.4	52.8	54.3	68.6
Resnet101Backbone	60.3M	51.3	49.7	53.0	65.5
NoGIoULoss	41.4M	55.8	52.8	58.8	67.3
DataAug+Resnet50+DIoU	41.4M	55.7	51.1	60.4	69.2
DataAug+Resnet50+CIoU	41.4M	60.4	54.5	66.3	68.4
Base(DataAug+Resnet50+GIoU)	41.4M	**64.8**	**58.3**	**71.4**	**75.3**

replace the original GIoU loss, the performance on mAP drops by 9.1% and 4.4%, respectively, indicating that even though DIoU and CIoU are improvements over GIoU, using GIoU loss still achieves the best performance. This may be due to the fact that both Backbone and Encoder-Decoder in MGTR are initialized using the parameters in DETR pretrained model that also uses GIoU loss. Therefore, using a consistent loss may give better results.

4.7   Qualitative Analysis

Different from the previous two-stage methods, MGTR simultaneously gives all head bounding boxes and mutual gaze labels in the scene in an end-to-end manner, which does not seem easy to understand. In this part, we will analyze the different roles of Encoder and Decoder of MGTR in different levels of image semantic understanding.

To study the different roles of Encoder and Decoder in MGTR, we visualize the last attention layer of Encoder and Decoder respectively, results can be seen in Fig. 7. It can be easily seen that the role of Encoder in MGTR is to find all the head bounding boxes in the social scene, because the head area of each person in the attention-map is given a larger attention weight. After the Encoder finds all the people in the scene, Decoder can find the relationship between different people. In the Decoder's attention-map, we can see that when the mutual gaze label is positive there will be two people's head regions with large attention weights. However, when the label is negative, only one person's head region will be focused. Therefore, we can conclude that the role of Decoder is to predict which pairs of people in the current scene are looking at each other so that model the relationship between different people.

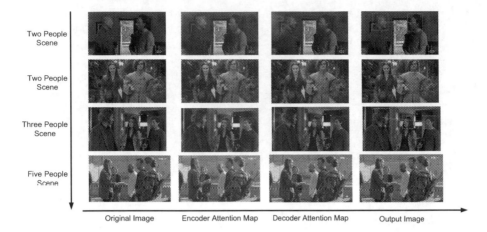

Original Image   Encoder Attention Map   Decoder Attention Map   Output Image

Fig. 7. Visualization of last attention layer in Encoder and Decoder and the predicted result by MGTR (from UCO-LAEO dataset). For each row, the first image is the original input image, the second image is the attention weight in Encoder, the third image is the attention weight in Decoder, and the last image is the predicted result by MGTR.

5 Conclusion

In this work, we propose a one-stage mutual gaze detection method called Mutual Gaze TRansformer or MGTR to directly predict mutual gaze instances in an end-to-end manner. Different from current two-stage mutual gaze detection methods, MGTR is the first work that integrates human head detection and mutual gaze recognition into one stage which simplifies the detection pipeline. Experiments on two mutual gaze datasets demonstrate that our proposed method can greatly accelerate inference process while improving performance. In the future, we will explore the application of incorporating mutual gaze information into the analysis of the interpersonal relationship, and the detected mutual gaze instance will serve as an important clue for social scene interpretation.

References

1. Abele, A.: Functions of gaze in social interaction: communication and monitoring. J. Nonverbal Behav. **10**(2), 83–101 (1986)
2. Admoni, H., Scassellati, B.: Social eye gaze in human-robot interaction: a review. J. Hum.-Robot Interact. **6**(1), 25–63 (2017)
3. Argyle, M., Cook, M.: Gaze and mutual gaze (1976)
4. Bello, I., Zoph, B., Vaswani, A., Shlens, J., Le, Q.V.: Attention augmented convolutional networks. In: Proceedings of the IEEE/CVF International Conference on Computer Vision, pp. 3286–3295 (2019)
5. Carion, N., Massa, F., Synnaeve, G., Usunier, N., Kirillov, A., Zagoruyko, S.: End-to-end object detection with transformers. In: Vedaldi, A., Bischof, H., Brox, T., Frahm, J.-M. (eds.) ECCV 2020. LNCS, vol. 12346, pp. 213–229. Springer, Cham (2020). https://doi.org/10.1007/978-3-030-58452-8_13

6. Doosti, B., Chen, C.H., Vemulapalli, R., Jia, X., Zhu, Y., Green, B.: Boosting image-based mutual gaze detection using pseudo 3D gaze. In: Proceedings of the AAAI Conference on Artificial Intelligence, vol. 35, pp. 1273–1281 (2021)

7. Fan, L., Chen, Y., Wei, P., Wang, W., Zhu, S.C.: Inferring shared attention in social scene videos. In: 2018 IEEE/CVF Conference on Computer Vision and Pattern Recognition, pp. 6460–6468. IEEE (2018)

8. Fan, L., Wang, W., Huang, S., Tang, X., Zhu, S.C.: Understanding human gaze communication by spatio-temporal graph reasoning. In: Proceedings of the IEEE/CVF International Conference on Computer Vision, pp. 5724–5733 (2019)

9. He, K., Zhang, X., Ren, S., Sun, J.: Deep residual learning for image recognition. In: Proceedings of the IEEE Conference on Computer Vision and Pattern Recognition, pp. 770–778 (2016)

10. Kleinke, C.L.: Gaze and eye contact: a research review. Psychol. Bull. **100**(1), 78 (1986)

11. Krizhevsky, A., Sutskever, I., Hinton, G.E.: Imagenet classification with deep convolutional neural networks. Commun. ACM **60**(6), 84–90 (2017)

12. Kuhn, H.W.: The Hungarian method for the assignment problem. Naval Res. Logist. Q. **2**(1–2), 83–97 (1955)

13. Lian, D., Yu, Z., Gao, S.: Believe it or not, we know what you are looking at! In: Jawahar, C.V., Li, H., Mori, G., Schindler, K. (eds.) ACCV 2018. LNCS, vol. 11363, pp. 35–50. Springer, Cham (2019). https://doi.org/10.1007/978-3-030-20893-6_3

14. Lin, T.Y., Goyal, P., Girshick, R., He, K., Dollár, P.: Focal loss for dense object detection. In: Proceedings of the IEEE International Conference on Computer Vision, pp. 2980–2988 (2017)

15. Lin, T.-Y., et al.: Microsoft COCO: common objects in context. In: Fleet, D., Pajdla, T., Schiele, B., Tuytelaars, T. (eds.) ECCV 2014. LNCS, vol. 8693, pp. 740–755. Springer, Cham (2014). https://doi.org/10.1007/978-3-319-10602-1_48

16. Liu, W., et al.: SSD: single shot MultiBox detector. In: Leibe, B., Matas, J., Sebe, N., Welling, M. (eds.) ECCV 2016. LNCS, vol. 9905, pp. 21–37. Springer, Cham (2016). https://doi.org/10.1007/978-3-319-46448-0_2

17. Loeb, B.K.: Mutual eye contact and social interaction and their relationship to affiliation (1972)

18. Loshchilov, I., Hutter, F.: Decoupled weight decay regularization. arXiv preprint arXiv:1711.05101 (2017)

19. Marín-Jiménez, M.J., Kalogeiton, V., Medina-Suárez, P., Zisserman, A.: LAEO-Net++: revisiting people looking at each other in videos. IEEE Trans. Pattern Anal. Mach. Intell. (2021). https://doi.org/10.1109/TPAMI.2020.3048482

20. Marin-Jimenez, M.J., Kalogeiton, V., Medina-Suarez, P., Zisserman, A.: LAEO-Net: revisiting people looking at each other in videos. In: Proceedings of the IEEE/CVF Conference on Computer Vision and Pattern Recognition, pp. 3477–3485 (2019)

21. Murray, N., Marchesotti, L., Perronnin, F.: AVA: a large-scale database for aesthetic visual analysis. In: 2012 IEEE Conference on Computer Vision and Pattern Recognition, pp. 2408–2415. IEEE (2012)

22. Parmar, N., et al.: Image transformer. In: International Conference on Machine Learning, pp. 4055–4064. PMLR (2018)

23. Redmon, J., Divvala, S., Girshick, R., Farhadi, A.: You only look once: unified, real-time object detection. In: Proceedings of the IEEE Conference on Computer Vision and Pattern Recognition, pp. 779–788 (2016)

24. Rezatofighi, H., Tsoi, N., Gwak, J., Sadeghian, A., Reid, I., Savarese, S.: Generalized intersection over union: a metric and a loss for bounding box regression. In: Proceedings of the IEEE/CVF Conference on Computer Vision and Pattern Recognition, pp. 658–666 (2019)
25. Vaswani, A., et al.: Attention is all you need. In: Advances in Neural Information Processing Systems, vol. 30 (2017)
26. Zhang, S., Zhu, X., Lei, Z., Shi, H., Wang, X., Li, S.Z.: S3FD: single shot scale-invariant face detector. In: Proceedings of the IEEE International Conference on Computer Vision, pp. 192–201 (2017)
27. Zheng, Z., Wang, P., Liu, W., Li, J., Ye, R., Ren, D.: Distance-IoU loss: faster and better learning for bounding box regression. In: Proceedings of the AAAI Conference on Artificial Intelligence, vol. 34, pp. 12993–13000 (2020)
28. Zhuang, N., et al.: Muggle: multi-stream group gaze learning and estimation. IEEE Trans. Circuits Syst. Video Technol. **30**(10), 3637–3650 (2019)
29. Zou, C., et al.: End-to-end human object interaction detection with HOI transformer. In: Proceedings of the IEEE/CVF Conference on Computer Vision and Pattern Recognition, pp. 11825–11834 (2021)

Is an Object-Centric Video Representation Beneficial for Transfer?

Chuhan Zhang[1(✉)], Ankush Gupta[2], and Andrew Zisserman[1]

[1] Visual Geometry Group, Department of Engineering Science,
University of Oxford, Oxford, UK
{czhang,az}@robots.ox.ac.uk
[2] DeepMind, London, UK
ankushgupta@google.com

Abstract. The objective of this work is to learn an *object-centric* video representation, with the aim of improving transferability to novel tasks, i.e., tasks different from the pre-training task of action classification. To this end, we introduce a new object-centric video recognition model based on a transformer architecture. The model learns a set of object-centric summary vectors for the video, and uses these vectors to fuse the visual and spatio-temporal trajectory 'modalities' of the video clip. We also introduce a novel trajectory contrast loss to further enhance object-ness in these summary vectors.

With experiments on four datasets—SomethingSomething-V2, SomethingElse, Action Genome and EpicKitchens—we show that the object-centric model outperforms prior video representations (both object-agnostic and object-aware), when: (1) classifying actions on unseen objects and unseen environments; (2) low-shot learning of novel classes; (3) linear probe to other downstream tasks; as well as (4) for standard action classification.

Keywords: Video action recognition · Object centric representations · Transfer learning

1 Introduction

Visual data is complicated—a seemingly infinite stream of events emerges from the interactions of a finite number of constituent *objects*. Abstraction and reasoning in terms of these entities and their inter-relationships—*object-centric reasoning*—has long been argued by developmental psychologists to be a *core* building block of infant cognition [1], and key for human-level common sense [2]. This object-centric understanding posits that objects exist [3], have permanence over time, and carry along physical properties such as mass and shape that govern their interactions

Supplementary Information The online version contains supplementary material available at https://doi.org/10.1007/978-3-031-26316-3_23.

with each other. Factorizing the environment in terms of these objects as recurrent entities allows for combinatorial generalization in novel settings [2]. Consequently, there has been a gradual growth in video models that embed object-centric inductive biases, e.g., augmenting the visual stream with actor or object bounding-box trajectories [4–7], graph algorithms on object nodes [8,9], or novel architectures for efficient discovery, planning and interaction [10–12].

The promise of object-centric representations is transfer across *tasks*. Due to the shared underlying physics across different settings, knowledge of object properties like shape, texture, and position can be repurposed with little or no modification for new settings [13], much like infants who learn to manipulate objects and understand their properties, and then apply these skills to new objects or new tasks [14,15] (Fig. 1).

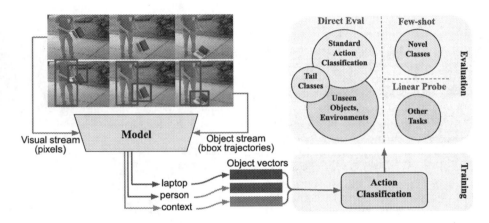

Fig. 1. Do Object-Centric video representations transfer better? To induce objectness into visual representations of videos, we learn a set of object-centric vectors—which are tied to specific objects present in the video, as well as context vectors—which reason about relations and context. The representation is built by fusing the two modality streams of the video—the visual stream, and the spatio-temporal object bounding-box stream. We train the model on the standard action recognition task, and show that using the object and context vectors can lead to SOTA results when evaluated for transfer to unseen objects, unseen environments, novel classes, other tasks and also standard action classification.

In this paper we investigate this promise by developing an *object-centric* video model and determining if it has superior task generalization performance compared to object-agnostic and other recent object-centric methods. In a similar manner to pre-training a classification model on ImageNet, and then using the backbone network for other tasks, we pre-train our object-centric model on the action recognition classification task, and then determine its performance on downstream tasks using a linear probe.

We consider a model to be object-centric if it learns a set of object summary vectors, that *explicitly* distil information about a specific object into a particular latent variable. In contrast, in object-agnostic [16–19] or previous state-of-the-art object-centric video models [4,5,20], the object information is de-localized throughout the representation.

To this end, we introduce a novel architecture based on a transformer [21] that achieves an object-centric representation through its design and its training. First, a bottleneck representation is learned, where a set of object query vectors [22] tied to specific constituent objects, cross-attend in the manner of DETR [22] to visual features and corresponding bounding-box trajectories. We demonstrate this cross-attention based fusion is an effective method for merging the two modality streams [4,5,23]—visual and geometric—complementing the individual streams. We call this 'modality' fusion module an *Object Learner*. Second, a novel *trajectory contrast* loss is introduced to further enhance object-awareness in the object summaries. Once learnt, this explicit set of object summary vectors are repurposed and refined for downstream tasks.

We evaluate the task generalization ability of the object-centric video representation using a number of challenging transfer tasks and settings:

1. **Unseen data:** Action classification with known actions (verbs), but novel objects (nouns) in the SomethingElse dataset [20]; Action classification with known actions (verbs and nouns), but unseen kitchens in EpicKitchens [24].
2. **Low-shot data:** Few-shot action classification in SomethingElse; Tail-class classification in EpicKitchens.
3. **Other downstream tasks:** Hand contact state estimation in SomethingElse, and human-object predicate prediction in ActionGenome [25].

Note, task 3 uses a linear probe on pre-trained representations for rigorously quantifying the transferability. In addition to evaluating the transferability as above, we also benchmark the learned object-centric representations on the standard task of action classification. In summary, our key contributions are:

1. A new object-centric video recognition model with explicit object representations. The object-centric representations are learned by using a novel cross-attention based module which fuses the visual and geometric streams, complementing the two individually.
2. The object-centric model sets a new record on a comprehensive set of tasks which evaluate transfer efficiency and accuracy on unseen objects, novel classes and new tasks on: SomethingElse, Action Genome and EpicKitchens.
3. Significant gains over the previous best results on standard action recognition benchmarks: 74.0%(+6.1%) on SomethingSomething-V2, 66.6%(+6.3%) on Action Genome, and 46.3%(+0.6%) top-1 accuracy on EpicKitchens.

2   Related Work

Object-Centric Video Models. Merging spatio-temporal object-level information and visual appearance for video recognition models has been explored extensively. These methods either focus solely on the human actors in the videos [6,7,26], or more generally model human-object interactions [27–30]. The dominant approach involves RoI-pooling [31,32] features extracted from a visual backbone using object/human bounding-boxes generated either from object detectors [33], or more generally using a region proposal network (RPN) [6,8,26,34–36] on each frame independently, followed by global aggregation using recurrent models [37]. The input to these methods is assumed to just be RGB pixels, and the object boxes are obtained downstream. A set of object-centric video models [4,5,20] assume object boxes as *input*, and focus on efficient fusion of the two streams; we follow this setting. Specifically, ORViT [4] is an object-aware vision transformer [38] which incorporates object information in two ways: (1) by attending to RoI-pooled object features, and (2) by merging encoded trajectories at several intermediate layers. STIN [20] encodes the object boxes and identity independently of the visual stream, and merges the two through concatenation before feeding into a classifier. STLT [5] uses a transformer encoder on object boxes, first across all objects in a given frame, and then across frames, before fusing with appearance features. We adopt STLT's hierarchical trajectory encoder, and develop a more performant cross-attention based fusion method.

Multi-modal Fusion. Neural network architectures which fuse multiple modalities, both within the visual domain, i.e., images and videos [39] with optical flow [23,40], bounding-boxes [6,8,26,34–36], as well as across other modalities, e.g., sound [41,42] and language [43–46], have been developed. The dominant approach was introduced in the classic two-stream fusion method [23] which processes the visual and optical flow streams through independent encoders before summing the final softmax predictions. Alternative methods [40] explore fusing at intermediate layers with different operations, e.g., sum, max, concatenation, and attention-based non-local operation [47]. We also process the visual and geometric streams independently, but fuse using a more recent cross-attention based transformer decoder [48] acting on object-queries [22]. An alternative to learning a single embedding representing all the input modalities, is to learn modality encoders which all map into the same joint vector space [43,49,50]; such embeddings are primarily employed for retrieval.

Benchmarks with Object Annotations. Reasoning at the *object* level lies at the heart of computer vision, where standard benchmarks for recognition [51], detection and segmentation [52,53], and tracking [54–57] are defined for categories of objects. Traditionally, bounding-box tracking of single [54,55] or multiple objects [56,57], or more spatially-precise video object segmentation [58–61] were the dominant benchmarks for object-level reasoning in videos. More recently, a number of benchmarks probe objects in videos in other ways, e.g., ActionGenome [25] augments the standard action recognition with human/object based scene-graphs, SomethingElse [20] tests for transfer of action recognition on novel objects, CATER [62] evaluates compositional reasoning over synthetic objects, and CLEVERER [63] for object-based counterfactual reasoning.

Object-Oriented Reasoning. There is a large body of work on building in and reasoning with object-level inductive biases across multiple domains and tasks. Visual recognition is typically *defined* at the object-level both in images

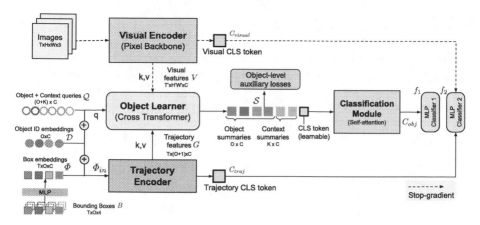

Fig. 2. The Object-Centric Video Transformer Model Architecture. The *Visual Encoder* module ingests the RGB video and produces a set of object agnostic spatio-temporal tokens. The *Trajectory Encoder* module ingests the bounding boxes, and labels them with object ID embedddings $\mathcal{D}$, to produce object-aware trajectory spatio-temporal tokens. The *Object Learner* module fuses the visual and trajectory streams by querying with the object IDs $\mathcal{D}$, and outputs object summaries which contains both visual and trajectory information. An object-level auxiliary loss is used to encourage each object summary vector to be tied to the object in the query. Finally, the *Classification* module ingests the outputs from the Object Learner to predict the class. The model is trained with cross-entropy losses applied to class predictions from the dual encoders and Object Learner, together with an auxiliary loss.

[53,64–66] and videos [25,33,67,68]. Learning relations, expressed as edges, between entities/particles, expressed as nodes in a graph has been employed for amortizing inference in simulators and modelling dynamics [11,69]. Such factorized dynamics models conditioned on structured object representations have been employed for future prediction and forecasting [70–72]. Object-conditional image and video decomposition [10,73–75] such as Monet [76] and Genesis [77] and generation [78–83] methods benefit from compositional generalization. Finally, object-level world-models have been used to constrain action-spaces, and states in robotics [84,85] and control domains [12,86,87].

3   An Object-Centric Video Action Transformer

We first describe the architecture of the object-centric video action recognition model for fusing visual and trajectory streams. We then describe the training objectives for action classification and for learning the object representations. Finally, we discuss our design choices, and the difference between our model and previous fusion methods, and explain its advantages.

3.1   Architecture

The model is illustrated in Fig. 2, and consists of four transformer-based modules. We briefly describe each module, with implementation details in Sect. 4.

Video Encoder. The encoder ingests a video clip F of RGB frames $F = (f_1, f_2, \ldots, f_t)$, where $f_i \in \mathbb{R}^{H \times W \times 3}$. The clip F is encoded by a Video Transformer [48] which tokenizes the frames by 3D patches to produce downsampled feature maps. These feature maps appended with a learnable CLS token are processed by self-attention layers to obtain the spatio-temporal visual representations $V \in \mathbb{R}^{T \times H'W' \times C}$ and video-level visual embedding C_{visual}. We take the representation V from the 6th self-attention layer to be the visual input of the Object Learner, and C_{visual} from the last layer to compute the final loss in Eq. (2).

Trajectory Encoder. The encoder ingests the bounding box coordinates $B^t = (b_1^t, b_2^t, \ldots, b_o^t)$ of $\mathcal{O}$ number of annotated objects in the tth frame, where each box b_i^t is in format $[x_1, y_1, x_2, y_2]$. These boxes are encoded into corresponding box embeddings $\Phi^t = (\Phi_1^t, \Phi_2^t, \ldots, \Phi_o^t)$ through an MLP. Object ID embeddings $\mathcal{D} = \{d_i\}_{i=1}^{O}$ are added to Φ to produce the sum Φ_{in}, This is to keep the ID information persistent throughout the video length T; Φ_{in} serves as the input to the Trajectory Encoder, which is a Spatial Temporal Layout Transformer (STLT) of [5]. STLT consists of two self-attention Transformers in sequence – a Spatial Transformer and a Temporal Transformer. First, the Spatial Transformer encodes boxes in every frame separately. It takes a learnable CLS token and box embeddings $\Phi_{in}^t \in \mathbb{R}^{O \times C}$ from frame t as the input into the self-attention layers, and output a frame-level representation $l^t \in \mathbb{R}^{1 \times C}$ and spatial-context-aware box embeddings $\Phi_{out}^t \in \mathbb{R}^{O \times C}$ respectively. The Temporal Transformer models trajectory information over frames, it applies self-attention on the frame-level embeddings $L = (l^1, l^2, l^3, \ldots, l^T)$ from the Spatial Transformer with another learnable CLS token. Its output are temporal-context-aware frame embeddings $L_{out} \in \mathbb{R}^{T \times C}$ and a video-level trajectory representation $C_{traj} \in \mathbb{R}^{1 \times C}$. C_{traj} is used to compute the final loss in eq. (2), while $L_{out} \in \mathbb{R}^{T \times 1 \times C}$ is concatenated with the $\Phi_{out} \in \mathbb{R}^{T \times O \times C}$ from the Spatial Transformer to be the spatio-temporal trajectory embeddings $G \in \mathbb{R}^{T \times (O+1) \times C}$, which are used as trajectory input to the Object Learner. (See Supp. for detailed architecture.)

Object Learner. The Object Learner module is a cross-attention Transformer [21] which has a query set $\mathcal{Q} = \{q_i\}_{i=1}^{O+K}$ made up of O learnable object queries and K learnable context queries. The same ID embeddings $\mathcal{D}$ from the Trajectory Encoder are added to the first O queries to provide object-specific identification, while the remaining K context queries can be learnt freely. We concatenate the visual feature maps $V \in \mathbb{R}^{T \times H'W' \times C}$ and trajectory embeddings $G \in \mathbb{R}^{T \times (O+1) \times C}$ as keys and values in the cross-attention layers. Note the query latents are video level (i.e., common across all frames), and attend to the features from the visual and trajectory encoders using cross-attention. The Object Learner outputs summary vectors $\mathcal{S} = \{s_i\}_{i=1}^{O+K}$, O of which are object centric, and the remaining K carry context information. The output is independent of the number of video frames, with the visual and trajectory information distilled into the summary vectors. Figure 3 presents a schematic of the module.

Classification Module. This is a light-weight cross-attention transformer that ingests the summary vectors output from the Object Learner, together with a learnable query vector C_{obj}. The vector output of this module is used for a linear classifier for the actions prediction.

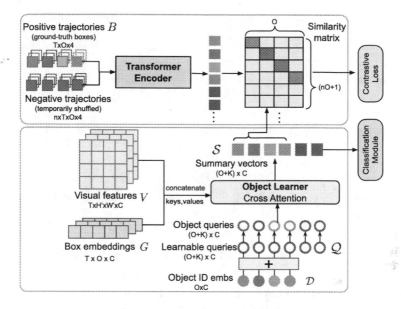

Fig. 3. Object Learner and Auxiliary Loss. The Object Learner is a cross-attention transformer, that outputs object-level summary vectors by attending to tokens from both the visual and trajectory encoders. These summary vectors are used for downstream tasks like classification. An auxiliary loss is added where the object summary vectors are tasked with distinguishing GT and shuffled trajectory embeddings.

3.2 Objectives

We apply two types of losses to train the model. One is an object-level auxiliary loss on the object summary vectors $S = (s_1, s_2, \ldots, s_o)$ to ensure object-centric information is learned in these vectors. The other is a standard cross-entropy loss on action category prediction.

Object-Level Trajectory Contrast Loss. The aim of this loss is to encourage the object specific latent queries in the Object Learner to attend to both the trajectory and visual tokens, and thereby fuse the information from the two input modalities. The key idea is to ensure that the video-level object queries do not ignore the identity, position encodings, and trajectory tokens. The loss is a contrastive loss which encourages discrimination between the correct trajectories and others that are randomly perturbed or from other video clips in the batch. This is implemented using an InfoNCE [88] loss, with a small transformer encoder used to produce vectors for each of the trajectories. This encoder only consists of two self-attention layers that encode the object trajectories into vectors.

In more detail, for an object j, the transformer takes its ground-truth trajectory $B_j \in \mathbb{R}^{T \times 4}$ and outputs the embedding $z_j \in \mathbb{R}^C$ as the positive to be matched against the summary vector $\hat{s}_j \in \mathbb{R}^C$. For negatives, other trajectories in the same batch as well as n new ones generated by temporally shuffling B_j encoded into $z_j^{shuffle}$ are used.

$$\mathcal{L}_{aux} = -\sum_j \left[\log \frac{\exp(\hat{s}_j^\top \cdot z_j)}{\sum_k \exp(\hat{s}_j^\top \cdot z_k) + \sum_k \exp(\hat{s}_j^\top \cdot z_k^{shuffle})} \right] \tag{1}$$

Final Objective. We use two MLPs as classifiers for the CLS vectors from visual and trajectory backbones and the Object Learner. The first classifier $f_1(.)$ is applied to concatenated C_{visual} and C_{traj} CLS vectors, and the second, $f_2(.)$, is applied to the CLS vector C_{obj} from the Object Learner. The total loss is the sum of the cross-entropy loss for the two classifiers and the auxiliary loss:

$$\mathcal{L}_{\text{total}} = \mathcal{L}_{CE}(f_1(C_{visual}; C_{traj}), gt) + \mathcal{L}_{CE}(f_2(C_{obj}), gt) + \mathcal{L}_{aux}, \tag{2}$$

The final class prediction is obtained by averaging the class probabilities from the two classifiers.

Discussion: Object Learner and Other Fusion Modules. Prior fusion methods can be categorized into three main types: (a) RoI-Pooling based methods like STRG [8], where visual features are pooled using boxes for the downstream tasks; (b) Joint training methods like ORViT [4] where the two modalities are encoded jointly from early stages; and (c) Two stream methods [5,20] with dual encoders for the visual and trajectory modalities, where fusion is in the last layer. The RoI-pooling based methods explicitly pool features inside boxes for downstream operations, omitting context outside the boxes. In contrast, our model allows the queries to attend to the visual feature maps freely. Joint training benefits from fine-grained communication between modalities, but this may not be as robust as the two-stream models under domain shift. Our method combines the two, by keeping the dual encoders for independence and having a bridging module to link the information from their intermediate layers. Quantitative comparisons are done in Sect. 5.

4 Implementation Details

Model Architecture. We use Motionformer [48] as the visual encoder, operating on 16 frames of size 224×224 pixels uniformly sampled from a video; the 3D patch size for tokenization is $2 \times 16 \times 16$. We use STLT [5] as the trajectory encoder which takes normalized bounding boxes from 16 frames as input. Our Object Learner is a Cross-Transformer with 6 layers and 8 heads. We adopt the trajectory attention introduced in [48] instead of the conventional joint spatio-temporal attention in the layers. The Classification Module has 4 self-attention layers with 6 heads. We set the number of context queries as 2 in all the datasets, and number of object queries as 6 in SomethingElse, SomethingSomething and EpicKitchens, 37 in ActionGenome.

Training. We train our models on 2 Nvidia RTX6k GPUs with the AdamW [89] optimizer. Due to the large model size and limited compute resources, we are not able to train the full model end-to-end with a large batch size. Instead, we first train the visual backbone for action classification with batch size 8, and then keep it frozen while we train the rest of the model with batch size 72. More details on architecture and training are in the supplementary material.

5   Experiments

We conduct experiments on four datasets, namely SomethingSomething-V2, SomethingElse, Action Genome and EpicKitchens. We first train and evaluate our model on the standard task of action recognition on these datasets, and then test its transferability on novel tasks/settings including action recognition on unseen objects, few-shot action recognition, hand state classification, and scene-graph predicate prediction. We first introduce the datasets and the metrics. Followed by a comparison of our method with other fusion methods, and then ablations on the design choices in the proposed Object Learner. Finally, we compare with SOTA models on different tasks and analyze the results.

5.1   Datasets and Metrics

SomethingSomething-V2 [90] is a collection of 220k labeled video clips of humans performing basic actions with objects, with 168k training videos and 24k validation videos. It contains 174 classes, these classes are object agnostic and named after the interaction, e.g., 'moving something from left to the right'. We use the ground-truth boxes provided in the dataset as input to our networks.

Something-Else [20] is built on the videos in SomethingSomething-V2 [90] and proposes new training/test splits for two new tasks testing for generalizability: compositional action recognition, and few-shot action recognition. The compositional action recognition is designed to ensure there is no overlap in object categories between 55k training videos and 58k validation videos. In the few-shot setting, there are 88 base actions (112k videos) for pre-training, 86 novel classes for fine-tuning. We use the ground-truth boxes provided in the dataset.

Action Genome [25] is a dataset which uses videos and action labels from Charades [91], and decomposes actions into spatio-temporal scene graphs by annotating human, objects and their relationship in them. It contains 10K videos (7K train/3k val) with 0.4M objects. We use the raw frames and ground-truth boxes provided for action classification over 157 classes on this dataset.

Epic-Kitchens [24] is a large-scale egocentric dataset with 100-hour activities recorded in kitchens. It provides 496 videos for training and 138 videos for val, each video has detected bounding boxes from [92]. We use an offline tracker [93] to build association between the boxes and use them as input. We use the detected boxes provided in the dataset as input to our networks.

5.2   Ablations

Comparison with Other Fusion Methods. For a fair comparison, we implement other fusion methods using our (i.e., the same) visual and trajectory backbones. We choose LCF (Late Concatenation Fusion) and CACNF (Cross-Attention CentralNet Fusion) as they are the two best methods among seven in the latest work [5]. We also compare against a baseline method where the class probabilities from the encoders trained independently are averaged.

We implement CACNF with the same number of cross-attention layers and attention heads as in our Object Learner. Table 1 summarizes the results. The results show that the above fusion methods work better than the baseline, and our model achieves better results than other parametric methods. Performance for Motionformer with other trajectory encoders, or STLT with other visual encoders, has been explored in previous works [4,5,20]—more comparisons are in Table 2.

Table 1. Different fusion methods with the same visual backbone. We show the performance of Motionformer and STLT alone on SthSth-V2, and compare the classification performance with different fusion methods on them, namely averaged class probabilities, LCF and CACNF and our Object Learner (OL).

Visual Enc.	Traj. Enc.	Fusion		SthSth-V2	
		Method	Parametric	Top1	Top5
Motionformer	-	-	-	66.5	90.1
-	STLT	-	-	57	85.2
Motionformer	STLT	Avg prob	✗	67.5	90.9
		CACNF [5]	✓	69.7	93.4
		LCF [5]	✓	73.1	**94.2**
		OL(ours)	✓	**74.0**	**94.2**

Ablating Trajectory Contrast Loss. We compare the performance of training with and without the trajectory contrast loss on different transfer tasks in Table 3. Having the object-level auxiliary loss (Eq. (1)) brings improvement in performance in 3 out of 4 tasks. The improvement is 1% in hand state classification in SomethingElse, 2% in scene-graph prediction and 1.7% on standard classification in Action Genome. The results show the auxiliary loss helps in both task transfer as well as standard action recognition. Figure 4 also shows the visualization of attention scores in the Object Learner – object queries trained with auxiliary loss are more object-centric when attending to the visual frames.

5.3   Results

We present the experiment results on a wide range of tasks organized into sections by the dataset used. In each dataset, we first compare the results from different models on standard action recognition, and then introduce the transfer tasks and discuss the performance.

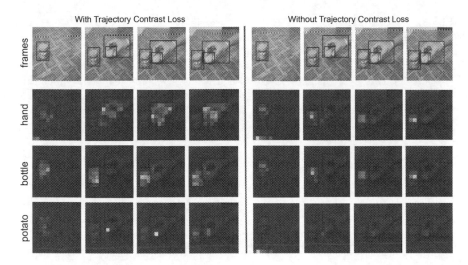

Fig. 4. Trajectory Contrast Loss Induces Object-Centric Attention. Object Learners's attention from various heads when trained with (left) and without (right) the trajectory contrast loss Eq. (1). In both cases, attention is on individual objects, indicating objectness in the model. However, there is a notable difference: without the trajectory contrast loss, there is no attention on the hands(first row). Hence, the trajectory contrast loss induces enhanced objectness in the object summary vectors.

5.3.1   SomethingSomething-V2 and SomethingElse

Standard Action Recognition. We evaluate on the regular train/val split on SomethingSomething-V2 for the performance on (seen) action recognition. The accuracy of our model is 5.9% higher than ORViT and 7.1% higher than STLT [5], showing the advantage is not only on transfer tasks but also on standard action classification.

Transfer to Unseen Objects. Compositional action recognition is a task in Something-Else where the actions are classified given unseen objects (i.e., objects not present in the training set). Thus it requires the model to learn appearance-agnostic information on the actions. Our object-centric model improves the visual Motionfomer by 10.9%, and outperforms the joint encoding ORViT model by a margin of 3.9 %, showing that keeping the trajectory encoding independent from the visual encoder can make the representations more generalizable.

Data-efficiency: Few-shot Action Recognition. We follow the experiment settings in [20] to freeze all the parameters except the classifiers in 5-shot and 10-shot experiments on SomethingElse. Again, models that are using both visual and trajectory modalities have an obvious advantage over visual only ones. The performance boost is more obvious in a low data regime, with a 4.7% and 0.3% improvement over R3D, STLT in 5-shot and 10-shot respectively. It's worth noting that while the raw classification results from the backbone and Object Learner classifiers only have a 0.1% difference, averaging the two together gives

Table 2. Comparison with SOTA models on Something-Else and SomethingSomething-V2. We report top1 and top5 accuracy on three action classification tasks, including compositional and few-shot action recognition on Something-Else, and action recognition on SomethingSomething-V2. From top to bottom: we show the performance of SOTA visual models, trajectory models, and the models which takes both modalities as input. In the last section we list the classification performance from the backbone baseline without an Object Learner(OL) and our model with an Object Learner(OL), Our model outperforms other methods by a clear margin on all the tasks.

Model	Video input	Box input	GFLOP	SthSth-V2		SomethingElse			
				Action recognition		Compositional action (unseen objects)		Few shot	
				Top 1	Top5	Top 1	Top 5	5 shot	10 shot
I3D [16]	✓	✗	28	61.7	83.5	46.8	72.2	21.8	26.7
SlowFast.R101 [39]	✓	✗	213	63.1	87.6	45.2	73.4	22.4	29.2
Motionformer [48]	✓	✗	369	66.5	90.1	62.8	86.7	28.9	33.8
STIN [20]	✗	✓	5.5	54.0	79.6	51.4	79.3	24.5	30.3
SFI [36]	✗	✓	-	-	-	44.1	74	24.3	29.8
STLT [5]	✗	✓	4	57.0	85.2	59.0	86	31.4	38.6
STIN+I3D [20]	✓	✓	33.5	-	-	54.6	79.4	28.1	33.6
STIN.I3D [20]	✓	✓	33.5	-	-	58.1	83.2	34.0	40.6
SFI [36]	✓	✓	-	-	-	61.0	86.5	35.3	41.7
R3D.STLT(CACNF) [5]	✓	✓	48	66.8	90.6	67.1	90.4	37.1	45.5
ORViT [4]	✓	✓	405	73.8	93.6	69.7	91	33.3	40.2
Motionformer+STLT(baseline)	✓	✓	373	72.8	94.1	72.0	92.3	38.9	44.6
Motionformer+STLT+OL(Ours)	✓	✓	383.3	**74.0**	**94.2**	**73.6**	**93.5**	**40.0**	**45.7**

Table 3. Ablate auxiliary loss and Object Learner (OL) on compositional action, hand state classification and predicate prediction. We show linear probe results on the backbone CLS token.

Method	Aux loss	Something-else		Hand contact state		Action genome		
		Compositional action				Action	Predicate	
		Top1	Top5	Per-video	Per-class	mAP	R@10	R@20
ORViT	-	69.7	91.0	70.2	66.0	-	-	-
MFormer+STLT (baseline)	-	72	93.2	66.8	43.3	66.0	78.3	83.5
MFormer+STLT+OL (ours)	✗	73.5	**93.5**	77.5	68.5	64.9	78.9	83.8
MFormer+STLT+OL (ours)	✓	**73.6**	**93.5**	**78.2**	**69.7**	**66.6**	**80.9**	**85.4**

more than 1% improvement. It suggests that our Object Learner has captured complementary information through combining the two streams.

Transfer to Hand State Classification. We further evaluate the object-level representations (pre-trained with standard action recognition) on hand contact state classification using a linear probe. We extract hand state labels using a pre-trained object-hand detector from [92] as ground truth, and design a 3-way classification task on SomethingElse. Specifically, the three classes are 'no hand contact', 'one hand contact' (one hand contacts with object) or 'two hands contact' (both hands contact with object). In our experiments, we average-pool the object summary vectors, train a linear classifier on the training set, and test on the validation set. We conduct the linear probe on summaries trained with and without the auxiliary loss in Eq. (1), and also on the baseline backbone classifier. Video-level top-1 accuracy and class-level top-1 accuracy are reported in Table 3. Our model is better than the baseline by 11.4 % in per-video accuracy and 26.4 % in per-class accuracy. Object summaries trained with auxiliary losses on trajectories outperform the one without by about 1%.

5.3.2   Epic-Kitchens

Standard Action Recognition. Table 4 shows the results of action recognition in Epic-Kitchens, With the Object-Learner, our model is 0.6–4.0% more accurate in action prediction than other methods that use both visual stream and trajectory stream as input, and 1.7% more accurate than the Late Concatenation Fusion (LCF) method without an Object-Learner.

Classification on Tail Classes and Unseen Kitchens. In Table 4 we also present the classification results on tail classes and videos from unseen kitchens. In average, Object-centric models are better than visual-only models by 4.8% on tail actions, and by 0.4% on unseen kitchens. Among all the models with objectness, our model with Object Learner achieves the best action classification accuracy on both tail classes and unseen kitchens.

5.3.3   Action Genome

Standard Action Recognition. In Action Genome, each action clip is labelled with object bounding boxes and their categories. We follow the experiment settings in [5], train and evaluate our model with RGB frames and ground truth trajectory as input. Table 5 shows the classification results. By using our Object Learner trained with auxiliary loss, we achieve the best result 66.6% mAP, outperforming other fusion methods using the same backbone. We also compare to SGFB [25], which uses scene graphs as input, our model is better by 6.3% without access to the relationship between objects.

Table 4. Action Classification results on Epic-Kitchens. Our model achieves the best results compared to other methods using the same backbone. MFormer uses 224×224 resolution input and MFormer-HR uses 336×336 resolution input.

Methods	Box input	Overall			Tail classes			Unseen kitchens		
		Action	Verb	Noun	Action	Verb	Noun	Action	Verb	Noun
SlowFast [39]	N	38.5	65.5	50.0	18.8	36.2	23.3	29.7	56.4	41.5
ViViT-L [94]	N	44.0	66.4	56.8	-	-	-	-	-	-
MFormer [48]	N	43.1	66.7	56.5	-	-	-	-	-	-
MFormer-HR [48]	N	44.5	67.0	58.5	19.7	34.2	28.4	34.8	58.0	46.6
MFormer-HR+STRG	Y	42.5	65.8	55.4	-	-	-	-	-	-
MFormer-HR+STRG+STIN	Y	44.1	66.9	57.8	24.7	**39.9**	34.4	34.8	59.5	48.1
MFormer-HR-ORVIT [4]	Y	45.7	68.5	57.9	-	-	-	-	-	-
MFormer-HR+STLT(baseline)	Y	44.6	67.4	58.8	23.3	38.5	34.1	35.1	**59.7**	**49.6**
MFormer-HR+STLT+OL(ours)	Y	**46.3**	**68.7**	**59.4**	**25.7**	**39.9**	**35.3**	**35.4**	**59.7**	48.3

Table 5. Action recognition and human-object predicate prediction results on Action Genome. In action classification, our model outperforms others with the same frame and boxes input, and even SGFB with scene graph (SG) input. When linear-probing the output features for predicate prediction, our Object Learner fuses the visual and trajectory streams in an efficient way and is 2.6% higher than baseline LCF in recall@10. We also show the object-centric representations learned with the auxiliary loss is better than the ones learned without the auxiliary loss in both tasks.

Backbone	Method	Boxes	SG	Aux loss	# Frames	Action CLS. mAP	Predicate Pred. R@10	Predicate Pred. R@20
I3D [5]	Avgpool	N	N	-	32	33.5	-	-
MFormer [48]	CLS token	N	N	-	16	36.5	76.4	82.6
STLT [5]	CLS token	Y	N	-	16	56.7	79.0	84.1
STLT [5]	CLS token	Y	N	-	32	60.0	-	-
I3D+STLT [5]	CACNF [5]	Y	Y	-	32	61.6	-	-
MFormer+STLT	CACNF [5]	Y	N	-	16	64.2	-	-
R101-I3D-NL [95]	SGFB [25]	Y	Y	-	32+	60.3	-	-
MFormer+STLT	LCF(baseline)	Y	N	-	16	66.0	78.3	83.5
MFormer+STLT	OL(ours)	Y	N	N	16	64.9	78.9	83.8
MFormer+STLT	OL(ours)	Y	N	Y	16	**66.6**	**80.9**	**85.4**

Transfer to Scene Graph Predicate Prediction. We transfer the trained model on action classification to scene graph predicate prediction by linear probing. In this task, the model has to predict the predicate (relationship) between human and object when the bounding boxes and categories are known. Given the object id, we concat one-hot object id vectors with the classification vector from the frozen models, and train a linear classifier to predict the predicate. As shown in Table 5, the result from object summaries trained with the auxiliary loss is 2.6% higher than linear probing the concatenated CLS tokens (LCF) from two backbones, and 2.0% higher than the one trained without auxiliary loss.

6   Conclusion

We set out to evaluate whether objectness in video representations can aid visual task transfer. To this end, we have developed an object-centric video model which fuses the visual stream with object trajectories (bounding-boxes) in a novel transformer based architecture. We indeed find that the object-centric representations learned by our model are more transferrable to novel tasks and settings in video recognition using a simple linear probe, i.e., they outperform both prior object-agnostic and object-centric representations on a comprehensive suite of transfer tasks. This work uses a very coarse geometric representation of objects, i.e., bounding-boxes, for inducing object awareness in visual representations. In the future more spatially precise/physically-grounded representations, e.g., segmentation masks or 3D shape, could further enhance the transferability.

Acknowledgements. This research is funded by a Google-DeepMind Graduate Scholarship, a Royal Society Research Professorship, and EPSRC Programme Grant VisualAI EP/T028572/1.

References

1. Spelke, E.S., Breinlinger, K., Macomber, J., Jacobson, K.: Origins of knowledge. Psychol. Rev. **99**, 605 (1992)
2. Tenenbaum, J.B., Kemp, C., Griffiths, T.L., Goodman, N.D.: How to grow a mind: statistics, structure, and abstraction. Science **331**, 1279–1285 (2011)
3. Grill-Spector, K., Kanwisher, N.: Visual recognition: as soon as you know it is there, you know what it is. Psychol. Sci. **16**, 152–160 (2005)
4. Herzig, R., et al.: Object-region video transformers. arXiv preprint arXiv:2110. 06915 (2021)
5. Radevski, G., Moens, M.F., Tuytelaars, T.: Revisiting spatio-temporal layouts for compositional action recognition. In: Proceedings of BMVC (2021)
6. Sun, C., Shrivastava, A., Vondrick, C., Murphy, K., Sukthankar, R., Schmid, C.: Actor-centric relation network. In: Ferrari, V., Hebert, M., Sminchisescu, C., Weiss, Y. (eds.) ECCV 2018. LNCS, vol. 11215, pp. 335–351. Springer, Cham (2018). https://doi.org/10.1007/978-3-030-01252-6_20
7. Zhang, Y., Tokmakov, P., Hebert, M., Schmid, C.: A structured model for action detection. In: Proceedings of CVPR (2019)
8. Wang, X., Gupta, A.: Videos as space-time region graphs. In: Ferrari, V., Hebert, M., Sminchisescu, C., Weiss, Y. (eds.) ECCV 2018. LNCS, vol. 11209, pp. 413–431. Springer, Cham (2018). https://doi.org/10.1007/978-3-030-01228-1_25
9. Chen, Y., Rohrbach, M., Yan, Z., Shuicheng, Y., Feng, J., Kalantidis, Y.: Graph-based global reasoning networks. In: Proceedings of CVPR (2019)
10. Locatello, F., et al.: Object-centric learning with slot attention. In: NeurIPS (2020)
11. Battaglia, P., Pascanu, R., Lai, M., Jimenez Rezende, D., et al.: Interaction networks for learning about objects, relations and physics. In: NeurIPS (2016)
12. Kulkarni, T.D., Gupta, A., Ionescu, C., Borgeaud, S., Reynolds, M., Zisserman, A., Mnih, V.: Unsupervised learning of object keypoints for perception and control. In: NeurIPS (2019)
13. Dubey, R., Agrawal, P., Pathak, D., Griffiths, T.L., Efros, A.A.: Investigating human priors for playing video games. In: Proceedings of ICML (2018)
14. Gopnik, A., Meltzoff, A.N., Kuhl, P.K.: The scientist in the crib: what early learning tells us about the mind. William Morrow Paperbacks (2000)
15. Smith, L.B., Jayaraman, S., Clerkin, E., Yu, C.: The developing infant creates a curriculum for statistical learning. Trends Cogn. Sci. **22**, 324–336 (2018)
16. Carreira, J., Zisserman, A.: Quo vadis, action recognition? a new model and the kinetics dataset. In: Proceedings of CVPR (2017)
17. Lin, J., Gan, C., Han, S.: TSM: temporal shift module for efficient video understanding. In: Proceedings of ICCV (2019)
18. Xie, S., Sun, C., Huang, J., Tu, Z., Murphy, K.: Rethinking spatiotemporal feature learning: speed-accuracy trade-offs in video classification. In: Ferrari, V., Hebert, M., Sminchisescu, C., Weiss, Y. (eds.) ECCV 2018. LNCS, vol. 11219, pp. 318–335. Springer, Cham (2018). https://doi.org/10.1007/978-3-030-01267-0_19
19. Yang, C., Xu, Y., Shi, J., Dai, B., Zhou, B.: Temporal pyramid network for action recognition. In: Proceedings of CVPR (2020)

20. Materzynska, J., Xiao, T., Herzig, R., Xu, H., Wang, X., Darrell, T.: Something-else: compositional action recognition with spatial-temporal interaction networks. In: Proceedings of CVPR (2020)

21. Vaswani, A., et al.: Attention is all you need. In: NeurIPS (2017)

22. Carion, N., Massa, F., Synnaeve, G., Usunier, N., Kirillov, A., Zagoruyko, S.: End-to-end object detection with transformers. In: Vedaldi, A., Bischof, H., Brox, T., Frahm, J.-M. (eds.) ECCV 2020. LNCS, vol. 12346, pp. 213–229. Springer, Cham (2020). https://doi.org/10.1007/978-3-030-58452-8_13

23. Simonyan, K., Zisserman, A.: Two-stream convolutional networks for action recognition in videos. In: NeurIPS (2014)

24. Damen, D., et al.: Scaling egocentric vision: the dataset. In: Ferrari, V., Hebert, M., Sminchisescu, C., Weiss, Y. (eds.) ECCV 2018. LNCS, vol. 11208, pp. 753–771. Springer, Cham (2018). https://doi.org/10.1007/978-3-030-01225-0_44

25. Ji, J., Krishna, R., Fei-Fei, L., Niebles, J.C.: Action genome: actions as compositions of spatio-temporal scene graphs. In: Proceedings of CVPR (2020)

26. Girdhar, R., Carreira, J., Doersch, C., Zisserman, A.: Video action transformer network. In: Proceedings of CVPR (2019)

27. Gao, C., Xu, J., Zou, Y., Huang, J.-B.: DRG: dual relation graph for human-object interaction detection. In: Vedaldi, A., Bischof, H., Brox, T., Frahm, J.-M. (eds.) ECCV 2020. LNCS, vol. 12357, pp. 696–712. Springer, Cham (2020). https://doi.org/10.1007/978-3-030-58610-2_41

28. Kato, K., Li, Y., Gupta, A.: Compositional learning for human object interaction. In: Ferrari, V., Hebert, M., Sminchisescu, C., Weiss, Y. (eds.) Computer Vision – ECCV 2018. LNCS, vol. 11218, pp. 247–264. Springer, Cham (2018). https://doi.org/10.1007/978-3-030-01264-9_15

29. Xu, B., Wong, Y., Li, J., Zhao, Q., Kankanhalli, M.S.: Learning to detect human-object interactions with knowledge. In: Proceedings of CVPR (2019)

30. Gkioxari, G., Girshick, R., Dollár, P., He, K.: Detecting and recognizing human-object interactions. In: Proceedings of CVPR (2018)

31. Ren, S., He, K., Girshick, R., Sun, J.: Faster R-CNN: towards real-time object detection with region proposal networks. In: NeurIPS (2015)

32. He, K., Gkioxari, G., Dollár, P., Girshick, R.: Mask R-CNN. In: Proceedings of ICCV (2017)

33. Gupta, A., Davis, L.S.: Objects in action: an approach for combining action understanding and object perception. In: Proceedings of CVPR (2007)

34. Baradel, F., Neverova, N., Wolf, C., Mille, J., Mori, G.: Object level visual reasoning in videos. In: Ferrari, V., Hebert, M., Sminchisescu, C., Weiss, Y. (eds.) ECCV 2018. LNCS, vol. 11217, pp. 106–122. Springer, Cham (2018). https://doi.org/10.1007/978-3-030-01261-8_7

35. Arnab, A., Sun, C., Schmid, C.: Unified graph structured models for video understanding. In: Proceedings of ICCV (2021)

36. Yan, R., Xie, L., Shu, X., Tang, J.: Interactive fusion of multi-level features for compositional activity recognition. arXiv preprint arXiv:2012.05689 (2020)

37. Ma, C.Y., Kadav, A., Melvin, I., Kira, Z., AlRegib, G., Graf, H.P.: Attend and interact: higher-order object interactions for video understanding. In: Proceedings of CVPR (2018)

38. Kolesnikov, A., et al.: An image is worth 16x16 words: Transformers for image recognition at scale. In: Proceedings of ICLR (2021)

39. Feichtenhofer, C., Fan, H., Malik, J., He, K.: Slowfast networks for video recognition. In: Proceedings of ICCV (2019)

40. Feichtenhofer, C., Pinz, A., Zisserman, A.: Convolutional two-stream network fusion for video action recognition. In: Proceedings of CVPR (2016)
41. Arandjelović, R., Zisserman, A.: Objects that sound. In: Ferrari, V., Hebert, M., Sminchisescu, C., Weiss, Y. (eds.) ECCV 2018. LNCS, vol. 11205, pp. 451–466. Springer, Cham (2018). https://doi.org/10.1007/978-3-030-01246-5_27
42. Owens, A., Efros, A.A.: Audio-visual scene analysis with self-supervised multisensory features. In: Ferrari, V., Hebert, M., Sminchisescu, C., Weiss, Y. (eds.) ECCV 2018. LNCS, vol. 11210, pp. 639–658. Springer, Cham (2018). https://doi.org/10.1007/978-3-030-01231-1_39
43. Aytar, Y., Vondrick, C., Torralba, A.: See, hear, and read: Deep aligned representations. arXiv preprint arXiv:1706.00932 (2017)
44. Frome, A., Corrado, G.S., Shlens, J., Bengio, S., Dean, J., Ranzato, M., Mikolov, T.: Devise: a deep visual-semantic embedding model. In: NeurIPS (2013)
45. Weston, J., Bengio, S., Usunier, N.: Wsabie: Scaling up to large vocabulary image annotation. In: Proceedings of IJCAI (2011)
46. Sun, C., Myers, A., Vondrick, C., Murphy, K., Schmid, C.: VideoBERT: a joint model for video and language representation learning. In: Proceedings of ICCV (2019)
47. Wang, X., Girshick, R., Gupta, A., He, K.: Non-local neural networks. In: Proceedings of CVPR (2018)
48. Patrick, M., Campbell, D., Asano, Y., Misra, I., Metze, F., Feichtenhofer, C., Vedaldi, A., Henriques, J.F.: Keeping your eye on the ball: Trajectory attention in video transformers. In: NeurIPS (2021)
49. Alayrac, J.B., et al.: Self-supervised multimodal versatile networks. In: NeurIPS (2020)
50. Nagrani, A., Albanie, S., Zisserman, A.: Learnable PINs: cross-modal embeddings for person identity. In: Ferrari, V., Hebert, M., Sminchisescu, C., Weiss, Y. (eds.) ECCV 2018. LNCS, vol. 11217, pp. 73–89. Springer, Cham (2018). https://doi.org/10.1007/978-3-030-01261-8_5
51. Deng, J., Dong, W., Socher, R., Li, L.J., Li, K., Fei-Fei, L.: Imagenet: a large-scale hierarchical image database. In: Proceedings of CVPR (2009)
52. Everingham, M., Eslami, S.M.A., Van Gool, L., Williams, C.K.I., Winn, J., Zisserman, A.: The PASCAL visual object classes challenge: a retrospective. Int. J. Comput. Vision 111(1), 98–136 (2014). https://doi.org/10.1007/s11263-014-0733-5
53. Lin, T.-Y., et al.: Microsoft COCO: common objects in context. In: Fleet, D., Pajdla, T., Schiele, B., Tuytelaars, T. (eds.) ECCV 2014. LNCS, vol. 8693, pp. 740–755. Springer, Cham (2014). https://doi.org/10.1007/978-3-319-10602-1_48
54. Kristan, M., et al.: The ninth visual object tracking vot2021 challenge results. In: Proceedings of ICCV (2021)
55. Huang, L., Zhao, X., Huang, K.: Got-10k: a large high-diversity benchmark for generic object tracking in the wild. In: IEEE PAMI (2019)
56. Dave, A., Khurana, T., Tokmakov, P., Schmid, C., Ramanan, D.: TAO: a large-scale benchmark for tracking any object. In: Vedaldi, A., Bischof, H., Brox, T., Frahm, J.-M. (eds.) ECCV 2020. LNCS, vol. 12350, pp. 436–454. Springer, Cham (2020). https://doi.org/10.1007/978-3-030-58558-7_26
57. Dendorfer, P., et al.: Motchallenge: a benchmark for single-camera multiple target tracking. IJCV. 129, 845–881 (2021)
58. Perazzi, F., Pont-Tuset, J., McWilliams, B., Van Gool, L., Gross, M., Sorkine-Hornung, A.: A benchmark dataset and evaluation methodology for video object segmentation. In: Proceedings of CVPR (2016)

59. Xu, N., et al.: Youtube-VOS: A large-scale video object segmentation benchmark. arXiv preprint arXiv:1809.03327 (2018)
60. Yang, L., Fan, Y., Xu, N.: Video instance segmentation. In: Proceedings of ICCV (2019)
61. Wang, W., Feiszli, M., Wang, H., Tran, D.: Unidentified video objects: a benchmark for dense, open-world segmentation. In: Proceedings of ICCV (2021)
62. Girdhar, R., Ramanan, D.: CATER: A diagnostic dataset for Compositional Actions and TEmporal Reasoning. In: ICLR (2020)
63. Yi, K., Gan, C., Li, Y., Kohli, P., Wu, J., Torralba, A., Tenenbaum, J.B.: Clevrer: Collision events for video representation and reasoning. arXiv preprint arXiv:1910.01442 (2019)
64. Krishna, R., et al.: Visual genome: connecting language and vision using crowd-sourced dense image annotations. IJCV **123**, 32–73 (2017)
65. Krishna, R., Chami, I., Bernstein, M., Fei-Fei, L.: Referring relationships. In: Proceedings of CVPR (2018)
66. Johnson, J., Karpathy, A., Fei-Fei, L.: Densecap: fully convolutional localization networks for dense captioning. In: Proceedings of CVPR (2016)
67. Saenko, K., et al.: Mid-level features improve recognition of interactive activities. Department of Electrical Engineering and Computer Science. University of California, Berkeley, Technical report (2012)
68. Xu, H., Yang, L., Sclaroff, S., Saenko, K., Darrell, T.: Spatio-temporal action detection with multi-object interaction. arXiv preprint arXiv:2004.00180 (2020)
69. Battaglia, P., et al.: Relational inductive biases, deep learning, and graph networks. arXiv preprint arXiv:1806.01261 (2018)
70. Ye, Y., Singh, M., Gupta, A., Tulsiani, S.: Compositional video prediction. In: Proceedings of ICCV (2019)
71. Liang, J., Jiang, L., Niebles, J.C., Hauptmann, A.G., Fei-Fei, L.: Peeking into the future: predicting future person activities and locations in videos. In: Proceedings of CVPR (2019)
72. Wu, Y., Gao, R., Park, J., Chen, Q.: Future video synthesis with object motion prediction. In: Proceedings of CVPR (2020)
73. Greff, K., et al.: Multi-object representation learning with iterative variational inference. In: Proceedings of ICML (2019)
74. Henderson, P., Lampert, C.H.: Unsupervised object-centric video generation and decomposition in 3d. In: NeurIPS (2020)
75. Yang, C., Lamdouar, H., Lu, E., Zisserman, A., Xie, W.: Self-supervised video object segmentation by motion grouping. In: Proceedings of ICCV (2021)
76. Burgess, C.P., et al.: Monet: Unsupervised scene decomposition and representation. arXiv preprint arXiv:1901.11390 (2019)
77. Engelcke, M., Kosiorek, A.R., Jones, O.P., Posner, I.: Genesis: Generative scene inference and sampling with object-centric latent representations. arXiv preprint arXiv:1907.13052 (2019)
78. Johnson, J., Gupta, A., Fei-Fei, L.: Image generation from scene graphs. In: Proceedings of CVPR (2018)
79. Herzig, R., Raboh, M., Chechik, G., Berant, J., Globerson, A.: Mapping images to scene graphs with permutation-invariant structured prediction. In: NeurIPS (2018)
80. Park, T., Liu, M.Y., Wang, T.C., Zhu, J.Y.: GauGAN: semantic image synthesis with spatially adaptive normalization. In: ACM SIGGRAPH 2019 Real-Time Live (2019)

81. Singh, K.K., Ojha, U., Lee, Y.J.: FineGAN: unsupervised hierarchical disentanglement for fine-grained object generation and discovery. In: Proceedings of CVPR (2019)
82. Yang, B., et al.: Learning object-compositional neural radiance field for editable scene rendering. In: Proceedings of ICCV (2021)
83. Herzig, R., Bar, A., Xu, H., Chechik, G., Darrell, T., Globerson, A.: Learning canonical representations for scene graph to image generation. In: Vedaldi, A., Bischof, H., Brox, T., Frahm, J.-M. (eds.) ECCV 2020. LNCS, vol. 12371, pp. 210–227. Springer, Cham (2020). https://doi.org/10.1007/978-3-030-58574-7_13
84. Ye, Y., Gandhi, D., Gupta, A., Tulsiani, S.: Object-centric forward modeling for model predictive control. In: Proceedings of CoRL (2020)
85. Devin, C., Abbeel, P., Darrell, T., Levine, S.: Deep object-centric representations for generalizable robot learning. In: Proceedings of International Conference on Robotics and Automation (2018)
86. Bapst, V., et al.: Structured agents for physical construction. In: Proceedings of ICML (2019)
87. Anand, A., Racah, E., Ozair, S., Bengio, Y., Côté, M.A., Hjelm, R.D.: Unsupervised state representation learning in Atari. In: NeurIPS (2019)
88. Oord, A.v.d., Li, Y., Vinyals, O.: Representation learning with contrastive predictive coding. arXiv preprint arXiv:1807.03748 (2018)
89. Loshchilov, I., Hutter, F.: Decoupled weight decay regularization. In: Proceedings of ICLR (2019)
90. Goyal, R., et al.: The "something something" video database for learning and evaluating visual common sense. In: Proceedings of ICCV (2017)
91. Sigurdsson, G.A., Varol, G., Wang, X., Farhadi, A., Laptev, I., Gupta, A.: Hollywood in homes: crowdsourcing data collection for activity understanding. In: Leibe, B., Matas, J., Sebe, N., Welling, M. (eds.) ECCV 2016. LNCS, vol. 9905, pp. 510–526. Springer, Cham (2016). https://doi.org/10.1007/978-3-319-46448-0_31
92. Shan, D., Geng, J., Shu, M., Fouhey, D.F.: Understanding human hands in contact at internet scale. In: Proceedings of the IEEE/CVF Conference on Computer Vision and Pattern Recognition, pp. 9869–9878 (2020)
93. Zhang, Y., et al.: ByteTrack: multi-object tracking by associating every detection box. arXiv preprint arXiv:2110.06864 (2021)
94. Arnab, A., Dehghani, M., Heigold, G., Sun, C., Lučić, M., Schmid, C.: Vivit: a video vision transformer. In: Proceedings of ICCV (2021)
95. Wu, C.Y., Feichtenhofer, C., Fan, H., He, K., Krahenbuhl, P., Girshick, R.: Long-term feature banks for detailed video understanding. In: Proceedings of the IEEE/CVF Conference on Computer Vision and Pattern Recognition, pp. 284–293 (2019)

DCVQE: A Hierarchical Transformer for Video Quality Assessment

Zutong Li and Lei Yang[✉]

Weibo R&D Limited, Palo Alto, USA
trilithy@gmail.com

Abstract. The explosion of user-generated videos stimulates a great demand for no-reference video quality assessment (NR-VQA). Inspired by our observation on the actions of human annotation, we put forward a Divide and Conquer Video Quality Estimator (DCVQE) for NR-VQA. Starting from extracting the frame-level quality embeddings (QE), our proposal splits the whole sequence into a number of clips and applies Transformers to learn the clip-level QE and update the frame-level QE simultaneously; another Transformer is introduced to combine the clip-level QE to generate the video-level QE. We call this hierarchical combination of Transformers the Divide and Conquer Transformer (DCTr) layer. An accurate video quality feature extraction can be achieved by repeating the process of this DCTr layer several times. Taking the order relationship among the annotated data into account, we also propose a novel correlation loss term for model training. Experiments on various datasets confirm the effectiveness and robustness of our DCVQE model.

1 Introduction

Recent years have witnessed a significant increase in user-generated content (UGC) on social media platforms like Youtube, Tiktok, and Weibo. Watching the UGC videos on computers or smartphones has even become part of our daily life. This trend stimulates a great demand for automatic video quality assessment (VQA), especially in popular video sharing/recommendation services.

UGC-VQA, also known as blind or No-Reference video quality assessment (NR-VQA), aims to evaluate in-the-wild videos without the corresponding pristine reference videos. Usually, UGC videos may suffer from complex distortions due to the diversity of capturing devices, uncertain shooting skills, compression, and poor editing process. Although many excellent algorithms have been proposed to evaluate video quality, it remains a challenging task to assess the quality of UGC videos accurately and consistently.

Z. Li—Work done when Z. Li was at Weibo. Z. Li is currently with Microsoft.

Supplementary Information The online version contains supplementary material available at https://doi.org/10.1007/978-3-031-26316-3_24.

L. Wang et al. (Eds.): ACCV 2022, LNCS 13844, pp. 398–416, 2023.
https://doi.org/10.1007/978-3-031-26316-3_24

Fig. 1. Three consecutive frames from Video B304 in LIVE-VQA dataset. An overall high mean opinion score (MOS) of 91.73 was annotated.

Besides the frame-level image information, temporal information is regarded as a critically important factor for video analysis tasks. Although many image quality assessment (IQA) models [14,22,33,35,60–62,64,67] can be applied to VQA base on a simple temporal pooling process [52], these models may not work very robustly because of the absence of proper time sequence aggregation. For instance, Fig. 1 shows three consecutive frames extracted from Video B304 in LIVE-VQA [48] dataset. As seen, the motion blur distortion appears on the actress's hand area. These frame images are most likely to be recognized as middle, even low quality when applying a sophisticated IQA method to them individually. However, the quality of this video was labeled as high by human annotators, because a very smooth movement of the actress can be observed when playing the video stream. RNNs [1,2,23] and 3D-CNNs [6,12,13,20,25] are potential models to integrate spatial and temporal information for NR-VQA. Though these algorithms perform well on many datasets, the difficulty of parallelizing RNN models and the non-negligible computational cost of 3D-CNNs make them infeasible to be applied to many internet applications that require quick responses, just like the online video sharing/recommendation services.

On the other hand, research on NR-VQA is often limited by the lack of sufficient training data, due to the tedious work to label the mean opinion score (MOS) for each video. Among the publicly available datasets, MCL-JCV [56], VideoSet [57], UGC-VIDEO [27], CVD-2014 [38], LIVE-Qualcomm [16] are generated in lab environments, while KoNViD-1k [18], LIVE-VQC [48], YouTube-UGC [58] and LSVQ [63] are collected in-the-wild. As mentioned above, one could consider VQA as a temporal extension of IQA task, thus can apply some frame-level distortion and ranking processes [28] to augment the small video datasets for algorithm development. However, in-the-wild videos are usually hard to synthesize, since they may suffer from compound distortions which cannot be exactly parameterized as a combination of certain distortion cases. Recently, a large-scale LSVQ dataset [63], which contains 39,075 annotated videos with authentic distortions is released for public research. We know that not many studies have been conducted so far based on this new dataset.

Additionally, most previous works take L1 or L2 loss [23,54,63,64,70] as the optimization criterion for model training. Since these criteria to some extent ignore the order relationship of quality scores of the training samples, the trained model may be not stable to quantify the perceptual differences between the videos with similar quality scores. For example, in our research we find that many existing NR-VQA models work well to identify both high and low quality videos, but struggle to distinguish the videos with middle quality scores. How to effectively quantify the difference between samples with similar perceptual scores therefore becomes the key to the success of a NR-VQA model.

To address the above problems, in this paper we put forward a new Divide and Conquer Video Quality Estimator (DCVQE) model for NR-VQA. We summarize our contributions as follows: (1) Inspired by our observation on the actions of human annotation, we propose a Divide and Conquer Transformer (DCTr) architecture to extract video quality features for NR-VQA. Our algorithm starts from extracting the frame-level quality representations. Regarded as a divide process, we split the input sequence into a number of clips and apply Transformers to learn the clip-level quality embeddings (QE) and update the frame-level QE simultaneously. Subsequently, a conquer process is conducted by using another Transformer to combine the clip-level QE to generate a video-level QE. After stacking several DCTr layers and topping with a linear regressor, our DCVQE model can be constructed to predict the quality value of the input video. (2) By taking the order relationship of the training samples into account, we propose a novel correlation loss to bring an additional order constraint of video quality to guide the training. Experiments indicate that the introduction of this correlation loss can consistently help to improve the performance of our DCVQE model. (3) We conduct plenty of experiments on different datasets and confirm that our DCVQE outperforms most other algorithms.

2   Related Works

Traditional NR-VQA Solutions: Many prior NR-VQA works are "distortion specific" because they are designed to identify different distortion types like blur [31], blockiness [39], or noise [49] in compressed videos or image frames. More recent and popular used models are deployed on natural video statistics (NVS) features, which are created by extending the highly regular parametric band-pass models of natural scene statistics (NSS) from IQA to VQA tasks. Among them, successful applications have explored in both frequency domain (BIQI [36], DIIVINE [37], BLINDS [44], BLINDS-II [45]) and spatial domain (NIQE [35], BRISQUE [33]). V-BLIINDS [46] combined spatio-temporal NSS with motion coherency models to estimate perceptual video quality. Inter-subband correlations, modeled by spatial domain statistical features in frame differences, were used to quantify the degree of distortion in VIIDEO [34]. 3D-DCT was applied on local space-time regions, to establish quality aware features in [26]. Based on hand-crafted features selection and combination, recent algorithms VIDEVAL [53] and TLVQM [21] demonstrated outstanding performance on many UGC datasets. The designation of these hand-crafted features is deliberate, though they are hard to be deployed in an end-to-end fashion for NR-VQA tasks.

Deep Learning Based NR-VQA Solutions: Deep neural networks have shown their superior abilities in many computer vision tasks. With the availability of perceptual image quality datasets [15,19,41,64], many successful applications have been reported in the past decade [28,29,50,64,70]. Combining with a convolutional neural aggregation network, DeepVQA [59] utilized the advantages of CNN to learn spatio-temporal visual sensitivity maps for VQA. Based on a weakly supervised learning and resampling strategy, Zhang et al. [69] proposed a general purpose NR-VQA framework which inherited the knowledge learned from full-reference VQA and can effectively alleviate the curse of inadequate training data. VSFA [23] used a pretrained CNN to abstract frame features, and introduced the gated recurrent unit (GRU) to learn the temporal dependencies. Following VSFA, MDTVSFA [24] proposed a mixed datasets training method to further improve VQA performance. Although the above methods performed well on synthetic distortion datasets, they may be unstable to analyze UGC videos with complex and diverse distortions. PVQ [63] reported a leading performance on the large-scale dataset LSVQ. For a careful study on the local and global spatio-temporal quality, spatial patch, temporal patch and spatio-temporal patch were introduced in [63]. RAPIQUE [54], by leveraging a set of NSS features concatenated with learned CNN features, shown the top performance on several public datasets, including KoNViD-1k, YouTube-UGC, and their combination.

Transformer Techniques in Computer Vision: Self-attention mechanism-based Transformer architecture shows its exceptional performance in natural language processing [9,55]. Recently, many researchers introduced Transformer to solve computer vision problems. ViT [10] directly run attention among image patches with positional embeddings for image classification. Detection Transformer (DETR) [5] reached a comparable performance with Faster-RCNN [43] by designing a new object detection systems based on Transformers and bipartite matching loss for direct set prediction. Through conducting contrastive learning on 400 million image-text pairs, CLIP [42] shown impressive performance to solve different zero-shot transfer learning problems. For IQA tasks, Transformer also shows its powerful strength. Inspired by ViT, TRIQ [66] connected Transformer with MLP head to predict perceptual image quality, where sufficient lengths of positional embeddings were set to analyze the images with different resolutions. IQT [7] achieved outstanding performance by applying Transformer encoder and decoder on the features of reference images and distorted images. Through introducing 1D CNN and Transformer to integrate short-term and long-term temporal information, a recent work LSCT [65] demonstrated excellent performance on VQA. Our proposal is also derived from Transformer, inspired by our observation on the actions of human annotation. Experiments on various datasets confirm the effectiveness and robustness of our method.

3 Divide and Conquer Video Quality Estimator (DCVQE)

Human judgements of video quality are usually content-dependent and affected by their temporal memory [3,11,32,47,51,57,68]. In our investigation, we notice that many human annotators like to give their opinions on the quality of a video after the following two actions: first, watch the video quickly (usually in the fast forward mode) to get an overall impression of its quality, then they may scroll mouse forward and backward to review some specific parts of the video for their final decisions. Inspired by this observation, we propose a hierarchical architecture, dubbed Divide and Conquer Video Quality Estimator (DCVQE) for NR-VQA. Our model is worked by extracting three levels of video quality representations from frames, video clips to whole video sequence progressively and repeatedly, somewhat similar to the reverse processes of human annotation. An additional correlation loss term is also presented to bring an additional order constraint of video quality to guide the training. We find that our method can effectively improve the performance of NR-VQA. We will describe our work in detail in the following paragraphs.

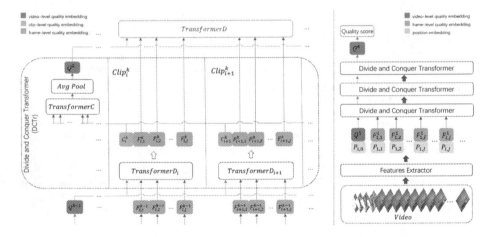

Fig. 2. The architectures of the proposed Divide and Conquer Transformer (DCTr) layer (left) and Divide and Conquer Video Quality Estimator (DCVQE) (right).

3.1 Overall Architecture

The left side of Fig. 2 represents the architecture of a key video quality analysis layer, Divide and Conquer Transformer (DCTr) in our proposal. In order to simulate the second action of human annotation mentioned above, we split the input sequence into a number of clips, and introduce a Transformer module $TransformerD$ to learn quality representations for each clip. As shown in Fig. 2,

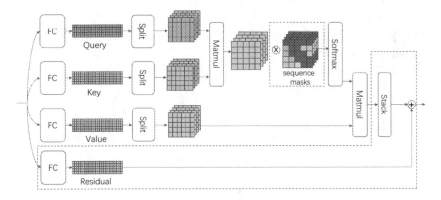

Fig. 3. The architecture of $TransformerD$ (include dashed box) and $TransformerC$ (exclude dashed box).

for the kth DCTr layer, we split the whole sequence into I clips, and each clip covers J frame-level quality representations generated by the previous layer. For the ith $(1 \leq i \leq I)$ clip C_i^k, a module $TransformerD_i$ is applied to combine the two levels of quality embeddings (QE) generated by the previous layer, that is, all the J frame-level QE $F_{i,j}^{k-1}(1 < j < J)$ and the video-level QE Q^{k-1}, to simultaneously learn the clip-level QE C_i^k and update the frame-level QE $F_{i,j}^k$ for the current layer. We further simulate the first action of human annotation by integrating the learned clip-level QE to generate a video-level QE. The other Transformer module $TransformerC$ with a topped average pooling layer are proposed to merge all clip-level quality representations C_i^k to predict the video-level QE Q^k for the current layer. In general, our proposed video quality analysis module is constructed in a divide and conquer format, so we coin it as a Divide and Conquer Transformer (DCTr) layer.

The overall architecture of our DCVQE model is shown in the right side of Fig. 2. As seen, we stack several DCTr layers (3 layers for our proposal, refer to supplementary material for details about this setting) to extract the final video-level QE Q^k for the input video. The final quality score is predicted by topping this embedding Q^k with a regressor. In practice, to improve the temporal sensitivity of our model, we progressively expand the coverage of video clips in the DCTr as the layers deepen. For example, supposing that each clip in the kth DCTr layer covers J frame-level QE from the previous layer, the coverage of each clip in the $(k+1)^{th}$ DCTr layer will be $2J$, which means two neighbor clips of the kth DCTr layer are combined to extract the clip-level QE in the $(k+1)^{th}$ DCTr layer. For the first DCTr layer, the input frame-level QE $F_{*,*}^1$ is generated by our feature extractor described in Subsect. 3.2, and the input video-level QE Q^1 is initialized randomly. Both of them are integrated with positional embeddings $P_{*,*}$ [9] for the following processes.

$TransformerD$ and $TransformerC$ are the two most important modules in our DCVQE. As shown in Fig. 3, they have similar architecture to the classic

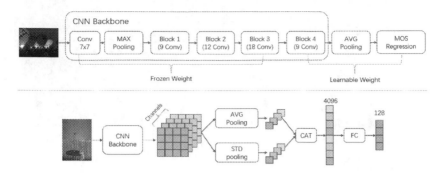

Fig. 4. Details of CNN backbone fine-tuning (top) and feature extraction (bottom).

Transformer [55]. Three fully connected layers are used to convert the input features into Query, Key, and Value tensors. Split operations are then adopted to group these three tensors into H heads, respectively. For the input frame-level feature with the shape of $B \times S \times D$, where B, S and D denote the batch size, sequence length and dimension of the frame embedding respectively, the shapes of its corresponding grouped Query, Key and Value tensors will be $H \times B \times S/H \times D$. The attention weights can be computed by the dot-product operations between the grouped Query and grouped Key tensors. To make $TransformerD$ module more concerned with the clip-level quality information, except the first position is reserved for the input video-level QE, we mask out the attention weights outside a predefined temporal range by using sequence masks (the dashed square plotted in Fig. 3) and adopt a Softmax layer to normalize the rests. The normalized weights are then applied to the grouped Value tensor to update the quality features. A stack operation is conducted on the head axis to recover the shape of the tensor. Here we also introduce the residual technique [17] to improve the robustness of $TransformerD$ module, so the final output of this module is the addition of the recovered features and residual. $TransformerC$ module is designed to update the video-level QE for each DCTr layer. The difference between $TransformerC$ and $TransformerD$ modules is that $TransformerC$ module does not contain sequence masks and residual block, as shown in Fig. 3.

3.2 Feature Extraction

To avoid the additional distortions that may be caused by applying some preprocessing steps to the input data, such as image resizing and filtering, we take the original full-size image frames as the input of our model. The pretrained Resnet-50 [17] is adopted as the CNN backbone of our feature extractor. To increase the sensitivity of the feature extractor to capture frame-level distortions, we first conduct a typical IQA task to fine-tune the CNN backbone. As shown in the top figure of Fig. 4, all CNN layers except the last group of CNN blocks are frozen for fine-tuning. We find this process can help to transfer the learned knowledge from ImageNet [8] to our application. It will be verified by our experiments.

Based on the fine-tuned backbone, as the bottom figure of Fig. 4 shows, the quality feature of each input frame is then extracted by adopting a concatenation operator to the outputs of global average (AVG) pooling and global standard deviation (STD) pooling processes [23,24]. Since both the dimensions of the outputs of AVG pooling and STD pooling are 2048, the frame-level QE for each frame will be a vector with a dimension of 4096. In practice, refer to [23], we further add a fully connected (FC) layer to reduce the dimension of the frame-level QE to 128. It is an efficient way to balance the accuracy of frame-level feature representation and overall computational cost.

3.3 Correlation Loss

Prior works usually use the following L1 loss to optimize the model:

$$L_1 = \frac{1}{N} \sum_{n=1}^{N} |p_n - g_n| \tag{1}$$

where $p_n, n \in [1, N]$ represents the nth predicted MOS, g_n denotes its corresponding ground truth, N is the total number of videos. Mathematically, this criterion to some extent ignores the relative order relationship of quality scores of the training samples in a batch, it sometimes may lead to limited ability to quantify the perceptual differences between the videos with similar quality scores. For example, considering the training processes of two NR-VQA models based on L1 loss criterion, for two video samples A and B with the MOS ground truths of 7.0 and 7.1, if their quality scores are predicted as $P_{A_1} = 7.1$, $P_{B_1} = 7.0$ and $P_{A_2} = 6.9$, $P_{B_2} = 7.0$ by the two models respectively (where P_{A_1} denotes the quality score of video A predicted by model 1, and so forth), the same strengths coming from the L1 losses (0.1) of these two training samples will be contributed to optimize the models correspondingly. Though it is hard to say which model is better than the other, in our opinion, it may be easier to optimize model 2 because of a positive correlation between the ground truths and its loss. Based on this idea, a new correlation loss, aiming to bring an additional order constraint of video quality to guide the training procedure, is proposed as follows:

$$L_c = \frac{1}{N} \sum_{n=1}^{N} \max \left(0, - \left(\sum_{m=1}^{N} (p_n - p_m) \right) \left(\sum_{m=1}^{N} (g_n - g_m) \right) \right) \tag{2}$$

where p_m represents the prediction of a training sample in a batch with size N, and p_n denotes the nth prediction which is picked to compare the ranking order with each element in the prediction set. g_m and g_n stand for the corresponding ground truth scores of p_m and p_n, respectively. This loss equation can be considered as that, we take the nth video in the batch as an anchor and compute the positive or negative correlations between predictions and ground truths for this anchor and any other one in the batch. Equation 2 can be further simplified as:

$$L_c = \frac{1}{N} \sum_{n=1}^{N} \max \left(0, -N^2 \left(p_n - \sum_{m=1}^{N} \frac{p_m}{N} \right) \left(g_n - \sum_{m=1}^{N} \frac{g_m}{N} \right) \right) \tag{3}$$

$$L_c = N \sum_{n=1}^{N} \max \left(0, -\left(p_n - \bar{p} \right) \left(g_n - \bar{g} \right) \right) \tag{4}$$

where $\bar{p}$ and $\bar{g}$ represent the mean values of predictions and ground truths, respectively. We find that this simplified L_c term is actually equivalent to the Spearman correlation coefficient without normalization. A total loss L, which combines the above two loss equations (1) and (4), is finally constructed to optimize our DCVQE model as

$$L = \alpha \times L_1 + \beta \times L_c \tag{5}$$

where α and β represent the weights of L_1 loss and L_c loss. We will discuss the optimal settings for these two weights in Subsect. 4.4.

4   Experiments

4.1   Datasets

We conduct experiments on four in-the-wild VQA datasets: KoNViD-1k [18], LIVE-VQC [48], YouTube-UGC [58], and LSVQ [63]. In these datasets, KoNViD-1k contains 1,200 unique contents with a duration of 8 s; LIVE-VQE, the smallest among them, consists of 585 video contents; YouTube-UGC contains 1,500 UGC video clips sampled from millions of YouTube videos; A recently released dataset LSVQ, containing 39,075 video samples with diverse durations and resolutions, is the largest publicly available UGC datasets for research right now. We also follow the suggestion given in [40,53] to select YouTube-UGC dataset as the anchor and map MOS values of KoNViD-1k and LIVE-VQC data onto a common scale to generate an All-Combined dataset for performance evaluation. The typical random 60-20-20 strategy [23] is applied to split data into three sets, i.e. 60% for training, 20% for evaluation, and the remaining 20% for testing. Especially, for LSVQ we follow the setting in [63] to first generate a Test-1080p set and then conduct the random 80–20 splitting on the rests to generate training and testing sets for experiments.

4.2   Evaluation Metrics

Here we adopt four commonly used criteria, Spearman Rank-Order Correlation Coefficient (SRCC), Kendall Rank-Order Correlation Coefficient (KRCC), Pearson Linear Correlation Coefficient (PLCC) and Root Mean Square Error (RMSE) to measure prediction monotonicity and prediction accuracy. For fair comparisons, we conduct each evaluation 100 times individually and report the median values of these four metrics as their final results.

4.3   Implementation Details

We first fine-tune ImageNet-pretrained Resnet-50 backbone on KonIQ-10k dataset [19] with random 80–20 splitting and maximum 20 epochs. The CNN backbone with the best performance of SRCC on the test set is selected to construct the feature extractor for our VQA task. The frame-level features are extracted using this feature extractor, where all frames of each video are considered without sampling. The temporal range of the sequence mask in the proposed $TransformerD$ (see Fig. 3) is empirically set to 15. For a tradeoff between GPU memory, running speed, and performance, we set the maximum length of each input video to 600 (i.e. only the first 600 frames of a video are considered if the video is longer than this maximum length) and the maximum training epoch to 75. An evaluation process will be conducted after each epoch, and the model with the lowest loss on the validation set will be stored.

4.4   Ablation Studies

In this subsection, we perform ablation studies to better understand the different components of the proposed DCVQE model.

CNN Backbone: Training of the CNN feature extraction backbone can also be regarded as a typical IQA task. Two series of architectures based on ResNet and ViT backbones are fine-tuned fully or partially in our study. Here we note the partial fine-tuning means that only the last blocks of these models, topped with average pooling and fully connected layers (see Fig. 4) are trainable. We determine that fully fine-tuned backbones act negatively with regard to the complexities of the networks. i.e. deeper structure results in worse performance, and partial fine-tuned backbones usually work better than fully fine-tuned ones for both two series of architectures. We also incorporate a new attention-based IQA architecture PHIQNet [65] to our study, and similar performance as partial fine-tuning on ResNet is observed (refer to supplementary material for details). In our application, for a tradeoff between model complexity and performance, also a fair comparison with previous work [23], we select the partially fine-tuned ResNet-50 as the feature extraction backbone to construct our DCVQE model.

Temporal Range of Sequence Mask: We introduce a group of experiments on KoNViD-1k dataset to study how the temporal range affects the DCVQE model. By directly replacing the proposed DCTr layers with the traditional Transformer layers [55], a baseline Transformer model is constructed for performance comparisons. Test results are plotted in Fig. 5. From these figures, we can see that the best temporal range for the baseline Transformer model is about 9, and the performance of this baseline model slightly decreases as the temporal range increases. Especially if we set the range to "all", which means that all the frame-level representations are involved in the self-attention processes [55] of the Transformer, the performance of this baseline model drops dramatically. Switching to our DCVQE, we can see that the overall performance stays at a high level. The best performance can be observed when the temporal range

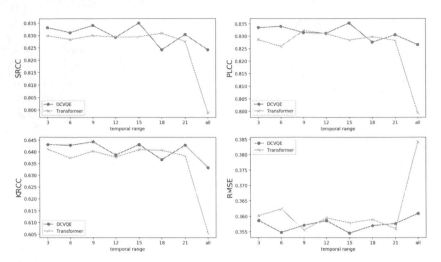

Fig. 5. Performance comparisons between a baseline Transformer model and our DCVQE with different settings of the temporal range.

Table 1. Performance comparisons of different weight combinations of α (for L1 loss term) and β (for proposed CL term) on KoNViD-1k dataset.

α	β	SRCC	PLCC	KRCC	RMSE
0.0	1.0	0.6980	0.6956	0.5149	3.0966
0.3	0.7	0.8259	0.8264	0.6360	0.3924
0.5	0.5	0.8349	0.8353	0.6431	0.3546
0.7	**0.3**	**0.8382**	**0.8375**	**0.6500**	**0.3515**
1.0	0.0	0.8278	0.8306	0.6396	0.3623

reaches 15. The results of the "all" setting for DCVQE, for which one frame only conducts self-attention with all frames in its own clip, also demonstrate the robust performance of our DCVQE model.

Correlation Loss: To show how the proposed Correlation Loss (CL) contributes to our VQA task, the performances of DCVQE with different weight settings of CL and L1 loss terms are reported in Table 1. The best performance can be identified when L1 Loss weight α reaches 0.7 and CL weight β reaches 0.3. Here we note that our CL does not intend to play a critical role in model optimization since it tries to describe the relative order relationships between videos, particularly for those videos with similar quality scores, so the overall improvement by adding this CL term may not look very significant. Overall, we confirm that our model can achieve better performance with the introduction of CL (Eq. 4). The rest experiments are conducted with the setting of $\alpha = 0.7$ and $\beta = 0.3$ if not specified.

Table 2. Component-by-component comparisons of several key parts in our DCVQE on KoNViD-1k dataset. Here we note that L1 loss is a default setting for all tests. Rows 3 and 4 apply AVG pooling (AVG) and $Transformer_C$ ($Trans_C$) respectively to conduct the "conquer" operation. "TR 15"s in rows 5, 6, 7, and 8 mean the temporal ranges of sequence masks are set to 15 for the tests.

Row #	Components	SRCC	PLCC	RMSE
1	Vanilla-Transformer	0.7988	0.7994	0.3842
2	Divide Only	0.7987	0.8032	0.3723
3	Divide + AVG	0.8025	0.8140	0.3644
4	Divide + $Trans_C$	0.8031	0.8121	0.3608
5	Divide + AVG + TR 15	0.8181	0.8267	0.3474
6	Divide + $Trans_C$ + AVG + TR 15	0.8278	0.8289	0.3453
7	Divide + $Trans_C$ + AVG + TR 15 + PW-RL	0.8307	0.8326	0.3542
8	Divide + $Trans_C$ + AVG + TR 15 + CL	0.8382	0.8375	0.3515

Component by Component Ablation: We further stack the key parts in our DCVQE component-by-component to analyze their impacts individually and gradually. To additionally study the effectiveness of our new CL term, a well-known pairwise ranking loss (PW-RL) [4], which is similar to our proposal, is also involved in our experiments. As shown in Table 2, compared with Vanilla-Transformer (row 1), only introduces the Divide operation (row 2) does not impact the system's performance. In row 3 we employ the AVG pooling operator to integrate clip-level embeddings to generate video-level embeddings, and the results show that this simple "conquer" operation can slightly improve the performance. In row 4 the conquer operation is switched to the proposed $Transformer_C$, where similar correlation scores with a lower RMSE are presented (compared with row 3). The temporal range of sequence mask 15 is introduced to tests in rows 5 and 6, where we can see that this change apparently benefits the system. Furthermore, the losses CL and PW-RL are compared directly in rows 7 and 8. The results confirm that the proposal CL provides the best performance for our VQA task.

4.5 Comparison with the State-of-the-Art Methods

We compare our method with 13 state-of-the-art VQA methods, which can be roughly divided into NR-IQA based methods: BRISQUE [33], GM-LOG [61], HIGRADE [22], FRIQUEE [14], CORNIA [62], HOSA [60], Koncept-512 [19], PaQ-2-PiQ [64], and NR-VQA based methods: V-BLIINDS [46], TLVQM [21], VSFA [23], VIDEVAL [53] and RAPIQUE [54]. For NR-IQA based methods, we abstract one frame feature per second and regard the average value of the features as the video representation to train a support vector regressor (SVR) for prediction [53, 54, 63].

Test results are reported in Table 3 and the best solution for each evaluation metric is marked in bold. Since it is hard to reproduce VIDEVAL and RAPIQUE methods, in Table 3 we directly cite the results given in [53,54] (marked as "*"). Different from our default 60-20-20 data splitting setting, these two methods used the random 80–20 strategy to split data for training and testing. For a fair comparison, we train an additional version of the DCVQE model under the same 80–20 data splitting setting. Test results of this additional DCVQE model are marked as "†". As seen from this table, our model (DCVQE†) significantly outperforms the current top method RAPIQUE on KoNViD-1K dataset by 4.37% SRCC and 2.45% PLCC, and greatly exceeds the second place method VIDEVAL on YouTube-UGC dataset by 5.62% SRCC and 5.67% PLCC. TLVQM method performs best on LIVE-VQC dataset. Looking into this dataset, we can see that most videos are captured using mobile devices, thus camera motion blurring is one of the common distortions affecting the video quality. TLVQM introduces several hand-crafted motion-related features to handle this issue, so it achieves the best performance, especially compared with all the deep learning based solutions. Nevertheless, besides TLVQM, our proposal performs the best among the remaining. The evaluation on the All-Combined dataset of KoNViD-1k, LIVE-VQC, YouTube-UGC also confirms the robustness of our model. As shown in the column "All Combined", our method (DCVQE†) surpasses the second place method (RAPIQUE) by 2.84% SRCC and 1.12% PLCC.

Table 3. Performance comparisons with the state-of-the-art methods.

| Dataset | KoNViD-1k | | | LIVE-VQC | | | YouTube-UGC | | | All combined | | |
model	SRCC	PLCC	RMSE	SRCC	PLCC	RMSE	SRCC	PLCC	RMSE	SRCC	PLCC	RMSE
BRISQUE	0.6567	0.6576	0.4813	0.5925	0.6380	13.100	0.3820	0.3952	0.5919	0.5695	0.5861	0.5617
GM-LOG	0.6578	0.6636	0.4818	0.5881	0.6212	13.223	0.3678	0.3920	0.5896	0.5650	0.5942	0.5588
HIGRADE	0.7206	0.7269	0.4391	0.6103	0.6332	13.027	0.7376	0.7216	0.4471	0.7398	0.7368	0.4674
FRIQUEE	0.7472	0.7482	0.4252	0.6579	0.7000	12.198	0.7652	0.7571	0.4169	0.7568	0.7550	0.4549
CORNIA	0.7169	0.7135	0.4486	0.6719	0.7183	11.832	0.5972	0.6057	0.5136	0.6764	0.6974	0.4946
HOSA	0.7654	0.7664	0.4142	0.6873	0.7414	11.353	0.6025	0.6047	0.5132	0.6957	0.7082	0.4893
KonCept-512	0.7349	0.7489	0.4260	0.6645	0.7278	11.626	0.5872	0.5940	0.5135	0.6608	0.6763	0.5091
PaQ-2-PiQ	0.6130	0.6014	0.5148	0.6436	0.6683	12.619	0.2658	0.2935	0.6153	0.4727	0.4828	0.6081
V-BLINDS	0.7101	0.7037	0.4595	0.6939	0.7178	11.765	0.5590	0.5551	0.5356	0.6545	0.6599	0.5200
TLVQM	0.7729	0.7668	0.4102	**0.7988**	**0.8025**	**10.145**	0.6693	0.6590	0.4849	0.7271	0.7342	0.4705
VSFA	0.7728	0.7754	0.4205	0.6978	0.7426	11.649	0.7611	0.7500	0.4269	0.7690	0.7862	0.4253
VIDEVAL*	0.7832*	0.7803*	0.4026*	0.7522*	0.7514*	11.100*	0.7787*	0.7733*	0.4049*	0.7960*	0.7939*	0.4268*
RAPIQUE*	0.8031*	0.8175*	**0.3623***	0.7548*	0.7863*	10.518*	0.7591*	0.7684*	0.4060*	0.8070*	0.8279*	0.3968*
DCVQE†	**0.8382†**	**0.8375†**	**0.3515†**	0.7620†	0.7858†	10.549†	**0.8225†**	**0.8172†**	**0.3770†**	**0.8299†**	**0.8372†**	**0.3824†**
DCVQE	**0.8206**	**0.8224**	0.3671	0.7479	0.7648	11.599	**0.8069**	**0.8050**	**0.3974**	**0.8239**	**0.8350**	0.3914

To demonstrate how well our model is learned from a given dataset, in Fig. 6 we also visualize the correlation between the video quality predictions and their ground truths (top row), as well as the video-level QE learned by the "conquer" operation of the last DCTr layer (bottom row). From the top row, we can see a strong correlation between our predictions and their ground truths. The figures in the bottom row plot the dimension-reduced video-level QE, where the gray value of each point is assigned with the normalized ground truth MOS score

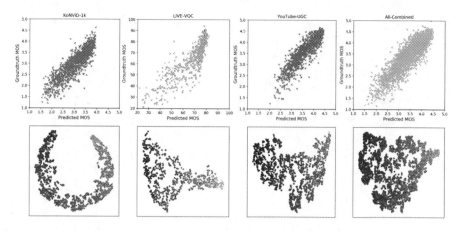

Fig. 6. Visualize test results. Top figures: Video quality predictions versus their ground truth MOS scores; Bottom figures: dimension-reduced video-level QE colored by ground truth MOS scores, where we apply t-SNE algorithm [30] to reduce the dimension of the embeddings to 2 for display.

of the corresponding video sample, that is, the darker a point, the greater its corresponding MOS score. From these figures, we can observe that the points with higher MOS scores scatter on one side while the points with lower MOS on the other. It indicates that the separability of quality for those video samples has been significantly enhanced by using our DCVQE model.

Table 4. Performance comparisons on the large-scale LSVQ dataset.

Dataset	Test		Test-1080p	
model	SRCC	PLCC	SRCC	PLCC
BRISQUE	0.579	0.576	0.497	0.531
TLVQM	0.772	0.774	0.589	0.616
VIDEVAL	0.794	0.783	0.545	0.554
VSFA	0.801	0.796	0.675	0.704
PVQ	0.827	0.828	0.711	0.739
DCVQE	**0.836**	**0.834**	**0.727**	**0.758**

Additionally, we compare our model with several others on the recently released large-scale LSVQ dataset. The PVQ model [63], which was proposed along with this dataset, is also involved in our evaluation. Test results listed in Table 4 confirm the excellent performance of our model again.

5   Conclusions

Inspired by our observation on the actions of human annotation, in this paper, we propose a new Divide and Conquer Transformer (DCTr) architecture to extract the video quality features for NR-VQA. Starting from extracting the frame-level quality embeddings (QE) of an input video, two types of Transformers are introduced to extract the clip-level QE and video-level QE progressively in our DCTr layer. Through stacking several DCTr layers and topping with a regressor, a hierarchical model, named Divide and Conquer Video Quality Estimator (DCVQE) is constructed to predict the quality score of the input video. We also put forward an additional correlation loss regarding the order relationship among the training data to guide the training. Experiments confirm that our proposal outperforms most other methods. What is more, our model is purely deep learning based, compared with other top methods where NSS/NVS features are more or less needed, so we believe our proposal is more practical.

References

1. Baccouche, M., Mamalet, F., Wolf, C., Garcia, C., Baskurt, A.: Sequential deep learning for human action recognition. In: Salah, A.A., Lepri, B. (eds.) HBU 2011. LNCS, vol. 7065, pp. 29–39. Springer, Heidelberg (2011). https://doi.org/10.1007/978-3-642-25446-8_4
2. Ballas, N., Yao, L., Pal, C., Courville, A.: Delving deeper into convolutional networks for learning video representations. CoRR (2016)
3. Bampis, C.G., Li, Z., Moorthy, A.K., Katsavounidis, I., Aaron, A., Bovik, A.C.: Study of temporal effects on subjective video quality of experience. IEEE Trans. Image Process. **26**(11), 5217–5231 (2017)
4. Cao, Z., Qin, T., Liu, T.Y., Tsai, M.F., Li, H.: Learning to rank: from pairwise approach to listwise approach, vol. 227, pp. 129–136, January 2007
5. Carion, N., Massa, F., Synnaeve, G., Usunier, N., Kirillov, A., Zagoruyko, S.: End-to-end object detection with transformers. In: Vedaldi, A., Bischof, H., Brox, T., Frahm, J.-M. (eds.) ECCV 2020. LNCS, vol. 12346, pp. 213–229. Springer, Cham (2020). https://doi.org/10.1007/978-3-030-58452-8_13
6. Carreira, J., Zisserman, A.: Quo vadis, action recognition? A new model and the kinetics dataset. In: 2017 IEEE Conference on Computer Vision and Pattern Recognition (CVPR), pp. 4724–4733 (2017)
7. Cheon, M., Yoon, S.J., Kang, B., Lee, J.: Perceptual image quality assessment with transformers. In: 2021 IEEE/CVF Conference on Computer Vision and Pattern Recognition Workshops (CVPRW), pp. 433–442 (2021)
8. Deng, J., Dong, W., Socher, R., Li, L.J., Li, K., Fei-Fei, L.: Imagenet: a large-scale hierarchical image database. In: 2009 IEEE Conference on Computer Vision and Pattern Recognition, pp. 248–255 (2009)
9. Devlin, J., Chang, M.W., Lee, K., Toutanova, K.: BERT: pre-training of deep bidirectional transformers for language understanding. In: NAACL-HLT (2019)
10. Dosovitskiy, A., et al.: An image is worth 16 × 16 words: Transformers for image recognition at scale. arXiv preprint arXiv:2010.11929 (2020)
11. Duanmu, Z., Ma, K., Wang, Z.: Quality-of-experience of adaptive video streaming: exploring the space of adaptations. In: Proceedings of the 25th ACM international conference on Multimedia, pp. 1752–1760 (2017)

12. Feichtenhofer, C.: X3d: expanding architectures for efficient video recognition. In: 2020 IEEE/CVF Conference on Computer Vision and Pattern Recognition (CVPR), pp. 200–210 (2020)
13. Feichtenhofer, C., Fan, H., Malik, J., He, K.: Slowfast networks for video recognition. In: 2019 IEEE/CVF International Conference on Computer Vision (ICCV), pp. 6201–6210 (2019)
14. Ghadiyaram, D., Bovik, A.: Perceptual quality prediction on authentically distorted images using a bag of features approach. J. Vision. **17**, 32 (2016)
15. Ghadiyaram, D., Bovik, A.C.: Massive online crowdsourced study of subjective and objective picture quality. IEEE Trans. Image Process. **25**(1), 372–387 (2016)
16. Ghadiyaram, D., Pan, J., Bovik, A.C., Moorthy, A.K., Panda, P., Yang, K.C.: In-capture mobile video distortions: a study of subjective behavior and objective algorithms. IEEE Trans. Circuits Syst. Video Technol. **28**(9), 2061–2077 (2018)
17. He, K., Zhang, X., Ren, S., Sun, J.: Deep residual learning for image recognition. In: 2016 IEEE Conference on Computer Vision and Pattern Recognition (CVPR), pp. 770–778 (2016)
18. Hosu, V., et al.: The Konstanz natural video database (Konvid-1k). In: 2017 Ninth International Conference on Quality of Multimedia Experience (QoMEX), pp. 1–6 (2017)
19. Hosu, V., Lin, H., Szirányi, T., Saupe, D.: KonIQ-10K: an ecologically valid database for deep learning of blind image quality assessment. IEEE Trans. Image Process. **29**, 1 (2020)
20. Karpathy, A., Toderici, G., Shetty, S., Leung, T., Sukthankar, R., Fei-Fei, L.: Large-scale video classification with convolutional neural networks. In: 2014 IEEE Conference on Computer Vision and Pattern Recognition, pp. 1725–1732 (2014)
21. Korhonen, J.: Two-level approach for no-reference consumer video quality assessment. IEEE Trans. Image Process. **28**(12), 5923–5938 (2019)
22. Kundu, D., Ghadiyaram, D., Bovik, A.C., Evans, B.L.: No-reference quality assessment of tone-mapped HDR pictures. IEEE Trans. Image Process. **26**(6), 2957–2971 (2017)
23. Li, D., Jiang, T., Jiang, M.: Quality assessment of in-the-wild videos. In: Proceedings of the 27th ACM International Conference on Multimedia, October 2019
24. Li, D., Jiang, T., Jiang, M.: Unified quality assessment of in-the-wild videos with mixed datasets training. Int. J. Comput. Vision **129**(4), 1238–1257 (2021)
25. Li, X., Wang, Y., Zhou, Z., Qiao, Y.: SmallBigNet: integrating core and contextual views for video classification. In: 2020 IEEE/CVF Conference on Computer Vision and Pattern Recognition (CVPR), pp. 1089–1098 (2020)
26. Li, X., Guo, Q., Lu, X.: Spatiotemporal statistics for video quality assessment. IEEE Trans. Image Process. **25**(7), 3329–3342 (2016)
27. Li, Y., Meng, S., Zhang, X., Wang, S., Wang, Y., Ma, S.: UGC-video: perceptual quality assessment of user-generated videos. In: 2020 IEEE Conference on Multimedia Information Processing and Retrieval (MIPR), pp. 35–38 (2020)
28. Liu, X., van de Weijer, J., Bagdanov, A.D.: RankIQA: learning from rankings for no-reference image quality assessment. In: 2017 IEEE International Conference on Computer Vision (ICCV), pp. 1040–1049 (2017)
29. Ma, K., Liu, W., Zhang, K., Duanmu, Z., Wang, Z., Zuo, W.: End-to-end blind image quality assessment using deep neural networks. IEEE Trans. Image Process. **27**(3), 1202–1213 (2018)
30. van der Maaten, L., Hinton, G.: Visualizing data using T-SNE. J. Mach. Learn. Res. **9**, 2579–2605 (2008)

31. Marziliano, P., Dufaux, F., Winkler, S., Ebrahimi, T.: A no-reference perceptual blur metric. In: Proceedings of International Conference on Image Processing, vol. 3 (2002)
32. Mirkovic, M., Vrgović, P., Stefanović, D., Anderla, A.: Evaluating the role of content in subjective video quality assessment. Sci. World J. **2014**, 625219 (2014)
33. Mittal, A., Moorthy, A.K., Bovik, A.C.: No-reference image quality assessment in the spatial domain. IEEE Trans. Image Process. **21**(12), 4695–4708 (2012)
34. Mittal, A., Saad, M.A., Bovik, A.C.: A completely blind video integrity oracle. IEEE Trans. Image Process. **25**(1), 289–300 (2016)
35. Mittal, A., Soundararajan, R., Bovik, A.C.: Making a "completely blind" image quality analyzer. IEEE Signal Process. Lett. **20**(3), 209–212 (2013)
36. Moorthy, A.K., Bovik, A.C.: A two-step framework for constructing blind image quality indices. IEEE Signal Process. Lett. **17**(5), 513–516 (2010)
37. Moorthy, A.K., Bovik, A.C.: Blind image quality assessment: from natural scene statistics to perceptual quality. IEEE Trans. Image Process. **20**(12), 3350–3364 (2011)
38. Nuutinen, M., Virtanen, T., Vaahteranoksa, M., Vuori, T., Oittinen, P., Häkkinen, J.: CVD 2014-a database for evaluating no-reference video quality assessment algorithms. IEEE Trans. Image Process. **25**(7), 3073–3086 (2016)
39. Pan, F., Lin, X., Rahardja, S., Ong, E.P., Lin, W.: Using edge direction information for measuring blocking artifacts of images. Multidimens. Syst. Signal Process. **18**, 297–308 (2007)
40. Pinson, M., Wolf, S.: Objective method for combining multiple subjective data sets. pp. 583–592 (2003)
41. Ponomarenko, N., et al.: Color image database tid2013: Peculiarities and preliminary results. In: European Workshop on Visual Information Processing (EUVIP), pp. 106–111 (2013)
42. Radford, A., et al.: Learning transferable visual models from natural language supervision. In: International Conference on Machine Learning, pp. 8748–8763. PMLR (2021)
43. Ren, S., He, K., Girshick, R., Sun, J.: Faster R-CNN: towards real-time object detection with region proposal networks. IEEE Trans. Pattern Anal. Mach. Intell. **39**(6), 1137–1149 (2017)
44. Saad, M.A., Bovik, A.C., Charrier, C.: A DCT statistics-based blind image quality index. IEEE Signal Process. Lett. **17**(6), 583–586 (2010)
45. Saad, M.A., Bovik, A.C., Charrier, C.: Blind image quality assessment: a natural scene statistics approach in the DCT domain. IEEE Trans. Image Process. **21**(8), 3339–3352 (2012)
46. Saad, M.A., Bovik, A.C., Charrier, C.: Blind prediction of natural video quality. IEEE Trans. Image Process. **23**(3), 1352–1365 (2014)
47. Siahaan, E., Hanjalic, A., Redi, J.: Semantic-aware blind image quality assessment. Signal Process. Image Commun. **60**, 237–252 (2017)
48. Sinno, Z., Bovik, A.C.: Large-scale study of perceptual video quality. IEEE Trans. Image Process. **28**(2), 612–627 (2019)
49. Tai, S.C., Yang, S.M.: A fast method for image noise estimation using Laplacian operator and adaptive edge detection. In: 2008 3rd International Symposium on Communications, Control and Signal Processing, pp. 1077–1081 (2008)
50. Talebi, H., Milanfar, P.: NIMA: neural image assessment. IEEE Trans. Image Process. **27**(8), 3998–4011 (2018)

51. Triantaphillidou, S., Allen, E., Jacobson, R.: Image quality comparison between JPEG and JPEG2000. II. Scene dependency, scene analysis, and classification. J. Imaging Sci. Technol. **51**, 259–275 (2007)

52. Tu, Z., Chen, C.J., Chen, L.H., Birkbeck, N., Adsumilli, B., Bovik, A.C.: A comparative evaluation of temporal pooling methods for blind video quality assessment. In: 2020 IEEE International Conference on Image Processing (ICIP), pp. 141–145 (2020)

53. Tu, Z., Wang, Y., Birkbeck, N., Adsumilli, B., Bovik, A.C.: UGC-VQA: benchmarking blind video quality assessment for user generated content. IEEE Trans. Image Process. **30**, 4449–4464 (2021)

54. Tu, Z., Yu, X., Wang, Y., Birkbeck, N., Adsumilli, B., Bovik, A.C.: RAPIQUE: rapid and accurate video quality prediction of user generated content. IEEE Open J. Signal Process. **2**, 425–440 (2021)

55. Vaswani, A., et al.: Attention is all you need. In: Advances in Neural Information Processing Systems, vol. 30 (2017)

56. Wang, H., et al.: MCL-JCV: A JND-based h.264/AVC video quality assessment dataset. In: 2016 IEEE International Conference on Image Processing (ICIP), pp. 1509–1513 (2016)

57. Wang, H., et al.: VideoSet: a large-scale compressed video quality dataset based on JND measurement. J. Visual Commun. Image Represent. **46**, 292–302 (2017)

58. Wang, Y., Inguva, S., Adsumilli, B.: YouTube UGC dataset for video compression research. In: 2019 IEEE 21st International Workshop on Multimedia Signal Processing (MMSP), pp. 1–5 (2019)

59. Woojae, K., Kim, J., Ahn, S., Kim, J., Lee, S.: Deep video quality assessor: from spatio-temporal visual sensitivity to a convolutional neural aggregation network. In: Ferrari, V., Hebert, M., Sminchisescu, C., Weiss, Y. (eds.) Computer Vision - ECCV 2018, pp. 224–241. Springer International Publishing, Cham (2018)

60. Xu, J., Ye, P., Li, Q., Du, H., Liu, Y., Doermann, D.: Blind image quality assessment based on high order statistics aggregation. IEEE Trans. Image Process. **25**(9), 4444–4457 (2016)

61. Xue, W., Mou, X., Zhang, L., Bovik, A.C., Feng, X.: Blind image quality assessment using joint statistics of gradient magnitude and Laplacian features. IEEE Trans. Image Process. **23**(11), 4850–4862 (2014)

62. Ye, P., Kumar, J., Kang, L., Doermann, D.: Unsupervised feature learning framework for no-reference image quality assessment. In: 2012 IEEE Conference on Computer Vision and Pattern Recognition, pp. 1098–1105 (2012)

63. Ying, Z., Mandal, M., Ghadiyaram, D., Bovik, A.: Patch-VQ: 'patching up' the video quality problem. In: Proceedings of the IEEE/CVF Conference on Computer Vision and Pattern Recognition, pp. 14019–14029 (2021)

64. Ying, Z., Niu, H., Gupta, P., Mahajan, D., Ghadiyaram, D., Bovik, A.: From patches to pictures (PAQ-2-PIQ): Mapping the perceptual space of picture quality. In: 2020 IEEE/CVF Conference on Computer Vision and Pattern Recognition (CVPR), pp. 3572–3582 (2020)

65. You, J.: Long short-term convolutional transformer for no-reference video quality assessment. pp. 2112–2120 (2021)

66. You, J., Korhonen, J.: Transformer for image quality assessment. In: 2021 IEEE International Conference on Image Processing (ICIP), pp. 1389–1393. IEEE (2021)

67. Zhang, L., Zhang, L., Bovik, A.C.: A feature-enriched completely blind image quality evaluator. IEEE Trans. Image Process. **24**(8), 2579–2591 (2015)

68. Zhang, R., Isola, P., Efros, A.A., Shechtman, E., Wang, O.: The unreasonable effectiveness of deep features as a perceptual metric. In: Proceedings of the IEEE Conference on Computer Vision and Pattern Recognition, pp. 586–595 (2018)
69. Zhang, Y., Gao, X., He, L., Lu, W., He, R.: Blind video quality assessment with weakly supervised learning and resampling strategy. IEEE Trans. Circuits Syst. Video Technol. **29**(8), 2244–2255 (2019)
70. Zhu, H., Li, L., Wu, J., Dong, W., Shi, G.: Metaiqa: Deep meta-learning for no-reference image quality assessment. In: 2020 IEEE/CVF Conference on Computer Vision and Pattern Recognition (CVPR). pp. 14131–14140 (2020)

Not End-to-End: Explore Multi-Stage Architecture for Online Surgical Phase Recognition

Fangqiu Yi, Yanfeng Yang, and Tingting Jiang[(✉)]

National Engineering Research Center of Visual Technology, School of Computer
Science, Peking University, Beijing, China
{chinayi,yangyanfeng,ttjiang}@pku.edu.cn

Abstract. Surgical phase recognition is of particular interest to computer assisted surgery systems, in which the goal is to predict what phase is occurring at each frame for a surgery video. Networks with multi-stage architecture have been widely applied in many computer vision tasks with rich patterns, where a predictor stage first outputs initial predictions and an additional refinement stage operates on the initial predictions to perform further refinement. Existing works show that surgical video contents are well ordered and contain rich temporal patterns, making the multi-stage architecture well suited for the surgical phase recognition task. However, we observe that when simply applying the multi-stage architecture to the surgical phase recognition task, the end-to-end training manner will make the refinement ability fall short of its wishes. To address the problem, we propose a new non end-to-end training strategy and explore different designs of multi-stage architecture for surgical phase recognition task. For the non end-to-end training strategy, the refinement stage is trained separately with proposed two types of disturbed sequences. Meanwhile, we evaluate three different choices of refinement models to show that our analysis and solution are robust to the choices of specific multi-stage models. We conduct experiments on two public benchmarks, the M2CAI16 Workflow Challenge and the Cholec80 dataset. The SOTA comparable results show that the multi-stage architecture holds the great potential to boost the performance of existing single-stage models. Code is available at https://github.com/ChinaYi/NETE.

Keywords: Surgical phase recognition · Surgical workflow segmentation · Multi-stage architecture

1 Introduction

Surgical phase recognition is of particular interest to computer assisted surgery systems, because it offers solutions to numerous demands of the modern operating room, such as monitoring surgical processes [1], scheduling surgeons [2]

F. Yi and Y. Yang—Contribute equally to this work.

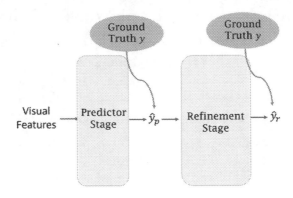

Fig. 1. Pipeline of multi-stage architecture.

and enhancing coordination among surgical teams [3]. This paper works on the online surgical phase recognition task, which requires to predict what phase is occurring at each frame without using the information of future frames.

Existing surgical phase recognition models can be divided into two groups. The first group is single-stage models which output prediction results with the input visual features, while the second group is the multi-stage models which additionally stack a refinement stage over the prediction results to perform a further refinement. In our opinion, the multi-stage architecture is the one which is well-suited for the surgical phase recognition task. First of all, networks with multi-stage architecture have been widely applied in many computer vision tasks with rich patterns, such as human pose estimation [4,5] and action segmentation [6]. Generally speaking, the idea of multi-stage architecture consists of a predictor stage and a refinement stage, as shown in Fig. 1. Sometimes, due to the hard-to-recognize visual features, the initial predictions output by the predictor stage may have errors that violate intrinsic patterns within the data. (*i.e.* The tiny spikes of over-segmentation errors for a continuous action or human pose estimation results that do not conform to the connections of human body joint.) The initial predictions are thus further refined by the refinement stage. Secondly, surgical video contents are well ordered and contain rich temporal patterns. Some works have been motivated by utilizing the rich temporal patterns to refine the predictions. The success in [7,8] shows that it is possible for the multi-stage architecture to rectify the misclassifications due to the ambiguous visual features in the predictor stage.

However, we observe that the improvement of multi-stage structure in the surgical phase recognition task is not as obvious as other tasks. Experiments in [9] show that the performance improvement of the additional refinement stage is very limited. This interesting phenomenon raises our concerns, *why* the multi-stage architecture does not work as well as we expected?

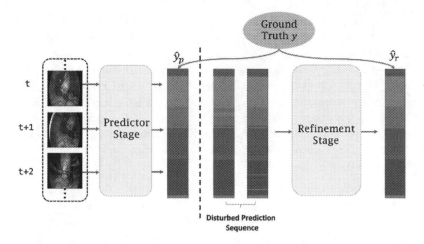

Fig. 2. Our non end-to-end training strategy where two stages are trained separately.

In this paper, we first answer the question of *why* with our analysis, and then further give a solution of *how* to make multi-stage architecture work better for surgical phase recognition task. The reason of *why* is two folds. Firstly, a common issue for the multi-stage architecture is that the refinement stage cannot actually learn how to refine with an end-to-end training manner. As shown in Fig. 1, the supervision signal is applied in both the predictor stage and refinement stage. After several epochs of training, the predictor stage will quickly converge to the ground truth, which means that the initial predictions $\hat{y}_p$ are almost 100% correct during the training process. However, in the inference process with the test data, $\hat{y}_p$ still remain lots of errors due to imperfect predictor. The huge gap between the inputs of the refinement stage during training and inference makes the refinement ability fall short of its wishes. Secondly, in the case of end-to-end training, the limited size of current datasets for surgery phase recognition cannot afford the training of refinement stage which brings additional parameters. This leads to a severe overfitting problem compared to the results of applying multi-stage architecture to other vision tasks.

With the answer of *why*, we propose a non end-to-end training strategy where the predictor stage and the refinement stage are trained separately to solve the above two issues simultaneously. As shown in Fig. 2, the predictor stage is trained with the raw video data. To reduce the gap between the inputs of the refinement stage during training and inference, two types of training sequences for refinement stage are carefully designed to simulate the real output of the predictor during inference, denoted as cross-validate type and mask-hard-frame type respectively. Meanwhile, as the refinement stage is separately trained with the two types of carefully designed training sequences, the over-fitting problem for surgical phase recognition can also be alleviated.

Besides the training strategy, we further explore the designs of multi-stage architecture by evaluating three different temporal models for the refinement stage, including TCN (offline) [10], causal TCN (online) [10] and GRU (online) [11]. As for the predictor stage, we use the causal TCN in [9] for its high efficiency and good performance. In principle, our solution can be applied to any single-stage predictor model. Extensive experiments are conducted on two public benchmarks, M2CAI16 [12] and Cholec80 [13]. Results show that all three refinement models trained with our strategy successfully boost the performance of the single predictor stage, demonstrating that our analysis and solution are robust to different choices of refinement models. And the SOTA comparable results show that the multi-stage architecture holds the great potential to boost the performance of existing single-stage models.

2   Related Work

Existing surgical phase recognition models can be divided into two groups. The first group is single-stage models which output prediction results with the input visual features. For example, a number of works utilized dynamic time warping [14,15], conditional random field [16], and variations of Hidden Markov Model (HMM) [17,18] over extracted visual features. [7] trained an end-to-end RNN model that first used a very deep ResNet to extract visual features for each frame and then applied a LSTM network to model the temporal dependencies of sequential frames. [19] proposed a LSTM-based temporal network structure that leveraged task-specific network representation to collect long-term sufficient statistics that were propagated by a sufficient statistics model. In addition, with the wide application of transformer in computer vision, there are also some single-stage models based on transformer. [20] used a novel attention regularization loss which encouraged the transformer model to focus on high-quality frames during training, and the attention weights could be utilized to identify characteristic high attention frames for each surgical phase, which could further help the surgery summarization. [21] proposed a hybrid embedding aggregation transformer model which used cleverly designed spatial and temporal embeddings by allowing for active queries based on spatial information from temporal embedding sequences.

The second group is the multi-stage models which additionally stack a refinement stage over the prediction results to perform a further refinement. [9] is the first one which brought in multi-stage architecture for surgical phase recognition task. They used a causal TCN [10] to output initial predictions over pre-extracted CNN features, and then appended another causal TCN [10] to refine the predictions. [22] proposed a multi-task multi-stage temporal convolutional network along with a multi-task convolutional neural network training setup to jointly predict the phases and steps and benefit from their complementarity to better evaluate the execution of the procedure.

In addition, some surgical phase recognition models used data augmentation techniques to improve performance, such as cross validation and hard frames detection. Cross validation is the simplest way to make part of training data unseen to the predictor model, so that the unseen part of data could be used to simulate the real predictions during the inference. Besides, the concept of hard frames was first proposed by [8], which denoted the frames that were not recognizable from their visual appearance. They observed that single-stage models usually made mistakes on hard frames, so they found out all the hard frames in the training videos and mapped them to corresponding phases separately.

3 Methods

The multi-stage architecture stacks a refinement stage over the predictor stage sequentially. We propose a training strategy where these two stages are trained separately and explore designs of multi-stage architecture. We first introduce the predictor stage and its training in Sec. 3.1, then describe the generation process of disturbed prediction sequences in Sec. 3.2, and finally discuss the refinement stage and its training in Sec. 3.3. It is worth noting that, although we train the multi-stage architecture in a non end-to-end manner, the inference process is still end-to-end as the normal multi-stage architecture.

3.1 Predictor Stage

We use causal TCN in [9] to get the initial predictions for its high efficiency and good performance. Instead of the general temporal convolutions which depend on both n past and n future frames, the causal temporal convolutions only rely on the current and previous frames and thus meet the demand of online surgical phase recognition. In principle, the predictor model can be any online model. The input of the causal TCN is the frame-wise extracted features from a pre-trained CNN. Denote the output prediction sequence as $\hat{y}_p \in \mathcal{R}^{C \times T}$. For each frame, the output is a vector of size C denoting the classification probability for each class. For the loss function $\mathcal{L}_p$, we use a combination of cross-entropy classification loss and a smoothing loss [23] by deploying a mean square error over the classification probabilities of every two adjacent frames. The loss function writes as

$$\mathcal{L}_p = \frac{1}{T} \sum_{t=1}^{T} -log(\hat{y}_{p(c,t)})$$
$$+ \frac{1}{TC} \sum_{m=1}^{C} \sum_{t=1}^{T-1} |\hat{y}_{p(m,t)} - \hat{y}_{p(m,t+1)}|^2 . \tag{1}$$

3.2 Disturbed Prediction Sequence Generation

The input of the refinement stage is the prediction results $\hat{y}_p$ output by the predictor stage. During the inference, $\hat{y}_p$ will remain lots of errors due to the imperfect predictor. In order to achieve better refinement results, we should minimize the distribution gap between the preliminary prediction results in training and inference. Thus, we design two types of disturbed prediction sequences by simulating the imperfect prediction results of the predictor model during the

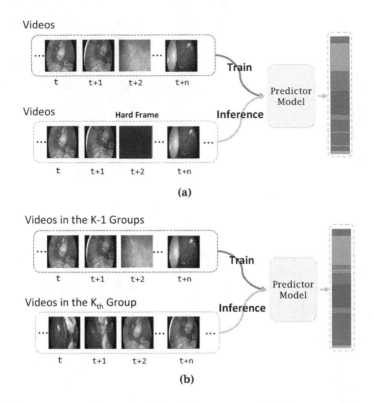

Fig. 3. (a) The generation of mask-hard-frame type. (b) The generation of cross-validate type.

inference. Note that the generation of both two types of disturbed prediction sequences are related to the predictor model, since we cannot directly obtain the prediction sequences from the raw video data.

Mask-Hard-Frame Type. The concept of hard frames was proposed by [8], which denoted the frames that were not recognizable from their visual appearance. Their work shows that single-stage models usually make mistakes on hard frames. Motivated by this, we seek to add perturbations to the prediction of these hard frames. We first train a predictor model with the normal video data. Then, we follow the rule of [8] to find out hard frames in the training set and add perturbations on those hard frames by using a black mask to cover the whole image. Finally, we pass the perturbed video data to the predictor model to get the disturbed prediction sequence. The workflow is shown in Fig. 3(a).

Cross-Validate Type. Cross validation is the simplest way to make part of training data unseen to the predictor model, so that the unseen part of data could be used to simulate the real predictions of the predictor model during

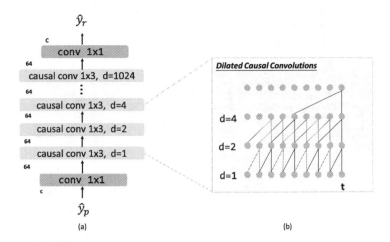

Fig. 4. (a) An overview of causal TCN, which can be performed online. The dilation factor is doubled at each causal layer, *i.e.* 1, 2, 4, ..., 1024. (b) The architecture of the causal convolutions with different dilation factors.

the inference. Specifically, we randomly partition the training videos into K groups of equal size. Each time, a single group is retained for validation, and the remaining $K - 1$ groups are used to train a predictor model. And then, we use the trained predictor model to obtain the predictions of the retained video to get the prediction sequence. The workflow is shown in Fig. 3(b). We set $K = 10$ in our experiments.

3.3 Refinement Stage

For a training set with N training videos, both two methods can generate N perturbed prediction sequences to train the refinement model. Both two types of disturbed sequences are used to train the refinement stage in following experiments except for special specification. For the loss function of the refinement stage, we use the cross-entropy loss. For the choice of specific refinement model, we evaluate three common temporal models, including TCN [10], causal TCN [10] and GRU [11]. We chose these three models as the refinement models because these are the three most common temporal models. In addition, stacking several predictors sequentially has shown significant improvements in many tasks like human pose estimation and action segmentation. The stacked architecture is composed of several models sequentially such that each model operates directly on the output of the previous one. So we also stack single GRU stages to form a stacked GRU for the refinement stage.

Causal TCN. The overview of causal TCN is illustrated in Fig. 4(a). The first layer of causal TCN is a 1×1 convolutional layer, that adjusts the dimension of

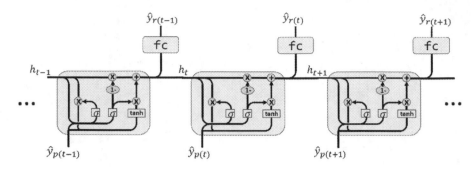

Fig. 5. The overview of single GRU network. The GRU unit employs three gates to modulate the interactions between the GRU cells and the environment. The dimension of the hidden state is set to 128.

the input features from 2048-d to 64-d. Then, this layer is followed by several layers of dilated causal convolutions, as shown in Fig. 4(b). The causal convolution is different from common temporal convolution, for each time step t and filter length d, it convolves from X_{t-d} to X_t, which meets the demands of online surgical phase recognition. Note that we do not use temporal pooling layer, because it might results in a loss of fine-grained information that is necessary for phase recognition. Instead, we use a dilation factor that is doubled at each causal layer, *i.e.* 1, 2, 4, ..., 1024, to enlarge the temporal receptive field. Finally, we apply a 1×1 convolution over the output of the last dilated causal convolution layer to get the probability vector dimension.

Single GRU. Gated Recurrent Unit (GRU) is a popular variant of Recurrent Neural Network (RNN). The architecture of GRU is shown in Fig. 5. The GRU unit employs three gates, *i.e.*, a reset gate r_t, an update gate z_t and a new gate n_t, to modulate the interactions between the GRU cells and the environment. The dimension of the hidden state is set to 128. At timestep t, given input p_t (probability vector of frame X_t belonging to each phase), hidden state h_{t-1}, the GRU unit updates with following equations:

$$
\begin{aligned}
r_t &= \sigma(W_{ir}p_t + b_{ir} + W_{hr}h_{(t-1)} + b_{hr}) , \\
z_t &= \sigma(W_{iz}p_t + b_{iz} + W_{hz}h_{(t-1)} + b_{hz}) , \\
n_t &= tanh(W_{in}p_t + b_{in} + r_t * (W_{hn}h_{(t-1)} + b_{hn})) , \\
h_t &= (1 - z_t) * n_t + z_t * h_{(t-1)} .
\end{aligned}
\tag{2}
$$

In above equations, σ is the sigmoid function, and $*$ is the Hadamard product. To get the refined predictions, we additionally apply a fully connected layer over the hidden state.

Stacked GRU. The refinement model takes an initial prediction as input, and then outputs a refined prediction. It is natural to come up with the idea that if we stack multiple refinement models sequentially, each model will successively operate on the refined predictions of the previous one. The effect of such composition is an incremental refinement of the predictions from the previous model. Therefore, in the architecture of stacked GRU, the input to the latter GRU is the refined result from the previous GRU, not the hidden state. The set of operations at each GRU in stacked GRU can be formally described as follows:

$$\mathcal{P}^0 = \hat{p}_{1:T} \ ,$$
$$\mathcal{P}^s = GRU(\mathcal{P}^{s-1}) \ . \tag{3}$$

In above equations, $\mathcal{P}^s$ is the refined predictions at sth GRU and GRU is the single-stage GRU discussed before. As for loss function, different from single-stage refinement models, we use the cross entropy loss on the refined probability sequence of each GRU.

3.4 Training Details

We employ the ResNet50 [24] as the visual feature extractor to extract off-the-shelf video features. Specifically, the ResNet50 is trained frame-wise without temporal information through cross-entropy loss. The dimension of the input ResNet50 features is 2048-d. With the pre-extracted video features, we train the predictor stage for 100 epochs with initial learning rate 1e-4 and Adam optimizer. For the refinement stage, we train the model for 40 epochs with two proposed disturbed sequences.

4 Experiment

4.1 Dataset

M2CAI16 Workflow Challenge Dataset. M2CAI16 dataset [12] contains 41 laparoscopic videos that are acquired at 25 fps of cholecystectomy procedures, and 27 of them are used for training and 14 videos are used for testing. These videos are segmented into 8 phases by experienced surgeons.

Cholec80 Dataset. Cholec80 dataset [13] contains 80 videos of cholecystectomy surgeries performed by 13 surgeons. The dataset is divided into training set (40 videos) and testing set (40 videos). These videos are segmented into 7 phases and are captured at 25 fps.

4.2 Metrics

To quantitatively analyze the performance of our method, we use three metrics [7] including the jaccard index ($JACC$), recall (Rec) and accuracy (Acc). Among them, Acc quantitatively evaluates the amount of correctly classified phases in the whole video, while Rec and $JACC$ evaluate the results for each individual phase.

Table 1. Comparison of multi-stage architectures with three different choices of refinement models under the end-to-end training strategy and ours on Cholec80 dataset. Predictor denotes single-stage predictor without the refinement stage.

Method	Acc	JACC	Rec
Predictor	88.8 ± 6.3	73.2 ± 9.8	84.9 ± 7.2
End-to-End+GRU	87.1 ± 7.8	69.7 ± 12.6	83.2 ± 9.4
End-to-End+causal TCN	87.7 ± 6.3	77.7 ± 11.2	84.3 ± 6.3
End-to-End+TCN	89.8 ± 6.6	75.8 ± 8.4	87.4 ± 7.5
Ours+GRU	90.8 ± 7.0	75.5 ± 11.1	85.6 ± 10.0
Ours+causal TCN	91.0 ± 5.2	74.2 ± 11.8	84.1 ± 9.6
Ours+TCN	**92.8 ± 5.0**	**78.7 ± 9.4**	**87.5 ± 8.3**

4.3   End-to-End VS. Not End-to-End

In this section, we evaluate the performance of multi-stage architectures with the end-to-end training strategy and our non end-to-end training strategy on the Cholec80 dataset. Results are shown in Table 1. We can observe that, with the end-to-end training strategy, the multi-stage architecture only achieves comparable results with the single-stage predictor. When we use causal TCN or GRU as the refinement model, performance is even slightly worse possibly due to the over-fitting problem. Such results are consistent with the results in [9]. Meanwhile, all three refinement models trained with our proposed disturbed sequences largely boost the performance of the predictor, which proves our previous analysis above the multi-stage architecture and shows that our solution is effective and not sensitive to the choices of different refinement models. Although TCN achieves the best performance as the refinement model, however, it does not meet the constraint of online surgical phase recognition because it needs information from future frames. So, we explore GRU as the refinement model for further experiments.

Figure 6 shows the qualitative results of GRU as the refinement model under the end-to-end training strategy and our non end-to-end training strategy on the Cholec80 dataset. We can observe that the results of single-stage predictor and end-to-end multi-stage model both contain a large number of short surgical phases, which do not meet the temporal continuity and destroy the smoothness of the predicted results. Besides, the result of end-to-end multi-stage model also contains more prediction errors than single-stage predictor. These results clearly highlight the ability of the multi-stage architecture trained with our solution which obtains consistent and smooth predictions.

4.4   Stacked Number of GRUs in Refinement Model

In this section, we use the stacked GRU as the refinement stage model. During training, the stacked GRU is trained as a whole with the two proposed sequences and the loss function is applied on the output of each GRU. Results of

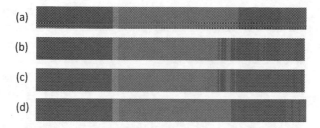

Fig. 6. Qualitative results on the Cholec80 dataset of the multi-stage architectures trained with end-to-end manner and ours. GRU is used for the refinement stage. (a) is the ground truth. (b) is the prediction from the single-stage predictor. (c) is the prediction from the multi-stage architecture trained with end-to-end manner. (d) is the prediction from the multi-stage architecture trained with our solution.

Table 2. Effects of the stacked number of GRUs on the Cholec80 dataset.

Method	Acc	JACC	Rec
Single GRU	90.8 ± 7.0	75.5 ± 11.1	85.6 ± 10.0
Stacked GRU(2 GRU)	91.3 ± 6.1	75.9 ± 11.2	84.4 ± 10.0
Stacked GRU(3 GRU)	$\mathbf{91.5 \pm 7.1}$	$\mathbf{77.2 \pm 11.2}$	$\mathbf{86.8 \pm 8.5}$
Stacked GRU(4 GRU)	90.8 ± 5.8	74.2 ± 15.9	83.1 ± 14.6

the stacked GRU with different numbers of GRUs on the Cholec80 dataset are shown in Table 2. We get the best results when stacking 3 GRUs sequentially on the Cholec80 dataset. In order to give an intuitive explanation about how the stacked refinement model works, we also show the qualitative results of a video in different stages for a stacked GRU with 3 GRUs in Fig. 7. Figure 7(a) shows the predictions at each stage, and Fig. 7(b) shows the probability sequences of the predictions. We can observe that adding more GRUs results in an incremental refinement of the predictions. We also conduct experiments on the M2CAI16 dataset. Results of the stacked GRU with different numbers of GRUs on the M2CAI16 dataset are shown in Table 3. we can observe that the experiments get the best results when stacking 2 GRUs, which is less than that of the Cholec80 dataset. This may due to the fact that the size of M2CAI16 is smaller.

Table 3. Effects of the stacked number of GRUs on the M2CAI16 dataset.

Method	Acc	JACC	Rec
Single GRU	86.2 ± 9.1	72.6 ± 11.6	90.0 ± 11.7
Stacked GRU(2 GRU)	$\mathbf{88.2 \pm 8.5}$	$\mathbf{75.1 \pm 10.6}$	$\mathbf{91.4 \pm 11.2}$
Stacked GRU(3 GRU)	86.9 ± 10.2	72.7 ± 11.1	89.9 ± 9.2
Stacked GRU(4 GRU)	87.0 ± 8.4	72.4 ± 11.0	89.0 ± 12.3

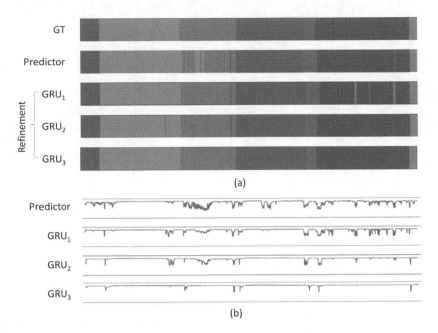

Fig. 7. (a) Qualitative results for the predictions in each GRU of the stacked GRU on the Cholec80 dataset. (b) Qualitative results for the probability sequences of the predictions in each GRU of the stacked GRU.

4.5 Impact of Disturbed Prediction Sequence

The disturbed prediction sequence for training the refinement stage is very important. In this section, we conduct ablative experiments by using different combinations of the disturbed prediction sequences and other augmentation techniques. For the multi-stage architecture, we use the stacked GRU with 3 GRUs as the refinement stage. Besides the mask-hard-frame type and the cross-validate type, we additionally designed the random-mask type and the normal-noise type. Different from mask-hard-frame type, the random-mask type randomly masks frames, no matter how important the masked frames are. The normal-noise type adds Gaussian noise with different standard deviations to the output of the predictor model to simulate the imperfect prediction results of the predictor model during the inference. Table 4 shows the results on the Cholec80 dataset. First, if we only use one type of disturbed sequence, the performances will drop due to the lack of training data. Meanwhile, we can observe that, the refinement stage trained with the random-mask type is not as good as the mask-hard-frame type. This demonstrates the importance of the location to add perturbations. Only when those frames that are not easily classified correctly during inference are masked, the sequence obtained is most similar to the real situation, and that's why the result of *mhf* is better than *rm*. In addition, when using one augmentation method, the way of adding noise performs better than the mask-hard-frame type and the random-mask type. And when combined with the other methods,

Table 4. Results of the multi-stage architectures trained with different augmentation techniques. Abbreviations: *cv* for the cross-validate type, *mhf* for the mask-hard-frame type, *rm* for the random-mask type, *nn* for the normal-noise type.

Method	Acc	JACC	Rec
nn($\sigma = 0.3$)	88.9 ± 6.3	72.8 ± 9.4	83.6 ± 9.6
nn($\sigma = 0.5$)	89.3 ± 6.4	74.2 ± 9.1	84.8 ± 8.2
nn($\sigma = 0.7$)	89.1 ± 6.4	73.2 ± 9.7	84.2 ± 9.2
cv	89.6 ± 5.6	70.4 ± 13.5	83.5 ± 13.2
mhf	88.7 ± 9.4	70.6 ± 9.4	81.8 ± 9.9
rm	86.5 ± 7.8	69.7 ± 10.9	82.4 ± 7.0
nn+cv	90.4 ± 5.3	72.9 ± 14.6	85.6 ± 10.6
nn+mhf	89.2 ± 7.2	71.9 ± 10.5	82.7 ± 9.7
nn+rm	87.4 ± 6.8	69.2 ± 11.5	82.6 ± 8.6
cv+mhf	91.5 ± 7.1	$\mathbf{77.2 \pm 11.2}$	86.8 ± 8.5
nn+cv+mhf	$\mathbf{92.0 \pm 5.3}$	77.1 ± 11.5	$\mathbf{87.0 \pm 7.3}$
cv+rm+mhf	91.0 ± 6.8	75.3 ± 13.4	85.4 ± 10.3
nn+cv+rm+mhf	91.7 ± 5.2	76.0 ± 12.6	86.4 ± 9.5

the normal-noise type improves the performance of other methods and achieves the SOTA performance. These results show that adding Gaussian noise to the output of the predictor model is also an effective perturbation, and it can play a complementary role to the disturbed prediction sequences.

4.6   Comparison with the SOTA Methods

In order to compare our solution with the SOTA methods, we use a stacked GRU with 3 GRUs and a stacked GRU with 2 GRUs as the refinement stage for the Cholec80 dataset and the M2CAI16 dataset, respectively. Table 5 and Table 6 show the results. Compared with the single-stage predictor causal TCN, multi-stage architecture trained with our solution largely boosts its performance. Noting that TeCNO [9] is also a multi-stage network, where causal TCN is used for both predictor stage and refinement stage. We can observe that, when simply applying the multi-stage network on surgical phase recognition task, the end-to-end training makes the refinement ability fall out of its wishes, TeCNO only achieves comparable results with the single-stage predictor causal TCN. However, when applying our strategy, the advantage of multi-stage network is revealed, which proves that our previous analysis above the multi-stage architecture and shows that our solution is effective. Besides, compared with these SOTA methods, our method is comparable to Trans-SVNet [21] on the Cholec80 dataset and much better than the other methods, while on the M2CAI16 dataset, our method outperforms all other methods.

Table 5. Comparison with the SOTA methods on the Cholec80 dataset.

Method	Acc	JACC	Rec
ResNet [8]	78.3 ± 7.7	52.2 ± 15.0	–
PhaseLSTM [25]	80.7 ± 12.9	64.4 ± 10.0	–
PhaseHMM [25]	71.1 ± 20.3	62.4 ± 10.4	-
EndoNet [13]	81.7 ± 4.2	–	79.6 ± 7.9
EndoNet-GTbin [13]	81.9 ± 4.4	–	80.0 ± 6.7
SV-RCNet [7]	85.3 ± 7.3	–	83.5 ± 7.5
OHFM [8]	87.0 ± 6.3	66.7 ± 12.8	–
TeCNO [9]	88.6 ± 2.7	–	85.2 ± 10.6
OperA [20]	85.8 ± 1.0	–	87.7 ± 0.7
Trans-SVNet [21]	90.3 ± 7.1	**79.3 ± 6.6**	**88.8 ± 7.4**
Causal TCN	88.8 ± 6.3	73.2 ± 9.8	84.9 ± 7.2
Ours	**92.0 ± 5.3**	77.1 ± 11.5	87.0 ± 7.3

Table 6. Comparison with the SOTA methods on the M2CAI16 dataset.

Method	Acc	JACC	Rec
ResNet [8]	76.3 ± 8.9	56.4 ± 10.4	–
PhaseLSTM [25]	72.5 ± 10.6	54.8 ± 8.9	–
PhaseHMM [25]	79.5 ± 12.1	64.1 ± 10.3	–
SV-RCNet [7]	81.7 ± 8.1	65.4 ± 8.9	81.6 ± 7.2
OHFM [8]	84.8 ± 8.0	68.5 ± 11.1	–
Trans-SVNet [21]	87.2 ± 9.3	74.7 ± 7.7	87.5 ± 5.5
Causal TCN	84.1 ± 9.6	69.8 ± 10.7	88.3 ± 9.6
Ours	**88.2 ± 8.5**	**75.1 ± 10.6**	**91.4 ± 11.2**

5   Conclusion and Future Work

In this paper, we observe that the end-to-end training manner makes the refinement ability of the multi-stage architecture fall out of its wishes. In order to solve the problem, we propose a new non end-to-end training strategy and explore different designs of multi-stage architectures for surgical phase recognition task. To minimize the distribution gap between the training and inference, we generate two types of disturbed sequences as the input of the refinement stage. In the future, we will explore other different predictor models, and apply our solution to other computer vision tasks where the multi-stage architecture is widely applied.

Acknowledgements. This work was partially supported by the Natural Science Foundation of China under contract 62088102. We also acknowledge the Clinical Medicine Plus X-Young Scholars Project, and High-Performance Computing Platform of Peking University for providing computational resources.

Author contributions. Fangqiu Yi and Yanfeng Yang : contribute equally to this work.

References

1. Bricon-Souf, N., Newman, C.R.: Context awareness in health care: a review. Int. J. Med. Inf. **76**, 2–12 (2007)
2. Bhatia, B., Oates, T., Xiao, Y., Hu, P.: Real-time identification of operating room state from video. Proc. Conf. Innov. Appl. Artif. Intell. **2**, 1761–1766 (2007)
3. Lin, H.C., Shafran, I., Murphy, T.E., Okamura, A.M., Yuh, D.D., Hager, G.D.: Automatic detection and segmentation of robot-assisted surgical motions. In: Medical Image Computing and Computer Assisted Intervention, pp. 802–810 (2005)
4. Newell, A., Yang, K., Deng, J.: Stacked hourglass networks for human pose estimation. In: European Conference on Computer Vision, pp. 483–499 (2016)
5. Wei, S.E., Ramakrishna, V., Kanade, T., Sheikh, Y.: Convolutional pose machines. In: IEEE Conference on Computer Vision and Pattern Recognition, pp. 4724–4732 (2016)
6. Farha, Y.A., Gall, J.: MS-TCN: multi-stage temporal convolutional network for action segmentation. In: IEEE Conference on Computer Vision and Pattern Recognition, pp. 3570–3579 (2019)
7. Jin, Y., et al.: SV-RCNet: workflow recognition from surgical videos using recurrent convolutional network. IEEE Trans. Med. Imaging **37**, 1114–1126 (2018)
8. Yi, F., Jiang, T.: Hard frame detection and online mapping for surgical phase recognition. In: Medical Image Computing and Computer Assisted Intervention (2019)
9. Czempiel, T., et al.: Tecno: surgical phase recognition with multi-stage temporal convolutional networks. In: Medical Image Computing and Computer Assisted Intervention, vol. 12263, pp. 343–352 (2020)
10. Lea, C., Flynn, M.D., Vidal, R., Reiter, A., Hager, G.D.: Temporal convolutional networks for action segmentation and detection. In: IEEE Conference on Computer Vision and Pattern Recognition, pp. 1003–1012 (2017)
11. Cho, K., van Merriënboer, B., Gulcehre, C., Bougares, F., Schwenk, H., Bengio, Y.: Learning phrase representations using RNN encoder-decoder for statistical machine translation. In: Proceedings of the Conference on Empirical Methods in Natural Language Processing, pp. 1724–1734 (2014)
12. Stauder, R., Ostler, D., Kranzfelder, M., Koller, S., Feußner, H., Navab, N.: The TUM lapchole dataset for the M2CAI 2016 workflow challenge. arxiv abs/1610.09278 (2016)
13. Twinanda, A.P., Shehata, S., Mutter, D., Marescaux, J., de Mathelin, M., Padoy, N.: EndoNet: a deep architecture for recognition tasks on laparoscopic videos. IEEE Trans. Med. Imaging **36**, 86–97 (2017)
14. Blum, T., Feußner, H., Navab, N.: Modeling and segmentation of surgical workflow from laparoscopic video. In: Medical Image Computing and Computer Assisted Intervention, pp. 400–407 (2010)
15. Padoy, N., Blum, T., Ahmadi, S.A., Feussner, H., Berger, M.O., Navab, N.: Statistical modeling and recognition of surgical workflow. Med. Image Anal. **16**, 632–641 (2012)
16. Tao, L., Zappella, L., Hager, G.D., Vidal, R.: Surgical gesture segmentation and recognition. In: Medical Image Computing and Computer Assisted Intervention, pp. 339–346 (2013)

17. Lalys, F., Riffaud, L., Morandi, X., Jannin, P.: Surgical phases detection from microscope videos by combining SVM and HMM. In: Menze, B., Langs, G., Tu, Z., Criminisi, A. (eds.) MCV 2010. LNCS, vol. 6533, pp. 54–62. Springer, Heidelberg (2011). https://doi.org/10.1007/978-3-642-18421-5_6

18. Padoy, N., Blum, T., Feussner, H., marie odile, B., Navab, N.: On-line recognition of surgical activity for monitoring in the operating room. In: Proceedings of the Conference on Innovative Applications of Artificial Intelligence, vol. 3, pp. 1718–1724 (2008)

19. Ban, Y., et al.: Aggregating long-term context for learning laparoscopic and robot-assisted surgical workflows. In: IEEE International Conference on Robotics and Automation, pp. 14531–14538 (2021)

20. Czempiel, T., Paschali, M., Ostler, D., Kim, S.T., Busam, B., Navab, N.: Opera: attention-regularized transformers for surgical phase recognition. In: Medical Image Computing and Computer Assisted Intervention, vol. 12904, pp. 604–614 (2021)

21. Gao, X., Jin, Y., Long, Y., Dou, Q., Heng, P.A.: Trans-svnet: accurate phase recognition from surgical videos via hybrid embedding aggregation transformer. Med. Image Comput. Comput. Assist. Interven. **12904**, 593–603 (2021)

22. Ramesh, S., et al.: Multi-task temporal convolutional networks for joint recognition of surgical phases and steps in gastric bypass procedures. Int. J. Comput. Assist. Radiol. Surg. **16**(7), 1111–1119 (2021). https://doi.org/10.1007/s11548-021-02388-z

23. Farha, Y.A., Gall, J.: MS-TCN: multi-stage temporal convolutional network for action segmentation. In: IEEE Conference on Computer Vision and Pattern Recognition, pp. 3575–3584 (2019)

24. He, K., Zhang, X., Ren, S., Sun, J.: Deep residual learning for image recognition. In: IEEE Conference on Computer Vision and Pattern Recognition (CVPR), pp. 770–778 (2016)

25. Twinanda, A.P., Mutter, D., Marescaux, J., Mathelin, M.D., Padoy, N.: Single- and multi-task architectures for surgical workflow challenge at M2CAI 2016. arXiv:1610.08844 (2016)

FunnyNet: Audiovisual Learning of Funny Moments in Videos

Zhi-Song Liu[1], Robin Courant[2(✉)], and Vicky Kalogeiton[2]

[1] Caritas Institute of Higher Education, Hong Kong, China
zhisong.liu@connect.polyu.hk
[2] VISTA, LIX, Ecole Polytechnique, Paris, IP, France
{robin.courant,vicky.kalogeiton}@lix.polytechnique.fr
http://www.lix.polytechnique.fr/vista/projects/2022_accv_liu

Abstract. Automatically understanding funny moments (i.e., the moments that make people laugh) when watching comedy is challenging, as they relate to various features, such as facial expression, body language, dialogues and culture. In this paper, we propose FunnyNet, a model that relies on cross- and self-attention for both visual and audio data to predict funny moments in videos. Unlike most methods that focus on text with or without visual data to identify funny moments, in this work in addition to visual cues, we exploit audio. Audio comes naturally with videos, and moreover it contains higher-level cues associated with funny moments, such as intonation, pitch and pauses. To acquire labels for training, we propose an unsupervised approach that spots and labels funny audio moments. We provide experiments on five datasets: the sitcoms TBBT, MHD, MUStARD, Friends, and the TED talk UR-Funny. Extensive experiments and analysis show that FunnyNet successfully exploits visual and auditory cues to identify funny moments, while our findings corroborate our claim that audio is more suitable than text for funny moment prediction. FunnyNet sets the new state of the art for laughter detection with audiovisual or multimodal cues on all datasets.

1 Introduction

We understand the world by using our senses, especially in multimedia area. All signals can stimulate one's feelings and reactions. Funniness is universal and timeless: in 1900 BC Sumerians wrote the first joke and it is still funny nowadays. However, whereas humans can easily understand funny moments, even from different cultures and eras, machines do not. Even though the number of interactions between humans and machines is growing fast, identifying funniness is still a brake on making these interactions spontaneous. Actually, understanding funny moments is a complex concept since they can be purely visual, purely auditory, or they can mix both cues: there is no recipe for the perfect joke.

Z.-S. Liu and R. Courant—These authors contributed equally to this work.

Supplementary Information The online version contains supplementary material available at https://doi.org/10.1007/978-3-031-26316-3_26.

L. Wang et al. (Eds.): ACCV 2022, LNCS 13844, pp. 433–450, 2023.
https://doi.org/10.1007/978-3-031-26316-3_26

Mia: Three tomatoes are walking down the street -- a poppa tomato, a momma tomato, and a little baby tomato. Baby tomato starts lagging behind. Poppa tomato gets angry, goes over to the baby tomato, and squishes him... and says, ... "Catch up."

Fig. 1. What is funny? Audio cue along with visual frame and facial data are a rich source of information for identifying funny moments in videos. Video scene from Pulp fiction, 1994, source video https://www.youtube.com/watch?v=4L5LjjYVsHQ

Recently, some works try to understand what is a joke, humor and funny moments [1,2]. These rely solely on text and only a couple of them use also videos [3,4]; in such cases, video is always combined with text. However, these works are limited as text (transcripts, subtitles) does not come naturally with videos, but it depends on external pipelines, which depend themselves on other factors such as accents, sound quality, simultaneous audios, language, annotation. Thus, such works lack flexibility and can hardly be used in the wild.

In contrast, audio comes naturally with videos, and it contains crucial and complementary cues, such as tones, pauses, pitch, pronunciation, background noises [5,6]. Indeed, when people talk, not only what they say matters, but also how they say it. In turn, the visual content is also very important. For instance, depending on the context, the very same phrase said by the same person can be funny or sad (Fig. 1). Yet, facial expressions, body gestures and scene context help better understand the sense of a phrase, thus impacting funniness.

Therefore, in this paper, we introduce FunnyNet, an audiovisual model for funny moment prediction. It consists of three encoders: (a) visual that encompasses the global context information of a scene, (b) face for representing the facial expressions of individuals, and (c) audio that captures voice and language effects; and the Cross Attention Fusion (CAF) module, i.e., a new module that learns cross-modality correlations hierarchically so that features from different modalities can be combined to form a unified feature for prediction. Thus, FunnyNet is trained to learn to embed all cross-attention features in the same space via self-supervised contrastive learning [7], in addition to classifying clips as funny or not-funny. To obtain labelled data, we exploit the laughter that naturally exists in sitcom TV shows. We define as 'funny-moment' any n-second clip followed by laughter; and 'not-funny' the clips not followed by laughter. To extract laughter, we propose an unsupervised labelling approach that clusters audio segments into laughter, music, voice and empty, based on their waveform difference[1]. Moreover, we enrich the Friends dataset with laughter annotations.

Our extensive experimentation and analysis show that combining visual with audio cues is suitable for funny-moment detection; specifically, our findings

---

[1] Note that we use the laughter solely as indicator for data labelling, but the laughter is not the included in the audio segments of FunnyNet. Once FunnyNet is trained, it can detect funny moments in any video, with or without laughter.

demonstrate that audio results in superior performances than language, hence revealing its superiority for the task and supporting our intuition that audio captures higher-level cues than subtitles. Moreover, we compare FunnyNet to the state of the art on five datasets including sitcoms (TBBT, MHD, MUStARD, and Friends) and TED talks (UR-Funny), and show that it outperforms all other methods for all metrics and input configurations. We also apply FunnyNet on data from other domains, i.e., movies, stand-up comedies, and audiobooks. For quantitative evaluation, we apply FunnyNet on a sitcom without canned laughter manually annotated. It shows that FunnyNet predicts funny moments without fine-tuning, revealing its flexibility for funny-moment detection in the wild.

Our contributions are: (1) We introduce FunnyNet, an audiovisual model for funny moment detection. It combines features from various modalities using the proposed CAF module relying on cross and self-attention; (2) Extensive experiments and analysis highlight that FunnyNet successfully exploits audiovisual cues, and show that audio is better suited than text for funny-moment detection; (3) FunnyNet achieves the new state of the art on five datasets, and we also demonstrate its flexibility by applying it to in-the-wild videos.

2   Related Work

Sarcasm and Humor Detection. Sarcasm and humor share similar styles (irony, exaggeration and twist) but also differ from each other in terms of representation. Sarcasm usually relates to dialogues; hence, most methods detect sarcasm by processing language using human efforts. For instance, [8] collect a speech dataset from social media using the hashtag and manual labeling, while others [9,10] study the acoustic patterns related to sarcasm, like slower speaking rates or higher volumes of voice. In contrast, a humorous moment is defined as the moment before laughter [6,11]. Hence, such methods [4,6,11–13] process audios to extract laughter for labeling. Nevertheless, for prediction, most such approaches focus solely on language models [1,2] or on multiple cues including text [11,13]. For instance, LaughMachine [4] propose vision and language attention mechanisms, while MSAM [3] combine self-attention blocks and LSTMs to encode vision and text. [14] use first an advanced BERT [15] model to process long-term textual correlation and then vision for the prediction. Following this, [16] propose a Multimodal Adaptation Gate to efficiently leverage textual cues to explore better representation for sentiment analysis. Few methods also explore audio. For instance, MUStARD [6] and URFUNNY [11] process text, audio and frames using LSTM to explore long-term correlations, while HKT [13] classifies language (context and punchline) and non-verbal cues (audio and frame) to learn cross-attention correlations for humor prediction. They combine audio with other information (video and texts) in a simple feature fusion process without investigating the inter-correlations in depth. Specifically, they stack multimodal features to learn the global weighting parameters without considering the biases in different domains. In contrast, we believe that funny scenes can be triggered by mutual signal from multimodalities; hence, we explore the cross-domain agreement of cues with contrastive training. Moreover, instead of text, FunnyNet relies solely on audiovisual cues, as audio comes naturally with videos, and it contains all essential cues for funny moment prediction.

Sound Event Detection detects which and when sound events happen in audios. Most attempts either rely on annotated data [17] or use source separation [18]. The input plays a crucial role, and most methods use Mel spectrograms [19–22] instead of audio waveforms. *Laughter detection* focuses on one specific event: laughter. For this, some methods rely on physiological sensors [23,24], while others [25,26] follow the supervised paradigm to train detectors. In contrast, our laughter detector is an unsupervised, robust and straightforward labelling method that exploits multichannel audio specificities.

Multimodal Signal is processed with models like LSTM [27], GRN [28], ViT [29] and VQVAE [30] and is studied for various tasks. For instance, [31,32] recognize the facial movements to separate the speaker's voice in the audio. [33,34] temporally align the audio and video using attention to locate the speaker. Several methods extend this to other applications, including speech recognition [35], audio-image retrieval [36,37], audiovisual generation [38,39], video-text retrieval [40], human replacement [41], visual question answering [42], and affect analysis [43]. Recently, Video Transformers models [44,45] showed improved accuracy on various video tasks, in particular for classification [44–47]. Their self/cross- attention operation provides a natural mechanism to connect multimodal signals. Thus, some works exploit this to account for multiple modalities, such as inter and intra cross-attention in [48], contrastive cross-attention in [49], iterative cross-attention in [47,50] or bottlenecks in [51] for various tasks, such as summarization [52], retrieval [53,54], audiovisual classification [51], predicting goals [55]. [48,49] iteratively apply self and cross-attention to explore correlations among modalities. Instead, FunnyNet both fuses all modalities and in parallel learns the cross-correlation among different modalities; this avoids any biases caused by one dominant modality.

3 Method

Here, we present FunnyNet, its training process and losses (Sects. 3.1–3.2). For training labels, we propose an unsupervised laughter detector (Sect. 3.3).

Overview. FunnyNet consists of (a) three encoders: visual with videos as input, face with face tracks as input, and audio with audios as input, and (b) the proposed Cross-Attention Fusion (CAF) module, which explores cross- and intra-modality correlations by using cross- and self-attentions in the encoders outputs. Then, the fused feature is fed to a binary classifier (Fig. 2). For comparison, FunnyNet can take text as extra input with a text encoder. It is trained to embed all modalities in the same space via self-supervised contrastive loss and to classify clips as funny or not. For training, we exploit laughter that naturally exists in TV Shows: we define as 'funny-moment' any audiovisual snippet followed by laughter; and 'not-funny' any audiovisual snippet not followed by laughter.

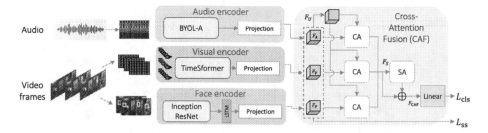

Fig. 2. FunnyNet. Given audio-visual clips, it predicts funny moments in videos. It consists of the audio (*blue*), visual (*green*), and face (*red*) encoders, whose outputs pass through the Cross Attention Fusion (*CAF*), which consists of cross-attention (*CA*) and self-attention (*SA*) for feature fusion. It is trained to embed all modalities in the same space via self-supervision (L_{ss}) and to classify clips as funny or not-funny (L_{cls}). (Color figure online)

3.1   FunnyNet Architecture

Audio Encoder. FunnyNet takes as input audio snippets X_{audio} in the form of Mel spectrogram[2]. It is fed into the audio encoder, i.e., BYOL-A [21] to obtain a 1D feature vector. Finally, we use a Projection module to map it to a 512-D vector for final prediction: $\mathbf{F}_A \in \mathbb{R}^{512}$.

Visual Encoder. It processes video frames with TimeSformer [45]. Its inputs are patches of size 16×16, partitioned from eight consecutive input frames X_{visual}. Unlike TimeSformer where the representation is obtained by the 'classification token', we obtain the representation by average pooling features from all patches, thus forming a 768-D vector; then, we use a Projection module to map it to a 512-D vector: $F_V \in \mathbb{R}^{512}$. Video context complements audio (or subtitles) to have richer content [11]. If there is no sound, hence no subtitle, visual cues can also provoke laughter.

Face Encoder. Face features capture local cues to enrich the visual representation. We use InceptionResNet [56,57] to extract up to eight faces per frame, that we then process with a LSTM to form a 512-D vector $\mathbf{F}_F \in \mathbb{R}^{512}$. Note, instead of more advanced models [58–60], we use InceptionResNet because of its robustness and efficiency [61].

Text Encoder. For a fair comparison to the state of the art, we also experiment with a text encoder that uses BERT [15] to extract key features and feeds them to a LSTM to model temporal correlations (more details in supplementary).

Projection Module. It consists of a linear layer followed by batch normalization, a tanH activation function and another linear layer. It takes input features from each encoder and projects them in a common 512-D feature space.

Cross-Attention Fusion (CAF) learns the cross-domain correlations among vision, audio and face (yellow box Fig. 2). It consists of (a) three cross-attention (CA) and (b) one self-attention (SA) modules, described below:

---

[2] Mel spectrogram is a 2D acoustic time-frequency representation of sound.

(a) **Cross-attention** is used in cross-domain knowledge transfer to learn across-cue correlations by attending the features from one domain to another [48,62,63].

In CAF, it models the relationship among vision, audio, and face features. We stack all features as $\mathbf{F}_U \in \mathbb{R}^{3 \times 512}$, and then feed $\mathbf{F}_U$ into three cross-attention modules to attend to vision, face, and audio, respectively (Fig. 2). Next, the scaled attention per modality is computed as $\sigma\left(\frac{\mathbf{Q}_U \mathbf{K}_i^T}{\sqrt{d}}\right) \mathbf{V}_i$, where $i = \{V, F, A\}$ for {vision, face, audio}, and σ the softmax. The query comes from the stacked features: $\mathbf{Q}_U = \mathbf{W}^{Q_U} \mathbf{F}_U$, while the key and value come from a single modality as $\mathbf{K}_i = \mathbf{W}^{K_i} \mathbf{F}_i$, and $\mathbf{V}_i = \mathbf{W}^{V_i} \mathbf{F}_i$. Next, we obtain three cross-attentions and sum them to a unified feature $\mathbf{F}_S$ as:

$$F_S = \sum_{i \in \{V,F,A\}} \sigma\left(\frac{\mathbf{Q}_U \mathbf{K}_i^T}{\sqrt{d}}\right) \mathbf{V}_i. \tag{1}$$

(b) **Self-attention** computes the intra-correlation of the F_S features, which are further summed with a residual F_S as:

$$F_{CAF} = \mathbf{F}_S + \sigma\left(\frac{\mathbf{Q}_S \mathbf{K}_S^T}{\sqrt{d}}\right) \mathbf{V}_S, \tag{2}$$

where $\mathbf{Q}_S = \mathbf{W}^{Q_S} \mathbf{F}_S$, $\mathbf{K}_S = \mathbf{W}^{K_S} \mathbf{F}_S$, $\mathbf{V}_S = \mathbf{W}^{V_S} \mathbf{F}_S$. Finally, we average F_{CAF} tokens and feed it to a classification layer.

Discussion. CAF differs to existing methods [48,62] in the computation of the cross attention. Using stacked features F_U to attend to each modality Q_U brings three benefits: (a) it is order-agnostic: for any modality pair we compute cross-attention once, instead of twice by interchanging queries and keys/values; this results in reduced computation; (b) each modality serves as a query to search for tokens in other modalities; this brings rich feature fusion; and (c) it generalizes to any number of modalities, resulting in scalability.

3.2   Training Model and Loss Functions

Positive and Negative Samples. To create samples, we exploit the laughter that naturally exists in episodes. We define as 'funny' any n-sec clip followed by laughter; 'not-funny' any n-sec clip not followed by laughter. More formally, given a laughter at timestep (t_s, t_e), we extract a n-sec clip at $(t_s - n, t_s)$ and we split it into audio and video. For each video, we sample n frames (1 FPS). For the audio, we resample it 16000 Hz and transform it to Mel spectrogram. Thus, each sample corresponds to n sec and consists of a Mel spectrogram for the audio and a n-frame long video. In practice, we use 8-s clips as the average time between two canned laughters, and it also leads to better performances (ablations of n-sec clips and n-frames per clip in supplementary). Note that we clip the audio based on the starting time of the laughter so the positive samples do not include any laughter.

Self-supervised Contrastive Loss. To capture 'mutual' audiovisual information, we solve a self-supervised synchronization task [64–66]: we encourage visual features to be correlated with true audios and uncorrelated with audios

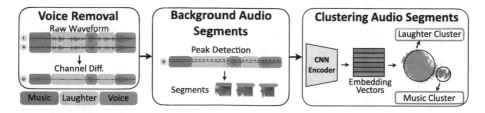

Fig. 3. Proposed laughter detector. It takes raw waveforms as input and consists of (i) removing voices by subtracting channels (here, the audio is stereo with 2 channels), (ii) detecting peaks, and (iii) clustering audios to music and laughter

from other videos. Given the i-th pair of visual v^i and true audio features a^i and N other audios from the same batch: $a_1, ..., a_N$ we minimize the loss [7,67,68]:

$$L_{\text{cotrs}} = -\log \frac{\exp(S(v^i, a^i)/\tau)}{\sum_{j=1}^{N} \exp(S(v^i, a^j)/\tau)}, \tag{3}$$

where S the cosine similarity and τ the temperature factor. Equation 3 accounts for audio and visual features. Here, we compute the contrastive loss between all three modalities, i.e., visual-audio, face-audio, and visual-face. Thus, our self-supervised loss is $L_{\text{ss}} = -\frac{1}{3}\left(L_{\text{cotrs}}^{v^i, a^i} + L_{\text{cotrs}}^{v^i, f^i} + L_{\text{cotrs}}^{f^i, a^i}\right)$.

Final Loss. FunnyNet is trained with a Softmax loss Y_{cls} to predict if the input is funny or not, and the L_{ss} to learn 'mutual' information across modalities. Thus, the final loss is: $L = \lambda_{\text{ss}}L_{\text{ss}} + \lambda_{\text{cls}}L_{\text{cls}}$, where λ_{ss}, λ_{cls} the weighting parameters that control the importance of each loss.

3.3   Unsupervised Laughter Detection

To detect funny moments automatically, we design an unsupervised laughter detector consisting of three steps (Fig. 3). **(i) Remove Voices.** Background audios include sounds, music, laughter; instead, voice (speech) is part of the foreground audio. We remove voices from audios by exploiting multichannel audio specificities. Given raw waveform audios, when the audio is stereo (two channels), the voices are centered and are common in both channels [69]; hence, by subtracting the channels, we remove the voice and keep the background audio. In surround tracks (six channels), we remove the voice channel [69] and keep the background ones. **(ii) Background Audios.** The waveforms from (i) are mostly empty with sparse peaks that correspond to audio: laughter and music. To split them into background and empty segments, we use an energy-based peak detector[3] that detects peaks based on the computed waveform energy. Then, we keep background segments and convert them to log-scaled Mel spectrograms. **(iii) Cluster Audio Segments.** For each laughter and music segment, we extract features using a self-supervised pre-trained encoder. Then, we cluster all audio segments using K-means to distinguish the laughter from the music ones.

---

[3] https://github.com/amsehili/auditok.

4 Datasets and Metrics

Datasets. We use five datasets (more details in suppl.). **The Big Bang Theory (TBBT)** dataset [4] contains 228 episodes of *TBBT* TV show: (183,23,22) for (train,val,test). All episodes come with video, audio and subtitles, labelled as humor (or non) if followed (or not) by laughter. **Multimodal Humor Dataset (MHD)** [3] contains episodes from *TBBT*, with 110 episodes split (84,6,20) for (train,val,test) (disjoint splits to TBBT). It contains multiple modalities; the subtitles are tagged as humor (or not). **MUStARD** [6] contains 690 segments from 4 TV shows with video-audio-transcript labelled as sarcastic or not. **UR-Funny** [11] contains 1866 TED-talk segments with video-audio-transcript labelled as funny or not. **Friends** [70,71] contains 25 episodes from the third season of *Friends* TV show, split as (1-15,16-20,21-25) for (train,val,test). Each episode contains video, audio, and face tracks. Here, we enrich it with manually annotated laughter time-codes, i.e., starting-ending time of laughter.

Metrics. To evaluate **FunnyNet**, we use classification accuracy (Acc) and F1 score (F1). For **laughter detector**, we use sample-scale at detection level and frame-scale at temporal level to compute precision (Pre), recall (Rec) and F1.

5 Experiments

We provide experiments for FunnyNet. In supplementary, we include more results on feature modalities, modules, impacts, time windows of inputs (n-sec inputs, n frames), losses, automatic/manual laughter, datasets, effect of fine-tuning, metrics, complexity, discussion of laughter detector, and videos.

Implementation Details. We train FunnyNet using Adam optimizer with a learning rate of 10^{-4}, batch size of 32 and Pytorch [72]. At training, we use data augmentation: for frames, we randomly apply rotation and horizontal/vertical flipping, and randomly set the sampling rate to 8 frames; for audios, we apply random forward/backward time shifts and random Gaussian noises. **Setting.** In our experiments, we train FunnyNet on Friends. For MUStARD/UR-Funny, we fine-tune FunnyNet on their respective train sets. For TBBT/MHD, we fine-tune it only with a subset of the training set from TBBT (32 random episodes).

5.1 Comparison to the State of the Art

Here, we evaluate FunnyNet on five datasets: TBBT, MHD, MUStARD, UR-Funny and Friends and compare it to the state of the art: MUStARD [6], MSAM [3], MISA [14], HKT [13] and LaughM [4]. Table 1 reports the results (including random, positive and negative baselines) for both metrics. We indicate the modalities each method uses as V: video, F: face, T: text and A: audio.

Overall, we observe that the proposed audiovisual FunnyNet (V+F+A) outperforms all methods for all metrics on all five datasets. Fsor TBBT it outperforms the LaughM by a notable margin of +5% for F1 and Acc, while for MHD it outperforms MSAM by 3% in F1 and 7% in Acc and LaughM by 3% in Acc.

Table 1. Comparison to the state of the art on five datasets. Modalities used per method A: audio, V: visual frames, F: Face, T: text. [†]Reproduced results: we use the exact model as in [4], pre-train it on Friends and fine-tune it on the other datasets

Method / Metrics	TBBT		MHD		MUStARD		UR-Funny		Friends	
	F1	Acc	F1	Acc	F1	Acc	F1	Acc	F1	Acc
Random	46.3	50.0	56.1	50.9	48.3	48.7	50.2	50.2	51.0	51.0
All positive	60.3	43.2	75.6	60.8	66.7	50.0	75.4	50.7	66.7	50.0
All negative	0.0	56.8	0.0	39.2	0.0	50.0	0.0	49.3	0.0	50.0
MUStARD 2019 (V+A+T) [6]	-	-	-	-	71.7	71.8	-	-	-	-
MSAM 2021 (V+T) [3]	-	-	81.3	72.4	-	-	-	-	-	-
MISA 2020 (V+A+T) [14]	-	-	-	-	-	66.2	-	69.8	-	-
HKT 2021 (V+A+T) [13]	-	-	-	-	-	79.4	-	77.4	-	-
LaughM[†] 2021 (T) [4]	64.2	70.5	**86.5**	76.3	68.6	68.7	71.9	67.6	74.7	59.8
FunnyNet: V+F+A	**69.6**	**74.0**	84.0	**79.3**	**81.4**	**81.0**	**83.7**	**78.0**	**86.8**	**84.8**
FunnyNet: V+A+T	73.8	75.8	83.4	78.6	79.5	79.9	84.1	79.9	88.2	85.8
FunnyNet: V+F+T	**76.0**	69.5	75.9	69.8	75.2	76.3	82.3	73.0	81.3	76.2
FunnyNet: V+F+A+T	75.9	**78.3**	85.2	**79.6**	**83.2**	**82.0**	**84.4**	**80.2**	**88.8**	**86.4**

For MUStARD and UR-Funny, the results are more conclusive as we compare against several methods that use different modalities; in all cases, FunnyNet outperforms MUStARD, MISA, HKT, LaughM by 10–12% in F1 and 2–15% in Acc for MUStARD and 20% in F1 and 1–10% in UF-Funny. For Friends, we observe similar patterns, where we outperform LaughM by 11% in F1 and 25% in Acc. These results confirm the effectiveness of FunnyNet compared to other methods.

Our remarks are: First, FunnyNet performs best among all methods that leverage audio (MUStARD, MISA, HKT), even without using text. Second, the performance in the out-of-domain UR-Funny is significantly high. Third, for TBBT and MHD our results are much less optimized than the ones from LaughM or MSAM, as we do not have access to the exact same test videos as either work, so inevitably there are some time shifts or wrong labels[4] and we use much fewer training data (32 vs 183 episodes in LaughM vs 84 episodes in MHD). These highlight that FunnyNet is an effective model for funny moment detection.

FunnyNet Using Text. For fair comparison, we explore a FunnyNet version that leverages subtitles in addition to audiovisual cues (FunnyNet: V+A+T). We compare it against MUStARD, MISA, and HKT that use the same modalities and observe that FunnyNet (V+A+T) outperforms them all by a large margin (1–14% for all metrics and datasets). This shows that the performance boost from FunnyNet stems from the superior architecture and the adequate modality fusion, rather than the difference in input modalities.

5.2 Analysis of Unsupervised Laughter Detector

We compare our laughter detector to the state-of-the-art LD [25] used in [6] and RLD [26]. Table 2 reports the results on Friends. We observe that overall, our

---

[4] The label time shift is 0.3–1s on TBBT and 0.3–2s on v2.

Table 2. Laughter detection evaluation on Friends. We compare 'Ours' to two audio feature extractors

	Temporal				Det IoU = 0.3			Det IoU = 0.7		
	Acc	Pre	Rec	F1	Pre	Rec	F1	Pre	Rec	F1
LD [25]	43.64	35.70	**98.95**	52.28	25.69	22.09	23.35	4.02	3.73	3.82
RLD [26]	74.46	58.91	61.98	59.69	66.15	53.71	59.10	18.45	15.04	16.52
Ours Wav2CLIP[73]	77.56	64.49	63.66	63.70	91.25	61.23	73.07	49.74	33.45	39.89
Ours BYOL-A[21]	**85.97**	**76.94**	79.38	**77.81**	**94.57**	**82.25**	**87.83**	**54.07**	**47.11**	**50.27**

Table 3. Ablation of modalities of FunnyNet on the test set of all five datasets (A: audio, V: visual frames, F: face)

Modalities			TBBT		MHD		MUStard		URFunny		Friends	
A	V	F	F1	Acc	F1	Acc	F1	Acc	F1	Acc	F1	Acc
✓	-	-	68.9	63.2	64.9	67.5	65.2	63.8	68.2	64.7	73.7	66.7
-	✓	-	65.0	58.9	66.3	66.5	64.6	60.2	67.5	60.4	72.9	63.4
-	-	✓	64.8	59.4	66.8	67.7	64.7	61.1	67.2	61.2	72.9	62.1
✓	✓	-	**70.3**	72.8	79.9	73.1	79.0	77.5	80.9	76.8	84.2	81.1
✓	-	✓	**70.3**	72.9	80.5	73.9	80.4	78.9	82.9	**79.4**	83.9	82.2
✓	✓	✓	69.6	**74.0**	**84.0**	**79.3**	**81.4**	**81.0**	**83.7**	78.0	**86.8**	**84.8**

Table 4. Ablation of CAF of FunnyNet on friends test set

CAF		A+V		A+F		A+V+F	
Self	Cross	F1	Acc	F1	Acc	F1	Acc
-	-	**84.51**	80.65	82.62	78.99	85.97	84.02
✓	-	84.10	80.64	**84.01**	81.01	86.57	84.13
-	✓	83.13	81.05	83.71	81.33	86.61	84.52
✓	✓	84.19	**81.14**	83.91	82.22	**86.79**	**84.75**
MMCA [48]		83.56	81.05	83.44	82.06	86.71	84.36
CoMMA [49]		83.71	81.08	83.79	**82.24**	86.69	84.63

detector outperforms both supervised ones. We also examine the efficiency of BYOL-A [21] and Wav2CLIP [73] encoders, where we observe that using BYOL-A outperforms Wav2CLIP, due to the richer audio representation capacity. From the laughters, we make three notes: (a) most false positives are unfiltered sounds not well separable by K-means; (b) most false negatives are intra-diegetic laughter, which is less loud, and hence, less detectable; (c) when the music is superposed with the laughter (e.g. party) the peak detector fails.

5.3   Ablation of FunnyNet

Modalities. Table 3 reports the ablation of all modalities of FunnyNet on five datasets. Using audio alone produces better results than any visual modality alone, underlying that audio is more suitable than visual cues for our task, as it encompasses the way of speaking (tone, pauses). Combining modalities outperforms using single ones: combining audio and visual increases the F1 by 3–12% and the Acc by 6–17%. This is expected as modalities bring complementarity and their combination helps discriminate funny moments. Audio+face leads to smaller boosts than audio+visual, as frames capture better than (low-level) faces global information. Overall, using all modalities achieves the best performance.

Cross-Attention Fusion (CAF). Table 4 reports results with various cross- and self-attention fusions in CAF. We observe that including either self- or cross-attention (second, third rows) brings improvements over not having any (first row), indicating that they enhance the feature representation. The fourth row shows that using them both for feature fusion leads to the best performance. For completeness, we also compare CAF against the state of the art MMCA [48] and

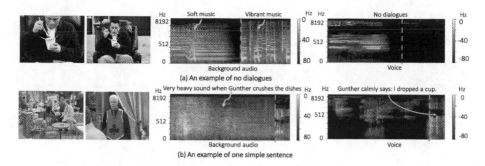

Fig. 4. Audio vs Text for funny-moment detection on Friends. Relying solely on voices (subtitles) fails when nobody is speaking; the audio, however, may succeed. (*a*) humorous background music without voices, (*b*) abrupt background sound (plates smashing) accompanied by a simple dialogue 'I dropped a cup'

CoMMA [49]. All CAF, MMCA and CoMMA use self and cross-attentions jointly for feature extraction. Their main difference is that both MMCA and CoMMA first use self-attention to individually process each modality, then concatenate all modalities together and process them using cross-attention to output the final feature representation. Instead, CAF uses cross-attention to gradually fuse one modality with the rest of modalities to fully explore cross-modal correlations. The results (fifth, last rows) show that CAF outperforms MMCA [48] and CoMMA [49] by 0.1~ 0.4 in F1 score and 0.03~ 0.2 in accuracy. This reveals the importance of the gradual modality fusion and hence the superiority of CAF.

6   Analysis of FunnyNet

6.1   Audio vs Subtitles

Instead of subtitles [4], FunnyNet relies on audio, as it (1) encodes mutual information to the text, (2) does not encode only words, but also the way of speaking (pauses, pitch and so on), and (3) can succeed when a scene is funny, yet nobody is speaking, by exploiting background sounds.

Quantitative Analysis. Table 1 reports FunnyNet results with the combination of text and audiovisual modalities. By comparing visual-textual to visual-audio features (V+F+T vs V+F+A) we observe that in most cases audiovisual cues perform better than the visual-textual ones (4–8% in both metrics for all datasets); this corroborates our claim that audio identifies better than text funny moments in videos. The last row reports results when combining all modalities (V+F+A+T). This further outperforms all other FunnyNet results, indicating that FunnyNet can efficiently exploit all sources of signals for funny prediction.

Qualitative Analysis. Figure 4 shows some scene frames from Friends and their audio, separated into background (middle) and vocal/voice (bottom) [74]. (a) Joey is trying to use chopsticks to pick up the food but end up dropping it. There is no voice, but vibrant and active music with a simpler rhythm to Joey's action.

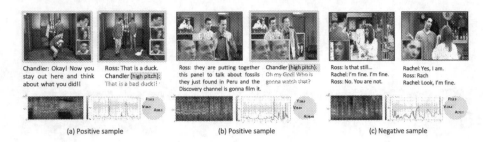

(a) Positive sample   (b) Positive sample   (c) Negative sample

Fig. 5. Visualization of (a,b) funny, (c) non-funny predictions on Friends. We show the audio and visual (frame and faces) inputs, the learned average weights of cross-attentions from CAF (*pie chart*), and the subtitles (for better understanding)

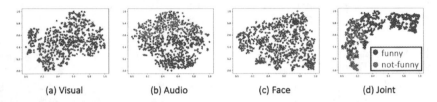

(a) Visual   (b) Audio   (c) Face   (d) Joint

Fig. 6. t-SNE visualization of embeddings on Friends for (a) visual, (b) audio, (c) face, (d) all modalities. We show positive (*blue*) and negative samples (*red*). (Color figure online)

FunnyNet correctly predicts the scene as funny by leveraging audio. (b) The extreme loud sound of several smashing dishes is followed by Gunther appearing and calmly saying "I drop A cup". Unlike text, the audio correctly detects the contradiction between the smashing sound and the calm words, hence predicting the scene as funny. In both cases, using text results in incorrect predictions, whereas audio successfully leverages no verbal cues and correctly predicts the scenes as funny, showing that audio better than text addresses such cases.

6.2 FunnyNet Architecture

Modality Impact. To visualize the impact of modalities, we compute the average attention values on the three CA modules (CA boxes in Fig. 2) and then, show the average weights for each modality in the pie chart of Fig. 5. For this, we show two positive and one negative samples on Friends with frames, face (on frames), and audio spectrogram (left) and pitch (right). Note, FunnyNet does not use subtitles; we show dialogues only for a better understanding. We observe that the contribution of each modality varies; the commonality though is that audio contributes more than half, followed by visual and finally face. When characters smile ('Chandler' in b), the contribution of face increases, indicating the importance of facial expressions, whereas the 'over the shoulder' shot of (c) shows that face posture play a small role. Moreover, the dramatic peaks of the audio pitch in (a,b) show that they are associated with funny punchline ('That is a bad duck'), whereas the smooth ones in (c) imply a 'normal scene'.

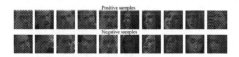

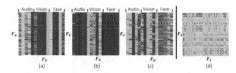

Fig. 7. Top-5 predicted positive and negative faces of 'Chandler' and 'Rachel' from Friends

Fig. 8. CAF attention maps on Friends. (a,b,c) CA between F_U and audio, vision and face; (d) SA on F_U

Fig. 9. Failure cases on Friends

Fig. 10. Funny moments in the wild

Feature Visualization. Figure 6 shows the t-SNE [75] visualization of features: (a-c) visual, audio, face, (d) all with blue for funny and red for not-funny samples. All single features, and in particular the visual and facial ones, scattered around the centre of the 2D space without clear boundaries between positives and negatives. However, the joint embedding shows clear separation between funny and not-funny, thus revealing the effectiveness of FunnyNet.

Impact of Face Encoder. To examine its effect, in Fig. 7 we depict the top-5 detected faces from positive (top) and negative (bottom) samples of two characters from Friends. We observe that all positives have rich expressions (yell, stare, smile, open mouth), while the negatives are either of bad quality (no useful features) or show neutral expression, thus indicating scenes without funniness.

Impact of CAF Module. To examine the effect of CAF, we visualize in Fig. 8 the learned attention maps: red indicates higher, and blue lower attention. (a,b,c) display the cross-attention between fused F_U and (a) audio, (b) visual, (c) face features. Since F_U is stacked from audio, vision, face, we observe that each modality highly attends to itself. We also observe in (a), both the visual and the face in F_U fire with the audio, thus indicating that FunnyNet captures correlations between audio and visual expressions or movements (e.g. character laughing). Then, in (c), all modalities attend to the face features, thus revealing their mutual information. Finally, (c) displays the self-attention map between F_S, where we observe that F_S attends to all tokens with different weights.

Failure Analysis. We note two groups of failure cases. First, when characters laugh sarcastically is not always funny; but, all modalities incorrectly, yet confidently tag them as funny. Figure 9-(a) shows this, where 'Rachel' laughs sarcastically, which is not funny (subs 'ha ha'). We wrongly predict it as positive. Second, visual cues fail in dark scenes; thus, we rely on audio. Figure 9-(b) shows a night scene with no clear faces and dark frames, where FunnyNet uses mostly the non-discriminative audio; hence, we wrongly predict the scene as negative.

6.3   Funny Scene Detection in the Wild

We show applications of FunnyNet in videos from other domains (more in suppl.).
1. Sitcoms without Canned Laughter. We collect 9 episodes of the first season (~180 min) of *Modern Family* (Lloyd and Levitan, 2009)[5] without canned laughter. We manually annotate as positive every punchline that could lead to laughter, resulting in 453 positives (we will make them available). We apply FunnyNet on the 8-s preceding funny moments, resulting in an accuracy of 55.4% vs 50% for random. Our remarks are: (i) FunnyNet is not fine-tuned on this data, and (ii), *Modern Family* differs from other sitcoms with live audience as characters are not reacting to punchlines. Thus, our results indicate that FunnyNet is capable of detecting funny-moments in out-of-domain cases. Figure 10 (a) shows a correctly predicted funny moment between two characters who vary their speech rhythm and tones. **2. Movies with Diverse Funny Styles.** Figure 10 (b) depicts such an example from the *Dumb and Dumber* film (Farrelly, 1994). FunnyNet correctly detects funny moments followed by silence or a speaker's change of tone. **3. Stand-Up Comedies** contain several punchlines that make audiences laugh. We experiment on the Jerry Seinfeld *23 h to Kill* stand-up comedy. Figure 10 (c) shows that FunnyNet detects funny moments correctly and confidently as Jerry is highly expressive (expressions, gestures). **4. Audio-Only.** As audio is the most discriminative cue, we examine its impact on out-of-domain audios: narrating jokes and reading books. FunnyNet detects funny punchlines from jokes, mostly when they are accompanied by a change of pitch or pause; for the audiobook, it successfully detects funny moments when the reader's voice imitates a character.

7   Conclusions

We introduced FunnyNet, an audiovisual model for funny moment detection. In contrast to works that rely on text, FunnyNet exploits audio that comes naturally with videos and contains high-level cues (pauses, tones, etc.). Our findings show audio is the dominant cue for signaling funny situations, while video offers complementary information. Extensive analysis and visualizations also support our finding that audio is better than text (in the form of subtitles) when it comes to scenes with no or simple dialogue but with hilarious acting or funny background sounds. Our results show the effectiveness of each component of FunnyNet, which outperforms the state of the art on the TBBT, MUStARD, MHD, UR-Funny and Friends. Future work includes analyzing the contribution of audio cues (pitch, tone, etc.).

Acknowledgements. We would like to thank Dim P. Papadopoulos and Xi Wang for proofreading and the anonymous reviewers for their feedback. This work was supported by a DIM RFSI grant, and the ANR-22-CE39-0016 and ANR-22-CE23-0007 projects.

---

[5] https://www.youtube.com/playlist?list=PL8v3aNB88WMM0iwOUeLpgFf3pHH9u
xz7_.

References

1. Annamoradnejad, I., Zoghi, G.: Colbert: Using BERT sentence embedding for humor detection. arXiv preprint arXiv:2004.12765 (2020)
2. Weller, O., Seppi, K.: The rJokes dataset: a large scale humor collection. In: LREC (2020)
3. Patro, B.N., Lunayach, M., Srivastava, D., Sarvesh, S., Singh, H., Namboodiri, V.P.: Multimodal humor dataset: predicting laughter tracks for sitcoms. In: WACV (2021)
4. Kayatani, Y., et al.: The laughing machine: predicting humor in video. In: WACV (2021)
5. Zadeh, A., Liang, P.P., Mazumder, N., Poria, S., Cambria, E., Morency, L.P.: Memory fusion network for multi-view sequential learning. In: AAAI (2018)
6. Castro, S., Hazarika, D., Pérez-Rosas, V., Zimmermann, R., Mihalcea, R., Poria, S.: Towards multimodal sarcasm detection (an Obviously perfect paper). In: ACL (2019)
7. Chen, T., Kornblith, S., Norouzi, M., Hinton, G.: A simple framework for contrastive learning of visual representations. In: Proceedings of ICML (2020)
8. Davidov, D., Tsur, O., Rappoport, A.: Semi-supervised recognition of sarcastic sentences in Twitter and Amazon. In: ACL (2010)
9. Rockwell, P.: Lower, slower, louder: vocal cues of sarcasm. J. Psycholinguist. Res. (2000)
10. Tepperman, J., Traum, D., Narayanan, S.S.: 'yeah right': sarcasm recognition for spoken dialogue systems. In: INTERSPEECH (2006)
11. Hasan, M.K., et al.: UR-FUNNY: a multimodal language dataset for understanding humor. In: EMNLP-IJCNLP (2019)
12. Bertero, D., Fung, P.: Deep learning of audio and language features for humor prediction. In: LREC (2016)
13. Hasan, M.K., et al.: Humor knowledge enriched transformer for understanding multimodal humor. In: AAAI (2021)
14. Hazarika, D., Zimmermann, R., Poria, S.: MISA: modality-invariant and-specific representations for multimodal sentiment analysis. In: Proceedings of the 28th ACM International Conference on Multimedia (2020)
15. Devlin, J., Chang, M.W., Lee, K., Toutanova, K.: BERT: Pre-training of deep bidirectional transformers for language understanding. In: NAACL (2019)
16. Rahman, W., et al.: Integrating multimodal information in large pretrained transformers. In: ACL (2020)
17. Mesaros, A., Heittola, T., Virtanen, T.: Tut database for acoustic scene classification and sound event detection. In: 24th European Signal Processing Conference (EUSIPCO) (2016)
18. Défossez, A., Usunier, N., Bottou, L., Bach, F.: Music source separation in the waveform domain. arXiv preprint arXiv:1911.13254 (2019)
19. Mesaros, A., Heittola, T., Benetos, E., Foster, P., Lagrange, M., Virtanen, T., Plumbley, M.D.: Detection and classification of acoustic scenes and events: outcome of the Dcase 2016 challenge. In: ACM Transactions on Audio Speech, and Language Processing (2017)
20. Wang, L., Luc, P., Recasens, A., Alayrac, J.B., Oord, A.: Multimodal self-supervised learning of general audio representations. arXiv preprint arXiv:2104.12807 (2021)

21. Niizumi, D., Takeuchi, D., Ohishi, Y., Harada, N., Kashino, K.: BYOL for audio: self-supervised learning for general-purpose audio representation. In: 2021 International Joint Conference on Neural Networks (IJCNN) (2021)
22. Saeed, A., Grangier, D., Zeghidour, N.: Contrastive learning of general-purpose audio representations. In: ICASSP (2021)
23. Barral, O., Kosunen, I., Jacucci, G.: No need to laugh out loud: predicting humor appraisal of comic strips based on physiological signals in a realistic environment. In: ACM Transactions on Computer-Human Interaction (TOCHI) (2017)
24. Shimasaki, A., Ueoka, R.: Laugh log: e-textile bellyband interface for laugh logging. In: Proceedings of the 2017 CHI Conference Extended Abstracts on Human Factors in Computing Systems (2017)
25. Ryokai, K., Durán López, E., Howell, N., Gillick, J., Bamman, D.: Capturing, representing, and interacting with laughter. In: Proceedings of the 2018 CHI Conference on Human Factors in Computing Systems (2018)
26. Gillick, J., Deng, W., Ryokai, K., Bamman, D.: Robust laughter detection in noisy environments. In: INTERSPEECH (2021)
27. Hochreiter, S., Schmidhuber, J.: Long short-term memory. Neural Comput. **9**, 1735–1780 (1997)
28. Gao, Y., Glowacka, D.: Deep gate recurrent neural network. In: ACCV (2016)
29. Dosovitskiy, A., et al.: An image is worth 16×16 words: transformers for image recognition at scale. In: ICLR (2020)
30. Walker, J., Razavi, A., Oord, A.: Predicting video with VQVAE. arXiv preprint arXiv:2103.01950 (2021)
31. Gabbay, A., Ephrat, A., Halperin, T., Peleg, S.: Seeing through noise: visually driven speaker separation and enhancement. In: ICASSP (2018)
32. Afouras, T., Chung, J.S., Zisserman, A.: The conversation: deep audio-visual speech enhancement. In: INTERSPEECH (2020)
33. Senocak, A., Oh, T.H., Kim, J., Yang, M.H., Kweon, I.S.: Learning to localize sound source in visual scenes. In: CVPR (2018)
34. Tian, Y., Shi, J., Li, B., Duan, Z., Xu, C.: Audio-visual event localization in unconstrained videos. In: Ferrari, V., Hebert, M., Sminchisescu, C., Weiss, Y. (eds.) ECCV 2018. LNCS, vol. 11206, pp. 252–268. Springer, Cham (2018). https://doi.org/10.1007/978-3-030-01216-8_16
35. Aytar, Y., Vondrick, C., Torralba, A.: See, hear, and read: Deep aligned representations. arXiv preprint arXiv:1706.00932 (2017)
36. Engel, J., Agrawal, K.K., Chen, S., Gulrajani, I., Donahue, C., Roberts, A.: Gansynth: adversarial neural audio synthesis. In: ICLR (2019)
37. Zhou, H., Sun, Y., Wu, W., Loy, C.C., Wang, X., Liu, Z.: Pose-controllable talking face generation by implicitly modularized audio-visual representation. In: CVPR (2021)
38. Owens, A., Isola, P., McDermott, J., Torralba, A., Adelson, E.H., Freeman, W.T.: Visually indicated sounds. In: CVPR (2016)
39. Zhou, H., Xu, X., Lin, D., Wang, X., Liu, Z.: Sep-stereo: visually guided stereophonic audio generation by associating source separation. In: Vedaldi, A., Bischof, H., Brox, T., Frahm, J.-M. (eds.) ECCV 2020. LNCS, vol. 12357, pp. 52–69. Springer, Cham (2020). https://doi.org/10.1007/978-3-030-58610-2_4
40. Dong, J., Li, X., Xu, C., Yang, X., Yang, G., Wang, X., Wang, M.: Dual encoding for video retrieval by text. IEEE TPAMI. **44**, 4065–4080 (2021)

41. Dufour, N., Picard, D., Kalogeiton, V.: Scam! Transferring humans between images with semantic cross attention modulation. In: In: Avidan, S., Brostow, G., Cissé, M., Farinella, G.M., Hassner, T. (eds) Computer Vision – 2022. ECCV 2022. LNCS, vol. 13674. Springer, Cham (2022). https://doi.org/10.1007/978-3-031-19781-9_41

42. Liang, Z., Jiang, W., Hu, H., Zhu, J.: Learning to contrast the counterfactual samples for robust visual question answering. In: EMNLP (2020)

43. Deng, D., Zhou, Y., Pi, J., Shi, B.E.: Multimodal utterance-level affect analysis using visual, audio and text features. arXiv preprint arXiv:1805.00625 (2018)

44. Arnab, A., Dehghani, M., Heigold, G., Sun, C., Lučić, M., Schmid, C.: VIVIT: a video vision transformer. In: ICCV (2021)

45. Bertasius, G., Wang, H., Torresani, L.: Is space-time attention all you need for video understanding? In: Proceedings of ICML (2021)

46. Ryoo, M.S., Piergiovanni, A., Arnab, A., Dehghani, M., Angelova, A.: TokenLearner: what can 8 learned tokens do for images and videos? arXiv preprint arXiv:2106.11297 (2021)

47. Jaegle, A., Gimeno, F., Brock, A., Zisserman, A., Vinyals, O., Carreira, J.: Perceiver: general perception with iterative attention. In: Proceedings of ICML (2021)

48. Wei, X., Zhang, T., Li, Y., Zhang, Y., Wu, F.: Multi-modality cross attention network for image and sentence matching. In: CVPR (2020)

49. Tan, R., Plummer, B.A., Saenko, K., Jin, H., Russell, B.: Look at what I'm doing: self-supervised spatial grounding of narrations in instructional videos. In: NeurIPS (2021)

50. Lee, J.T., Jain, M., Park, H., Yun, S.: Cross-attentional audio-visual fusion for weakly-supervised action localization. In: ICLR (2020)

51. Nagrani, A., Yang, S., Arnab, A., Jansen, A., Schmid, C., Sun, C.: Attention bottlenecks for multimodal fusion. In: NeurIPS (2021)

52. Narasimhan, M., Rohrbach, A., Darrell, T.: Clip-it! Language-guided video summarization. In: NeurIPS (2021)

53. Gabeur, V., Sun, C., Alahari, K., Schmid, C.: Multi-modal transformer for video retrieval. In: Vedaldi, A., Bischof, H., Brox, T., Frahm, J.-M. (eds.) ECCV 2020. LNCS, vol. 12349, pp. 214–229. Springer, Cham (2020). https://doi.org/10.1007/978-3-030-58548-8_13

54. Bain, M., Nagrani, A., Varol, G., Zisserman, A.: Frozen in time: a joint video and image encoder for end-to-end retrieval. In: ICCV (2021)

55. Epstein, D., Vondrick, C.: Learning goals from failure. In: CVPR (2021)

56. Schroff, F., Kalenichenko, D., Philbin, J.: FaceNet: A unified embedding for face recognition and clustering. In: CVPR (2015)

57. Szegedy, C., Vanhoucke, V., Ioffe, S., Shlens, J., Wojna, Z.: Rethinking the inception architecture for computer vision. In: CVPR (2016)

58. Wang, C., Fang, H., Zhong, Y., Deng, W.: MLFW: a database for face recognition on masked faces. arXiv preprint arXiv:2109.05804 (2021)

59. Chrysos, G.G., Moschoglou, S., Bouritsas, G., Deng, J., Panagakis, Y., Zafeiriou, S.P.: Deep polynomial neural networks. In: IEEE TPAMI (2021)

60. Chou, H.R., Lee, J.H., Chan, Y.M., Chen, C.S.: Data-specific adaptive threshold for face recognition and authentication. In: 2019 IEEE Conference on Multimedia Information Processing and Retrieval (MIPR) (2019)

61. Du, H., Shi, H., Zeng, D., Zhang, X.P., Mei, T.: The elements of end-to-end deep face recognition: a survey of recent advances. ACM Comput. Surv. (CSUR) **54**, 1–42 (2020)

62. Mohla, S., Pande, S., Banerjee, B., Chaudhuri, S.: FusATnet: dual attention based spectrospatial multimodal fusion network for hyperspectral and lidar classification. In: CVPRW (2020)
63. Nam, H., Ha, J.W., Kim, J.: Dual attention networks for multimodal reasoning and matching. In: CVPR (2017)
64. Chung, J.S., Zisserman, A.: Out of time: automated lip sync in the wild. In: ACCV (2016)
65. Korbar, B.: Co-training of audio and video representations from self-supervised temporal synchronization. CoRR (2018)
66. Owens, A., Efros, A.A.: Audio-visual scene analysis with self-supervised multisensory features. In: Ferrari, V., Hebert, M., Sminchisescu, C., Weiss, Y. (eds.) ECCV 2018. LNCS, vol. 11210, pp. 639–658. Springer, Cham (2018). https://doi.org/10.1007/978-3-030-01231-1_39
67. Chung, S.W., Chung, J.S., Kang, H.G.: Perfect match: improved cross-modal embeddings for audio-visual synchronisation. In: ICASSP (2019)
68. Oord, A., Li, Y., Vinyals, O.: Representation learning with contrastive predictive coding. arXiv preprint arXiv:1807.03748 (2018)
69. Huber, D.M., Runstein, R.: Modern Recording Techniques. Routledge Milton Park (2012)
70. Kalogeiton, V., Zisserman, A.: Constrained video face clustering using 1nn relations. In: BMVC (2020)
71. Brown, A., Kalogeiton, V., Zisserman, A.: Face, body, voice: video person-clustering with multiple modalities. In: ICCV (2021)
72. Paszke, A., et al.: PyTorch: an imperative style, high-performance deep learning library. In: NeurIPS (2019)
73. Wu, H.H., Seetharaman, P., Kumar, K., Bello, J.P.: Wav2clip: learning robust audio representations from clip. arXiv preprint arXiv:2110.11499 (2021)
74. Hennequin, R., Khlif, A., Voituret, F., Moussallam, M.: Spleeter: a fast and efficient music source separation tool with pre-trained models. J. Open Source Softw. **5**, 5124 (2020)
75. Hinton, G., Roweis, S.: Stochastic neighbor embedding. In: NeurIPS (2002)

ConTra: (Con)text (Tra)nsformer
for Cross-Modal Video Retrieval

Adriano Fragomeni, Michael Wray, and Dima Damen[(⊠)]

Department of Computer Science, University of Bristol, Bristol, UK
Dima.Damen@bristol.ac.uk

Abstract. In this paper, we re-examine the task of cross-modal clip-sentence retrieval, where the clip is part of a longer untrimmed video. When the clip is short or visually ambiguous, knowledge of its local temporal context (i.e. surrounding video segments) can be used to improve the retrieval performance. We propose **Con**text **Tra**nsformer (ConTra); an encoder architecture that models the interaction between a video clip and its local temporal context in order to enhance its embedded representations. Importantly, we supervise the context transformer using contrastive losses in the cross-modal embedding space.

We explore context transformers for video and text modalities. Results consistently demonstrate improved performance on three datasets: YouCook2, EPIC-KITCHENS and a clip-sentence version of ActivityNet Captions. Exhaustive ablation studies and context analysis show the efficacy of the proposed method.

1 Introduction

Millions of hours of video are being uploaded to online platforms every day. Leveraging this wealth of visual knowledge relies on methods that can understand the video, whilst also allowing for videos to be searchable, e.g. via language. Methods can query the entire video [22,49,67] or the individual segments, or clips, that make up a video [14, 41]. In this work, we focus on the latter problem of clip-sentence retrieval, specifically from long untrimmed videos. This is particularly beneficial to retrieve all instances of the same step (e.g. folding dough or jacking up a car) from videos of various procedures.

In Fig. 1, we compare current clip-sentence retrieval approaches (e.g. [3,7,11,39, 46,58]) to our proposed context transformer. We leverage local temporal context clues, readily available in long videos, to improve retrieval performance. Local sequences of actions often use similar objects or include actions towards the same goal, which can enrich the embedded clip representation.

We emphasise the importance of learnt **local temporal clip context**, that is the *few* clips surrounding (i.e. before and after) the clip to be embedded. Our model, ConTra, learns to attend to relevant neighbouring clips by using a transformer encoder, differing from previous works which learn context over frames [20] or globally across the entire video [22] (see Video-Paragraph Retrieval, Fig. 1 top). We supervise ConTra with cross-modal contrastive losses and a proposed neighbouring loss that ensures the embedding is distinct across overlapping contexts.

Supplementary Information The online version contains supplementary material available at https://doi.org/10.1007/978-3-031-26316-3_27.

L. Wang et al. (Eds.): ACCV 2022, LNCS 13844, pp. 451–468, 2023.
https://doi.org/10.1007/978-3-031-26316-3_27

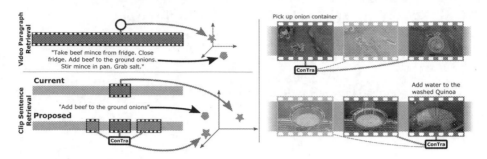

Fig. 1. Left: We compare video-paragraph retrieval (top) to current and proposed clip-sentence retrieval (bottom) in long videos. In ConTra, we propose to attend to local context of neighbouring clips. **Right**: Examples where ConTra can enrich the clip representation from next/previous clips, observing the onion (top) or that the quinoa has already been washed (bottom). Line thickness/brightness represents attention weights.

Our contributions are summarised as follows: (i) we explore the task of cross-modal clip-sentence retrieval when using local context in clip, text or in both modalities simultaneously (ii) we propose ConTra, a transformer based architecture that learns to attend to local temporal context, supervised by a multi-term loss that is able to distinguish consecutive clips by the introduction of a neighbouring contrastive loss (iii) we demonstrate the added value of local context by conducting detailed experiments on three datasets.

2   Related Works

In this section, we split video-text retrieval works into those which primarily focus on either Clip-Sentence or Video-Paragraph retrieval before presenting works that use temporal context for other video understanding tasks.

Clip-Sentence Retrieval: Most works primarily rely on two-stream (i.e. dual) approaches [3,23,32,40,42,46,58,60,65,70], using multiple text embeddings [9,11, 15,43,61], video experts [36,37,41], or audio [1,2,7,58]. Recently, single stream cross-modal encoders have also been used [38,39,52,64,72], improving inter-modality modelling at the cost of increased computational complexity. In ConTra, we use a dual stream model with separate branches for the visual and the textual components.

Temporal modelling of frames *within a clip* is a common avenue for retrieval approaches [20,58,72]. Gabeur et al. [20] use multiple video experts with a multi-modal transformer to better capture the temporal relationships between modalities. Wang et al. [58] learn an alignment between words and frame features alongside the clip-sentence alignment. ActBert [72] also models alignment between clip and word-level features using self-supervised learning. Bain et al. [3] adapt a ViT [16] model, trained with a curriculum learning schedule, to gradually attend to more frames within each clip. MIL-NCE [40] alleviates noise within the automated captions, matching clips to neighbouring sentences. However, the learned representation does not go beyond the clip extent. VideoCLIP [65] creates positive clips by sampling both the centre-point (within narration timestamp) and the clip's duration to better align clips and sentences, foregoing the reliance on explicit start/end times. In our work, we go beyond temporal modelling of the clip itself to using local context outside the clip.

Other works improve modelling of the textual representation [15,46,61]. Patrick et al. [46] introduce a generative task of cross-instance captioning to alleviate false negatives. They create a support set of relevant captions and learn to reconstruct a sample's text representation as a weighted combination of a support-set of video representations from the batch. However, whilst they use information from other sentences, there is no notion of those which are temporally related. Instead, we propose to explore relationships between neighbouring sentences using local context within the same video.

Video-Paragraph Retrieval: Another retrieval task is video-paragraph retrieval [11, 22,34,36,49,51,58,60,67], where videos and paragraphs describing the full videos are embedded in their entirety. There are two main approaches: using hierarchical representations between the paragraph/video and constituent sentences/clips [22,36,67] or jointly modelling the entire video/paragraph with a cross-modal transformer [34,51].

COOT [22] models the interactions between levels of granularity for each modality by using a hierarchical transformer. The video and paragraph embeddings are obtained via a combination of their clips and sentences. ClipBERT [34] inputs both text and video to a single transformer encoder after employing sparse sampling, where only a single or a few sampled clips are used at each training step. In this work, we focus on clip-sentence retrieval, but take inspiration from video-paragraph works in how they relate clips within a video. Importantly, we focus on local context, which is applicable to long videos with hundreds of clips.

Temporal Context for Video Understanding: We also build on works that successfully used *local temporal context* for other video understanding tasks such as: action recognition [6,30,62,68]; action anticipation [19,48]; object detection and tracking [4,5]; moment localisation [69]; and Content-Based Retrieval [50]. Bertasius and Torresani [5] use local context for mask propagation to better segment and track occluded objects. Kazakos et al. [30] use the context of neighbouring clips for action recognition using a transformer encoder along with a language model to ensure the predicted sequence of actions is realistic. Feichtenhofer et al. [62] allow for modelling features beyond the short clips. They use a non-local attention block and find that using context from up to 60 s can help recognise actions. Shao et al. [50] use a self-attention mechanism to model long-term dependencies for content-based video retrieval. They use a supervised contrastive learning method that performs automatic hard negative mining and utilises a memory bank to increase the capacity of negative samples.

To the best of our knowledge, ours is the first work to explore using neighbouring clips as context for cross-modal clip-sentence retrieval in long untrimmed videos.

3   Context Transformer (ConTra)

We first explicitly define the task of clip-sentence retrieval in Sect. 3.1 before extending the definition to incorporate *local* clip context, in untrimmed videos. We then present our clip context embedding in Sect. 3.2 where we provide details of our architecture followed by the training losses in Sect. 3.3. We then extend ConTra to context in both modalities in Sect. 3.4. An overview of our approach can be seen in Fig. 2.

3.1   Definitions

We begin with a set of untrimmed videos, $v_i \in V$. These are broken down further into ordered clips, $c_{ij} \in v_i$, each with a corresponding sentence/caption, t_{ij}, describing the

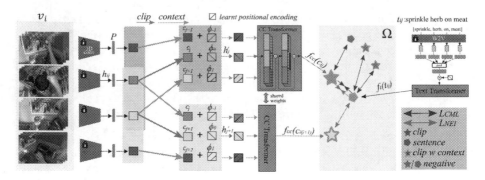

Fig. 2. Overview of ConTra: Given a video v_i split into clips c_{ij}, we encode clips into features h_{ij}, projected by P and tagged with a learnt position encoding ϕ *relative* to the centre clip. A Clip Context (CC) encoder learns an enriched representation of the centre clip (cyan arrow) attending to context clips (orange arrows). The embedding space Ω is learnt with cross-modal (CML) and neighbouring (NEI) losses. NEI pushes overlapping contexts further apart—shown for $f_{cc}(c_{ij})$ and $f_{cc}(c_{i(j+1)})$. (Color figure online)

action within the clip. Querying by the sentence t_{ij} aims to retrieve the corresponding clip c_{ij} and vice versa for cross-modal retrieval.

Learning a dual stream retrieval model focuses on learning two projection functions, $f_c : c \longrightarrow \Omega \subseteq \mathbb{R}^d$ and $f_t : t \longrightarrow \Omega \subseteq \mathbb{R}^d$, which project the video/text modalities respectively into a common d-dimensional embedding space, Ω, where c_{ij} and t_{ij} are close. The weights of these embedding functions can be collectively trained using contrastive-based losses including a triplet loss [47,54–56] or noise-contrastive estimation [24,29,40].

Instead of using the clip solely, we wish to utilise its *local* temporal context to enrich the embedded representation. We define the temporal Clip Context (CC) using m, around the clip c_{ij}, as follows:

$$CC_m(c_{ij}) = (c_{i(j-m)}, \cdots, c_{ij}, \cdots, c_{i(j+m)}) \qquad (1)$$

There are $2m$ clips that are temporally adjacent to c_{ij} in the same video v_i. Note that the length of the adjacent clips governed by m differs per dataset and video. Importantly, we still aim to retrieve the corresponding sentence t_{ij} for the centre clip c_{ij}, but utilise the untrimmed nature of the video around the clip to enrich c_{ij}'s representation.

In Table 1, we differentiate between existing tasks in Sect. 2 and our proposed settings. Note that in Video-Paragraph retrieval, models cannot be used to retrieve individual clips. Different from the standard Clip-Sentence setting, we utilise neighbouring clips to enrich the clip representation, the sentence representation or both. Next, we describe our Clip Context Transformer.

Table 1. Comparison of tasks with/without using context in video and text.

Task	Clip?	Video	Text
Video-paragraph	×	All clips	All sentences
Clip-sentence			
No context	✓	Clip	Sentence
Clip context	✓	Clip+context	Sentence
Text context	✓	Clip	Sent.+context
Context in both	✓	clip+context	sent.+context

3.2 Clip Context Transformer

We learn the embedding function $f_{cc} : CC_m(c) \longrightarrow \Omega$, using the local clip context in Eq. 1. We consider each clip as a linear projection of its features $P(h_j)$. We drop the video index i here for simplicity. We learn $2m + 1$ distinct positional embeddings, $(\phi_{-m}, \cdots, \phi_0, \cdots, \phi_m)$, that are added such that $h'_{j+\alpha} = P(h_{j+\alpha}) + \phi_{0+\alpha}$, where $-m \leq \alpha \leq m$ and ϕ_0 is the positional embedding of the centre clip. Note that the positional embeddings emphasise the order of the clip within the context window rather than the full video, thus reflecting the *relative* position of the context to the centre clip c_{ij}, and are identical across contexts. We showcase this on two neighbouring clips in Fig. 2. We form the input to the encoder transformer as:

$$H' = [h'_{j-m}, \cdots, h'_j, \cdots, h'_{j+m}] \tag{2}$$

From H', we aim to learn the embedding of the centre clip. We use a multi-headed attention block [53] with the standard self-attention heads and residual connections. The output of the rth attention head is thus computed as,

$$A_r = \sigma \left(\frac{(\theta_r^Q H')(\theta_r^K H')^\top}{\sqrt{d}} \right) (\theta_r^W H') \tag{3}$$

where θ_r^Q, θ_r^K, $\theta_r^W \in \mathbb{R}^{(2m+1) \times d/R}$ are learnable projection matrices. The output of the multi head attention is then calculated as the concatenation of all R heads:

$$A = [A_1, \ldots, A_R] + H' \tag{4}$$

For the clip embedding, we focus on the output from A corresponding to the centre clip j, such that:

$$f_{cc}(c_{ij}) = g(A_j) + A_j \tag{5}$$

where g is one or more linear layers with ReLU activations, along with another residual connection. Note that the size of f_{cc} is d, independent of the context length m. f_{cc} can be extended with further multi-head attention layers. We discuss how we train the ConTra model next.

3.3 Training ConTra

Cross-Modal Loss. For both training and inference, we calculate the cosine similarity $s(c_{ij}, t_{kl})$ between the embeddings of the context-enriched clip $f_{cc}(c_{ij})$ and a sentence $f_t(t_{kl})$. Cross-modal losses are regularly used in retrieval works such as the triplet loss [9,22,36,42] and the Noise-Contrastive Estimation (NCE) loss [1,7,35,64]. We use NCE as our cross-modal loss (L_{CML}) [24,29]:

$$L_{CML} = \frac{1}{|B|} \sum_{(c_{ij}, t_{ij}) \in B} -\log \left(\frac{e^{s(c_{ij}, t_{ij})/\tau}}{e^{s(c_{ij}, t_{ij})/\tau} + \sum_{(c', t') \sim \mathcal{N}'} e^{s(c', t')/\tau}} \right) \tag{6}$$

where B is a set of corresponding clip-captions pairs, i.e. (c_{ij}, t_{ij}) and τ is the temperature parameter. We construct the negative set $\mathcal{N}'$ in each case from the batch by combining $(c_{ij}, t_{lk})_{ij \neq lk}$ as well as $(c_{lk}, t_{ij})_{ij \neq lk}$, considering negatives for both clip and sentence across elements in the batch.

Uniformity Loss. The uniformity loss is less regularly used, but was proposed in [57] and used in [10, 17] works, for image retrieval. It ensures that the embedded representations preserve maximal information, i.e. feature vectors are distributed uniformly on the unit hypersphere. We use the uniformity loss (L_{UNI}) such as:

$$L_{UNI} = \log \left(\frac{1}{|B|} \sum_{u,u' \in U \times U} e^{-2\|u-u'\|_2^2} \right) \tag{7}$$

where $U = \{c_1, t_1, ..., c_B, t_B\}$, are all the clips and sentences in the batch. This loss term is applied to all the clips and sentences in a batch.

Neighbouring Loss. We additionally propose a neighbour-contrasting loss (L_{NEI}) to ensure that the embeddings of context items are well discriminated. Indeed, one of the challenges of introducing local temporal context is the overlap between contexts of neighbouring clips. Consider two neighbouring clips in the same video, say c_{ij} and $c_{i(j+1)}$ (see Fig. 2), the context windows $CC(c_{ij})$ and $CC(c_{i(j+1)})$ share $2m$ clips. While the positional encoding of the clips differ, distinguishing between the embedded neighbouring clips can be challenging. This can be considered as a special case of hard negative mining, as in [18, 27], however our usage of it, where only neighbouring clips are considered as negatives is novel.

Accordingly, we define the L_{NEI} using the NCE loss:

$$L_{NEI} = \frac{1}{|B|} \sum_{(c_{ij},t_{ij}) \in B} - \log \left(\frac{e^{s(c_{ij},t_{ij})/\tau}}{e^{s(c_{ij},t_{ij})/\tau} + e^{s(c_{i(j+\alpha)},t_{ij})/\tau}} \right) \tag{8}$$

where α is randomly sampled from $[-m, m]$ subject to $t_{ij} \neq t_{i(j+\alpha)}$. We thus randomly sample a neighbouring clip, avoiding neighbours where the sentences are matching (e.g. the sentence, "mix ingredients" might be repeated in consecutive clips).

In practice, the neighbouring loss is calculated by having another batch of sampled neighbouring clips of size B. We use a single negative neighbour per clip to keep the batch size to B regardless of the length m, though we do ablate differing numbers of sampled negatives in Sect. 4.2.

We optimize our ConTra model by minimizing the overall loss function L:

$$L = L_{CML} + L_{NEI} + L_{UNI} \tag{9}$$

We keep the weights between the three losses the same in all experiments and datasets showcasing that we outperform other approaches without hyperparameter tuning. In supplementary, we report results when tuning the weights, to ablate these.

Once the model is trained, it can be used for both sentence-to-clip and clip-to-sentence retrieval. The clip is enriched with the context, whether used in the gallery set (in sentence-to-clip) or in the query (in clip-to-sentence). When performing sentence-to-clip retrieval, our query consists of only one sentence as usually done in other approaches, and is thus comparable to these. During inference, the gallery of clips is always given, and thus all approaches have access to the same information.

3.4 Multi-modal Context

In previous sections (Sects. 3.1–3.3), we motivated our approach by focusing on local clip context—i.e. context in the visual modality. However, ConTra could similarly be applied to the local context of the text modality. As an example, given steps of a recipe, these can be utilised to build a Text Context (TC), such that:

$$TC_m(t_{ij}) = (t_{i(j-m)}, \cdots, t_{ij}, \cdots, t_{i(j+m)}) \tag{10}$$

To give an example for clip-to-sentence retrieval, a single clip is used as the query, but the gallery is constructed of captions that have had their representation enriched via sentences of neighbouring clips (e.g. "Add the mince to the pan" has attended to "Take the beef mince out of its wrapper" and "Fry until the mince is browned"). This contextual text knowledge could come from video narrations or steps in a recipe.

We also assess the utilisation of context in both modalities. This setup assumes access to local context in both clip and sentence. L_{NEI} is thus applied to both neighbouring clip contexts and text contexts, using one negative for each case. The architecture for both f_{cc} and f_{tc} are identical, but are learned as two separate embedding functions with unique weights and positional embeddings[1].

4 Results

We first present our experimental settings and the choice of untrimmed datasets in Sect. 4.1. We then focus on clip context results including comparison with state-of-the-art (SOTA) methods in Sect. 4.2 before exploring text context and context in both modalities in Sect. 4.3. Finally, we discuss limitations and avenues for future work.

4.1 Experimental Settings

Datasets. Video datasets commonly used for cross-modal retrieval can be split into two groups: trimmed and untrimmed. In trimmed datasets, such as MSRVTT [66], MSVD [8] and VATEX [59], the full video is considered as a single clip and thus no context can be utilised. In Table 2, we compare the untrimmed datasets for their size and the number of clips per video. Datasets with 1–2 clips per video on average limit the opportunity to explore long or local temporal context. While we include QuerYD [44] in the table, this dataset does not allow for context to be explored as clips from the same video are split between the train and test sets. We choose to evaluate our method on three untrimmed datasets, whose average number of clips/video is greater than 3. We describe the notion of context in each:

YouCook2 [71] contains YouTube cooking videos. On average, training videos contain 7.75 clips, each associated with a sentence. The dataset has been evaluated for clip-sentence retrieval [23,40,52,70] as well as video-paragraph retrieval [22]. We focus on clip-sentence retrieval, utilising the local context, which represents previous/follow-up steps in a recipe. Given YouCook2's popularity, we use it for all ablation experiments.

ActivityNet Captions [33] consists of annotated YouTube videos from ActivityNet [26]. The dataset has only been evaluated for video-paragraph retrieval [22,34,67]

---

[1] We experimented with sharing these embeddings but similar to previous approaches [28], this performed worse, see supplementary.

Table 2. Comparing untrimmed video datasets by size and number of clips per video. *QuerYD videos are split across train and test so we report overall clips/video.

	#clips		#videos		#clips per video	
Datasets	Train	Test	Train	Test	Train	Test
Charades-STA [21]	5,657	1,596	5338	1334	1.60	1.20
DiDeMo [27]	21,648	2,650	8511	1037	2.21	2.21
QuerYD* [44]	9,118	1,956	1,283	794	8.3	
YouCook2 [71]	10,337	3,492	1,333	457	7.75	7.64
ActivityNet CS [33]	37,421	17505	10,009	4,917	3.74	3.56
EPIC-KITCHENS 100 [12]	67,217	9,668	495	138	135.79	70.06

where all clips in the same video are concatenated, and all corresponding captions are also concatenated to form the paragraph. Instead, we consider the *val_1* split, and introduce an **ActivityNet Clip-Sentence (CS)** variant using all the individual clips and their corresponding captions/sentences. We emphasise that this evaluation *cannot* be compared to published results on video-paragraph retrieval and instead evaluate two methods to act as baselines for comparison.

EPIC-KITCHENS-100 [12] offers a unique opportunity to explore context in significantly longer untrimmed videos. On average, there are 135.8 clips per video of kitchen-based actions, shot from an egocentric perspective. We use the train/test splits for the multi-instance retrieval benchmark, but evaluate on the single-instance retrieval task using the set of unique narrations.

Evaluation Metrics. We report the retrieval performance, for both clip-to-sentence and sentence-to-clip tasks, using the two standard metrics of: Recall at $K = \{1, 5, 10\}$ (R@K) and median rank (MR). We also report the sum of cross-modal R@K as RSum to demonstrate overall performance. Where figures are plotted, tables of exact results are given in supplementary.

Visual Features. To be comparable to prior work, we use the same features as recent methods per dataset. For YouCook2, we use the S3D backbone provided by [40] pre-trained on [42], extracting 1024-d features. We uniformly sample 32 frames from each clip with a 224×224 resolution. For ActivityNet CS, we use frame features provided by [67]. These frame features are combined into clip features using a single transformer, trained with shared weights across clips, as proposed in [22], obtaining 384-d features. Note that this transformer is trained for clip-sentence alignment, using the code from [22], without the global context and contextual transformer. For EPIC-KITCHENS-100, we use the publicly available 3072-d features from [31].

Text Features. For YouCook2 and EPIC-KITCHENS-100, we take a maximum of 16 words without removing stopwords from each sentence and we extract 2048-d feature vectors using the text branch in [40] pre-trained on [42]. This consists of a linear layer with a ReLU activation applied independently to each word embedding followed by max pooling and a randomly initialised linear layer to reduce dimensionality. We fine-tune the text branch layer, to accommodate missing vocabulary[2]. For ActivityNet CS, we feed

---

[2] We add 174 and 104 missing words from the model in [42] for YouCook2 and EPIC-KITCHENS-100 respectively.

the sentences into a pretrained BERT-Base Uncased model [13] and use the per-token outputs of the last 2 layers to train a sentence transformer, as in [22], and obtain 384-d text features.

Architecture Details. The number of layers and heads in the ConTra encoder differs depending on the dataset size. For the small-scaled YouCook2, we use 1 layer and $R = 2$ heads to avoid overfitting. For the larger two datasets we use 2 layers and $R = 8$ heads. The inner dimension of the transformers is 2048-d. The learnt positional encoding matches the feature dimension: 512-d for YouCook2 and EPIC-KITCHENS-100, and 384-d for ActivityNet CS, initialised from $\mathcal{N}(0, 0.001)$. Due to the small dimension of the features of ActivityNet CS, we remove the linear projection P from our architecture. We apply dropout of 0.3 at h_j and h'_j.

Implementation Details. We use the Adam optimizer with a starting learning rate of 1×10^{-4} and decrease it linearly after a warmup period of 1300 iterations. The size of the batch is fixed to 512. The temperature τ in Eq. 6 and 8 is set to 0.07 as in [25, 45, 63], and the dimension of the common embedding space Ω is set to 512 for YouCook2 and EPIC-KITCHENS-100, and 384 for ActivityNet CS. We ablate these values in supplementary. If the clip does not have sufficient temporal context (i.e. is at the start/end of the video), we pad the input by duplicating the first/last clip to obtain a fixed-length context. *Our code is available at* https://github.com/adrianofragomeni/ConTra.

4.2   Clip Context Results

Context Length Analysis. We analyse the effect of clip context length by varying m from its no-context baseline, $m = 0$, to $m = 5$. Results are presented in Fig. 3. In all three datasets, the largest improvement is obtained comparing $m = 0$, i.e. no context, to $m = 1$, i.e. introducing the smallest context, where the RSum increases by 8.5, 17.6 and 23.2 for Youcook2, ActivityNet CS, and EPIC-KITCHENS-100, respectively. This highlights that neighbouring clips are able to improve the retrieval performance. Moreover, every $m > 0$ outperforms $m = 0$ on all datasets. We obtain the best performance on YouCook2 at $m = 3$. ActivityNet CS also obtained best performance when using $m = 3$, and EPIC-KITCHENS-100 when $m = 4$. Although ConTra introduces a new hyperparameter, m, Fig. 3 shows that RSum saturates when $m \geq 3$ across all datasets. Using a larger context does not further improve the performance.

We show the attention weights learned by the multi-headed attention layers in Fig. 4 averaged over all videos, per layer (left), and for specific examples (right). YouCook2 focuses more on the later clips due to the recipes being more recognisable when ingredients are brought together. The attention weights for EPIC-KITCHENS-100 and ActivityNet CS are higher for earlier clips, with ActivityNet's first layer and EPIC-KITCHENS-100's second layer attending to past clips. EPIC-KITCHENS-100 specifically has higher attention weights

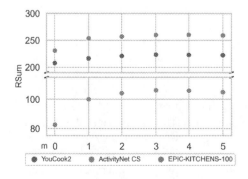

Fig. 3. Analysis of clip context (CC) with differing m across YouCook2, ActivityNet CS and EPIC-KITCHENS-100.

Fig. 4. Left: average attention weights over videos as m changes, per dataset and layer. Right: qualitative examples with clip attention.

on directly neighbouring clips. From the examples in Fig. 4 (right), ConTra uses local context to discriminate objects which may be occluded, such as chicken in YouCook2, the contents of the salad bag in EPIC-KITCHENS-100, or the brush in ActivityNet CS.

In Fig. 5, we analyse how individual words are affected by using context. On a word-by-word basis we find all captions that contain a given word and count the number of times the rank of those captions improved/worsened after adding context. E.g., the word 'pan' in YouCook2 is present in 373 captions, in which 182 captions improve their rank with context while 126 captions worsen their rank resulting in a delta of 56. In YouCook2, 'salt', 'sauce', and 'oil' all see a large improvement when using context, likely due to easily being obscured by their containers. In comparison, for EPIC-KITCHENS-100, verbs benefit the most from context—these actions tend to be very short so surrounding context can help discriminate them.

Overall, these results showcase that using local Clip Context (CC) enhances the clip embedding representation and consistently results in a boost in performance.

Comparison with State of the Art. The most commonly used untrimmed dataset in prior work is YouCook2. For fair comparison, we split the SOTA methods on this dataset into blocks according to the pre-training and fine-tuning datasets: (i) training only on YouCook2, (ii) training only on other large-scale datasets, (iii) pre-training on large-scale datasets then fine-tuning on YouCook2; note that this is where ConTra lies and (iv) additionally, pre-training with proxy tasks on large-scale datasets. Table 3 compares ConTra with the SOTA on YouCook2.

Overall, ConTra outperforms all directly-comparable SOTA works [22, 23,42]. ConTra outperforms COOT [22] that trains for video-paragraph retrieval

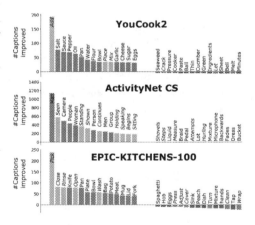

Fig. 5. 15 most improved/hindered words per dataset when context is used. Verbs are lighter/italicised. Best viewed in colour.

on the full video. Distinct from COOT [22], we only use clip context, i.e. single sentences in training and inference, and local context.

Table 3. Sentence-to-Clip comparison with SOTA on YouCook2 test set. PX: pre-train end-to-end with proxy tasks. FT: fine-Tuning on YouCook2. −: unreported results.

	Method	PX	FT	R@1	R@5	R@10	MR	RSum
(i)	HGLMM [32]	×	✓	4.6	14.3	21.6	75	40.5
	UniVL (FT-joint) [38]	×	✓	7.7	23.9	34.7	21	66.3
(ii)	ActBert [72]	✓	×	9.6	26.7	38.0	19	74.3
	MMV FAC [2]	×	×	11.7	33.4	45.4	13	90.5
	VATT-MBS [1]	×	×	-	-	45.5	13	-
	MCN [7]	×	×	18.1	35.5	45.2	-	98.8
	MIL-NCE [40]	×	×	15.1	38.0	51.2	10	104.3
(iii)	HowTo100M [42]	×	✓	8.2	24.5	35.3	24	68.0
	GRU+SSA [23]	×	✓	10.9	28.4	-	-	-
	COOT [22]	×	✓	16.7	40.2	52.3	9	109.2
	MIL-NCE (from [70])	×	✓	15.8	40.3	54.1	8	110.2
	ConTra (ours)	×	✓	16.7	42.1	55.2	8	114.0
(iv)	CUPID [70]	✓	✓	17.7	43.2	57.1	7	117.9
	DeCEMBERT [52]	✓	✓	17.0	43.8	59.8	9	120.6
	UniVL (FT-joint) [38]	✓	✓	22.2	52.2	66.2	5	140.6
	VLM [64]	✓	✓	27.0	56.9	69.4	4	153.3
	VideoCLIP [65]	✓	✓	32.2	62.6	75.0	-	169.8

Table 4. Comparison on ActivityNet CS for Clip-Sentence Retrieval. *Our reproduced results. †: Transformer fine-tuned first—clip/sentence features match those from COOT [22] −g.

	Method	FT	Sentence-to-clip				Clip-to-sentence				RSum
			R@1	R@5	R@10	MR	R@1	R@5	R@10	MR	
(ii)	MIL-NCE [40]*	×	2.4	6.8	10.0	460	2.1	6.2	9.2	543	36.7
(iii)	HowTo100M [42]*	✓	3.8	12.5	18.9	68	3.6	11.4	17.3	78	67.5
	COOT [22] −g*	✓	3.7	11.9	18.6	67	3.7	12.0	18.8	64	68.7
	ConTra (ours)	✓†	5.9	18.4	27.6	38	6.4	19.3	28.5	37	106.1
	COOT [22]*	✓	6.2	18.8	28.4	33	6.3	19.0	28.4	32	107.0

Table 5. Comparison with baseline on EPIC-KITCHENS-100. *Our reproduced results.

	Method	FT	Sentence-to-clip				Clip-to-sentence				RSum
			R@1	R@5	R@10	MR	R@1	R@5	R@10	MR	
(ii)	MIL-NCE [40]*	×	3.2	10.4	15.4	188	2.1	7.8	12.3	194	51.2
(iii)	JPoSE [61]*	✓	2.5	7.5	11.6	13	4.4	17.4	27.2	17	70.6
	ConTra (ours)	✓	22.2	43.4	53.4	9	28.2	52.0	61.1	5	260.3

The last block of Table 3 includes works that are not directly comparable to ConTra, as these models are pre-trained *end-to-end* on HowTo100M with additional proxy tasks, e.g. masked language modelling, whereas ConTra is initialised randomly. Although DeCEMBERT [52] is not directly comparable, ConTra is less complex with $9.5M$ parameters compared to DeCEMBERT's $115.0M$ and our results are only marginally lower.

To the best of our knowledge, no prior work has evaluated on ActivityNet CS for clip-sentence retrieval. For comparison, we evaluate [40,42] on ActivityNet CS for clip-sentence retrieval using public code. We also run the code from COOT [22], trained on video-paragraph retrieval, and obtain their results on clip-sentence retrieval. We then replace the text input with sentence-level representa-

Table 6. Ablation of loss function terms: Neighbouring Loss (L_{NEI}), Cross Modal Loss (L_{CML}), Hard Triplet Mining ($L_{HardMining}$), and Uniformity Loss (L_{UNI}).

Loss	Sentence-to-clip				Clip-to-sentence				
	R@1	R@5	R@10	MR	R@1	R@5	R@10	MR	RSum
L_{NEI}	6.4	18.3	27.3	39	4.3	15.5	23.6	43	95.4
L_{CML}	15.7	39.9	53.4	9	14.5	38.5	52.3	9	214.3
$L_{CML}+L_{HardMining}$	15.7	39.8	53.5	9	14.2	39.0	51.9	10	214.1
$L_{CML}+L_{NEI}$	16.2	41.4	54.1	9	**14.8**	39.3	52.6	9	218.4
$L_{CML}+L_{NEI}+L_{UNI}$	**16.7**	**42.1**	**55.2**	**8**	**14.8**	**40.5**	**53.9**	9	**223.2**

tions, and remove their global alignment to produce the COOT$-g$ baseline reported above. Table 4 shows that ConTra outperforms MIL-NCE [40], HowTo100M [42] and COOT$-g$ by a considerable margin. Our RSum is only marginally lower than COOT, where global context is considered for both modalities during training. Note that methods that train for global context cannot be used for datasets with hundreds of clips per video, like EPIC-KITCHENS-100. In Table 5, we compare ConTra to JPoSE [61] and our reproduced results of MIL-NCE [40] on EPIC-KITCHENS-100, outperforming on all metrics by a large margin. We cannot train or evaluate COOT on EPIC-KITCHENS-100 which has 136 clips per video on average. Additionally, clips in EPIC-KITCHENS-100 are significantly shorter, increasing the benefits of attending to local context.

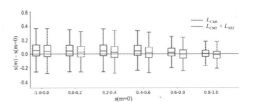

Fig. 6. Comparison between similarities to neighbouring clips, $s(f_{cc}(j+1), f_s(j))$, with and without using L_{NEI}. Without L_{NEI} ConTra gives higher similarities to neighbouring clips.

Ablation studies. We ablate ConTra on YouCook2. Ablations on the other two datasets are in supplementary.

Loss Function. In Table 6, we first test the formulation of the two losses L_{NEI} and L_{CML} individually, using the NCE loss [1,7,35,64].

On its own, L_{NEI} learns with limited variety of only neighbouring clips as negatives.

Then, we compare L_{NEI} to the standard hard mining approach proposed in [18]. L_{NEI} consistently outperforms hard mining. Our proposed loss L, with its 3 terms, performs the best, improving RSum by 4.8 when adding the uniformity loss L_{UNI} which allows preserving maximal information and so obtains better embeddings.

We further demonstrate the benefits of L_{NEI} in Fig. 6. We bin the neighbouring clips $j \pm 1$ based on their similarity to the sentence j along the x-axis. We then calculate the difference between this similarity with and without context, and provide the average and extent of these differences in a box plot over all datasets. When this difference, on the y-axis, is > 0, the context transformer would have increased the similarity between the neighbouring clip and the sentence. Without L_{NEI}, the similarity is increased further, particularly for clips and sentences with low cosine similarity, depicted on the x-axis.

Number of negatives neighbouring clips. Table 7 shows how the performance changes when we consider more than one negative for our neighbouring loss L_{NEI}. Increasing the negatives from 1 to 2 improves the retrieval results marginally. RSum remains the same when increasing the number of negatives further to 3. We keep the number of negatives equal to 1 in all the experiments.

Table 7. Analysis of the performance varying the number of negative in L_{NEI}.

	Sentence-to-clip				Clip-to-sentence				
#Negatives	R@1	R@5	R@10	MR	R@1	R@5	R@10	MR	RSum
1	16.7	42.1	55.2	8	14.8	40.5	**53.9**	9	223.2
2	16.9	42.2	**55.7**	8	**15.6**	40.5	**53.9**	9	224.8
3	**17.3**	**42.4**	55.6	8	14.9	**40.7**	**53.9**	9	224.8

Aggregate Context. As explained in Sect. 3.2, we select the middle output of the transformer encoder as our clip embedding. In order to justify this design choice, we compare to other aggregation approaches. These are of two types based on where local context is aggregated, i.e. the visual features h_j or the outputs of the clip transformer encoder f_{cc}. Moreover we experimented with two aggregation techniques, Maximum and Average.

Table 8 shows that aggregating features has a poor performance. Moreover, using the middle output outperforms the other two aggregation techniques, as the model is enriching the embedding of the anchor clip from its contextual neighbours.

Table 8. Comparing aggregation approaches.

h_j	f_{cc}	Sentence-to-clip				Clip-to-sentence				
		R@1	R@5	R@10	MR	R@1	R@5	R@10	MR	RSum
Avg	-	4.9	16.3	26.3	40	4.3	16.0	25.8	41	93.6
Max	-	3.1	13.1	22.0	46	3.4	13.4	21.7	49	76.7
-	Avg	15.6	40.1	53.9	9	14.0	38.6	51.6	10	213.8
-	Max	**16.9**	41.7	54.8	8	14.7	40.1	53.7	9	221.9
-	Mid	16.7	**42.1**	**55.2**	8	**14.8**	**40.5**	**53.9**	9	223.2

4.3 Results of Modality Context

In Figs. 7 and 8 we provide comparable analysis as m increases from no context ($m = 0$) up to $m = 5$, for text context (Fig. 7) and context in both modalities (Fig. 8). Results consistently demonstrate context to be helpful in both cases, $m = 1$ outperforms $m = 0$ by a large margin in every case. For some datasets, e.g. ActivityNet CS, performance saturates and drops slightly for $m > 3$. For long videos, e.g. EPIC-KITCHENS-100, performance continues to improve with larger context.

In Fig. 9 we show qualitative examples from models trained with clip context and context in both modalities. Clip context (left) shows additional visual context helps. For example, in row 1, the previous clip includes potatoes before being mashed—a more recognisable shape compared to their mashed state. When using context in both modalities (right), these benefits are combined, leading to the model discriminating between difficult examples in which very similar clips are described.

Limitations. We find that local context is certainly beneficial for Clip-Sentence retrieval, but also acknowledge there are cases in which it is detrimental. An example of this is in Fig. 9 (row 3, left—"add water"), context drops the rank of the correct clip from 4 to 15. Studying the correct video, we note that neighbouring clips are not

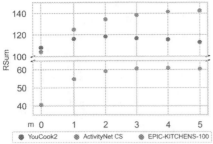

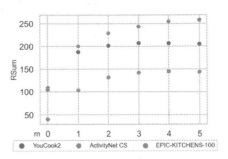

Fig. 7. Analysis of temporal text context (TC), reporting RSum in S2C.

Fig. 8. Analysis of temporal both context (BC), reporting RSum in S2C.

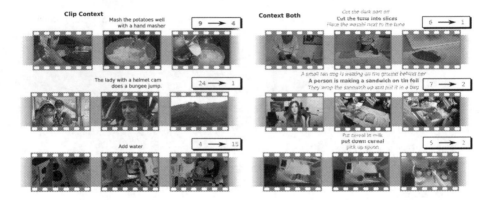

Fig. 9. Qualitative clip-to-sentence results for clip context (left) and context in both modalities (right) from 3 datasets: YouCook2 (top), ActivityNet CS (middle), and EPIC-KITCHENS-100 (bottom). The change in rank of retrieved video from no context (cyan) to using context (orange), e.g. from rank 9 to rank 4 when using context. (Color figure online)

always related to the main action, i.e. —"turn on the cooker". The enriched clip representation is thus less similar to the query sentence. In Fig. 5, we also show specific words harmed by clip context.

5   Conclusions

In this work, we introduce the notion of local temporal context, for clip-sentence cross-modal retrieval in long videos. We propose an attention-based deep encoder, which we term Context Transformer (ConTra), that is trained using contrastive losses from the embedding space. We demonstrate the impact of ConTra on individual modalities as well as both modalities in cross-modal retrieval. We ablate our method to further show the benefit of each component. Our results indicate, both qualitatively and by comparing to other approaches, that local context in retrieval decreases the ambiguity in clip-sentence retrieval on three video datasets.

Acknowledgements. This work used public dataset and was supported by EPSRC UMPIRE (EP/T004991/1) and Visual AI (EP/T028572/1).

References

1. Akbari, H., et al.: VATT: Transformers for multimodal self-supervised learning from raw video, audio and text. In: Conference on Neural Information Processing Systems (NeurIPS) (2021)
2. Alayrac, J., et al.: Self-supervised multimodal versatile networks. In: Conference on Neural Information Processing Systems (NeurIPS) (2020)
3. Bain, M., Nagrani, A., Varol, G., Zisserman, A.: Frozen in time: a joint video and image encoder for end-to-end retrieval. In: International Conference on Computer Vision (ICCV) (2021)
4. Beery, S., Wu, G., Rathod, V., Votel, R., Huang, J.: Context R-CNN: long term temporal context for per-camera object detection. In: Conference on Computer Vision and Pattern Recognition (CVPR) (2020)
5. Bertasius, G., Torresani, L.: Classifying, segmenting, and tracking object instances in video with mask propagation. In: Conference on Computer Vision and Pattern Recognition (CVPR) (2020)
6. Cartas, A., Radeva, P., Dimiccoli, M.: Modeling long-term interactions to enhance action recognition. In: International Conference on Pattern Recognition (ICPR) (2021)
7. Chen, B., et al.: Multimodal clustering networks for self-supervised learning from unlabeled videos. In: International Conference on Computer Vision (ICCV) (2021)
8. Chen, D.L., Dolan, W.B.: Collecting highly parallel data for paraphrase evaluation. In: Association for Computational Linguistics (ACL/IJCNLP) (2011)
9. Chen, S., Zhao, Y., Jin, Q., Wu, Q.: Fine-grained video-text retrieval with hierarchical graph reasoning. In: Conference on Computer Vision and Pattern Recognition (CVPR) (2020)
10. Chun, S., Oh, S.J., de Rezende, R.S., Kalantidis, Y., Larlus, D.: Probabilistic embeddings for cross-modal retrieval. In: Conference on Computer Vision and Pattern Recognition (CVPR) (2021)
11. Croitoru, I., et al.: TeachText: crossmodal generalized distillation for text-video retrieval. In: International Conference on Computer Vision (ICCV) (2021)
12. Damen, D., et al.: Rescaling egocentric vision: collection, pipeline and challenges for epic-kitchens-100. Int. J. Comput. Vision (IJCV) **130**, 33–55 (2021)
13. Devlin, J., Chang, M., Lee, K., Toutanova, K.: BERT: pre-training of deep bidirectional transformers for language understanding. In: Conference of the North American Chapter of the Association for Computational Linguistics: Human Language Technologies, (NAACL-HLT) (2019)
14. Dong, J., Li, X., Snoek, C.G.: Word2visualvec: Image and video to sentence matching by visual feature prediction. CoRR, abs/1604.06838 (2016)
15. Dong, J., et al.: Dual encoding for video retrieval by text. Trans. Pattern Anal. Mach. Intell. (TPAMI) **44**, 4065–4080 (2021)
16. Dosovitskiy, A., et al.: An image is worth 16×16 words: transformers for image recognition at scale. In: International Conference on Learning Representations (ICLR) (2021)
17. El-Nouby, A., Neverova, N., Laptev, I., Jégou, H.: Training vision transformers for image retrieval. CoRR (2021)
18. Faghri, F., Fleet, D.J., Kiros, J.R., Fidler, S.: VSE++: improving visual-semantic embeddings with hard negatives. In: British Machine Vision Conference (BMVC) (2018)

19. Furnari, A., Farinella, G.M.: What would you expect? Anticipating egocentric actions with rolling-unrolling LSTMS and modality attention. In: International Conference on Computer Vision (ICCV) (2019)
20. Gabeur, V., Sun, C., Alahari, K., Schmid, C.: Multi-modal transformer for video retrieval. In: Vedaldi, A., Bischof, H., Brox, T., Frahm, J.-M. (eds.) ECCV 2020. LNCS, vol. 12349, pp. 214–229. Springer, Cham (2020). https://doi.org/10.1007/978-3-030-58548-8_13
21. Gao, J., Sun, C., Yang, Z., Nevatia, R.: TALL: temporal activity localization via language query. In: International Conference on Computer Vision (ICCV) (2017)
22. Ging, S., Zolfaghari, M., Pirsiavash, H., Brox, T.: COOT: cooperative hierarchical transformer for video-text representation learning. In: Conference on Neural Information Processing Systems (NeurIPS) (2020)
23. Guo, X., Guo, X., Lu, Y.: SSAN: separable self-attention network for video representation learning. In: Conference on Computer Vision and Pattern Recognition (CVPR) (2021)
24. Gutmann, M., Hyvärinen, A.: Noise-contrastive estimation of unnormalized statistical models, with applications to natural image statistics. J. Mach. Learn. Res. **13**, 1–55 (2012)
25. He, K., Fan, H., Wu, Y., Xie, S., Girshick, R.B.: Momentum contrast for unsupervised visual representation learning. In: Conference on Computer Vision and Pattern Recognition (CVPR) (2020)
26. Heilbron, F.C., Escorcia, V., Ghanem, B., Niebles, J.C.: ActivityNet: a large-scale video benchmark for human activity understanding. In: Conference on Computer Vision and Pattern Recognition (CVPR) (2015)
27. Hendricks, L.A., Wang, O., Shechtman, E., Sivic, J., Darrell, T., Russell, B.C.: Localizing moments in video with temporal language. In: Conference on Empirical Methods in Natural Language Processing (EMNLP) (2018)
28. Jaegle, A., Gimeno, F., Brock, A., Vinyals, O., Zisserman, A., Carreira, J.: Perceiver: general perception with iterative attention. In: International Conference on Machine Learning (ICML) (2021)
29. Józefowicz, R., Vinyals, O., Schuster, M., Shazeer, N., Wu, Y.: Exploring the limits of language modeling. CoRR, abs/1602.02410 (2016)
30. Kazakos, E., Huh, J., Nagrani, A., Zisserman, A., Damen, D.: With a little help from my temporal context: Multimodal egocentric action recognition. In: British Machine Vision Conference (BMVC) (2021)
31. Kazakos, E., Nagrani, A., Zisserman, A., Damen, D.: Epic-fusion: audio-visual temporal binding for egocentric action recognition. In: International Conference on Computer Vision (ICCV) (2019)
32. Klein, B., Lev, G., Sadeh, G., Wolf, L.: Associating neural word embeddings with deep image representations using fisher vectors. In: Conference on Computer Vision and Pattern Recognition (CVPR) (2015)
33. Krishna, R., Hata, K., Ren, F., Fei-Fei, L., Niebles, J.C.: Dense-captioning events in videos. In: International Conference on Computer Vision (ICCV) (2017)
34. Lei, J., et al.: Less is more: Clipbert for video-and-language learning via sparse sampling. In: Conference on Computer Vision and Pattern Recognition (CVPR) (2021)
35. Liu, S., Fan, H., Qian, S., Chen, Y., Ding, W., Wang, Z.: Hit: hierarchical transformer with momentum contrast for video-text retrieval. In: Proceedings of the IEEE/CVF International Conference on Computer Vision (ICCV) (2021)
36. Liu, Y., Albanie, S., Nagrani, A., Zisserman, A.: Use what you have: Video retrieval using representations from collaborative experts. In: British Machine Vision Conference (BMVC) (2019)
37. Liu, Y., Chen, Q., Albanie, S.: Adaptive cross-modal prototypes for cross-domain visual-language retrieval. In: Conference on Computer Vision and Pattern Recognition (CVPR) (2021)

38. Luo, H., et al.: UNIVILM: a unified video and language pre-training model for multimodal understanding and generation. CoRR, abs/2002.06353 (2020)
39. Miech, A., Alayrac, J., Laptev, I., Sivic, J., Zisserman, A.: Thinking fast and slow: Efficient text-to-visual retrieval with transformers. In: Conference on Computer Vision and Pattern Recognition (CVPR) (2021)
40. Miech, A., Alayrac, J., Smaira, L., Laptev, I., Sivic, J., Zisserman, A.: End-to-end learning of visual representations from uncurated instructional videos. In: Conference on Computer Vision and Pattern Recognition (CVPR) (2020)
41. Miech, A., Laptev, I., Sivic, J.: Learning a text-video embedding from incomplete and heterogeneous data. CoRR, abs/1804.02516 (2018)
42. Miech, A., Zhukov, D., Alayrac, J., Tapaswi, M., Laptev, I., Sivic, J.: How to 100 m: learning a text-video embedding by watching hundred million narrated video clips. In: International Conference on Computer Vision (ICCV) (2019)
43. Mithun, N.C., Li, J., Metze, F., Roy-Chowdhury, A.K.: Learning joint embedding with multimodal cues for cross-modal video-text retrieval. In: International Conference on Multimedia Retrieval (ICMR) (2018)
44. Oncescu, A., Henriques, J.F., Liu, Y., Zisserman, A., Albanie, S.: QUERYD: a video dataset with high-quality text and audio narrations. In: International Conference on Acoustics, Speech and Signal Processing (ICASSP) (2021)
45. Patrick, M., et al.: Multi-modal self-supervision from generalized data transformations. In: International Conference on Computer Vision (ICCV) (2021)
46. Patrick, M., et al.: Support-set bottlenecks for video-text representation learning. In: International Conference on Learning Representations (ICLR) (2021)
47. Schroff, F., Kalenichenko, D., Philbin, J.: FaceNet: a unified embedding for face recognition and clustering. In: Conference on Computer Vision and Pattern Recognition (CVPR) (2015)
48. Sener, F., Singhania, D., Yao, A.: Temporal aggregate representations for long-range video understanding. In: Vedaldi, A., Bischof, H., Brox, T., Frahm, J.-M. (eds.) ECCV 2020. LNCS, vol. 12361, pp. 154–171. Springer, Cham (2020). https://doi.org/10.1007/978-3-030-58517-4_10
49. Shao, D., Xiong, Yu., Zhao, Y., Huang, Q., Qiao, Yu., Lin, D.: Find and focus: retrieve and localize video events with natural language queries. In: Ferrari, V., Hebert, M., Sminchisescu, C., Weiss, Y. (eds.) ECCV 2018. LNCS, vol. 11213, pp. 202–218. Springer, Cham (2018). https://doi.org/10.1007/978-3-030-01240-3_13
50. Shao, J., Wen, X., Zhao, B., Xue, X.: Temporal context aggregation for video retrieval with contrastive learning. In: Winter Conference on Applications of Computer Vision (WACV) (2021)
51. Sun, C., Myers, A., Vondrick, C., Murphy, K., Schmid, C.: VideoBERT: A joint model for video and language representation learning. In: International Conference on Computer Vision (ICCV) (2019)
52. Tang, Z., Lei, J., Bansal, M.: DecemBERT: learning from noisy instructional videos via dense captions and entropy minimization. In: Conference of the North American Chapter of the Association for Computational Linguistics: Human Language Technologies (NAACL-HLT) (2021)
53. Vaswani, A., et al.: Attention is all you need. In: Conference on Neural Information Processing Systems (NeurIPS) (2017)
54. Wang, J., et al.: Learning fine-grained image similarity with deep ranking. In: Conference on Computer Vision and Pattern Recognition (CVPR) (2014)
55. Wang, L., Li, Y., Huang, J., Lazebnik, S.: Learning two-branch neural networks for image-text matching tasks. Trans. Pattern Anal. Mach. Intell. (TPAMI) **41**, 394–407 (2018)
56. Wang, L., Li, Y., Lazebnik, S.: Learning deep structure-preserving image-text embeddings. In: Conference on Computer Vision and Pattern Recognition (CVPR) (2016)

57. Wang, T., Isola, P.: Understanding contrastive representation learning through alignment and uniformity on the hypersphere. In: International Conference on Machine Learning (ICML) (2020)
58. Wang, X., Zhu, L., Yang, Y.: T2VLAD: global-local sequence alignment for text-video retrieval. In: Conference on Computer Vision and Pattern Recognition (CVPR) (2021)
59. Wang, X., Wu, J., Chen, J., Li, L., Wang, Y., Wang, W.Y.: VATEX: a large-scale, high-quality multilingual dataset for video-and-language research. In: International Conference on Computer Vision (ICCV) (2019)
60. Wei, J., Xu, X., Yang, Y., Ji, Y., Wang, Z., Shen, H.T.: Universal weighting metric learning for cross-modal matching. In: Conference on Computer Vision and Pattern Recognition (CVPR) (2020)
61. Wray, M., Csurka, G., Larlus, D., Damen, D.: Fine-grained action retrieval through multiple parts-of-speech embeddings. In: International Conference on Computer Vision (ICCV) (2019)
62. Wu, C., Feichtenhofer, C., Fan, H., He, K., Krähenbühl, P., Girshick, R.B.: Long-term feature banks for detailed video understanding. In: Conference on Computer Vision and Pattern Recognition (CVPR) (2019)
63. Wu, Z., Xiong, Y., Yu, S.X., Lin, D.: Unsupervised feature learning via non-parametric instance discrimination. In: Conference on Computer Vision and Pattern Recognition (CVPR) (2018)
64. Xu, H., et al.: VLM: task-agnostic video-language model pre-training for video understanding. In: Association for Computational Linguistics (ACL/IJCNLP) (2021)
65. Xu, H., et al.: VideoCLIP: contrastive pre-training for zero-shot video-text understanding. In: Conference on Empirical Methods in Natural Language Processing (EMNLP) (2021)
66. Xu, J., Mei, T., Yao, T., Rui, Y.: MSR-VTT: a large video description dataset for bridging video and language. In: Conference on Computer Vision and Pattern Recognition (CVPR) (2016)
67. Zhang, B., Hu, H., Sha, F.: Cross-modal and hierarchical modeling of video and text. In: Ferrari, V., Hebert, M., Sminchisescu, C., Weiss, Y. (eds.) ECCV 2018. LNCS, vol. 11217, pp. 385–401. Springer, Cham (2018). https://doi.org/10.1007/978-3-030-01261-8_23
68. Zhang, C., Gupta, A., Zisserman, A.: Temporal query networks for fine-grained video understanding. In: Conference on Computer Vision and Pattern Recognition (CVPR) (2021)
69. Zhang, Z., Han, X., Song, X., Yan, Y., Nie, L.: Multi-modal interaction graph convolutional network for temporal language localization in videos. IEEE Trans. Image Process. **30**, 8265–8277 (2021)
70. Zhou, L., Liu, J., Cheng, Y., Gan, Z., Zhang, L.: CUPID: adaptive curation of pre-training data for video-and-language representation learning. CoRR, abs/2104.00285 (2021)
71. Zhou, L., Xu, C., Corso, J.J.: Towards automatic learning of procedures from web instructional videos. In: Conference on Artificial Intelligence (AAAI) (2018)
72. Zhu, L., Yang, Y.: ActBERT: learning global-local video-text representations. In: Conference on Computer Vision and Pattern Recognition (CVPR) (2020)

A Compressive Prior Guided Mask Predictive Coding Approach for Video Analysis

Zhimeng Huang[1], Chuanmin Jia[2]([✉]), Shanshe Wang[1], and Siwei Ma[1]

[1] National Engineering Research Center of Visual Technology, Peking University, Beijing 100871, China
[2] Wangxuan Institute of Computer Technology, Peking University, Beijing 100871, China
cmjia@pku.edu.cn

Abstract. In real-world scenarios, video analysis algorithms are conducted for visual signals after compression and transmission. Generally speaking, most codecs introduce irreversible distortion due to coarse quantization during compression. The distortion may lead to significant perception degradation in terms of video analysis performance. To tackle this problem, we propose an efficient plug-and-play approach to preserve the essential semantic information in video sequences explicitly. The proposed approach could boost the video analysis performance with a little extra bit cost. Specifically, we employ the proposed approach on an emerging video analysis task, video object segmentation (VOS). Massive experimental results prove that the our work outperforms the existing coding approaches over multiple VOS datasets. Concretely, it could improve the analysis performance by up to 13% at similar bitrates. Additional experiments also verifies the flexibility of our scheme because there is no dependency on any specific VOS model or encoding method. Essentially, the proposed approach provides novel insights for the emerging Video Coding for Machine (VCM) standard.

1 Introduction

In recent years, videos has become the dominant component of the internet traffic. Considering the data volume of video big data, it is necessary to develop high efficient video compression from analysis-friendly perspective. However, in earlier studies [1], the target of video compression is to simply keep the signal fidelity.

Supported in part by the National Natural Science Foundation of China under grant 62072008, 62025101, 61931014, 62101007, and in part by the High Performance Computing Platform of Peking University, which are gratefully acknowledged.

Supplementary Information The online version contains supplementary material available at https://doi.org/10.1007/978-3-031-26316-3_28.

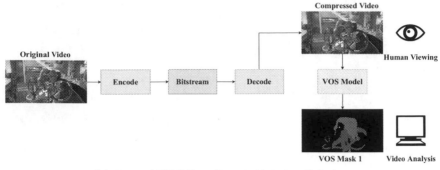

(a) Typical VOS Paradigm in Existing Solutions

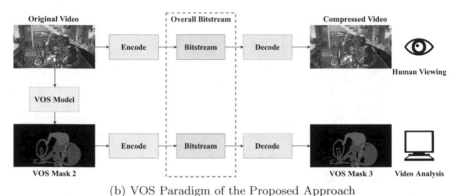

(b) VOS Paradigm of the Proposed Approach

Fig. 1. The comparison between the paradigm of vision tasks in real-world application and that in our proposed approach. Our work uses an additional bitstream to significantly improve the VOS performance.

Methods following this target tend to ignore other useful information in compressed domain, which may cause severe performance degradation in video analysis tasks. To discuss this problem, we regard video object segmentation, a popular video analysis task recently, as the representation of video analysis tasks. Semi-supervised VOS targets segmenting particular objects throughout the entire video sequence, given only the object mask of the first frame. Existing state-of-the-art (SOTA) solutions [2–4] could achieve high accuracy by fully utilizing the semantic information of the input videos. Nevertheless, the performance of these high efficient methods sharply decrease when dealing with video sequences with compression distortions, especially at low bitrate. Therefore it is critical to boost the performance of the existing video analysis methods on compressed videos.

For better comprehension, we summarize two kinds of paradigms for VOS. As shown in Fig. 1a, the VOS algorithm are conducted after signal capture, signal encoding, bitstream transmitting, and bitstream decoding. In other words, VOS is more like a downstream task for video compression. To avoid analytical performance degradation caused by video compression, we propose a compressive prior guided mask predictive coding approach. The proposed scheme is shown in Fig. 1b. Different from the traditional one, the proposed work could improve the analysis performance by an additional bitstream. Following the proposed approach, the VOS methods could be deployed on the raw videos[1] directly rather than compressed video sequences with severe semantic distortion.

Specifically, our framework could be divided into three parts, Motion Estimation (ME), Feature Compression (FC), and Motion Comprehension (MC). For the first part, we utilize the existing masks as a reference to extract a motion feature of the current mask. Then the motion feature is compressed by an end-to-end autoencoder for transmission. Finally, the compressed motion feature and the reference masks are fed into a convolution neural network (CNN) based comprehension network to generate the predicted masks. Experiments are deployed on two VOS baseline models and three standard VOS datasets. The consistently superior performances to the baseline codec demonstrate the effectiveness and generality of the proposed framework. The contributions are summarized as follows:

- We propose a novel approach to improve VOS performance on compressed video sequences. To the best of our knowledge, this is the first work who jointly considers the bitrate and the corresponding VOS performance.
- The proposed framework is high-efficient, generalizable, and flexible. It could be transferred to any VOS methods or datasets without any fine-tuning.
- Experimental results prove that the proposed framework outperforms traditional codecs over two VOS models and three VOS datasets. Exhaustive experiments also demonstrate the robustness of our framework. Our method provides novel possibilities the Video Coding for Machine (VCM) research.

2   Related Works

2.1   Video Object Segmentation (VOS)

VOS has two sub-tasks, semi-supervised VOS and unsupervised VOS. The difference between them is whether an initial mask is provided. In this paper, we mainly consider the former one. Current semi-supervised VOS methods fall into one of two categories. Following the first category, VOS method [5–7] generates the masks by fine-tuning the provided initial mask. Regarding the second category, these approaches [8–10] adopt propagating the mask from the previous frame using optical flow and then refining these estimates using a fully convolutional network. However, both VOS methods are facing difficulties when dealing with compressed video sequences due to the discriminative feature damage caused by quantization error.

Moreover, some VOS approaches utilize compressed videos as an extra supplementary information for better accuracy or efficiency. To increase VOS performance, several algorithms [11,12] convert the input video sequences into the compressed domain. In [13], the sparse motion vector are utilized for object

---

[1] Note that we omit the encoding distortions caused by signal capturing tools such as cameras. Namely, we assume that the video sequences pre-processed in the dataset are all pristine videos.

segmentation. To realize high efficient processing, many researches [14] use bit-stream of compressed video sequences to accelerate existing VOS methods. A plug-and-play acceleration framework is proposed in [15] by propagating the motion vectors extracted from the HEVC [16] bitstream. However, existing approaches for accuracy and efficiency pay little attention to the critical problem, the bitrate analysis of the compressed videos and their associated VOS performance. For example, if the bandwidth for compressed videos is unlimited, the common codecs allow almost lossless compression at an extremely high bitrate, which is unrealistic in practical scenarios. Therefore in this paper, we regard bitrate as another dimension of discussions, which is different from works mentioned in this subsection.

2.2   Image/Video Compression for Vision Tasks

Most of the existing compression frameworks for visual tasks aim at image tasks such as image classification, object detection, and semantic segmentation. Therefore we will introduce some image compression methods which could improve analysis performance for vision tasks. Benefited from emerging neural image/video compression approaches [17–19], most existing machine vision oriented methods mainly utilize a task-related learning objective to optimize the entire framework [20–23]. Chamain $et\ al.$ [24] formulate a detection loss to optimize existing end-to-end image compression framework. Moreover, a content-adaptive end-to-end compression approach [25] is proposed for instance segmentation task. However, these methods rely heavily on analytical models and datasets. Therefore, they are less effective in practical application scenarios. To overcome this problem, we aim to design a framework that is effective for different datasets and analytical algorithms.

3   Methods

The proposed method is elaborated in this section. We first demonstrate the schematic illustration of our method with terminology definition. Subsequently, the detailed descriptions of each module are provided and analyzed. The notations and preliminary concepts are shown in Table 1.

3.1   Overview

The overall flowchart of the proposed approach in illustrated in Fig. 2. Denote $\mathcal{X}$ as the original video sequence with length T. For each frame $X_t \in \mathcal{X}$ at time step t, we generate the corresponding VOS mask $M_t = F(X_t)$ by VOS model F. After the generation of all of the masks, we get the mask sequence $\mathcal{M} = \{M_1, M_2, ...M_t, ...\}$. Note that the first mask M_1 is provided by the self-supervised VOS task. For other masks M_t at time step t ($t \in [2, T]$), we deconstruct it into N_t binary masks, in which N_t indicates the number of objects in

Table 1. Notations and descriptions of the proposed scheme

Notations	Descriptions
$\mathcal{X} = \{X_1, X_2, \ldots, X_t, \ldots\}$	a sequence of video frames with timestep t
$\mathcal{M} = \{M_1, M_2, \ldots, M_t, \ldots\}$	a sequence of masks
M_t	original mask at t
N_t	number of objects at t
$m_{t,k}$	binary mask of object k at t
$v_{t,k}$	motion feature of object k at t
$\hat{v}_{t,k}$	compressed motion feature of object k at t
$\hat{m}_{t,k}$	the prediction of $m_{t,k}$
P_t	the prediction of M_t
$z_{t,k}$	bitstream of $v_{t,k}$
$b_{t,k}$	bitrates of compressed $v_{t,k}$
B_t	bitrates of compressed motion features at t
F	baseline VOS method
Θ_τ	trainable parameters of module τ

M_t. Denote the binary map of the kth object in M_t as $m_{t,k}$, which could be formulated as,

$$m_{t,k}(i,j) = \begin{cases} 1 & M_t(i,j) = k \\ 0 & others, \end{cases} \tag{1}$$

in which (i,j) represents the coordinate of each pixel. The reason for doing so lies in that objects with different labels will essentially disturb the motion estimation between each other. The related analysis will be conduced in Sect. 4.4. Note that the $m_{t-1,k}$ is not available at the decoder, thus for each binary mask $m_{t,k}$, we utilize the previous reconstructed binary mask $\hat{m}_{t-1,k}$ as the reference mask to ensure encoder-decoder consistency. When $t = 2$, the reference mask is $m_{1,k}$ given by the labeled data. Then the binary mask $m_{t,k}$ and its reference is fed into ME module to estimate the changes from time step $t-1$ to t. The changes are represented by a motion feature $v_{t,k} = ME(m_{t,k}, \hat{m}_{t-1,k})$. After that, we encode the motion feature into a more compact representation $z_{t,k}$ by **FC** for transmission. We utilize $b_{t,k}$, the size of $z_{t,k}$, to evaluate the bit cost to compress each motion feature. Moreover, given all of the $b_{t,k}$ at time step t, the bit cost for the entire mask B_t is calculated by,

$$B_t = \sum_{i=1}^{N_t} b_{t,k}. \tag{2}$$

The compressed motion feature $\hat{v}_{t,k}$, together with the reference mask $\hat{m}_{t-1,k}$, are utilized to generate the predicted binary mask $\hat{m}_{t,k}$ by MC. Specifically, MC module consists of two parts. Firstly, $\hat{v}_{t,k}$ is warped on $\hat{m}_{t-1,k}$ to obtain a coarse prediction for $m_{t,k}$. Then the warped mask is further refined by a CNN-based network. Finally, we merge all of the predicted binary masks into the predicted

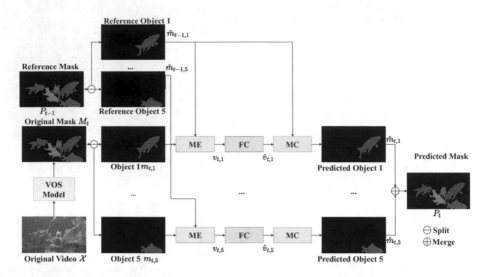

Fig. 2. The overall flowchart of the proposed approach. **ME, FC, MC** respectively denote the *Motion Estimation, Feature Compression,* and *Motion Comprehension* module. Firstly, the mask M_t is generated by a VOS model. Then the masks are split into several binary masks. For each binary mask $m_{t,k}$, we utilize the responding predicted mask $\hat{m}_{t-1,k}$ as the reference mask to extract the motion feature by ME. Then the motion feature $v_{t,k}$ is compressed by the FC module for transmission and decompression. After decompression, the compressed motion feature $\hat{v}_{t,k}$, together with $\hat{m}_{t-1,k}$, are utilized to predict $\hat{m}_{t,k}$, the predicted binary mask by MC. Finally, all of the binary masks are merged to the final reconstructed mask P_t. Note that we color some binary masks (including $m_{t,k}$, $\hat{m}_{t-1,k}$ and $\hat{m}_{t,k}$) for better comprehension.

mask P_t at time step t.

$$P_t(i,j) = \begin{cases} k & \hat{m}_{t,k} = 1 \\ 0 & others. \end{cases} \tag{3}$$

3.2   Detailed Architecture

ME. In our proposed framework, we employ a masked CNN-based optical flow estimation approach [26] to estimate the motion between the temporal adjacent binary mask. The visualization of ME module is shown in Fig. 3. The splitted binary mask $m_{t,k} \in \{0,1\}^{W \times H}$ is fed into the optical flow estimation model to generate $o_{t,k} \in \mathbb{R}^{W \times H \times 2}$, which denotes the motion displacement between $\hat{m}_{t-1,k}$ and $m_{t,k}$. Then $o_{t,k}$ is element-wise multiplied with $\hat{m}_{t-1,k}$ to generate the motion feature $v_{t,k} \in \mathbb{R}^{W \times H \times 2}$. Instead of directly deploying the estimated flow, the element-wise multiplication have two advantages. One is that the multiplication could remove the disturbs of other regions. Note that the optical flow responding of the black region are the noises of the estimation. The other is the multiplication could make the motion feature easier to compress. which

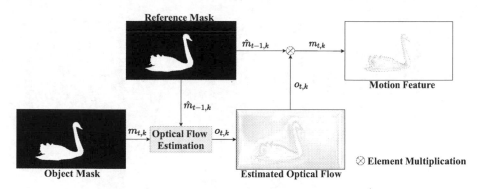

Fig. 3. Network of ME module. The example is the video *"swan"* in DAVIS 2017. The optical flow estimation module follows the implementation in [26].

means the bit cost of the proposed approach decreases. Compared to the optical flow or motion vector in traditional codecs, the proposed ME module could be end-to-end optimized with more flexibility.

FC. Since the motion feature extracted by ME module could not be transmitted directly. We deploy a modified hyperprior guided autoencoder proposed by [27], which is utilized to compress images with three channels. Therefore we modify the shape of the inputs and outputs to meet that of $v_{t,k}$. And the output of FC is the compressed motion feature $\hat{v}_{t,k}$, the bitstream $z_{t,k}$ and the bits of the bitstream $b_{t,k}$. The architecture of FC module is shown in Fig. 4. Every $v_{t,k} \in \mathbb{R}^{W \times H \times 2}$ is compressed into $z^1 \in \mathbb{R}^{\frac{W}{16} \times \frac{H}{16} \times 96}$ and $z^2 \in \mathbb{R}^{\frac{W}{64} \times \frac{H}{64} \times 64}$. z^1 denotes the representation of $v_{t,k}$. z^2 represents the parameters of the distribution to recover $\hat{v}_{t,k} \in \mathbb{R}^{W \times H \times 2}$.[2] Compared to other traditional compression methods, the proposed FC module is more efficient. Because the trainable FC module can be optimized to fit the signal distribution of $v_{t,k}$, which is different from that of images.

MC. The MC module is deployed to generate the predicted binary mask $\hat{m}_{t,k}$ by the compressed motion feature $\hat{v}_{t,k}$ and the reference mask $\hat{m}_{t-1,k}$. The flowchart of MC is shown in Fig. 5. Firstly, we warp the compressed motion feature $\hat{v}_{t,k}$ on the reference binary mask $\hat{m}_{t-1,k}$. However, the warped mask is far from an accurate prediction. Thus we deploy a CNN based refinement network after the warp operation. The network takes the reference mask and the warped mask as the inputs to generate the final predicted binary mask. Limited by the length of the paper, more comparisons between the warped masks before refinement and after refinement are shown in the supplementary materials.

---

[2] $z_{t,k}^1$ and $z_{t,k}^2$ for completeness but we drop the subscript (t, k) for simplicity.

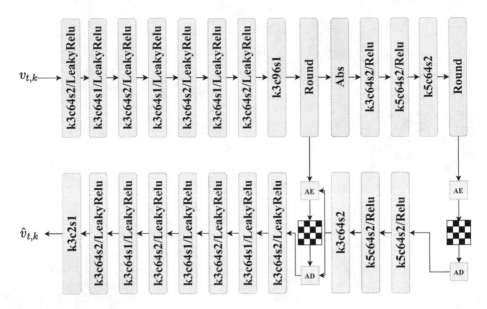

Fig. 4. Network architecture of FC module. The notations on the blocks denote the hyper-parameters. For example, "k3c64s2" indicates a convolution or deconvolution layer with 64 channels, the stride step is 2, and the kernel size is 3. The blocks in green represent convolution layers, and the blocks in blue denote deconvolution layers. The parameter of all of the LeakyRelu is set to be 0.1. (Color figure online)

4 Experimentation

4.1 Training Details

Loss Function. The goal of the proposed framework is to leverage the accuracy of the predicted masks and minimize bits required for transmission simultaneously. Therefore, the optimization problem at time t could be formulated as:

$$\mathcal{L}_t = \alpha R_t + \lambda D_t$$
$$= \alpha \sum_{k=1}^{N_t} H(\hat{v}_{t,k}) + \lambda \sum_{k=1}^{N_t} d(m_{t,k}, \hat{m}_{t,k}), \tag{4}$$

in which $d(m_{t,k}, \hat{m}_{t,k})$ denotes the distortions between $m_{t,k}$ and $\hat{m}_{t,k}$. In practice, we use the mean square error (MSE) in our experiments. $H(\cdot)$ represents the number of bits utilized to compress the motion feature. Actually, the bitrate estimation is a very complicated problem in end-to-end image compression. However, it is beyond the scope of this work. Therefore, we directly utilize the implementation of hyper-prior based entropy model [27] denoted by function $H(\cdot)$. λ indicates the Lagrange multiplier to adjust the trade-off between R_t and D_t. α is a binary parameter to remove the loss from R_t.

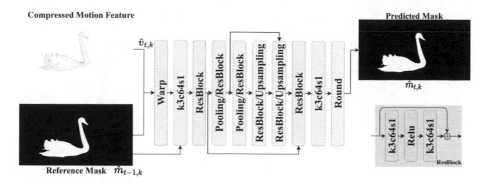

Fig. 5. Network architecture of the MC module. The notations on the blocks follows the rules mentioned in Fig. 4. All of the ResBlocks are deployed with the structure at the bottom right corner.

Table 2. Detailed training configuration

	Components trained	α	λ	Learning rate	Batch size	Epochs
Stage I	ME, MC	0	1	1e−3	6	40
Stage II	FC	1	1e−5	1e−3	4	50
Stage III	FC,ME,MC	1	1e−5	1e−3	2	200

Training Details. The training details of each stage is shown in Table 2. And the optical flow estimation net in ME module is initialized with a pre-train model[3]. The optimizer is AdamW with decay, ϵ equalling to 5e−5 and 1e−8, respectively. Although the approach is set to optimize for 200 epochs.

The entire scheme is implemented by using PyTorch 1.11.0 with CUDA 11.3. The simulation environment is based on Ubuntu 18.04 with one NVIDIA 3080ti graphic card. We train distinct models for different bitrate points to realize optimal performance.

4.2   Experimental Settings

Datasets and Preparation. The experiments are deployed on three VOS benchmarks: DAVIS 2016 [28], DAVIS 2017 [29], and YouTube-VOS [30]. DAVIS 2016 and DAVIS 2017 are small datasets with 50 and 120 video sequences, respectively. YouTube-VOS is a large-scale dataset with 3945 video sequences. All the video sequences are converted from RGB to YUV420 format, which is widely employed color space for traditional codecs such as x265 and VVEnc[4] to compress.

For the test on three datasets, we use the same model trained by DAVIS 2017 dataset. Note that our framework does not use enormous videos for training. Only 60 video sequences are utilized. This configuration proves that our work is simple and robust for unseen VOS datasets.

---

[3] https://github.com/zacjiang/GMA.

[4] VVEnc [31] is a light-weighted implementation of the reference software of VVC [32].

Table 3. Experimental results on DAVIS2016 and DAVIS2017

Dataset	VOS model	Method	Bitrate↓	$\mathcal{J}_m$↑	$\mathcal{F}_m$↑	$(\mathcal{J}\&\mathcal{F})_m$↑
DAVIS 2016 [28]	AOT [33]	Original	–	0.9014	0.9217	0.9117
		x265(baseline)	0.0198	0.7944	0.8118	0.8031
		x265+Ours	0.0168	0.8728	0.9044	0.8886
	STCN [34]	Original	–	0.9042	0.9305	0.9172
		x265(baseline)	0.0198	0.7517	0.7848	0.7683
		x265+Ours	0.0158	0.8814	0.9148	0.8981
DAVIS 2017 [29]	AOT [33]	Original	–	0.8251	0.8791	0.8521
		x265(baseline)	0.0209	0.7165	0.7608	0.7386
		x265+Ours	0.0178	0.8056	0.8652	0.8354
	STCN [34]	Original	–	0.8200	0.8862	0.8531
		x265(baseline)	0.0209	0.6734	0.7161	0.6947
		x265+Ours	0.0158	0.7890	0.8641	0.8265

Metrics. For the accuracy on VOS, we follows the standard criteria from [28]: Jaccard Index $\mathcal{J}$ and $\mathcal{F}$-scores, which represent the region similarity and contour accuracy, respectively. Additionally, we also report some detailed performance for each dataset: {Mean↑, Recall↑, Decay↓}x{$\mathcal{J}$,$\mathcal{F}$} for DAVIS 2017 and {Seen, Unseen}x{$\mathcal{J}$,$\mathcal{F}$} for YouTube-VOS. The experimental results of these metrics are provided in supplementary materials

In addition to the traditional VOS evaluation metrics, we have added the evaluation metric: bitrate. Bitrate denotes the compression ratio of codecs. Generally speaking, the method with less bitrate is better given the same performance. Specifically, we utilize bits-per-pixel (bpp) as the evaluation of bitrate in the experiments. As the proposed approach is an additional module to existing codecs. Thus the bitrate of our work is the sum of it and the baseline codec for a fair comparison.

Base Video Codecs. We choose two conventional codecs (x265 and VVEnc) as the baselines. Considering the efficiency and effectiveness, we choose the x265 library in FFmpeg with *veryfast* preset for main experiments. And VVEnc is utilized to compress videos for the supplementary experiments. The compressed sequences will be shared to encourage others to research compressed video segmentation.

Base VOS Models. We choose the AOT and STCN as the base model in our framework.[5] AOT employs a Long Short-Term Transformer to construct hierarchical matching and propagation. STCN combines the computational advantages of temporal convolutional networks with the representational power and robustness of stochastic latent spaces.

4.3   Experimental Results

In this subsection, the experiments are deployed with different VOS models and codecs for comparison. Please refer to supplementary material for more details

---

[5] AOT and STCN are representative VOS models with codes and models available.

Table 4. Experimental results on YouTube-VOS dataset.

VOS model	Method	Bitrate↓	$\mathcal{J}_s$↑	$\mathcal{F}_s$↑	$\mathcal{J}_u$↑	$\mathcal{F}_u$↑	$\mathcal{G}$↑
AOT [33]	Original	–	0.8387	0.7990	0.8880	0.8848	0.8526
	x265(baseline)	0.0159	0.7966	0.8429	0.7530	0.8450	0.8094
	x265+Ours	0.0121	0.8265	0.8811	0.7782	0.8742	0.8400
STCN [34]	Original	–	0.8259	0.8695	0.7946	0.8772	0.8418
	x265(baseline)	0.0159	0.7867	0.8264	0.7372	0.8200	0.7867
	x265+Ours	0.0122	0.8149	0.8634	0.7726	0.8656	0.8291

and commands about the x265/VVEnc settings. Furthermore, the concrete VOS models are also provided in it.

The performance of the proposed framework compared to x265 codec is reported in Table 3 and Table 4. On DAVIS 2016, our work achieves 8% and 13% accuracy($\mathcal{J}\&\mathcal{F}$) improvement for AOT and STCN with 35% and 25% bitrate saving, respectively. On DAVIS 2017, our work achieves 10% and 12% accuracy($\mathcal{J}\&\mathcal{F}$) improvement for AOT and STCN with 14% and 24% bitrate saving, respectively. On YouTube-VOS, we achieve 3.1% and 4.3% $\mathcal{G}$($\mathcal{G}$ denotes the score calculated by the evaluation server[6]) improvement for AOT and STCN with 20% and 23% bitrate saving. The performance improvement over all of the VOS models and datasets reveals the generality of our framework.

Rate-Performance Curves. We report the Rate-Performance (RP) curves in Fig. 6. The QP set of both x265 and x265+Ours are $\{32, 37, 42, 47\}$. Taking **DAVIS 2016+AOT** in Fig. 6(a) as an example, the line in orange denotes the RP curve of video sequences compressed by x265 codec without the proposed framework. The line in blue indicates the RP curve of video sequences compressed by x265 with the proposed framework. From these curves, several conclusions could be drawn. First of all, the proposed framework outperforms x265 at every bitrate. Secondly, the accuracy loss of our method is minimal compared to the results on the original videos. Last but not the least, the proposed framework could achieve remarkable performance for different datasets and VOS methods.

4.4   Ablation Studies and Modular Analysis

FC Module. We utilize a hyperprior guided auto-encoder model to compress the underlying motion feature. Actually, it is also reasonable to deploy a traditional codec to encode the feature. Thus we investigate the efficiency of FC module by using x265 codec for comparison. As shown in Table 5, experimental results prove that the proposed FC module could efficiently compress the motion vector. It is because that the learned parameters in FC module were optimized to fit the distribution of motion features during training.

MC Module. In our framework, we propose a CNN-based MC model to refine the predicted mask after the warp operation. Another alternative is to utilize the

---

[6] https://competitions.codalab.org/competitions/20127#participate-submit_results.

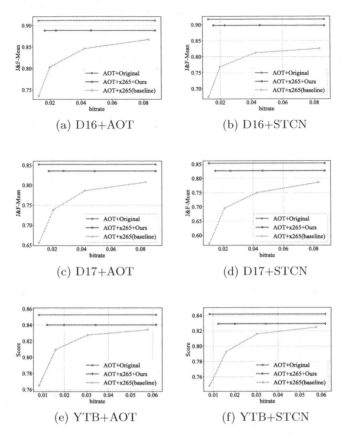

Fig. 6. Rate-Performance curves of our approach. D16, D17, and YTB represent DAVIS 2016, DAVIS 2017 and YouTube-VOS, respectively. Original denotes that the videos are uncompressed.

warped mask directly as the prediction results. Thus we conduct a set of ablation study to verify the effectiveness of the MC module. From the visualization in Table 6, it is obvious that the propose MC module significantly improve the quality of the predicted mask. Then the experimental results shows that the $(\mathcal{J}\&\mathcal{F})_m$ of the approach without MC module will drop about 30% on DAVIS-2017 dataset.

More Codecs. We conduct experiments on more codecs to verify the generality of our framework. The experimental results on are shown in Table 7. Although VVEnc is the most efficient conventional codecs, the proposed framework still achieve 9.1% improvement on DAVIS 2017 with 32% bits saving.

More Bitrates. By adjusting the λ in the loss function ($\lambda = 10^{-a}$, a = 2, 3, 4, 5), the proposed framework are trained for different bitrates. To valid the performance at different bitrates, we conduct experiments on DAVIS 2017. Note that the bitrate is calculated without that of codecs. As shown in Table 8, with the

Table 5. Ablation study for FC module on DAVIS 2017 dataset.

Method	Bitrate↓	$\mathcal{J}_m\uparrow$	$\mathcal{F}_m\uparrow$	$(\mathcal{J}\&\mathcal{F})_m\uparrow$
x265+Ours(x265)	0.0209	0.8052	0.8655	0.8353
x265+Ours(FC)	0.0178	0.8056	0.8652	0.8354

Table 6. Ablation study for MC module on DAVIS 2017 dataset.

Method	Bitrate↓	$\mathcal{J}_m\uparrow$	$\mathcal{F}_m\uparrow$	$(\mathcal{J}\&\mathcal{F})_m\uparrow$
x265+Ours(Warp)	0.0178	0.4725	0.5358	0.5041
x265+Ours(MC)	0.0178	0.8056	0.8652	0.8354

increase of the bitrate, the difference compared to the analysis performance of the original videos become smaller and smaller. This experimental result demonstrates that our model could be adapted to different application scenarios by adjusting the bitrate.

Direct Comparison to Codecs. Rethinking the paradigm illustrated in Fig. 1b, it is feasible to utilize existing compression codecs to compress the VOS masks to replace our approach. Therefore we conduct a comparative experiment. We utilize x265 as the mask codec to compress the masks generated by AOT on DAVIS 2017. The experimental results are shown in Table 9. It is obvious that the masks encoded by x265 are not accurate for VOS anymore. Then main reason to the degradation is the noise caused by codec at extremely low bitrate.

5 Discussions

Extensibility. The proposed framework can be easily extended to other mask-based vision tasks. Mask based vision tasks denotes tasks where the output is a mask related to the semantics of the video such as video instance segmentation and multi object tracking. To run our work on other tasks, simply replace the VOS model with the model for the corresponding task.

Robustness. As reported in Table 8, the proposed framework could utilize an extremely low bitrate to achieve 95% performance compared to that on original video sequences. It proves that our work could handle practical scenarios with low network.

Generality. The generality of the proposed framework is manifested in two aspects. Firstly, the proposed framework does not rely on any VOS methods.

Table 7. Experimental results for VVEnc on DAVIS 2017 dataset.

Method	Bitrate↓	$\mathcal{J}_m\uparrow$	$\mathcal{F}_m\uparrow$	$(\mathcal{J}\&\mathcal{F})_m\uparrow$
Original	-	0.8251	0.8791	0.8521
VVEnc(baseline)	0.0179	0.7235	0.7646	0.7440
VVEnc+Ours	0.0121	0.8056	0.8652	0.8354

Table 8. Experimental results for different $\lambda(\lambda = 10^{-a})$

a	Bitrate↓	$\mathcal{J}_m$↑	$\mathcal{F}_m$↑	$(\mathcal{J}\&\mathcal{F})_m$↑
Original	-	0.8251	0.8791	0.8521
2	0.0262	0.8174	0.8706	0.8440
3	0.0162	0.8135	0.8690	0.8413
4	0.0069	0.8056	0.8652	0.8354
5	0.0023	0.7752	0.8502	0.8127

Table 9. Experimental results for direct comparison to codecs

Method	Bitrate↓	$\mathcal{J}_m$↑	$\mathcal{F}_m$↑	$(\mathcal{J}\&\mathcal{F})_m$↑
Original	–	0.8251	0.8791	0.8521
x265	0.0072	0.5856	0.6488	0.6172
Ours	0.0069	0.8056	0.8652	0.8354

Therefore, our framework can also be adapted to new VOS methods with higher performance. Secondly, the input of the proposed frameworks are masks rather than objects, which means the category information of the objects are not utilized in our work. Thus it could be easily transferred to other datasets without fine-tuning.

6   Conclusion

In this paper, we propose a simple and effective framework to improve the performance of semi-supervised VOS on compressed video sequences by preserving the semantic information during compression procedure. Such a framework could effectively diminish the degradation of VOS performance during lossy compression. Moreover, the proposed is a plug-and-play framework which means that it does not rely on a specific VOS method or a specific codec. Concretely, it could improve the analysis performance by up to 13% at similar bitrates. As VCM has been an emerging research topic recently, we provide an promising solution for efficient and high performance compressed VOS.

References

1. Wiegand, T., Sullivan, G.J., Bjontegaard, G., Luthra, A.: Overview of the h. 264/AVC video coding standard. IEEE Trans. Circuits Syst. Video Technol. **13**, 560–576 (2003)
2. Luiten, J., Voigtlaender, P., Leibe, B.: PReMVOS: proposal-generation, refinement and merging for video object segmentation. In: Jawahar, C.V., Li, H., Mori, G., Schindler, K. (eds.) ACCV 2018. LNCS, vol. 11364, pp. 565–580. Springer, Cham (2019). https://doi.org/10.1007/978-3-030-20870-7_35

3. Maninis, K.K., et al.: Video object segmentation without temporal information. IEEE Trans. Pattern Anal. Mach. Intell. **41**, 1515–1530 (2018)
4. Oh, S.W., Lee, J.Y., Xu, N., Kim, S.J.: Video object segmentation using space-time memory networks. In: IEEE International Conference on Computer Vision, pp. 9226–9235 (2019)
5. Caelles, S., Maninis, K.K., Pont-Tuset, J., Leal-Taixé, L., Cremers, D., Van Gool, L.: One-shot video object segmentation. In: IEEE Conference on Computer Vision and Pattern Recognition, pp. 221–230 (2017)
6. Robinson, A., Lawin, F.J., Danelljan, M., Khan, F.S., Felsberg, M.: Learning fast and robust target models for video object segmentation. In: IEEE Conference on Computer Vision and Pattern Recognition, pp. 7406–7415 (2020)
7. Voigtlaender, P., Leibe, B.: Online adaptation of convolutional neural networks for the 2017 davis challenge on video object segmentation. In: The 2017 DAVIS Challenge on Video Object Segmentation-CVPR Workshops, vol. 5 (2017)
8. Li, X., Loy, C.C.: Video object segmentation with joint re-identification and attention-aware mask propagation. In: Ferrari, V., Hebert, M., Sminchisescu, C., Weiss, Y. (eds.) ECCV 2018. LNCS, vol. 11207, pp. 93–110. Springer, Cham (2018). https://doi.org/10.1007/978-3-030-01219-9_6
9. Li, X., et al.: Video object segmentation with re-identification. arXiv preprint arXiv:1708.00197 (2017)
10. Perazzi, F., Khoreva, A., Benenson, R., Schiele, B., Sorkine-Hornung, A.: Learning video object segmentation from static images. In: IEEE Conference on Computer Vision and Pattern Recognition, pp. 2663–2672 (2017)
11. Jamrozik, M.L., Hayes, M.H.: A compressed domain video object segmentation system. In: International Conference on Image Processing, vol. 1, I-I. IEEE (2002)
12. Porikli, F., Bashir, F., Sun, H.: Compressed domain video object segmentation. IEEE Trans. Circuits Syst. Video Technol. **20**, 2–14 (2009)
13. Babu, R.V., Ramakrishnan, K., Srinivasan, S.: Video object segmentation: a compressed domain approach. IEEE Trans. Circuits Syst. Video Technol. **14**, 462–474 (2004)
14. Tan, Z., et al.: Real time video object segmentation in compressed domain. IEEE Trans. Circuits Syst. Video Technol. **31**, 175–188 (2020)
15. Xu, K., Yao, A.: Accelerating video object segmentation with compressed video. In: IEEE Conference on Computer Vision and Pattern Recognition, pp. 1342–1351 (2022)
16. Sullivan, G.J., Ohm, J.R., Han, W.J., Wiegand, T.: Overview of the high efficiency video coding (HEVC) standard. IEEE Trans. Circuits Syst. Video Technol. **22**, 1649–1668 (2012)
17. Lu, G., Zhang, X., Ouyang, W., Chen, L., Gao, Z., Xu, D.: An end-to-end learning framework for video compression. IEEE Trans. Pattern Anal. Mach. Intell. **43**, 3292–3308 (2020)
18. Hu, Z., Lu, G., Xu, D.: FVC: a new framework towards deep video compression in feature space. In: IEEE Conference on Computer Vision and Pattern Recognition, pp. 1502–1511 (2021)
19. Li, J., Li, B., Lu, Y.: Deep contextual video compression. Adv. Neural. Inf. Process. Syst. **34**, 18114–18125 (2021)
20. Chen, Z., He, T.: Learning based facial image compression with semantic fidelity metric. Neurocomputing **338**, 16–25 (2019)
21. Wang, S., et al.: Towards analysis-friendly face representation with scalable feature and texture compression. IEEE Trans. Multimedia **24**, 3169–3181 (2021)

22. Le, N., Zhang, H., Cricri, F., Ghaznavi-Youvalari, R., Rahtu, E.: Image coding for machines: an end-to-end learned approach. In: IEEE International Conference on Acoustics, Speech and Signal Processing, pp. 1590–1594 (2021)
23. Huang, Z., Jia, C., Wang, S., Ma, S.: HMFVC: a human-machine friendly video compression scheme. IEEE Trans. Circuits Syst. Video Technol. (2022)
24. Chamain, L.D., Racapé, F., Bégaint, J., Pushparaja, A., Feltman, S.: End-to-end optimized image compression for machines, a study. In: IEEE Data Compression Conference, pp. 163–172 (2021)
25. Le, N., Zhang, H., Cricri, F., Ghaznavi-Youvalari, R., Tavakoli, H.R., Rahtu, E.: Learned image coding for machines: a content-adaptive approach. In: IEEE International Conference on Multimedia and Expo, pp. 1–6 (2021)
26. Jiang, S., Campbell, D., Lu, Y., Li, H., Hartley, R.: Learning to estimate hidden motions with global motion aggregation. In: International Conference on Computer Vision, pp. 9772–9781 (2021)
27. Minnen, D., Ballé, J., Toderici, G.D.: Joint autoregressive and hierarchical priors for learned image compression. In Advances in Neural Information Processing Systems, vol. 31 (2018)
28. Perazzi, F., Pont-Tuset, J., McWilliams, B., Van Gool, L., Gross, M., Sorkine-Hornung, A.: A benchmark dataset and evaluation methodology for video object segmentation. In: IEEE Conference on Computer Vision and Pattern Recognition, pp. 724–732 (2016)
29. Pont-Tuset, J., et al.: The 2017 davis challenge on video object segmentation. arXiv preprint arXiv:1704.00675 (2017)
30. Xu, N., et al.: YouTube-VOS: a large-scale video object segmentation benchmark. arXiv preprint arXiv:1809.03327 (2018)
31. Wieckowski, A., et al.: VVENC: an open and optimized VVC encoder implementation. In: IEEE International Conference on Multimedia & Expo Workshops, pp. 1–2. IEEE (2021)
32. Bross, B., Chen, J., Ohm, J.R., Sullivan, G.J., Wang, Y.K.: Developments in international video coding standardization after AVC, with an overview of versatile video coding (VVC). Proc. IEEE **109**, 1463–1493 (2021)
33. Yang, Z., Wei, Y., Yang, Y.: Associating objects with transformers for video object segmentation. Adv. Neural. Inf. Process. Syst. **34**, 2491–2502 (2021)
34. Aksan, E., Hilliges, O.: STCN: stochastic temporal convolutional networks. In: International Conference on Learning Representations (2019)

BaSSL: Boundary-aware Self-Supervised Learning for Video Scene Segmentation

Jonghwan Mun[1(✉)], Minchul Shin[1], Gunsoo Han[1], Sangho Lee[2], Seongsu Ha[2], Joonseok Lee[2(✉)], and Eun-Sol Kim[3(✉)]

[1] Kakao Brain, Seongnam, South Korea
{jason.mun,craig.starr,coco.han}@kakaobrain.com
[2] Graduate School of Data Science, Seoul National University, Seoul, South Korea
{sangho.lee,sha17,joonseok}@snu.ac.kr
[3] Department of Computer Science, Hanyang University, Seoul, South Korea
eunsolkim@hanyang.ac.kr

Abstract. Self-supervised learning has drawn attention through its effectiveness in learning in-domain representations with no ground-truth annotations; in particular, it is shown that properly designed pretext tasks bring significant performance gains for downstream tasks. Inspired from this, we tackle video scene segmentation, which is a task of temporally localizing scene boundaries in a long video, with a self-supervised learning framework where we mainly focus on designing effective pretext tasks. In our framework, given a long video, we adopt a sliding window scheme; from a sequence of shots in each window, we discover a moment with a maximum semantic transition and leverage it as pseudo-boundary to facilitate the pre-training. Specifically, we introduce three novel boundary-aware pretext tasks: 1) Shot-Scene Matching (SSM), 2) Contextual Group Matching (CGM) and 3) Pseudo-boundary Prediction (PP); SSM and CGM guide the model to maximize intra-scene similarity and inter-scene discrimination by capturing contextual relation between shots while PP encourages the model to identify transitional moments. We perform an extensive analysis to validate effectiveness of our method and achieve the new state-of-the-art on the MovieNet-SSeg benchmark. The code is available at https://github.com/kakaobrain/bassl.

Keywords: Video scene segmentation · Self-supervised learning

1 Introduction

Understanding long videos such as movies, for an AI system, has been viewed as an extremely challenging task [1]. In contrast, for humans, as studies in cognitive science [49] tell us it is naturally achieved by breaking down a video into

J. Mun and M. Shin—Equal contribution.

Supplementary Information The online version contains supplementary material available at https://doi.org/10.1007/978-3-031-26316-3_29.

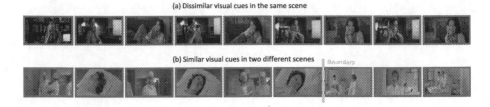

Fig. 1. Examples of the video scene segmentation. In each row, we visualize the shots including similar visual cues (*e.g.*, characters, places, etc.) with the same colored border.

meaningful units (*e.g.*, event) and reasoning about these units and their relation [42]. From this point of view, dividing a long video into a series of shorter temporal segments can be considered as an essential step towards the high-level video understanding. Motivated by this, in this paper, we tackle the video scene segmentation task, temporally localizing scene boundaries from a long video; the term scene is widely used in filmmaking and scene (a series of semantically cohesive shots) is considered as a basic unit for understanding the story of movies.

One of the biggest challenges in video scene segmentation is that it is not achieved simply by detecting changes in visual cues. As shown in Fig. 1(a), we present an example of nine shots, all of which belong to the same scene, where two characters are talking on the phone; the overall visual cues within the scene do not stay the same but rather change repeatedly when each character appears. On the other hand, Fig. 1(b) shows two different scenes which contain visually similar shots (highlighted in blue) where the same character appears in the same place. Thus, it is expected that two adjacent scenes which share shots with similar visual cues need to be contextually discriminated. From this observation, it is important for the video scene segmentation task to model contextual relationship between shots by maximizing 1) *intra-scene similarity* (*i.e.*, the shots in the same scene should be close to each other) and 2) *inter-scene discrimination* across two adjacent scenes (*i.e.*, shots across the scene boundary should be distinguishable).

Supervised learning approaches (*e.g.*, [34]) are clearly limited due to the lack of large-scale datasets with reliable ground-truth annotations; in addition, collecting boundary annotations from long videos is extremely expensive. Recently, self-supervision [5,9,17,37] is spotlighted through its effectiveness in learning in-domain representation without relying on costly ground-truth annotations. The self-supervised learning methods [11,14,33] in the video domain have been proposed to learn spatio-temporal patterns in a short term; inspired by this, ShotCoL [8] proposed shot-level representation pre-training algorithm based on contrastive prediction task. Although ShotCoL shows the remarkable performance, such shot-level representation learned without being aware of the semantic transition is insufficient for video scene segmentation. This is because the task requires not only a good representation for individual shots but also contextual representation considering neighboring shots at a higher level as observed in Fig. 1. Thus, we set our main goal to design effective pre-text tasks for video scene segmentation so that the model can learn the contextual relationship between shots across semantic transition during pre-training.

We introduce a novel **B**oundary-aware **S**elf-**S**upervised **L**earning (BaSSL) framework where we learn boundary-aware contextualized representation effective in capturing semantic transtion during pre-training and adapt the learned representation for precise scene boundary detection through fine-tuning. The main idea during pre-training is identifying a moment with a maximum semantic transition and using it as psuedo-boundary. Then, we propose three boundary-aware pretext tasks that are beneficial to the video scene segmentation task as follows: 1) Shot-Scene Matching (SSM) matching shots with their associated scenes, 2) Contextual Group Matching (CGM) aligning shots whether they belong to the same scene or not and 3) Pseudo-boundary Prediction (PP) capturing semantic changes. SSM and CGM encourage the model to maximize intra-scene similarity and inter-scene discrimination, while PP enables the model to learn the capability of identifying transitional moments. In addition, we perform Masked Shot Modeling task inspired by CBT [46] to further learn temporal relationship between shots. The comprehensive analysis demonstrates the effectiveness of the boundary-aware pre-training compared to shot-level pre-training as well as the contribution of the individual proposed components (*i.e.*, pseudo-boundary discovery algorithm and boundary-aware pretext tasks).

Our main contributions are summarized as follows: (*i*) we introduce a novel boundary-aware pre-training framework which leverages pseudo-boundaries to learn contextual relationship between shots during the pre-training; (*ii*) we propose three boundary-aware pretext tasks, which are carefully designed to learn essential capabilities required for the video scene segmentation task; (*iii*) we perform extensive ablations to demonstrate the effectiveness of the proposed framework; (*iv*) we achieve the new state-of-the-art on the MovieNet-SSeg benchmark with large margins compared to existing methods.

2   Related Work

Video Scene Segmentation approaches formulate the task as a problem of temporal grouping of shots. In this formulation, the optimal grouping can be achieved by clustering-based [7,35,36,40], dynamic programming-based [16,39,48] or multi-modal input-based [30,43] methods. However, the aforementioned methods have been trained and evaluated on small-scale datasets such as OVSD [38] and BBC [3] which can produce a poorly generalized model. Recently, [19] introduce a large-scale video scene segmentation dataset (*i.e.*, MovieNet-SSeg) that contains hundreds of movies. Training with large-scale data, [34] proposes a strong supervised baseline model that performs a shot-level binary classification followed by grouping using the prediction scores. [8] proposes a shot contrastive pre-training method that learns shot-level representation. We found ShotCoL [8] to be the most similar to our method. However, our method is different from ShotCoL in that we focus on learning contextual representations by considering the relationship between shots through boundary-aware pre-text tasks.

Action Segmentation in Videos is one of the related works for video scene segmentation, which identifies action labels of individual frames, thus can divide a video into a series of action segments. Supervised methods [13,24] proposed CNN-based architectures to effectively capture temporal relationship between frames in order to address an over-segmentation issue. As frame-level annotations are prohibitively costly to acquire, weakly supervised methods [6,15,26,27,41,44,59] have been suggested to use an ordered list of actions occurring in a video as supervision. Most of the methods are trained to find (temporal) semantic alignment between frames and a given action list using an HMM-based architecture [21], a DP-based assignment algorithm [15] or a DTW-based temporal alignment method [6]. Recently, unsupervised methods [22,23,28,51,54] have been further proposed; in a nutshell, clustering-based prototypes (corresponding to one of the actions) are discovered from unlabeled videos, then the methods segment the videos by assigning prototypes into frames. Contrary to action segmentation localizing segments each of which represents a single action within an activity, video scene segmentation requires localizing more complex segments each of which may be composed of more than two actions (or activities).

Self-supervised Learning in Videos has been actively studied for the recent years with approaches proposing various pretext tasks such as future frame prediction [2,45,52], temporal ordering of frames [25,31,55], geometric transformations prediction [20], colorization of videos [53], multimodal correspondence [57] and contrastive prediction [11,14,33]. In addition, CBT [46,47] proposes a pretext task of masked frame modeling to learn temporal dependency between frames (or clips). Note that since most of those methods are proposed for the classification task, they would be sub-optimal to the video scene segmentation task. On the other hand, BSP [56] proposes a pre-training algorithm based on pseudo-boundary synthesis for temporal localization tasks. However, the method still requires video-level class labels to synthesize pseudo-boundaries thus is not applicable to videos such as movies that are hard to define semantic labels. Also, note that we empirically show that pseudo-boundaries identified by our method are more effective for pre-training than synthesized pseudo-boundaries.

3   Preliminary

Terminologies. A long video (*e.g.*, documentaries, TV episodes and movies) is assumed to have a hierarchical structure at three-level semantics: frame, shot and scene. A shot is a series of frames physically captured by the same camera during an uninterrupted period of time. A scene is a series of semantically cohesive shots and serves as a semantic unit for making a story. Note that, in this paper, our focus is on finding scene-level boundaries.

Video Scene Segmentation Task. Given a long video, which contains a series of N shots $\{\mathbf{s}_1, ..., \mathbf{s}_N\}$ with class labels $\{y_1, ..., y_N\}$ where $y_i \in \{0, 1\}$ indicating if it is the last shot of a scene, the video scene segmentation task is formulated as a simple binary classification task at an individual shot level. Leveraging the local context from the neighbor shots, existing methods [8,34] adopt a sliding window scheme. For n^{th} shot $\mathbf{s}_n$, the window is defined by $\mathbf{S}_n = \{\mathbf{s}_{n-K}, ..., \mathbf{s}_n, ..., \mathbf{s}_{n+K}\}$

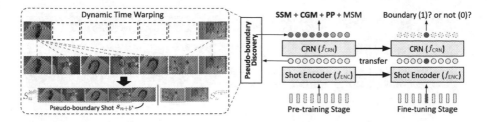

Fig. 2. Overall pipeline of our proposed framework, BaSSL.

containing a sequence of $2K + 1$ shots where K is the number of neighbor shots before and after $\mathbf{s}_n$. Then, supervised learning methods typically train a parameterized (θ) model by maximizing the expected log-likelihood:

$$\theta^* = \arg\max_{\theta} \mathbb{E}\left[\log p_{\theta}(y_n | \mathbf{S}_n)\right]. \tag{1}$$

Note that each shot $\mathbf{s}$ is given by a set of N_k key-frames, resulting in a tensor with size of (N_k, C, H, W) where C, H and W are the RGB channels, the height and the width, respectively.

Model Architecture. The model (θ) consists of two main components: 1) *shot encoder* embedding a shot by capturing its spatio-temporal patterns, and 2) *contextual relation network* (CRN) capturing contextual relation between shots. Taking a window $\mathbf{S}_n = \{\mathbf{s}_{n-K}, ..., \mathbf{s}_n, ..., \mathbf{s}_{n+K}\}$ centered at $\mathbf{s}_n$ as an input, two-level representations are extracted as follows:

$$\mathbf{e}_n = f_{\text{ENC}}(\mathbf{s}_n) \quad \text{and} \quad \mathbf{C}_n = f_{\text{CRN}}(\mathbf{E}_n), \tag{2}$$

where $f_{\text{ENC}}\colon \mathbb{R}^{N_k \times C \times H \times W} \to \mathbb{R}^{D_e}$ and $f_{\text{CRN}}\colon \mathbb{R}^{(2K+1) \times D_e} \to \mathbb{R}^{(2K+1) \times D_c}$ represent the shot encoder and CRN while D_e and D_c mean dimensions of encoded and contextualized features, respectively. $\mathbf{e}_n$ is an encoding of shot $\mathbf{s}_n$ by f_{ENC} while $\mathbf{E}_n = \{\mathbf{e}_{n-K}, ..., \mathbf{e}_n, ..., \mathbf{e}_{n+K}\}$ and $\mathbf{C}_n = \{\mathbf{c}_{n-K}, ..., \mathbf{c}_n, ..., \mathbf{c}_{n+K}\}$ correspond to the input and output feature sequence for f_{CRN}, respectively. In addition, the model employs additional pretext-specific heads for pre-training or a scene boundary detection head for fine-tuning.

Shot-level Self-supervised Learning. ShotCoL [8] proposes a shot-level contrastive self-supervised learning algorithm for video scene segmentation, which learns to make representation of visually similar nearby shots—highly likely to belong to the same scene—similar. However, the method has following two limitations. First, since ShotCoL pre-trains a model without explicitly identifying semantic boundaries during pre-training, it may fail to properly maximize intra-scene similarity and inter-scene dissimilarity. For example, the visually similar shots in different scenes may be learned indistinguishable. Second, the method learns shot representation given by the shot encoder (f_{ENC}) only and does not learn temporal and contextual relation between shots given by the contextual relation network (f_{CRN}). Contrary to such shot-level self-supervised learning, we propose boundary-aware self-supervised learning that trains both f_{ENC} and f_{CRN}

Fig. 3. An example in each row shows an input window sampled from the same scene where there exists no ground-truth scene-level boundary. Our method finds a pseudo-boundary shot (highlighted in red) that divides a sequence into two pseudo-scenes (represented by green and orange bars, respectively) so that semantics (*e.g.*, places, characters) maximally changes. (Color figure online)

while capturing the desired contextual relation between shots across semantic change. More detailed comparisons are given in appendix.

4 Boundary-aware Self-supervised Learning (BaSSL)

4.1 Overview

As illustrated in Fig. 2, our framework BaSSL is based on two-stage training following common practice [8]: pre-training on large-scale unlabeled data with self-supervision and fine-tuning on relatively small labeled data. Our main focus is in the pre-training stage, designing effective pretext tasks for video scene segmentation. Furthermore, we aim to train both shot encoder and contextual relation network while maximizing intra-scene similarity and inter-scene discrimination across semantic transition.

During pre-training, given an input window $\mathbf{S}_n$, BaSSL finds a shot across which the semantic transition becomes maximum and uses it as a pseudo-boundary to self-supervise the model. To be specific, we leverage the dynamic time warping technique to divide the shots in a window into two semantically disjoint sub-sequences, thus yielding a pseudo-boundary (Sect. 4.2). Then, we pre-train a model θ using three boundary-aware pre-text tasks and the masked shot modeling task adopted from CBT [46] to maximize intra-scene similarity and inter-scene dissimilarity (Sect. 4.3). After pre-trained with the four pretext tasks, the model is fine-tuned with labeled scene boundaries (Sect. 4.4).

4.2 Pseudo-boundary Discovery

The goal of our pre-training is to learn a capability of capturing semantic change before and after a semantic transition moment, thereby leading to higher performance in video scene segmentation. Specifically, we leverage a pseudo-boundary as a clue for self-supervision. However, extracting scene-level pseudo-boundaries from an input window is challenging. This is because there may be no scene boundary and it is also difficult to determine how many boundaries there are.

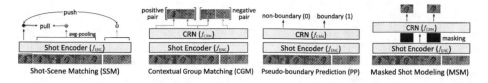

Fig. 4. Illustration of four pre-training pretext tasks.

Therefore, given an input window, we adopt a simple approach of finding a single moment where the semantics is maximally transitioning, and use it as pseudo-boundary. Although such identified moment may not correspond to scene-level boundary, it is still effective in learning a capability to capture semantic transition and the capability can be adapted to detect scene-level transition via fine-tuning. Figure 3 shows identified pseudo-boundaries from input windows having no ground-truth scene boundary; we observe that the resulting two sub-sequences are still cognitively distinguishable. More examples are presented in appendix.

The process, dividing an input window $\mathbf{S}_n$ into two continuous, non-overlapping sub-sequences $\mathbf{S}_n^{\text{left}}$ and $\mathbf{S}_n^{\text{right}}$ with maximum semantic transition, can be seen as a temporal alignment problem between $\mathbf{S}_n$ and $\mathbf{S}_n^{\text{slow}}$; specifically, observing the first shot should belong to $\mathbf{S}_n^{\text{left}}$ and the last one to $\mathbf{S}_n^{\text{right}}$, we define $\mathbf{S}_n^{\text{slow}} = \{\mathbf{s}_{n-K}, \mathbf{s}_{n+K}\}$, which can be seen as a same video with $\mathbf{S}_n$ with lower sampling frequency. Then, the problem becomes aligning intermediate shots either to the first shot $\mathbf{s}_{n-K}$ or the last shot $\mathbf{s}_{n+K}$ while preserving continuity.

Under the problem setting, we adopt dynamic time warping (DTW) [4] to find the optimal alignment between $\mathbf{S}_n$ and $\mathbf{S}_n^{\text{slow}}$. DTW solves the following optimization problem using dynamic programming to maximize semantic coherence of the resulting two sub-sequences among all possible boundary candidates:

$$b^* = \underset{b=-K+1,\ldots,K-1}{\arg\max} \left(\frac{1}{b+K} \sum_{i=-K+1}^{b} \text{sim}(\mathbf{e}_{n-K}, \mathbf{e}_{n+i}) \right.$$
$$\left. + \frac{1}{K-b-1} \sum_{j=b+1}^{K-1} \text{sim}(\mathbf{e}_{n+K}, \mathbf{e}_{n+j}) \right), \qquad (3)$$

where b and b^* are the candidate and optimal boundary offsets, respectively. $\text{sim}(\mathbf{x}, \mathbf{y}) = \frac{\mathbf{x}^\top \mathbf{y}}{\|\mathbf{x}\|\|\mathbf{y}\|}$ computes cosine similarity between two given shot encodings. Two sub-sequences are inferred as $\mathbf{S}_n^{\text{left}} = \{\mathbf{s}_{n-K}, \ldots, \mathbf{s}_{n+b^*}\}$ and $\mathbf{S}_n^{\text{right}} = \{\mathbf{s}_{n+b^*+1}, \ldots, \mathbf{s}_{n+K}\}$. $\mathbf{s}_{n+b^*}$ is the pseudo-boundary shot, which is the last shot of $\mathbf{S}_n^{\text{left}}$. The results are used for learning boundary-aware pretext tasks, which will be described Sect. 4.3.

Discussion on Single Pseudo-boundary. One might question if identifying multiple pseudo-boundaries is more reasonable, since there may exist more than two semantic transitions in an input window. However, we emphasize that the goal of our boundary-aware pre-training is learning a capability of capturing semantic transition, not lying on perfectly capturing all semantic transitions (or scene boundaries) at once; the capability to capture all scene-level boundaries is adapted via fine-tuning. In experiments, we verify that pre-training with one semantically strongest pseudo-boundary brings remarkable performance gain.

4.3   Pre-training Objectives

As shown in Fig. 4, we pre-train a model with three novel boundary-aware pre-text tasks—1) shot-scene matching ($\mathcal{L}_{\text{ssm}}$), 2) contextual group matching ($\mathcal{L}_{\text{cgm}}$) and 3) pseudo-boundary prediction ($\mathcal{L}_{\text{pp}}$)—and an additional one, masked shot modeling ($\mathcal{L}_{\text{msm}}$) as follows:

$$\mathcal{L}_{\text{pretrain}} = \mathcal{L}_{\text{ssm}} + \mathcal{L}_{\text{cgm}} + \mathcal{L}_{\text{pp}} + \mathcal{L}_{\text{msm}}. \tag{4}$$

Shot-Scene Matching (SSM). The objective of this task is to make the representations of a shot and its associated scene similar to each other, while the representations of the shot and other scenes dissimilar. In other words, SSM encourages the model to maximize intra-scene similarity, while minimizing inter-scene similarity. Considering the splitted two sub-sequences ($\mathbf{S}_n^{\text{left}}$ and $\mathbf{S}_n^{\text{right}}$) as pseudo-scenes, we train the model using the InfoNCE loss [32]:

$$\mathcal{L}_{\text{ssm}} = \mathcal{L}_{\text{nce}}\left(h_{\text{ssm}}(\mathbf{e}_{n-K}), h_{\text{ssm}}(\mathbf{r}_n^{\text{left}})\right) + \mathcal{L}_{\text{nce}}\left(h_{\text{ssm}}(\mathbf{e}_{n+K}), h_{\text{ssm}}(\mathbf{r}_n^{\text{right}})\right), \tag{5}$$

$$\mathcal{L}_{\text{nce}}(\mathbf{e}, \mathbf{r}) = -\log \frac{e^{\text{sim}(\mathbf{e},\mathbf{r})/\tau}}{e^{\text{sim}(\mathbf{e},\mathbf{r})/\tau} + \sum_{\bar{\mathbf{e}} \in \mathcal{N}_e} e^{\text{sim}(\bar{\mathbf{e}},\mathbf{r})/\tau} + \sum_{\bar{\mathbf{r}} \in \mathcal{N}_r} e^{\text{sim}(\mathbf{e},\bar{\mathbf{r}})/\tau}}, \tag{6}$$

where h_{ssm} is a SSM head of a linear layer, τ is a temperature hyperparameter and $\mathbf{r}_n^{\text{left}}$ means a scene-level representation, which is defined by the averaged encoding of shots in the sub-sequence $\mathbf{S}_n^{\text{left}}$. $\mathcal{N}_e$ and $\mathcal{N}_r$ in Eq. (6) are constructed using other shots and pseudo-scenes in a mini-batch, respectively.

Contextual Group Matching (CGM). Since directly matching representations of shots and scenes would not be effective when the scenes are composed of visually dissimilar shots, CGM is introduced to bridge this gap. Similar to SSM, CGM is also designed to maximize intra-scene similarity and inter-scene discrimination. However, CGM measures semantic coherence of the shots rather than comparing visual cues. With CGM, the model learns to decide if the given two shots belong to the same group (*i.e.*, scene) or not. In detail, we use the center shot $\mathbf{s}_n$ in the input sequence as the anchor and construct a triplet of ($\mathbf{s}_n$, $\mathbf{s}_{\text{pos}}$, $\mathbf{s}_{\text{neg}}$). We sample each shot from $\mathbf{S}_n^{\text{left}}$ and $\mathbf{S}_n^{\text{right}}$; the one sampled within the same sub-sequence with $\mathbf{s}_n$ is used as the positive shot $\mathbf{s}_{\text{pos}}$, while the other as the negative $\mathbf{s}_{\text{neg}}$. CGM loss is defined using binary cross-entropy by

$$\mathcal{L}_{\text{cgm}} = -\log\left(h_{\text{cgm}}(\mathbf{c}_n, \mathbf{c}_{\text{pos}})\right) - \log\left(1 - h_{\text{cgm}}(\mathbf{c}_n, \mathbf{c}_{\text{neg}})\right), \tag{7}$$

where h_{cgm} is a CGM head taking two shots as input and predicting a matching score. $\mathbf{c}_n$, $\mathbf{c}_{\text{pos}}$ and $\mathbf{c}_{\text{neg}}$ are the contextualized features by f_{CRN} for the center, positive and negative shots, respectively.

Pseudo-boundary Prediction (PP). Through the above two pretext tasks, our model learns the contextual relationship between shots. In addition to these, we design an extra pretext task, PP, which is more directly related to boundary detection; PP makes the model have a capability of identifying transitional moments that semantic changes. Using the pseudo-boundary shot and one randomly sampled non-boundary shot, the PP loss is defined as a binary cross-entropy loss:

$$\mathcal{L}_{\text{pp}} = -\log\left(h_{\text{pp}}(\mathbf{c}_{n+b^*})\right) - \log\left(1 - h_{\text{pp}}(\mathbf{c}_{\bar{b}})\right), \tag{8}$$

where h_{pp} is a PP head that projects the contextualized shot representation to a probability distribution over binary class. $\mathbf{c}_{n+b^*}$ and $\mathbf{c}_{\bar{b}}$ indicate the contextualized representation from f_{CRN} for the pseudo-boundary shot $\mathbf{s}_{n+b^*}$ and randomly sampled non-boundary shot $\mathbf{s}_{\bar{b}}$, respectively.

Masked Shot Modeling (MSM). Inspired by masked frame modeling [46,47], we adopt the MSM task whose goal is to reconstruct the representation of masked shots based on their surrounding shots. In this task, given a set of encoded shot representations, we randomly apply masking each of them with a probability of 15%. For a set $\mathcal{M}$ of masked shot offsets, we learn to regress the output on each masked shot to its encoded shot representation:

$$\mathcal{L}_{\text{msm}} = \sum_{m \in \mathcal{M}} \|\mathbf{e}_m - h_{\text{msm}}(\mathbf{c}_m)\|_2^2, \tag{9}$$

where h_{msm} is a MSM head to match the dimension of contextualized shot representation with that of encoded one. $\mathbf{e}_m$ and $\mathbf{c}_m$ denote the encoded and contextualized features by f_{ENC} and f_{CRN} for a masked shot $\mathbf{s}_m$, respectively.

4.4   Fine-tuning for Scene Boundary Detection

Recall that we formulate the video scene segmentation as a binary classification task to identify contextual transition across the scene. Different from the pretraining stage, given an input window $\mathbf{S}_n$, we employ a scene boundary detection head h_{sbd} to infer a prediction from the contextualized representation $(\mathbf{c}_n)$ for the center shot $\mathbf{s}_n$. Following ShotCoL [8], we freeze the parameters of the shot encoder and then train only CRN and the scene boundary detection head using a binary cross-entropy loss with the ground truth label y_n as follows:

$$\mathcal{L}_{\text{finetune}} = -y_n \log(h_{\text{sbd}}(\mathbf{c}_n)) + (1 - y_n)\log(1 - h_{\text{sbd}}(\mathbf{c}_n)). \tag{10}$$

With a sidling window scheme, each shot is decided to be a scene boundary when its prediction score is higher than a pre-defined threshold (set to 0.5).

5   Experiment

5.1   Experimental Settings

Dataset. For evaluation, we use the MovieNet-SSeg dataset [19] containing 1,100 movies with 1.6M shots. Only 318 out of 1,100 movies have scene boundary annotations, which are divided into 190, 64, and 64 movies for training, validation,

Table 1. Comparison with other algorithms. † and ‡ denote that the numbers are copied from [34] and [19], respectively. * indicates the methods exploiting additional information (*e.g.*, audio, place, cast, transcript). The best numbers are in bold.

Method	AP (↑)	mIoU (↑)	AUC-ROC (↑)	F1 (↑)
Supervised Learning				
Siamese [3]‡	35.80	39.60	–	–
MS-LSTM [19]‡*	46.50	46.20	–	–
LGSS [34]†*	47.10	48.80	–	–
Unsupervised Learning				
GraphCut [36]†	14.10	29.70		–
SCSA [7]†	14.70	30.50	–	–
DP [16]†	15.50	32.00	–	–
Story Graph [48]†	25.10	35.70	–	–
Grouping [39]‡*	33.60	37.20		–
BaSSL w/o fine-tuning (10 epochs)	31.55	39.36	71.67	32.55
Self-supervised Learning				
ShotCoL [8]	53.40			–
BaSSL (10 epochs)	56.26 ± 0.04	49.50 ± 0.11	90.27 ± 0.02	45.70 ± 0.24
BaSSL (40 epochs)	**57.40 ± 0.08**	**50.69 ± 0.45**	**90.54 ± 0.03**	**47.02 ± 0.87**

and test split, respectively. Following ShotCoL [8], we use the entire 1,100 movies with no ground truth labels for the pre-training and fine-tune the model on the training split. The performance is measured on the test split.

Metric. Following [19], we compare algorithms using AP and mIoU. Also, we adopt F1 score[1] and AUC-ROC as additional evaluation metrics. We also report Meta-Sum metric inspired by [10,29] for easy and straightforward comparison.

Implementation Details. We employ ResNet-50 [18] and Transformer [50] as shot encoder and CRN, respectively. We cross-validate the number of neighbor shots among $K = \{4, 8, 12, 16\}$ and $K = 8$ is selected due to its good performance and computational efficiency. In all experiments, we report mean and std from 5 fine-tuned models with random seeds. More details are presented in appendix.

5.2   Comparison with State-of-the-Art Methods

We compare BaSSL with 1) supervised ones: LGSS [34], Siamese [3], MS-LSTM [19] and, 2) unsupervised ones: GraphCut [36], SCSA [7], DP [16], Story-Graph [48] and Grouping [39], and 3) self-supervised ones: ShotCoL [8]. Without fine-tuning, BaSSL can be seen as an unsupervised model in that it is trained to predict the pseudo-boundary by the PP task. Table 1 summarizes comparison against competing methods. BaSSL without fine-tuning shows competitive or outperforming performance based only on the basic visual cue compared to competing unsupervised ones. Furthermore, fine-tuning BaSSL with ground-truth scene boundaries improves AP by 24.71%p and BaSSL outperforms all other algorithms. Finally, through longer pre-training (40 epochs), BaSSL surpasses

---

[1] Contrary to the previous works [8,34] that report recall, we use F1 score to consider for balanced comparison between precision and recall.

Table 2. Average precision (AP) comparison with pre-training baselines. Note that SimCLR (NN) corresponds to our reproduced ShotCoL using SimCLR.

Method		Pre-training f_{ENC}	f_{CRN}	Transfer f_{ENC}	f_{CRN}	Architecture of f_{CRN} during fine-tuning MLP	MS-LSTM	Transformer
Supervised pre-training using image dataset								
M1	ImageNet	✓		✓		43.12 ± 0.14	45.10 ± 0.55	47.13 ± 1.04
M2	Places365	✓		✓		43.82 ± 0.10	45.87 ± 0.40	48.71 ± 0.50
Shot-level pre-training								
M3	SimCLR (instance)	✓		✓		45.60 ± 0.07	49.09 ± 0.24	51.51 ± 0.31
M4	SimCLR (temporal)	✓		✓		45.55 ± 0.11	49.24 ± 0.26	50.05 ± 0.78
M5	SimCLR (NN)	✓		✓		45.99 ± 0.13	50.73 ± 0.19	51.17 ± 0.69
Boundary-aware pre-training								
M6	BaSSL	✓	✓	✓		46.53 ± 0.11	50.58 ± 0.14	50.82 ± 0.69
M7	BaSSL	✓	✓	✓	✓	−	−	56.26 ± 0.04
M8	M5+M7	✓	✓	✓	✓	−	−	56.86 ± 0.01

the previous state-of-the-art method (*i.e.*, ShotCoL) by a large margin (4.00%p in AP).

5.3   Comparison with Pre-training Baselines

We perform extensive experiments to compare BaSSL with the pre-training baselines learning shot-level representation by f_{ENC} only. In the experiments, we compare the following three types of pre-training approaches. The first group (M1-2) trains f_{ENC} using image-level supervision on ImageNet [12] or place labels on Places365 [58]. The second group (M3-5) trains f_{ENC} through shot-level contrastive learning (*i.e.*, SimCLR [9]) with different positive pair sampling strategies. Specifically, *instance* (M3) takes an instance of the center shot with different augmentation, *temporal* (M4) takes one randomly sampled neighbor shot as positive pair in local temporal window, and *NN* (M5) takes the most visually similar shot among the neighbor shots as positive pair, which is also known as ShotCoL [8]. The last group (M6-8) learns both f_{ENC} and f_{CRN} through boundary-aware pretext tasks proposed in this paper. Given pre-trained representations of f_{ENC}, we train a video scene segmentation model with three different types of f_{CRN} including MLP [8], MS-LSTM [19][2] and Transformer. For fair comparison, all pre-training methods employ ResNet-50 as f_{ENC} and we pre-train the models for 10 epochs.

In Table 2, we found the following observations. First, when transferring pretrained shot representation, employing MS-LSTM and Transformer as f_{CRN} is more effective than using MLP, as they are favorably designed to capture contextual relation between shots (see M1-6). Second, BaSSL (M7) outperforms all competing baselines (M1-5), which shows the importance of boundary-aware pretraining. Third, it turns out that transferring the representation through f_{CRN} is important for the boundary detection task where it leads to a performance gain of 5.44%p in AP (see M6-7). Finally, learning shot-level and contextual

---

[2] https://github.com/AnyiRao/SceneSeg/tree/master/lgss

Table 3. Ablation study on varying combinations of pretext tasks for pre-training. The best scores are highlighted in bold.

	Pretext tasks			Evaluation metric				
	SSM	CGM	PP	MSM	AP	mIoU	AUC-ROC	F1	Sum
P1	✓				42.57 ±0.29	40.12 ±0.50	84.11 ±0.15	30.83 ±0.79	197.63
P2		✓			36.76 ±0.02	40.59 ±0.18	82.06 ±0.04	30.94 ±0.32	190.35
P3			✓		36.55 ±0.04	39.58 ±0.05	81.36 ±0.03	29.96 ±0.04	187.45
P4				✓	13.33 ±0.23	29.80 ±0.39	64.65 ±0.98	18.68 ±0.39	126.45
P5	✓	✓			55.77 ±0.05	48.19 ±0.21	90.19 ±0.03	43.17 ±0.39	237.32
P6	✓		✓		56.04 ±0.08	49.00 ±0.16	90.13 ±0.02	44.74 ±0.29	239.91
P7		✓	✓		38.09 ±0.03	41.25 ±0.10	82.85 ±0.01	32.24 ±0.24	195.43
P8	✓			✓	54.39 ±0.07	47.54 ±0.18	89.72 ±0.03	42.48 ±0.22	234.13
P9		✓		✓	39.49 ±0.04	41.71 ±0.12	83.27 ±0.02	32.85 ±0.20	197.32
P10			✓	✓	38.53 ±0.07	40.85 ±0.15	82.78 ±0.04	31.47 ±0.16	193.63
P11		✓	✓	✓	41.02 ±0.07	40.89 ±0.10	83.79 ±0.02	31.53 ±0.18	197.23
P12	✓		✓	✓	56.10 ±0.08	49.10 ±0.17	90.09 ±0.03	45.42 ±0.30	240.71
P13	✓	✓		✓	56.20 ±0.06	48.00 ±0.17	90.13 ±0.01	43.24 ±0.27	237.57
P14	✓	✓	✓		**56.26** ±0.02	48.42 ±0.33	90.25 ±0.01	43.98 ±0.58	238.91
P15	✓	✓	✓	✓	**56.26** ±0.04	**49.50** ±0.11	**90.27** ±0.02	**45.70** ±0.24	**241.73**

representations is complementary to each other; that is, incorporating ShotCoL (M5) and our framework (M7) provides further improved performance (M8).

5.4 Ablation Studies

Impact of Individual Pretext Tasks. We investigate the contribution of individual pretext tasks. In this experiment, we train models by varying the combinations of the pretext tasks. From Table 3, we can obtain following two observations. First, among models trained by a single pretext task (P1-4), the MSM leads to the worst performance compared to the others. This indicates that boundary-aware pretext tasks (*i.e.*, SSM, CGM and PP) are indeed important for scene boundary detection. Second, the more pretext tasks are used, the better the performance is, and the best one is obtained from all tasks (P15). This means all tasks are complementary to each other, contributing to improvement.

Pseudo-boundary Discovery Method. To check the effectiveness of DTW-based pseudo-boundary discovery, we train three models with different pseudo-boundary decision strategies—1) *Random* defining one randomly sampled shot in the input window as a pseudo-boundary, 2) *Fixed* always taking the center shot as a pseudo-boundary, and 3) *Synthesized*, inspired by BSP [56], synthesizing the input window by concatenating two sub-sequences sampled from different movies and using the last shot of the first sub-sequence as a pseudo-boundary. Table 4(a) summarizes the results. Our approach to adopting DTW to find pseudo-boundaries achieves the best performance. It is notable that BaSSL with *Synthesized* pseudo-boundaries also outperforms the pre-training baselines in Table 2, which shows the importance of boundary-aware pre-training.

Table 4. Ablations to check the impact of pseudo-boundary discovery strategies, the number of neighboring shots (K) and longer pre-training. The best scores are in bold.

Pseudo-boundary	AP
Random	46.64 ± 0.37
Fixed	49.53 ± 0.32
Synthesized	54.61 ± 0.03
DTW (ours)	**56.26 ± 0.04**

(a) Pseudo-boundary discovery methods.

# Neighbors	AP
4	55.98 ± 0.10
8	56.26 ± 0.04
12	**56.29 ± 0.03**
16	55.31 ± 0.04

(b) The number of neighbor shots.

Epochs	AP
10	56.26 ± 0.04
20	56.74 ± 0.04
30	56.74 ± 0.07
40	**57.40 ± 0.08**
50	57.15 ± 0.08

(c) The number of pre-training epochs.

Table 5. Scene clustering quality measured by normalized mutual information (NMI).

Model	Scene length			$\Delta \downarrow$ (Short → Long)
	Short ($N_c = 8$)	Medium ($N_c = 16$)	Long ($N_c = 32$)	
ImageNet	67.50	61.60	56.25	−16.67%
SimCLR (temporal)	82.40	81.65	78.99	−4.14%
SimCLR (NN)	83.54	83.17	81.25	−2.75%
BaSSL (ours)	**86.22**	**86.72**	**85.63**	**−0.68%**

Hyperparameters. We analyze the impact of two hyperparameters: the number of neighbor shots K and pre-training epochs. Table 4(b) shows that we achieve higher performance with more neighbor shots, saturating around $K = 12$. Table 4(c) shows the impact of longer pre-training. We find that performance increases until certain numbers (40 epochs) and decrease afterward. We conjecture that this is partly due to overfitting to noise from incorrect pseudo-boundaries.

5.5   Analysis on Pre-trained Shot Representation Quality

We analyze the quality of pre-trained shot representations using normalized mutual information (NMI) to measure the clustering quality. Specifically, we randomly sample 100 scenes from the test split of MovieNet-SSeg while we vary the length of scenes $N_c \in \{8, 16, 32\}$ (the number of shots included in a single scene). Then, we perform K-Means clustering on $N_c \times 100$ shot representations extracted by the pre-trained model with the number of classes (=100). This intends to form a single cluster for each scene, assuming that high-quality representation would locate the shot embeddings within the same scene close to each other. Considering the randomness in the K-Means clustering and scene sampling, we report the averaged score from 5 trials.

Table 5 shows the NMI score for different pre-trained models. BaSSL outperforms the shot-level pre-training baselines and the model pre-trained using ImageNet. With respect to different scene lengths (N_c), we found BaSSL is more robust than the others. When the visual diversity across the shots increases as the scenes become longer (N_c=8 → 32), the performance of BaSSL drops only

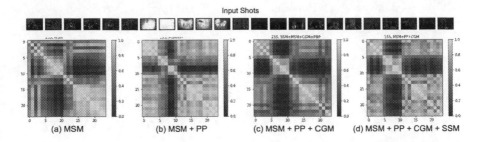

Fig. 5. Visualization of similarity (below) between shot representations in randomly sampled consecutive shots (above). We observe that the shot representations are clearly clustered as adding pretext tasks one by one.

-0.68% while the other baselines suffer from severe degradation. This implies the effectiveness of BaSSL in maximizing intra-scene similarity.

5.6 Qualitative Analysis

To qualitatively check the effect of individual pretext tasks, we visualize the matrix of cosine similarity between shot representations from the randomly sampled 16 consecutive shots in Fig. 5. The shot representations are computed by models without the fine-tuning in order to solely focus on the behavior of each pretext task at the pretraining stage. When the MSM is used only, approximately three clusterings are shown, but similarity around boundaries is smoothed. Next, when we add PP, dissimilarities around the boundaries are to be sharpened. Then, with additional CGM, the clusters are more clearly obtained. Finally, adding SSM makes the similarity of shots within the same cluster higher (*i.e.*, more yellow ones). On the other hand, we present more qualitative analysis for discovered pseudo-boundaries and scene boundary predictions in supplementary material.

6   Conclusion

We present BaSSL, a novel self-supervised framework for video scene segmentation, especially designed to learn contextual relationship between shots. Through the pseudo-boundary discovery, we can define and conduct boundary-aware pretext tasks that encourage the model to learn the contextual relational representation and a capability of capturing transitional moments. Comprehensive experiments demonstrate the effectiveness of our framework and we achieve outstanding performance in the MovieNet-SSeg dataset.

Acknowledgements. This work was supported by Kakao Brain and partly by Korea research grant from NRF (2021H1D3A2A03038607/10%, 2022R1C1C1010627/10%), and IITP (2021-0-01778/10%, 2022-0-00264/40%, 2022-0-00951/10%, 2022-0-00612/10%, 2020-0-01373/10%).

References

1. Abu-El-Haija, S., et al.: YouTube-8M: a large-scale video classification benchmark. arXiv:1609.08675 (2016)
2. Ahsan, U., Sun, C., Essa, I.: DiscrimNet: semi-supervised action recognition from videos using generative adversarial networks. arXiv:1801.07230 (2018)
3. Baraldi, L., Grana, C., Cucchiara, R.: A deep Siamese network for scene detection in broadcast videos. In: ACM MM (2015)
4. Berndt, D.J., Clifford, J.: Using dynamic time warping to find patterns in time series. In: SIGKDD Workshop (1994)
5. Caron, M., Misra, I., Mairal, J., Goyal, P., Bojanowski, P., Joulin, A.: Unsupervised learning of visual features by contrasting cluster assignments. arXiv:2006.09882 (2020)
6. Chang, C.Y., Huang, D.A., Sui, Y., Fei-Fei, L., Niebles, J.C.: D3TW: discriminative differentiable dynamic time warping for weakly supervised action alignment and segmentation. In: CVPR (2019)
7. Chasanis, V.T., Likas, A.C., Galatsanos, N.P.: Scene detection in videos using shot clustering and sequence alignment. IEEE Trans. Multimedia $11(1)$, 89–100 (2008)
8. Chen, S., Nie, X., Fan, D., Zhang, D., Bhat, V., Hamid, R.: Shot contrastive self-supervised learning for scene boundary detection. In: CVPR (2021)
9. Chen, T., Kornblith, S., Norouzi, M., Hinton, G.: a simple framework for contrastive learning of visual representations. In: ICML (2020)
10. Chen, Y.C., Li, L., Yu, L., El Kholy, A., Ahmed, F., Gan, Z., Cheng, Y., Liu, J.: UNITER: UNiversal Image-TExt representation learning. In: ECCV (2020)
11. Dave, I., Gupta, R., Rizve, M.N., Shah, M.: TCLR: temporal contrastive learning for video representation. arXiv:2101.07974 (2021)
12. Deng, J., Dong, W., Socher, R., Li, L.J., Li, K., Fei-Fei, L.: ImageNet: a large-scale hierarchical image database. In: CVPR (2009)
13. Farha, Y.A., Gall, J.: MS-TCN: multi-stage temporal convolutional network for action segmentation. In: CVPR (2019)
14. Feichtenhofer, C., Fan, H., Xiong, B., Girshick, R., He, K.: A large-scale study on unsupervised spatiotemporal representation learning. In: CVPR (2021)
15. Fried, D., Alayrac, J.B., Blunsom, P., Dyer, C., Clark, S., Nematzadeh, A.: Learning to segment actions from observation and narration. arXiv:2005.03684 (2020)
16. Han, B., Wu, W.: Video scene segmentation using a novel boundary evaluation criterion and dynamic programming. In: IEEE International Conference on Multimedia and Expo (2011)
17. He, K., Fan, H., Wu, Y., Xie, S., Girshick, R.: Momentum contrast for unsupervised visual representation learning. In: CVPR (2020)
18. He, K., Zhang, X., Ren, S., Sun, J.: Deep residual learning for image recognition. In: CVPR (2016)
19. Huang, Q., Xiong, Yu., Rao, A., Wang, J., Lin, D.: MovieNet: a holistic dataset for movie understanding. In: Vedaldi, A., Bischof, H., Brox, T., Frahm, J.-M. (eds.) ECCV 2020. LNCS, vol. 12349, pp. 709–727. Springer, Cham (2020). https://doi.org/10.1007/978-3-030-58548-8_41
20. Jing, L., Tian, Y.: Self-supervised spatiotemporal feature learning by video geometric transformations. arXiv preprint arXiv:1811.11387 (2018)
21. Kuehne, H., Richard, A., Gall, J.: A hybrid RNN-HMM approach for weakly supervised temporal action segmentation. IEEE Trans. Pattern Anal. Mach. Intell. $42(4)$, 765–779 (2018)

22. Kukleva, A., Kuehne, H., Sener, F., Gall, J.: Unsupervised learning of action classes with continuous temporal embedding. In: CVPR (2019)
23. Kumar, S., Haresh, S., Ahmed, A., Konin, A., Zia, M.Z., Tran, Q.H.: Unsupervised activity segmentation by joint representation learning and online clustering. arXiv:2105.13353 (2021)
24. Lea, C., Reiter, A., Vidal, R., Hager, G.D.: Segmental spatiotemporal CNNs for fine-grained action segmentation. In: Leibe, B., Matas, J., Sebe, N., Welling, M. (eds.) ECCV 2016. LNCS, vol. 9907, pp. 36–52. Springer, Cham (2016). https://doi.org/10.1007/978-3-319-46487-9_3
25. Lee, H.Y., Huang, J.B., Singh, M., Yang, M.H.: Unsupervised Representation Learning by Sorting Sequences. In: ICCV (2017)
26. Li, J., Lei, P., Todorovic, S.: Weakly Supervised Energy-based Learning for Action Segmentation. In: ICCV (2019)
27. Li, J., Todorovic, S.: Set-constrained Viterbi for set-supervised action segmentation. In: CVPR (2020)
28. Li, J., Todorovic, S.: Action shuffle alternating learning for unsupervised action segmentation. In: CVPR (2021)
29. Li, L., et al.: VALUE: a multi-task benchmark for video-and-language understanding evaluation. In: NeurIPS (2021)
30. Liang, C., Zhang, Y., Cheng, J., Xu, C., Lu, H.: A novel role-based movie scene segmentation method. In: Pacific-Rim Conference on Multimedia (2009)
31. Misra, I., Zitnick, C.L., Hebert, M.: Shuffle and learn: unsupervised learning using temporal order verification. In: Leibe, B., Matas, J., Sebe, N., Welling, M. (eds.) ECCV 2016. LNCS, vol. 9905, pp. 527–544. Springer, Cham (2016). https://doi.org/10.1007/978-3-319-46448-0_32
32. Oord, A.v.d., Li, Y., Vinyals, O.: Representation learning with contrastive predictive coding. arXiv:1807.03748 (2018)
33. Qian, R., et al.: Spatiotemporal contrastive video representation learning. In: CVPR (2021)
34. Rao, A., et al.: A local-to-global approach to multi-modal movie scene segmentation. In: CVPR (2020)
35. Rasheed, Z., Shah, M.: Scene detection in Hollywood movies and TV shows. In: CVPR (2003)
36. Rasheed, Z., Shah, M.: Detection and representation of scenes in videos. IEEE Trans. Multimedia 7(6), 1097–1105 (2005)
37. Roh, B., Shin, W., Kim, I., Kim, S.: Spatially consistent representation learning. In: CVPR (2021)
38. Rotman, D., Porat, D., Ashour, G.: Robust and efficient video scene detection using optimal sequential grouping. In: IEEE International Symposium on Multimedia (ISM) (2016)
39. Rotman, D., Porat, D., Ashour, G.: Optimal sequential grouping for robust video scene detection using multiple modalities. Int. J. Seman. Comput. 11(02), 193–208 (2017)
40. Rui, Y., Huang, T.S., Mehrotra, S.: Exploring video structure beyond the shots. In: Proceedings of the IEEE International Conference on Multimedia Computing and Systems (1998)
41. Shen, Y., Wang, L., Elhamifar, E.: Learning to segment actions from visual and language instructions via differentiable weak sequence alignment. In: CVPR (2021)
42. Shou, M.Z., Lei, S.W., Wang, W., Ghadiyaram, D., Feiszli, M.: Generic event boundary detection: a benchmark for event segmentation. In: ICCV (2021)

43. Sidiropoulos, P., Mezaris, V., Kompatsiaris, I., Meinedo, H., Bugalho, M., Trancoso, I.: Temporal Video Segmentation to Scenes using High-level Audiovisual Features. IEEE Trans. Circuits Syst. Video Technol. **21**(8), 1163–1177 (2011)

44. Souri, Y., Fayyaz, M., Minciullo, L., Francesca, G., Gall, J.: Fast weakly supervised action segmentation using mutual consistency. IEEE Trans. Pattern Anal. Mach. Intell. **44**, 6196–6208 (2021)

45. Srivastava, N., Mansimov, E., Salakhudinov, R.: Unsupervised learning of video representations using LSTMs. In: ICML (2015)

46. Sun, C., Baradel, F., Murphy, K., Schmid, C.: Learning video representations using contrastive bidirectional transformer. arXiv:1906.05743 (2019)

47. Sun, C., Myers, A., Vondrick, C., Murphy, K., Schmid, C.: VideoBERT: a joint model for video and language representation learning. In: ICCV (2019)

48. Tapaswi, M., Bauml, M., Stiefelhagen, R.: StoryGraphs: visualizing character interactions as a timeline. In: CVPR (2014)

49. Tversky, B., Zacks, J.M.: Event perception. Oxford Handbook of Cognitive Psychology (2013)

50. Vaswani, A., et al.: Attention is all you need. In: NIPS (2017)

51. VidalMata, R.G., Scheirer, W.J., Kukleva, A., Cox, D., Kuehne, H.: Joint visual-temporal embedding for unsupervised learning of actions in untrimmed sequences. In: WACV (2021)

52. Vondrick, C., Pirsiavash, H., Torralba, A.: Generating videos with scene dynamics. NIPS (2016)

53. Vondrick, C., Shrivastava, A., Fathi, A., Guadarrama, S., Murphy, K.: Tracking emerges by colorizing videos. In: Ferrari, V., Hebert, M., Sminchisescu, C., Weiss, Y. (eds.) ECCV 2018. LNCS, vol. 11217, pp. 402–419. Springer, Cham (2018). https://doi.org/10.1007/978-3-030-01261-8_24

54. Wang, Z., et al.: Unsupervised action segmentation with self-supervised feature learning and co-occurrence parsing. arXiv:2105.14158 (2021)

55. Xu, D., Xiao, J., Zhao, Z., Shao, J., Xie, D., Zhuang, Y.: Self-supervised spatiotemporal learning via video clip order prediction. In: CVPR (2019)

56. Xu, M., et al.: Boundary-sensitive pre-training for temporal localization in videos. arXiv:2011.10830 (2020)

57. Zhang, B., et al.: A hierarchical multi-modal encoder for moment localization in video corpus. arXiv:2011.09046 (2020)

58. Zhou, B., Lapedriza, A., Khosla, A., Oliva, A., Torralba, A.: Places: a 10 million image database for scene recognition. IEEE Trans. Pattern Anal. Mach. Intell. **40**, 1452–1464 (2017)

59. Zhukov, D., Alayrac, J.B., Cinbis, R.G., Fouhey, D., Laptev, I., Sivic, J.: Cross-task weakly supervised learning from instructional videos. In: CVPR (2019)

HaViT: Hybrid-Attention Based Vision Transformer for Video Classification

Li Li[1,2], Liansheng Zhuang[1(✉)], Shenghua Gao[3], and Shafei Wang[2]

[1] University of Science and Technology of China, Hefei 230026, China
lili1234@mail.ustc.edu.cn, lszhuang@ustc.edu.cn
[2] Peng Cheng Laboratory, Shenzhen 518000, China
[3] ShanghaiTech University, Shanghai 201210, China

Abstract. Video transformers have become a promising tool for video classification due to its great success in modeling long-range interactions through the self-attention operation. However, existing transformer models only exploit the patch dependencies within a video when doing self-attention, while ignoring the patch dependencies across the different videos. This paper argues that external patch prior information is beneficial to the performance of video transformer models for video classification. Motivated by this assumption, this paper proposes a novel Hybrid-attention based Vision Transformer (HaViT) model for video classification, which explicitly exploits both internal patch dependencies within a video and external patch dependencies across videos. Different from existing self-attention, the hybrid-attention is computed based on internal patch tokens and an external patch token dictionary which encodes external patch prior information across the different videos. Experiments on Kinetics-400, Kinetics-600, and Something-Something v2 show that our HaViT model achieves state-of-the-art performance in the video classification task against existing methods. Moreover, experiments show that our proposed hybrid-attention scheme can be integrated into existing video transformer models to improve the performance.

1 Introduction

The task of video classification is to understand the visual and audio features to assign one or more relevant tags to the video. With the rapid increase of video content, this task is critical for many applications such as video retrieval [1] and video surveillance [2]. Compared with image classification, video classification is more challenging due to the temporal dimension, which increases the overall size of the input and variations in sequence. Though many methods have been proposed to model spatial relationships for image classification, it is still an open problem to jointly model spatial and temporal features in a video.

The remarkable progress of the transformer [3] in natural language processing (NLP) has inspired researchers to investigate its adaptation to image classification. The transformer is notable for its use of multi-head self-attention to model long-range dependencies, which are often modeled by the large receptive fields

L. Wang et al. (Eds.): ACCV 2022, LNCS 13844, pp. 502–517, 2023.
https://doi.org/10.1007/978-3-031-26316-3_30

formed by deep stacks of convolutional operations. However, convolutional operations can only capture local neighborhood in images, and the deep stack strategies are inherently limited in capturing long-range dependencies by means of aggregation of shorter-range information [4]. Conversely, the self-attention operation attends to all elements in the input sequence, and thus can capture both local as well as global spatial relationships on non-overlapping image patches in images. Recently, a pure transformer-based architecture with the Vision Transformer (ViT) [5] has been proposed to replace convolutions completely, and outperformed its convolution counterparts in image classification [5]. Inspired by the fact that attention-based architecture is an intuitive choice for modelling long-range contextual relationships in video, several transformer-based models have been proposed for video classification [6–11]. Some models apply self-attention on top of convolutional layers [6], while others use self-attention as the exclusive building block in the video classification models [7, 8].

The natural extension of Vision Transformers to 3-dimensional video signal is challenging. Specially, each encoder of a transformer contains heavy computations such as pair-wise self-attention. Meanwhile, a video has a longer sequential representation than an image due to the additional temporal axis. Consequently, it is not economical or easy to optimize if directly applying the joint space-time attention to flattened video sequences. To reduce the computation costs, some efforts [7, 8] have factorized the spatial and temporal domains via a factorized encoder or factorized self-attention, and have achieved a good speed-accuracy trade-off. Though achieving promising results in video classification, all these transformer models only exploit internal patch dependencies across the spatial and temporal dimensions within a video, while ignoring external patch dependencies across the different videos. In fact, external patch dependencies across the different images or videos which capture the external patch prior information plays an important role in many low-level vision tasks such as image restore [12] and video super-resolution [13]. This paper argues that external patch prior information is beneficial to transformer-based models for video classification.

Motivated by the above assumptions, this paper introduces a novel Hybrid-attention based Vision Transformer (HaViT) for video classification, which explicitly exploits both internal patch dependencies within a video and external patch dependencies across videos. The main operation performed in this architecture is hybrid-attention, which is computed on a sequence of spatio-temporal tokens extracted from a video and an external token dictionary extracted from extra videos. The spatio-temporal tokens encode internal patch information within a video, while the external token dictionary encodes external patch information across the different videos. HaViT uses the hybrid-attention instead of self-attention to model both internal patch dependencies within a video and external patch dependencies across the different videos. To improve the model performance, HaViT inserts the class token later in the transformer. This choice eliminates the discrepancy on the first layer of the transformer, which ithus used to perform hybrid-attention between patches and an external token dictionary only. Extensive experiments on three public datasets (Kinetics-400 [14],

Kinetics-600 [15] and Something-Something v2 [16]) show that HaViT achieves competitive results on video classification against existing state-of-the-art models.

In summary, our main contributions are as follows:

- A new Hybrid-attention based Vision Transformer (HaViT) is proposed for video classification, which is mainly built on the hybrid-attention module. To our best knowledge, this is the first work on transformer architecture for video classification which explicitly exploits external patch dependencies across videos.
- A new hybrid-attention mechanism is introduced to model both long-range patch dependencies within a video and external patch dependencies across the different videos, which is easily integrated into existing transformer models to improve their performance.
- Extensive experiments on public datasets demonstrate that our proposed HaViT model outperforms most existing video transformer models in most cases. When combined with existing video transformer models, the hybrid-attention does improve their model performance on video classification.

2 Related Work

Early works on video classification use hand-crafted features to encode appearance and motion information [17,18]. With the success of AlexNet in image classification [19], deep learning increasingly dominates visual modeling for video classification. Previously for convolutional models, backbone architectures for the video were adapted from those for images simply by extending the modeling through the temporal axis. Consequently, 3D convolution neural networks (3D-CNNs) have become a de-facto standard for video classification [20–24]. Compared with their image counterparts, 3D-CNNs have significantly more parameters and thus require more computation. To alleviate this, a large body of works (such as P3D [25], R(2+1)D [26], and S3D [27]) factorize convolutions across spatial and temporal dimensions to achieve a better speed-accuracy trade-off. However, the potential of convolution based approaches is limited by the small receptive field of the convolution operator. With a self-attention mechanism, the receptive field can be broadened with fewer parameters and lower computation costs, which leads to better performance. In [4], non-local network introduces self-attention on top of CNNs. Further, CBA-QSA CNN [28] extends self-attention with compact bilinear mapping for fine-grained action classification.

With the success of Vision Transformer (ViT) in image classification [5], a shift in backbone architectures is currently underway for video classification, from Convlutional Neural Networks (CNNs) to attention-based transformers [7–11]. Attention-based transformers use self-attention blocks at each layer to understand a frame's role with respect to other frames in the video. Since performing full spatio-temporal attention is computationally prohibitive, many efforts have been devoted to reducing computation costs via factorizing temporal and spatial domains. In TimeSformer [8], the authors propose applying spatial and temporal attention in an alternating manner reducing the complexity of calculating attention weights. Similarly, ViViT [7] explores several methods of space-time factorization. In addition, they also proposed to adapt the patch embedding process from [5] to 3D data. However, all existing works only exploit internal patch dependencies information within a video via self-attention, while ignoring external patch dependencies information across the different video. In fact, external patch dependencies have been proven to be important for many vision tasks. This paper will try to exploit the external information in transformer model for video classification.

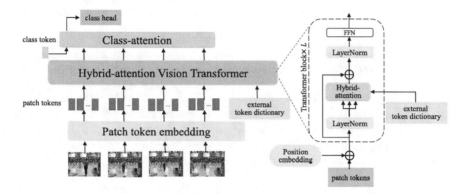

Fig. 1. Diagram of HaViT. The patches extracted from the frames are linearly projected into the token embedding, and position embedding is added. The patches and the class token are fed into the transformer with L hybrid-attention layers.

3 Our Method

3.1 The Overall Architecture

The core idea of our proposed method is to introduce an external patch token dictionary to represent the texture feature space of all image patches. It uses the similarity of image patches in videos to calculate the internal attention and uses the similarity of image patches in current videos and the external token dictionary to calculate the external attention. In the calculation of hybrid-attention, the external token dictionary is used to expand the elements involved in the attention for the encoding of prior information. Specifically, when using the attention mechanism for feature embedding, the embedding of one image patch is not only linearly combined by the image patch features of the current video, but also by the elements of the external patch token dictionary. As shown in Fig. 1, the overall architecture includes a patch token embedding module, a video transformer module with hybrid-attention, and a class-attention layer. In the following, we elaborate on the processing flow: Firstly, F frame images are sampled from the video sequence to form a multidimensional tensor $x \in R^{H \times W \times C \times F}$ as the model input, where H, W and C denote the height, the width and the number of channels of each frame, respectively. Then, each frame in the video is divided into a fixed number of non-overlapping image patches and the image patches are reshaped into a flatten vector $x_{p,t}$, with p $= 1, ..., N$ denoting the spatial locations and t $= 1, ..., F$ denoting the index of frames. Then, we get the image patch feature sequence $z_{p,t}^{0}$ through the patch token embedding module. Next, we input the image patch feature sequence to the video transformer with hybrid-attention. The module uses the image patch feature sequence and the external token dictionary to calculate the feature representation of image patches through the multi-head hybrid-attention mechanism. Finally, the feature for classification is calculated using the class-attention layer, combining the feature representation of image patches with the class token.

3.2 Hybrid-attention Vision Transformer

The hybrid-attention module is designed to model the relationship between image patches in a video and the relationship across the different videos. It consists of the internal attention based on patch tokens within a video and the external attention based on an external token dictionary D^{token}. In the following, we describe the hybrid-attention module and its several variants.

Internal Attention. The internal attention is designed to model the internal patch dependencies in a video. Each patch representation is projected into the query, key, and value vector. The vectors are computed from $z_{p,t}^{(l-1)}$ encoded by the preceding block:

$$
\begin{aligned}
q_{p,t}^{(l,a)} &= W_Q^{(l,a)} f_{LN}\left(z_{p,t}^{(l-1)}\right) \in R^{d_h}, \\
k_{p,t}^{(l,a)} &= W_K^{(l,a)} f_{LN}\left(z_{p,t}^{(l-1)}\right) \in R^{d_h}, \\
v_{p,t}^{(l,a)} &= W_V^{(l,a)} f_{LN}\left(z_{p,t}^{(l-1)}\right) \in R^{d_h}, \\
& \text{p,t} \in \{p', t' | \begin{matrix} p' = 1, ..., N \\ t' = 1, ..., F \end{matrix} \} \cup \{0, 0\}
\end{aligned}
\tag{1}
$$

where $a = 1, 2, ..., A$ is the index over the multiple attention heads, and $d_h = d/A$ is the hidden dim of each head. $W_Q, W_K, W_V \in R^{d_h \times d}$ are learnable parameter matrices for projecting the queries, keys and values. Next, the attention weights $\alpha_{p,t}^{(l,a)} \in R^{NF+1}$ are computed via the dot products of the query $q_{p,t}^{(l,a)}$ with all keys:

$$\alpha_{p,t}^{(l,a)} = \sigma \left(\frac{q_{p,t}^{(l,a)^T}}{\sqrt{d_h}} \cdot \left[k_{0,0}^{(l,a)} \{k_{p',t'}^{(l,a)}\} \begin{matrix} p' = 1, ..., N \\ t' = 1, ..., F \end{matrix} \right] \right), \tag{2}$$

where σ denotes the activation function. Above attention weights are used as coefficients in a weighted summation over value vectors to obtain the result of each attention head $s_{p,t}^{(l,a)}$. Then, these outputs from each attention head are concatenated and passed through embedding matrix W_O and the feed-forward network (FFN) which contains two MLP layers with GeLU activation:

$$\tilde{z}_{p,t}^{(l)} = W_O \begin{bmatrix} s_{p,t}^{(l,1)} \\ \vdots \\ s_{p,t}^{(l,A)} \end{bmatrix} + z_{p,t}^{(l-1)} \tag{3}$$

In (2), the attention coefficient is calculated by using the query vector of the image patch and the key vector of all image patches in the video. The internal attention is jointly computed by the spatial and the temporal dimension. A reduction in computation can be achieved by disentangling the spatial and the temporal dimension. For the spatial dimension, only $N+1$ query-key comparisons are made, using keys from the same frame as the query patch token exclusively:

$$\alpha_{p,t}^{(l,a)space} = \sigma \left(\frac{q_{p,t}^{(l,a)^T}}{\sqrt{d_h}} \cdot \left[k_{0,0}^{(l,a)} \{k_{p',t}^{(l,a)}\}_{p'=1,...,N} \right] \right). \tag{4}$$

If we only consider the space-attention, the model becomes the ViT which encodes the feature from each frame, and the classifier vector is the global average of the features of all the frames. The baseline of time dimensional dependencies are proposed by TimeSformer [8], only making $F+1$ query-key comparisons and using the patches from the other frames in the same location as the query patch:

$$\alpha_{p,t}^{(l,a)time} = \sigma \left(\frac{q_{p,t}^{(l,a)^T}}{\sqrt{d_h}} \cdot \left[k_{0,0}^{(l,a)} \{k_{p,t'}^{(l,a)}\}_{t'=1,...,F} \right] \right) \tag{5}$$

The internal attention with divided space-time dimension can effectively reduce the computation costs, but it will increase the number of parameters of the model. In (4) and (5), the query vector and key vector are used twice, which are calculated with two different parameters. Compared with joint space-time attention mechanism, this method increases the number of parameters but reduces the costs of computation.

External Attention. The external attention is to model the external patch dependencies across the different videos. Since the number of external patches is huge, this paper proposes to use an external patch token dictionary to encode the external patch information. In this way, each patch is represented by the combination of the element of the dictionary. Corresponding attention coefficients encode the dependencies between internal patches and external patches in different videos. The external token dictionary includes two trainable parts, the value set $\{v_1^e, v_2^e, ..., v_n^e\}$ and its corresponding key set $\{k_1^e, k_2^e, ..., k_n^e\}$. Given a patch embedding $z_{p,t}^{(l-1)}$ from the layer, the attention coefficient is calculated via the dot products with the key set of the external token dictionary. Then the value set of the external token dictionary is linearly weighted to obtain the external feature $\hat{z}_{p,t}^{(l)}$:

$$\alpha_i = \sigma \left(\frac{q_{p,t}^{(l)^T}}{\sqrt{d}} \cdot \{k_i^e\}_{i=1,...,n} \right),$$

$$\hat{z} = \sum_{i=1}^{n} \alpha_i v_i^e, \tag{6}$$

$$\hat{z}_{p,t}^{(l)} = f_{FFN} \left(f_{LN} \left(\hat{z} \right) \right) + \hat{z}.$$

where $q_{p,t}^{(l)}$ is the query vector of $z_{p,t}^{(l-1)}$. Note that the external token dictionary in the external attention aims to learn general visual features from the dictionary. For the sake of simplicity, we only give the formula of the attention. We can also use the multi-head attention to calculate values of the external attention (Fig. 2).

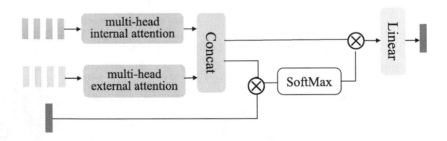

Fig. 2. Diagram of the combined-correlation hybrid-attention mechanism. The green block represents the image patches in the video and the yellow block represents the external patch token dictionary. (Color figure online)

Hybrid Schemes for Hybrid-attention. First of all, it is necessary to explain how to model the external prior information of videos using the external token dictionary. This paper assumes that the features of image patches are composed of two parts. The one is a linear combination of features of image patches in the video, utilizing the self-attention mechanism (i.e., the internal attention mechanism introduced earlier); The other is constructed from shared visual features instead of the current video. We use a learnable dictionary to construct shared visual features, which are related to all videos.

The keys and values calculated from the external dictionary and internal patches can be spliced together for subsequent calculation. In internal attention, the feature information of all image patches in the video is aggregated by calculating attention. It should be noted that not only the key vectors of all image patches but also all key sets from the external token dictionary are used to calculate the attention coefficient. In this way, an expanded attention coefficient is calculated as follows:

$$\alpha_{\mathrm{p,t}}^{l,a} = \sigma \left(\frac{q_{\mathrm{p,t}}^{l,a}}{\sqrt{d_h}} \cdot \{k_i^{l,a}\} \right), \tag{7}$$

where i is the index of the image patch feature sequence and the external token dictionary. In this scheme, the model splices the attention coefficient matrix between the internal and external attention of a query patch to form an extended correlation matrix (called the combined-correlation type). With the attention coefficients calculated by (7), the final output of hybrid-attention is obtained by using the weighted summation over corresponding value vectors.

3.3 Class-attention

As we all know, the class token has two functions: guiding the learning of attention weights between patches and aggregating overall information to the linear classifier for classification [29]. Recent work has shown that separating two functions is beneficial to the classification. In this paper, we will explore whether this method influences the performance of video classification. Our implementation includes two stages: the hybrid-attention stage and the class-attention stage. In the hybrid-attention attention stage, we get the space-temporal feature of patches without the class token. In the class-attention stage, we only update the class token embedding while keeping patch features frozen.

Specifically, we first calculate the query, key, and value vectors:

$$
\begin{aligned}
q_{0,0}^{(l,a)} &= W_Q^{(l,a)} f_{\mathrm{LN}} \left(z_{0,0}^{(l-1)} \right), \\
k_{\mathrm{p,t}}^{(l,a)} &= W_K^{(l,a)} f_{\mathrm{LN}} \left(z_{\mathrm{p,t}}^{(l-1)} \right), \\
v_{\mathrm{p,t}}^{(l,a)} &= W_V^{(l,a)} f_{\mathrm{LN}} \left(z_{\mathrm{p,t}}^{(l-1)} \right).
\end{aligned}
\tag{8}
$$

Next, the attention weights are given by

$$\alpha_{0,0}^{(l,a)} = \sigma \left(\frac{q_{0,0}^{(l,a)^T}}{\sqrt{d_h}} \cdot \left[k_{\mathrm{p',t'}}^{(l,a)} \right]_{\substack{\mathrm{p'}=1,...,N \\ \mathrm{t'}=1,...,F}} \right) \tag{9}$$

Then, we use the (3) to calculate the $z_{0,0}^{(l)}$ as the output of class-attention block. Finally, HaViT has a hybrid-attention module (which combines the internal attention and the external attention) and several class-attention layers (at which only the values of class tokens are updated).

4　Experiment

4.1　Setup

Datasets. All the experiments in this paper were conducted on the following datasets. The Kinetics dataset contains short clips sampled from YouTube. Since some videos on YouTube have been deleted or privatized, the dataset versions used in this paper include about 260k clips of Kinetics-400 and 397k clips of Kinetics-600. Note that, these numbers are lower than the original dataset and thus might induce a negative performance bias when compared with previous works. The Something-Something v2 (SSv2) consists of about 220k short videos with a length of 2 to 6 s, depicting human beings performing predefined basic actions on daily objects. Since the objects and backgrounds in the videos are consistent across the different action classes, this dataset tends to require stronger temporal modeling.

Network Architecture. The backbone modules closely follow the ViT architecture. Most of the experiments were performed using the HaViT-B/16($L = 12$, $A = 12$, $d = 768$, $P = 16$)and HaViT-S/16($L = 12$, $A = 6$, $d = 384$, $P = 16$), where L, A, d, P denotes the number of transformer layers, the number of heads, the embedding dimension, and the patch size.

Training and Inference. Unless otherwise stated, we sample frames uniformly across the video. For the training stage, we resize the smaller dimension of each frame to a value $\in [256, 320]$ and take a random crop of size 224×224 from the same location for all frames of the same video. In the inference, we give the accuracy results for 4×3 views (4 temporal clips and 3 spatial crops). The models are implemented by python and pytorch, and were trained on a DGX-v1 server (Table 1).

Table 1. Training hyperparamters for experiments.

Config	Value
Optimizer	AdamW [30]
Momentum	$\beta_1, \beta_2 = 0.9, 0.999$
Batch size	64
Learning rate	$2.5e^{-4}$
Weight decay	0.05
Learning rate schedule	Cosine with linear warmup
Linear warmup epochs	3
Epochs	30
Dropout	0.1

Table 2. Comparison with state-of-the-art methods on the Kinetics-400 dataset. T ×
frames are used in our experiments.

Method	Top-1	Top-5	Views	TFLOPs
blVNet [31]	73.5	91.2	3 × 3	0.84
STM [32]	73.7	91.6	10 × 3	2.01
TEA [33]	76.1	92.5	10 × 3	2.10
CorrNet-101 [35]	79.2	–	10 × 3	6.70
ip-CSB-152 [23]	79.3	93.8	10 × 3	3.27
LGD-R101 [36]	79.4	94.4	10 × 3	–
SlowFast [24]	79.8	93.9	10 × 3	7.02
X3D-XXL [22]	80.4	94.6	10 × 3	5.82
TimeSformer-L [8]	80.7	94.7	1 × 3	7.14
ViViT-L/16 × 2 [7]	80.6	94.7	4 × 3	17.35
Swin-B [9]	80.6	94.6	4 × 3	3.38
Our model(8×)	80.6	94.3	4 × 3	6.96
Our model(16×)	**81.7**	**95.2**	4 × 3	13.12

4.2 Comparison with State-of-the-Art

In this subsection, we compare our HaViT model with state-of-the-art models on
three mentioned datasets. The results are shown in the Table 2-Table 4. Unless
otherwise stated, we report the results on all the datasets using the 4 × 3 views.

Table 2 gives a comparison with the state-of-the-art on Kinetics-400, including convolution based networks and transformer-based networks. Compared with
transformer-based model TimeSformer-L, the performance of our proposed structure is largely improved, and the classification accuracy is improved by 1.1%.
Compared with the most advanced convolution network X3d, our model also
improves the classification accuracy by 1.3% and uses fewer temporal views.

Table 3. Comparison with state-of-the-art on the Kinetics-600.

Method	Top-1	Top-5	TFLOPs
AttentionNAS [6]	79.8	94.4	1.03
LGD-R101 [36]	81.5	95.6	–
SlowFast [24]	81.8	95.1	7.02
X3D-XL [22]	81.9	95.5	1.05
TimeSformer [8]	82.4	95.3	5.11
ViViT-L/16 × 2 [7]	82.5	–	17.35
Swin-B [9]	84.0	**96.5**	3.38
Our model(16×)	**84.5**	96.1	13.12

Table 3 shows the comparison between our model and the state-of-the-art on Kinetics-600. The classification accuracy is much higher than the previous convolution network based method (+2.6%) and transformer-based method (+0.5%). Compared with the Kinetics-400, the size of the Kinetics-600 dataset is 0.6 times larger, which also shows that the performance of the transformer model will be improved when the dataset is large.

Table 4. Comparison with state-of-the-art on the SSv2.

Method	Top-1	Top-5	TFLOPs
TRN [37]	48.8	77.6	–
SlowFast [24]	61.7	–	7.02
TSM [34]	63.4	88.5	0.95
STM [32]	64.2	89.7	2.01
TEA [33]	65.1	–	2.10
blVNet [31]	65.2	90.3	0.84
TimeSformer-L [8]	62.5	–	7.14
ViViT-L/16 × 2 [7]	65.4	89.8	17.35
Our model(16×)	**67.3**	**90.5**	13.12

Table 4 compares our model with the state-of-the-art on the Something-Something v2 dataset. In terms of classification accuracy, our model is 2.1% higher than the previous convolution network blvnet, However, it is 1.9% higher than the previous transformer model ViViT-L/16.

4.3 Ablation Studies

This subsection studies the impact of different components on the HaViT performance. For all experiments in this subsection, we use a lightweight model HaViT-S/16 with the model dim of 384 and adopt the Kinetics-400 dataset.

Table 5. Effect of different hybrid schemes for hybrid-attention.

Hybrid scheme	Top-1	Top-5
Simplified	77.1	92.5
Multi-view	77.5	92.7
Combined-correlation	**78.2**	**93.1**

Hybrid Schemes for Hybrid-attention. First, we consider the effect of different hybrid schemes for hybrid-attention on the final performance. Three schemes of hybrid-attention are discussed, including simplified scheme, multi-view scheme, and combined-correlation scheme. The baseline is the simplified scheme, which directly adds the internal attention result and the external attention results. As shown in Table 5, compared to the simplified scheme, the multi-view scheme is more flexible and achieves better performance. For combined-correlation scheme, each head of attention is influenced by internal attention and external attention weights, and each query patch's feature is decided by the attention which influences it most. There's a trade-off between the internal attention and the external attention in the combined-correlation scheme. So, it's not surprising that the combined-correlation scheme has the best performance. Our model also adopts the combined-correlation scheme. Here, we give the visualization results of HaViT model. It can be seen from the Fig. 3 that compared with the existing divided space-time self-attention scheme, the proposed hybrid-attention scheme can better reflect the attention mechanism to the related objects in the video.

(a) drinking   (b) eating hot dog

Fig. 3. Here are the results of attention visualization of the two models. "drinking" and "eating hot dog" represent two categories of data respectively; three of them represent the visualization results of the original video image, divided space-time self-attention and hybrid-attention respectively

Effect of Attention Realization. Previously, we introduced different ways to achieve attention, and here we will give the experimental results of different ways to achieve attention. The models using joint space-time self-attention and divided space-time self-attention are pre-trained on the image classification dataset Imagenet. From the Table 6, it is not difficult to find that the parameters of the model using the divided self-attention mechanism are more than those of the implementation of joint space-time self-attention. There are three modeling methods of joint space-time self-attention: the original attention method and two linear computational complexity methods (linear activation and cosine re-weighting) [38]. The accuracy of the proposed linear activation method is lower than that of the original joint space-time method (-2.7%), but the inference speed is about 5 times faster. After using cosine re-weighting technology, the performance of the model is improved ($+2.2\%$), but the classification accuracy is still not as good as the original joint space-time attention method. Cosine re-weighting technology attaches a larger weight of the attention coefficient to the query value vector, so that the query image patches pay more attention to the surrounding image patches, so the classification effect of this linear compu-tational complexity method is better. Although the space-time joint method of linear computational complexity introduced above is fast, there are also unsta-ble problems in the training process. Compared with the joint space-time self-attention method, the accuracy of divided space-time self-attention method is improved by about 1%, and the calculation speed of the model is about 3 times faster. Therefore, divided space-time attention is used in internal attention mod-eling.

Table 6. Effect of internal attention realization.

Internal attention	Top-1(%)	Parameters(M)
Joint space-time	77.3	26.4
Linear activation	74.6	26.4
Cosine re-weighting	76.8	26.4
Divided space-time	**78.1**	34.5

4.4   With Different Vision Transformers

To verify that the hybrid-attention scheme proposed in this chapter can be com-bined with different transformer models, different vision transformers are used for experiments. Firstly, different vision transformer models are extended to 3-dimensional space to get the corresponding video transformer model. Then different models are used to experiment on the Kinetics-400 dataset and test the classification accuracy. Finally, the models obtained by combining different video transformers with hybrid-attention schemes are trained and tested. The classification accuracy results are shown in Table 7. It is not difficult to find that in the three different transformer models, the hybrid-attention scheme obtained by using the external token dictionary can achieve better results. The proposed hybrid-attention scheme can be combined with other vision transformer models and improve the performance of its models.

Table 7. Effect of hybrid schemes combined with different vision transformers

Transformer model	External token dictionary	Top-1(%)
ViT-S [5]	without	77.6
	with	**78.2**↑
CvT [39]	without	76.6
	with	**77.4**↑
Swin-T [9]	without	78.4
	with	**78.9**↑

5   Conclusion

In this paper, we propose a new hybrid-attention based vision transformer model for video classification, which explicitly exploits external patch dependencies across videos. Instead of using self-attention, it uses hybrid-attention to model both long-range patch dependencies within a video as well as external patch dependencies across videos. Compared to existing vision transformers, our model achieves competitive or better performance on public datasets including Kinetics-400/600 and SSv2. Experiments also show that hybrid-attention can be integrated into existing transformer models and improve their performance.

Acknowledgements. This work was supported in part to Dr. Liansheng Zhuang by NSFC under contract No. U20B2070 and No. 61976199.

References

1. Dong, Y., Li, J.: Video retrieval based on deep convolutional neural network. In: Proceedings of the 3rd International Conference on Multimedia Systems and Signal Processing, pp. 12–16 (2018)
2. Muhammad, K., Khan, S., Elhoseny, M., Ahmed, S.H., Baik, S.W.: Efficient fire detection for uncertain surveillance environment. IEEE Trans. Industr. Inf. **15**, 3113–3122 (2019)
3. Vaswani, A., et al.: Attention is all you need (2017)
4. Wang, X., Girshick, R., Gupta, A., He, K.: Non-local neural networks. In: Proceedings of the IEEE Conference on Computer Vision and Pattern Recognition, pp. 7794–7803 (2018)
5. Dosovitskiy, A., et al.: An image is worth 16x16 words: transformers for image recognition at scale (2020)
6. Wang, X., et al.: AttentionNAS: spatiotemporal attention cell search for video classification. In: Vedaldi, A., Bischof, H., Brox, T., Frahm, J.-M. (eds.) ECCV 2020. LNCS, vol. 12353, pp. 449–465. Springer, Cham (2020). https://doi.org/10. 1007/978-3-030-58598-3_27
7. Arnab, A., Dehghani, M., Heigold, G., Sun, C., Lučić, M., Schmid, C.: Vivit: a video vision transformer. In: Proceedings of the IEEE/CVF International Conference on Computer Vision, pp. 6836–6846 (2021)

8. Bertasius, G., Wang, H., Torresani, L.: Is space-time attention all you need for video understanding? In: ICML, p. 4 (2021)
9. Liu, Z., et al.: Video swin transformer. arXiv preprint arXiv:2106.13230 (2021)
10. Zhang, Y., et al.: VidTr: video transformer without convolutions. In: Proceedings of the IEEE/CVF International Conference on Computer Vision, pp. 13577–13587 (2021)
11. Neimark, D., Bar, O., Zohar, M., Asselmann, D.: Video transformer network. In: Proceedings of the IEEE/CVF International Conference on Computer Vision, pp. 3163–3172 (2021)
12. Chen, F., Zhang, L., Yu, H.: External patch prior guided internal clustering for image denoising. In: Proceedings of the IEEE/CVF International Conference on Computer Vision (2015)
13. Ma, Z., Liao, R., Tao, X., Xu, L., Jia, J., Wu, E.: Handling motion blur in multi-frame super-resolution. In: 2015 IEEE Conference on Computer Vision and Pattern Recognition (CVPR), pp. 5224–5232 (2015)
14. Kay, W., et al.: The kinetics human action video dataset. arXiv preprint arXiv:1705.06950 (2017)
15. Carreira, J., Noland, E., Banki-Horvath, A., Hillier, C., Zisserman, A.: A short note about kinetics-600 (2018)
16. Goyal, R., et al.: The "something something" video database for learning and evaluating visual common sense. In: Proceedings of the IEEE International Conference on Computer Vision, pp. 5842–5850 (2017)
17. Laptev, I.: On space-time interest points. Int. J. Comput. Vision 64, 107–123 (2005)
18. Wang, H., Kläser, A., Schmid, C., Liu, C.L.: Dense trajectories and motion boundary descriptors for action recognition. Int. J. Comput. Vision 103, 60–79 (2013)
19. He, K., Zhang, X., Ren, S., Sun, J.: Deep residual learning for image recognition. In: Proceedings of the IEEE Conference on Computer Vision and Pattern Recognition, pp. 770–778 (2016)
20. Tran, D., Bourdev, L., Fergus, R., Torresani, L., Paluri, M.: Learning spatiotemporal features with 3D convolutional networks. In: Proceedings of the IEEE International Conference on Computer Vision, pp. 4489–4497 (2015)
21. Carreira, J., Zisserman, A.: Quo vadis, action recognition? a new model and the kinetics dataset. In: Proceedings of the IEEE Conference on Computer Vision and Pattern Recognition (CVPR) (2017)
22. Feichtenhofer, C.: X3D: expanding architectures for efficient video recognition. In: Proceedings of the IEEE/CVF Conference on Computer Vision and Pattern Recognition (CVPR) (2020)
23. Tran, D., Wang, H., Torresani, L., Feiszl, M.: Video classification with channel-separated convolutional networks. In: Proceedings of the IEEE/CVF International Conference on Computer Vision (ICCV) (2019)
24. Feichtenhofer, C., Fan, H., Malik, J., He, K.: Slowfast networks for video recognition. In: Proceedings of the IEEE/CVF International Conference on Computer Vision (ICCV) (2019)
25. Qiu, Z., Yao, T., Mei, T.: Learning spatio-temporal representation with pseudo-3D residual networks. In: Proceedings of the IEEE Conference on Computer Vision (ICCV) (2017)
26. Tran, D., Wang, H., Torresani, L., Ray, J., LeCun, Y., Paluri, M.: A closer look at spatiotemporal convolutions for action recognition. In: Proceedings of the IEEE/CVF Conference on Computer Vision and Pattern Recognition (CVPR) (2018)

27. Xie, S., Sun, C., Huang, J., Tu, Z., Murphy, K.: Rethinking spatiotemporal feature learning: speed-accuracy trade-offs in video classification. In: Ferrari, V., Hebert, M., Sminchisescu, C., Weiss, Y. (eds.) ECCV 2018. LNCS, vol. 11219, pp. 318–335. Springer, Cham (2018). https://doi.org/10.1007/978-3-030-01267-0_19

28. Hao, Y., Zhang, H., Ngo, C.W., Liu, Q., Hu, X.: Compact bilinear augmented query structured attention for sport highlights classification. In: Proceedings of the 28th ACM International Conference on Multimedia (2020)

29. Touvron, H., Cord, M., Sablayrolles, A., Synnaeve, G., Jégou, H.: Going deeper with image transformers. arXiv preprint arXiv:2103.17239 (2021)

30. Loshchilov, I., Hutter, F.: Decoupled weight decay regularization. In: International Conference on Learning Representations (2018)

31. Fan, Q., Chen, C.F.R., Kuehne, H., Pistoia, M., Cox, D.: More is less: learning efficient video representations by big-little network and depthwise temporal aggregation. In: Wallach, H., Larochelle, H., Beygelzimer, A., d' Alché-Buc, F., Fox, E., Garnett, R. (eds.) Advances in Neural Information Processing Systems, Curran Associates, Inc. (2019)

32. Jiang, B., Wang, M., Gan, W., Wu, W., Yan, J.: STM: spatiotemporal and motion encoding for action recognition. In: Proceedings of the IEEE/CVF International Conference on Computer Vision (ICCV) (2019)

33. Li, Y., Ji, B., Shi, X., Zhang, J., Kang, B., Wang, L.: Tea: temporal excitation and aggregation for action recognition. In: Proceedings of the IEEE/CVF Conference on Computer Vision and Pattern Recognition, pp. 909–918 (2020)

34. Lin, J., Gan, C., Han, S.: TSM: temporal shift module for efficient video understanding. In: Proceedings of the IEEE/CVF International Conference on Computer Vision (ICCV) (2019)

35. Wang, H., Tran, D., Torresani, L., Feiszli, M.: Video modeling with correlation networks. In: Proceedings of the IEEE/CVF Conference on Computer Vision and Pattern Recognition (CVPR) (2020)

36. Qiu, Z., Yao, T., Ngo, C.W., Tian, X., Mei, T.: Learning spatio-temporal representation with local and global diffusion. In: Proceedings of the IEEE/CVF Conference on Computer Vision and Pattern Recognition (CVPR) (2019)

37. Zhou, B., Andonian, A., Oliva, A., Torralba, A.: Temporal relational reasoning in videos. In: Ferrari, V., Hebert, M., Sminchisescu, C., Weiss, Y. (eds.) ECCV 2018. LNCS, vol. 11205, pp. 831–846. Springer, Cham (2018). https://doi.org/10.1007/978-3-030-01246-5_49

38. Qin, Z., et al.: cosformer: Rethinking softmax in attention. In: International Conference on Learning Representations (2021)

39. Wu, H., Xiao, B., Codella, N., Liu, M., Dai, X., Yuan, L., Zhang, L.: CVT: introducing convolutions to vision transformers. In: Proceedings of the IEEE/CVF International Conference on Computer Vision, pp. 22–31 (2021)

Vision and Language

From Sparse to Dense: Semantic Graph Evolutionary Hashing for Unsupervised Cross-Modal Retrieval

Yang Zhao[1], Jiaguo Yu[1], Shengbin Liao[2], Zheng Zhang[3],
and Haofeng Zhang[1(✉)]

[1] School of Computer Science and Engineering, Nanjing University of Science and Technology, Nanjing 210094, China
{zhao_yang,yujiaguo,zhanghf}@njust.edu.cn
[2] National Engineering Research Center for E-learning, Huazhong Normal University, Wuhan 430079, China
[3] School of Computer Science and Technology, Harbin Institute of Technology, Shenzhen 518055, China

Abstract. In recent years, cross-modal hashing has attracted an increasing attention due to its fast retrieval speed and low storage requirements. However, labeled datasets are limited in real application, and existing unsupervised cross-modal hashing algorithms usually employ heuristic geometric prior as semantics, which introduces serious deviations as the similarity score from original features cannot reasonably represent the relationships among instances. In this paper, we study the unsupervised deep cross-modal hash retrieval method and propose a novel Semantic Graph Evolutionary Hashing (SGEH) to solve the above problem. The key novelty of SGEH is its evolutionary affinity graph construction method. To be concrete, we explore the sparse similarity graph with clustering results, which evolve from fusing the affinity information from code-driven graph on intrinsic data and subsequently extends to dense hybrid semantic graph which restricts the process of hash code learning to learn more discriminative results. Moreover, the batch-inputs are chosen from edge set rather than vertexes for better exploring the original spatial information in the sparse graph. Experiments on four benchmark datasets demonstrate the superiority of our framework over the state-of-the-art unsupervised cross-modal retrieval methods. Code is available at: https://github.com/theusernamealreadyexists/SGEH.

Keywords: Cross-modal hashing · Visual-text retrieval · Sparse affinity graph · Semantic graph evolution

1 Introduction

Cross-modal retrieval aims to search the related results of other different modalities from a query term of one modal, *e.g.*, using a caption to retrieve the related

---

Supplementary Information The online version contains supplementary material available at https://doi.org/10.1007/978-3-031-26316-3_31.

L. Wang et al. (Eds.): ACCV 2022, LNCS 13844, pp. 521–536, 2023.
https://doi.org/10.1007/978-3-031-26316-3_31

pictures in database. With the explosive growth of multimedia data, hashing technology, which encodes continuous features into common hash space where relative samples have similar binary codes, is widely used in cross-modal retrieval technology due to its few storage, low Hamming distance computational complexity and fast retrieval speed [6–8, 10, 14, 17, 26, 29, 30].

Recently, [24] proposes an unsupervised deep cross-modal hashing method that learns hash codes via Laplacian constraint in objective function to preserve the neighborhood information from code-driven dense semantic graph. However, a significant shortcoming is that it only preserves the original relationship from different modalities, while integrating the affinity information from instances into a unified structure in advance can improve affinity relation construction, which brings a better performance. Then [31] adopts Generative Adversarial Network (GAN) to train cross-modal hash training. With the intention to preserve the correlation from inter-modal and intra-modal in the latent hash space, this method maintains a manifold structure across the attributes of different modalities. In their algorithm, hash code plays a significant role for both the generator and the discriminator. Based on CycleGAN, [25, 33] proposed a new method to learn hash codes via unpaired instances. [20] proposes Deep Joint-Semantics Reconstructing Hashing (DJSRH) which fuses the semantic similarities into a unified matrix to explore the latent relevance for the input multi-modal instances. Though impressive progress the above methods have made, several challenges still exist in this task. Therefore, our study is motivated according to the following issues.

(1) The difficulty of excavating non-label relationships from intra-modal and inter-modal. Unsupervised hashing cross-modal methods usually have no access to accurate relationship among instances. Based on co-occurrence information, recent unsupervised techniques [10, 11, 20, 22, 27, 31, 32] usually adopt Attention Mechanism or GAN to generate affinity graph structure which aims to aggregate neighborhood information.

(2) The semantic gap of different modality. Given that each modality has its own geometric prior, each modal have its own affinity code-driven graph. the different between multi-graph may confuse the training process. Thus, coming up with a semantic-unified graph is necessary for conducting the task.

(3) The impossibility to utilize the adjacency matrix as the feature space of large graph. Graph embedding methods can be utilized to fix out the huge storage consumption of large adjacency matrix. However, graph embedding loses a lot of original information during dimension reduction process, which may lead to a sub-optimal performance and fail to preserve the similarity information.

In this paper, to tackle the aforementioned issues, we propose a novel unsupervised method called Semantic Graph Evolutionary Hashing (SGEH). We define a graph evolutionary module which extends the sparse affinity graph to a dense semantic graph, the core idea of which is illustrated in Fig. 1. Specifically, we first take the code-driven similarity graph of both visual modality and textual modality built upon geometric prior and fuse them in a automatically updated weight. Then, after keeping updating the fused graph, we generate a sparse semantic graph which can be shared by both image modal and text modal

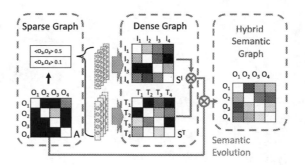

Fig. 1. The process of semantic graph evolution. Sparse graph evolves with the similarity information on geometrical priority. O means object which composes of visual information and caption, and O_i means i-th object in train set.

and relief the problem of lacking label. Our fuse method is inspired by [21] and preserves similarity information hidden in original data from both two modalities. Subsequently, to maximize the use of the spatial information in the sparse graph, we randomly select samples from the connected node pairs in the sparse graph to construct the input of the convolutional neural network. In addition, on one hand, it is impossible to employ the huge adjacency matrix as the feature space for the input of deep network; on the other hand, focusing on neighboring sample pairs can avoid the interference of non-neighboring sample pairs. Hence, we evolve the spares affinity graph to a dense manner by taking the local geometric characters. To map data from different modalties into one common latent hash space, we try to reduce the distance between binary hash codes of the same instance from two modalities, which is reflected in our objective function. In summary, this method has the following main contributions:

- We propose a novel SGEH method by evolving the joint sparse graph obtained by cross-modal clustering into a dense semantic graph that can be used for mini-batch deep learning, which enables hash learning to obtain rich semantic associations among examples.
- We propose a graph evolutionary mechanism to learn a hybrid semantic graph structure from sparse to dense. Instead of directly using dense graphs constructed according to the geometric characteristics of the samples, the graph evolutionary mechanism first learn a sparse affinity graph, and subsequently extend it to a dense form by fusing the dense relation built upon the geometric graph.
- Comprehensive experiments are conducted on four popular datasets, and the results show the priority of the proposed model.

2 Methodology

2.1 Preliminaries

The overall pipeline of SGEH is shown in Fig. 2. We first introduce several definitions in our methods. Let $\boldsymbol{F}^I \in \mathbb{R}^{m \times d_I}$ and $\boldsymbol{F}^T \in \mathbb{R}^{m \times d_V}$ denote m training

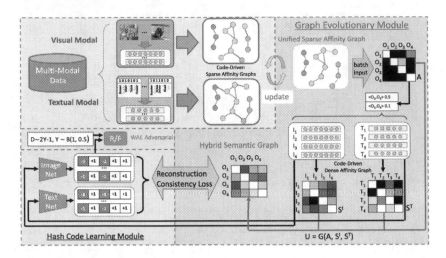

Fig. 2. The framework of our proposed Semantic Graph Evolutionary Hashing (SGEH).

visual and textual instance features respectively. m means the amount of instance in the whole training set. With n equals to the amount of instances in each batch, the visual and textual features in each batch are denoted as $\boldsymbol{X}^I \in \mathbb{R}^{n \times d_I}$ and $\boldsymbol{X}^T \in \mathbb{R}^{n \times d_V}$ respectively. Here d_I and d_T represent the dimensions of image and caption features respectively. Furthermore, we aim to generate binary hash codes $\boldsymbol{B}^I$ and $\boldsymbol{B}^T$ by embedding continuous features into common latent hash space, where $\boldsymbol{B}^H \in \mathbb{R}^{n \times b}, (H \in \{I, T\})$ and b is hash code length.

Utilizing hash code that preserve the neighborhood information can greatly improve the performance of retrieval task. Specifically, previous methods can be grouped into two categories in terms of how to guide hashing learning based on original features. The first category methods preserve the information of original features and use them to learning hash codes directly, they share the following common loss function:

$$\mathcal{L}_H = \left\| \boldsymbol{X}^I \boldsymbol{W}^I - \boldsymbol{B}^I \right\|_F^2 + \left\| \boldsymbol{X}^T \boldsymbol{W}^T - \boldsymbol{B}^T \right\|_F^2,$$
$$s.t. \boldsymbol{B}^g \in \{+1, -1\}^{n \times l}, (\boldsymbol{B}^g)^T \boldsymbol{B}^g = n\boldsymbol{I}, \tag{1}$$

where $\boldsymbol{W}^g$ is learning parameter matrix and $g \in \{I, T\}$. Equation 1 aims to reduce the gap between features and hash codes. The second constraint $(\boldsymbol{B}^g)^T \boldsymbol{B}^g = n\boldsymbol{I}$ aims to generate mutually independent hash codes.

Evolved from the first category, the second category typically generates similarity structure from the features in two modalities, and further combines with the method of graph optimization or graph fusion methods to obtain joint-semantic affinity matrices. Both the design of construing matrices and the strategy of employing the matrices in training stage have an impact on the final performance. To be specific, the common formulation is as following:

$$\mathcal{L}_G = \|\boldsymbol{S} - \boldsymbol{Q}\|_F^2,$$
$$s.t. \boldsymbol{S} = g_1(\boldsymbol{X}^I, \boldsymbol{X}^T) \in \{+1, -1\}^{n \times n}, \boldsymbol{Q} = g_2(\boldsymbol{B}^I, \boldsymbol{B}^T) \in \{+1, -1\}^{n \times n}, \tag{2}$$

where m is the batch size, S preserves the affinity information of samples in each batch and is used for hash learning. Q represent the affinity matrix generated by hash codes. g_1 and g_2 means two similarity calculating functions. However, Graph structure is a typical non-euclidean structure data. Generally, building adjacency matrix based on euclidean distance not only causes similarity information loss but also calculates unrealistic similarity which confuses the training process. For instance, image feature $\boldsymbol{X}_i^I$ and $\boldsymbol{X}_j^I$ are unrelated but may got a few similarity score in a mini-batch, which misleads the process of hash code training. As demonstrated in [22], the sparse graph distilling knowledge from geometric and semantic information of the whole training set can save the most useful similarity information hidden in the data samples to avoid useless or interfering information. Then the unified semantic graph is kept updated based on these features from two modalities, which is defined as $Z = f(\boldsymbol{F}^I, \boldsymbol{F}^T)$, where $f(\cdot)$ is neighborhood information fusion function, the solution of which will be demonstrated in Subsect. 2.2, from Eq. 18 to Eq. 23.

2.2 Unified Sparse Affinity Graph

Clustering results provided by pre-computed global graph can solve the problem of lacking label information to a certain degree. Therefore, in this stage, we need to generate two similarity-induced graphs from both visual and textual modalities which are used to learn a fusion semantic graph $\boldsymbol{Z} \in [0, 1]^{m \times m}$. And $\boldsymbol{Z}$ should keep samples with smaller distance responding to a larger similarity score, and ones with larger distance responding to a smaller similarity score. Finally, $\boldsymbol{Z}$ need to produce clustering results for better hash learning. What's more, it is desirable to update the similarity-induced graphs and fusion semantic graph at the same time for strengthening the accuracy of clustering result. As demonstrated in [21, 22], sparse structure has strong anti-noise ability and friendly to storage. Thus, we first use Gaussian Kernel $S(\boldsymbol{F}_i^H, \boldsymbol{F}_j^H) = exp(\dfrac{-\left\|F_i^H - F_j^H\right\|_2^2}{2\sigma^2})$ to define the weight of two instances from same modal, where σ is the width parameter of the function and controls the radial range of the function, and lately keep k nearest neighborhoods of each vertex to keep graph sparse. Thus, we get $\boldsymbol{S}^I \in [0, 1]^{m \times m}$ and $\boldsymbol{S}^T \in [0, 1]^{m \times m}$, where $\boldsymbol{S}^H$ is the similarity-induced graph and $H \in \{I, T\}$. SGEH mimics GMC [21] loss function and translated it into double-modal expression.

$$\min_{\{\boldsymbol{S}^I, \boldsymbol{S}^T\}} \sum_{H \in \{I,T\}} \sum_{i,j=1}^m \left\|\boldsymbol{F}_i^H - \boldsymbol{F}_j^H\right\|_2^2 \boldsymbol{S}_{ij}^H + \lambda_1 \sum_{H \in \{I,T\}} \sum_i \left\|\boldsymbol{S}_i^H\right\|_2^2, \tag{3}$$
$$s.t. \boldsymbol{S}_{ii}^H = 0, \boldsymbol{S}_{ij}^H \geq 0, \boldsymbol{1}^T \boldsymbol{S}_i^H = 1,$$

and the optimal solution of $\boldsymbol{S}^I$ and $\boldsymbol{S}^T$ can be written in closed-form as Eq. (4) with Lagrange Multiplier Method, which is proved in [21].

$$
\boldsymbol{S}_{ij}^H = \begin{cases} \dfrac{\boldsymbol{b}_{i,p+1} - \boldsymbol{b}_{ij}}{p\boldsymbol{b}_{i,p+1}, -\sum_{h=1}^{p} \boldsymbol{b}_{ih}} & j \le p, \\ 0 & j > p, \end{cases} \tag{4}
$$

where p is a hyper parameter which adjusts the number of neighbors kept in graph and $\left\| \boldsymbol{F}_i^H - \boldsymbol{F}_j^H \right\|_2^2$ is simplified as $\boldsymbol{b}_{ij}$.

Then we need to learn a sparse fusion semantic graph $\boldsymbol{Z}$ to represent the similarity connection in both visual and textual modalities. We design the loss function describing the distance between $\boldsymbol{Z}$ and $\boldsymbol{S}^H$ as:

$$
\min_{\boldsymbol{Z}} \sum_{H \in \{I,T\}} w^H \left\| \boldsymbol{Z} - \boldsymbol{S}^H \right\|_F^2, s.t. \boldsymbol{Z}_{ij} \ge 0, \boldsymbol{1}^T \boldsymbol{Z}_i = 1, \tag{5}
$$

where w^H is the weight of similarity-induced graph. As deliberated in Sect. 1, it is desirable to weight both visual modal and textual modal automatically.

The last problem we need to figure out is how to produce clustering result directly on $\boldsymbol{Z}$. This can be tackled be adding an rank constraint on the graph Laplacian matrix of $\boldsymbol{Z}$ as

$$
\min_{\boldsymbol{D}} Tr \left(\boldsymbol{D}^T \boldsymbol{L} \boldsymbol{D} \right), s.t. \boldsymbol{D}^T \boldsymbol{D} = \boldsymbol{I}, \tag{6}
$$

where $\boldsymbol{L}$ is the Laplacian matrix of $\boldsymbol{Z}$ and $\boldsymbol{D} \in \mathbb{R}^{m \times c}$ is the embedding matrix composed by cluster center vectors. It comes from a theorem that if a matrix is non-negative, then the dimension of the nullspace of Laplacian matrix of the graph of this matrix is the number of connected components of the graph. Thus, let c denotes the number of connected components of $\boldsymbol{Z}$, if rank(L)=$m - c$, the vertexes in $\boldsymbol{Z}$ can be divided into C clusters. However, rank($\boldsymbol{L}$)=$m - c$ is difficult to achieve. Given that $\boldsymbol{L}$ is positive semi-definite, the constraint rank($\boldsymbol{L}$)=$m - c$ can be achieved if the summation of top-c eigenvalue of $\boldsymbol{L}$ equals to zero. Ky Fan's Theorem [4] told us $\sum_{i=1}^{c} v_i = \min Tr \left(\boldsymbol{D}^T \boldsymbol{L} \boldsymbol{D} \right), s.t. \boldsymbol{D}^T \boldsymbol{D} = \boldsymbol{I}$, where v_i is the ith smallest eigenvalue of $\boldsymbol{L}$. Then based on the above fact, this problem can be further tackled by restricting $Tr \left(\boldsymbol{D}^T \boldsymbol{L} \boldsymbol{D} \right) = 0$. Mathematically, we got the loss function of producing sparse fusion semantic graph U as:

$$
\mathcal{L}^A = \sum_{H \in \{I,T\}} \sum_{i,j=1}^{m} \left\| \boldsymbol{F}_i^H - \boldsymbol{F}_j^H \right\|_2^2 \boldsymbol{S}_{ij}^H + \lambda_1 \sum_{H \in \{I,T\}} \sum_i^m \left\| \boldsymbol{S}_i^H \right\|_2^2
$$
$$
+ \sum_{H \in \{I,T\}} w^H \left\| \boldsymbol{Z} - \boldsymbol{S}^H \right\|_F^2 + 2\lambda_2 Tr \left(\boldsymbol{D}^T \boldsymbol{L} \boldsymbol{D} \right), \tag{7}
$$
$$
s.t. \boldsymbol{S}_{ii}^H = 1, \boldsymbol{S}_{ij}^H \ge 0, \boldsymbol{1}^T \boldsymbol{S}_i^H = 1, \boldsymbol{Z}_{ij} \ge 0, \boldsymbol{1}^T \boldsymbol{Z}_i = 1, \boldsymbol{D}^T \boldsymbol{D} = \boldsymbol{I},
$$

where $H \in \{I,T\}$, $\boldsymbol{S}_i \in \mathbb{R}^{n \times 1}$, $\boldsymbol{U}_i \in \mathbb{R}^{n \times 1}$, w^H is the weight of similarity-induced graph, $\boldsymbol{L}$ is the Laplacian matrix of Z and $\boldsymbol{D} \in \mathbb{R}^{m \times c}$ is the clustering center matrix. The optimization of above problem will be solved in Sect. 2.5.

2.3   Semantic Graph Evolution

Although the matrix we obtained at this time can more accurately reflect the relationship between instances, we need to use mini-batch method in the neural network model, which means that in each batch, similarity matrix composed of randomly selected samples tends to be sparse for each instance only have p neighbors. In the hash learning stage, the semantic similarity matrix is utilized to constrain the hash similarity matrix generated by the hash code. However, sparse matrices are incompetent to guide the hash learning process. For example, two pairs of data whose similarity is 0 in the sparse semantic matrix are not also 0 in the hash similarity matrix. At the same time, the sparse matrix has the problem of information loss. Thus, in order to solve the problem of sparse, we need to evolve the similarity matrix into a denser form.

To be specific, suppose there are t edges in ε_Z, which is the edge set of $\boldsymbol{Z}$, and n $(t+1 < n \leq 2t)$ related vertexes for each batch. Then we use V and E to represent its vertex set and edge set respectively. And subsequently we constructs a local sparse graph $\boldsymbol{A} = (V, E)$ on batch-inputs. The features of vertexes in V can be denoted as $\boldsymbol{X}^I$ and $\boldsymbol{X}^T$, which are defined above respectively. We define the similarity matrices in mini batch from visual and textual modalities as:

$$\boldsymbol{S}_{ij}^H = \boldsymbol{X}_i^H (\boldsymbol{X}_j^H)^T / (\|\boldsymbol{X}_i^H\| \|\boldsymbol{X}_j^H\|), \tag{8}$$

where $\boldsymbol{S}_{ij}^H \in [-1, +1]$, $H \in \{I, T\}$. $\boldsymbol{X}_i^H$ means the i-th row in $\boldsymbol{X}^H$ and $\boldsymbol{X}_j^H$ stands for the j-th row in $\boldsymbol{X}^H$. We employ $\boldsymbol{S}^I$ and $\boldsymbol{S}^T$ to integrate the original similarity information in image and text modal. Then unified sparse affinity graph A is evolved from spare to dense by fusing the information from $\boldsymbol{S}^I$ and $\boldsymbol{S}^I$. Then we get hybrid semantic affinity matrix $\boldsymbol{U} = \mathcal{G}(\boldsymbol{A}, \boldsymbol{S}^I, \boldsymbol{S}^T) \in [-1, +1]^{n \times n}$ to describe the affinity structure in both two modalities, with $\boldsymbol{U}_{ij}$ describing the captured fusion semantic affinity information between the input samples $\boldsymbol{e}_i$ and $\boldsymbol{e}_j$. The hybrid semantic affinity matrix is calculated as:

$$\boldsymbol{U} = \mathcal{G}_1(\boldsymbol{A}, \boldsymbol{S}^I, \boldsymbol{S}^T) = (1 - \lambda_3)[\lambda_4 \boldsymbol{S}_I + (1 - \lambda_4) \boldsymbol{S}_T] + \lambda_3 \boldsymbol{A}, \tag{9}$$

where λ_3 adjust the importance of clustering result and λ_4 balances the weights between affinity structure information of visual modality and textual modality. The manner of constructing $\boldsymbol{U}$ combines the similarity information across both clustering result and original affinity structure in two modalities. Given that samples selected in batch are highly related, $\boldsymbol{U}$ refines the affinity more accurate than randomly training samples in mini-batch, which makes it capturing more effective latent common similarity relationship over multi-modal perspective. In another word, $\boldsymbol{U}$ reflects the original affinity connection among input samples, after which we can subsequently learn binary hash code that are employed to achieve cross-modal retrieval task. Alternatively, following the form of combination in [20], we can make $\boldsymbol{A}$ evolved in the form as:

$$\overline{\boldsymbol{S}} = \lambda_4 \boldsymbol{S}_I + (1 - \lambda_4) \boldsymbol{S}_T, \boldsymbol{U} = \mathcal{G}_2(\boldsymbol{A}, \boldsymbol{S}^I, \boldsymbol{S}^T) = \lambda_3 \boldsymbol{A} + (1 - \lambda_3)(\frac{\overline{\boldsymbol{S}}\,\overline{\boldsymbol{S}}^T}{n}). \tag{10}$$

2.4   Hash-Code Learning

In this subsection, we utilize the accurate semantic matrix U, which represents the original affinity relations of the input instances, to restrict the generation stage of hash code. In latent hash hypercube, adjacent vertices share small Hamming distance and more similar hash codes. Thus, hash codes can be understood as discrete features. To calculate the similarity with neighborhoods in Hamming space, the similarity function can be defined as:

$$\mathcal{Z}(\boldsymbol{B}_i^H, \boldsymbol{B}_j^H) = \boldsymbol{B}_i^H (\boldsymbol{B}_j^H)^T / (\|\boldsymbol{B}_i^H\| \|\boldsymbol{B}_j^H\|), \tag{11}$$

where $H \in \{I, T\}$, $\boldsymbol{B}_i^H$ means the i-th row in $\boldsymbol{B}^H$ and $\boldsymbol{B}_j^H$ means the j-th row in $\boldsymbol{B}^H$. The result of Eq.(11) is the cosine affinity score which representing the angular connection among discrete features. We adopt manner of Eq.(2) that minimize the reconstruction error between the similarity matrix of hash code and the affinity matrix U of continuous features to keep their similarity consistency. Therefore, we define pairwise cosine similarity matrices as Q and $\boldsymbol{Q}_{ij}^{HH} = \mathcal{Z}(\boldsymbol{B}_i^H, \boldsymbol{B}_j^H)$. Then, we employ

$$\mathcal{L}^{B_{HH}} = \min_{\boldsymbol{B}_H} \left\| \alpha U - \mathcal{Z}(\boldsymbol{B}_i^H, \boldsymbol{B}_j^H) \right\|_F^2, \tag{12}$$

as the formulation to compute the difference between U and $\boldsymbol{Q}_{ij}^{HH}$. In Eq. (12), α is a hyper-parameter which makes reconstruction more flexible, as discussed in [20]. Given that $U \in [-1, +1]^{n \times n}$, $\alpha U \in [-\alpha, +\alpha]^{n \times n}$. For example, supposed that $U_{ij} = 0.8$, which means that ith instance and jth instance got 0.8 similarity score, then the similarity score of corresponding hash codes calculated from Hamming space need to be close to 0.8. $\alpha > 1$ means the similarity score of hash codes pair need to lager than 0.8 and accordingly make the nodes in Hamming space dense, while $\alpha < 1$ means the similarity score of hash codes pair need to smaller than 0.8 and accordingly make the nodes in Hamming space sparse. We empirically find that it is beneficial to threshold $\alpha > 1$. And this phenomenon can be attributed to the fact that cosine similarity measures the similarity between two vectors by measuring the cosine of the angle between them. The result is not related to the length of the vector, but only related to the direction of the vector. Setting $\alpha > 1$ means we force the binary hash code close to the latent clustering center in Hamming space in direction than it should be, which bring a better performance as we are trying to Increase the distance between categories and reduce the distance within category.

Given that each instance is still represented by hash codes from two modalities, we need to restrict the reconstruction in the manner of intro-modal and inter-modal. Specifically, we employ $\boldsymbol{Q}^{II}$ and $\boldsymbol{Q}^{TT}$ as the intro-modal reconstruction for image modal and text modal respectively, and $\boldsymbol{Q}^{IT}$ is engaged as inter-modal reconstruction. Finally, the consistency loss between binary hash code and continues original features can be summarized as:

$$
\begin{aligned}
\mathcal{L}^B =\mathcal{L}^{B_{II}} + \eta_1 \mathcal{L}^{B_{TT}} + \eta_2 \mathcal{L}^{B_{IT}} &= \min_{B_I} \left\| \alpha U - \mathcal{Z}(\boldsymbol{B}_i^I, \boldsymbol{B}_j^I) \right\|_F^2 \\
&+ \eta_1 \min_{B_T} \left\| \alpha U - \mathcal{Z}(\boldsymbol{B}_i^T, \boldsymbol{B}_j^T) \right\|_F^2 + \eta_2 \min_{B_I, B_T} \left\| \alpha U - \mathcal{Z}(\boldsymbol{B}_i^I, \boldsymbol{B}_j^I) \right\|_F^2,
\end{aligned}
\tag{13}
$$

where η_1 and η_2 are the trade-off parameters to balance the reconstruction of inter-modal and intro-modal in latent hash space.

To avoid wasting bits and align representation distributions, it is worth generating mutually independent hash codes as the constraint $\boldsymbol{B}_g^T \boldsymbol{B}_g = m\boldsymbol{I}$ in Eq. 1. However, the above loss cannot tackle this problem. Following [18], we regularize the latent discrete features with auxiliary discriminator $d(\cdot, \zeta)$. To be concrete, we assume the each row in $\boldsymbol{B}$ comes from a distribution $\mathcal{D} = 2\mathcal{Y} - 1, \mathcal{Y} \sim \mathcal{B}(1, 0.5)$, which maximizes the code entropy. We suppose that each binary code is prioired by a binomial distribution which is $\mathcal{D}$. Then, to adversarially regularize the latent variables, we utilize auxiliary discriminator d which involves two fully-connected layers successively with ReLu and sigmoid non-linearities. In a word, it is to balance the amount of zeros and ones in each binary code and further maximize the code entropy. In this end, we can employ the following discriminator d to balance -1 and $+1$ in a binary hash code:

$$
d(\boldsymbol{B}_i, \zeta) \in (-1, +1); d(\boldsymbol{y}^{\boldsymbol{B}_i}, \zeta) \in (-1, +1),
\tag{14}
$$

where $\boldsymbol{y}^{\boldsymbol{B}_i}$ obeys the same distribution as $\boldsymbol{B}_i$ for implicit regularizing $\boldsymbol{B}_i$. Therefore, our final loss can be written as:

$$
\mathcal{L} = \mathcal{L}^A + \mathcal{L}^B - \frac{\eta_4}{b} \sum_{H \in I, T} \sum_{i=1}^{b} (log(1 - d(\boldsymbol{B}_i, \zeta)) + log\, d(\boldsymbol{y}^{\boldsymbol{B}_i}, \zeta))).
\tag{15}
$$

2.5   Optimization

In this method, the global sparse graph is employed to solve the problem of missing label information and we first need to optimize Eq. 7.

Sparse Affinity Graph. In this stage, we basically refer to the alternating rules used in [21] to optimize Eq. 7. As there are four variables in total and are coupled with each other, the problem is split into four step. It is beneficial to get detailed information from the above method. Here, we directly give the close-form solution of the variables $\boldsymbol{Z}, \boldsymbol{S}, \boldsymbol{D}, \boldsymbol{w}^H$:

$step1$: Fix $\boldsymbol{Z}, \boldsymbol{D}$ and w^H, update $\boldsymbol{S}^H$. When $\boldsymbol{Z}, \boldsymbol{D}$ and w^H fixed, the last item of Eq. 7 is constant and accordingly original problem is translated into following pattern:

$$
\begin{aligned}
\min_{\boldsymbol{S}^H} \sum_{H \in \{I,T\}} \sum_{i,j=1}^{m} \left\| \boldsymbol{F}_i^H - \boldsymbol{F}_j^H \right\|_2^2 \boldsymbol{S}_{ij}^H + \lambda_1 \sum_{H \in \{I,T\}} \sum_i^m \left\| \boldsymbol{S}_i^H \right\|_2^2 \\
+ \sum_{H \in \{I,T\}} w^H \left\| \boldsymbol{Z} - \boldsymbol{S}^H \right\|_F^2, \quad s.t. \boldsymbol{S}_{ii}^H = 1, \boldsymbol{S}_{ij}^H \geq 0, \boldsymbol{1}^T \boldsymbol{S}_i^H = 1,
\end{aligned}
\tag{16}
$$

Updating both two view is independent, we update each S^H in the following way:

$$\min_{S^H} \sum_{i,j=1}^{m} \left\| F_i^H - F_j^H \right\|_2^2 S_{ij}^H + \lambda_1 \sum_i^m \left\| S_i^H \right\|_2^2 + w^H \left\| Z - S^H \right\|_F^2,$$

$$s.t. S_{ii}^H = 1, S_{ij}^H \geq 0, \mathbf{1}^T S_i^H = 1, \tag{17}$$

For simplicity, we suppose that a feature of sample is similar to its neighbours and accordingly can update the representation using its p neighbor data points, where p is the number of neighbors. We employ the solution from [21] and give the final solution as follows:

$$S_{ij}^{H*} = \begin{cases} \dfrac{b_{i,p+1} - b_{ij} + 2w^H Z_{ij} - 2w^H Z_{i,p+1}}{pb_{i,p+1}, -\sum_{h=1}^{p} b_{ih} - 2pw^H Z_{i,p+1} + 2\sum_{h=1}^{p} w^H Z_{ih}} & j \leq p, \\ 0 & j > p, \end{cases} \tag{18}$$

$step2$: Fix Z, D and S^H, update w^H. In this step, fixing problem 7 is the same way to solve the problem (5).

Theorem. *If the weights w^H are fixed, solving problem 5 is equivalent to solving the following probelm:*

$$\min_{Z} \sum_{H \in \{I,T\}} \sqrt{\left\| Z - S^H \right\|_F^2}, s.t. Z_{ij} \geq 0, \mathbf{1}^T Z_i = 1, \tag{19}$$

Proof. The Lagrange function of Eq (19) is:

$$\sum_{H \in \{I,T\}} \sqrt{\left\| Z - S^H \right\|_F^2} + \Theta(\Lambda, Z), s.t. Z_{ij} \geq 0, \mathbf{1}^T Z_i = 1, \tag{20}$$

where Λ is the Lagrange multiplier, and $\Theta(\Lambda, Z)$ is the formalized term derived from constraints. Taking the derivative of Eq. (20) with respect to Z and setting the derivative to zero, we get the following equation:

$$w^{H*} = \frac{1}{\sqrt[2]{\left\| Z - S^H \right\|_F^2}}. \tag{21}$$

$step3$: Fix all the other variables except Z, and it can be proved that solving Eq. 7 is equivalent to solving the following problem:

$$\min_{Z_i} \sum_{H \in \{H,I\}} \left\| Z_i - S_i^H + \frac{\lambda_1}{4w^H} d_i \right\|_2^2, s.t. Z_{ij} \geq 0, \mathbf{1}^T Z_i = 1, \tag{22}$$

where Z_i means the i-th row in Z and d_{ij} means the similarity score between S_i^H and Z_i. The problem in Eq. 22 can be solved with Lagrange Multiplier Method as proved in [21] with several steps:

$$\boldsymbol{q}^H = \boldsymbol{S}_i^H - \frac{\lambda_1}{4w^H}\boldsymbol{d}_i, \quad \boldsymbol{p} = \frac{\sum_{H \in \{I,T\}} \boldsymbol{q}^H}{2} + \frac{1}{m} - \frac{\mathbf{1}^T \boldsymbol{q}^H \mathbf{1}}{2m},$$
$$f(t) = \frac{1}{m} \sum_{j=1}^{m} (t - \boldsymbol{p}_j)_+ - t, \quad \boldsymbol{Z}_{ij}^* = (\boldsymbol{p}_j - t^*)_+, \tag{23}$$

where t^* makes $f(t^*) = 0$ and $(\cdot)_+ = max(\cdot, 0)$. In summary, the produce for solving the proposed problem in Eq. 7 can be found in the supplementary material.

   $step4$:Fix all the other variables except $\boldsymbol{D}$, optimizing problem (7) is equivalent to problem (6), which is formed by the c eigenvectors of $\boldsymbol{L}$ corresponding to the c smallest eigenvalues.

   Deep Hash Learning. In traditional hash methods [5,18], the process of mapping continuous features to discrete space causes huge quantization loss stem from the fact that the sign function, which outputs +1 for positive number and -1 for negative number, can not be derived. To handle this problem, we follow [20] to adopt a scaled tanh function:

$$\boldsymbol{B} = tanh(\beta \boldsymbol{Y}) \in [-1, +1]^{m \times d}, \beta \in \mathbb{R}^+, \tag{24}$$

where $\boldsymbol{Y}$ represent that final output of Convolutional Neural Network. It is noticed that β is kept increasing during deep training stage. To be noted that it is motivated by a crucial fact that $\lim_{\alpha \to \infty} tanh(\beta y) = sgn(y)$.

3   Experiments

3.1   Datasets

Four datasets, including **Wiki** [15], **NUS-WIDE** [2], **MIRFlickr-25K** [9] and **MSCOCO** [12], are employed to evaluate the proposed methods, more details about the four datasets can be found in the supplementary material.

3.2   Implementation Details

For all of our experiments, we follow previous methods to employ the VGG-16 fc7 to extract the 4,096-dimensional deep features $\boldsymbol{X}^I \in \mathbb{R}^{n \times 4096}$ from original images, while for original textual features we utilize the universal sentence encoder [1] to represent final textual features $\boldsymbol{X}^T$ whose dimension is 512. Besides, considering the computational burden in the solution process of $\boldsymbol{Z}$, we randomly pick up 20,000 instances from NUS-WIDE and MSCOCO dataset. It is worth noting that to calculate the consistency loss as the manner of Eq. 2, we need to force the items in the ranges. However, the cosine similarity ranges -1 from +1 while the affinity value elements in $\boldsymbol{Z}$ are non-negative, which can be obtained by Eq. 23. Therefore, as $\boldsymbol{A}$ is the batch-input of $\boldsymbol{Z}$, we preprocess

Table 1. The mAP@all results on image query text $(I \rightarrow T)$ and text query image $(T \rightarrow I)$ retrieval at various encoding lengths and datasets. The best performances are shown in bold.

Task	Method	WIKI			MIRFlicker-25K			MSCOCO			NUS-WIDE		
		16bit	32bit	64bit	16bit	32bit	64bit	16bit	32bit	64bit	16bit	32bit	64bit
$I \rightarrow T$	CMFH	0.173	0.169	0.184	0.580	0.572	0.554	0.442	0.423	0.492	0.381	0.429	0.416
	PDH	0.196	0.168	0.184	0.544	0.544	0.545	0.442	0.423	0.492	0.368	0.368	0.368
	IMH	0.151	0.145	0.133	0.557	0.565	0.559	0.416	0.435	0.442	0.349	0.356	0.370
	QCH	0.159	0.143	0.131	0.579	0.565	0.554	0.496	0.470	0.441	0.401	0.382	0.370
	DJSRH	0.274	0.304	0.350	0.649	0.662	0.669	0.561	0.585	0.585	0.496	0.529	0.528
	DGCPN	0.226	0.326	0.410	0.651	0.670	0.702	0.469	0.586	0.630	0.517	0.553	0.567
	DSAH	0.249	0.333	0.381	0.654	0.693	0.700	0.518	0.595	0.632	0.539	0.566	0.576
	JDSH	0.253	0.289	0.325	0.665	0.681	0.697	0.571	0.613	0.624	0.545	0.553	0.572
	SGEH	**0.396**	**0.422**	**0.441**	**0.665**	**0.695**	**0.703**	**0.578**	**0.617**	**0.634**	**0.565**	**0.584**	**0.579**
$T \rightarrow I$	CMFH	0.176	0.170	0.179	0.583	0.566	0.556	0.453	0.435	0.499	0.394	0.451	0.447
	PDH	0.344	0.293	0.251	0.544	0.544	0.546	0.437	0.440	0.440	0.366	0.366	0.367
	IMH	0.236	0.237	0.218	0.560	0.569	0.563	0.560	0.561	0.520	0.350	0.356	0.371
	QCH	0.341	0.289	0.246	0.585	0.567	0.556	0.505	0.478	0.445	0.405	0.385	0.372
	DJSRH	0.246	0.287	0.333	0.658	0.660	0.665	0.563	0.577	0.572	0.499	0.530	0.536
	DGCPN	0.186	0.297	0.522	0.648	0.676	0.703	0.474	0.594	0.634	0.509	0.556	0.574
	DSAH	0.249	0.315	0.393	0.678	0.700	0.708	0.533	0.590	0.630	0.546	0.572	0.578
	JDSH	0.256	0.303	0.320	0.660	0.692	0.710	0.565	0.619	0.632	0.545	0.566	0.576
	SGEH	**0.452**	**0.510**	**0.530**	**0.687**	**0.706**	**0.711**	**0.578**	**0.626**	**0.635**	**0.570**	**0.588**	**0.595**

the $\boldsymbol{A}$ with $\boldsymbol{A} \leftarrow 2\boldsymbol{A} - 1$. Additionally, we fix the batch size as 8 and employ the SGD optimizer with 0.9 momentum and 0.0005 weight decay. We experimentally take $\alpha = 1.5$ and $\lambda_3 = 0.4$ for all four datasets. Then we set $c = 5$, $p = 10000$, $\lambda_4 = 0.6$, $\eta_1 = \eta_2 = 0.1$ for NUM-WIDE, $c = 5$, $p = 3000$, $\lambda_4 = 0.9$, $\eta_1 = \eta_2 = 0.1$ for MIRFlicker, $c = 5$, $p = 1000$, $\lambda_4 = 0.3$, $\eta_1 = \eta_2 = 0.3$ for Wiki and $c = 5$, $p = 3000$, $\lambda_4 = 0.6$, $\eta_1 = \eta_2 = 0.1$ for MSCOCO.

3.3   Retrieval Performance

Baselines. Previous methods can be categorized into two kinds according to whether takes the whole retrieved points into consideration or not. Hence, in order to prove that our method has superior performance under different evaluation indicators, we conduct experiments on two aspects. Specifically, on the one hand, we compare the mAP results with IMH [19], CMFH [3], PDH [16], QCH [23], DJSRH [20], DSAH [28], JDSH [13], DGCPN [30] conducted on Wiki, MIRFlicker, MSCOCO and NUS-WIDE datasets, with the whole retrieved points occupied (*i.e.*, mAP@all), and the results are shown in Tab. 1. All the compared method are conducted according to their released codes or description in their original papers. The retrieval performance on mAP@50 can be found in the supplementary material.

Quantitative Results. It can be observed that the proposed SGEH outperforms all of other unsupervised cross-modal hashing methods in Table 1 on all

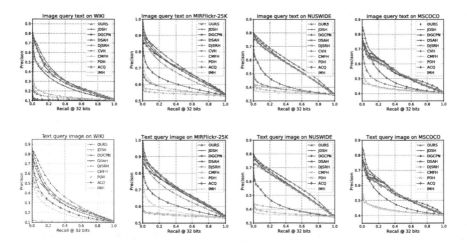

Fig. 3. Results of Precision VS Recall Curves of various unsupervised hashing methods on datasets WIKI, MIRFLickr-25K, MSCOCO and NUS-WIDE with 32-bit codes.

four datasets regardless of the cross-modal retrieval tasks and code lengths, which demonstrates the effectiveness of the proposed methods. Specifically, our image query for text retrieval performance on Wiki dataset improves a lot compared with other deep methods on three kinds of hash codes, especially on 16 bits and 32 bits, while the text query for image retrieval performance outperforms them more than 10.8%, 19.5%, 0.8% on 16 bits, 32 bits, and 64 bits respectively. While improvements on NUS-WIDE and MSCOCO are related lower, which is stemmed from that we only using 20,000 samples as training set. The corresponding Precision-Recall (P-R) curves of represented methods are also retorted in Fig. 3, which can further prove the effectiveness of our method. In particular, our curves for Wiki are all located above those of the other methods, which means that the precision of our approach can significantly surpass that of the other works at the same recall rates. As for the multi-label datasets, Our P-R curves on MSCOCO and NUS-WIDE are also higher than the other, but not as obviously as the curves on Wiki. On the MIRFlickr-25K, we can obtain that the results are slightly worse than DSAH for 32 bits when the recall rate is higher than 0.14 when image queries text and 0.12 when text queries image. However, taking text query image for instance, it can be seen that our curve get (recall = 0.05, precision = 0.81), which means that our method can obtain images with 81% accuracy among the $0.05 \times 20,000 = 1000$ return images.

3.4 Ablation Study

To further demonstrate the effectiveness of each part in SGEH, we design several variants to evaluate the performance when adding the proposed each components. Following the introduction order in Sect. 2, SGEH-1 and SGEH-2 are the basic variants which respectively only employ A as similarity matrix and

Table 2. The mAP@all on MIRFlickr-25K to evaluate the value of each component.

Model	Configuration	32bits		64bits	
		$I \to T$ $T \to I$		$I \to T$ $T \to I$	
SGEH-1	$U = A$	0.658	0.679	0.650	0.678
SGEH-2	$U = \lambda_4 S^I + (1 - \lambda_4)S^T$	0.684	0.695	0.680	0.701
SGEH-3	$U = \mathcal{G}_1(A, S^I, S^T)$	0.685	0.694	0.692	0.699
SGEH-4	$U = \mathcal{G}_2(A, S^I, S^T)$	0.688	0.692	0.693	0.702
SGEH	$\eta_4 = 0.001$	0.695	0.706	0.703	0.711

only employ S^I with S^T as similarity matrix. SGEH is the variant that merges the affinity metrics in the manner of $\mathcal{G}(A, S^I, S^T) = (1 - \lambda_3)[\lambda_4 S_I + (1 - \lambda_4)S_T] + \lambda_3 A$. SGEH is the variant based on SGEH-4 which further supplements the loss of discriminator. The results are shown in Table 2, and from which we can discover that each component of our proposed method has its own effect. Table 2 suggests that removing any component of our final framework leads to performance degradation. Specially, compared with the results of SGEH-1 and SGEH-2, the better performance of SGEH-3 and SGEH-4 shows illustrate the effectiveness of the proposed fusion strategy Eq. (9). The combination of clustering information from total dataset and neighborhood information in each mini-batch can much more accurately define the similarity relationship, impelling to learn more consistent hash codes and accordingly achieving better performance. What's more, SGEH demonstrate the important role of hashcode regularization. It facilitates the proposed method for the end-to-end batch-wise training which better refine the similarity relationship by combining the clustering results and mini-batch neighborhood information than previous mini-batch pattern.

4   Conclusion

This paper proposed Semantic Graph Evolutionary Hashing (SGEH) for unsupervised cross-modal retrieval. SGEH first employs sparse affinity graph to update the unified sparse affinity graph, which is shared by both visual modal and textual modal. And subsequently the sparse graph is evolved from sparse to dense by fusing code-driven similarity information. Consequently, the sparse graph extends to the Hybrid Semantic Graph which is utilized to restrict hash code learning. The key novelty of this method is the graph evolution scheme. Then hash code can be learned via construction consistence loss with a more effective feature space. Extensive experiments demonstrate the superiority of our proposed method and detailed ablation study shows the effect of each module utilized in our method.

Acknowledgements. This work was supported in part by the National Natural Science Foundation of China (NSFC) under Grants No. 61872187, No. 62072246 and No.

62077023, in part by the Natural Science Foundation of Jiangsu Province under Grant No. BK20201306, and in part by the "111" Program under Grant No. B13022.

References

1. Cer, D., et al.: Universal sentence encoder. arXiv preprint arXiv:1803.11175 (2018)
2. Chua, T.S., Tang, J., Hong, R., Li, H., Luo, Z., Zheng, Y.: Nus-wide: a real-world web image database from National University of Singapore. In: Proceedings of the ACM International Conference on Image and Video Retrieval, pp. 1–9 (2009)
3. Ding, G., Guo, Y., Zhou, J.: Collective matrix factorization hashing for multimodal data. In: Proceedings of the IEEE Conference on Computer Vision and Pattern Recognition, pp. 2075–2082 (2014)
4. Fan, K.: On a theorem of weyl concerning eigenvalues of linear transformations i. Proc. Natl. Acad. Sci. United States America **35**(11), 652 (1949)
5. Gong, Y., Lazebnik, S., Gordo, A., Perronnin, F.: Iterative quantization: a procrustean approach to learning binary codes for large-scale image retrieval. IEEE Trans. Pattern Anal. Mach. Intell. **35**(12), 2916–2929 (2012)
6. He, L., Xu, X., Lu, H., Yang, Y., Shen, F., Shen, H.T.: Unsupervised cross-modal retrieval through adversarial learning. In: ICME, pp. 1153–1158 (2017)
7. Hu, H., Xie, L., Hong, R., Tian, Q.: Creating something from nothing: Unsupervised knowledge distillation for cross-modal hashing. In: Proceedings of the IEEE Conference on Computer Vision and Pattern Recognition, June 2020
8. Hu, M., Yang, Y., Shen, F., Nie, N., Hong, R., Shen, H.: Collective reconstructive embeddings for cross-modal hashing. IEEE IEEE Trans. Image Process. **28**(6), 2770–2784 (2019)
9. Huiskes, M.J., Lew, M.S.: The mir flickr retrieval evaluation. In: Proceedings of the 28th ACM International Conference on Multimedia, pp. 39–43 (2008)
10. Li, C., Deng, C., Li, N., Liu, W., Gao, X., Tao, D.: Self-supervised adversarial hashing networks for cross-modal retrieval. In: Proceedings of the IEEE Conference on Computer Vision and Pattern Recognition, pp. 4242–4251, June 2018
11. Li, C., Deng, C., Wang, L., Xie, D., Liu, X.: Coupled cyclegan: unsupervised hashing network for cross-modal retrieval. In: Proceedings of the AAAI Conference on Artificial Intelligence, pp. 176–183 (2019)
12. Lin, T.-Y., et al.: Microsoft COCO: common objects in context. In: Fleet, D., Pajdla, T., Schiele, B., Tuytelaars, T. (eds.) ECCV 2014. LNCS, vol. 8693, pp. 740–755. Springer, Cham (2014). https://doi.org/10.1007/978-3-319-10602-1_48
13. Liu, S., Qian, S., Guan, Y., Zhan, J., Ying, L.: Joint-modal distribution-based similarity hashing for large-scale unsupervised deep cross-modal retrieval. In: Proceedings of the 43rd International ACM SIGIR Conference on Research and Development in Information Retrieval, pp. 1379–1388 (2020)
14. Lu, X., Zhu, L., Li, J., Zhang, H., Shen, H.T.: Efficient supervised discrete multiview hashing for large-scale multimedia search. IEEE Trans. Multimedia **22**(8), 2048–2060 (2020). https://doi.org/10.1109/TMM.2019.2947358
15. Rasiwasia, N., et al.: A new approach to cross-modal multimedia retrieval. In: Proceedings of the ACM International Conference on Multimedia, pp. 251–260 (2010)
16. Rastegari, M., Choi, J., Fakhraei, S., Hal, D., Davis, L.: Predictable dual-view hashing. In: Proceedings of International Conference on Machine Learning, pp. 1328–1336. PMLR (2013)

17. Shen, H.T., Liu, L., Yang, Y., Xu, X., Huang, Z., Shen, F., Hong, R.: Exploiting subspace relation in semantic labels for cross-modal hashing. IEEE Trans. Knowl. Data Eng. (2020)
18. Shen, Y., et al.: Auto-encoding twin-bottleneck hashing. In: Proceedings of the IEEE Conference on Computer Vision and Pattern Recognition, pp. 2818–2827 (2020)
19. Song, J., Yang, Y., Yang, Y., Huang, Z., Shen, H.T.: Inter-media hashing for large-scale retrieval from heterogeneous data sources. In: Proceedings of the International Conference on Management of Data, pp. 785–796 (2013)
20. Su, S., Zhong, Z., Zhang, C.: Deep joint-semantics reconstructing hashing for large-scale unsupervised cross-modal retrieval. In: Proceedings of the International Conference on Computer Vision, pp. 3027–3035 (2019)
21. Wang, H., Yang, Y., Liu, B.: Gmc: graph-based multi-view clustering. IEEE Trans. Knowl. Data Eng. **32**(6), 1116–1129 (2019)
22. Wang, W., Shen, Y., Zhang, H., Yao, Y., Liu, L.: Set and rebase: determining the semantic graph connectivity for unsupervised cross modal hashing. In: Proceedings of the International Joint Conference on Artificial Intelligence, pp. 853–859 (2020)
23. Wu, B., Yang, Q., Zheng, W.S., Wang, Y., Wang, J.: Quantized correlation hashing for fast cross-modal search. In: Proceedings of the International Joint Conference on Artificial Intelligence, pp. 3946–3952. Citeseer (2015)
24. Wu, G., Lin, Z., Han, J., Liu, L., Ding, G., Zhang, B., Shen, J.: Unsupervised deep hashing via binary latent factor models for large-scale cross-modal retrieval. In: Proceedings of the International Joint Conference on Artificial Intelligence, pp. 2854–2860 (2018)
25. Wu, L., Wang, Y., Shao, L.: Cycle-consistent deep generative hashing for cross-modal retrieval. IEEE Trans. Image Process. **28**(4), 1602–1612 (2018)
26. Xie, L., Shen, J., Zhu, L.: Online cross-modal hashing for web image retrieval. In: Proceedings of the AAAI Conference on Artificial Intelligence (2016)
27. Xu, R., Li, C., Yan, J., Deng, C., Liu, X.: Graph convolutional network hashing for cross-modal retrieval. In: Proceedings of the International Joint Conference on Artificial Intelligence, pp. 982–988 (2019)
28. Yang, D., Wu, D., Zhang, W., Zhang, H., Li, B., Wang, W.: Deep semantic-alignment hashing for unsupervised cross-modal retrieval. In: Proceedings of the 2020 International Conference on Multimedia Retrieval, pp. 44–52 (2020)
29. Yang, E., Deng, C., Liu, W., Liu, X., Tao, D., Gao, X.: Pairwise relationship guided deep hashing for cross-modal retrieval. In: Proceedings of the AAAI Conference on Artificial Intelligence, vol. 31 (2017)
30. Yu, J., Zhou, H., Zhan, Y., Tao, D.: Deep graph-neighbor coherence preserving network for unsupervised cross-modal hashing. In: Proceedings of the AAAI Conference on Artificial Intelligence (2021)
31. Zhang, J., Peng, Y., Yuan, M.: Unsupervised generative adversarial cross-modal hashing. In: Proceedings of the AAAI Conference on Artificial Intelligence, vol. 32 (2018)
32. Zhang, X., Lai, H., Feng, J.: Attention-aware deep adversarial hashing for cross-modal retrieval. In: Proceedings of European Conference on Computer Vision, September 2018
33. Zhu, J.Y., Park, T., Isola, P., Efros, A.A.: Unpaired image-to-image translation using cycle-consistent adversarial networks. In: Proceedings of the International Conference on Computer Vision, pp. 2223–2232 (2017)

SST-VLM: Sparse Sampling-Twice Inspired Video-Language Model

Yizhao Gao[1,2] and Zhiwu Lu[1,2(✉)]

[1] Gaoling School of Artificial Intelligence, Renmin University of China, Beijing, China
{gaoyizhao,luzhiwu}@ruc.edu.cn
[2] Beijing Key Laboratory of Big Data Management and Analysis Methods, Beijing, China

Abstract. Most existing video-language modeling methods densely sample dozens (or even hundreds) of video clips from each raw video to learn the video representation for text-to-video retrieval. This paradigm requires high computational overload. Therefore, sparse sampling-based methods are proposed recently, which only sample a handful of video clips with short time duration from each raw video. However, they still struggle to learn a reliable video embedding with fragmented clips per raw video. To overcome this challenge, we present a novel video-language model called SST-VLM inspired by a Sparse Sampling-Twice (SST) strategy, where each raw video is represented with only two holistic video clips (each has a few frames, but throughout the entire video). For training our SST-VLM, we propose a new Dual Cross-modal MoCo (Dual X-MoCo) algorithm, which includes two cross-modal MoCo modules to respectively model the two clip-text pairs (for each video-text input). In addition to the classic cross-modal contrastive objective, we devise a clip-level alignment objective to obtain more consistent retrieval performance by aligning the prediction distributions of the two video clips (based on the negative queues of MoCo). Extensive experiments show that our SST-VLM achieves new state-of-the-art in text-to-video retrieval.

1 Introduction

Video-language modeling has drawn great attention in recent years, because it is applicable to a wide variety of practical downstream tasks, including text-to-video retrieval [1–4], video captioning [1,5,6], and video question answering [7–9]. In this paper, we focus on text-to-video retrieval, and hopefully our work can bring some inspirations to other video-language tasks. Since raw videos consist of a series of image frames, processing these frames acquires tremendous computation cost and resource consumption. Therefore, how to efficiently and effectively utilize/integrate the video frames to obtain informative video representation has become a great challenge in text-to-video retrieval.

Existing approaches [10–19] address this challenge mainly by encoding each raw video with multiple sampled video clips. Most of them [10–14,16] sample video clips with short time duration (e.g., 1 s for each clip) from the raw video. Since such local clips can hardly represent the holistic content of the raw

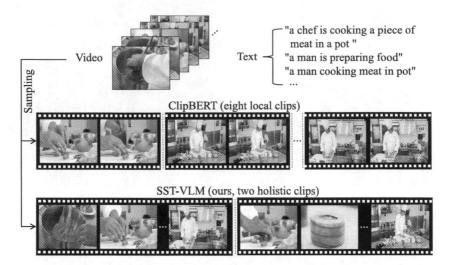

Fig. 1. Comparison among different sparse sampling strategies for text-to-video retrieval. We draw one video-text pair from the original dataset (1st row). Note that the video content of 'meat' (that appears in several captions) fails to be sampled in ClipBERT [10] (2nd row), but it is correctly sampled in our SST-VLM (3rd row).

video, these methods often sample them densely (i.e., sample a large number of local clips per raw video). Unlike these dense sampling methods, ClipBERT [10] applies a sparse sampling strategy to each raw video (i.e., only a handful of local clips are sampled), which has been reported to be effective. However, it still has limitations: the sampled local clips with short time duration are separately matched with the query text to obtain the clip-level predictions (before aggregated into the final video-level prediction), and thus the video content of some important concepts may be ignored by such sparse sampling strategy (see Fig. 1). Therefore, matching sampled local clips to the whole video description is not reliable for video-language modeling.

To overcome these limitations, in this work, we propose a new video sampling strategy named '**Sparse Sampling-Twice (SST)**' for text-to-video retrieval, which sparsely and holistically samples two video clips from each raw video. Our sampling strategy has two key characteristics: (1) *Sparse Random Sampling* – we first subdivide a raw video into a handful of equal segments and then randomly sample a single frame from each segment, resulting in a holistic video clip. (2) *Sampling-Twice* – since sampling only one holistic clip may ignore some key information of the raw video and make the video-text prediction unreliable, we propose to sample two holistic clips by imposing the same sparse random sampling strategy twice on each raw video. Note that we can easily sample more clips per raw video, but in this work, we focus on sampling-twice due to the GPU resource restriction. The detailed comparison between the sparse sampling strategies used in ClipBERT [10] and our SST strategy is shown in Fig. 1.

Inspired by our SST, we present a novel video-language model termed SST-VLM for text-to-video retrieval (see Fig. 2). To train our model, we propose a new

Dual Cross-modal MoCo (**Dual X-MoCo**) algorithm, which includes two cross-modal MoCo [20] modules to respectively model the two clip-text pairs for each video-text input. For the video clip, we employ a 2D image encoder (i.e., ViT-base [21]) to embed the sampled frames and obtain the video embedding by a Transformer [22] module. For the text description, we employ a text encoder (i.e., BERT-base [23]) to obtain its embedding. Note that making retrieval prediction with only one sparsely sampled clip is not reliable and the model's performance varies significantly across different sampled clips per raw video. Therefore, in addition to the classic cross-modal contrastive objective, we devise a new clip-level alignment objective to obtain more consistent retrieval performance based on the two video clips sampled by SST (per raw video). Concretely, in each training step, the retrieval distributions of the two video clips are aligned by minimizing the Kullback-Leibler (KL) divergence between them. Since the retrieval distributions have actually been computed during obtaining the cross-modal contrastive loss, our alignment objective almost requires no extra computation cost. Overall, our clip-level alignment objective enables our SST-VLM to achieve more consistent performance in text-to-video retrieval. Importantly, we find that it is even effective without using more frames per raw video (see the ablation study in Table 1).

Our main contributions are three-fold: (1) We present a novel video-language contrastive learning framework termed SST-VLM for text-to-video retrieval, which is inspired by the 'Sparse Sampling-Twice (SST)' strategy. Different from ClipBERT [10], our SST sparsely and holistically samples two video clips from each raw video. (2) We propose a new Dual X-MoCo algorithm for training our SST-VLM. It is seamlessly integrated with the SST strategy so that our model can achieve more stable as well as better performance. (3) Extensive experiments show that our SST-VLM achieves new state-of-the-art in text-to-video retrieval.

2   Related Work

Text-to-Video Retrieval. Text-to-video retrieval has recently become a popular video-language modeling task. Classic approaches [24, 24–30] pre-extract the video features using expert models, including those trained on other tasks such as object recognition, and action classification. They also pre-extract the text features using pre-trained language models [23, 31]. The major drawback of this paradigm is the lack of cross-modal interaction during feature pre-extraction. To tackle this problem, a number of works [10–14, 16, 18, 32] have proposed to train video-language models without using pre-extracted features. Most of them [11–14, 16] embed each raw video with densely sampled video clips. Different from such costly dense sampling, ClipBERT [10] introduces a sparse sampling strategy, which samples a handful of video clips with short-time duration to learn the video representation. In this work, instead of sampling locally multiple times (like ClipBERT), our SST-VLM proposes a Sparse Sampling-Twice strategy which sparsely and holistically samples two video clips from each raw video. Importantly, we choose to align the clip-level prediction distributions in the retrieval task to obtain more reliable video embeddings.

Contrastive Learning. Contrastive learning has achieved great success in visual recognition [20, 33–39]. The infoNCE loss [33] has been widely used for contrastive learning, where a large number of negative samples are proven to be crucial for better performance. There are two ways of collecting negative samples: (1) SimCLR [37] utilizes the augmented view of each sample to be the positive sample and all other samples in the current batch to be negative ones. (2) MoCo [20] and its variant [40] introduce a momentum mechanism to maintain a large negative queue. Since MoCo can decouple the number of negative samples from the batch size, MoCo-based models are applicable to the setting with a small total batch size (less GPUs are needed for training). In this work, we thus choose to employ MoCo for video-language modeling. Interestingly, we have also explored BYOL [39] and SimSiam [41] in text-to-video retrieval, but found that they fail without using negative samples.

Note that contrastive learning has already been applied to text-to-video retrieval in the latest works [28, 30]. Concretely, TACo [28] adopts token-aware cascade contrastive learning (enhanced with hard negative mining), and HiT [30] introduces cross-modal MoCo based on both feature-level and semantic-level matching. They both utilize pre-extracted features as their inputs, leading to suboptimal results. Different from them, in this work, we focus on learning better video/text embeddings directly from *raw* videos/texts by proposing a new Dual X-MoCo algorithm, which includes two cross-modal MoCo modules with both cross-modal contrastive loss and clip-level alignment loss.

3 Methodology

3.1 Framework Overview

The flowchart of our SST-VLM model for text-to-video retrieval is illustrated in Fig. 2. Concretely, we are given a training set of N video-text pairs $\mathcal{D} = \{V_i, T_i\}_{i=1}^{N}$, where each video V_i has S_i frames of resolution $H \times W$ and each text T_i is represented by the natural language in English. Our model aims to learn a video encoder f_{θ_v} and a text encoder f_{θ_t} to project each video and its paired text into the joint embedding space so that they can be aligned with each other. The video and text encoders of our model are presented in Sect. 3.2, and our proposed Dual X-MoCo algorithm for model training is given in Sect. 3.3.

3.2 Video and Text Encoders

Video Encoder. Given a raw video V_i, we randomly and sparsely sample $N_c = 2$ video clips $\{c_{i,r}\}_{r=1}^{N_c}$, with s ($s < S_i$) frames per video clip (see Sect. 3.3 for more details). For each sampled video clip $c_{i,r}$, we first extract the visual embeddings $F_{i,r}^v \in \mathbb{R}^{s \times D_v}$ of all frames through an pre-trained image encoder f^v (e.g., the ViT-base model [21]) with an output dimension D_v:

$$F_{i,r}^v[k] = f^v(c_{i,r}[k]), k = 1, \cdots, s, \tag{1}$$

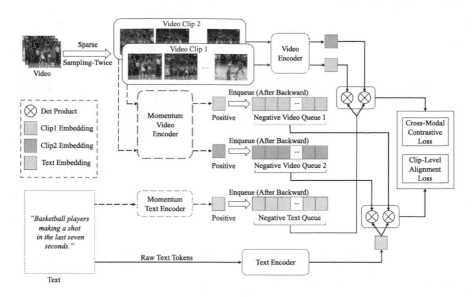

Fig. 2. Schematic illustration of our SST-VLM model. We sparsely and randomly sample two video clips from each raw video. Inspired by this 'Sparse Sampling-Twice' strategy, we devise a Dual X-MoCo algorithm to train our model. The video encoder consists of a ViT-base model followed by a Transformer module, and the text encoder is a BERT-base model. The momentum video/text encoders marked with dotted lines are initialized by the video/text encoders and updated by Eqs. (8) and (9).

where $c_{i,r}[k]$ denotes the k-th frame of the video clip $c_{i,r}$ and $F_{i,r}^v[k]$ denotes the k-th row of $F_{i,r}^v$. We then utilize a Transformer [22] module f_{att} to capture the temporal correlation across all frame embeddings:

$$\widehat{F}_{i,r}^v = f_{att}(F_{i,r}^v[1], F_{i,r}^v[2], \cdots, F_{i,r}^v[s]), \tag{2}$$

where $\widehat{F}_{i,r}^v \in \mathbb{R}^{s \times D_v}$ are the embeddings output by the Transformer. We finally obtain the embedding $F_{i,r}^c \in \mathbb{R}^D$ of video clip $c_{i,r}$ by averaging all frame embeddings, which is followed by a linear projection layer:

$$F_{i,r}^c = \text{Linear}(\text{Avg}(\widehat{F}_{i,r}^v[1], \widehat{F}_{i,r}^v[2], \cdots, \widehat{F}_{i,r}^v[s])), \tag{3}$$

where $\widehat{F}_{i,r}^v[k]$ is the k-th row of $\widehat{F}_{i,r}^v$ and D is the common dimension for video and text embeddings. Linear($\cdot$) denotes a linear projection layer and Avg($\cdot$) denotes the average pooling function that computes the mean of input embeddings. Overall, our video encoder f_{θ_v} (with parameters θ_v) encodes video clips by Eqs. (1)–(3).

Text Encoder. For each text T_i, we first tokenize it into a sequence of tokens $[t_1^i, t_2^i, \cdots, t_{l_i}^i]$, where l_i denotes the length of T_i. A pre-trained language model f^t (e.g., the BERT-base model [23]) is then used to map the sequence of tokens to a D_t-dimensional embedding $F_i^t \in \mathbb{R}^{D_t}$:

$$F_i^t = f^t(t_1^i, t_2^i, \cdots, t_{l_i}^i). \tag{4}$$

Similar to Eq. (3), we obtain the final text embedding by:

$$F_i^t = \text{Linear}(F_i^t), \tag{5}$$

where $\text{Linear}(\cdot)$ is a linear layer with the output dimension D. Overall, our text encoder f_{θ_t} (with parameters θ_t) encodes raw texts by Eqs. (4) and (5).

3.3 Dual X-MoCo

Sparse Sampling-Twice. In this work, to overcome the limitations of Clip-BERT [10] (see Fig. 1), we propose to sparsely and holistically sample two video clips from each raw video, noted as the 'Sparse Sampling-Twice' strategy. Specifically, given a raw video V_i with S_i frames from a mini-batch $\mathcal{B} = \{V_i, T_i\}_{i=1}^B$, we first subdivide it into s equal segments, where $s = 4$ in our implementation. We then randomly sample one frame from each segment to form the first video clip $c_{i,1}$. The second video clip $c_{i,2}$ is obtained by the same random sampling strategy. Finally, we feed the two sparsely sampled video clips $c_{i,1}, c_{i,2}$ and paired text T_i separately into the video encoder and text encoder to obtain video clip embeddings $F_{i,1}^c$, $F_{i,2}^c$ and text embedding F_i^t:

$$F_{i,1}^c = f_{\theta_v}(c_{i,1}), F_{i,2}^c = f_{\theta_v}(c_{i,2}), F_i^t = f_{\theta_t}(T_i). \tag{6}$$

Cross-Modal Contrastive Loss. As shown in Fig. 2, our Dual X-MoCo includes two cross-modal MoCo [20] modules to respectively model the sampled two clip-text pairs for each video-text input. Concretely, in a mini-batch $\mathcal{B} = \{V_i, T_i\}_{i=1}^B$, we first sample two video clips $c_{i,1}$ & $c_{i,2}$ from each video V_i by our SST strategy. Further, we encode the video clips $c_{i,1}$, $c_{i,2}$ and paired text T_i by Eq. (6). After that, we obtain the key embeddings $K_{i,1}^c$, $K_{i,2}^c$, K_i^t of $c_{i,1}$, $c_{i,2}$, T_i by the momentum encoders $f_{\theta_v^m}$ and $f_{\theta_t^m}$:

$$K_{i,1}^c = f_{\theta_v^m}(c_{i,1}), K_{i,2}^c = f_{\theta_v^m}(c_{i,2}), K_i^t = f_{\theta_t^m}(T_i), \tag{7}$$

where $f_{\theta_v^m}$ (with parameters θ_v^m) is initialized by the video encoder f_{θ_v} and $f_{\theta_t^m}$ (with parameters θ_t^m) is initialized by the text encoder f_{θ_t}. During training, the parameters of the momentum encoders $f_{\theta_v^m}$ and $f_{\theta_t^m}$ are updated by the momentum-update rule as follows:

$$\theta_v^m = m \cdot \theta_v^m + (1 - m) \cdot \theta_v, \tag{8}$$
$$\theta_t^m = m \cdot \theta_t^m + (1 - m) \cdot \theta_t, \tag{9}$$

where m is a momentum coefficient. After loss calculation, with the earliest B momentum embeddings popped out, we push $\{K_{i,1}^c\}_{i=1}^B$, $\{K_{i,2}^c\}_{i=1}^B$, and $\{K_i^t\}_{i=1}^B$ respectively into queues Q_1^c, Q_2^c, and Q^t, where $Q_1^c = \{n_{j,1}^c\}_{j=1}^{N_q}$, $Q_2^c = \{n_{j,2}^c\}_{j=1}^{N_q}$, and $Q^t = \{n_j^t\}_{j=1}^{N_q}$. Note that the queue size N_q is decoupled from B.

As a result, for each query text embedding F_i^t, we have positive video clip embeddings $K_{i,1}^c$, $K_{i,2}^c$ and negative video clip embeddings $n_{j,1}^c \in Q_1^c$, $n_{j,2}^c \in Q_2^c$.

We thus define the text-to-video contrastive loss L_{t2v} as an InfoNCE-based loss:

$$P_{i,1} = e^{F_i^t \cdot K_{i,1}^c / \tau}, \quad P_{i,2} = e^{F_i^t \cdot K_{i,2}^c / \tau}, \tag{10}$$

$$N_{i,1} = \sum_{j=1}^{N_q} e^{F_i^t \cdot n_{j,1}^c / \tau}, \quad N_{i,2} = \sum_{j=1}^{N_q} e^{F_i^t \cdot n_{j,2}^c / \tau}, \tag{11}$$

$$L_{t2v} = -\frac{1}{B} \sum_{i=1}^{B} (\log \frac{P_{i,1}}{P_{i,1} + N_{i,1}} + \log \frac{P_{i,2}}{P_{i,2} + N_{i,2}}), \tag{12}$$

where τ is the temperature, $P_{i,1}$ (or $P_{i,2}$) is the similarity score between positive clip embedding $K_{i,1}^c$ (or $K_{i,2}^c$) and query text embedding F_i^t. Additionally, $N_{i,1}$ (or $N_{i,2}$) is the summed similarity score between negative clip embeddings $n_{j,1}^c$ (or $n_{j,2}^c$) and query text embedding F_i^t. Similarly, for query video clips with embeddings $F_{i,1}^c$, $F_{i,2}^c$ computed by Eq. (6) and their positive/negative text embeddings K_i^t, n_j^t, we define the video-to-text contrastive loss L_{v2t} as follows:

$$\widehat{P}_{i,1} = e^{F_{i,1}^c \cdot K_i^t / \tau}, \quad \widehat{P}_{i,2} = e^{F_{i,2}^c \cdot K_i^t / \tau}, \tag{13}$$

$$\widehat{N}_{i,1} = \sum_{j=1}^{N_q} e^{F_{i,1}^c \cdot n_j^t / \tau}, \quad \widehat{N}_{i,2} = \sum_{j=1}^{N_q} e^{F_{i,2}^c \cdot n_j^t / \tau}, \tag{14}$$

$$L_{v2t} = -\frac{1}{B} \sum_{i=1}^{B} (\log \frac{\widehat{P}_{i,1}}{\widehat{P}_{i,1} + \widehat{N}_{i,1}} + \log \frac{\widehat{P}_{i,2}}{\widehat{P}_{i,2} + \widehat{N}_{i,2}}). \tag{15}$$

The total cross-modal contrastive loss L_{cl} is given by:

$$L_{cl} = L_{t2v} + L_{v2t}. \tag{16}$$

Clip-Level Alignment Loss. Note that different video clips sampled from the same video should correspond to the same ground-truth text. Therefore, a good model should have consistent retrieval performance over different video clips sampled from each video. To this end, in addition to the cross-modal contrastive loss, we propose to enhance the performance consistency of our SST-VLM model by minimizing a clip-level alignment loss defined with the Symmetrical Kullback-Leibler (Sym-KL) divergence.

Given a video V_i with two video clips $c_{i,r}$ ($r = 1, 2$) and its paired text T_i, the video-to-text retrieval probability distribution of the query video clip is denoted as $\widehat{U}_{i,r} = [\hat{u}_{i,r}^0, \cdots, \hat{u}_{i,r}^{N_q}] \in \mathbb{R}^{N_q+1}$. Similarly, the text-to-video retrieval probability distribution of the query text is $U_{i,r} = [u_{i,r}^0, \cdots, u_{i,r}^{N_q}] \in \mathbb{R}^{N_q+1}$. Concretely, the j-th element ($j = 0, \cdots, N_q$) of $\widehat{U}_{i,r}$ or $U_{i,r}$ is defined as:

$$\hat{u}_{i,r}^j = \frac{\widehat{P}_{i,r}}{\widehat{P}_{i,r} + \widehat{N}_{i,r}}, \quad u_{i,r}^j = \frac{P_{i,r}}{P_{i,r} + N_{i,r}}, (j = 0), \tag{17}$$

$$\hat{u}_{i,r}^j = \frac{e^{F_{i,r}^c \cdot n_j^t / \tau}}{\widehat{P}_{i,r} + \widehat{N}_{i,r}}, \quad u_{i,r}^j = \frac{e^{F_i^t \cdot n_{j,r}^c / \tau}}{P_{i,r} + N_{i,r}}, (j > 0), \tag{18}$$

where $\widehat{P}_{i,r}$, $\widehat{N}_{i,r}$, $P_{i,r}$, $N_{i,r}$ have been computed in Eqs. (13), (14), (10) and (11). We then define the clip-level alignment loss L_{al} with the Sym-KL divergence:

$$L_{\hat{u}} = \frac{1}{B} \sum_{i=1}^{B} \sum_{j=0}^{N_q} (\hat{u}_{i,1}^j \log \frac{\hat{u}_{i,1}^j}{\hat{u}_{i,2}^j} + \hat{u}_{i,2}^j \log \frac{\hat{u}_{i,2}^j}{\hat{u}_{i,1}^j}), \tag{19}$$

$$L_u = \frac{1}{B} \sum_{i=1}^{B} \sum_{j=0}^{N_q} (u_{i,1}^j \log \frac{u_{i,1}^j}{u_{i,2}^j} + u_{i,2}^j \log \frac{u_{i,2}^j}{u_{i,1}^j}), \tag{20}$$

$$L_{al} = L_{\hat{u}} + L_u. \tag{21}$$

Total Loss. Our SST-VLM model is trained by minimizing both the cross-modal contrastive loss and clip-level alignment loss. We thus have the total loss:

$$L_{total} = L_{cl} + \lambda * L_{al}, \tag{22}$$

where λ is the weight hyper-parameter.

3.4 Model Pre-training

Note that our model can be readily applied to the image-text retrieval task when the temporal Transformer module is removed. Therefore, similar to Clip-BERT [10], our model (excluding the temporal Transformer module) is pre-trained on a widely-used image-text dataset (with overall 5.3M image-text pairs), which consists of CC3M [42], VisGenome [43], SBU [44], COCO [45], and Flickr30k [46]. In this work, we do not pre-train our model on a large-scale external video-text dataset like HowTo100M [12] due to the limited computation resources. Although only pre-trained on an image-text dataset rather than a large-scale video-text dataset, our model still achieves new state-of-the-art on several benchmark datasets for text-to-video retrieval (see Table 3).

4 Experiments

4.1 Datasets and Settings

Datasets. We evaluate our SST-VLM on three benchmarks: (1) **MSR-VTT** [1] contains 10k videos with 200k descriptions. We first follow recent works [18,24, 47], using the 1k-A split of 9k training videos and 1k test videos. Further, we also adopt the split in [10,28] (called 7k-1k split in our work), having 7k training and 1k test videos. (2) **MSVD** [2] consists of 80k English descriptions for 1,970 videos from YouTube, and each video has around 40 captions. As in [18,29,47], we use the standard split: 1,200 videos for training, 100 videos for validation, and 670 ones for test. (3) **VATEX** [3] includes 25,991 videos for training, 3000 videos for validation, and 6000 ones for test. Since the original test set is private, we follow [29,48] to randomly split the original val set into two equal parts with 1500 videos for validation and the other 1500 videos for test.

Table 1. Ablation study for our SST-VLM model. Text-to-video retrieval results are reported on the MSR-VTT 1K-A test set.

L_{cl}	L_{al}	Frames	R@1 ↑	R@5 ↑	R@10 ↑
×	×	4	31.4	59.5	69.8
×	×	8	31.3	59.3	70.1
✓	×	4 + 4	32.1	61.5	71.8
✓	✓	4 + 4	**33.4**	**62.5**	**73.5**

Evaluation Metrics. We evaluate the text-to-video retrieval performance with the widely-used evaluation metrics in information retrieval, including Recall at K (shortened as R@K with K = 1, 5, 10) and Median Rank (shortened as MedR). R@K refers to the percentage of queries that are correctly retrieved in the top-K results. MedR measures the median rank of correct answers in the retrieved ranking list, where lower score indicates better performance.

Implementation Details. We adopt ViT-base [21] as the frame feature extractor of our video encoder and BERT-base [23] as the text encoder. For visual augmentation at the training stage, we apply random-crop, gray-scaling, horizontal-flip, and color-jitter to the input video frames that are resized to 384 × 384 (but only frame-resizing and center-crop are deployed at the evaluation stage). Due to the computation constraint, we empirically set the hyperparameters as: $\tau = 1$, $\lambda = 0.1$, and the initial learning rate is 5e−5. We only update the last 8 layers of the video and text encoders (but the other layers are frozen) during training. It takes about 2 h per epoch to train our model on MSR-VTT with 8 T V100 GPUs. In addition, different from the SST strategy (4 frames per clip) used for model training, we sample two video clips with 8 frames from each video V_i at the evaluation stage. With two clip embeddings $F_{i,1}^c$, $F_{i,2}^c$ obtained by Eq. (6), we have the final embedding of V_i for evaluation by averaging $F_{i,1}^c$ and $F_{i,2}^c$.

4.2 Ablation Study

Contributions of Contrastive and Alignment Losses. We analyze the contributions of cross-modal contrastive loss L_{cl} and clip-level alignment loss L_{al} used in our SST-VLM. The obtained ablative results are shown in Table 1. The baseline model (1st row) is formed with a single cross-modal MoCo framework where only one video clip (with *frames* = 4) is sampled for each video. Based on our SST strategy (with *frames* = 4 + 4), we obtain another baseline method by removing the clip-level alignment loss L_{al} from our Dual X-MoCo (3rd row). To further demonstrate the effectiveness of our full model, we train a baseline model (2nd row) based on single cross-modal MoCo with *frames* = 8. We can observe that: (1) Sampling one video clip with 4 or 8 frames leads to comparable performance (2nd row vs. 1st row). (2) With our SST strategy (4 + 4 frames per video), our model achieves significant improvements (3rd row vs. 1st/2nd row). This suggests that our SST-VLM model is even effective without using more frames per raw video. (3) The clip-level alignment loss can further improve

Table 2. Comparative results obtained by different alignment methods used in Eq. (21). Text-to-video retrieval results are reported on the MSR-VTT 1K-A test set.

Alignment method	R@1 ↑	R@5 ↑	R@10 ↑
NC	32.8	62.0	71.7
L2	32.4	62.2	71.5
Asym-KL	33.2	62.2	72.1
Sym-KL (ours)	**33.4**	**62.5**	**73.5**

the performance of our SST-VLM model, yielding around 1% improvement on all R@K (K = 1, 5, 10) results over our SST-VLM model with only contrastive loss (4th row vs. 3rd row).

Comparison to Alternative Alignment Methods. We further analyze the impact of alternative methods used for our clip-level alignment loss in Table 2. Note that the alignment loss L_{al} in Eq. (16) is defined with the Symmetric Kullback-Leiber (Sym-KL) divergence. This Sym-KL distance can be replaced by the negative cosine similarity (NC), L2 distance, or Asymmetric KL (Asym-KL) divergence. Concretely, the alternative alignment losses are defined by:

$$L_{al}^{NC} = -\frac{1}{B}\sum_{i=1}^{B}\left(\frac{\widehat{U}_{i,1}\cdot\widehat{U}_{i,2}}{||\widehat{U}_{i,1}||_2||\widehat{U}_{i,2}||_2} + \frac{U_{i,1}\cdot U_{i,2}}{||U_{i,1}||_2||U_{i,2}||_2}\right), \tag{23}$$

$$L_{al}^{L2} = \frac{1}{B}\sum_{i=1}^{B}(||\widehat{U}_{i,1} - \widehat{U}_{i,2}||_2 + ||U_{i,1} - U_{i,2}||_2), \tag{24}$$

$$L_{al}^{Asym} = \frac{1}{B}\sum_{i=1}^{B}\sum_{j=0}^{N_q}\left(\hat{u}_{i,1}^{j}\log\frac{\hat{u}_{i,1}^{j}}{\hat{u}_{i,2}^{j}} + u_{i,1}^{j}\log\frac{u_{i,1}^{j}}{u_{i,2}^{j}}\right), \tag{25}$$

where $\widehat{U}_{i,r}$, $U_{i,r}$, $\hat{u}_{i,r}$, $u_{i,r}$ are defined in Eqs. (17) and (18). We can find that SST-VLM with NC, L2 or Asym-KL leads to slightly lower performance on R@1 and R@5, and nearly 2% performance degradation on R@10, as compared with our SST-VLM using Sym-KL. We thus choose Sym-KL in this work.

4.3   Comparative Results

Table 3 shows the comparative results for text-to-video retrieval on MSR-VTT. We compare our SST-VLM with a wide range of representative/state-of-the-art methods including those [27–29] pre-trained on HowTo100M and those [10,18] pre-trained on image-text datasets. For extensive comparison, we also include methods [24,30,47] that utilize pre-extracted expert features. Although our SST-VLM is pre-trained on the smallest dataset with only 5.3M image-text pairs, it still achieves the best performance under both 7k-1k and 1k-A splits, demonstrating the effectiveness of our Dual X-MoCo for video-language modeling. Concretely, under the 7k-1k split, our SST-VLM outperforms the second best competitor by 5.0% on R@1, 5.9% on R@5, and 4.2% on R@10. It also leads to the

Table 3. Comparison to the state-of-the-art results for text-to-video retrieval on the MSR-VTT test set. **w/o PE**: methods trained without using multi-modal pre-extracted features. **VLM PT**: datasets for pre-training visual-language models. **VL Pairs**: the number of visual-language pairs in the pre-training datasets.

Method	w/o PE	VLM PT	VL Pairs	R@1 ↑	R@5 ↑	R@10 ↑	MedR ↓
7k-1k Split							
JSFusion [11]	✓	–	–	10.2	31.2	43.2	13.0
HT MIL-NCE [12]	✓	HowTo100M	>100M	14.9	40.2	52.8	9.0
ActBERT [13]	✓	HowTo100M	>100M	16.3	42.8	56.9	10.0
HERO [14]	✓	HowTo100M	>100M	16.8	43.4	57.7	–
VidTranslate [16]	✓	HowTo100M	>100M	14.7	–	52.8	–
NoiseEstimation [26]		HowTo100M	>100M	17.4	41.6	53.6	8.0
UniVL [27]		HowTo100M	>100M	21.2	49.6	63.1	6.0
ClipBERT [10]	✓	COCO, VisGenome	5.6M	22.0	46.8	59.9	6.0
TACo [28]		HowTo100M	>100M	24.8	52.1	64.5	5.0
SST-VLM (Ours)	✓	CC3M, Others	5.3M	**29.8**	**58.0**	**68.7**	**3.0**
1k-A Split							
CE [47]		–	–	20.9	48.8	62.4	6.0
AVLnet [25]		HowTo100M	>100M	27.1	55.6	66.6	4.0
MMT [24]		HowTo100M	>100M	26.6	57.1	69.6	4.0
Support Set [29]		HowTo100M	>100M	30.1	58.5	69.3	3.0
HiT [30]		HowTo100M	>100M	30.7	60.9	73.2	**2.6**
TACo [28]		HowTo100M	>100M	28.4	57.8	71.2	4.0
Frozen in Time [18]	✓	CC3M, WebVid-2M	5.5M	31.0	59.5	70.5	3.0
SST-VLM (ours)	✓	CC3M, Others	5.3M	**33.4**	**62.5**	**73.5**	3.0

Table 4. Comparison to the state-of-the-arts on MSVD for text-to-video retrieval.

Method	R@1 ↑	R@5 ↑	R@10 ↑	MedR ↓
VSE [49]	12.3	30.1	42.3	14.0
VSE++ [50]	15.4	39.6	53.0	9.0
Multi. Cues [51]	20.3	47.8	61.1	6.0
CE [47]	19.8	49.0	63.8	6.0
Support Set [29]	28.4	60.0	72.9	4.0
Frozen in Time [18]	33.7	64.7	76.3	3.0
SST-VLM (ours)	**36.2**	**66.4**	**76.9**	**2.0**

best MedR = 3.0. Moreover, under the 1k-A split (with more training data than the 7k-1k split), our SST-VLM outperforms the latest state-of-the-arts [18,30] by 2.4% on R@1 and 1.6% on R@5. In particular, as compared with HiT [30], our SST-VLM achieves better results on R@1 and R@5, and obtains competitive results on R@10 and MedR. This is still impressive and remarkable, given that HiT not only is pre-trained on the much larger dataset HowTo100M but also utilizes numerous pre-extracted expert features.

Table 4 shows the comparative results on MSVD. Our SST-VLM outperforms all competitors, especially yielding 2.5% margin on R@1 against the latest stat-of-the-art [18]. The results on VATEX in Table 5 further demonstrate that our

Table 5. Comparison to the state-of-the-arts on VATEX for text-to-video retrieval.

Method	R@1 ↑	R@5 ↑	R@10 ↑	MedR ↓
VSE [49]	28.0	64.3	76.9	3.0
VSE++ [50]	33.7	70.1	81.0	2.0
Dual [51]	31.1	67.4	78.9	3.0
HGR [47]	35.1	73.5	83.5	2.0
HANet [19]	36.4	74.1	84.1	2.0
Support Set [29]	45.9	82.4	90.4	**1.0**
SST-VLM (ours)	**53.4**	**85.3**	**92.0**	**1.0**

Table 6. Comparison to the state-of-the-arts on MSR-VTT (1k-A split) for video-to-text retrieval.

Method	R@1 ↑	R@5 ↑	R@10 ↑	MedR ↓
CE [47]	20.9	48.8	62.4	6.0
AVLnet [25]	28.5	54.6	65.2	4.0
MMT [51]	28.0	57.5	69.7	3.7
Support Set [29]	28.5	58.6	71.6	**3.0**
SST-VLM (ours)	**33.2**	**61.2**	**72.0**	**3.0**

Table 7. Comparison to the state-of-the-arts on MSVD for video-to-text retrieval.

Method	R@1 ↑	R@5 ↑	R@10 ↑	MedR ↓
VSE++ [50]	21.2	43.4	52.2	9.0
Multi. Cues [51]	31.5	51.0	61.5	5.0
Support Set [29]	34.7	59.9	70.0	3.0
SST-VLM (ours)	**47.3**	**72.0**	**78.0**	**2.0**

SST-VLM achieves 7.5% improvement on R@1, 2.9% improvement on R@5, and 1.6% improvement on R@10 over the second best competitor [29].

For comprehensive comparison, we also provide video-to-text retrieval results on the MSR-VTT 1k-A split in Table 6, in addition to the text-to-video retrieval results. Our SST-VLM outperforms the latest state-of-the-art (i.e., Support Set [29] pre-trained on Howto100M [12]) by 4.7% on R@1, 2.6% on R@5, and 0.4% on R@10. It also achieves the best MedR = 3.0. Moreover, we present the results for video-to-text retrieval on the MSVD [2] test set in Table 7. Our SST-VLM outperforms the second best competitor [29] by 12.6% on R@1, 12.1% on R@5, and 8.0% on R@10. It also leads to the best MedR = 2.0. These results indicate that our SST-VLM is effective for video-language modeling on both video-to-text and text-to-video retrieval tasks.

4.4 Visualization Results

Retrieval Rank Distribution. To show the stability of our SST-VLM, we visualize the text-to-video retrieval rank results on MSR-VTT 1k-A test set in

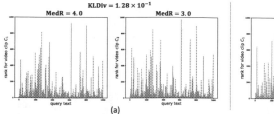
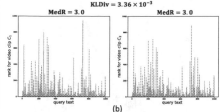

Fig. 3. Visualization results of text-to-video retrieval on the MSR-VTT 1k-A test set. (a) The results by our SST-VLM without using the clip-level alignment loss (but with cross-modal contrastive loss); (b) The results by our full SST-VLM.

Fig. 4. Attention visualization for our SST-VLM. The (attention) heatmaps are shown for two video-text pairs sampled from the MSR-VTT test set. Texts in *red* denote key objects for each video caption. (Color figure online)

Fig. 3. Note that we still sample two video clips (as in the training stage) from each raw video in the test set, resulting in two video clip sets $C_1 = \{c_{i,1}\}_{i=1}^{1,000}$ and $C_2 = \{c_{i,2}\}_{i=1}^{1,000}$. To show the effectiveness of our clip-level alignment loss L_{al} in enhancing stability, we evaluate two related models: model in Fig. 3(a) is trained without L_{al} (but with cross-modal contrastive loss L_{cl}), while model in Fig. 3(b) is exactly our full SST-VLM model. For each model, we visualize the retrieval rank (range from 1 to 1,000) distribution between video clip set C_1 (or C_2) and the same set of text queries. In addition, we report the MedR results for each distribution and KL divergence (KLDiv) between distributions of differently sampled video clips for each model. We find that: (1) The MedR results in Fig. 3(b) are equal to 3.0 for both C_1 and C_2, while those in Fig. 3(a) are different (4.0 for C_1 and 3.0 for C_2). (2) The KL divergence in Fig. 3(b) is two orders of magnitude smaller than that in Fig. 3(a). Therefore, the clip-level alignment loss L_{al} indeed leads to more stable results.

Attention Visualization. To further show that our SST-VLM has learned to understand the semantic content in videos, we adopt a recent Transformer visualization method [52] to highlight the relevant regions of the input frames according to the input texts. In this work, different from the original visualization method that computes the gradients directly from the total loss backward, we compute the separate gradients of each input frame and visualize the attention

Fig. 5. Text-to-video retrieval examples obtained by our SST-VLM on the MSR-VTT 1k-A test set. For each query text, we visualize the top-2 retrieved videos (with 3 frames shown per video). We also present the *original paired text* under each video.

maps of all frames. Concretely, as shown in Fig. 4, we present two video-text pairs (and their visualization results) sampled from the MSR-VTT test set. The left part presents a 4-frame video clip with text 'Penguins wander around'. The attention visualization shows that penguins in all frames have actually been noticed by our model. Moreover, the right part presents a 4-frame video clip with a longer text 'A rocket is launching. Smoke is emerging from the base of the rocket'. The attention visualization is rather interesting: with the rocket launching, our model pays more attention to the rocket and its smoke. Overall, these visualization results indicate that our SST-VLM has actually learned to understand the semantic content in videos.

Text-to-Video Retrieval Examples. Figure 5 shows the text-to-video retrieval qualitative results obtained by our SST-VLM on the MSR-VTT 1k-A test set. We visualize the top-2 videos (with 3 frames show per video) for each query text. Concretely, the left part of Fig. 5 consists of a query text 'a little girl does gymnastics' and the retrieved top-2 videos (with their original paired texts) shown below, while the right part of Fig. 5 is organized similarly. For each query text, we have the following observations: (1) The ground-truth video is correctly retrieved at the first place. (2) The texts of the second retrieved videos are also similar to the query text, which means that the semantic content of these videos is still consistent with the query text. Overall, these qualitative results indicate that our SSL-VLM has indeed aligned the video and text embeddings well in the learned joint space (which is crucial for video-text retrieval).

5   Conclusion

In this paper, we propose a novel video-language model called SST-VLM inspired by the Sparse Sampling-Twice (SST) strategy that sparsely and holistically samples two video clips from each raw video. For training our SST-VLM, we devise a new Dual X-MoCo algorithm, which includes both cross-modal contrastive and clip-level alignment losses to enhance the performance stability of our model.

Extensive results on several benchmarks show that our SST-VLM achieves new state-of-the-art in text-to-video retrieval. The ablation study and attention visualization further demonstrate the effectiveness of our SST-VLM. In our ongoing research, we will apply our SST-VLM to other video-language understanding tasks such as video captioning and video question answering.

References

1. Xu, J., Mei, T., Yao, T., Rui, Y.: MSR-VTT: a large video description dataset for bridging video and language. In: CVPR, pp. 5288–5296 (2016)
2. Chen, D., Dolan, W.B.: Collecting highly parallel data for paraphrase evaluation. In: ACL, pp. 190–200 (2011)
3. Wang, X., Wu, J., Chen, J., Li, L., Wang, Y., Wang, W.Y.: VaTeX: a large-scale, high-quality multilingual dataset for video-and-language research. In: ICCV, pp. 4580–4590 (2019)
4. Hendricks, A.L., Wang, O., Shechtman, E., Sivic, J., Darrell, T., Russell, B.: Localizing moments in video with natural language. In: ICCV, pp. 5804–5813 (2017)
5. Rohrbach, A., Rohrbach, M., Tandon, N., Schiele, B.: A dataset for movie description. In: CVPR, pp. 3202–3212 (2015)
6. Zhou, L., Xu, C., Corso, J.: Towards automatic learning of procedures from web instructional videos. AAA I, 7590–7598 (2018)
7. Antol, S., et al.: VQA: visual question answering. In: ICCV, pp. 2425–2433 (2015)
8. Lei, J., Yu, L., Bansal, M., Berg, T.L.: TVQA: Localized, compositional video question answering. In: EMNLP, pp. 1369–1379 (2018)
9. Jang, Y., Song, Y., Yu, Y., Kim, Y., Kim, G.: TGIF-QA: toward spatio-temporal reasoning in visual question answering. In: CVPR, 1359–1367 (2017)
10. Lei, J., Li, L., Zhou, L., Gan, Z., Berg, T.L., Bansal, M., Liu, J.: Less is more: ClipBERT for video-and-language learning via sparse sampling. CVPR, pp. 7331–7341 (2021)
11. Yu, Y., Kim, J., Kim, G.: A joint sequence fusion model for video question answering and retrieval. In: ECCV, pp. 487–503 (2018)
12. Miech, A., Zhukov, D., Alayrac, J.B., Tapaswi, M., Laptev, I., Sivic, J.: HowTo100M: learning a text-video embedding by watching hundred million narrated video clips. In: ICCV, pp. 2630–2640 (2019)
13. Zhu, L., Yang, Y.: ActBERT: learning global-local video-text representations. In: CVPR, pp. 8743–8752 (2020)
14. Li, L., Chen, Y.C., Cheng, Y., Gan, Z., Yu, L., Liu, J.: HERO: hierarchical encoder for video+ language omni-representation pre-training. In: EMNLP, pp. 2046–2065 (2020)
15. Feng, Z., Zeng, Z., Guo, C., Li, Z.: Exploiting visual semantic reasoning for video-text retrieval. IJCA I, 1005–1011 (2020)
16. Korbar, B., Petroni, F., Girdhar, R., Torresani, L.: Video understanding as machine translation. arXiv preprint arXiv:2006.07203 (2020)
17. Li, Z., Guo, C., Yang, B., Feng, Z., Zhang, H.: A novel convolutional architecture for video-text retrieval. In: ICME, pp. 1–6 (2020)
18. Bain, M., Nagrani, A., Varol, G., Zisserman, A.: Frozen in time: a joint video and image encoder for end-to-end retrieval. arXiv preprint arXiv:2104.00650 (2021)
19. Wu, P., He, X., Tang, M., Lv, Y., Liu, J.: HANet: hierarchical alignment networks for video-text retrieval. In: ACM-MM, pp. 3518–3527 (2021)

20. He, K., Fan, H., Wu, Y., Xie, S., Girshick, R.B.: Momentum contrast for unsupervised visual representation learning. In: CVPR, pp. 9726–9735 (2020)
21. Dosovitskiy, A., et al.: An image is worth 16×16 words: transformers for image recognition at scale. In: ICLR (2021)
22. Vaswani, A., et al.: Attention is all you need. In: NeurIPS, pp. 5998–6008 (2017)
23. Devlin, J., Chang, M., Lee, K., Toutanova, K.: BERT: pre-training of deep bidirectional transformers for language understanding. In: NAACL-HLT, pp. 4171–4186 (2019)
24. Gabeur, V., Sun, C., Alahari, K., Schmid, C.: Multi-modal transformer for video retrieval. In: Vedaldi, A., Bischof, H., Brox, T., Frahm, J.-M. (eds.) ECCV 2020. LNCS, vol. 12349, pp. 214–229. Springer, Cham (2020). https://doi.org/10.1007/978-3-030-58548-8_13
25. Rouditchenko, A., et al.: AVLnet: learning audio-visual language representations from instructional videos. arXiv preprint arXiv:2006.09199 (2020)
26. Amrani, E., Ben-Ari, R., Rotman, D., Bronstein, A.: Noise estimation using density estimation for self-supervised multimodal learning. AAA I, 6644–6652 (2021)
27. Luo, H., et al.: UniVL: a unified video and language pre-training model for multimodal understanding and generation. arXiv preprint arXiv:2002.06353 (2020)
28. Yang, J., Bisk, Y., Gao, J.: TACo: token-aware cascade contrastive learning for video-text alignment. arXiv preprint arXiv:2108.09980 (2021)
29. Patrick, M., et al.: Support-set bottlenecks for video-text representation learning. In: ICLR (2021)
30. Liu, S., Fan, H., Qian, S., Chen, Y., Ding, W., Wang, Z.: HiT: hierarchical transformer with momentum contrast for video-text retrieval. arXiv preprint arXiv:2103.15049 (2021)
31. Sanh, V., Debut, L., Chaumond, J., Wolf, T.: DistilBERT, a distilled version of BERT: smaller, faster, cheaper and lighter. arXiv preprint arXiv:1910.01108 (2019)
32. Wang, X., Zhu, L., Yang, Y.: T2VLAD: global-local sequence alignment for text-video retrieval. In: CVPR, pp. 5079–5088 (2021)
33. van den Oord, A., Li, Y., Vinyals, O.: Representation learning with contrastive predictive coding. arXiv preprint arXiv:1807.03748 (2018)
34. Wu, Z., Xiong, Y., Yu, S.X., Lin, D.: Unsupervised feature learning via non-parametric instance discrimination. In: CVPR, pp. 3733–3742 (2018)
35. Tian, Y., Krishnan, D., Isola, P.: Contrastive multiview coding. In: Vedaldi, A., Bischof, H., Brox, T., Frahm, J.-M. (eds.) ECCV 2020. LNCS, vol. 12356, pp. 776–794. Springer, Cham (2020). https://doi.org/10.1007/978-3-030-58621-8_45
36. Khosla, P., et al.: Supervised contrastive learning. In: NeurIPS, pp. 18661–18673 (2020)
37. Chen, T., Kornblith, S., Norouzi, M., Hinton, G.E.: A simple framework for contrastive learning of visual representations. In: ICML, pp. 1597–1607 (2020)
38. Chen, T., Kornblith, S., Swersky, K., Norouzi, M., Hinton, G.E.: Big self-supervised models are strong semi-supervised learners. In: NeurIPS, pp. 22243–22255 (2020)
39. Grill, J., et al.: Bootstrap your own latent - a new approach to self-supervised learning. In: NeurIPS, pp. 21271–21284 (2020)
40. Chen, X., Xie, S., He, K.: An empirical study of training self-supervised vision transformers. arXiv preprint arXiv:2104.02057 (2021)
41. Chen, X., He, K.: Exploring simple siamese representation learning. In: CVPR, pp. 15750–15758 (2021)
42. Sharma, P., Ding, N., Goodman, S., Soricut, R.: Conceptual captions: a cleaned, hypernymed, image alt-text dataset for automatic image captioning. In: ACL, pp. 2556–2565 (2018)

43. Krishna, R., et al.: Visual genome: Connecting language and vision using crowd-sourced dense image annotations. IJCV, pp. 32–73 (2017)
44. Ordonez, V., Kulkarni, G., Berg, T.: Im2Text: describing images using 1 million captioned photographs. In: NeurIPS, pp. 1143–1151 (2011)
45. Chen, X., et al.: Microsoft COCO captions: data collection and evaluation server. arXiv preprint arXiv:1504.00325 (2015)
46. Plummer, B.A., Wang, L., Cervantes, C.M., Caicedo, J.C., Hockenmaier, J., Lazebnik, S.: Flickr30k entities: collecting region-to-phrase correspondences for richer image-to-sentence models. In: ICCV, pp. 2641–2649 (2015)
47. Liu, Y., Albanie, S., Nagrani, A., Zisserman, A.: Use what you have: video retrieval using representations from collaborative experts. In: BMVC, p. 279 (2019)
48. Chen, S., Zhao, Y., Jin, Q., Wu, Q.: Fine-grained video-text retrieval with hierarchical graph reasoning. In: CVPR, pp. 10635–10644 (2020)
49. Kiros, R., Salakhutdinov, R., Zemel, R.S.: Unifying visual-semantic embeddings with multimodal neural language models. arXiv preprint arXiv:1411.2539 (2014)
50. Faghri, F., Fleet, D.J., Kiros, J.R., Fidler, S.: VSE++: improving visual-semantic embeddings with hard negatives. In: BMVC, p. 12 (2018)
51. Mithun, N.C., Li, J., Metze, F., Roy-Chowdhury, A.K.: Learning joint embedding with multimodal cues for cross-modal video-text retrieval. In: ICMR, pp. 19–27 (2018)
52. Chefer, H., Gur, S., Wolf, L.: Transformer interpretability beyond attention visualization. In: CVPR, pp. 782–791 (2021)

PromptLearner-CLIP: Contrastive Multi-Modal Action Representation Learning with Context Optimization

Zhenxing Zheng[1,2,3], Gaoyun An[2,3(✉)], Shan Cao[2,3], Zhaoqilin Yang[2,3], and Qiuqi Ruan[2,3]

[1] School of Communications and Information Engineering, Xi'an University of Posts and Telecommunications, Xi'an 710121, China
zhxzheng@xupt.edu.cn
[2] Institute of Information Science, Beijing Jiaotong University, Beijing 100044, China
[3] Beijing Key Laboratory of Advanced Information Science and Network Technology, Beijing 100044, China
{gyan,18112001,19112010,qqruan}@bjtu.edu.cn

Abstract. An action contains rich multi-modal information, and current methods generally map the action class to a digital number as supervised information to train models. However, numerical labels cannot describe the semantic content contained in the action. This paper proposes PromptLearner-CLIP for action recognition, where the text pathway uses PromptLearner to automatically learn the text content of prompt as the input and calculates the semantic features of actions, and the vision pathway takes video data as the input to learn the visual features of actions. To strengthen the interaction between features of different modalities, this paper proposes a multi-modal information interaction module that utilizes Graph Neural Network(GNN) to process both the semantic features of text content and the visual features of a video. In addition, the single-modal video classification problem is transformed into a multi-modal video-text matching problem. Multi-modal contrastive learning is used to disclose the feature distance of the same but different modalities samples. The experimental results showed that PromptLearner-CLIP could utilize the textual semantic information to significantly improve the performance of various single-modal backbone networks on action recognition and achieved top-tier results on Kinetics400, UCF101, and HMDB51 datasets. Code is available at https://github.com/ZhenxingZheng/PromptLearner.

1 Introduction

With the development of mobile devices and communication networks, video has become the main carrier of information. It is of great practical significance to understand and analyze human actions in the video. As an essential branch of video understanding, action recognition aims to analyze and recognize human actions in videos by analyzing video data and using specific algorithms.

L. Wang et al. (Eds.): ACCV 2022, LNCS 13844, pp. 554–570, 2023.
https://doi.org/10.1007/978-3-031-26316-3_33

Different from image processing tasks, action recognition needs to analyze not only the appearance information but also the semantics of an action. How to effectively encode the feature of an action remains a fundamental problem to be solved. An action contains rich multi-modal information, and current methods generally map the action class to a digital number as supervised information to train models. However, numerical labels cannot describe the semantic content contained in the action. The sample **playing tennis** on Youtube provides the corresponding text description of **A 12-year-old boy playing tennis**, which not only describes the class of the action but also includes the action subject. Therefore, the text provides rich semantic information and the visual feature can be enhanced by relevant text content.

Although some samples on Youtube are accompanied by detailed text descriptions, most samples contain a lot of information unrelated to the video content. Recently, in the field of natural language processing, researchers proposed a new paradigm: "pretrain, prompt, predict", where according to the downstream task, a template is designed such that the model can fit the task of pre-training when predicting. Based on this, CLIP [29] constructed the text input by designing a variety of natural language description templates and filled the image label text into the blank positions of templates. The experiments showed that different prompts have an important impact on the model, and subtle differences may lead to changes in performance. CoOp [52] used continuous representations to represent prompt whose parameters are optimized in an end-to-end fashion.

Based on the above analyses, this paper proposes a multi-modal semantic-guided network PromptLearner-CLIP for action recognition. In the training phase, the text labels of the samples in a batch are filled in the prompt template and then processed by the text encoder to extract text features. At the same time, the visual features of the videos in the batch are extracted by the visual encoder. Finally, the similarities of visual features and text features are computed, resulting in the similarity matrix that is used for optimization. In the inference phase, all labels are filled in the prompt template and the feature similarity scores between each video in the test dataset and all prompts are computed. The label with the highest similarity score is assigned to the video. Our contributions are summarized as follows: (1) PromptLearner is used to learn the text content of prompt as the input to the text pathway and its parameters are optimized together with the backbone network; (2) GNN is used to process both the semantic features of text content and the visual features of the video and strengthen the interaction between semantic features and visual features; (3) Finally, the Kullback-Leibler(KL) loss and supervised contrastive loss are used to disclose the feature distance of the same but different modalities samples.

2   Related Work

Single-Modal Action Recognition. C3D [33] stacked 3D convolutional layers to learn spatial-temporal features. ARTNet [37] designed appearance and relation branches to perform spatial modeling and relation modeling in a parallel way. R(2+1)D [34] decomposed the 3D convolution kernel into a 2D spatial convolution kernel and a 1D temporal convolution kernel. Because action recognition needs to process multiple frames of a video, it has large computational complexity. Based on the fact that adjacent frames of a video have redundant information, TSN [39] proposed a sparse sampling strategy and a feature aggregation module to model long-term temporal relationships of an action. AdaScan [15] pooled the video frames containing important information and discarded the video frames with less information. To complete effective temporal modeling for actions, TSM [24] shifted the feature map by a position along the temporal dimension, so that the convolutional feature map of the current frame obtains the information of adjacent frames.

Cross-Modal Action Recognition. PoTion [5] encoded the displacement of key points of the human body on a color image that was then processed by CNN to obtain action features containing pose motion information. Multistream network [43] used three streams to process an RGB image, multiple optical flow images, and spectrograms to model appearance features, short-term motion features, and sound features of actions respectively. In addition to sound information, the text as a rich expression can describe the semantic content of a video. CLIPBERT [19] fused the feature of each video clip and the feature of text to model the multi-modal feature by Transformer [36]. ActBERT [53] learned joint video-text feature representations to capture global and local visual cues from each pair of the video clip and text description.

Contrastive Learning. Contrastive learning as an unsupervised representation learning method has been successfully applied in the field of computer vision. MoCo [12] built a dynamic dictionary with a queue composed of previous sample features and set an instance discrimination task for contrastive unsupervised learning. VideoMoCo [28] built a generator to mask partial video frames and used a discriminator to distinguish the full video sequence features from the masked video sequence features. Inspired by the mask prediction task, MaskCo [50] masked a specific region of an enhanced image while keeping the other enhanced image unchanged, and then calculated the region-level feature contrastive loss of two images to implement the contrastive mask prediction task for visual representation learning. In the training of a network, a batch of samples may contain multiple samples belonging to the same class. SupCon [17] incorporated the label information into contrastive loss, which considers multiple positive samples of the same class for each anchor point so that the features from the same class are closer than the features of different classes.

3   Method

This paper proposes a multi-modal semantic-guided action recognition network PromptLearner-CLIP for action recognition, as shown in Fig. 1. The vision pathway uses ViT [7] as the backbone network to process video frames and obtains frame features and video features. The text pathway uses Transformer to process text content and obtains word features and sentence features. To construct valid text input, PromptLearner is used to automatically learn text content as the input to the text pathway. After extracting text features and visual features, a multi-modal information interaction module is used to interact with text features and visual features. Finally, multi-modal contrastive learning is used to disclose the feature distance of the videos belonging to the same class.

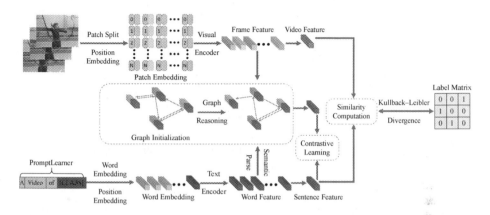

Fig. 1. Illustration of PromptLearner-CLIP consisting of vision feature extraction, text feature extraction, multi-modal information interaction, and contrastive learning.

3.1 Text Pathway

Multi-modal learning aims to use specific algorithms to learn the complementarity and eliminate the redundancy between different modalities. However, most of the datasets for action recognition only provide action classes without corresponding text descriptions. Recently, a new paradigm of prompt has been proposed in the field of natural language processing. By setting different fill-in-the-blank templates, the downstream task is adjusted to the form similar to the pre-training task, which can effectively solve the downstream task. PromptLearner proposed in this paper uses a text template to process action label text by expanding the label text into the sentence with certain semantic content as text descriptions of a video, which is used as the input to the text pathway to extract semantic information. PromptLearner uses a continuous vector to represent the content of text, and its parameters are updated together with the backbone network in an end-to-end fashion, represented as follows:

$$t^i = [V_1^i][V_2^i][V_3^i]...[V_M^i][CLASS^i], \qquad (1)$$

where $[V_m](m \in [1, M])$ represents the context token, $[CLASS^i]$ is the i-th label text of the action, t^i is the learnable text content, $i \in [1, I]$, and M and I represent the number of context tokens and action classes respectively. Class-specific context token is used in this paper.

After the learnable prompt is constructed, Transformer is used to process the text content. Transformer adopts an encoder-decoder structure and the text pathway only uses the Transformer encoder to extract the features of text content. The encoder consists of multiple encoding layers consisting of the self-attention layer, LayerNorm layer, and feed-forward layer. The structure of an encoder is shown in Fig. 2.

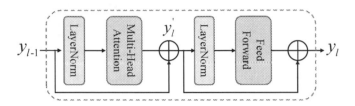

Fig. 2. Structure of Transformer encoder consisting of the self-attention layer, Layer-Norm layer, and feed-forward layer.

The overall process of the Transformer encoder extracting text features is represented as follows:

$$y_0 = [V_1 + PE_1, \ldots, V_M + PE_M, V_{class} + PE_{class}], \tag{2}$$

$$q_l = k_l = v_l = \text{LayerNorm}(y_{l-1}), \tag{3}$$

$$y_l' = \text{MSA}(q_l, k_l, v_l) + y_{l-1}, \tag{4}$$

$$y_l = \text{FFN}(\text{LayerNorm}(y_l')) + y_l', \quad l = 1, \ldots, L \tag{5}$$

$$[s_1, s_2, \ldots, s_M, s_{class}] = y_L, \tag{6}$$

where l denotes the l-th encode layer, V_m and PE_m denote the m-th context token and positional embedding, MSA denotes the multi-head self-attention layer, FFN denotes the feed-forward layer, and L is the number of Transformer encoder layers. Transformer uses the self-attention layer to effectively capture dependencies between features at any location and learn text features. After the text content is processed by the Transformer encoder, the word features $S = \{s_m\}_{m=1}^{M+1}$ and sentence features s_0 are obtained.

3.2   Vision Pathway

The original input to Transformer is a sequence of tokens. To satisfy the requirements of Transformer, ViT first pre-processes the frame to obtain the input token sequence. Given a frame $x \in R^{H \times W \times C}$, ViT first splits the frame into many non-overlapped patches of the same size, and then these frame patches are flattened into the 1D vectors composed of pixel values, denoted as $x_p \in R^{N \times (P^2 \cdot C)}$:

$$x_p = [x_p^1, x_p^2, \ldots, x_p^N], \tag{7}$$

where H and W denote the height and width of the frame respectively, C is the number of channels, (P, P) is the resolution of a patch, x_p^n represents the flatten vector from the n-th patch, and $N = H \cdot W/P^2$ is the number of split patches. Through linear projection $\mathbf{E}$ consisting of fully-connected layers, the vector consisting of pixels is transformed to the patch embedding feature with the dimension of d_{model}. At the start of the patch sequence, we prepend a learnable embedding z_{cls} and its state at the output layer denotes the frame feature. A learnable 1D position embedding is used to retrain the position information of each patch and is added to the patch embedding feature, as shown as follows:

$$z_0 = [z_{cls}, \mathbf{E}x_p^1, \mathbf{E}x_p^2, \ldots, \mathbf{E}x_p^N] + \mathbf{p}, \tag{8}$$

The overall process of ViT extracting the frame feature is summarized as follows:

$$q_l = k_l = v_l = \text{LayerNorm}(z_{l-1}), \tag{9}$$

$$z_l' = \text{MSA}(q_l, k_l, v_l) + z_{l-1}, \tag{10}$$

$$z_l = \text{FFN}(\text{LayerNorm}(z_l')) + z_l'. \quad l = 1, \ldots, L \tag{11}$$

Finally, Transformer is used to process the video frame by frame and we obtained the sequence of frame features $V = \{v_k\}_{k=1}^K$. These features are averaged as the video-level feature v_0.

3.3   Multi-Modal Information Interaction Module

After extracting vision features and text features, the model obtains frame-level features $V = \{v_k\}_{k=1}^K$, the video-level feature v_0, word-level features $S = \{s_m\}_{m=1}^{M+1}$, and the sentence-level feature s_0. Inspired by Dynamic Graph Attention Network [47], the multi-modal information interaction module represents the sentence as multiple soft distributions over words and parses the language structure of the sentence gradually. Firstly, the sentence-level feature s_0 is linearly projected to the question feature q^t at the t-th step and is concatenated with the results of the previous step, resulting in u^t:

$$q^t = W^t \times s_0 + b^t, \tag{12}$$

$$u^t = [q^t, y^{t-1}], \tag{13}$$

where W^t and b^t denote the learnable parameters, y^{t-1} denotes the results at the $(t-1)$-th step. Then, the semantic parsing module computes the similarity between u^t and each word-level feature to predict the visual reasoning processing, obtaining the soft distribution over all words $R^t = \{r_m^t\}_{m=1}^{M+1}$:

$$s^t = \delta(W_u \times u^t + b_u), \tag{14}$$

$$a_m^t = W_{s2} \times [\tanh(W_{s0} \times s^t + W_{s1} \times s_m)], \tag{15}$$

$$r_m^t = \frac{\exp(a_m^t)}{\sum\limits_{m=1}^{M+1} \exp(a_m^t)}, \tag{16}$$

where W_u, b_u, W_{s0}, W_{s1}, and W_{s2} are parameter matrices and are shared at different visual reasoning steps, and δ denotes ReLU. Finally, the output at the t-step is calculated as follows:

$$y^t = \sum_{m=1}^{M+1} r_m^t \cdot s_m. \tag{17}$$

Then, the semantic parsing feature y^t and the frame-level features $V = \{v_k\}_{k=1}^K$ are jointly fed into GNN for multi-modal feature interaction. To effectively embed textual semantic information into visual features, a question-guided graph attention mechanism is used to dynamically assign higher weights to frame features related to textual content. In this paper, and all frame features are concatenated with the output of the semantic parsing module as the vertices of the graph. The edges connecting the vertices represent the relationship between frames, and the vertex features are represented as follows:

$$v_k' = [v_k, y^t], \quad \text{for} \quad k = 1, ..., K. \tag{18}$$

Next, the self-attention layer is used to calculate the correlation between the feature of any vertex in the graph and the features of all neighboring vertices, the neighboring vertex information is aggregated to update the vertex feature. Feature correlation is calculated as follows:

$$\alpha_{ij}^v = (W_q \times v_i') \times (W_k \times v_j')^T, \tag{19}$$

where W_q and W_k are parameter matrices used for projecting the vertex feature into the feature subspace in which the correlations between all other vertex features and the i-th vertex feature are computed. The correlation α^v is normalized by the softmax function as a weight to aggregate the information of other vertices:

$$\alpha_{ij} = \frac{\exp(\alpha_{ij}^v)}{\sum\limits_{j=1,j\neq i}^{K} \exp(\alpha_{ij}^v)}, \tag{20}$$

$$v_i^* = \delta(v_i' + \sum_{j,j\neq i} \alpha_{ij} \cdot v_j'), \tag{21}$$

where α_{ij} denotes the weight between the i-th and j-th vertex features.

Finally, the multi-modal fusion method BUTD [1] is used to obtain the multi-modal representation:

$$\boldsymbol{J} = f(v^*, y^t; W_{fuse}), \tag{22}$$

where W_{fuse} denotes the parameter of the fusion method, and $\boldsymbol{J}$ is the resulted multi-modal feature.

3.4   Cross-Modal Contrastive Learning

Cross-modal contrastive learning plays an important role in image retrieval by learning a shared feature space for image-text matching. Therefore, the matching loss between the multi-modal feature similarity matrix and the label matrix in ActionCLIP [40] is used to pull the pairwise text and visual features close to each other:

$$\mathcal{L}^{KL} = \frac{1}{2}\mathbb{E}_{(s,v)\sim\mathcal{D}}[\mathrm{KL}(\boldsymbol{p}^{s2v}(\boldsymbol{s}), \boldsymbol{q}^{s2v}(\boldsymbol{s})) + \mathrm{KL}(\boldsymbol{p}^{v2s}(\boldsymbol{v}), \boldsymbol{q}^{v2s}(\boldsymbol{v}))], \qquad (23)$$

where $\boldsymbol{q}^{s2v}(\boldsymbol{s})$ and $\boldsymbol{q}^{v2s}(\boldsymbol{v})$ are label matrices where the position of pairwise video and text is set to 1 and other positions are set to 0. $\boldsymbol{p}^{s2v}(\boldsymbol{s})$ and $\boldsymbol{p}^{v2s}(\boldsymbol{v})$ are multi-modal feature similarity matrices where cosine distance is used to measure the feature similarity.

Although KL loss can disclose the difference between the multi-modal feature similarity matrix and the label matrix, there may be multiple positive sample pairs belonging to the same class in a batch of samples. If the label information is included in contrastive learning, the feature encoder will produce the features at a closer distance. The supervised contrastive learning is calculated as follows:

$$\mathcal{L}^{sup} = \sum_{i\in\mathcal{D}}\mathcal{L}_i^{sup} = \sum_{i\in\mathcal{D}}\frac{-1}{|P(i)|}\sum_{p\in P(i)}\log\frac{\exp(\boldsymbol{J}_i\times\boldsymbol{s}_0^p/\tau)}{\sum_{a\in A(i)}\exp(\boldsymbol{J}_i\times\boldsymbol{s}_0^a/\tau)}, \qquad (24)$$

where $P(i) \equiv \{p \in P(i) : y_p = y_i\}$ is the set of indices of all positives to the i-th sample in a batch, $|P(i)|$ is its cardinality, $A(i) \equiv I \setminus \{i\}$, and $\mathcal{D}$ denotes a batch of samples. The overall loss is represented as follows:

$$\mathcal{L} = \mathcal{L}^{KL} + \mathcal{L}^{sup}. \qquad (25)$$

4   Experiments

4.1   Datasets

The training set of Kinetics400 [16] has 240K videos and the validation set has 20K videos. Kinetics400 is divided into 400 categories, each of which contains at least 400 samples. Each sample is obtained by cropping Youtube videos and lasts about 10 s.

Mini-Kinetics-200 [44] is a subset of the Kinetics400 dataset and contains 200 categories. There are 400 samples and 25 samples for each category in the training set and the validation set respectively.

The UCF101 [31] dataset contains 13 320 videos with a total of 101 action categories. The HMDB51 [18] dataset contains 51 categories of daily actions with a total of 6 766 videos. UCF101 and HMDB51 datasets provide three splits of training sets and testing sets, and researchers need to compare the average accuracy of the three splits to verify the effectiveness of the method.

4.2   Implementation Details

The text encoder adopts Transformer with 12 layers, where the self-attention layer contains 8 heads and the number of neurons in the hidden layer is set to 512. For the visual feature encoder, this paper uses ViT that also has a 12-layer Transformer. Two types of feature encoders are initialized from CLIP [29]. The number of learnable context tokens of PromptLearner is set to 16. The initial template is "a video of action X", where X represents the label text of the action and the context vector is randomly initialized with mean 0 and variance 0.02. For the multi-modal information interaction module, the number of neurons in the hidden layer of GNN is set to 512, and the dimension of the output composite feature is set to 512.

The AdamW optimizer is used to optimize the model's parameters, where the initial learning rates of the encoders and the remaining module parameters are set to 5e-6 and 5e-5 respectively, and the weight decay is set to 0.2. The batch size is set to 64, and the total training epoch is set to 50. The first 5 epochs use the warm-up strategy and the remaining 45 epochs use the half-cosine decay strategy. RandAugment is used to crop the region with 224×224 size from each frame. This paper adopts a sampling method to randomly sample 8 or 16 frames from a video. During the testing, 10 groups of frame sequences were randomly sampled from each video, and the average of the 10 groups of similarity scores is calculated to predict the action category. After the model is trained on Kinetics400, PromptLearner-CLIP is transferred to UCF101 and HMDB51 datasets, keeping the training and testing strategies unchanged.

4.3   Ablation Study

The loss function mainly consists of three parts, the video-text supervised contrastive loss, the video-text KL loss, and the text-video KL loss. The numbers in the first column of Table 1 represent the coefficients of corresponding loss functions, and the second and third columns report Top-1 and Top-5 accuracies on mini-Kinetics-200 respectively.

Table 1. Analysis of loss functions on mini-Kinetics-200 (%)

Contrastive loss:video-text loss:text-video loss	Top-1	Top-5
1:0:1	67.15	88.89
1:1:0	84.70	97.29
0:1:1	85.10	97.39
1:1:1	85.34	97.15

The experimental results in Table 1 show that when three loss functions are used to optimize parameters, PromptLearner-CLIP achieves the best experimental results on mini-Kinetics-200, and the Top-1 accuracy is 85.34%.

When the video-text KL loss is removed, the Top-1 accuracy drops to 67.15%. At the same time, it can be seen from the third row of the table that when the text-video KL loss is removed, the Top-1 accuracy of the model drops to 84.70%. Finally, observing the experimental results in the fourth row when removing the multi-modal supervised contrastive loss, the model has a little drop in the Top-1 accuracy, which confirms that the multi-modal supervised contrastive loss can further increase the similarity of samples belonging to the same class and improve the performance of the model.

Table 2 builds multiple models to study the influence of different semantic information on the multi-modal information interaction module. The average feature in the first column means that the average vector of word features encoded by Transformer is used as the input, and the Transformer feature means that the output at [EOS] position of the highest layer is used as the input. The models from the fourth row to the seventh row represent the text semantic parsing features with different semantic parsing steps.

Table 2. Analysis of the multi-modal information interaction module on mini-Kinetics-200 (%)

Textual information	Top-1	Top-5
Average feature	85.16	97.11
Transformer feature	85.04	97.49
One-step semantic parse	84.74	97.27
Two-step semantic parse	85.02	97.17
Three-step semantic parse	85.26	97.05
Four-step semantic parse	85.06	97.19

Table 2 shows that when the average feature and Transformer feature are used as the inputs, the model achieves similar performance, which demonstrates that both features can effectively represent the input without parsing text semantics. The experimental results in the fourth row to the seventh row processed by different semantic parsing steps show that the accuracy of the model is gradually increased, and the model with three parsing steps achieves the highest value, rising from 84.74% to 85.26%, which indicates that parsing the text content can provide more detailed semantic information to guide the learning process of visual features.

Table 3 conducts ablation analyses of PromptLearner-CLIP on mini-Kinetics-200. For a fair comparison, a baseline model was set up to use ViT to process visual input. ActionCLIP uses the single-modal model ViT as the backbone network for visual feature extraction and builds a multi-modal learning framework to process visual data and text data simultaneously. The fourth, fifth, and sixth rows of Table 3 represent the results removing the multi-modal information interaction module, PromptLearner initialization, and video-text supervised contrastive loss respectively.

Table 3. Ablation study of PromptLearner-CLIP on mini-Kinetics-200 (%)

Interaction module	Prompt initializaiton	Contrastive learning	Top-1	Top-5
–	–	–	83.70	96.53
–	–	–	84.02	96.85
–	✓	✓	84.78	97.41
✓	–	✓	84.82	97.27
✓	✓	–	85.10	97.39
✓	✓	✓	85.26	97.05

When textual information is incorporated, ActionCLIP improves the classification results of ViT on mini-Kinetics-200, which demonstrates that text content can provide semantic clues for action recognition. From the fourth to seventh rows in the table, when the main modules in PromptLearner-CLIP are removed one by one, the performance decreases to different degrees, indicating that each module in the model has a certain contribution to improving the performance. When the model removes the multi-modal information interaction module, the accuracy in Top-1 drops to 84.78%. Compared with the experimental results of the fifth and sixth rows, the performance drops the most.

4.4 Pathway Finetune

PromptLearner-CLIP contains text and vision pathways to extract features of different modalities. Table 4 summarizes the results of fine-tuning the backbone network parameters of different pathways on mini-Kinetics-200. Table 4 shows that the model finetuning two pathways(the fifth row) is significantly higher than the model freezing the parameters of two pathways(the second row) in the Top-1 accuracy, and the Top-1 and Top-5 accuracies are increased by 3.71% and 0.86% respectively, which demonstrates that it is still necessary to finetune the model parameters on the target dataset and learn the dataset-specific features.

Table 4. Analysis of finetuning different pathways on mini-Kinetics-200 (%)

Text pathway	Vision pathway	Top-1	Top-5
–	–	81.55	96.19
✓	–	81.73	95.93
–	✓	85.04	97.29
✓	✓	85.26	97.05

4.5 Different Backbone Networks

In this section, the visual backbone networks have completed the training on Kinetics400 and the parameters of them are frozen. Table 5 shows the accuracies of PromptLearner-CLIP using three visual backbone networks on Kinetics400. In the comparison of each group of the same backbone network, PromptLearner-CLIP achieves performance improvement on ActionCLIP. The experimental results in the third and fifth rows of the table show that reducing the size of frame patches will bring more computation, but the model will learn more detailed relationships between frame regions and discriminative features. The experimental results in the fifth and seventh rows show that more video frames can help the model to obtain more complete action information and improve the accuracy of the model.

Table 5. Analysis of different visual backbone networks on Kinetics400 (%)

Model	Top-1	Top-5
ActionCLIP(ViT-32-8f)	77.49	93.88
PromptLearner-CLIP(ViT-32-8f)	77.92	94.51
ActionCLIP(ViT-16-8f)	80.32	95.41
PromptLearner-CLIP(ViT-16-8f)	80.86	95.61
ActionCLIP(ViT-16-16f)	81.12	95.73
PromptLearner-CLIP(ViT-16-16f)	81.60	95.86

4.6 Comparison with State-of-the-Art Methods

Finally, PromptLearner-CLIP and current state-of-the-art methods are compared on the Kinetics400 dataset. Table 6 summarizes Top-1 and Top-5 accuracies on Kinetics400 of different methods.

First, the classification accuracies of Transformer-based methods in the table, such as MViT-B [8], TimeSformer-L [42], and ViT-B-VTN [27] on Kinetics400 are higher than 2D CNN-based [23] and 3D CNN-based [10] methods. PromptLearner-CLIP uses the same backbone network and achieves higher experimental results, although the number of input frames to the model is less than these three models. ViViT-L [2] uses a deeper ViT as the backbone network and achieves 80.6% Top-1 accuracy on Kinetics400, which is lower than our model by 1.0%. When ViViT-L initializes ViT-B/16 parameters with the model pre-trained on the large-scale dataset JFT, ViViT-L achieves the best experimental results in the table. Deeper models, more video frames, larger image resolutions, and larger pre-training datasets will stimulate the potential of the model to achieve a higher action recognition accuracy.

4.7 Finetune on Small Datasets

Finally, the PromptLearner-CLIP pre-trained on Kinetics400 is transferred to HMDB51 and UCF101, the results are shown in Table 7. MSM-ResNets [54] takes RGB images, optical flow images, and action saliency images as inputs,

Table 6. Comparison with state-of-the-art methods on Kinetics400 (%)

Model	Source	Top-1	Top-5
TEA-ResNet50 [23]	CVPR2020	76.1	92.5
TEINet [25]	AAAI2020	76.2	92.5
SmallBigNet [22]	CVPR2020	77.4	93.3
SVT [30]	CVPR2022	78.1	–
TPN-R101 [46]	CVPR2020	78.9	93.9
TANet [26]	ICCV2021	79.3	94.1
TDN [38]	CVPR2021	79.4	93.9
SlowFast [11]	ICCV2019	79.8	93.9
ViT-B-VTN [27]	ICCV2021	79.8	94.2
X3D-XXL [10]	CVPR2020	80.4	94.7
TokenShift [48]	ACM2021	80.4	94.5
BEVT [41]	CVPR2022	80.6	–
ViViT-L [2]	ICCV2021	80.6	94.7
TimeSformer-L [3]	ICML2021	80.7	94.7
MViT-B [8]	ICCV2021	81.2	95.1
DirecFormer [35]	CVPR2022	82.8	94.9
ViViT-L(JFT) [2]	ICCV2021	**82.8**	95.3
PromptLearner-CLIP	-	81.6	**95.9**

which obtains 66.7% on HMDB51 and 93.5% on UCF101. Since MSM-ResNets is not pre-trained on large-scale video datasets, the performance of the model is significantly lower than the current state-of-the-art methods. PoTion [5] extracts the motion information of human pose from the video to learn pose motion features. Two-stream I3D [4] extracts appearance features and action features from RGB images and optical flow images respectively. When the spatial-temporal features of I3D are fused with the PoTion features, the accuracies are improved by 0.7% on HMDB51 and 0.3% on UCF101. However, I3D+PoTion is lower than PromptLearner-CLIP+I3D(Flow) on both datasets, revealing that the multi-modal learning framework proposed in this paper can effectively utilize the action clues contained in other modal information to improve the performance. The comparison results of different methods in the table show that PromptLearner-CLIP achieves competitive results with state-of-the-art methods.

Table 7. Comparison with state-of-the-art methods on HMDB51 and UCF101 (%)

Model	HMDB51	UCF101
MSM-ResNets [54]	66.7	93.5
SVT [30]	67.2	93.7
two-stream TSN [39]	68.5	94.0
Temporal Squeeze Network [13]	71.5	95.2
TVNet+IDT [9]	72.6	95.4
TokenShift [48]	–	96.8
TEA-ResNet50 [23]	73.3	96.9
VidTr [49]	74.4	96.7
Dense Dilated Network [45]	74.5	96.9
Dynamic Network [51]	75.5	96.8
ActionCLIP [40]	76.2	97.1
BQN [14]	77.6	97.6
S3D-G [44]	78.2	96.8
MARS+RGB [6]	79.5	97.6
SIFP+SlowFast [20]	80.1	96.9
two-stream I3D [4]	80.2	97.9
PoTion+I3D [5]	80.9	98.2
STRM [32]	81.3	98.1
STA-MARS [21]	**81.4**	98.4
PromptLearner-CLIP+I3D(Flow)	81.3	**98.5**

5   Conclusion

This paper proposes a multi-modal semantic-guided action recognition network PromptLearner-CLIP that utilizes textual information to enhance the representation ability of features. Experiments on Kinetics400, UCF101, and HMDB51 demonstrate that PromptLearner can automatically learn the text content of prompt and provide semantic clues for action recognition. Besides, by multi-modal information interaction module, features of different modalities pass information and disclose the difference of multi-modal features effectively. And the supervised contrastive loss is used to further reduce the feature distance between samples of the same class but different modalities. PromptLearner-CLIP achieves highly competitive accuracies on these three action recognition datasets with state-of-the-art methods. In future work, we will study the methods that incorporate the visual content of a video into prompt to better learn semantic information.

Acknowledgements. This work was supported in part by the National Key Research and Development Program of China under Grant 2021YFE0110500 and in part by the National Natural Science Foundation of China under Grant 62006015 and Grant 62072028.

References

1. Anderson, P., et al.: Bottom-up and top-down attention for image captioning and visual question answering. In: CVPR, pp. 6077–6086. IEEE, Salt Lake City, UT, USA (2018)
2. Arnab, A., Dehghani, M., Heigold, G., Sun, C., Lučić, M., Schmid, C.: ViViT: a video vision transformer. In: ICCV, pp. 6836–6846. IEEE, Montreal, Canada (2021)
3. Bertasius, G., Wang, H., Torresani, L.: Is space-time attention all you need for video understanding? In: ICML, pp. 813–824. ACM, Virtual (2021)
4. Carreira, J., Zisserman, A.: Quo vadis, action recognition? a new model and the kinetics dataset. In: CVPR, pp. 4724–4733. IEEE, Honolulu, HI, USA (2017)
5. Choutas, V., Weinzaepfel, P., Revaud, J., Schmid, C.: PoTion: pose motion representation for action recognition. In: CVPR, pp. 7024–7033. IEEE, Salt Lake City (2018)
6. Crasto, N., Weinzaepfel, P., Alahari, K., Schmid, C.: MARS: motion-augmented RGB stream for action recognition. In: CVPR, pp. 7874–7883. IEEE, Long Beach, CA, USA (2019)
7. Dosovitskiy, A., et al.: An image is worth 16x16 words: transformers for image recognition at scale. In: ICLR, pp. 1–21. Virtual (2021)
8. Fan, H., et al.: Multiscale vision transformers. In: ICCV, pp. 6824–6835. IEEE, Montreal, Canada (2021)
9. Fan, L., Huang, W., Gan, C., Ermon, S., Gong, B., Huang, J.: End-to-end learning of motion representation for video understanding. In: CVPR, pp. 6016–6025. IEEE, Salt Lake City, UT, USA (2018)
10. Feichtenhofer, C.: X3D: expanding architectures for efficient video recognition. In: CVPR, pp. 200–210. IEEE, Seattle, WA, USA (2020)
11. Feichtenhofer, C., Fan, H., Malik, J., He, K.: Slowfast networks for video recognition. In: CVPR, pp. 6202–6211. IEEE, Seoul, Korea (2019)
12. He, K., Fan, H., Wu, Y., Xie, S., Girshick, R.: Momentum contrast for unsupervised visual representation learning. In: CVPR, pp. 9726–9735. IEEE, Seattle, WA, USA (2020)
13. Huang, G., Bors, A.G.: Learning spatio-temporal representations with temporal squeeze pooling. In: ICASSP, pp. 2103–2107. IEEE, Barcelona, Spain (2020)
14. Huang, G., Bors, A.G.: Busy-quiet video disentangling for video classification. In: WACV, pp. 1341–1350. IEEE, Waikoloa, HI, USA (2022)
15. Kar, A., Rai, N., Sikka, K., Sharma, G.: AdaScan: adaptive scan pooling in deep convolutional neural networks for human action recognition in videos. In: CVPR, pp. 5699–5708. IEEE, Honolulu, HI, USA (2017)
16. Kay, W., et al.: The kinetics human action video dataset (2017)
17. Khosla, P., et al.: Supervised contrastive learning. In: NeurIPS, pp. 18661–18673. MIT Press, Virtual (2021)
18. Kuehne, H., Jhuang, H., Garrote, E., Poggio, T., Serre, T.: HMDB: a large video database for human motion recognition. In: ICCV, pp. 2556–2563. IEEE, Barcelona, Spain (2011)

19. Lei, J., et al.: Less is more: clipBERT for video-and-language learning via sparse sampling. In: CVPR, pp. 7331–7341. IEEE, Virtual (2021)
20. Li, J., Wei, P., Zhang, Y., Zheng, N.: A slow-i-fast-p architecture for compressed video action recognition. In: ACM MM, pp. 2039–2047. ACM, Seattle, WA, USA (2020)
21. Li, J., Liu, X., Zhang, W., Zhang, M., Song, J., Sebe, N.: Spatiotemporal attention networks for action recognition and detection. IEEE Trans. Multimedia **22**(11), 2990–3001 (2020)
22. Li, X., Wang, Y., Zhou, Z., Qiao, Y.: SmallBigNet: integrating core and contextual views for video classification. In: CVPR, pp. 1092–1101. IEEE, Seattle, WA, USA (2020)
23. Li, Y., Ji, B., Shi, X., Zhang, J., Kang, B., Wang, L.: Tea: temporal excitation and aggregation for action recognition. In: CVPR, pp. 906–915. IEEE, Seattle, WA, USA (2020)
24. Lin, J., Gan, C., Han, S.: TSM: temporal shift module for efficient video understanding. In: ICCV, pp. 7082–7092. IEEE, Seoul, Korea (2019)
25. Liu, Z., et al.: TEINet: towards an efficient architecture for video recognition. In: AAAI, pp. 11669–11676. AAAI, New York, USA (2020)
26. Liu, Z., Wang, L., Wu, W., Qian, C., Lu, T.: TAM: temporal adaptive module for video recognition. In: ICCV, pp. 13708–13718. IEEE, Montreal, Canada (2021)
27. Neimark, D., Bar, O., Zohar, M., Asselmann, D.: Video transformer network. In: ICCV, pp. 3163–3172. IEEE, Montreal, Canada (2021)
28. Pan, T., Song, Y., Yang, T., Jiang, W., Liu, W.: VideoMoCo: contrastive video representation learning with temporally adversarial examples. In: CVPR, pp. 11200–11209. IEEE, Virtual (2021)
29. Radford, A., et al.: Learning transferable visual models from natural language supervision. In: ICML, pp. 8748–8763. ACM, Virtual (2021)
30. Ranasinghe, K., Naseer, M., Khan, S., Khan, F.S., Ryoo, M.: Self-supervised video transformer. In: CVPR, pp. 2874–2884. IEEE, New Orleans, Louisiana, USA (2022)
31. Soomro, K., Zamir, A.R., Shah, M.: UCF101: a dataset of 101 human actions classes from videos in the wild (2012)
32. Thatipelli, A., Narayan, S., Khan, S., Anwer, R.M., Khan, F.S., Ghanem, B.: Spatio-temporal relation modeling for few-shot action recognition. In: CVPR, pp. 19958–19967. IEEE, New Orleans, Louisiana, USA (2022)
33. Tran, D., Bourdev, L., Fergus, R., Torresani, L., Paluri, M.: Learning spatiotemporal features with 3d convolutional networks. In: ICCV, pp. 4489–4497. IEEE, Santiago, Chile (2015)
34. Tran, D., Wang, H., Torresani, L., Ray, J., LeCun, Y., Paluri, M.: A closer look at spatiotemporal convolutions for action recognition. In: CVPR, pp. 6450–6459. IEEE, Salt Lake City, UT, USA (2018)
35. Truong, T.D., et al.: DirecFormer: a directed attention in transformer approach to robust action recognition. In: CVPR, pp. 20030–20040. IEEE, New Orleans, Louisiana, USA (2022)
36. Vaswani, A., et al.: Attention is all you need. In: NeurIPS, pp. 5998–6008. MIT Press, Long Beach, CA, USA (2017)
37. Wang, L., Li, W., Li, W., Van Gool, L.: Appearance-and-relation networks for video classification. In: CVPR, pp. 1430–1439. IEEE, Salt Lake City, UT, USA (2018)
38. Wang, L., Tong, Z., Ji, B., Wu, G.: TDN: temporal difference networks for efficient action recognition. In: CVPR, pp. 1895–1904. IEEE, Virtual (2021)

39. Wang, L., et al.: Temporal segment networks for action recognition in videos. IEEE Trans. Pattern Anal. Mach. Intell. **41**(11), 2740–2755 (2019)
40. Wang, M., Xing, J., Liu, Y.: ActionCLIP: a new paradigm for video action recognition (2021)
41. Wang, R., et al.: BEVT: BERT pretraining of video transformers. In: CVPR, pp. 14733–14743. IEEE, New Orleans, Louisiana, USA (2022)
42. Wang, X., Gupta, A.: Videos as space-time region graphs. In: Ferrari, V., Hebert, M., Sminchisescu, C., Weiss, Y. (eds.) ECCV 2018. LNCS, vol. 11209, pp. 413–431. Springer, Cham (2018). https://doi.org/10.1007/978-3-030-01228-1_25
43. Wu, Z., Jiang, Y.G., Wang, X., Ye, H., Xue, X.: Multi-stream multi-class fusion of deep networks for video classification. In: ACM MM, pp. 791–800. ACM, Amsterdam, Netherlands (2016)
44. Xie, S., Sun, C., Huang, J., Tu, Z., Murphy, K.: Rethinking spatiotemporal feature learning: speed-accuracy trade-offs in video classification. In: Ferrari, V., Hebert, M., Sminchisescu, C., Weiss, Y. (eds.) ECCV 2018. LNCS, vol. 11219, pp. 318–335. Springer, Cham (2018). https://doi.org/10.1007/978-3-030-01267-0_19
45. Xu, B., Ye, H., Zheng, Y., Wang, H., Luwang, T., Jiang, Y.: Dense dilated network for video action recognition. IEEE Trans. Image Process. **28**(10), 4941–4953 (2019)
46. Yang, C., Xu, Y., Shi, J., Dai, B., Zhou, B.: Temporal pyramid network for action recognition. In: CVPR, pp. 588–597. IEEE, Seattle, WA, USA (2020)
47. Yang, S., Li, G., Yu, Y.: Dynamic graph attention for referring expression comprehension. In: ICCV, pp. 4643–4652. IEEE, Seoul, Korea (2019)
48. Zhang, H., Hao, Y., Ngo, C.W.: Token shift transformer for video classification. In: ACM MM, pp. 917–925. ACM, Chengdu, China (2021)
49. Zhang, Y., et al.: VidTr: video transformer without convolutions. In: ICCV, pp. 13577–13587. IEEE, Montreal, Canada (2021)
50. Zhao, Y., Wang, G., Luo, C., Zeng, W., Zha, Z.J.: Self-supervised visual representations learning by contrastive mask prediction. In: ICCV, pp. 10160–10169. IEEE, Virtual (2021)
51. Zheng, Y., Liu, Z., Lu, T., Wang, L.: Dynamic sampling networks for efficient action recognition in videos. IEEE Trans. Image Process. **29**, 7970–7983 (2020)
52. Zhou, K., Yang, J., Loy, C.C., Liu, Z.: Learning to prompt for vision-language models. Int. J. Comput. Vision **130**, 2337–2348 (2022)
53. Zhu, L., Yang, Y.: ActBERT: learning global-local video-text representations. In: CVPR, pp. 8746–8755. IEEE, Seattle, WA, USA (2020)
54. Zong, M., Wang, R., Chen, X., Chen, Z., Gong, Y.: Motion saliency based multi-stream multiplier ResNets for action recognition. Image Vis. Comput. **107**, 104108 (2021)

Causal Property Based Anti-conflict Modeling with Hybrid Data Augmentation for Unbiased Scene Graph Generation

Ruonan Zhang[1,2] and Gaoyun An[1,2(✉)]

[1] Institute of Information Science, Beijing Jiaotong University, Beijing 100044, China
{21120318,gyan}@bjtu.edu.cn
[2] Beijing Key Laboratory of Advanced Information Science and Network Technology, Beijing 100044, China

Abstract. Scene Graph Generation (SGG) aims to detect visual triplets of pairwise objects based on object detection. There are three key factors being explored to determine a scene graph: visual information, local and global context, and prior knowledge. However, conventional methods balancing losses among these factors lead to conflict, causing ambiguity, inaccuracy, and inconsistency. In this work, to apply evidence theory to scene graph generation, a novel plug-and-play Causal Property based Anti-conflict Modeling (CPAM) module is proposed, which models key factors by Dempster-Shafer evidence theory, and integrates quantitative information effectively. Compared with the existing methods, the proposed CPAM makes the training process interpretable, and also manages to cover more fine-grained relationships after inconsistencies reduction. Furthermore, we propose a Hybrid Data Augmentation (HDA) method, which facilitates data transfer as well as conventional debiasing methods to enhance the dataset. By combining CPAM with HDA, significant improvement has been achieved over the previous state-of-the-art methods. And extensive ablation studies have also been conducted to demonstrate the effectiveness of our method.

Keywords: Scene graph generation · D-S evidence theory · Data augmentation

1 Introduction

Scene Graph Generation (SGG) is an important task in computer vision, which can bridge low-level visual tasks such as object detection [26] and high level visual tasks such as visual question answering [44], 3D scene synthesis [25], image caption [38], etc. Today's SGG [3, 30, 32, 42] is considered as a deterministic task, recognizing relationships between pairwise objects. Recent work has made steady progress on SGG and provides powerful models to encode both visual and linguistic context of the scene. As illustrated in Fig. 1, a scene graph is composed of visual triplets in the form of $\langle subject - predicate - object \rangle$.

© The Author(s), under exclusive license to Springer Nature Switzerland AG 2023
L. Wang et al. (Eds.): ACCV 2022, LNCS 13844, pp. 571–587, 2023.
https://doi.org/10.1007/978-3-031-26316-3_34

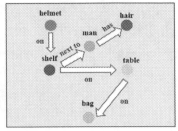

Fig. 1. An example of Visual Genome [14] dataset and its scene graph. Obviously, most of the predicates in this image are trivial, and the same is true for the dataset.

However, due to the long-tailed distribution of annotations, SGG is far from practical. In Visual Genome [14], there are 50 predicate classes, yet more than 100K samples are top 5 predicate classes [43]. BA-SGG [12] divided all predicates into two categories: informative and common. The frustrating fact is that annotators prefer common predicates, which are exactly the "head" predicates in the dataset. Therefore, we should not blame the model generates trivial and less informative predicates. To address this problem, we propose a Hybrid Data Augmentation (HDA) to deal with it.

Meanwhile, as Fig. 2 shown, counterfactual inference by causal graph [29] is also adopted to eliminate this highly-skewed long-tailed bias. A causal graph summarizes three key factors to determine a scene graph: visual appearance feature, context feature, and prior knowledge. Specifically, visual features are extracted from paired objects, context features are encoded by RNN or GNN for each object and prior knowledge denotes the predicate class distribution in the dataset after the categories of subject and object are identified. Then the predicate confidence scores are predicted by these features separately. Finally, these scores are fused to get a final result, usually by adding or gating.

Though models based on causal graphs outperform the conventional ones, we find an extra bias introduced by casual graphs. As Fig. 3 shown, the output distributions of branches in casual graph may conflict each other. For the triplet of $\langle helmet - on - shelf \rangle$, if only context branch is considered, the final prediction is "on", while visual branch tends to wrongly predict "in". Meanwhile, background is also regarded as a highest probable prediction by prior knowledge. What's worse, after these three predicted distributions are fused together, the final result is misdirected to "in", although one branch performs correctly. To eliminate this extra bias, we propose a Causal Property Anti-conflict Modeling (CPAM) module, direct modeling these three branches via Dempster-Shafer (D-S) evidence theory [40], which is first applied in expert system [20] to handle uncertain information. In D-S evidence theory, evidences are mutually exclusive probability distributions which represent all possible answers to a question. By transferring this theory to causal graph based SGG task, the predicted confidence scores of these three branches can be explicitly modeled as evidences. Then the fused scores can be computed by Dempster combination rules, and thus this additional bias may be removed. By the way, CPAM is a plug-and-play module.

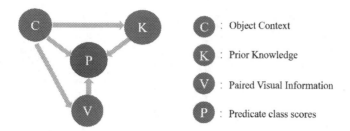

Fig. 2. The causal graph [29] in SGG. The final confidence scores of predicates come for three branches: context feature encoded by RNN or GNN from each proposal, visual appearance feature extracted by paired proposals, and prior knowledge distribution after the categories of subject and object are identified. Meanwhile, as input of the context branch, object visual features are also passed to the other two branches.

To sum up, the main contributions of our work contain: (1) We systematically reveal the long-tailed bias which limits SGG's overall performance and an extra bias introduced by multibranch prediction of a causal graph. (2) We propose a novel Causal Property Anti-conflict Modeling (CPAM) module, which applies Dempster-Shafer evidence theory and can serve as a plug-and-play module. (3) By combining data transfer and conventional debiasing method, we propose a Hybrid Data Augmentation (HDA) method to balance data distribution in Visual Genome. (4) Experimental results demonstrate the effectiveness of the proposed CPAM and HDA, which achieve state-of-the-art performance under existing evaluation metrics on Visual Genome and may improve the mean recall metric significantly.

2   Related Work

2.1   Scene Graph Generation

Scene Graph Generation has drawn widespread attention in computer vision community since Lu *et al.* [24] formalized SGG as a visual task to recognize relationships between objects. After the large-scale image semantic understanding dataset Visual Genome [14] is proposed, SGG has become a popular task in computer vision gradually. However, more and more researchers recognize the highly-skewed long-tailed distribution in the dataset which limits SGG performance seriously. To address this issue, TDE [29] introduced counterfactual causal analysis in the inference stage. BA-SGG [12] transferred common predicates to informative predicates by semantic adjustment. BGNN [16] proposed a bi-level data sampling strategy to sample instances of different entities and predicates. SHA-GCL [9] first grouped all the predicates and then employed a median-resampling method to make a balanced distribution. We, therefore, also propose an effective strategy HDA to reduce the impact of long-tailed bias and better results are shown by experiments.

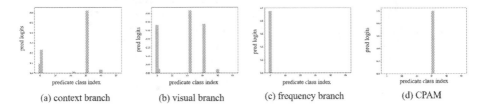

(a) context branch   (b) visual branch   (c) frequency branch   (d) CPAM

Fig. 3. This is an example of $\langle helmet - on - shelf \rangle$ in Fig. 1. Predicate probability distributions arise from different causal properties in re-implemented MOTIFS [42] as shown in (a) to (c). Predictions generated from prior knowledge in (c) are intuitive guesses obtained from the probability distribution in the dataset, after the categories of subject and object are known. It is obvious that predictions of these branches cause ambiguity because maximum logits of (a), (b), and (c) are not all the same. After CPAM, the wrong prediction is corrected.

The existing SGG models can be roughly divided into two categories: two-stage methods [22,35,45] and one-stage methods. One-stage methods [17–19] predict object labels and relationships at last simultaneously, which loses the assistance of language semantics during training. The co-occurrence of object pairs and predicates in the dataset shows that knowing the categories of subjects and objects are quite helpful to make a correct prediction, such as the predicate between "person" and "horse" is probably "riding on". Specifically, two-stage methods can provide extra semantic information of object labels, while the one-stage methods are not able to do this. Currently, two-stage methods are in the slight majority, and we also use this method to build our pipeline.

2.2   Evidence Theory

In 1967, Dempster proposed evidence theory and applied it to statistical problems [4,5]. After that, Shafer published the first monograph of evidence theory in 1976. By introducing the concept of belief function, Shafer developed and improved evidence theory, namely Dempster Shafer (D-S) theory, marking the birth of evidence theory [27].

Currently, evidence theory has been rapidly developed because of its strong ability in uncertain inference and this is why we apply it to model the ambiguity brought by different predictions of casual graphs. Evidence theory is widely used in pattern recognition [6], information fusion [10], artificial intelligence [1], expert systems [36], etc. Researchers in the community of computer vision are trying to introduce evidence theory to model the uncertainty [8,15,31,33]. More detail background knowledge can be found in [11,13,37,40].

3   Approach

3.1   Overview of Approach

Conventional SGG models can be defined as $SG = (B, O, R)$, where B, O and R represent the bounding box, object and relational prediction model respectively. The whole process of SGG can be expressed in the following form conventionally. I is the given image.

$$P(SG|I) = P(B|I)P(O|B,I)P(R|O,B,I) \tag{1}$$

The Faster R-CNN [26] is adopted to return coordinate $b_i \in \mathbb{R}^4$ of each proposal, visual feature $v_i \in \mathbb{R}^{4096}$ and object classification confidence scores $c_i \in \mathbb{R}^O$ of each node. O is the number of object categories. The visual feature x_i of each node is formed by concatenating semantic embedding feature, spatial embedding feature and extracted feature by Faster R-CNN. Then the global context is encoded and object-level context and edge-level context are obtained. Furthermore, paired boxes feature of i-th and j-th proposals is extracted, denoting as $u_{ij} \in \mathbb{R}^{4096}$. As Fig. 4(c) shown, to better capture the connection of properties in a causal graph and avoid conflict, our proposed CPAM module is adopted, which employs uncertainty modeling instead of a discriminative manner. Conventional causal properties fusion method is replaced with CPAM. As Fig. 4(a) shown, our proposed HDA method is also adopted to relieve long-tail distribution.

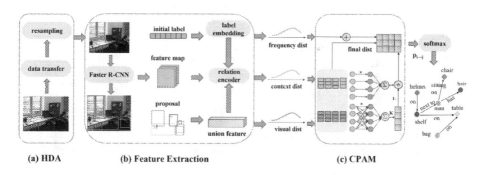

(a) HDA   (b) Feature Extraction   (c) CPAM

Fig. 4. The pipeline of our proposed method. In this work, we utilize Faster R-CNN as an object detector, and divide the method into three parts: (a) Enhancing the dataset to balance data distribution. (b) Extracting all the features used in the framework, including initial object relabel generated by Faster R-CNN, semantics embedding feature and encoded global context, etc. (c) Modeling data distributions of different branches based on the causal graph.

3.2 D-S Evidence Theory

D-S evidence theory models uncertainty by applying Dempster combination rules [40]. Some basic concepts of evidence theory are introduced below.

Defination 1. Frame of Discernment. The $\Theta = \{H_1, H_2, \ldots, H_N\}$ is defined as a frame of discernment which consists of a set of mutually exclusive non-empty events. Namely, $\forall i, j \in [1, N]$ and $i \neq j$, $H_i \cap H_j = \Phi$. Φ is the empty set and H_i denotes i-th event. In SGG task, events are the predicate class distribution predicted by the three branches of a casual graph.

Defination 2. Mass Function. Basic Probability Assignment (BPA) is the output of the mass function. The mass function is a mapping function and denotes as follows. In our method, n is the number of predicate categories and $m(H_i)$ is the confidence score corresponding to each predicate class.

$$m(\Phi) = 0, \sum_{i=1}^{n} m(H_i) = 1 \tag{2}$$

Defination 3. Belief Function and Plausibility Function. $Bel(\cdot)$ and $Pl(\cdot)$ are basic concepts in evidence theory to measure the confidence of evidence. $Bel(A)$ denotes the sum of the BPAs of all subsets of proposition A, while $Pl(A)$ denotes the sum of BPA of all subsets intersecting proposition A.

$$Bel(A) = \sum_{B \subseteq A} m(B) \tag{3}$$

$$Pl(A) = \sum_{B \cap A \neq \Phi} m(B) \tag{4}$$

Defination 4. Dempster Combination Rules. Suppose there are two independent and completely reliable evidence, corresponding to BPAs $m1$ and $m2$ respectively. $\forall A \subseteq \Theta$, the Dempster Combination Rules (DCR) are defined by:

$$m(A) = \begin{cases} 0 & , A = \Phi \\ \frac{1}{1-K} \sum_{B \cap C = A} m_1(B) m_2(C) & , A \neq \Phi \end{cases} \tag{5}$$

with

$$K = \sum_{B \cap C = \Phi} m_1(B) m_2(C) \tag{6}$$

where K is the conflict coefficient between m_1 and m_2. Note that some works denote $1 - K$ as the conflict coefficient, and the effect is the same. Predicted distribution of visual branch is modeled as B, while C represents the result from context branch. The intersection of B and C are confidence scores corresponding to the same predicate class.

3.3   Causal Property Anti-conflict Modeling

To reduce the possibility of conflict between causal properties, a Casual Property Anti-Conflict Modeling module is proposed. To apply D-S evidence theory to SGG, we should figure out which are the frame of discernment and exclusive events. Take "person" as the subject and "motorbike" as the object for example, the events will be: "The predicate between them is riding", or "The predicate between them is sitting on". There are 50 classes of predicate in Visual Genome. For each pair of objects, the discriminant results of the predicate category are events in the frame of discernment. To capture the intrinsic uncertainty of a causal graph, we try to model the output distribution of each branch as evidence explicitly.

First, visual feature v_i and context feature c_i are projected into a subspace of the same dimension. $W_1, W_2 \in \mathbb{R}^{4096 \times 512}$ are linear transformation matrices. To satisfy Eq. (2), we normalize the projected features as:

$$v' = softmax(W_1^T v) \tag{7}$$

$$c' = softmax(W_2^T c) \tag{8}$$

To obtain conflict coefficient K, an auxiliary matrix A is introduced. All elements of A are 1 except that the diagonal elements are 0. N is the number of predicate categories. $I \subset \mathbb{R}^{N \times N}$ is identity matrix. After getting the conflict coefficient K, the combined probability assignment is calculated as m, which is taken as the weighted coefficient of visual dist v' and context dist c' with the frequency dist of each pair of objects added. We call this approach binary causal attribute fusion. Also, a triple fusion is introduced among all of the three properties. These strategies are compared in Sect. 4.4. The classification score vector of relationships can be obtained as follows:

$$K - \sum_{i=1}^{N} \sum_{j=1}^{N} v'(i) \times c'(j) \times A(i,j) \tag{9}$$

$$m = \frac{1}{1-K} \sum_{i=1}^{N} \sum_{j=1}^{N} v'(i) \times c'(j) \times I(i,j) \tag{10}$$

$$p = m \times v'(i) + m \times c'(i) + frequency \tag{11}$$

The category of relationship between a pair of nodes is predicted by:

$$r = \arg\max_{r \in N}(p) \tag{12}$$

3.4   Hybrid Data Augmentation

There are many semantically ambiguous triplets in the dataset. Some predicates share similar semantic spaces like "holding" and "carrying", while some predicates can reasonably describe relations at the same time, such as "on" and "standing on".

To deal with the highly-skewed long-tailed distribution, a Hybrid Data Augmentation method is proposed. Based on [43], internal and external data transfer are adopted to deal with semantic ambiguity and long-tailed problems. Internal data transfer tries to transform general predicates into informative predicates, and external data transfer takes advantage of negative samples. Specifically, negative samples are relabeled to generate a more diverse training set.

Furthermore, to get a more enhanced dataset, the conventional resampling strategy [2] is introduced. Predicate categories in the tail of data distribution were up-sampled according to sample fraction during training. $Count(\cdot)$ denotes the number of training samples in the dataset. We set $p = 3.0$ as default, and the sampling rate φ_i is calculated as below.

$$\varphi_i = \begin{cases} 1.0 \;, if & \frac{\sum\limits_{k=1}^{M} Count(p_k)}{Count(p_i)} < m \\ p \;\;, if & \frac{\sum\limits_{k=1}^{M} Count(p_k)}{Count(p_i)} \geq m \end{cases} \quad (13)$$

4   Experiment

In this section, a series of comprehensive experiments are conducted to validate the effectiveness of our method. The generalizability of our method is demonstrated by plugging into different baseline models. Below introduces implementation details while training. Then we show experimental analysis and do ablation studies on Visual Genome.

4.1   Evaluation Settings

Dataset. The popular Visual Genome [14] is used to train and evaluate our method, and followed the popular split VG-150 [32,42] benchmark. VG-150 is composed of 108k images with the most frequent 150 object categories and 50 predicate categories. The whole dataset is divided into a training set and a testing set, which includes 70% images and 30% images separately. We also preserve 5k images for validation.

Tasks. In SGG, three widely-used subtasks are supposed to implement. (1) Predicate Classification (PredCls) classifies predicate categories with ground truth bounding boxes and labels. (2) Scene Graph Classification (SGCls) classifies object categories and predicate categories only with correct localization. (3) Scene Graph Detection (SGDet) requires a model to localize objects and recognize both objects and predicate classes. Namely, SGDet asks the model to detect scene graphs from scratch.

Metrics. Following the previous works [3, 29, 30], Mean Recall@K (mR@K) metric is used as our evaluation metrics. mR@K could evaluate the performance of a model more fairly, because it treats each predicate category as equal, and does not give more importance to "head" predicates due to the number of samples. mR@K first calculates Recall@K (R@K) of each predicate, and then averages them for all predicates. Furthermore, Zero-Shot Recall@K (zR@k) metric is introduced to better evaluate model performance, which was firstly proposed by [24]. zR@K aims to evaluate the generalization ability of the model when encountering triplets unseen in the training set. In other words, zR@K reports only the R@K of unseen triples.

4.2   Implementation Details

Object Detector. Following previous works [32, 42], a two-stage method is applied to build the overall model. For object detector, we employ a pre-trained Faster R-CNN and adopt ResNeXt-101-FPN as backbone. To reduce the computation cost, the parameters of the object detector are frozen while training. The object detector has 38.52 mAP on the training set and 28.14 mAP on the testing set.

Scene Graph Generation. In the training process, a SGD optimizer with an initial learning rate of 5e-3 is applied. We do not decay the learning rate while training but apply a warmup strategy to make the whole training process steady. The batch size of the three subtasks was set to be 8. Originally, the time complexity of all candidate pair boxes is $O(n^2)$. For HDA, all sampling rates are set to 1.0 for SGDet. To limit the number of candidate pair boxes, candidate bounding boxes are sorted in descending order by confidence scores and only choose 256 candidate pairs, thus a lot of time and computation can be saved. Furthermore, the ground truth triplets are added while training in case some gt boxes are missing only for SGDet. And during preparing candidate pair of objects, overlapping boxes are not required, because there is no need for a spatial overlap between the subject and the object, like $\langle person - throw - ball\rangle$ or $\langle girl - looking\,at - kite\rangle$.

4.3   Comparisons with State-of-the-Art Methods

We evaluate three models on VG-150 dataset: MOTIFS [42], VCTree [30] and VTransE [45] to demonstrate the generalization ability of our proposed method. As Table 1 shown, our proposed method can boost all metrics in three subtasks (PredCls, SGCls, SGDet) and achieve state-of-the-art. The model architecture includes LSTM, TreeLSTM, and translation embedding. Furthermore, the training process can be divided into supervised learning and reinforcement learning. For the MOTIFS model, our method achieves 25.0% higher on mR@100 for PredCls. Compared with other plug-and-play methods, our method outperforms all of them in nearly all metrics. For VTransE model, in particular, which has the

Table 1. Performance (%) of our method and other state-of-the-art models on VG-150. The re-implemented models under codebase of [29] are denoted by †.

Models	Predicate classification			Scene graph classification			Scene graph detection		
	mR@20	mR@50	mR@100	mR@20	mR@50	mR@100	mR@20	mR@50	mR@100
FC-SGG [23]	4.9	6.3	7.1	2.9	3.7	4.1	2.7	3.6	4.2
KERN [3]	-	17.7	19.2	-	9.4	10.0	-	6.4	7.3
GBNet [41]	-	22.1	24.0	-	12.7	13.4	-	7.1	8.5
BA-SGG [12]	26.7	31.9	34.2	15.7	18.5	19.4	11.4	14.8	17.1
PCPL [34]	-	35.2	37.8	-	18.6	19.6	-	9.5	11.7
BGNN [16]	-	30.4	32.9	-	14.3	16.5	-	10.7	12.6
GPS-Net [22]	-	19.2	21.4	-	11.7	12.5	-	7.4	9.5
MOTIFS† [42]	11.5	14.6	15.8	6.5	8.0	8.5	4.1	5.5	6.8
TDE [20]	18.5	25.5	20.1	0.8	13.1	14.9	5.8	8.2	9.8
-CogTree [39]	20.9	26.4	29.0	12.1	14.9	16.1	7.9	10.4	11.8
-CPAM+HDA (ours)	29.3	37.3	40.8	17.3	20.8	22.2	9.7	12.2	13.7
VCTree† [30]	11.7	14.9	16.1	6.2	7.5	7.9	4.2	5.7	6.9
-TDE [29]	18.4	25.4	28.7	8.9	12.2	14.0	6.9	9.3	11.1
-CogTree [39]	22.0	27.6	29.7	15.4	18.8	19.9	7.8	10.4	12.1
-CPAM+HDA (ours)	25.9	33.5	38.2	17.7	22.9	26.0	9.7	11.7	13.8
VTransE† [45]	11.6	14.7	15.8	6.7	8.2	8.7	3.7	5.0	6.0
-TDE [29]	18.9	25.3	28.4	9.8	13.1	14.7	6.0	8.5	10.2
-CPAM+HDA (ours)	25.9	34.6	38.7	15.1	18.9	21.4	10.3	14.6	17.1

weakest performance for SGDet in baseline, our method has made tremendous improvement with 65.0% on mR@100.

Meanwhile, compared with other strong baselines, our method still achieve competitive performance when compared with state-of-the-art KERN [3], GBNet [41], BA-SGG [12], PCPL [34], BGNN [16] and GPS-Net [22]. Considering SGDet, our method is slightly lower than BA-SGG, while performs best in Pred-Cls and SGCls.

On the other hand, model performance on zR@K is also evaluated as Table 2 shown. The focal loss [21], reweight [34] and resample [7, 16] are chosen as conventional plug-and-play debiasing methods, while EBM [28] as a debiasing method using deep learning model. Compared with these debiasing methods, our method outperforms them. Specifically, for MOTIFS, our method performs almost five times better than the baseline, changing the embarrassing result that almost no unseen triplets are recognized on zR@20.

4.4 Ablation Studies

Model Components. CPAM is proposed to solve the conflict problem among causal attributes, and HDA to deal with the imbalance distribution in VG-150. To prove the effectiveness of the above components, ablation studies are performed to get a clear sense of how these different components affect final performance. Experiments on VCTree [30] are performed, and the results are reported in Table 3. w/o-HDA represents models without HDA, and only CPAM works. We observe that all the metrics are improved, which means the existence of

Table 2. Performance (%) of Zero-Shot Recall (zR@K) of baseline model and model using our method on VG-150.

Models	Scene graph detection		
	zR@20	zR@50	zR@100
MOTIFS [42]	0.0	0.1	0.2
- Focal [21]	-	0.1	0.3
- Resample	-	0.1	0.3
- Reweight	-	0.0	0.0
- EMB [28]	0.2	0.3	-
- CPAM+HDA (ours)	0.4	0.6	1.0
VCTree [30]	0.2	0.5	0.7
- EMB [28]	0.3	0.6	-
- CPAM+HDA (ours)	0.6	1.1	1.5

causal property conflict while training and our proposed CPAM can effectively eliminate it. Meanwhile, CPAM improves the performances of all metrics. w/o-CPAM represents models evaluating only with HDA. We also find the performance outperforms the baseline with a large margin, which means the imbalance distribution (i.e. long-tailed bias) impairs model performance extremely.

To sum up, the necessity of all components are demonstrated. Only when these two components are applied at the same time, the model may achieve the best performance.

Fusion Types for CPAM. As aforementioned, the two fusion types for CPAM are proposed: binary fusion and triple fusion. Binary fusion fuses visual distribution and context distribution, with the frequency distribution added. Triple fusion fuses all of the three causal attribute distributions simultaneously. As shown in Table 4, the performance of these two fusion types is compared on several baseline models.

Intuitively, the more causal attributes fuse, the better performance a model could behave. However, we observe an opposite result: triple fusion is generally weaker than binary fusion. Therefore, the reasons are shown as follows: (1) As prior knowledge, the frequency distribution is a fixed attribute, while visual and context distribution is uncertain. If certain attributes and uncertain attributes are forced to combine, it will violate the premise of D-S evidence theory, introducing extra bias at the same time. (2) CPAM is applied at the end of a model to calculate the joint distribution of the three attributes directly, and this will lead to a slight gradient vanishing through backward propagation. The fact is that the fusion of three pieces of evidence is more computationally intensive than the fusion of two, so the problem of gradient vanishing will be enlarged.

In a word, the performance of binary fusion with triple fusion is compared and we conclude that binary fusion is better. Also, the experimental results are analyzed and the reasons for them are summarized.

Table 3. Ablation study of model components.

	PredCls		SGCls		SGDet	
	mR@50	mR@100	mR@50	mR@100	mR@50	mR@100
Baseline	14.9	16.1	7.5	7.9	5.7	6.9
w/o-HDA	17.4	18.9	11.1	11.9	7.3	8.7
w/o-CPAM	30.3	33.9	16.5	18.1	11.5	14.0
CPAM+HDA	33.5	38.2	22.9	26.0	11.7	14.1

Table 4. Ablation study of CPAM using different fusion types.

		PredCls		SGCls		SGDet	
		mR@50	mR@100	mR@50	mR@100	mR@50	mR@100
MOTIFS [42]	Triple	15.1	16.4	8.0	9.0	6.7	8.1
	Binary	15.6	16.9	9.0	9.6	6.9	8.2
VTransE [45]	Triple	15.6	16.8	8.0	8.5	6.0	7.1
	Binary	16.7	18.0	8.9	9.5	6.4	7.6
VCTree [30]	Triple	15.7	17.8	9.1	9.7	5.6	6.5
	Binary	17.4	18.9	11.1	11.9	7.3	8.7

4.5   Qualitative Studies

Several testing examples of VTransE on the PredCls subtask are visualized. As Fig. 5 shown, our method generates more fine-grained relationships such as "people-near-train" v.s. "people-looking at-train" in the first row and "tree-near-building" v.s. "tree-in front of-building" in the second row. Take predicate *on*, for example, it is divided into *painted on, parked on, walking on*, etc., and the latter predicates are more semantically informative instead of trivial and common. Head predicates such as *has, of, in,* and *near* rarely appear in predictions and the performance of tail predicates is promoted.

Furthermore, R@100 of each predicate category is calculated for baseline and our method as Fig. 6 shown. It can proved our inference proposed before that long-tailed distribution is alleviated. Specifically, we observe the drop of head predicate and substantial improvement of the tail.

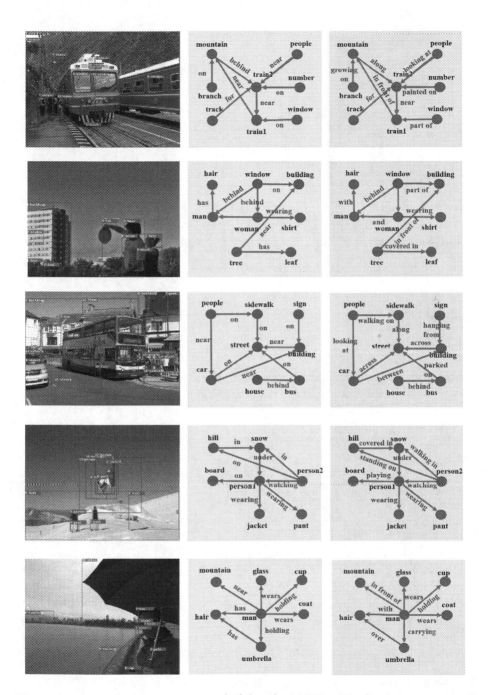

Fig. 5. Qualitative results of VTransE [45] (gray) and VTransE adopting our proposed method (orange). (Color figure online)

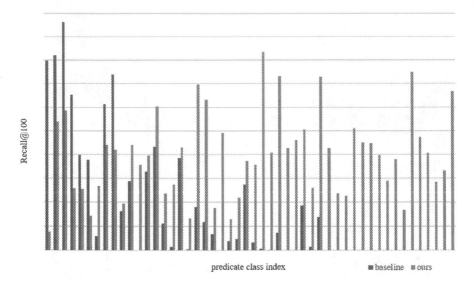

Fig. 6. R@100 of all the predicate classes of baseline and our method on VG-150.

5   Conclusion

In this work, the problems are analyzed in SGG, which could be attributed to long-tailed bias, and an extra bias caused by conflicts of different causal branches. To address the problems mentioned above, a novel plug-and-play CPAM module is proposed, applying the Dempster-Shafer evidence theory to eliminate conflict among causal attributions. The predicted distribution of each branch is regarded as evidence and Dempster combination rules are employed to model the uncertainty. Meanwhile, a HDA method is introduced to get an enhanced dataset by integrating IETrans and conventional debiasing strategies. Through extensive comparative experiments, our method achieved state-of-the-art performance under the existing evaluation metric mR@K of the three subtasks (PredCls, SGCls, SGDet), either model-agnostic baselines or specific baselines.

Furthermore, comprehensive ablation studies are conducted to demonstrate the universal effectiveness of CPAM and HDA, and the effectiveness of both model components are validated. For CPAM, we do more detailed experiments. Two fusion types of CPAM are proposed and the performance of binary fusion type and triple fusion type is compared. Then we show scene graphs generated by baseline and our method, and the results demonstrate our method can perform more fine-grained predictions and relieve long-tailed distribution effectively.

Acknowledgement. This work was supported by the National Key R&D Program of China (2019YFB2204200) and the National Natural Science Foundation of China (62006015 and 62072028).

References

1. Barnett, J.A.: Computational methods for a mathematical theory of evidence. In: Yager, R.R., Liu, L. (eds.) Classic Works of the Dempster-Shafer Theory of Belief Functions, pp. 197–216. Springer, Heidelberg (2008). https://doi.org/10.1007/978-3-540-44792-4_8
2. Burnaev, E., Erofeev, P., Papanov, A.: Influence of resampling on accuracy of imbalanced classification. In: Eighth International Conference on Machine Vision (ICMV 2015), vol. 9875, pp. 423–427. SPIE (2015)
3. Chen, T., Yu, W., Chen, R., Lin, L.: Knowledge-embedded routing network for scene graph generation. In: Proceedings of the IEEE/CVF Conference on Computer Vision and Pattern Recognition, pp. 6163–6171 (2019)
4. Dempster, A.P.: A generalization of Bayesian inference. J. Roy. Stat. Soc. Ser. B (Methodol.) **30**(2), 205–232 (1968)
5. Dempster, A.P.: Upper and lower probabilities induced by a multivalued mapping. In: Yager, R.R., Liu, L. (eds.) Classic Works of the Dempster-Shafer Theory of Belief Functions, pp. 57–72. Springer, Heidelberg (2008). https://doi.org/10.1007/978-3-540-44792-4_3
6. Denœux, T., Zouhal, L.M.: Handling possibilistic labels in pattern classification using evidential reasoning. Fuzzy Sets Syst. **122**(3), 409–424 (2001)
7. Desai, A., Wu, T.Y., Tripathi, S., Vasconcelos, N.: Learning of visual relations: the devil is in the tails. In: Proceedings of the IEEE/CVF International Conference on Computer Vision, pp. 15404–15413 (2021)
8. Dong, M., Peng, J., Ding, S., Wang, Z.: Vision and EMG information fusion based on DS evidence theory for gesture recognition. In: Deng, Z. (ed.) Proceedings of 2021 Chinese Intelligent Automation Conference. LNEE, vol. 801, pp. 492–501. Springer, Singapore (2022). https://doi.org/10.1007/978-981-16-6372-7_55
9. Dong, X., Gan, T., Song, X., Wu, J., Cheng, Y., Nie, L.: Stacked hybrid-attention and group collaborative learning for unbiased scene graph generation. In: Proceedings of the IEEE/CVF Conference on Computer Vision and Pattern Recognition, pp. 19427–19436 (2022)
10. Florea, M.C., Jousselme, A.L., Bossé, É., Grenier, D.: Robust combination rules for evidence theory. Inf. Fusion **10**(2), 183–197 (2009)
11. Gordon, J., Shortliffe, E.H.: A method for managing evidential reasoning in a hierarchical hypothesis space. Artif. Intell. **26**(3), 323–357 (1985)
12. Guo, Y., et al.: From general to specific: informative scene graph generation via balance adjustment. In: Proceedings of the IEEE/CVF International Conference on Computer Vision, pp. 16383–16392 (2021)
13. Jiang, W.: A correlation coefficient for belief functions. Int. J. Approx. Reason. **103**, 94–106 (2018)
14. Krishna, R., et al.: Visual genome: connecting language and vision using crowd-sourced dense image annotations. Int. J. Comput. Vision **123**(1), 32–73 (2017)
15. Li, B., Han, Z., Li, H., Fu, H., Zhang, C.: Trustworthy long-tailed classification. In: Proceedings of the IEEE/CVF Conference on Computer Vision and Pattern Recognition, pp. 6970–6979 (2022)
16. Li, R., Zhang, S., Wan, B., He, X.: Bipartite graph network with adaptive message passing for unbiased scene graph generation. In: Proceedings of the IEEE/CVF Conference on Computer Vision and Pattern Recognition, pp. 11109–11119 (2021)
17. Li, Y., Ouyang, W., Wang, X., Tang, X.: VIP-CNN: visual phrase guided convolutional neural network. In: Proceedings of the IEEE Conference on Computer Vision and Pattern Recognition, pp. 1347–1356 (2017)

18. Li, Y., Ouyang, W., Zhou, B., Shi, J., Zhang, C., Wang, X.: Factorizable net: an efficient subgraph-based framework for scene graph generation. In: Proceedings of the European Conference on Computer Vision (ECCV), pp. 335–351 (2018)

19. Li, Y., Ouyang, W., Zhou, B., Wang, K., Wang, X.: Scene graph generation from objects, phrases and region captions. In: Proceedings of the IEEE International Conference on Computer Vision, pp. 1261–1270 (2017)

20. Liao, S.H.: Expert system methodologies and applications-a decade review from 1995 to 2004. Expert Syst. Appl. **28**(1), 93–103 (2005)

21. Lin, T.Y., Goyal, P., Girshick, R., He, K., Dollár, P.: Focal loss for dense object detection. In: Proceedings of the IEEE International Conference on Computer Vision, pp. 2980–2988 (2017)

22. Lin, X., Ding, C., Zeng, J., Tao, D.: GPS-Net: graph property sensing network for scene graph generation. In: Proceedings of the IEEE/CVF Conference on Computer Vision and Pattern Recognition, pp. 3746–3753 (2020)

23. Liu, H., Yan, N., Mortazavi, M., Bhanu, B.: Fully convolutional scene graph generation. In: Proceedings of the IEEE/CVF Conference on Computer Vision and Pattern Recognition, pp. 11546–11556 (2021)

24. Lu, C., Krishna, R., Bernstein, M., Fei-Fei, L.: Visual relationship detection with language priors. In: Leibe, B., Matas, J., Sebe, N., Welling, M. (eds.) ECCV 2016. LNCS, vol. 9905, pp. 852–869. Springer, Cham (2016). https://doi.org/10.1007/978-3-319-46448-0_51

25. Qi, S., Zhu, Y., Huang, S., Jiang, C., Zhu, S.C.: Human-centric indoor scene synthesis using stochastic grammar. In: Proceedings of the IEEE Conference on Computer Vision and Pattern Recognition, pp. 5899–5908 (2018)

26. Ren, S., He, K., Girshick, R., Sun, J.: Faster R-CNN: towards real-time object detection with region proposal networks. In: Advances in Neural Information Processing Systems, vol. 28 (2015)

27. Shafer, G.: A mathematical theory of evidence. In: A Mathematical Theory of Evidence. Princeton University Press, Princeton (2021)

28. Suhail, M., et al.: Energy-based learning for scene graph generation. In: Proceedings of the IEEE/CVF Conference on Computer Vision and Pattern Recognition, pp. 13936–13945 (2021)

29. Tang, K., Niu, Y., Huang, J., Shi, J., Zhang, H.: Unbiased scene graph generation from biased training. In: Proceedings of the IEEE/CVF Conference on Computer Vision and Pattern Recognition, pp. 3716–3725 (2020)

30. Tang, K., Zhang, H., Wu, B., Luo, W., Liu, W.: Learning to compose dynamic tree structures for visual contexts. In: Proceedings of the IEEE/CVF Conference on Computer Vision and Pattern Recognition, pp. 6619–6628 (2019)

31. Wang, X., Wang, T.: Research on face recognition algorithm based on ds evidence theory and local domain pattern. In: 2021 International Conference on Intelligent Computing, Automation and Applications (ICAA), pp. 261–266. IEEE (2021)

32. Xu, D., Zhu, Y., Choy, C.B., Fei-Fei, L.: Scene graph generation by iterative message passing. In: Proceedings of the IEEE Conference on Computer Vision and Pattern Recognition, pp. 5410–5419 (2017)

33. Xu, Z., Zhang, B., Fu, H., Yue, X., Lv, Y.: Multi-branch recurrent attention convolutional neural network with evidence theory for fine-grained image classification. In: Denœux, T., Lefèvre, E., Liu, Z., Pichon, F. (eds.) BELIEF 2021. LNCS (LNAI), vol. 12915, pp. 177–184. Springer, Cham (2021). https://doi.org/10.1007/978-3-030-88601-1_18

34. Yan, S., et al.: PCPL: predicate-correlation perception learning for unbiased scene graph generation. In: Proceedings of the 28th ACM International Conference on Multimedia, pp. 265–273 (2020)
35. Yang, G., Zhang, J., Zhang, Y., Wu, B., Yang, Y.: Probabilistic modeling of semantic ambiguity for scene graph generation. In: Proceedings of the IEEE/CVF Conference on Computer Vision and Pattern Recognition, pp. 12527–12536 (2021)
36. Yang, J.B., Liu, J., Wang, J., Sii, H.S., Wang, H.W.: Belief rule-base inference methodology using the evidential reasoning approach-RIMER. IEEE Trans. Syst. Man Cybern.-Part A: Syst. Hum. **36**(2), 266–285 (2006)
37. Yang, J.B., Xu, D.L.: Evidential reasoning rule for evidence combination. Artif. Intell. **205**, 1–29 (2013)
38. Yang, X., Tang, K., Zhang, H., Cai, J.: Auto-encoding scene graphs for image captioning. In: Proceedings of the IEEE/CVF Conference on Computer Vision and Pattern Recognition, pp. 10685–10694 (2019)
39. Yu, J., Chai, Y., Wang, Y., Hu, Y., Wu, Q.: Cogtree: cognition tree loss for unbiased scene graph generation. arXiv preprint arXiv:2009.07526 (2020)
40. Zadeh, L.A.: On the validity of Dempster's rule of combination of evidence. Infinite Study (1979)
41. Zareian, A., Karaman, S., Chang, S.-F.: Bridging knowledge graphs to generate scene graphs. In: Vedaldi, A., Bischof, H., Brox, T., Frahm, J.-M. (eds.) ECCV 2020. LNCS, vol. 12368, pp. 606–623. Springer, Cham (2020). https://doi.org/10.1007/978-3-030-58592-1_36
42. Zellers, R., Yatskar, M., Thomson, S., Choi, Y.: Neural motifs: scene graph parsing with global context. In: Proceedings of the IEEE Conference on Computer Vision and Pattern Recognition, pp. 5831–5840 (2018)
43. Zhang, A., et al.: Fine-grained scene graph generation with data transfer. arXiv preprint arXiv:2203.11654 (2022)
44. Zhang, C., Chao, W.L., Xuan, D.: An empirical study on leveraging scene graphs for visual question answering. arXiv preprint arXiv:1907.12133 (2019)
45. Zhang, H., Kyaw, Z., Chang, S.F., Chua, T.S.: Visual translation embedding network for visual relation detection. In: Proceedings of the IEEE Conference on Computer Vision and Pattern Recognition, pp. 5532–5540 (2017)

gScoreCAM: What Objects Is CLIP Looking At?

Peijie Chen[1(✉)], Qi Li[1], Saad Biaz[1], Trung Bui[2], and Anh Nguyen[1]

[1] Auburn University, Auburn, USA
peijiechen@gmail.com, {qzl0019,biazsaa}@auburn.edu
[2] Adobe Research, San Jose, USA
bui@adobe.com

Abstract. Large-scale, multimodal models trained on web data such as OpenAI's CLIP are becoming the foundation of many applications. Yet, they are also more complex to understand, test, and therefore align with human values. In this paper, we propose gScoreCAM—a state-of-the-art method for visualizing the main objects that CLIP is looking at in an image. On zero-shot object detection, gScoreCAM performs similarly to ScoreCAM, the best prior art on CLIP, yet 8 to 10 times faster. Our method outperforms other existing, well-known methods (HilaCAM, RISE, and the entire CAM family) by a large margin, especially in multi-object scenes. gScoreCAM sub-samples $k = 300$ channels (from 3,072 channels—i.e. reducing complexity by almost 10 times) of the highest gradients and linearly combines them into a final "attention" visualization. We demonstrate the utility and superiority of our method on three datasets: ImageNet, COCO, and PartImageNet. Our work opens up interesting future directions in understanding and de-biasing CLIP.

1 Introduction

Large-scale, multimodal neural networks trained on web data are becoming important "foundation models" [1] for academic research, industry applications, and the wider society. Within only one year since release, OpenAI's CLIP [2], which learns to match captions and images, has powered a multitude of applications [3], including text-to-image synthesis [4–6], video retrieval [7,8], visual question answering [9], and image editing [10,11]. As foundation models are scaled up larger and becoming more ubiquitous, it is imperative to understand how they work internally to ensure safe deployment, avoid unexpected harms due to biases [12–14], and also further improve the models' functionality [15]. However, the inner-workings of foundation models (here, CLIP) are still largely unknown. For example, it is intriguing why the highly-knowledgeable CLIP model is fooled by a simple piece of paper with text [16] (see Fig. 4). In a complex, many-object scene (Fig. 1), which objects are *important* to CLIP?

Existing ViT-based interpretability methods [17–19] visualize the similarities between image and text tokens of vision-language models. Yet, they only

---

Supplementary Information The online version contains supplementary material available at https://doi.org/10.1007/978-3-031-26316-3_35.

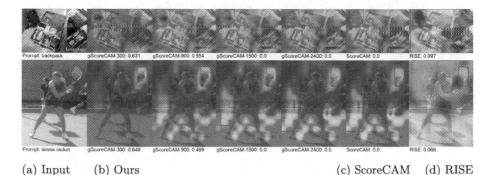

(a) Input   (b) Ours   (c) ScoreCAM   (d) RISE

Fig. 1. In a complex COCO scene, SotA feature importance methods often produce noisy heatmaps for CLIP RN50x16, questioning what objects are the most important to CLIP. Here, RISE [24] heatmap covers both suitcases and the "baggage" text (top row) while ScoreCAM [27] highlights both the racket and the tennis court (bottom row), yielding a poor 0.0 IoU between the highlighted region ☐ and the ground truth box ☐. By using only the top-300 channels of the highest gradients, we (1) localize *the most important* objects in a complex scene (e.g., here, racket at IoU of 0.649); and (2) produce a SotA zero-shot, open-vocabulary, object localization method for COCO and PartImageNet. (Color figure online)

perform well on single-object images [17,18,20] or require cross-attention [21], which does not exist in CLIP ViTs [2]. On the other hand, applying well-known saliency methods [22–24] for convolutional networks (CNNs) to CLIP ResNets often yields noisy heatmaps (see Figs. 1 and 3) perhaps because CLIP neurons are highly multifaceted [16,25] and responsive to a wide variety of information in an image, including text [16,26]. We find ScoreCAM [27] to be the highest-accuracy prior art in CLIP-based, zero-shot object localization (Table 4) but also prohibitively slow. For an input image, ScoreCAM requires $\geq 3{,}072$ forward passes through CLIP RN50x16 to compute the CLIP scores, i.e., the coefficients for linearly combining 3,072 layer4 channels into a heatmap. This overhead is even higher for CLIP RN50x64, which has 4,096 layer4 channels.

To overcome this problem, we propose gScoreCAM, a simple-yet-effective technique for reducing ScoreCAM's theoretical time complexity by 10× by using only the top $k = 300$ most important (i.e., highest-gradient) channels instead of all 3,072 layer4 channels of CLIP RN50x16. By using the 10% most important channels, gScoreCAM serves as (1) an interpretability technique for localizing the *most important* objects in a complex scene (Fig. 1; backpack); (2) a state-of-the-art (SotA) *zero*-shot, open-vocabulary, object localization method that does not rely on any CLIP finetuning or object detectors. We find that:[1]

1. Compared to the CAM-based family, which proposes alternative sets of coefficients for linearly combining all layer4 channels into a single heatmap, gScore-CAM is the most accurate, zero-shot object localization method for CLIP on 2017 COCO (Sect. 4).

---

[1] Code and an interactive demo are at https://github.com/anguyen8/gScoreCAM.

2. gScoreCAM performs comparably to ScoreCAM, the best prior art on zero-shot object detection using CLIP, but is ~8× faster in practice (Sect. 5.1).
3. gScoreCAM is around 2 to 4× more accurate than the other methods (excluding ScoreCAM) on three different datasets: ImageNet [28], COCO [29], and PartImageNet [30] (Sect. 5.1). In particular, our method performs better in the harder cases, i.e., localizing an object in multi-object scenes or localizing a part of an object (Sect. 5.2).
4. For both RN50x4 and RN50x16 versions of CLIP, gScoreCAM is a SotA localization method that does not require finetuning CLIP or any specific training to object localization (Sect. 5.3).

2   Related Work

Visualizing Multimodal Networks. As vision-language Transformers become increasingly more popular, many recent methods [17–19] were proposed to visualize the attention of these models or the similarity between image and text tokens [21]. Some methods [21] require cross-attention, which does not exist in both CLIP ViTs and ResNets because cross-modal attention does not allow caching of image or text embeddings and therefore admits a much slower retrieval speed in practice. Other interpretability methods that do not require cross-attention only perform well on single-object image crops [17,18,20,31]. That is, applying the SotA ViT feature importance method (i.e., HilaCAM [17]) on CLIP ViT-B/32 yields a localization accuracy 1.5× *worse* than when applying gScoreCAM on CLIP RN50x16 (12.82 vs. 20.83; Table 4).

Feature Importance Methods for CNNs. As ViT visualization methods are either not applicable to or performing poorly on CLIP ViTs, an alternative technique for interpreting CLIP is applying feature importance methods for CNNs [22,23,32] to CLIP ResNet-50 models. These methods can be grouped into two categories: **(1) white-box** i.e., using gradients or activation maps (a.k.a. channels) or both to derive an attribution map e.g.[22,23,27,32]; and **(2) black-box**, i.e. relying on perturbation-based analysis to compute the attribution of each input feature [24,33,34]. In the white-box group, visualizing the image gradients alone often yields noisy heatmaps [35,36]. CAM-based methods linearly combine the channels at the last convolutional layer in CNNs into a single "attention" heatmap. Because CLIP ResNets do not have a global average pooling (GAP) layer followed by the last classification layer, the original CAM [32] is not applicable to CLIP ResNets. Instead, one needs to resort to other CAM-based methods that compute the channel coefficients differently, e.g.using gradients [22,23] or confidence scores [27]. Compared to the CAM-based family, our method is the SotA in CLIP-based zero-shot, object localization and uses both gradients and scores for linearly combining the channels. Like ScoreCAM [27], our gScoreCAM also uses CLIP scores generated for masked images and a prompt to compute a channel's importance weight; however, we apply ScoreCAM on only the top-10% channels, discarding the rest. RISE [24] is a black-box version of ScoreCAM where it generates a random mask instead of directly leveraging an activation map as a mask. Our gScoreCAM outperforms RISE in both efficiency and localization accuracy.

3   Methods

We first describe the original Class Activation Map (CAM) method [32] and then ScoreCAM [27] before introducing our gScoreCAM, which extends ScoreCAM.

3.1   Revisiting CAM and ScoreCAM

CAM. [32] is applied to all N channels $\{A_i\}^N$ at the last convolutional layer (e.g.layer4 at ResNet-50 [37]) that is followed by a GAP layer and then a linear 1000-output classification layer (whose weight matrix $\in \mathbb{R}^{N \times 1000}$). That is, for each ImageNet class c, CAM uses the N corresponding weights $\{w_i^c\}^N$ to linearly combine N channels to create a saliency map M_{CAM}^c:

$$M_{\mathrm{CAM}}^c = \sum_i^N A_i \times softmax(w_i^c) \tag{1}$$

ScoreCAM. [27] is the same as CAM but uses the confidence scores of the CNN in place of the weights w_i^c in Eq. 1. Specifically, to explain why an input image $\boldsymbol{x}$ belongs to a target class c w.r.t. a CNN $f_c(.)$, the ScoreCAM algorithm is:

1. Upsample all N channels at the last convolutional layer to the input-image size using bilinear interpolation, yielding a set of upsampled channels $\{A_i^{\mathrm{up}}\}^N$, which serve as masks in the next step.
2. Element-wise multiply each mask A_i^{up} with all color channels of the input image $\boldsymbol{x}$ and feed the resultant (masked) images to the CNN $f_c(.)$ to obtain an output confidence score corresponding to the target class c.
3. Use the N confidence scores obtained in place of the w_i^c to compute M_{CAM}^c following Eq. 1.

In sum, the ScoreCAM saliency map is computed by:

$$M_{\mathrm{ScoreCAM}}^c = \sum_i^N A_i \times softmax(f_c(\boldsymbol{x} \odot A_i^{\mathrm{up}})) \tag{2}$$

where $\odot$ is the Hadamard product.

3.2   Proposed Method: Gradient-Guided ScoreCAM (gScoreCAM)

While performing fairly accurately with CLIP CNNs (Table 4), a major drawback of ScoreCAM is its prohibitively slow runtime as it requires many (e.g., $N = 3{,}072$) forward passes to generate a single saliency map when CLIP RN50x16 has $N = 3{,}072$ layer4 channels. A second problem is that since the convolutional channels inside CLIP CNNs tend to react to a diverse variety of details (including both graphical and textual ones) present in the input image [16], using *all* the channels (as in CAM and ScoreCAM) often yields noisy heatmaps.

To address these two problems, we propose to apply ScoreCAM but only to the top-k (e.g., $k = 300$ i.e. only $\sim 10\%$) of the channels of the CLIP's image encoder. This effectively reduces the runtime by almost $10\times$ and yields more object-focused heatmaps. We choose the top-k channels by ranking them using their mean gradients over three color channels, which we empirically find to be the best proxy for *importance* among common channel-ranking criteria (Sect. 4.1).

Our full algorithm for running gScoreCAM on CLIP is:

Algorithm 1. Explaining CLIP (for both ResNet- & ViT-based CLIP encoders)

Input: Input image $\boldsymbol{x} \in \mathbb{R}^{W \times H \times C}$, a text prompt P, and a target layer L (in CNNs, L is the last convolutional layer) whose activation maps are $\in \mathbb{R}^{W' \times H'}$.

Output: A heatmap $M \in \mathbb{R}^{W' \times H'}$.

1: Run 1 forward pass through CLIP $(\boldsymbol{x}, P)$ to get N channels $\{A_i\}^N$ at layer L
2: Run 1 backward pass to get the N channel gradients at layer L
3: Take the top-k largest-gradient channels and use them as masks (as in ScoreCAM) to generate k masked images, i.e. $\{\boldsymbol{x}_i^*\}_k$ where $k \ll N$ and $\boldsymbol{x}_i^* = \boldsymbol{x} \odot A_i^{\mathrm{up}}$.
4: Run k forward passes through CLIP $(\boldsymbol{x}_i^*, P)$ to obtain k CLIP scores.
5: Use the k CLIP scores as the coefficients in place of w_i^c in Eq. 1 to linearly combine the k channels into a single heatmap (i.e., discarding all N-k other channels).

For example, in CLIP RN50x16, L is the last convolutional layer whose the output volume is $12 \times 12 \times 3,072$ (i.e. $W' = H' = 12$ and $N = 3,072$). For ViT-B/32, L is the penultimate layer of size $7 \times 7 \times 768$ (more details in Sect. 3.5).

3.3 Evaluation Datasets

To thoroughly understand our method's localization capability, we test it on three localization benchmarks of increasing granularity: (1) localizing the main object in a single-object image (ImageNet-v2 [38]); (2) localizing one object in multi-object scenes (2017 MS COCO [29]); and (3) localizing an object part in a single-object image (PartImageNet [30]).

ImageNet-v2 is a re-make version of the original 1000-class ImageNet [28] attempting to understand how well existing algorithms overfit to the common ImageNet and generalize to a new reproduction of it. We use all 10,000 images in **train-fullsup** set (a.k.a **val2** in the data preparation script) annotated by Choe et al. [39].

COCO is an 80-class dataset designed for object detection and segmentation. COCO images often include multiple objects of different sizes and locations, making object detection more complex than ImageNet-v2. Therefore, we expect a significant contrast between different methods in the object detection of COCO. We use all 50,000 images in **val2017** set for our evaluation.

PartImageNet is a variance of ImageNet, where the selected images are further labeled at the part level. That is, it has a hierarchy label. e.g., in the bird class, the bird will be further labeled to part level: head, body, wing, foot, and tail. It consists of 24,000 images with 11 super-class. The object detection task in PartImageNet is challenging; it requires the attribution method to capture precisely the model's attention for each part of the object. We use all 4,598 images in **test** set for evaluation.

3.4 Evaluation Metrics

To assess how well a saliency method localizes an object or an object part in an image, we perform **Z**ero-**S**hot object **D**etection (ZSD) on the three datasets described in Sect. 3.3. Specifically, we first binarize the saliency map, derive a predicted bounding box, and compare it with the ground truth bounding box under two common metrics: BoxAcc [40] and MaxBoxAccV2 [39].

Inferring Bounding Box from a Heatmap.[2] A common method for deriving the bounding box from a heatmap is to maximize task performance by grid searching for binarization threshold. An alternative approach, Otsu's method, is a non-task-specific binarization method, which tries to maximize foreground and background contrast after binarization. Following Chefer et al. [17], we use Otsu's method to binarize the heatmap in COCO and ParImageNet evaluation and use the grid search approach in ImageNet-v2, which is recommended by Choe et al. [39]. Note that the choice of binarization method is trivial to the ZSD results (see Sec. A2 for more details).

BoxAcc Metric. For COCO and PartImageNet, we use BoxAcc [40], which measures the following accuracy:

$$BoxAcc(\tau, \delta) := \frac{1}{M} \sum_m 1_{IoU(box(h,\tau)_m, B_m) \geq \delta} \tag{3}$$

where $\tau \in [0, 1]$ is the binarization threshold, h is the saliency generated by the model, $box(h, \tau)$ is the tightest box around the largest-area connected component of the binarized saliency with threshold τ, B is the ground truth box, $m \in M$ is the box index, and δ is the IoU threshold, we use $\delta = 0.5$ in our experiments. Note that the binarization threshold τ will be determined by Otsu's method in our evaluation.

MaxBoxAccV2 Metric. MaxBoxAccV2 [39] is defined as:

$$MaxBoxAccV2(\delta) := \frac{1}{3} \sum_\delta \max_\tau \left(BoxAcc(\tau, \delta)\right) \tag{4}$$

where $\delta \in \{0.3, 0.5, 0.7\}$, $\tau \in [0 : 0.05 : 0.95]$ is the binarization threshold to binarize saliency maps for evaluation against the binary ground truth masks. We use the default hyperparameters from the authors.

---

[2] We provide a detailed description for Otsu-based bounding box inferencing in Sec. A1.

Intersection over Union (IoU) and Area Under the Curve (AUC). We measure the IoU at threshold $\delta = 0.5$ (Sect. 5.2) and the AUC of the IoU over different binarization thresholds (Sect. 4.2) to better understand the differences between multiple visualization methods.

Computation Resources. We conduct our experiments on 8 Nvidia RTX 1080Ti and 3 Nvidia Tesla T4 GPUs.

3.5   CLIP Networks

Model and Methods. We conduct all our experiments based on the CLIP [2] model. We use the pre-trained model **RN50x16** for all CNN-based methods (except for the experiment in Sect. 5.3, which also includes **RN50x4**) as it provides the best performance among the convolutional-based variances available. For comparison, we also apply the CNN-based saliency method to CLIP **ViT-B/32** by reshaping the embedding of the target layer.

Target Layers for Visualization. Similar to the idea in the CAM-based family, for interpretability, in both ResNet and ViT versions of the image encoder, we choose the valid channel closest to the CLIP prediction layer:

- RN50x16 and RN50x4: We use relu3 of the last BottleNeck in layer4, which is the last layer of the image encoder in CLIP. RN50x16 has 3072 channels with spatial dimension 12×12. RN50x4 has 2560 channels with spatial dimension 9×9.
- ViT-B/32: We use the second-last ResidualAttentionBlock in VisionTransformer. The output dimension is: $50 \times 1 \times 768$, we exclude the [CLS] vector then reshape into $7 \times 7 \times 768$ for CAM-based visualizations.

We choose the second-to-last layer in ViT-B/32 because the gradients in the last layer are zero except for the [CLS] vector, and, only the [CLS] in the last layer embedding is used for the final prediction.

The implementation of our attribution methods for CNNs is based on PyTorchCAM [41]. The CLIP models we used are from OpenAI [42]. Our Hila-CAM implementation is from the code released by [17].

Model Hyperparameters. We list some key hyperparameters of the models in Table S2. More hyperparameters can be found in Table 19 of Radford et al. [2].

Prompts of CLIP. For ImageNet-v2 and COCO, we directly use "{class name}" (without quotation marks) as the prompt. For PartImageNet, we use "{class name} {part name}" as the prompt.

4   Design of gScoreCAM

In this section, we first introduce two ablation studies (Sect. 4.1) to explain the choices of hyperparameter k and the dimension reduction technique of the gradients. Secondly, we study the weight quality of CAM-based methods by measuring the level of overlap between the target and the activation maps (Sect. 4.2).

We find that gScoreCAM has the best weight quality among the methods that directly weigh the activation maps (Table 3). Lastly, we measure the noise level of the weighted activation map with Total Variation [43] (Sect. 4.3). Note that the noise level can not directly reflect the performance of the object localization but instead indicates the *confidence* of the visualizing method.

4.1 Ablation Study of gScoreCAM

We first introduce why we set $k = 300$ in our proposed method by studying how the hyperparameter k affects Zero-Shot object Detection (ZSD) performance. We then compare the performance of ZSD using different pooling methods to rank the channels.

Effects on the Number of Channels. From Table 1, we find that gScoreCAM reaches its peak performance as the number of channels increases to 500. We choose $k = 300$ to conduct most of our experiments since it performs similarly to $k = 500$ but only needs 60% of its run-time.

Table 1. $k = 300$ is the smallest number of channels that yield a high localization accuracy on ImageNet-v2 and COCO.

Number of channels	ImageNet-v2 [38] (MaxBoxAccV2)	COCO [29] (BoxAcc)
300 (random)	55.12	18.55
20	49.83	15.72
100	54.97	19.25
200	53.75	20.50
300	**56.61**	**20.83**
400	56.60	20.89
500	**57.38**	**20.89**
600	56.55	20.89

Table 2. Taking the average of the channel-wise gradients yields the highest localization accuracy when $k = 300$.

Pooling method	ImageNet-v2 [38] (MaxBoxAccV2)	COCO [29] (BoxAcc)
Average	**56.61**	**20.83**
Max-abs	56.46	20.62
Average-abs	54.57	20.35

Effects on the Choice of Gradient Dimension Reduction. To use the gradients to guide us in choosing important channels, we first reduce the dimensions of the gradients from $\mathbb{R}^{c \times w \times h}$ to $\mathbb{R}^c$. Here, we study some of the most common methods, Average-Pooling, Average-Pooling over the **absolute** of the gradients, and Max-Pooling over the **absolute** of the gradients[3]. As shown in Table 2, the best method is simple Average-Pooling. This result coincides interestingly with the choice of GradCAM [22].

4.2 gScoreCAM Is the Best Weighting System among the Candidates

Experiment. We design an experiment to measure the quality of the weighting systems by measuring the level of overlap of the weighted activation maps and

---

[3] Method with **-abs** means operate over absolute value of the gradients.

the ground truth. Precisely, we first measure the AUC of the IoU over different binary thresholds (e.g., $\tau \in [0.0 : 0.05 : 0.95]$) for each activation map that CAM-based methods use. We then compute the weighted sum of the corresponding AUC with the weights given by the method. Finally, we average the weighted AUC of all testing samples. Note that in some methods (GradCAM, xGradCAM, GradCAM++), in which the weights are not summed to one, we first discard the negative values and divide the remaining by its sum. This measurement will provide us with the upper bound of the mean weighted AUC. We set a baseline by assuming that all activation maps have equal weights.

Results. We find that the gScoreCAM weighting is 2× better than the upper bound of GradCAM and slightly better than ScoreCAM (Table 3). The upper bound of the mean weighted AUC of GradCAM (in ImageNet-v2) is similar to the baseline, which explains why the ZSD performance of GradCAM is worse than the Gaussian noise baseline in [39].

Interestingly, we find a large gap between the IoU score of a random channel and the IoU score of the best channel (0.145 vs. 0.76). That is, a random channel may often not capture the content of the target class. It indicates that choosing the correct channels plays a vital role in visualizing the model's decision.

Table 3. We directly evaluate the weighting of different CAM-based methods and find that gScoreCAM is the best among them. The uniform baseline simply averages all the activation maps. The total variation of the heatmap provides information about how noisy the heatmap is. The lower total variation means that the heatmap tends to be less noisy.

Baseline	Mean weighted AUC	Mean total variation
	0.0380	1745 ± 268
GradCAM [22]	0.0390	885 ± 484
xGradCAM [44]	0.0421	1090 ± 657
GradCAM++ [23]	0.0357	1500 ± 458
ScoreCAM [27]	0.0881	1363 ± 391
gScoreCAM (ours)	**0.0936**	1301 ± 422

4.3   gScoreCAM is Less Noisy Compared to ScoreCAM

Experiment. We compute the mean total variation of the heatmaps generated by different methods.

Results. The total variation provides a statistical view of the noise level of the resulting heatmap from different methods. We find that gScoreCAM is statistically less noisy than ScoreCAM in Table 3. An interesting result is that Grad-CAM is the least noisy method. Although we want the resulting heatmap to be less noisy, the noise level itself can not guarantee better localization performance.

Why GradCAM is the Least Noisy Method? As discussed above, Grad-CAM provides a less noisy, lower coverage heatmap than other methods. We study the weighted activation maps of GradCAM over 4,000 random samples and find that the average resulting heatmap is entirely negative. Statistically, about 47% of the weights in GradCAM are negative. On the other hand, all the weights of gScoreCAM and ScoreCAM are positive, leading to higher total variation and a noisier heatmap. Note that the activation maps are after ReLU; therefore, all the activation maps are positive.

5 Experiments and Results

To directly compare performance between different methods, we use Zero-Shot object Detection (ZSD) to evaluate the heatmaps generated by different methods (Sect. 5.1). This evaluation provides information on how accurate the heatmap is with respect to the object. Our experiments find that gScoreCAM is the best method in COCO and PartImageNet (see Table 4 for details). To better understand the ZSD results, we further study how they perform in different scenarios, e.g., different object sizes and the number of objects in the test image (Sect. 5.2). Lastly, we conduct the same experiment as in Sect. 5.1 but for a different model (RN50x4), which reveals that our method is better than others without extra hyperparameter tuning (Sect. 5.3).

5.1 Zero-Shot Object Detection Results

Since CAM-based methods aim to detect the corresponding area of a given class or prompt, measuring the performance of ZSD will be a direct measurement of visualization results.

Table 4. CLIP zero-shot object detection results with different CAM variances. Score-CAM and gScoreCAM are similar overall, while gScoreCAM is faster. The gScoreCAM on RN-50x16 is substantially better than HilaCAM [31], a state-of-the-art method for CLIP ViT-B/32. In contrast, gScoreCAM performs on par with HilaCAM on ViT-B/32. The results are for CLIP RN50x16 unless noted (ViT-B/32).

	ImageNet-v2 [38] (MaxBoxAccV2)	COCO [29] (BoxAcc)	PartImageNet [30] (BoxAcc)	Run time (s)	Number of forward passes	Number of backward passes
GradCAM [22]	38.90	11.59	10.91	0.21	1	1
xGradCAM [44]	24.24	5.60	2.93	0.24	1	1
GradCAM++ [23]	44.15	9.68	6.57	0.39	1	1
LayerCAM [45]	43.70	9.19	12.42	0.82	1	1
GroupCAM [46]	50.85	13.06	6.16	1.99	96	1
RISE [24]	41.39	7.26	8.69	166.57	8001	0
HilaCAM [17] (ViT-B/32)	47.79	12.82	11.80	0.26	1	1
gScoreCAM (ViT-B/32)	45.26	12.73	10.67	0.84	301	1
ScoreCAM [27]	**57.78**	20.43	15.76	55.75	3073	0
gScoreCAM (ours)	56.61	**20.83**	**16.34**	7.40	301	1

ScoreCAM and gScoreCAM are the Best Methods Among the Tests.
ScoreCAM and gScoreCAM outperform other methods in the object localization
tests (Table 4). In particular, they are about 1.5 to 4× better in COCO and 1.2
to 2× better in PartImageNet compared to other methods.

gScoreCAM Runs 8 Times Faster than ScoreCAM. Since the computation of the visualization methods is model-dependent, we measure the computation overhead by its approximate elapsed time and the forward and backward
passes required by these methods. We measure the average run-time (in seconds)
for each image-prompt pair under different CAM methods and report in Table 4.
The average run-time is measured by averaging the elapsed time of 200 samples
on a single Nvidia 1080Ti GPU.

The required forward passes of ScoreCAM are 10× more than gScoreCAM,
which means the run-time is 10× longer in theory. In our approximate experiments, the actual run-time of gScoreCAM is about 8× less than ScoreCAM.

gScoreCAM Performs Better on Complex Tasks. The object localization
task on ImageNet-v2 is relatively "simple" because most test images are object-
centric, and a center-gaussian baseline reaches 52.5% accuracy, as reported in
[39]. However, the same tasks on COCO and PartImageNet are much harder due
to the variety of target sizes, shapes, and locations. Interestingly, gScoreCAM
performs better on these more complex tasks compared to ScoreCAM.

Apply gScoreCAM to ViT-Based CLIP. Although gScoreCAM is designed
for the CNN-based model, we also apply it to the ViT-based model by reshaping the embedding as discussed in Sect. 3.5. As shown in Table 4, we achieve a
similar performance as HilaCAM, which is the state-of-the-art method in ViT
visualization. One interesting note is that HilaCAM uses only attention in generating heatmaps, and our method uses only activation. Despite the enormous
difference in the approaches, both techniques end up showing similar results.

5.2   Why Does gScoreCAM Perform Better in COCO and PartImageNet?

Table 4 shows that gScoreCAM is slightly worse in ImageNet-v2 but better in
COCO and PartImageNet than ScoreCAM. We conduct two sets of controlled
experiments to find out why gScoreCAM is better in these two datasets.

5.2.1   gScoreCAM Performs Better in Multi-object Scenes Experi-
ment. We conduct a controlled experiment based on the number of objects in
an image. Specifically, we measure the average IoU of different methods when
the number of classes varies. For diversity, we select images that only have
one instance per class. This experiment uses a union of the COCO and LVIS
labels because it provides more labels for each image. We measure the mean IoU
because it directly measures the level of overlap with the object. Based on the
number of classes in the images, we split the test images into three groups: (1–3,
4–6, 7–9) classes with (1150, 2790, 874) samples correspondingly.

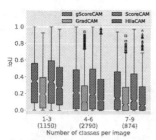 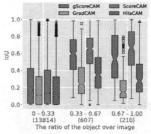 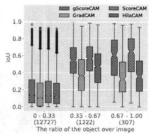

(a) Controlled experiment results of object number on COCO.

(b) Controlled experiment results of object ratio on COCO.

(c) Controlled experiment results of object ratio on PartImageNet.

Fig. 2. Controlled experiments on COCO and PartImageNet. Figure 2a shows how IoU changes with different methods when the number of classes per image is different. Figures 2b and c show how the object ratio affects the methods' IoU. The number in parenthesis of x-axis on each plot is the number of samples in that group. In sum, **gScoreCAM is more accurate than other methods when a scene contains more objects (a) and object size (measured as the ratio between the object size and the image size) is smaller (b–c).**

Results. We find that as the number of classes per image increases, the IoU of all methods decreases. gScoreCAM has the lowest IoU drop, resulting in better performance when the number of classes per image is greater than or equal to 4. The median IoU of gScoreCAM is approximately 0.03 higher than ScoreCAM and 0.07 to 0.10 higher than GradCAM and HilaCAM, as shown in Fig. 2a. This advantage makes gScoreCAM performs better in COCO because about 61% of the COCO validation images have more than three objects.

5.2.2 gScoreCAM Can Better Localize Small Objects Experiment.

Similar to Sect. 5.2.1, we conduct another controlled experiment on COCO and PartImagenet based on the size of the target part or object. This experiment divides the test samples into three groups according to the target ratio. The target ratio is measured by the area of the target part or object over the full image. We divide the images into three groups: small (0–0.33), medium (0.33–0.67), and large (0.67–1). COCO and PartImageNet have samples (13811, 607, 210) and (12727, 1222, 307) in the corresponding group.

Results. As Figs. 2b and c show that gScoreCAM consistently has a higher IoU in the small object groups, while ScoreCAM always has a higher IoU in the large groups. Combined with the results in Sect. 4.3, we find that ScoreCAM tends to generate a large and possibly noisy heatmap, resulting in better performance in the ZSD task when the object is large. On the other hand, gScoreCAM can better locate small objects. It is a critical capability in interpretability since we want the resulting heatmap to be as accurate as possible. Interestingly, when the target is a large scene (e.g., road, sea), gScoreCAM is still the best method (see Sec. A4 for details).

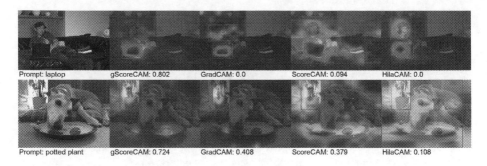

Fig. 3. In complex scenes (i.e. not ImageNet-v2), gScoreCAM outperforms other methods, yielding more precise localization and cleaner heatmaps. IoU scores between the ground truth (☐) and inferred box (☐) are shown next to each method name. More examples in Figs. S2 and S3. (Color figure online)

5.3 gScoreCAM Performs Better on Different CLIP Models

We repeat the ZSD experiments in Sect. 5.1 for RN50x4 to confirm that our proposed method can generally be applied to different CLIP variations. For generality (i.e., without further hyperparameter tuning), we use the same hyperparameter $k = 300$ as in Sect. 5.1.

Results. Our proposed method has the best ZSD performance, as shown in Table 5. The accuracy of gScoreCAM is around 2 to 4× higher than other methods (except ScoreCAM). It suggests that gScoreCAM can be applied to other CLIP variations without hyperparameter tuning.

Table 5. For both CLIP RN50x16 and RN50x4, gScoreCAM has the best overall ZSD performance in the CAM-based family.

	ImageNet-v2 [38] (MaxBoxAccV2)		COCO [29] (BoxAcc)		PartImageNet [30] (BoxAcc)	
	RN50x16	RN50x4	RN50x16	RN50x4	RN50x16	RN50x4
GradCAM [22]	38.9	32.61	11.59	9.86	10.91	9.60
xGradCAM [44]	24.24	18.94	5.6	6.11	6.57	5.01
GradCAM++ [23]	44.15	46.23	9.68	10.68	2.93	8.00
LayerCAM [45]	43.7	47.01	9.19	9.87	12.42	13.36
GroupCAM [46]	50.85	22.77	13.06	1.29	6.16	3.01
ScoreCAM [27]	**57.78**	57.99	20.43	21.31	15.67	15.39
gScoreCAM (ours)	56.61	**58.76**	**20.83**	**22.17**	**16.34**	**16.22**

5.4 Qualitative Study via CLIP

We first study a progressing plot that shows how the heatmaps change with the number of channels used by gScoreCAM. We then visually study the CLIP heatmap from different methods. We find that our proposed method provides a more accessible model explanation from the visual studies.

The Heatmap is Getting Noisier as the Number of Channels Used by gScoreCAM Increasing. Figure 1 shows a progressing plot when the number of channels used by gScoreCAM increases from 300 to 3072 (ScoreCAM) and the heatmap generated by RISE (last column). We find a clear trend that the gradient-guided ranking successfully ranks the channels by the heatmaps' contribution to the target. This visualization further confirms the result in Sect. 4.1 that we only need the top k channels to localize the target.

Visual Comparison of gScoreCAM to Other Methods. It turns out that gScoreCAM can generate a very accurate heatmap when the target object is tiny, as shown in Fig. 3. But other methods result in noisier or incorrect heatmaps. These accurate heatmaps allow us to study what the model is looking at and can be a helpful tool for studying the model. See Figs. S4 to S10 for more examples.

6   Discussion and Conclusions

Limitations. One major limitation of our proposed method is that although it is 10× faster than its predecessor; it is still not comparable to methods that do not require multiple forward passes. Our proposed method is CNN-based; it does not generalize well on popular transformer-based networks. One last thing is that our proposed method introduces a hyperparameter (k).

Fig. 4. While Goh et al. [16] reported that CLIP is easily fooled by *typographic attacks*, our gScoreCAM visualizations reveal interesting insights that CLIP indeed was able to distinguish the objects between apple, iPod and even the background. The misclassification was merely due to the fact that there are multiple objects in the scene (i.e., ill-posed, single-label, image classification task).

Conclusions. In this paper, we propose gScoreCAM, a gradient-guided CAM method to visualize and explain multimodal models. Our design is generic such that it can be easily applied to visualize and explain other CNN-based networks. In a systematic analysis of different visualization methods, our method performs the best in explaining and visualizing CLIP. We also find that our method can solve a common problem, as shown in Fig. 4: the text in the image misleads CLIP while giving the prediction. To the best of our knowledge, our proposed gScoreCAM is the best method to visualize and explain current large CNN-based models like CLIP. Therefore, we believe that gScoreCAM can help the community better understand recent foundation models and make improvements.

References

1. Bommasani, R., et al.: On the opportunities and risks of foundation models. arXiv preprint arXiv:2108.07258 (2021)
2. Radford, A., et al.: Learning transferable visual models from natural language supervision. In: International Conference on Machine Learning, pp. 8748–8763. PMLR (2021)
3. yzhuoning: yzhuoning/awesome-clip: Awesome list for research on clip (contrastive language-image pre-training) (2022). https://github.com/yzhuoning/Awesome-CLIP. Accessed 18 May 2022
4. Patashnik, O., Wu, Z., Shechtman, E., Cohen-Or, D., Lischinski, D.: Styleclip: text-driven manipulation of stylegan imagery. In: Proceedings of the IEEE/CVF International Conference on Computer Vision, pp. 2085–2094 (2021)
5. nerdyrodent: nerdyrodent/vqgan-clip: Just playing with getting vqgan+clip running locally, rather than having to use colab (2022). https://github.com/nerdyrodent/VQGAN-CLIP. Accessed 18 May 2022
6. Kim, G., Ye, J.C.: Diffusionclip: text-guided image manipulation using diffusion models. arXiv preprint arXiv:2110.02711 (2021)
7. Luo, H., et al.: Clip4clip: an empirical study of clip for end to end video clip retrieval. arXiv preprint arXiv:2104.08860 (2021)
8. Lei, J., et al.: Less is more: clipbert for video-and-language learning via sparse sampling. In: Proceedings of the IEEE/CVF Conference on Computer Vision and Pattern Recognition, pp. 7331–7341 (2021)
9. Song, H., Dong, L., Zhang, W.N., Liu, T., Wei, F.: Clip models are few-shot learners: empirical studies on VQA and visual entailment. arXiv preprint arXiv:2203.07190 (2022)
10. Kwon, G., Ye, J.C.: Clipstyler: image style transfer with a single text condition. arXiv preprint arXiv:2112.00374 (2021)
11. Vinker, Y., et al.: Clipasso: semantically-aware object sketching. arXiv preprint arXiv:2202.05822 (2022)
12. Sheng, E., Chang, K.W., Natarajan, P., Peng, N.: The woman worked as a babysitter: on biases in language generation. arXiv preprint arXiv:1909.01326 (2019)
13. Verge, T.: What a machine learning tool that turns obama white can (and can't) tell us about ai bias - the verge (2022). www.theverge.com/21298762/face-depixelizer-ai-machine-learning-tool-pulse-stylegan-obama-bias. Accessed 19 May 2022
14. Li, Q., Mai, L., Alcorn, M.A., Nguyen, A.: A cost-effective method for improving and re-purposing large, pre-trained GANs by fine-tuning their class-embeddings. In: Proceedings of the Asian Conference on Computer Vision (2020)
15. Phillips, P.J., Hahn, C.A., Fontana, P.C., Broniatowski, D.A., Przybocki, M.A.: Four principles of explainable artificial intelligence. Gaithersburg, Maryland (2020)
16. Goh, G., et al.: Multimodal neurons in artificial neural networks. Distill 6, e30 (2021)
17. Chefer, H., Gur, S., Wolf, L.: Generic attention-model explainability for interpreting bi-modal and encoder-decoder transformers. In: Proceedings of the IEEE/CVF International Conference on Computer Vision (ICCV), pp. 397–406 (2021)
18. Subramanian, S., Merrill, W., Darrell, T., Gardner, M., Singh, S., Rohrbach, A.: Reclip: a strong zero-shot baseline for referring expression comprehension. arXiv preprint arXiv:2204.05991 (2022)
19. Aflalo, E., et al.: VL-interpret: an interactive visualization tool for interpreting vision-language transformers. arXiv preprint arXiv:2203.17247 (2022)

20. vijishmadhavan: vijishmadhavan/crop-clip: Crop using clip (2022). https://github.com/vijishmadhavan/Crop-CLIP. Accessed 23 May 2022

21. Kim, W., Son, B., Kim, I.: ViLT: vision-and-language transformer without convolution or region supervision. In: International Conference on Machine Learning, pp. 5583–5594. PMLR (2021)

22. Selvaraju, R.R., Cogswell, M., Das, A., Vedantam, R., Parikh, D., Batra, D.: Gradcam: visual explanations from deep networks via gradient-based localization. In: Proceedings of the IEEE International Conference on Computer Vision, pp. 618–626 (2017)

23. Chattopadhay, A., Sarkar, A., Howlader, P., Balasubramanian, V.N.: Grad-CAM++: generalized gradient-based visual explanations for deep convolutional networks. In: 2018 IEEE Winter Conference on Applications of Computer Vision (WACV), pp. 839–847. IEEE (2018)

24. Petsiuk, V., Das, A., Saenko, K.: Rise: randomized input sampling for explanation of black-box models. arXiv preprint arXiv:1806.07421 (2018)

25. Nguyen, A., Yosinski, J., Clune, J.: Multifaceted feature visualization: uncovering the different types of features learned by each neuron in deep neural networks. arXiv preprint arXiv:1602.03616 (2016)

26. Materzyńska, J., Torralba, A., Bau, D.: Disentangling visual and written concepts in clip. In: Proceedings of the IEEE/CVF Conference on Computer Vision and Pattern Recognition, pp. 16410–16419 (2022)

27. Wang, H., et al.: Score-CAM: score-weighted visual explanations for convolutional neural networks. In: Proceedings of the IEEE/CVF Conference on Computer Vision and Pattern Recognition Workshops, pp. 24–25 (2020)

28. Russakovsky, O., et al.: Imagenet large scale visual recognition challenge. Int. J. Comput. Vision **115**, 211–252 (2015)

29. Lin, T.-Y., et al.: Microsoft COCO: common objects in context. In: Fleet, D., Pajdla, T., Schiele, B., Tuytelaars, T. (eds.) ECCV 2014. LNCS, vol. 8693, pp. 740–755. Springer, Cham (2014). https://doi.org/10.1007/978-3-319-10602-1_48

30. He, J., et al.: Partimagenet: a large, high-quality dataset of parts. arXiv preprint arXiv:2112.00933 (2021)

31. Chefer, H., Gur, S., Wolf, L.: Transformer interpretability beyond attention visualization. In: Proceedings of the IEEE/CVF Conference on Computer Vision and Pattern Recognition, pp. 782–791 (2021)

32. Zhou, B., Khosla, A., Lapedriza, A., Oliva, A., Torralba, A.: Learning deep features for discriminative localization. In: CVPR (2016)

33. Ribeiro, M.T., Singh, S., Guestrin, C.: "Why should i trust you?" explaining the predictions of any classifier. In: Proceedings of the 22nd ACM SIGKDD International Conference on Knowledge Discovery and Data Mining, pp. 1135–1144 (2016)

34. Agarwal, C., Nguyen, A.: Explaining image classifiers by removing input features using generative models. In: Proceedings of the Asian Conference on Computer Vision (2020)

35. Simonyan, K., Vedaldi, A., Zisserman, A.: Deep inside convolutional networks: visualising image classification models and saliency maps. arXiv preprint arXiv:1312.6034 (2013)

36. Nourelahi, M., Kotthoff, L., Chen, P., Nguyen, A.: How explainable are adversarially-robust CNNs? arXiv preprint arXiv:2205.13042 (2022)

37. He, K., Zhang, X., Ren, S., Sun, J.: Deep residual learning for image recognition. In: Proceedings of the IEEE Conference on Computer Vision and Pattern Recognition, pp. 770–778 (2016)

38. Recht, B., Roelofs, R., Schmidt, L., Shankar, V.: Do imagenet classifiers generalize to imagenet? In: International Conference on Machine Learning, pp. 5389–5400. PMLR (2019)
39. Choe, J., Oh, S.J., Lee, S., Chun, S., Akata, Z., Shim, H.: Evaluating weakly supervised object localization methods right. In: Proceedings of the IEEE/CVF Conference on Computer Vision and Pattern Recognition, pp. 3133–3142 (2020)
40. Gupta, T., Vahdat, A., Chechik, G., Yang, X., Kautz, J., Hoiem, D.: Contrastive learning for weakly supervised phrase grounding. In: Vedaldi, A., Bischof, H., Brox, T., Frahm, J.-M. (eds.) ECCV 2020. LNCS, vol. 12348, pp. 752–768. Springer, Cham (2020). https://doi.org/10.1007/978-3-030-58580-8_44
41. Gildenblat, J., contributors: Pytorch library for cam methods (2021). https://github.com/jacobgil/pytorch-grad-cam
42. OpenAI: openai/clip: Contrastive language-image pretraining (2022). https://github.com/openai/CLIP. Accessed 06 July 2022
43. Radin, L., Osher, S., Fatemi, E.: Non-linear total variation noise removal algorithm. Phys. D **60**, 259–268 (1992)
44. Fu, R., Hu, Q., Dong, X., Guo, Y., Gao, Y., Li, B.: Axiom-based grad-CAM: towards accurate visualization and explanation of CNNs. arXiv preprint arXiv:2008.02312 (2020)
45. Jiang, P.T., Zhang, C.B., Hou, Q., Cheng, M.M., Wei, Y.: Layercam: exploring hierarchical class activation maps for localization. IEEE Trans. Image Process. **30**, 5875–5888 (2021)
46. Zhang, Q., Rao, L., Yang, Y.: Group-CAM: group score-weighted visual explanations for deep convolutional networks. arXiv preprint arXiv:2103.13859 (2021)

From Within to Between: Knowledge Distillation for Cross Modality Retrieval

Vinh Tran$^{(\boxtimes)}$, Niranjan Balasubramanian, and Minh Hoai

Stony Brook University, Stony Brook, NY 11790, USA
{tquangvinh,niranjan,minhhoai}@cs.stonybrook.edu

Abstract. We propose a novel loss function for training text-to-video and video-to-text retrieval networks based on knowledge distillation. This loss function addresses an important drawback of the max-margin loss function often used in existing cross-modality retrieval methods, in which a fixed margin is used in training to separate matching video-and-caption pairs from non-matching pairs, treating all non-matching pairs the same and failing to account for the different degrees of non-matching. We address this drawback by introducing a novel loss for the non-matching pairs; this loss leverages the knowledge within one domain to train a better network for matching between two domains. This proposed loss does not require extra annotation. It is complementary to the existing max-margin loss, and it can be integrated into the training pipeline of any cross-modality retrieval method. Experimental results on four cross-modal retrieval datasets namely MSRVTT, ActivityNet, DiDeMo, and MSVD show the effectiveness of the proposed method. Code is available at: https://github.com/tqvinhcs/CrossKD.

Keywords: Text-video retrieval · Knowledge distillation

1 Introduction

Given that videos have become a big part of our lives with hundred of thousands of video hours produced, uploaded, and consumed every day, a problem of growing importance in computer vision is to index and search for videos based on their content. In this paper, we tackle two important cross modal retrieval problems: text-to-video and video-to-text. In the former problem, the input is a text query, and the system has to retrieve a list of videos with relevant content [15,35]. In the latter, the input is a video, and the system has to rank a list of captions based on how likely it is for the captions to be used to describe the content of the video [15,35]. This video-to-text retrieval task is useful for automatically captioning a video based on its content. Hereafter, for brevity, we will refer to these two problems as Text-and-Video Retrieval, or TVR for short.

TVR can be tackled by projecting text and video into a joint embedding space and learning a similarity scoring function for text and video embedding

Supplementary Information The online version contains supplementary material available at https://doi.org/10.1007/978-3-031-26316-3_36.

vectors. At test time, the retrieved candidates, either text or video, are ranked based on their similarity with respect to the input query. Usually, an encoder such as an LSTM/RNN [30,57,58,60] or a language model such as BERT [11,15] is used for encoding the caption. Whereas, the video representation is often a composition of multiple types of features such as faces, scene, motion, and sound. A feature aggregation method, such as NetVLAD [40] or a Transformer [15], is then used for encoding the video features into a single representation.

The key question in this setting is how the joint embedding space and the similarity scoring function are learned. Usually, one can assume that there is labeled training data containing *matching* video-caption pairs i.e., the video and the captions associated with it. Non-matching video-caption pairs can be constructed by pairing a random video with a random caption of another video. A common approach in cross-modal retrieval is to treat matching and non-matching pairs as positive and negative samples respectively, and a binary classifier is trained to separate the two sets using the well-known max-margin loss. One major problem of this approach is that it treats all non-matching pairs the same, demanding a fixed separating margin for all of them. However, not all non-matching pairs are alike. A caption C_A of video A might be similar to a caption C_B of video B, so the pair (C_A, B) would be a noisy negative sample. This severely affects the performance of the learned embedding space and scoring function.

In this work, we propose a simple but effective technique to address this problem. We use the similarity of the captions (or videos) as a better indicator for the degree of matching and non-matching. In other words, we use the knowledge of one domain (either text or video) to guide the training of neural networks that connect between two domains. This is a type of knowledge distillation, where knowledge *within* one domain is distilled to the *between* of two domains. The main rationale behind this approach is as follows: we can more reliably learn a within domain similarity function than a cross-domain similarity function, since there is often larger amounts of within-domain training data. This suggests that the former can be used as a teacher similarity function to train the latter. Figure 1 illustrates this idea.

The proposed within-to-between knowledge distillation can be implemented as a loss function and added to the existing training loss of any TVR framework to learn the joint embedding of text and video. Experimental results on four TVR benchmarks, MSRVTT [57], Activity Net [30], DiDeMo [20], and MSVD [5] show that the proposed knowledge distillation loss improves the performance of two cross modal retrieval frameworks.

2   Related Work

Text-Video Retrieval (TVR) is an emerging research area [15,17,30,35,39, 41,58,60]. Early works use language model such as LSTM [23] to represent text for capturing the sequential properties in a sentence [6,30,58,60]. Before feeding to the language model to extract sentence features, Word2Vec [42] is often used for representing word embedding. Recent methods use more advanced architectures such as OpenAI-GPT [35], BERT [15,61] or GPT2-XL and GPT2-XL-F [9] for text encoding. The video side is more complex where multiple

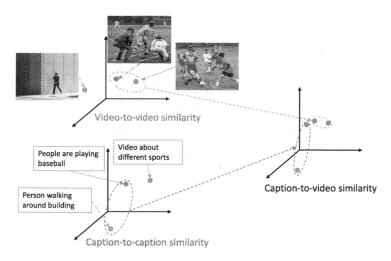

Fig. 1. Knowledge distillation for learning similarity scoring function between a caption and a video. The distillation, illustrated as dash lines, can be drawn from either the text domain or video domain. Our idea is based on the observation that the similarity scoring function in the same domain is more reliable than the similarity scoring function between two domains. Hence, the former can be used as the teacher to train the latter.

modalities are used for extracting video features. Then a multi-expert model such as NetVLAD [40], Collaborative Experts [35], or Transformer [15] is used for representing video features. However, previous works mainly focus on the features aggregation. Instead, we propose a new learning objective for better learning the similarity of text and video across modalities.

Knowledge distillation [22] has widely been used to transfer knowledge learned between different deep learning models. This method was originally introduced for decision tree simplification [2] and model compression [3]. This method has been proven to be useful in many visual recognition problems [7,27,29,31,32,38,44,48]. Conventionally, the knowledge is transferred from the teacher to the student model by considering each data sample individually, but Park *et al.* [45] propose a method that distills the mutual relation by penalizing structural differences of data samples. To some extent, we also utilize the relational knowledge from captions to captions and from videos to videos, but we perform knowledge distillation across the domains of text and video. In recent works, knowledge distillation has also been employed in the problem of cross modal learning [7]. However, this work focuses on aligning audio, image, and video representations, which is different from the TVR task. Moreover, their proposed method is semi-supervised and relies on class label prediction of human action for distillation. Whereas our method is unsupervised and is trained without using any class label. Recently, Croitoru *et al.* [9] proposed TEACHTEXT, a generalized distillation method utilizing multiple teachers to improve the retrieval performance. Our knowledge distillation method differs from this by not using external teachers during training. Alternatively, our distillation comes directly from the available structure of the video and caption domains, which is very efficient to compute and does not require an external model for distillation. Along with our work, [8,55] shows that related captions can also help improve the result embedding.

Contrastive Learning for Cross Modal Retrieval. To learn a joint embedding space for cross modal retrieval, a common approach is to use contrastive learning [26,33,38–41,51,61]. This learning method aims to maximize the similarity of text and video representations extracted from the same instance while maximizing the difference between text and video representations from different instances. To this end, bidirectional max-margin loss is used for training [6,7,9,15,35,41]. By pushing dissimilar text-video pair away, the model implicitly learns the ranking of video and text model based on their correspondent similarity. Beside max-margin loss, recent works also use normalized softmax loss [59] for contrastive learning in a similar fashion [1]. Instead of learning ranking function, some previous works also directly learn the similarity scores between the two models using regression loss [54].

3   Knowledge Distillation for Cross Modal Retrieval

3.1   Framework for Cross Modal Retrieval

Our method is applicable to different retrieval frameworks, including Multimodal Transformer (MMT) [15] and TEACHTEXT [9]. For brevity, we will describe our method with the MMT framework in this section, but we will demonstrate its benefits for both MMT, CLIP4Clip and TEACHTEXT in the experiment section.

Video Encoder. MMT [15] uses a Transformer architecture that combines multiple video embedding experts, with each expert corresponding to one type of video feature among motion [56], audio [21], scene [25], OCR [10,36], face [19,34], speech [42], and appearance [24]. The output of this video encoder is a representation that consists of N different embeddings, denoted by $\Psi(v) = \{\psi_i\}_{i=1}^{N}$. Please refer to [15] for the implementation details of the multi-modal experts.

Text Encoder. MMT uses a pre-trained BERT [11] model for encoding the caption text. BERT is a transformer-based model that has been shown to produce effective text representations for a wide variety of tasks [11]. Each caption is represented by BERT's output vector for the "[CLS]" token. Subsequently, gated embedding modules [40] are used to generate N different embeddings of this caption representation corresponding to N video experts. The caption embedding is denoted by $\Phi(c) = \{\phi_i\}_{i=1}^{N}$, where ϕ_i is the embedding vector of the ith expert.

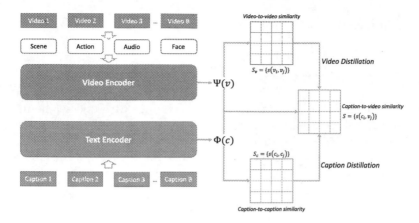

Fig. 2. The proposed distillation methods for cross modal text-video retrieval. We use a BERT model for caption representation and Multi-modal transformer for video representation. The video representation is the aggregation of multiple experts. Using these output representations, the caption-to-caption and video-to-video similarity can be computed efficiently within a training batch. The compatibility function $s(c, v)$ between caption and video can be trained with the additional training signal.

The caption-video similarity is taken as the weighted sum of the experts' caption-video similarity values:

$$s(c, v) = \sum_{i=1}^{N} w_i \langle \phi_i, \psi_i \rangle, \tag{1}$$

where $\langle \cdot, \cdot \rangle$ denotes the dot product, and w_i is the weight for the ith expert. Given a query item in one domain, these similarity scores are then used to rank items in the other domain. Figure 2 depicts the processing pipeline of our cross modal retrieval framework.

3.2   Bidirectional Max-Margin Ranking Loss

To encode the relationships and to measure similarity between a caption and a video, existing methods use bidirectional max-margin loss to separate matching caption-video pairs (called positive pairs) from the non-matching ones (called negative pairs). Positive and negative pairs of caption-video are created within each input training batch. A positive pair is the one where the caption and the video come from the same training data instance. Whereas, a negative caption-video pair consists of a caption and a video from different training instances. An instance is a pair of a caption and a video in each training batch. For each training instance, the similarity $s_{ij} = s(c_i, v_j)$ between a caption i and a video j is computed. Then, a bidirectional max-margin loss is used to train the compatibility function as:

$$\mathcal{L}_{\text{margin}} = \frac{1}{B} \sum_{i=1}^{B} \sum_{j \neq i} [\max(0, s_{ij} - s_{ii} + m) + \max(0, s_{ji} - s_{ii} + m)], \quad (2)$$

where B is the number of instances in each training batch and m the margin. The bidirectional max-margin loss function requires that the compatibility of a caption and a video from each positive pair to be higher than those from a negative pair by at least a margin m. In other frameworks [1,37], normalized softmax loss function is used for contrastive learning instead of bidirectional max-margin ranking loss.

3.3 Knowledge Distillation Loss from Caption and Video

Since videos and captions come from two different domains, learning a compatibility function based solely on the max-margin loss may be sub-optimal. We propose a complementary loss function that exploits the relationship between data instances in the same domain and distill this knowledge between domains. This is to use the similarity between captions (or between videos) from different training samples as a guidance for training the compatibility function between captions and videos. To this end, we compute the caption similarity of a given caption to the other captions in an input training batch. For a caption c that is represented by embedding vectors $\{\phi_i\}_{i=1}^{N}$ and a caption c' represented by embedding vectors $\{\phi_i'\}_{i=1}^{N}$, the caption-to-caption similarity is computed as:

$$s(c, c') = \sum_{i=1}^{N} w_i \langle \phi_i, \phi_i' \rangle. \quad (3)$$

Given a training batch B consisting of multiple caption and video pairs $\{(c_j, v_j)\}$, and a particular caption c_i, we can compute the similarity between c_i to other c_j's, which can then be normalized to get a proper probability distribution:

$$\mathcal{P}_{ij}^{cc} = \frac{\exp(s(c_i, c_j)/\tau)}{\sum_{l=1}^{B} \exp(s(c_i, c_l)/\tau)}, \forall j \in B, \quad (4)$$

where $\tau > 0$ is a temperature parameter controlling the smoothness of the distribution. Similarly, we can also compute the cross-domain caption-to-video similarity probability distribution as:

$$\mathcal{Q}_{ij}^{cv} = \frac{\exp(s(c_i, v_j)/\tau)}{\sum_{l=1}^{B} \exp(s(c_i, v_l)/\tau)}, \forall j \in B, \quad (5)$$

Given the above two distributions, we use the knowledge distillation loss to measure the dissimilarity between the caption-to-caption distribution $\mathcal{P}$ and the caption-to-video distribution $\mathcal{Q}$ using Kullback-Leibler divergence as:

$$\mathcal{L}_{\text{cap\_distill}} = \frac{1}{B} \sum_{i=1}^{B} KL(\mathcal{P}_{i,:}^{cc} || \mathcal{Q}_{i,:}^{cv}). \tag{6}$$

Analogously, the similarity between videos features can be used as another training signal. The video-to-video and video-to-caption similarity can also be written in form of two distributions as:

$$\mathcal{P}_{ij}^{vv} = \frac{\exp(s(v_i, v_j)/\tau)}{\sum_{l=1}^{B} \exp(s(v_i, v_l)/\tau)}, \forall j \in B, \tag{7}$$

and

$$\mathcal{Q}_{ij}^{vc} = \frac{\exp(s(c_j, v_i)/\tau)}{\sum_{l=1}^{B} \exp(s(c_l, v_i)/\tau)}, \forall j \in B. \tag{8}$$

Again, the video distillation loss is computed as the Kullback-Leibler divergence between the two distributions:

$$\mathcal{L}_{\text{vid\_distill}} = \frac{1}{B} \sum_{i=1}^{B} KL(\mathcal{P}_{i,:}^{vv} || \mathcal{Q}_{i,:}^{vc}). \tag{9}$$

3.4 Composition Loss for Training Retrieval Model

We add the two new distillation loss functions in Eq. (6) and Eq. (9) to the standard bidirectional max-margin ranking loss for training the compatibility function between a caption and a video. The distillation loss for the captions is given by:

$$\mathcal{L}_{\text{cap\_compose}} = \mathcal{L}_{\text{margin}} + \mathcal{L}_{\text{cap\_distill}}. \tag{10}$$

In the same way, the knowledge distillation for videos is given by:

$$\mathcal{L}_{\text{vid\_compose}} = \mathcal{L}_{\text{margin}} + \mathcal{L}_{\text{vid\_distill}}. \tag{11}$$

In our experiments, we train with both losses and also show an ablation study for each loss function. During development, we explored several ways for combining the two losses and find that treating both of them equally yields the best results.

4   Experimental Results

4.1   Text-Video Retrieval Frameworks

Retrieval Frameworks. One advantage of the proposed method is its ability to be used with different TVR frameworks. To demonstrate this, we experiment with three recent TVR frameworks: Multi-modal Transformer (MMT) [15],

Table 1. Benefits of knowledge distillation for the MMT framework on the MSRVTT and ActivityNet datasets. Both type of distillations effectively improve the performances of text-to-video and video-to-text retrieval on all datasets. Caption Distillation is slightly better than Video Distillation

Datasets & Methods	Text → Video			Video → Text		
MSRVTT 1k-A	R@1↑	R@5↑	R@10↑	R@1↑	R@5↑	R@10↑
MMT [15]	26.6	57.1	69.6	**27.0**	57.5	69.7
MMT + Caption Distillation	**27.8**	58.4	70.4	**27.0**	**58.8**	70.2
MMT + Video Distillation	26.7	**59.0**	**71.8**	26.4	58.1	**71.7**
MSRVTT 1k-B	R@1↑	R@5↑	R@10↑	R@1↑	R@5↑	R@10↑
MMT [15]	20.3	49.1	63.9	21.1	49.4	63.2
MMT + Caption Distillation	20.8	**52.4**	**66.5**	**22.7**	52.5	**67.4**
MMT + Video Distillation	**21.3**	51.4	66.2	21.3	**52.5**	66.3
ActivityNet	R@1↑	R@5↑	R@50↑	R@1↑	R@5↑	R@50↑
MMT [15]	22.7	54.2	93.2	22.9	54.8	93.1
MMT + Caption Distillation	24.3	57.1	**94.1**	**26.5**	**58.7**	**94.1**
MMT + Video Distillation	**24.4**	**58.0**	94.0	25.7	58.0	93.3

CLIP4Clip [37], and TEACHTEXT [9]. Both MMT and TEACHTEXT are based on Collaborative Experts (CE) architecture [35] with multiple experts for video encoding. Meanwhile, CLIP4Clip [37] is the most recent framework for TVR with a single video encoder based on vision transformer.

Video Encoders. MMT [15] uses seven pretrained experts for video encoding, namely: **Motion** from S3D [56], **Audio** from VGGish [21], **Scene** from DenseNet [25], **OCR** from the output of text detector embedded with Word2Vec [42], **Face** from ResNet50 [18] trained on VGGFace2, **Speech** using Google Cloud Speech, and **Appearance** from SENet-154 [24].

TEACHTEXT [9] also uses seven experts for video encoding. Most of these video experts are similar to that of MMT, except for the **Motion** expert which comprises of two action experts *Action(KN)* and *Action(IG)*. The first action expert, *Action(KN)*, is an I3D model trained on Kinetics [4], and the second expert, *Action(IG)*, is a 34-layer R(2+1)D model [52] pretrained on IG-65m dataset [16]. TEACHTEXT establishes a stronger baseline than CE (denoted as CE+) by using the more powerful text embeddings from GPT2-XL [50]. The full TEACHTEXT framework (denoted as TT-CE+) is trained with additional knowledge given by multiple text encoders as teachers including Word2Vec [42], GPT2-XL [50], and GPT2-XL-F.

CLIP4Clip [37] uses the pretrained CLIP [49] model as a single video encoder. The encoder is based on Vision Transformer (ViT) [12]. In our experiment, we use the pretrained CLIP (ViT-B/16) [49] as our backbone to encode video. The visual encoder has 12 layers and patch size of 16. Unfortunately, due to our limited

Table 2. Text-to-video retrieval performance for CLIP4CLIP with and without knowledge distillation on the MSRVTT, MSVD and DiDeMo datasets. Both type of distillations effectively improve the performances on all datasets. All the results are obtained using ViT-B/16 as video encoder. Caption Distillation is slightly better than Video Distillation

Datasets and methods	R@1↑	R@5↑	R@10↑	MnR↓
MSRVTT 1k-A				
CLIP4Clip [37]	43.4	72.3	80.5	15.4
CLIP4Clip + Caption Distillation	**46.2**	**73.1**	**81.2**	**13.2**
CLIP4Clip + Video Distillation	44.7	72.8	81.1	13.8
MSVD				
CLIP4Clip [37]	48.6	78.6	87.2	9.0
CLIP4Clip + Caption Distillation	48.8	**79.2**	**87.5**	**8.7**
CLIP4Clip + Video Distillation	**48.9**	79.1	87.4	9.0
DiDeMo				
CLIP4Clip [37]	42.0	69.1	78.1	18.8
CLIP4Clip + Caption Distillation	**43.2**	**69.7**	79.2	**17.5**
CLIP4Clip + Video Distillation	**43.2**	69.2	**79.3**	17.9

computational resources, we have to keep the first 6 layers of the ViT-B/16 frozen during training. Furthermore, we can only train our model with relatively smaller batch sizes in comparison to [37]. The batch sizes are set to be 24 for MSRVTT and MSVD datasets. For DiDeMo, we can only use the batch size of 6, since the videos in this dataset are relatively long and the video encoder requires a long observation window of 64 frames. As a result, our reproduced results for CLIP4CLIP on DiDeMo are not as good as the previously reported results [37].

Training Details. Our training procedure is based on the implementation of MMT[1], TEACHTEXT[2], and CLIP4Clip[3] We train all the models on PyTorch [46] with Adam optimizer [28]. The bidirectional max-margin ranking loss is used for MMT and TEACHTEXT (CE+ and TT-CE+), while the normalized softmax loss is used in CLIP4Clip for contrastive learning. If not otherwise specified, all training parameters are the same as reported in MMT [15], TEACHTEXT [9], and CLIP4Clip [37].

4.2 Datasets

We perform experiments on four challenging TVR benchmarks: MSRVTT [57], ActivityNet [30], DiDeMo [20], and MSVD [5]. We report performances on both

---

[1] https://github.com/gabeur/mmt.
[2] https://github.com/albanie/collaborative-experts.
[3] https://github.com/ArrowLuo/CLIP4Clip.

text-to-video and video-to-text retrieval tasks. For experiments using TEACH-TEXT and CLIP4Clip, we follow [9,37] to report only text-to-video performance.

MSRVTT. [57] is a large-scale dataset for video understanding, especially for TVR. The dataset contains 10,000 video clips crawled from web. Each video clip is associated with 20 natural sentences annotated by AMT workers. In the MMT framework, we follow [15,40,58] and perform experiments on Split 1k-A and Split 1k-B. Split 1k-A [58] uses 9000 videos for training and 1000 for testing. Meanwhile, Split 1k-B [40] uses 6656 videos for training and 1000 for testing. In addition, for direct comparison with prior work, we also provide the retrieval performance on the full split of this dataset with TEACHTEXT framework (6513 videos for training, 497 for validation, and 2990 for testing).

ActivityNet. [30] contains 20,000 Youtube videos amounting to 849 video hours with temporally annotated sentence descriptions. Following [60] and [15], we concatenate all the sentence descriptions for each video to form a paragraph. There are 10009 instances in the training set. Following the same paragraph-video retrieval setup in [15,35,60], we perform evaluation on the "val1" split (4917 videos) of this dataset.

DiDeMo. [20] stands for Distinct Describable Moments. This dataset contains unedited 10,464 personal videos in multiple content such as sports, concerts, and pets. Each video in the dataset has three to five captions. We follow [9,35,60] and use 8392 videos for training, 1065 for validation, and 1004 for testing.

MSVD. [5] dataset has 1970 video clips associated with 80K English captions. The setup is similar to [9,35], where 1200 clips are used for training, 100 for validation, and 670 for testing. Since MSVD videos do not contain sound, audio-based features are not used for this dataset.

Evaluation Metrics. The performance of all models are evaluated with recall at rank N (R@N), a standard retrieval metric. A better model should achieve higher recall. We also report median rank (MdR) and mean rank (MnR) of the correct results. For median rank and mean rank, a lower number indicates better performance. Similar to [9], we also report the geometric mean of R@1, R@5, and R@10 for conciseness. The geometric mean summarizes the overall retrieval performance at multiple recall ranking steps. Since each TVR framework is performed on a different subset of datasets, we will perform our experiments following the setups in previous works [9,15,37].

4.3 Knowledge Distillation for the MMT Retrieval Framework

We first evaluate the benefits of knowledge distillation for MMT on the MSRVTT and ActivityNet datasets. As can be seen from Table 1, training retrieval models with knowledge distillation improves the retrieval performance. Our models yield significant improvement over its direct baseline MMT on both text-to-video and video-to-text retrieval tasks. For Split 1k-A, Caption Distillation improves the text-to-video retrieval performance from 26.6% to 27.8% at R@1. For R@5 and R@10, the performance gain brought by Video Distillation are even higher, from

Table 3. Text-to-video retrieval performance of TEACHTEXT [9] without and with Caption Distillation on four datasets

Methods	R@1↑	R@5↑	R@10↑	MdR↓	Methods	R@1↑	R@5↑	R@10↑	MdR↓
MSRVTT									
CE+ [9]	13.8	36.5	49.4	11.0	TT-CE+ [9]	14.6	37.9	50.9	**10.0**
+Our Distillation	**14.7**	**37.8**	**50.6**	**10.0**	+Our Distillation	**14.7**	**38.1**	**51.1**	**10.0**
ActivityNet									
CE+ [9]	19.4	49.3	65.4	6.0	TT-CE+ [9]	23.5	57.2	**73.6**	4.0
+Our Distillation	**20.6**	**50.6**	**66.9**	**5.0**	+Our Distillation	**23.9**	**57.3**	73.5	4.0
MSVD									
CE+ [9]	25.1	56.5	70.9	4.0	TT-CE+ [9]	25.1	56.8	71.2	4.0
+Our Distillation	**26.0**	**58.3**	**72.9**	4.0	+Our Distillation	**25.5**	**57.1**	**71.7**	4.0
DiDeMo									
CE+ [9]	18.2	43.9	57.1	7.9	TT-CE+ [9]	21.6	48.6	**62.9**	6.0
+Our Distillation	**20.2**	**45.2**	**58.8**	**7.0**	+Our Distillation	**21.7**	**49.2**	62.4	**5.7**

Table 4. Text-to-video retrieval performance for TEACHTEXT methods with and without knowledge distillation. The performance measure is the geometric mean of R@1, R@5 and R@10. The left columns on each dataset are the base models of CE+ and TT-CE+ from TEACHTEXT. The right columns are the results obtained by adding our distillation loss. Our proposed method improves the performance on all base models in all datasets

Method	MSRVTT		ActivityNet		DiDeMo		MSVD	
	Base	Ours	Base	Ours	Base	Ours	Base	Ours
CE+ [9]	29.2	**30.4**	39.7	**41.1**	35.8	**37.7**	46.5	**47.9**
TT-CE+ [9]	30.4	**30.6**	46.3	**46.5**	40.4	**40.5**	46.6	**47.1**

57.1% to 59.0% and 69.6% to 71.8%, respectively. On the ActivityNet dataset, the performance gaps are even wider, clearly demonstrating the benefits of our proposed distillation loss. Our models perform better than the direct baseline MMT at every recall step. Both types of knowledge distillation provide benefits for training the retrieval system. Between the two, Caption Distillation performs slightly better than Video Distillation. There are two possible reasons that can explain this. One reason is that caption similarity starts with a pretrained BERT model, whereas video similarity is trained from scratch. The other is that the text similarity computation is arguably simpler in that it is done over a single representation, while the video similarity is computed over a composition of features from multiple experts.

4.4 Knowledge Distillation for the CLIP4Clip Retrieval Framework

We also consider a recent TVR framework CLIP4Clip [37]. For this framework, the bidirectional max-margin loss is replaced by the normalized softmax loss for contrastive learning. Hence, the composition losses $\mathcal{L}_{norm\_softmax} + \mathcal{L}_{cap\_distill}$

Table 5. Comparison to other methods on MSRVTT 1k-A dataset. Results are obtained by applying caption distillation on CLIP4Clip [37] framework.

Method	R@1↑	R@5↑	R@10↑	MnR↓
ActBERT [61]	8.6	23.4	33.1	-
MIL-NCE [39]	9.9	24.0	32.4	-
JSFusion [58]	10.2	31.2	43.2	-
HT [41]	12.1	35.0	48.0	-
HT-pretrained [41]	14.9	40.2	52.8	-
CE [35]	20.9	48.8	62.4	28.2
CLIP [49]	22.5	43.3	53.7	61.7
MMT [15]	24.6	54.0	67.1	-
MMT-pretrained [15]	26.6	57.1	69.6	24.0
TT-CE+ [9]	29.6	61.6	74.2	-
SSB [47]	30.1	58.5	69.3	-
Frozen [1]	31.0	59.5	70.5	-
MDMMT [13]	38.9	69.0	79.7	16.5
CLIP4Clip [37]	44.5	71.4	**81.6**	15.3
Ours	**46.2**	**73.1**	81.2	**13.2**

and $\mathcal{L}_{\text{norm\_softmax}} + \mathcal{L}_{\text{vid\_distill}}$ are used instead. As can be seen from Table 2, both proposed distillation losses, especially the Caption Distillation loss, improve the retrieval performance on the three datasets.

4.5   Knowledge Distillation for the TEACHTEXT Framework

We also evaluate the benefits of knowledge distillation for the TEACHTEXT [9] framework. As before, the distillation losses are added to the training loss of this method. Since caption distillation is slightly better than video distillation, we perform further experiments with caption distillation only. The experiments are conducted for two training settings: *(1) without external teachers (CE+)*, and *(2) with external teachers (TT-CE+)*. Specifically, we use the loss $\mathcal{L}_{\text{margin}} + \mathcal{L}_{\text{cap\_distill}}$ when applying our method on the CE+ setting, and the loss $\mathcal{L}_{\text{margin}} + \mathcal{L}_{\text{d}} + \mathcal{L}_{\text{cap\_distill}}$ when applying our method on the TT-CE+ setting. Here, $\mathcal{L}_{\text{d}}$ is the distillation loss from external teachers [9].

Following [9], we perform experiments on the four TVR benchmarks. For a fair comparison with most previous methods, we do not use the denoising trick [9] when training on the crowd-sourced datasets MSRVTT and MSVD.

Performance Without Using External Teachers (CE+). Tables 3 and 4 show the results of this experiment. On all four datasets, the CE+ method with the proposed caption distillation outperforms the baseline CE+ method without any knowledge distillation. On average, caption distillation brings a

Table 6. Comparison with the other methods on the MSVD dataset.

Method	R@1↑	R@5↑	R@10↑	MnR↓
VSE++ [14]	15.4	39.6	53.0	-
M-Cues [43]	20.3	47.8	61.1	-
MEE [40]	21.1	52.0	66.7	-
CE [9]	21.5	52.3	67.5	-
TT-CE [9]	22.1	52.2	67.2	-
CE+ [9]	25.1	56.5	70.9	-
TT-CE+ [9]	25.1	56.8	71.2	-
SSB [47]	28.4	60.0	72.9	-
Frozen [1]	33.7	64.7	76.3	-
CLIP4Clip [37]	46.2	76.1	84.6	10.0
Ours	**48.8**	**79.2**	**87.5**	**8.7**

Fig. 3. Qualitative retrieval performance on DiDeMo dataset with CE+ baseline. On the left is the query. The first column is the correct video clip. The next two columns show the top 1 retrieved clips by CE+ and our proposed method respectively.

1.4% gain over the direct baseline CE+. We also show some qualitative results of our proposed method in Fig. 3. Notably, the proposed method is a form of self-distillation and does not use any external information for training, while the full TT-CE+ employs the external teachers for distillation.

Performance Using External Teachers (TT-CE+). As can be also seen in Tables 3 and 4, caption distillation is still beneficial when combining with the external teachers of TEACHTEXT. However, the additional benefit is not as large as when training without external teachers (i.e., with the direct CE+ baseline). This is likely due to the saturation of information, since we already have the extra distillation from the three external teachers. A similar saturation phenomenon was also reported in [9]. The performance of TEACHTEXT plateaued after three external teachers had been used for distillation, and no significant further improvement was observed.

Table 7. Comparison with other methods on the DiDeMo dataset.

Method	R@1↑	R@5↑	R@10↑	MnR↓
S2VT [53]	11.9	33.6	-	-
FSE [60]	13.9	36.0	-	-
MEE [40]	16.1	41.2	55.2	43.7
CE [9]	17.1	41.9	56.0	-
TT-CE [9]	21.0	47.5	61.9	-
CE+ [9]	18.2	43.9	57.1	-
ClipBERT [33]	20.4	48.0	60.8	-
TT-CE+ [9]	21.6	48.6	62.9	-
Frozen [1]	34.6	65.0	74.7	-
MDMMT [13]	38.9	69.0	79.7	-
CLIP4Clip [37] (reported in [37])	**43.4**	**70.2**	**80.6**	**17.5**
CLIP4Clip-rerun (frozen layers + smaller batches)	42.0	69.1	78.1	18.8
CLIP4Clip-rerun + Caption Distillation (Ours)	43.2	69.7	79.2	**17.5**

4.6   Comparison to Other Methods

Using the proposed caption distillation loss with CLIP4CLIP, we achieve better text-to-video retrieval performance than the previous state-of-the-art results, as can be seen in Table 5 for MSRVTT 1k-A and Table 6 for MSVD datasets. The results on the DiDeMo dataset is shown in Table 7, and our method is not better than the current state-of-the-art due to the lack of computational resources to follow the recommended experiment setting. As explained in Sect. 4.1, for DiDeMo, we has to freeze some network layers and use much smaller batch size. This method is denoted as CLIP4CLIP-rerun in Table 7, and our method outperforms this direct baseline.

5   Conclusions

In this paper, we proposed a novel knowledge distillation loss for cross modal text-to-video and video-to-text retrieval. The new loss exploits the information from features of the same domain as knowledge to guide the similarity learning for the cross domain matching. This information does not require any external data or additional annotation and can be drawn directly from either the text or video features for distillation. This loss function can be combined with the original loss function of a retrieval framework. More importantly, our proposed knowledge distillation loss is framework agnostic, and is applicable to any retrieval framework. Extensive experiments on three retrieval frameworks and four large-scale datasets for cross modal retrieval show the benefits of our method.

Acknowledgement. This material is based on research that is supported in part by the Air Force Research Laboratory (AFRL), DARPA, for the KAIROS program under agreement number FA8750-19-2-1003.

References

1. Bain, M., Nagrani, A., Varol, G., Zisserman, A.: Frozen in time: a joint video and image encoder for end-to-end retrieval. In: Proceedings of the International Conference on Computer Vision (2021)
2. Breiman, L., Shang, N.: Born again trees. University of California, Berkeley, Berkeley, CA, Technical report, vol. 1, no. 2 (1996)
3. Buciluva, C., Caruana, R., Niculescu-Mizil, A.: Model compression. In: Proceedings of the 12th ACM SIGKDD International Conference on Knowledge Discovery and Data Mining, pp. 535–541 (2006)
4. Carreira, J., Zisserman, A.: Quo vadis, action recognition? A new model and the kinetics dataset. In: Proceedings of the IEEE Conference on Computer Vision and Pattern Recognition (2017)
5. Chen, D., Dolan, W.B.: Collecting highly parallel data for paraphrase evaluation. In: Proceedings of the 49th Annual Meeting of the Association for Computational Linguistics: Human Language Technologies (2011)
6. Chen, S., Zhao, Y., Jin, Q., Wu, Q.: Fine-grained video-text retrieval with hierarchical graph reasoning. In: Proceedings of the IEEE Conference on Computer Vision and Pattern Recognition (2020)
7. Chen, Y., Xian, Y., Koepke, A., Shan, Y., Akata, Z.: Distilling audio-visual knowledge by compositional contrastive learning. In: Proceedings of the IEEE Conference on Computer Vision and Pattern Recognition (2021)
8. Chun, S., Oh, S.J., De Rezende, R.S., Kalantidis, Y., Larlus, D.: Probabilistic embeddings for cross-modal retrieval. In: Proceedings of the IEEE Conference on Computer Vision and Pattern Recognition, pp. 8415–8424 (2021)
9. Croitoru, I., et al.: Teachtext: crossmodal generalized distillation for text-video retrieval. In: Proceedings of the International Conference on Computer Vision (2021)
10. Deng, D., Liu, H., Li, X., Cai, D.: Pixellink: detecting scene text via instance segmentation. In: Proceedings of AAAI Conference on Artificial Intelligence (2018)
11. Devlin, J., Chang, M.W., Lee, K., Toutanova, K.: Bert: pre-training of deep bidirectional transformers for language understanding. In: Proceedings of the Association for Computational Linguistics (2018)
12. Dosovitskiy, A., et al.: An image is worth 16x16 words: transformers for image recognition at scale. In: Proceedings of International Conference on Learning and Representation (2021)
13. Dzabraev, M., Kalashnikov, M., Komkov, S., Petiushko, A.: MDMMT: multidomain multimodal transformer for video retrieval. In: Proceedings of the IEEE Conference on Computer Vision and Pattern Recognition (2021)
14. Faghri, F., Fleet, D.J., Kiros, J.R., Fidler, S.: VSE++: improving visual-semantic embeddings with hard negatives. In: Proceedings of the British Machine Vision Conference (2018)
15. Gabeur, V., Sun, C., Alahari, K., Schmid, C.: Multi-modal transformer for video retrieval. In: Proceedings of the European Conference on Computer Vision (2020)

16. Ghadiyaram, D., Tran, D., Mahajan, D.: Large-scale weakly-supervised pre-training for video action recognition. In: Proceedings of the IEEE Conference on Computer Vision and Pattern Recognition (2019)

17. Ging, S., Zolfaghari, M., Pirsiavash, H., Brox, T.: COOT: cooperative hierarchical transformer for video-text representation learning. In: Advances in Neural Information Processing Systems, vol. 33, pp. 22605–22618 (2020)

18. He, K., Zhang, X., Ren, S., Sun, J.: Deep residual learning for image recognition. In: Proceedings of the IEEE Conference on Computer Vision and Pattern Recognition (2016)

19. He, K., Zhang, X., Ren, S., Sun, J.: Identity mappings in deep residual networks. In: Leibe, B., Matas, J., Sebe, N., Welling, M. (eds.) ECCV 2016. LNCS, vol. 9908, pp. 630–645. Springer, Cham (2016). https://doi.org/10.1007/978-3-319-46493-0_38

20. Hendricks, L.A., Wang, O., Shechtman, E., Sivic, J., Darrell, T., Russell, B.: Localizing moments in video with temporal language. In: Proceedings of the International Conference on Computer Vision (2017)

21. Hershey, S., et al.: CNN architectures for large-scale audio classification. In: Proceedings of IEEE International Conference on Acoustics, Speech and Signal Processing (2017)

22. Hinton, G., Vinyals, O., Dean, J.: Distilling the knowledge in a neural network. arXiv preprint arXiv:1503.02531 (2015)

23. Hochreiter, S., Schmidhuber, J.: Long short-term memory. Neural Comput. $9(8)$, 1735–1780 (1997)

24. Hu, J., Shen, L., Sun, G.: Squeeze-and-excitation networks. In: IEEE Transactions on Pattern Analysis and Machine Intelligence, pp. 7132–7141 (2018)

25. Huang, G., Liu, Z., Van Der Maaten, L., Weinberger, K.Q.: Densely connected convolutional networks. In: Proceedings of the IEEE Conference on Computer Vision and Pattern Recognition (2017)

26. Huang, Z., et al.: Learning with noisy correspondence for cross-modal matching. In: Ranzato, M., Beygelzimer, A., Dauphin, Y., Liang, P., Vaughan, J.W. (eds.) Advances in Neural Information Processing Systems, vol. 34, pp. 29406–29419 (2021)

27. Jha, A., Kumar, A., Banerjee, B., Namboodiri, V.: SD-MTCNN: self-distilled multi-task CNN. In: Proceedings of the British Machine Vision Conference (2020)

28. Kingma, D.P., Ba, J.: Adam: a method for stochastic optimization. arXiv preprint arXiv:1412.6980 (2014)

29. Koepke, A., Wiles, O., Zisserman, A.: Visual pitch estimation. In: Proceedings of the Sound and Music Computing Conference (2019)

30. Krishna, R., Hata, K., Ren, F., Fei-Fei, L., Niebles, J.C.: Dense-captioning events in videos. In: Proceedings of the International Conference on Computer Vision (2017)

31. Le, Q.V., et al.: Building high-level features using large scale unsupervised learning. In: Proceedings of the International Conference on Machine Learning (2012)

32. Lee, S., Song, B.C.: Graph-based knowledge distillation by multi-head attention network. In: Proceedings of the British Machine Vision Conference (2019)

33. Lei, J., et al.: Less is more: clipbert for video-and-language learning via sparse sampling. In: Proceedings of the IEEE Conference on Computer Vision and Pattern Recognition (2021)

34. Liu, W., et al.: SSD: single shot MultiBox detector. In: Leibe, B., Matas, J., Sebe, N., Welling, M. (eds.) ECCV 2016. LNCS, vol. 9905, pp. 21–37. Springer, Cham (2016). https://doi.org/10.1007/978-3-319-46448-0_2

35. Liu, Y., Albanie, S., Nagrani, A., Zisserman, A.: Use what you have: video retrieval using representations from collaborative experts. In: Proceedings of the British Machine Vision Conference (2019)

36. Liu, Y., Wang, Z., Jin, H., Wassell, I.: Synthetically supervised feature learning for scene text recognition. In: Proceedings of the European Conference on Computer Vision (2018)

37. Luo, H., et al.: CLIP4Clip: an empirical study of clip for end to end video clip retrieval. arXiv preprint arXiv:2104.08860 (2021)

38. Miech, A., Alayrac, J.B., Laptev, I., Sivic, J., Zisserman, A.: Thinking fast and slow: efficient text-to-visual retrieval with transformers. In: Proceedings of the IEEE Conference on Computer Vision and Pattern Recognition, pp. 9826–9836 (2021)

39. Miech, A., Alayrac, J.B., Smaira, L., Laptev, I., Sivic, J., Zisserman, A.: End-to-end learning of visual representations from uncurated instructional videos. In: Proceedings of the IEEE Conference on Computer Vision and Pattern Recognition, pp. 9879–9889 (2020)

40. Miech, A., Laptev, I., Sivic, J.: Learning a text-video embedding from incomplete and heterogeneous data. arXiv preprint arXiv:1804.02516 (2018)

41. Miech, A., Zhukov, D., Alayrac, J.B., Tapaswi, M., Laptev, I., Sivic, J.: HowTo100M: learning a text-video embedding by watching hundred million narrated video clips. In: Proceedings of the International Conference on Computer Vision (2019)

42. Mikolov, T., Chen, K., Corrado, G., Dean, J.: Efficient estimation of word representations in vector space. In: ICLR Workshop (2013)

43. Mithun, N.C., Li, J., Metze, F., Roy-Chowdhury, A.K.: Learning joint embedding with multimodal cues for cross-modal video-text retrieval. In: ACM International Conference on Multimedia Retrieval (2018)

44. Nguyen, N., et al.: Dictionary-guided scene text recognition. In: Proceedings of the IEEE Conference on Computer Vision and Pattern Recognition (2021)

45. Park, W., Kim, D., Lu, Y., Cho, M.: Relational knowledge distillation. In: Proceedings of the IEEE Conference on Computer Vision and Pattern Recognition (2019)

46. Paszke, A., et al.: Pytorch: an imperative style, high-performance deep learning library. In: Advances in Neural Information Processing Systems, vol. 32, pp. 8026–8037 (2019)

47. Patrick, M., et al.: Support-set bottlenecks for video-text representation learning. In: Proceedings of International Conference on Learning and Representation (2021)

48. Phuong, M., Lampert, C.H.: Distillation-based training for multi-exit architectures. In: Proceedings of the International Conference on Computer Vision (2019)

49. Radford, A., et al.: Learning transferable visual models from natural language supervision. arXiv preprint arXiv:2103.00020 (2021)

50. Radford, A., Wu, J., Child, R., Luan, D., Amodei, D., Sutskever, I., et al.: Language models are unsupervised multitask learners. OpenAI Blog 1(8), 9 (2019)

51. Tang, Z., Lei, J., Bansal, M.: Decembert: learning from noisy instructional videos via dense captions and entropy minimization. In: Proceedings of the Conference of the North American Chapter of the Association for Computational Linguistics: Human Language Technologies, pp. 2415–2426 (2021)

52. Tran, D., Wang, H., Torresani, L., Ray, J., LeCun, Y., Paluri, M.: A closer look at spatiotemporal convolutions for action recognition. In: Proceedings of the IEEE Conference on Computer Vision and Pattern Recognition (2018)

53. Venugopalan, S., Xu, H., Donahue, J., Rohrbach, M., Mooney, R., Saenko, K.: Translating videos to natural language using deep recurrent neural networks. arXiv preprint arXiv:1412.4729 (2014)
54. Wang, L., Li, Y., Huang, J., Lazebnik, S.: Learning two-branch neural networks for image-text matching tasks. IEEE Trans. Pattern Anal. Mach. Intell. **41**(2), 394–407 (2018)
55. Wray, M., Doughty, H., Damen, D.: On semantic similarity in video retrieval. In: Proceedings of the IEEE Conference on Computer Vision and Pattern Recognition, pp. 3650–3660 (2021)
56. Xie, S., Sun, C., Huang, J., Tu, Z., Murphy, K.: Rethinking spatiotemporal feature learning: Speed-accuracy trade-offs in video classification. In: Proceedings of the European Conference on Computer Vision (2018)
57. Xu, J., Mei, T., Yao, T., Rui, Y.: MSR-VTT: a large video description dataset for bridging video and language. In: Proceedings of the IEEE Conference on Computer Vision and Pattern Recognition (2016)
58. Yu, Y., Kim, J., Kim, G.: A joint sequence fusion model for video question answering and retrieval. In: Proceedings of the European Conference on Computer Vision (2018)
59. Zhai, A., Wu, H.Y.: Classification is a strong baseline for deep metric learning. In: Proceedings of the British Machine Vision Conference (2018)
60. Zhang, B., Hu, H., Sha, F.: Cross-modal and hierarchical modeling of video and text. In: Proceedings of the European Conference on Computer Vision (2018)
61. Zhu, L., Yang, Y.: ActBERT: learning global-local video-text representations. In: Proceedings of the IEEE Conference on Computer Vision and Pattern Recognition, pp. 8746–8755 (2020)

Thinking Hallucination for Video Captioning

Nasib Ullah[1(✉)] and Partha Pratim Mohanta[2]

[1] LIVIA, Department of Systems Engineering, ÉTS, Montreal, Canada
nasibullah.nasibullah.1@ens.etsmtl.ca
[2] ECSU, Indian Statistical Institute, Kolkata, India
ppmohanta@isical.ac.in

Abstract. With the advent of rich visual representations and pre-trained language models, video captioning has seen continuous improvement over time. Despite the performance improvement, video captioning models are prone to hallucination. Hallucination refers to the generation of highly pathological descriptions that are detached from the source material. In video captioning, there are two kinds of hallucination: object and action hallucination. Instead of endeavoring to learn better representations of a video, in this work, we investigate the fundamental sources of the hallucination problem. We identify three main factors: (i) inadequate visual features extracted from pre-trained models, (ii) improper influences of source and target contexts during multi-modal fusion, and (iii) exposure bias in the training strategy. To alleviate these problems, we propose two robust solutions: (a) the introduction of auxiliary heads trained in multi-label settings on top of the extracted visual features and (b) the addition of context gates, which dynamically select the features during fusion. The standard evaluation metrics for video captioning measures similarity with ground truth captions and do not adequately capture object and action relevance. To this end, we propose a new metric, COAHA (caption object and action hallucination assessment), which assesses the degree of hallucination. Our method achieves state-of-the-art performance on the MSR-Video to Text (MSR-VTT) and the Microsoft Research Video Description Corpus (MSVD) datasets, especially by a massive margin in CIDEr score.

1 Introduction

Video captioning is the translation of a video into a natural language description. It has many potential applications, including video retrieval, human-computer interface, assisting visually challenged, and many more. The encoder-decoder based sequence to sequence architecture (initially proposed for machine translation) has helped video captioning exceptionally. Pre-trained vision models

This work was done while Nasib Ullah was a PLP at ECSU, ISI Kolkata, India. The code is available at: https://github.com/nasib-ullah/THVC.

Supplementary Information The online version contains supplementary material available at https://doi.org/10.1007/978-3-031-26316-3_37.

Object Hallucination

MARN: A man is riding a horse.
BLEU4: 76.0, METEOR: 42.4, ROUGE_L: 83.3, COAHA: 8.12

Action Hallucination

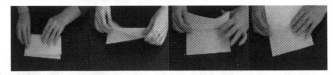

MARN: A woman is drawing a paper.
BLEU4: 0.008, METEOR: 37.5, ROUGE_L: 83.3, COAHA: 10.7

Fig. 1. Video Captioning models suffer from two types of hallucination, Object hallucination (above) and action hallucination (below). Unfortunately, the standard metrics fail to adequately capture hallucination, whereas our proposed COAHA metric is more relevant to hallucination. The results are generated using the MARN [23] model. For COAHA, lower is better.

extract the frame features at the encoder section, and the decoder is a conditional language model. Recent improvements happened in broadly three areas, better visual feature extraction [20,23], better and support of external pre-trained language models [21,47] and, better strategy [9,29] to sample informative frames.

Despite the improvements, a significant problem remains that is known as a hallucination [24,27,38] in the literature. Unlike image captioning, there are two types of hallucination in the case of video captioning: object hallucination occurs when the model describes objects that are not present in the video, and action hallucination occurs when the model describes an action that has not been performed in the video as depicted in Fig. 1. Another evidence to understand that hallucination is a significant problem can be seen from the token position vs. model confidence plot in Fig. 2. Generally, the object and action words occur towards the middle of the caption. Thus, we can see that the model has low confidence while generating tokens related to objects and actions.

In this work, we point out three fundamental sources of hallucination for video captioning: (i) inadequate visual features from pre-trained models, (ii) improper influence of features during intra-modal and multi-modal fusion, and (iii) exposure bias in the model training strategy. We propose two robust and straightforward techniques that alleviate the problem. The exposure bias problem acts mainly under domain shift [19,38], so we will not be focusing on that. Standard evaluation metrics only measure similarity with ground truth caption and may not fully capture the object and action relevance. To this end,

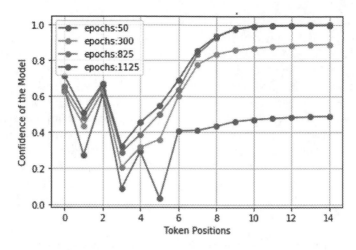

Fig. 2. The plot of the model confidence at the different token positions. The plot is done by averaging confidence over validation set on MSVD data and using the model MARN [23].

we propose a new metric COAHA (caption object and action hallucination assessment) to assess the hallucination rate. The CHAIR metric [27] has been proposed for image captioning. However, it considers only object hallucination and relies on object segmentation annotation unavailable for video captioning datasets. Unlike CHAIR, our metric considers both object and action hallucination and leverage existing captions instead of relying on segmentation annotations.

Finally, we show that our solutions outperform the state-of-the-art models in standard metrics, especially by a large margin in CIDEr-D on Microsoft Research Video Description Corpus (MSVD) and MSR-Video to Text (MSR-VTT) datasets.

2   Related Work

2.1   Video Captioning

The main breakthrough in video captioning happened with the encoder-decoder based sequence to sequence architecture. Although MP-LSTM [35] applied this first, they did not capture the temporal information between frames. S2VT [34] and SA-LSTM [43] capture the temporal dependencies between frames. The former shares a single LSTM network for both encoder and decoder, whereas the later uses attention weights over frames and 3D HOG features. Recent methods have improved upon SA-LSTM [43]. For example, RecNet [2] has added reconstruction loss based on backward flow from decoder output to the video feature construction. MARN [23] has used external memory to capture various visual contexts corresponding to each word. Both MARN [23] and M3 [39] utilize motion and appearance features. However, unlike MARN [23], M3 [39] uses heterogeneous memory to model long-term visual-text dependency. OA-BTG [45] and STG-KD [20] both use object features along with appearance and motion features. OA-BTG [45] leverage the temporal trajectory features of salient objects in the video. In contrast, STG-KD [20] uses a spatio-temporal graph network for object interaction features and a Transformer network for the language model instead of recurrent neural networks. ORG-TRL [47] also uses object interaction features, but unlike STG-KD [20], it utilizes an external language model which guides the standard recurrent neural network based language model. Another line of work focuses on a better sampling strategy to choose informative video frames. PickNet [9] samples informative frames based on reward-based objectives, whereas SGN [29] uses partially decoded information to sample frames. Regardless of the improvements, hallucination is still one of the major problems in existing models. In this work, we will be focusing on significant sources of hallucination, how to mitigate them, and finally, a metric to measure the rate of hallucination.

2.2   Hallucination Problems in Natural Language Generation

Koehn et al. [16] pointed out the problem of hallucination among the six challenges in neural machine translation. Müller et al. [19] and Wang et al. [38] linked the problem of hallucination with exposure bias. Wang et al. [38] have used minimum risk training (MRT) instead of Maximum likelihood (MLE) to reduce the exposure bias and concluded the improvement mostly happens under domain shift. Lee et al. [17] has claimed that NMT models hallucinate under source perturbation. Tu et al. [31] has used the context gates to control the source and target contribution, reducing the hallucination for neural machine translation. In the case of image captioning, Anna et al. [27] discussed object hallucination and proposed CHAIR metric to assess the degree of object hallucination. Unlike neural machine translation and image captioning, video captioning is more complex because of the spatio-temporal nature of the input data. Although state-of-art video captioning models suffer from hallucination, there is no pre-attempt. To the best of our knowledge, this is the first attempt to address hallucination for video captioning.

3   Methodology

As shown in Fig. 3, the components of our proposed system are (a) Visual Encoder responsible for the extraction of different visual features from the input video, (b) Auxiliary heads to reduce the hallucination that occurs from inadequate visual features, (c) Context gates to dynamically control the importance of features during intra-modal and multi-modal fusion and, (d) Decoder which is a conditional language model. We utilize running visual and language memory along with these components to act as an external short-term memory. We have used multi-label Auxiliary head loss and Coherent loss along with the standard cross-entropy loss for training.

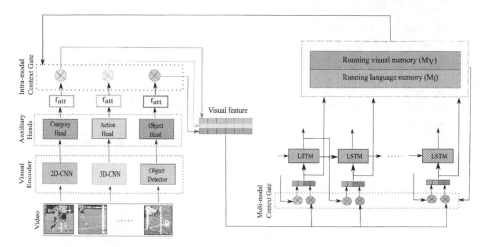

Fig. 3. The architecture of our proposed model. The components are (1) Visual Encoder for visual feature extraction, (2) auxiliary heads, and (3) context gates to reduce hallucination, and finally, (4) Decoder for sentence generation.

3.1   Visual Encoder

From a given input video, we uniformly sample N frames $\{f_i\}_{i=1}^N$ and clips $\{c_i\}_{i=1}^N$, where each c_i is a sequence of frames around f_i. We extract appearance features $\{a_i\}_{i=1}^N$ and motion features $\{m_i\}_{i=1}^N$ using pre-trained 2D CNN ϕ^a [10] and 3D CNN ϕ^m [13] respectively, where $a_i = \phi^a(f_i)$ and $m_i = \phi^m(c_i)$. Apart from appearance and motion, we also extract object features $\{o_i\}_{i=1}^N$ using a pre-trained object detection module ϕ^o [26] where $o_i = \phi^o(f_i)$. For each frame, we extract salient objects based on the objectiveness threshold v and average the features of those salient objects. Appearance ($\{a_i\}_{i=1}^N$) and motion ($\{m_i\}_{i=1}^N$) feature helps to understand the global context and the motion information of the video. In contrast, the object features ($\{o_i\}_{i=1}^N$) are more localized and helps to understand fine-grained information.

3.2   Auxiliary Heads: Category, Action, and Object

Unlike image captioning, the sequence of frames in video captioning does make the training of the feature extraction module computationally extensive. Existing methods use pre-trained models for different visual feature extraction and consider these features as the input to the model. However, the pre-trained models are trained on different data distribution, and extracted features are not adequate, which leads to the hallucination problem. We hypothesize that inadequate object and appearance features are responsible for object hallucination, whereas inadequate motion features lead to action hallucination.

We propose specialized classification heads on top of each type of visual feature, and they are trained in multi-label settings. The supervision is generated by leveraging existing captions, and the features from the heads are more suitable and semantically relevant. We include object head ψ^h_{Obj} on top of the object features $\{o_i\}^N_{i=1}$, where the training objective is to predict visual objects present in the caption for that input video. Similarly, action head ψ^h_{Act} applies to motion features $\{m_i\}^N_{i=1}$ to predict actions present in the caption. The ground truth for both the action head and object head are extracted using Named Entity Recognition [7,12] and Parts-of-speech tagging [37] methods from NLP literature. To refine the appearance features $\{a_i\}^N_{i=1}$, we add categorical head ψ^h_{Cat} to predict the video category, which is available for the MSR-VTT dataset but not for MSVD. Although we have used three specific objectives, this framework can be applied with other objectives, such as predicting attributes on top of the appearance feature. Finally, we use the features $\{a^r_i\}^N_{i=1}$, $\{m^r_i\}^N_{i=1}$ and $\{o^r_i\}^N_{i=1}$ from the specific auxiliary head as the input to the model, where $a^r_i = \psi^h_{Cat}(a_i)$, $m^r_i = \psi^h_{Act}(m_i)$ and $o^r_i = \psi^h_{Obj}(o_i)$. The Fig. 4 shows the modeling of the action head.

3.3   Context Gates: Balancing Intra-modal and Multi-modal Fusion

There are two types of fusion of features in the existing video captioning framework: intra-modal fusion and inter-modal or multi-modal fusion. In intra-modal fusion, different visual (appearance, motion, and object) features are merged, and thus one feature might dominate other vital features. For example, motion features might dominate appearance and object features, which lead to object hallucination, and action hallucination occurs due to dominance of object and appearance features over motion features. Similarly, in the case of multi-modal fusion at the decoder, the dominant language prior leads to hallucination [31,36], whereas the dominant source context leads to a not fluent caption.

We have included two separate context gate units for intra-modal and multi-modal fusion of features, which mitigate the hallucination. The gates are responsible for dynamically selecting the importance of each type of feature. Tu et al. [31] have used the context gate before to balance the source and target context in neural machine translation. Nevertheless, unlike Tu et al. [31], we have used separate gates for intra-modal and multi-modal fusion. Also, instead of depending on the decoder memory, our gates are conditioned on running visual (M_v) and language (M_l) memory as shown in Fig. 3. Empirical studies have shown better performance with separate running memories rather than decoder memory. One possible reason might be the decoder memory is contaminated by a mixture of source and target features and noisy LSTM gates.

At the encoder, the combined visual feature $v_{m,t}$ is,

$$v_{m,t} = [CG^E_a \circ a^r_t; CG^E_m \circ m^r_t; CG^E_o \circ o^r_t] \tag{1}$$

where,

$$a^r_t = \sum_{i=1}^{N} \alpha_{i,t} a^r_{i,t}, m^r_t = \sum_{i=1}^{N} \alpha_{i,t} m^r_{i,t}, o^r_t = \sum_{i=1}^{N} \alpha_{i,t} o^r_{i,t} \tag{2}$$

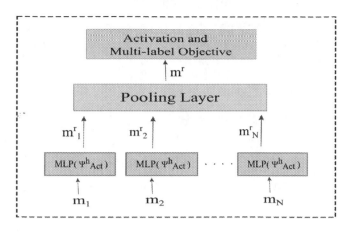

Fig. 4. Diagram of action auxiliary head. $\{m_i\}_{i=1}^N$ are the motion features from visual encoder. The MLP is shared over clip features. The same architecture follows for object and categorical auxiliary heads.

$$CG_{a/m/o}^E = f(a_t^r, m_t^r, o_t^r, M_v, M_l) \qquad (3)$$

$CG_{a/m/o}^E$ are the context gates corresponding to appearance, motion, and object features, respectively, and $\alpha_{i,t}$ are the attention weights [43]. We share the attention network over the object, motion, and appearance features for the regularization purpose.

At the decoder, the multi-modal feature C_t is,

$$C_t = [CG_S^D \circ v_{m,t}; CG_T^D \circ E(y_{t-1})] \qquad (4)$$

where,

$$CG_{S/T}^D = f(v_{m,t}, E(y_{t-1}), M_v, M_l) \qquad (5)$$

$CG_{S/T}^D$ are the context gates corresponding to source and target context, respectively, and $E(y_{t-1})$ is the embedding feature of the previous token. f is realized using MLP, [;] denotes concatenation, and $\circ$ denotes Hadamard product.

3.4 Decoder

The decoder is a language model conditioned on merged visual and language prior features C_t. We have used the LSTM network because of its superior sequential modeling performance. At t time steps during the sentence generation, the hidden memory of LSTM can be expressed as,

$$h_t = LSTM(C_t, h_{t-1}) \qquad (6)$$

where h_{t-1} is the hidden memory at previous time steps, and C_t is the combined multi-modal feature as shown in Eq. 4. Although transformer-based [32] models have outperformed LSTM in most NLP benchmarks, LSTM and transformer models performed almost equally in our case. Finally, the probability distribution over words is generated using a fully connected layer followed by a softmax layer.

$$P(y_t|V, y_1, y_2, .., y_{t-1}) = softmax(V_h h_t + b_h) \tag{7}$$

where V_h and b_h are learnable parameters.

3.5   Parameter Learning

Along with the standard cross-entropy loss, our model is trained using two additional losses, Auxiliary head loss, and Coherent loss.

Attention Based Decoder. Negative log-likelihood (or cross-entropy) is the standard objective function for the video captioning models. The loss for a mini-batch is

$$L_{CE} = -\sum_{i=1}^{B}\sum_{t=1}^{T} \log p(y_t|V, y_1, y_2, .., y_{t-1}; \theta) \tag{8}$$

where θ is learnable parameters, V is the video feature, y_t is the tth word in the sentence of length T, and B is the mini-batch size.

Auxiliary Heads. Auxiliary heads are trained in multi-label settings. For all three heads, we have used binary cross-entropy loss over sigmoid activation. The total auxiliary loss is,

$$L_{AH} = \lambda_C L_{AH}^C + \lambda_A L_{AH}^A + \lambda_O L_{AH}^O \tag{9}$$

where λ_C, λ_A and λ_O are hyperparameters corresponding to Category head loss L_{AH}^C, Action head loss L_{AH}^A, and Object head loss L_{AH}^C, respectively. Each loss is calculated as

$$L_{AH}^C = \Upsilon(Y^C, \psi_{Cat}^h(a_{j\,j=1}^N)), L_{AH}^A = \Upsilon(Y^A, \psi_{Act}^h(m_{j\,j=1}^N))$$

$$L_{AH}^O = \Upsilon(Y^O, \psi_{Obj}^h(o_{j\,j=1}^N))$$

where,

$$\Upsilon(y,p) = \sum_{i=1}^{B}\sum_{m=1}^{M} [y_{i,m} \log p_{i,m} + (1 - y_{i,m}) \log(1 - p_{i,m})] \tag{10}$$

Y^C, Y^A and Y^O are category, action and object ground truth for the given batch (B) of data.

Coherent Loss. The consecutive frames in a video are highly repetitive. So encoding of consecutive frames should be similar. We apply the coherent loss to restrict the embedding of consecutive frames to be similar. Coherent loss [23] has been used before to regularize attention weights, but unlike Pei et al. [23], we apply coherent loss on appearance, motion, and object features. It is debatable whether to apply coherent loss on motion features, but empirically we got a good result, and the possible reason might be shorter clip length. The total coherent loss for a mini-batch is,

$$L_{CL} = \lambda_{fcl} L_{CL}^a + \lambda_{mcl} L_{CL}^m + \lambda_{ocl} L_{CL}^o + \lambda_{acl} L_{CL}^\alpha \tag{11}$$

where λ_{fcl}, λ_{mcl}, λ_{ocl} and λ_{acl} are hyperparameters corresponding to appearance coherent loss L_{CL}^a, motion coherent loss L_{CL}^m, object coherent loss L_{CL}^o and attention coherent loss L_{CL}^α respectively.
The individual coherent losses are calculated as, $L_{CL}^a = \Phi(a_i^r)$, $L_{CL}^m = \Phi(m_i^r)$, $L_{CL}^o = \Phi(o_i^r)$ and $L_{CL}^\alpha = \Phi(\alpha_i)$ where,

$$\Phi(f) = \sum_{i=1}^{B} \sum_{t=1}^{T} \sum_{n=2}^{N} |f_{n,t}^{(i)} - f_{n-1,t}^{(i)}| \tag{12}$$

The overall loss function of our model is

$$L = L_{CE} + L_{AH} + L_{CL} \tag{13}$$

3.6   The COAHA Metric

In order to measure the degree of hallucination, we propose COAHA (caption object and action hallucination assessment) metric. The metric is defined for per instance and is given by,

$$COAHA = OH + AH \tag{14}$$

where,

$$OH = \frac{\sum_{h_i \in H_O} d_i^h}{T}, \quad AH = \frac{\sum_{a_i \in H_A} d_i^a}{T} \tag{15}$$

OH measures the degree of object hallucination, and AH measures the degree of action hallucination. T is the average ground truth caption length and N_O and N_A are the set of objects and actions mentioned in the video. H_O is the set of hallucinated objects, and H_A is the set of hallucinated actions. d_i^h measures the average semantic distance between the current hallucinated object and all objects in N_O. Similarly, d_i^a measures the average semantic dissimilarity of hallucinated action. The higher d_i^h or higher cardinality of H_O will lead to a higher value of OH, and the same follows for AH. So, if the hallucinated object or action is a synonym, then the semantic distance would be small, which will lead to a low hallucination value. Finally, we have taken the addition of OH and AH.

We have applied the same approach mentioned in the Auxiliary Head section to extract the list of objects and actions from each video (from the ground truth caption). For measuring the semantic distance (d_i^h and d_i^a) we have used pre-trained Fasttext [6] embedding model. Mathematically,

$$d_i^h = \frac{1}{|N_O|} \sum_{w_k \in N_O} \vartheta(w_k, h_i) \tag{16}$$

where, ϑ is the cosine similarity on top of the Fasttext [6] embedding. Similarly the formulae follows for d_i^a.

4   Experimental Results

We evaluate our proposed model on two most popular benchmark datasets: MSVD and MSR-VTT. To compare with the state-of-the-art results, we have used the four most popular metrics: CIDEr [33], METEOR [3], ROUGE-L [18] and BLEU-4 [22].

4.1   Datasets

MSVD. Microsoft Video Description (MSVD) dataset [8] consists of 1970 open domain videos of single activity and is described by 40 captions generated by Amazon Mechanical Turk. For a fair comparison, we have followed the standard split [35,43] of 1200 videos for training, 100 for validation, and 670 for testing.

MSR-VTT. MSR Video-to-Text (MSR-VTT) dataset [42] is a large scale benchmark datasets with 10k videos and 20 categories. Each video is annotated with 20 captions with an average length of 20 s. We have followed the standard benchmark split [42] of 6513 for training, 497 for validation, and 2990 for testing.

4.2   Implementation Details

Feature Extraction and Decoding. We uniformly sample 28 frames per video, and sentences longer than 30 words are truncated. The 1024-D appearance features of each frame are extracted by ViTL [10] pre-trained on Imagenet [28]. For the 2048-D motion features, we use C3D [13] with ResNeXt-101 [41] and pre-trained on the Kinetic-400 dataset. To extract the object features, we apply Faster-RCNN [26] with ResNet-101 and FPN backbone pre-trained on MSCOCO [40]. Appearance, motion, and objects features are embedded to 512-D before sending to the next stage. At the decoder LSTM, we use a hidden layer size of 512, and the word embedding size is fixed to 512. The vocabulary is built by words with at least 5 occurrences. During testing, we use greedy decoding to generate the sentences.

Auxiliary Heads and Context Gates Related Details. The running visual and language memory size is set to 512. We use the Spacy [14] module pre-trained with transformers [32] for object and action extraction from ground truth captions. We have also used Porter Stemmer [5] to work with the root

Table 1. Performance comparison on MSVD and MSR-VTT benchmarks. B4, M, R, and C denote BLEU-4, METEOR, ROUGE L, and CIDEr, respectively.

Models	MSVD				MSR-VTT			
	B@4	M	R	C	B@4	M	R	C
SA-LSTM [43]	45.3	31.9	64.2	76.2	36.3	25.5	58.3	39.9
h-RNN [44]	44.3	31.1	-	62.1	-	-	-	-
hLSTMat [30]	53.0	33.6	-	73.8	38.3	26.3	-	-
RecNet [2]	52.3	34.1	69.8	80.3	39.1	26.6	59.3	42.7
M3 [39]	52.8	33.3	-	-	38.1	26.6	-	-
PickNet [9]	52.3	33.3	69.6	76.5	41.3	27.7	59.8	44.1
MARN [23]	48.6	35.1	71.9	92.2	40.4	28.1	60.7	47.1
GRU-EVE [1]	47.9	35.0	71.5	78.1	38.3	28.4	60.7	48.1
POS+CG [37]	52.5	34.1	71.3	88.7	42.0	28.2	61.6	48.7
OA-BTG [45]	**56.9**	36.2	-	90.6	41.4	28.2	-	46.9
STG-KD [20]	52.2	**36.9**	73.9	93.0	40.5	28.3	60.9	47.1
SAAT [46]	46.5	33.5	69.4	81.0	40.5	28.2	60.9	49.1
ORG-TRL [47]	54.3	36.4	73.9	95.2	**43.6**	28.8	**62.1**	50.9
SGN [29]	52.8	35.5	72.9	94.3	40.8	28.3	60.8	49.5
Ours	53.3	36.5	**74.0**	**99.9**	41.1	**28.9**	61.9	**51.7**

form of words. Finally, to measure the semantic distance between words, we have used the pre-trained Fasttext [6] model from Gensim [25].

Other Details. We apply Adam [15] with a fixed learning rate of 1e-4 and a gradient clip value of 5. Our model is trained for 900 epochs with a batch size of 100. The coherent loss weights λ_{acl}, λ_{fcl}, λ_{mcl}, and λ_{ocl} are set as 0.01, 0.1, 0.01, and 0.1, respectively. All the experiments are done in a single Titan X GPU.

4.3   Quantitative Results

To proclaim the effectiveness of our approach, we compare the performance of our model with state-of-the-art models.

The quantitative results in Table 1 show that our model performs significantly better than other methods, especially in CIDEr. CIDEr is specially designed for captioning tasks and is more similar to human judgment than other metrics. Compared to OA-BTG [45] and ORG-TRL [47] our model lags in the BLEU-4 score, which is more suitable and designed for machine translation evaluation. Also, ORG-TRL [47] and OA-BTG [45] have utilized object interaction and object trajectory features, respectively, whereas we have only taken mean localized object features for simplicity.

4.4   Ablation Studies

We perform the quantitative evaluation to investigate the effectiveness of the components we proposed. We conduct ablation studies that start with a primary

Table 2. Ablation studies of the context gates and auxiliary heads on MSVD benchmark. C.Gates denotes context gates, and A.Heads denotes auxiliary heads.

Methods						
A.Heads	C.Gates	B@4	M	R	C	COAHA
✗	✗	50.1	34.8	72.9	91.1	10.57
✓	✗	52.6	36.2	73.4	96.6	7.90
✗	✓	52.5	36.3	73.6	97.1	7.96
✓	✓	53.3	36.5	74.0	99.9	7.02

decoder and add auxiliary heads and context gates incrementally. The Table 2 shows the results.

Table 3. Semantic significance of COAHA metric. *OH* and *AH* represent object and action hallucination scores respectively. For *OH*, *AH* and COAHA, lower is better.

Caption	OH	AH	COAHA
A man is riding a motorcycle	0	0	0
A man is riding a **horse**	8.12	0	8.12
A man is riding a **car**	5.10	0	5.10
A **woman** is riding a motorcycle	4.67	0	4.67
An **animal** is riding a motorcycle	9.58	0	9.58
A man is **playing** with a motorcycle	0	9.52	9.52
A man is **eating** a motorcycle	0	10.4	10.4
A man is **playing** with a **toy**	6.45	9.52	15.97

4.5 Validation of Context Gates

To validate that the context gates are helping in dynamically selecting the feature importance, we have shown the values of context gates during multi-modal fusion. The Fig. 5(a) shows that the source context gate value is high during the generation of visual object words. In contrast, the target context gate value is high during the generation of non-visual words.

4.6 Effect of Exposure Bias on Hallucination

In the case of machine translation, it is proven [19,38] that exposure bias affects mainly under domain shift. In order to validate that exposure bias does not affect in same domain cases, we have shown the model confidence (Fig. 5(b)) and COAHA values (Table 4) at different training strategies. Although the scheduled sampling [4] and the professor forcing [11] are designed to mitigate exposure bias, they do not affect hallucination significantly.

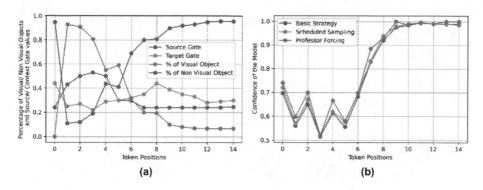

Fig. 5. (a) Contribution of context gates during the generation of visual and non-visual words. (b) The model confidence vs. token positions under different training strategies.

Table 4. COAHA scores at different training strategies.

Training strategy	COAHA
Teacher forcing	7.02
Scheduled sampling	7.01
Professor forcing	6.98

4.7   Significance of COAHA

In order to show the semantic significance of our proposed metric, we have calculated OH (Object hallucination rate), AH (Action hallucination rate), and COAHA for different semantically perturbed captions, as shown in Table 3.

4.8   Qualitative Results

We have compared the captions generated by our model and by Pei et al. [23]. From Fig. 6, we can see that our model is less prone to hallucination and better identify objects and actions in the video.

Groundtruth: A man rides a bike through beach.
MARN: A man is riding a horse.
Ours: A man is riding a motorcycle.

(a)

Groundtruth: A man is taking a right turn in a car.
MARN: A car is driving a car.
Ours: A man is driving a car.

(b)

Groundtruth: A man is eating spaghatti out of a large bowl.
MARN: A man is pouring a chicken.
Ours: A man is eating a bowl of food.

(c)

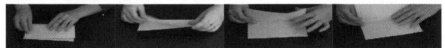

Groundtruth: Someone is folding a piece of paper.
MARN: A woman is drawing a paper.
Ours: A man is folding a piece of paper.

(d)

Fig. 6. Captions generated by our model and MARN [23].

5   Conclusion

This work shows that hallucination is a significant problem for video captioning and points out three significant hallucination sources in the existing framework. We propose two robust and straightforward strategies to mitigate hallucination. We have got new state-of-the-art on two popular benchmark datasets by applying our solutions, especially by a significant margin in CIDEr score. Furthermore, we have shown that exposure bias is not a significant issue in same domain cases. Finally, we propose a new metric to measure the hallucination rate that the standard metrics cannot capture. Qualitative and quantitative experimental results show our model's effectiveness in overcoming the hallucination problem in video captioning.

References

1. Aafaq, N., Akhtar, N., Liu, W., Gilani, S.Z., Mian, A.: Spatio-temporal dynamics and semantic attribute enriched visual encoding for video captioning. In: IEEE Conference on Computer Vision and Pattern Recognition, CVPR 2019, pp. 12487–12496 (2019)
2. Bairui, W., Lin, M., Wei, Z., Wei, L.: Reconstruction network for video captioning. In: Proceedings of the IEEE/CVF Conference on Computer Vision and Pattern Recognition (2018)
3. Banerjee, S., Lavie, A.: METEOR: an automatic metric for MT evaluation with improved correlation with human judgments. In: Proceedings of the Workshop on Intrinsic and Extrinsic Evaluation Measures for Machine Translation and/or Summarization@ACL 2005, Ann Arbor, Michigan, USA, 29 June 2005, pp. 65–72. Association for Computational Linguistics (2005). https://aclanthology.org/W05-0909/
4. Bengio, S., Vinyals, O., Jaitly, N., Shazeer, N.: Scheduled sampling for sequence prediction with recurrent neural networks. In: Advances in Neural Information Processing Systems 28: Annual Conference on Neural Information Processing Systems 2015, Montreal, Quebec, Canada (2015)
5. Bird, S., Klein, E., Loper, E.: Natural Language Processing with Python. O'Reilly, Sebastopol (2009). http://www.oreilly.de/catalog/9780596516499/index.html
6. Bojanowski, P., Grave, E., Joulin, A., Mikolov, T.: Enriching word vectors with subword information. Trans. Assoc. Comput. Linguist. 5, 135–146 (2017)
7. Chen, D., Manning, C.D.: A fast and accurate dependency parser using neural networks. In: Proceedings of the 2014 Conference on Empirical Methods in Natural Language Processing, EMNLP 2014, Qatar (2014)
8. Chen, D.L., Dolan, W.B.: Collecting highly parallel data for paraphrase evaluation. In: The 49th Annual Meeting of the Association for Computational Linguistics: Human Language Technologies, Proceedings of the Conference, 19–24 June 2011, Portland, Oregon, USA, pp. 190–200. The Association for Computer Linguistics (2011). https://aclanthology.org/P11-1020/
9. Chen, Y., Wang, S., Zhang, W., Huang, Q.: Less is more: picking informative frames for video captioning. In: Proceedings of the European Conference on Computer Vision (2018)
10. Dosovitskiy, A., et al.: An image is worth 16x16 words: transformers for image recognition at scale. In: 9th International Conference on Learning Representations, ICLR 2021, Virtual Event, Austria (2021)
11. Goyal, A., Lamb, A., Zhang, Y., Zhang, S., Courville, A.C., Bengio, Y.: Professor forcing: a new algorithm for training recurrent networks. In: Advances in Neural Information Processing Systems 29: Annual Conference on Neural Information Processing Systems 2016, Barcelona, Spain (2016)
12. Guillaume, L., Miguel, B., Sandeep, S., Kazuya, K., Chris, D.: Neural architectures for named entity recognition. In: Proceedings of the 2016 Conference of the North American Chapter of the Association for Computational Linguistics: Human Language Technologies (2016)
13. Hara, K., Kataoka, H., Satoh, Y.: Can spatiotemporal 3D CNNs retrace the history of 2D CNNs and imagenet? In: Proceedings of the IEEE Conference on Computer Vision and Pattern Recognition (2018)
14. Honnibal, M., Montani, I.: spaCy 2: Natural language understanding with Bloom embeddings, convolutional neural networks and incremental parsing (2017, to appear)

15. Kingma, D.P., Ba, J.: Adam: a method for stochastic optimization. In: 3rd International Conference on Learning Representations, ICLR 2015, San Diego, CA, USA (2015)
16. Koehn, P., Knowles, R.: Six challenges for neural machine translation. In: Proceedings of the First Workshop on Neural Machine Translation, NMT@ACL 2017, Vancouver, Canada, 4 August 2017, pp. 28–39. Association for Computational Linguistics (2017). https://doi.org/10.18653/v1/w17-3204
17. Lee, K., Firat, O., Agarwal, A., Fannjiang, C., Sussillo, D.: Hallucinations in neural machine translation. In: NIPS 2018 Interpretability and Robustness for Audio, Speech and Language Workshop (2018)
18. Lin, C.: Rouge: a package for automatic evaluation of summaries. In: Text Summarization Branches Out: Proceedings of the ACL 2004 Workshop, Barcelona, Spain, vol. 8 (2004)
19. Müller, M., Rios, A., Sennrich, R.: Domain robustness in neural machine translation. In: Proceedings of the 14th Conference of the Association for Machine Translation in the Americas, AMTA 2020, Virtual, 6–9 October 2020, pp. 151–164. Association for Machine Translation in the Americas (2020). https://aclanthology.org/2020.amta-research.14/
20. Pan, B., et al.: Spatio-temporal graph for video captioning with knowledge distillation. In: Proceedings of the IEEE/CVF Conference on Computer Vision and Pattern Recognition (2020)
21. Pan, P., Xu, Z., Yang, Y., Wu, F., Zhuang, Y.: Hierarchical recurrent neural encoder for video representation with application to captioning. In: Proceedings of the IEEE Conference on Computer Vision and Pattern Recognition (2016)
22. Papineni, K., Roukos, S., Ward, T., Zhu, W.: Bleu: a method for automatic evaluation of machine translation. In: Proceedings of the 40th Annual Meeting of the Association for Computational Linguistics, 6–12 July 2002, Philadelphia, PA, USA, pp. 311–318. ACL (2002). https://doi.org/10.3115/1073083.1073135. https://aclanthology.org/P02-1040/
23. Pei, W., Zhang, J., Wang, X., Ke, L., Shen, X., Tai, Y.: Memory-attended recurrent network for video captioning. In: IEEE Conference on Computer Vision and Pattern Recognition, CVPR 2019, Long Beach, CA, USA, 16–20 June 2019, pp. 8347–8356. Computer Vision Foundation/IEEE (2019)
24. Raunak, V., Menezes, A., Junczys-Dowmunt, M.: The curious case of hallucinations in neural machine translation. In: Proceedings of the 2021 Conference of the North American Chapter of the Association for Computational Linguistics: Human Language Technologies, NAACL-HLT 2021, Online, 6–11 June 2021, pp. 1172–1183. Association for Computational Linguistics (2021). https://doi.org/10.18653/v1/2021.naacl-main.92
25. Řehůřek, R., Sojka, P.: Software framework for topic modelling with large corpora. In: Proceedings of the LREC 2010 Workshop on New Challenges for NLP Frameworks, Valletta, Malta, pp. 45–50. ELRA (2010). http://is.muni.cz/publication/884893/en
26. Ren, S., He, K., Girshick, R.B., Sun, J.: Faster R-CNN: towards real-time object detection with region proposal networks. IEEE Trans. Pattern Anal. Mach. Intell. (2017)
27. Rohrbach, A., Hendricks, L.A., Burns, K., Darrell, T., Saenko, K.: Object hallucination in image captioning. In: Proceedings of the 2018 Conference on Empirical Methods in Natural Language Processing, Brussels, Belgium, 31 October–4 November 2018, pp. 4035–4045. Association for Computational Linguistics (2018). https://doi.org/10.18653/v1/d18-1437

28. Russakovsky, O., et al.: Imagenet large scale visual recognition challenge. Int. J. Comput. Vis. **115**, 211–252 (2015)
29. Ryu, H., Kang, S., Kang, H., Yoo, C.D.: Semantic grouping network for video captioning. In: Thirty-Fifth AAAI Conference on Artificial Intelligence, AAAI 2021 (2021)
30. Song, J., Gao, L., Guo, Z., Liu, W., Zhang, D., Shen, H.T.: Hierarchical LSTM with adjusted temporal attention for video captioning. In: Proceedings of the 26th International Joint Conference on Artificial Intelligence (2017)
31. Tu, Z., Liu, Y., Lu, Z., Liu, X., Li, H.: Context gates for neural machine translation. Trans. Assoc. Comput. Linguist. **5**, 87–99 (2017). https://transacl.org/ojs/index.php/tacl/article/view/948
32. Vaswani, A., et al.: Attention is all you need. In: Advances in Neural Information Processing Systems (2017)
33. Vedantam, R., Zitnick, C.L., Parikh, D.: Cider: consensus-based image description evaluation. In: IEEE Conference on Computer Vision and Pattern Recognition, CVPR 2015, Boston, MA, USA, 7–12 June 2015, pp. 4566–4575. IEEE Computer Society (2015). https://doi.org/10.1109/CVPR.2015.7299087
34. Venugopalan, S., Rohrbach, M., Donahue, J., Mooney, R.J., Darrell, T., Saenko, K.: Sequence to sequence - video to text. In: 2015 IEEE International Conference on Computer Vision, ICCV 2015, Santiago, Chile, 7–13 December 2015, pp. 4534–4542. IEEE Computer Society (2015). https://doi.org/10.1109/ICCV.2015.515
35. Venugopalan, S., Xu, H., Donahue, J., Rohrbach, M., Mooney, R.J., Saenko, K.: Translating videos to natural language using deep recurrent neural networks. In: NAACL HLT 2015, The 2015 Conference of the North American Chapter of the Association for Computational Linguistics: Human Language Technologies, Denver, Colorado, USA, 31 May–5 June 2015, pp. 1494–1504. The Association for Computational Linguistics (2015). https://doi.org/10.3115/v1/n15-1173
36. Voita, E., Sennrich, R., Titov, I.: Analyzing the source and target contributions to predictions in neural machine translation. In: Proceedings of the 59th Annual Meeting of the Association for Computational Linguistics and the 11th International Joint Conference on Natural Language Processing, ACL/IJCNLP 2021, (Volume 1: Long Papers), Virtual Event, 1–6 August 2021, pp. 1126–1140. Association for Computational Linguistics (2021). https://doi.org/10.18653/v1/2021.acl-long.91
37. Wang, B., Ma, L., Zhang, W., Jiang, W., Wang, J., Liu, W.: Controllable video captioning with POS sequence guidance based on gated fusion network. In: The IEEE International Conference on Computer Vision (ICCV) (2019)
38. Wang, C., Sennrich, R.: On exposure bias, hallucination and domain shift in neural machine translation. In: Proceedings of the 58th Annual Meeting of the Association for Computational Linguistics, ACL 2020, Online, 5–10 July 2020, pp. 3544–3552. Association for Computational Linguistics (2020). https://doi.org/10.18653/v1/2020.acl-main.326
39. Wang, J., Wang, W., Huang, Y., Wang, L., Tan, T.: M3: multimodal memory modelling for video captioning. In: Proceedings of the IEEE Conference on Computer Vision and Pattern Recognition (2018)
40. Wu, Y., Kirillov, A., Massa, F., Lo, W.Y., Girshick, R.: Detectron2 (2019). https://github.com/facebookresearch/detectron2
41. Xie, S., Girshick, R.B., Dollár, P., Tu, Z., He, K.: Aggregated residual transformations for deep neural networks. In: IEEE Conference on Computer Vision and Pattern Recognition, CVPR 2017, pp. 5987–5995 (2017)

42. Xu, J., Mei, T., Yao, T., Rui, Y.: MSR-VTT: a large video description dataset for bridging video and language. In: 2016 IEEE Conference on Computer Vision and Pattern Recognition, CVPR 2016, Las Vegas, NV, USA, 27–30 June 2016, pp. 5288–5296. IEEE Computer Society (2016). https://doi.org/10.1109/CVPR.2016.571

43. Yao, L., et al.: Describing videos by exploiting temporal structure. In: 2015 IEEE International Conference on Computer Vision, ICCV 2015, Santiago, Chile, 7–13 December 2015, pp. 4507–4515. IEEE Computer Society (2015). https://doi.org/10.1109/ICCV.2015.512

44. Yu, H., Wang, J., Huang, Z., Yang, Y., Xu, W.: Video paragraph captioning using hierarchical recurrent neural networks. In: Proceedings of the IEEE Conference on Computer Vision and Pattern Recognition (2016)

45. Zhang, J., Peng, Y.: Object-aware aggregation with bidirectional temporal graph for video captioning. In: Proceedings of the IEEE Conference on Computer Vision and Pattern Recognition (2019)

46. Zheng, Q., Wang, C., Tao, D.: Syntax-aware action targeting for video captioning. In: Proceedings of the IEEE/CVF Conference on Computer Vision and Pattern Recognition (2020)

47. Ziqi, Z., et al.: Object relational graph with teacher- recommended learning for video captioning. In: Proceedings of the IEEE/CVF Conference on Computer Vision and Pattern Recognition (2020)

Boundary-Aware Temporal Sentence Grounding with Adaptive Proposal Refinement

Jianxiang Dong and Zhaozheng Yin[✉]

Stony Brook University, New York, USA
jianxiang.dong@stonybrook.edu, zyin@cs.stonybrook.edu

Abstract. Temporal sentence grounding (TSG) in videos aims to localize the temporal interval from an untrimmed video that is relevant to a given query sentence. In this paper, we introduce an effective proposal-based approach to solve the TSG problem. A Boundary-aware Feature Enhancement (BAFE) module is proposed to enhance the proposal feature with its boundary information, by imposing a new temporal difference loss. Meanwhile, we introduce a Boundary-aware Feature Aggregation (BAFA) module to aggregate boundary features and propose a Proposal-level Contrastive Learning (PCL) method to learn query-related content features by maximizing the mutual information between the query and proposals. Furthermore, we introduce a Proposal Interaction (PI) module with Adaptive Proposal Selection (APS) strategies to effectively refine proposal representations and make the final localization. Extensive experiments on Charades-STA, ActivityNet-Captions and TACoS datasets show the effectiveness of our solution. Our code is available at https://github.com/DJX1995/BAN-APR.

1 Introduction

Temporal sentence grounding (TSG) in videos has been an important but challenging problem in Vision-Language understanding area [1], which requires techniques from both Computer Vision and Natural Language Processing. It aims to localize a temporal video moment[1] in an untrimmed video that semantically matches a given query sentence, as shown in Fig. 1(a). In addition, TSG can be widely used in many applications such as question answering in videos, video event captioning and human computer interaction [2–4].

Most TSG methods can be categorized into two groups: (1) proposal-based approaches and (2) proposal-free methods. The former one borrows techniques from object detection. A set of proposals with start and end timestamps are proposed, then visual-language interaction is performed to generate proposal representations. Afterwards, it computes the similarity between each proposal and the given query sentence and selects the proposal with the highest matching

---

[1] We define a moment to be an interval in the video with the start and end timestamps.

© The Author(s), under exclusive license to Springer Nature Switzerland AG 2023
L. Wang et al. (Eds.): ACCV 2022, LNCS 13844, pp. 641–657, 2023.
https://doi.org/10.1007/978-3-031-26316-3_38

score as the localized video moment. The latter one also performs complex visual-language interaction but it directly predicts the start and end timestamps based on the feature representation.

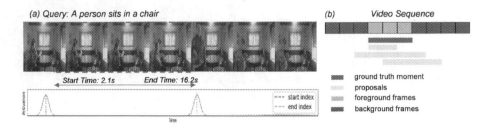

Fig. 1. (a) An illustration of the temporal sentence grounding in videos. *Bottom:* We expect the learned boundary-aware feature representation to have high activations of temporal difference around the start and end timestamps. (b) Proposal-based methods aggregate video clips within each proposal to generate the proposal feature. However, these proposals may include video clips (background frames) outside the ground truth moment (foreground frames), which will affect the quality of the learned proposal representations.

Despite the success of previous methods, there are three challenges remaining unsolved in TSG:

- *Feature discrimination* around temporal boundaries. Most of previous methods usually apply the pooling operation on cross-modal representations, e.g., average pooling or RoI (Region of Interest) pooling [5,6] to generate proposal features, which neglects the discriminative temporal boundary information. Without this information, it is difficult to distinguish overlapping or nearby proposals correctly because the pooling results of overlapped or neighbouring regions are very similar.
- *Proposal construction* with precise feature representations. Another challenging but crucial problem in TSG task is how to extract the query-related visual information when constructing proposal representations. For example, in Fig. 1(b), some proposals include the whole ground truth moment and some have an overlap with the ground truth moment. When aggregating proposal representations based on video clips within proposals' start and end timestamps, we introduce a lot of noises from the unrelated background frames (frames outside the ground truth moment), which may affect the quality of the generated proposal representations.
- *Proposal Interaction.* The proposal-free methods do not construct proposal representations and also neglect the informative proposal relationship. Proposal-based approaches usually have a large number of proposals and use a complex proposal interaction module (e.g., a 2D CNN [7] or graph neural network [4]) to model the proposal relationship, which requires a lot of computations during training and inference.

To address the above challenges, we develop a novel Boundary-aware Network with Adaptive Proposal Refinement (BANet-APR) for temporal sentence grounding:

- Firstly, we design a Boundary-aware Feature Enhancement (BAFE) module to extract the start and the end boundary information. A temporal difference loss is applied onto the boundary-aware feature to ensure that the feature representation has a high activation of temporal difference around the start and end boundary positions (The concept is shown in Fig. 1(a) and the validation is shown in Fig. 6 in the experiment section).
- Secondly, we introduce a Boundary-aware Feature Aggregation (BAFA) module to generate discriminative and informative proposal representations where the boundary feature and semantic content feature are both considered when constructing proposal representations. Meanwhile, we design a Proposal-level Contrastive Learning (PCL) method to implicitly enforce the semantic content feature to be query-related. Unlike previous methods [8,9] which either adopt frame-based contrastive learning or predefine an IoU threshold to construct positive and negative sets, we enforce proposals that include the whole ground truth moment to be close to the query while proposals that have no overlap with the ground truth moment to be away from the query. The boundary-aware feature and the semantic content feature are complement to each other when generating proposal representations to learn more discriminative (from boundary-aware feature) and informative (from semantic content feature) proposal features for the TSG task.
- Finally, we propose a refinement module which makes coarse predictions, selects k proposals via Adaptive Proposal Selection (APS), performs Proposal Interaction (PI) among the selected proposals and then localizes the moment. This refinement module only models the interaction between the selected k confident and representative proposals instead of all proposals as used in many previous works [4,7,10], which is more effective and efficient.

Our main contributions are fourfold: 1) We design a Boundary-aware Feature Enhancement (BAFE) module to extract boundary information and design a temporal difference loss that is directly applied on the feature representation to learn more discriminative proposal features; 2) We propose a novel Boundary-aware Feature Aggregation (BAFA) with Proposal-level Contrastive Learning (PCL) method to learn more discriminative and informative proposal features; 3) We propose a Proposal Interaction (PI) module with Adaptive Proposal Selection (APS) strategies to effectively refine proposal representations and make the final localization prediction; and 4) Compared to the latest state-of-the-art, the experiments on three datasets (Charades-STA [11], ActivityNet Captions [12] and TACoS [13]) show the effectiveness of our proposed BANet-APR.

2   Related Work

Temporal sentence grounding (TSG) is a new task introduced in the computer vision community recently [11,14,15]. Formally, it aims to retrieve a video moment with start and end timestamps from an untrimmed video using a query sentence. Current existing methods can be roughly grouped into two categories, namely propose-based and proposal-free methods [10,14,16–22]. Proposal-based methods first pre-define a set of video moment proposals and solve the problem by choosing the best matched proposal [4,7,11]. CTRL [11] adopts a sliding window to generate moment proposals and rank the proposals based on their similarity scores to the query sentence. 2D-TAN [7] and RaNet [4] enumerate all possible proposals and introduce complex proposal interaction methods using convolution and graph neural network, respectively. Proposal-free methods directly regress the start and end timestamps or classify which frames are the start and end boundary frames of the matched video moment [1,23–25].

Extracting informative boundary feature is very important and has been explored by some previous works [4,26–28]. RaNet [4] adopts two branches, e.g., a start branch and an end branch, to capture the start and end boundary information to learn more discriminative proposal representations. In addition to boundary branches, De-VLTrans-MSA [26] attaches a MLP as a classifier on top of the feature and defines a downstream task to predict the probability of a frame being the start or end stage to help boundary feature learning. CBP [27] designs an anchor submodule and a boundary submodule which take the feature vector at every temporal position as input and predict the probability of a proposal ending at that position. However, the classifier for the downstream tasks is easy to overfit to the statistics of the dataset and we may train a good classifier instead of good feature representations. Unlike previous methods which either have no boundary-related constraints on the feature, or use a classifier to predict the probability of a frame being the start and end boundary, our paper directly imposes a novel temporal difference loss on the boundary feature to enforce it to be boundary-aware. The loss is calculated only based on boundary regions instead of all temporal positions (so non-boundary regions with high temporal difference will not get punished). It allows the model itself to determine the high-salient temporal locations, which reduces the side effect of overfitting.

Proposal-based methods are intuitive and follow similar spirits of anchor-based approaches in object detection, but it suffers from the redundant computation cost. 2D-TAN [7] proposes a sparse sampling strategy to sparsely sample long proposals and densely sample short proposals, which is adopted by many other works. However, the proposal number is still too large for modeling dense proposal relationships. APGN [29] proposes a two-step model which first regresses a small number of (start, end) tuples as proposals and then ranks the proposed proposals. LPNet [30] proposes a novel model with a fixed set of learnable proposals where proposals' start and end positions serve as model parameters and are updated at every iteration. However, there is no constraint on the generated proposals, which may lead to repeated and useless proposals. Therefore, we propose to adaptively select unrepeated and representative proposals for further refinement.

Contrastive learning [31–34] methods are commonly used in self-supervised learning (SSL) research to learn high quality representations in an unsupervised manner. In image representation learning, it brings the representation of different transformations of the same image closer and push the representation of transformations from different images apart [35]. Contrastive learning is a MI (mutual information)-based approach and, in practice, contrastive learning models are usually trained by maximizing an estimation of MI between different transformations of data [36]. In the TSG task, IVG-DCL [8] adopts the contrastive learning concept in their model for the TSG task in which they apply contrastive loss on each element in the video sequence (clip-level) and on each video among all videos within a batch (video-level). In our PCL, we perform proposal-level contrastive learning which directly helps the discriminative proposal representation learning. Moreover, unlike previous methods [8,9] which construct positive and negative set based on IOUs, we do not manually set an IoU threshold for positive and negative splitting, which can reduce the side effect of confusing proposals and enforce the proposal encoder focusing more on the query-related information.

3 Methodology

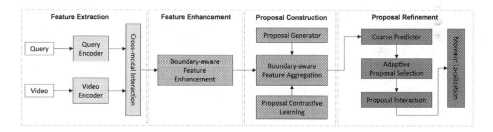

Fig. 2. An overview of our proposed model architecture for TSG.

Figure 2 depicts the overall framework of our BANet-APR, which consists of feature extraction, feature enhancement, proposal construction and proposal refinement: 1) Given a video and a query sentence, we use two encoders to generate the visual feature and the language feature, which are then fed into a cross-modal interaction module; 2) The output cross-modal feature will pass through a Boundary-aware Feature Enhancement module to strengthen the boundary information; 3) Then, we aggregate the enhanced feature and generate proposal representations by a proposal construction module; and 4) Finally, we design a Proposal Refinement module to adaptively select confident and informative proposals and refine their feature representations for the final moment localization.

3.1 Feature Extraction

Given an untrimmed video $\mathcal{V}$ and a sentence query $\mathcal{Q}$, we denote the video as $\mathcal{V} = \{v_t\}_{t=1}^{\hat{T}}$, where v_t is the t-th frame and $\hat{T}$ is the total number of frames. Similarly, each query sentence is represented by $\mathcal{Q} = \{w_i\}_{i=1}^{N}$, where w_i is the i-th word and N is the total number of words.

Feature Encoding. We use a pre-trained video feature extractor, such as C3D [37] or I3D [38], to obtain video features $\widetilde{\boldsymbol{V}} = \{\widetilde{\boldsymbol{v}}_i\}_{i=1}^T \in \mathbb{R}^{T \times d_v}$, where $\widetilde{\boldsymbol{v}}_i$ is the i-th video feature, d_v refers to the video feature dimension, $T = \hat{T}/s$ is the total number of extracted features from the video and s is the number of frames of a video clip in C3D or I3D models. Afterwards, we feed the video feature into a stacked Bi-LSTM [39] network as the visual encoder to further aggregate its contextual information over the temporal domain as:

$$\boldsymbol{V} = \text{VisualEncoder}(\widetilde{\boldsymbol{V}}), \tag{1}$$

where $\boldsymbol{V} = \{\boldsymbol{v}_i\}_{i=1}^T \in \mathbb{R}^{T \times d}$ is the final encoded visual feature and d is the dimension of the visual feature.

Similarly, we use the GloVe [40] to encode each word in the sentence. The encoded feature of the input sentence is denoted as $\widetilde{\boldsymbol{S}} = \{\widetilde{\boldsymbol{s}}_i\}_{i=1}^N \in \mathbb{R}^{N \times d_w}$, where d_w is the GloVe embedding dimension. Similarly, we use a stacked Bi-LSTM network as the lauguage encoder to further aggregate its sequential context:

$$\boldsymbol{Q} = \text{LanguageEncoder}(\widetilde{\boldsymbol{S}}), \tag{2}$$

where $\boldsymbol{Q} = \{\boldsymbol{q}_i\}_{i=1}^N \in \mathbb{R}^{N \times d}$ is the final encoded query feature for N words in the query, each of which has the same feature dimension of the visual feature.

Cross-modal Interaction. After getting the visual feature and the language feature, we adopt the Context-Query Attention Layer (CQA) [41] to model the interaction between the visual and the language feature. It takes the visual and the language feature as input and outputs the fused feature that provides rich cross-modal information for localization. The output will then be fed into another BiLSTM($\cdot$) to capture the sequential relationship.

$$\mathbf{F} = \text{BiLSTM}(\text{CQA}(\boldsymbol{V}, \boldsymbol{Q})) \in \mathbb{R}^{T \times d}, \tag{3}$$

where $\mathbf{F} \in \mathbb{R}^{T \times d}$ is the cross-modal representation.

3.2   Feature Enhancement

The generated cross-modal representation $\mathbf{F}$ contains both the visual and language information for the TSG task. However, it does not have the precise query-related boundary information, which is crucial for accurate moment localization.

Therefore, we design a Boundary-aware Feature Enhancement (BAFE) module which consists of two branches to 1) strengthen the start and end boundary information and 2) enforce the proposal construction to pay attention on the query-related elements in cross-modal features as well. The process is shown in Fig. 3. Given the cross-modal representation $\mathbf{F}$, we first pass it through two independent BiLSTMs to generate the boundary-aware feature ($\widetilde{\mathbf{F}}_b$) and the semantic content feature ($\widetilde{\mathbf{F}}_c$):

$$\widetilde{\mathbf{F}}_b = \text{BiLSTM}(\mathbf{F}), \tag{4}$$

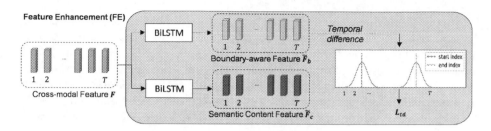

Fig. 3. Feature Enhancement. We have two branches to generate the boundary-aware feature and semantic content feature.

$$\widetilde{\mathbf{F}}_c = \text{BiLSTM}(\mathbf{F}), \tag{5}$$

where $\widetilde{\mathbf{F}}_{j \in \{b,c\}} = \{\widetilde{\boldsymbol{f}}_j^i\}_{i=1}^T \in \mathbb{R}^{T \times d}$ are the enhanced cross-modal feature.

We impose a temporal difference loss on the boundary-aware feature $\widetilde{\mathbf{F}}_b$ by calculating its temporal difference:

$$\Delta \widetilde{\boldsymbol{f}}_b = \{\Delta \widetilde{f}_b^i\}_{i=1}^T, \text{ where } \Delta \widetilde{f}_b^i = \|\widetilde{\boldsymbol{f}}_b^i - \widetilde{\boldsymbol{f}}_b^{i-1}\|_2 + \|\widetilde{\boldsymbol{f}}_b^i - \widetilde{\boldsymbol{f}}_b^{i+1}\|_2 \tag{6}$$

$$\boldsymbol{d} = Softmax(\Delta \widetilde{\boldsymbol{f}}_b) \tag{7}$$

where $\boldsymbol{d} = \{d_i\}_{i=1}^T$, d_i is a scalar value representing the temporal feature difference at position i, and $\| \cdot \|_2$ refers to l_2-norm.

For each video-query pair with the start and end ground truth timestamps $\hat{t}_s$ and $\hat{t}_e$, we calculate the ground truth temporal difference regarding boundaries:

$$\tilde{d}_i = \frac{1}{\sqrt{2\pi\sigma^2}}(e^{-\frac{(i-\hat{t}_s)^2}{2\sigma^2}} + e^{-\frac{(i-\hat{t}_e)^2}{2\sigma^2}}), \tag{8}$$

where i is the position index in the temporal domain, σ is the standard deviation of a Gaussian distribution (In this paper, $\sigma = \alpha(\hat{t}_e - \hat{t}_s)$ where α is a hyperparameter), $\hat{t}_s$ and $\hat{t}_e$ are the ground truth start and end positions. We expect the temporal difference of $\widetilde{\boldsymbol{F}}_b$ to have high activation values at the ground truth start and end boundaries as $\tilde{d}_i$. Thus, we define a temporal difference loss L_{td} as below.

$$L_{td} = -\sum_{i=1}^T \hat{d}_i \log(d_i) \tag{9}$$

where $\hat{d}_i = \tilde{d}_i / \sum_{i=1}^T \tilde{d}_i$ is the normalized ground truth temporal difference. By minimizing L_{td}, $\widetilde{\boldsymbol{F}}_b$ is forced to be a boundary-aware feature.

3.3 Proposal Construction

The general idea of proposal construction is to leverage both the discriminative boundary-aware feature and the informative semantic content feature to construct high quality proposal representations.

Proposal Generator. As illustrated in Fig. 4(a), in the proposal generator, the vertical and horizontal axes in the proposal generator denote the start and end indices of a proposal, respectively. Each block indicates a (start, end) pair and blocks in blue color are proposals with the valid (start, end) pairs, where the start index is smaller than the end index. The proposal generator will enumerate all possible valid pairs. Then, we densely sample short length proposals and sparsely sample longer length proposals so that proposals with large overlaps will not be selected [7].

Boundary-aware Feature Aggregation (BAFA). As shown in Fig. 4(b), given a proposal with the start and end timestamps (s, e) from the proposal generator, a BAFA module is designed to aggregate the boundary information from the boundary-aware feature, $\widetilde{\mathbf{F}}_b$, and the content information from the semantic content feature, $\widetilde{\mathbf{F}}_c$, to construct the proposal feature representation.

$$\mathbf{R} = \{r_m\}_{m=1}^{M}, \text{ where } r_m = MLP([\bar{f}_c^m; \widetilde{f}_b^{s_m}; \widetilde{f}_b^{e_m}]) \in \mathbb{R}^{1 \times d}, \qquad (10)$$

where $\mathbf{R} \in \mathbb{R}^{M \times d}$ are proposal features, M is the number of sampled proposals, $\widetilde{f}_b^{s_m}$ and $\widetilde{f}_b^{e_m}$ are elements at the start and end timestamps in the boundary-aware feature and $\bar{f}_c^m = maxpool(\widetilde{f}_c^{s_m} : \widetilde{f}_c^{e_m})$ is maxpooling the features in the semantic content feature sequence from the start to the end position of the mth proposal.

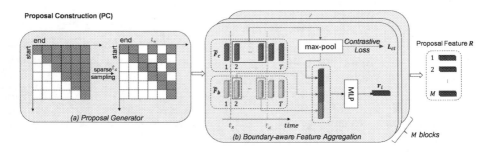

Fig. 4. (a)Proposal Constructor. We enumerate all possible proposals and sparsely sample long proposals. (b)Boundary-aware Feature Aggregation. We aggregate video clips in the boundary-aware feature and semantic content feature and pass it through a MLP layer to generate the final proposal feature.

Proposal-level Contrastive Learning. We impose a contrastive loss on proposal representations aggregated from the semantic content feature (shown in Fig. 4(b)), which serves as an additional supervision to guide the BAFA module to generate query-related proposals. Given all proposals P, we treat proposals that include the whole ground truth interval as positive sample set P_+, and the ones that have no overlap with the ground truth interval as negative sample set P_-. Then, we use the MIL-NCE loss [42] for our contrastive loss:

$$L_{cl} = -\log \frac{\sum\limits_{p \in P_+} e^{h_t(q)^\top h_v(p)}}{\sum\limits_{p \in P_+} e^{h_t(q)^\top h_v(p)} + \sum\limits_{p \in P_-} e^{h_t(q)^\top h_v(p)}}, \tag{11}$$

where $q = AvgPool(Q)$ is the global query feature representation and p (equivalent to the $\widetilde{f}_c^m$ in the BAFA module) is the proposal feature generated via max pooling over the semantic content feature from video clips. Similar to the Sim-CLR approach [34], we add two learnable projectors $h_v(\cdot)$ and $h_t(\cdot)$ to project proposal feature and text feature to a compatible embedding spaces, respectively, and apply the contrastive loss on the projected features.

3.4 Proposal Refinement

The proposal refinement module aims to effectively select a subset of proposals and refine their feature representations for the final localization.

Coarse Predictor. After getting all proposals (shown in Fig. 5(a)) and their corresponding proposal features, we impose a predictor to predict the matching score of each proposal regarding the query. The coarse predictor takes the proposal feature as the input and predicts the matching score of that proposal:

$$p_i = Sigmoid(MLP(\boldsymbol{r}_i)) \tag{12}$$

where $\boldsymbol{r}_i$ and p_i are the feature representation and the predicted matching score of the ith proposal, respectively.

Adaptive Proposal Selection. After obtaining the coarse matching scores of all proposals, we select k proposals out of them. However, instead of directly selecting proposals with the *top k* highest matching scores (shown in Fig. 5(b)), we design adaptive selection strategies which consider both the global diversity and local compaction. Specifically, we first perform Non-maximum Suppression to select *top m* confident anchor proposals with small overlaps. Then, for each anchor proposal, we select *top n* confident nearby proposals that have large overlaps with the corresponding anchor proposal ($k = m \times (n+1)$). Therefore, we are able to model both the global (between anchors) and local (between an anchor and its neighbors) relationships in the next proposal interaction step. Figure 5(c) shows the effectiveness of the adaptive selection. By comparing Fig. 5(b) and (c), we can observe that using adaptive proposal selection strategies has a larger chance to include the ground truth moment in the pool of proposals for further refinement.

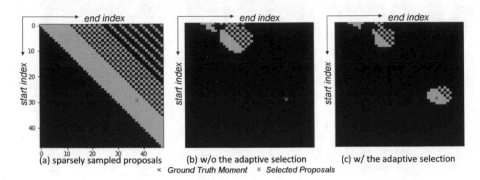

Fig. 5. Effectiveness of the adaptive proposal selection. Each cell denotes a possible proposal and the x axis and y axis represent the end and start indices of the proposal, respectively. Cells in green are selected proposals, and cells in black are not selected. The red cross represents the ground truth temporal interval.

Proposal Interaction. Given a proposal with the start and end positions, we embed it with its positional information based on sine and cosine functions of different frequencies [43]. Then, we adopt an MLP to project it into a new space:

$$\hat{r}_i = MLP([r_i; f_s^{pos}; f_e^{pos}])\tag{13}$$

where r_i is the ith proposal feature and f_s^{pos} and f_e^{pos} are the positional embeddings for the start and end positions, respectively.

Then, we adopt a Graph Neural Network (GNN) with Edge Convolution [29] to perform proposal interactions to refine proposal features where all the selected proposals are densely connected:

$$\tilde{R} = GNN(\hat{R}), \hat{R} = \{\hat{r}_i\}_{i=1}^k\tag{14}$$

where $\tilde{R} = \{\tilde{r}_i\}_{i=1}^k$ is the refined proposal representations.

Moment Localization. After getting the refined proposal feature, we impose two predictors to predict the matching score of a proposal and the boundary offsets of that proposal.

$$\hat{p}_i = Sigmoid(MLP(\tilde{r}_i))\tag{15}$$

$$(\Delta t_s^i, \Delta t_e^i) = MLP(\tilde{r}_i),\tag{16}$$

where $\hat{p}_i$ is the final predicted matching score and Δt_s^i and Δt_e^i are the predicted start and end boundary offsets, respectively. The refined proposal representations enable the newly predicted matching scores in Eq. 15 to be more precise than the matching scores from the coarse predictor.

We use the truncated IoU value [7] as the ground truth matching score and adopt a binary cross entropy loss for both the coarse and final matching score prediction losses L_{mc} and L_{mf}, respectively. And we use smooth $L1$ loss for the boundary offsets loss L_b:

$$L_{mc} = -\frac{1}{M}\Sigma_{i=1}^{M}(\hat{y}_i \log p_i + (1 - \hat{y}_i) \log(1 - p_i)) \qquad (17)$$

$$L_{mf} = -\frac{1}{k}\Sigma_{i=1}^{k}(\hat{y}_i \log \hat{p}_i + (1 - \hat{y}_i) \log(1 - \hat{p}_i)) \qquad (18)$$

$$L_b = SmoothL1(\Delta\hat{t}_s - \Delta t_s) + SmoothL1(\Delta\hat{t}_e - \Delta t_e) \qquad (19)$$

where $\hat{y}_i$ is the ground truth matching score, k is the number of selected proposals, M is the total number of sampled proposals and $\Delta\hat{t}_s$ and $\Delta\hat{t}_e$ are the ground truth boundary offsets.

3.5   Training and Inference

The overall training objective consists of the temporal difference loss L_{td} (Eq. 9), contrastive loss L_{cl} (Eq. 11), matching score prediction losses L_{mc} (Eq. 17) and L_{mf} (Eq. 18) and boundary offset loss L_b (Eq. 19):

$$L = \alpha_{td}L_{td} + \alpha_{cl}L_{cl} + \alpha_{mc}L_{mc} + \alpha_{mf}L_{mf} + \alpha_b L_b, \qquad (20)$$

where α_{td}, α_{cl}, α_{mc}, α_{mf} and α_b are loss weights to balance different loss contributions and are determined by the validation set.

During inference, we pass the video sequence and query sentence into the model and get a set of proposals together with their corresponding matching scores. We then select the proposal with the highest matching score to generate the final localization.

4   Experiments

We evaluate the proposed BANet-APR method and compare to the state-of-the-art approaches on Charades-STA, ActivityNet Captions and TACoS dataset.

4.1   Dataset and Evaluation

Charades-STA. Charades-STA is built on the Charades dataset [44] for indoor activities and is extended by [11] with temporal annotation of text descriptions. In total, it has 6,672 videos and 16,128 moment-query pairs, where 12408 pairs and 3720 pairs for training and testing, respectively.

ActivityNet-Captions. ActivityNet Captions is first introduced by [12] and there are more than 20k videos. Following the given split, we use val_1 as the validation set and val_2 as the testing set, resulting in 37417, 17505, and 17031 samples for training, validation and testing, respectively.

TACoS. TACoS [13] dataset is based on 127 indoor cooking videos in around 7 min on average. We follow the same splits as in [11] and have 10146, 4589 and 4083 moment-query pairs for training, validation and testing.

Evaluation Metric. For a fair comparison, we adopt the recall 1 at various thresholds of the Intersection over Union, $R1@IoU=m$, following previous works [11,45] to measure the percentage of the predicted proposals that have IoU with the ground truth annotation larger than m, where $m \in \{0.3, 0.5, 0.7\}$.

4.2 Implementation Details

For fair comparisons with state-of-the-art, we adopt the C3D [37] feature for ActivityNet and TACoS dataset, and use the I3D [38] feature for the Charades-STA dataset videos. We utilize a pre-trained Glove 840B 300d [40] to encode query sentences. In the experiment, we set the number of hidden units in Bi-LSTM to 256 and the feature dimension d to 512. In the contrastive loss, the dimension of the projected feature space is set to 128. In the adaptive proposal selection, we set the number of anchors $m = 16$ and the number of neighbors $n = 4$. The video feature sequence length for Charades-STA, ActivityNet and TACoS is set to 48, 64 and 128, respectively. The batch size is set to 32. We train our model using the Adam optimizer with a learning rate of 1×10^{-3}.

4.3 Comparison with State-of-the-Arts

In Table 1 we compare our BANet-APR with recent state-of-the-art methods [4,7,8,21,23,24,29,30,46,47]. Our model achieves the highest score in terms of $R1@IoU=0.7$ on Charades-STA and ActivityNet. In terms of $R1@IoU=0.5$, we have the highest score on Charades-STA dataset and the second best score on both ActivityNet and TACoS dataset. Although our model only achieves the second best scores on TACoS dataset, it has a large improvement compared with all other methods (except CPN [48]), and it gives much higher performance on the other two datasets compared with CPN.

4.4 Ablation Study

Effectiveness of Model Components. As shown in Table 2, we perform in-depth ablation studies on the effectiveness of different components in our model, including Boundary-aware Feature Enhancement (BAFE) in Feature Enhancement, Proposal-level Contrastive Learning (PCL) in Proposal Construction and Proposal Interaction (PI) and Adaptive Proposal Selection (APS) in Proposal Refinement modules based on the Charades-STA dataset. We can observe that the removal of any component will decrease the performance. Moreover, we can also observe from the last three rows that introducing a proposal interaction can help moment localization and using adaptive selection strategies can further improve the performance.

More Ablations on the Boundary-Aware Feature Enhancement. Table 3 illustrates the results of different designs of the boundary-aware feature enhancement module, from which we can observe the followings. First, simply

Table 1. Comparisons with other state-of-the-art methods. Bold and underline denote the best and the second best results, respectively. For a fair comparison, we only compare with models using the I3D or C3D feature.

Method	ActivityNet captions R1@		Charades-STA R1@		TACoS R1@	
	0.5	**0.7**	**0.5**	**0.7**	**0.3**	**0.5**
DRN [24]	45.45	24.36	53.09	31.75	–	23.17
LGI [23]	41.51	23.07	59.46	35.48	–	–
CPN [48]	45.10	28.10	59.77	36.67	**48.29**	**36.58**
IVG-DCL [8]	43.84	27.10	–	–	38.84	29.07
DeNet [49]	43.79	–	59.70	38.52	–	–
DRFT [45]	42.37	25.23	60.79	36.72	–	–
CBLN [47]	<u>48.12</u>	27.60	61.13	38.22	38.98	27.65
2D-TAN [7]	44.51	26.54	–	–	37.29	25.32
RaNet [4]	45.59	<u>28.67</u>	60.40	<u>39.65</u>	43.34	33.54
LPNet [30]	45.92	25.39	54.33	34.03	–	–
APGN [29]	**48.92**	28.64	<u>62.58</u>	38.86	39.34	28.34
Ours	<u>48.12</u>	**29.67**	**63.68**	**42.28**	<u>48.24</u>	<u>33.74</u>

taking the max-pooling results from the semantic-content feature to construct proposal representations gives the worst result. Secondly, introducing boundary-aware feature gives better results on the higher IoU metric. Finally, the combination of them gives the best performance which verifies that they are complement to each other.

4.5 Qualitative Results

Figure 6 demonstrates how the boundary-aware feature helps the localization. The bottom row in Fig. 6 displays the temporal difference values of the

Table 2. Effectiveness of different components in our model on Charades-STA dataset

Component				R1@	
BAFE	PCL	PI	APS	**0.5**	**0.7**
✗	✗	✗	✗	58.01	34.35
✓	✗	✗	✗	61.75	39.68
✓	✓	✗	✗	62.23	40.78
✓	✓	✓	✗	63.12	41.34
✓	✓	✓	✓	**63.68**	**42.28**

Table 3. Ablations of the boundary-aware feature enhancement on Charades-STA dataset. $\tilde{F}_c$ is the semantic content feature and $\tilde{F}_b$ is the boundary-aware feature.

Component		R1@	
$\tilde{F}_c$	$\tilde{F}_b$	**0.5**	**0.7**
✓	✗	62.35	39.07
✗	✓	61.96	40.45
✓	✓	**63.68**	**42.28**

boundary-aware feature $\tilde{F}_b$ after softmax normalization. We can observe that the boundary-aware feature has high activation values around the start and end positions, aiding the temporal localization accuracy.

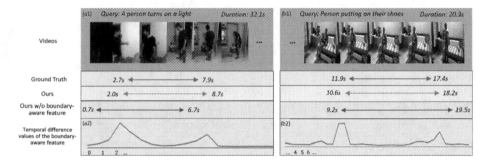

Fig. 6. Qualitative examples on the boundary-aware feature.

5   Conclusion

In this paper, we propose a novel Boundary-aware Network with Adaptive Proposal Refinement (BANet-APR) for the TSG task. Specifically, we design a Boundary-aware Feature Enhancement (BAFE) module to extract the boundary information and introduce a Boundary-aware Feature Aggregation (BAFA) module where the boundary-aware feature and the semantic content feature work together to construct discriminative and informative proposal representations. Moreover, a Proposal-level Contrastive Learning (PCL) method is proposed to enforce the semantic content feature to be query-related. Finally, we propose to adaptively select a subset of proposals and perform proposal interactions to refine their feature representations for the final localization. We conduct experiments on three benchmark datasets and show the effectiveness of our model.

Acknowledgement. Jianxiang Dong and Zhaozheng Yin have been supported by National Science Foundation via National Robotics Initiative grant CMMI-1954548 and Human Technology Frontier grant ECCS-2025929.

References

1. Zhang, H., Sun, A., Jing, W., Zhou, J.T.: Span-based localizing network for natural language video localization. arXiv preprint arXiv:2004.13931 (2020)
2. Zhu, F., Zhu, Y., Chang, X., Liang, X.: Vision-language navigation with self-supervised auxiliary reasoning tasks. In: Proceedings of the IEEE/CVF Conference on Computer Vision and Pattern Recognition, pp. 10012–10022 (2020)
3. Huang, D., Chen, P., Zeng, R., Du, Q., Tan, M., Gan, C.: Location-aware graph convolutional networks for video question answering. Proceed. AAAI Conf. Artif. Intell. **34**, 11021–11028 (2020)
4. Gao, J., Sun, X., Xu, M., Zhou, X., Ghanem, B.: Relation-aware video reading comprehension for temporal language grounding. arXiv preprint arXiv:2110.05717 (2021)
5. Girshick, R.: Fast R-CNN. In: Proceedings of the IEEE International Conference on Computer Vision, pp. 1440–1448 (2015)
6. Xu, H., Das, A., Saenko, K.: R-C3D: Region convolutional 3D network for temporal activity detection. In: Proceedings of the IEEE International Conference on Computer Vision, pp. 5783–5792 (2017)
7. Zhang, S., Peng, H., Fu, J., Luo, J.: Learning 2D temporal adjacent networks for moment localization with natural language. Proceed. AAAI Conf. Artif. Intell. **34**, 12870–12877 (2020)
8. Nan, G., et al.: Interventional video grounding with dual contrastive learning. In: Proceedings of the IEEE/CVF Conference on Computer Vision and Pattern Recognition, pp. 2765–2775 (2021)
9. Wang, Z., Wang, L., Wu, T., Li, T., Wu, G.: Negative sample matters: a renaissance of metric learning for temporal grounding. Proceed. AAAI Conf. Artif. Intell. **36**, 2613–2623 (2022)
10. Zhang, D., Dai, X., Wang, X., Wang, Y.F., Davis, L.S.: MAN: moment alignment network for natural language moment retrieval via iterative graph adjustment. In: Proceedings of the IEEE/CVF Conference on Computer Vision and Pattern Recognition, pp. 1247–1257 (2019)
11. Gao, J., Sun, C., Yang, Z., Nevatia, R.: TALL: temporal activity localization via language query. In: Proceedings of the IEEE International Conference on Computer Vision, pp. 5267–5275 (2017)
12. Krishna, R., Hata, K., Ren, F., Fei-Fei, L., Carlos Niebles, J.: Dense-captioning events in videos. In: Proceedings of the IEEE International Conference on Computer Vision, pp. 706–715 (2017)
13. Regneri, M., Rohrbach, M., Wetzel, D., Thater, S., Schiele, B., Pinkal, M.: Grounding action descriptions in videos. Trans. Assoc. Comput. Linguist. **1**, 25–36 (2013)
14. Hendricks, L.A., Wang, O., Shechtman, E., Sivic, J., Darrell, T., Russell, B.: Localizing moments in video with natural language. In: Proceedings of the IEEE International Conference on Computer Vision (ICCV), pp. 5803–5812 (2017)
15. Dong, J., et al.: Dual encoding for video retrieval by text. IEEE Trans. Patt. Anal. Mach. Intell. **44**, 4065–4080 (2021)
16. Ge, R., Gao, J., Chen, K., Nevatia, R.: MAC: mining activity concepts for language-based temporal localization. In: IEEE Winter Conference on Applications of Computer Vision (WACV), pp. 245–253 (2019)
17. Liu, M., Wang, X., Nie, L., He, X., Chen, B., Chua, T.S.: Attentive moment retrieval in videos. In: Proceedings of the 41nd International ACM SIGIR Conference on Research and Development in Information Retrieval (SIGIR), pp. 15–24 (2018)

18. Chen, J., Chen, X., Ma, L., Jie, Z., Chua, T.S.: Temporally grounding natural sentence in video. In: Proceedings of the 2018 Conference on Empirical Methods in Natural Language Processing (EMNLP), pp. 162–171 (2018)

19. Zhang, Z., Lin, Z., Zhao, Z., Xiao, Z.: Cross-modal interaction networks for query-based moment retrieval in videos. In: Proceedings of the 42nd International ACM SIGIR Conference on Research and Development in Information Retrieval (SIGIR), pp. 655–664 (2019)

20. Liu, M., Wang, X., Nie, L., Tian, Q., Chen, B., Chua, T.S.: Cross-modal moment localization in videos. In: Proceedings of the 26th ACM international conference on Multimedia, pp. 843–851 (2018)

21. Yuan, Y., Ma, L., Wang, J., Liu, W., Zhu, W.: Semantic conditioned dynamic modulation for temporal sentence grounding in videos. In: Advances in Neural Information Processing Systems (NIPS), pp. 534–544 (2019)

22. Xu, H., He, K., Plummer, B.A., Sigal, L., Sclaroff, S., Saenko, K.: Multilevel language and vision integration for text-to-clip retrieval. Proceed. AAAI Conf. Artif. Intell. **33**, 9062–9069 (2019)

23. Mun, J., Cho, M., Han, B.: Local-global video-text interactions for temporal grounding. In: Proceedings of the IEEE/CVF Conference on Computer Vision and Pattern Recognition, pp. 10810–10819 (2020)

24. Zeng, R., Xu, H., Huang, W., Chen, P., Tan, M., Gan, C.: Dense regression network for video grounding. In: Proceedings of the IEEE/CVF Conference on Computer Vision and Pattern Recognition (CVPR), pp. 10287–10296 (2020)

25. Cao, M., Chen, L., Shou, M.Z., Zhang, C., Zou, Y.: On pursuit of designing multimodal transformer for video grounding. arXiv preprint arXiv:2109.06085 (2021)

26. Zhang, M., et al.: Multi-stage aggregated transformer network for temporal language localization in videos. In: Proceedings of the IEEE/CVF Conference on Computer Vision and Pattern Recognition, pp. 12669–12678 (2021)

27. Wang, J., Ma, L., Jiang, W.: Temporally grounding language queries in videos by contextual boundary-aware prediction. Proceed. AAAI Conf. Artif. Intell. **34**, 12168–12175 (2020)

28. Xu, M., et al.: Boundary-sensitive pre-training for temporal localization in videos. In: Proceedings of the IEEE/CVF International Conference on Computer Vision, pp. 7220–7230 (2021)

29. Liu, D., Qu, X., Dong, J., Zhou, P.: Adaptive proposal generation network for temporal sentence localization in videos. arXiv preprint arXiv:2109.06398 (2021)

30. Xiao, S., Chen, L., Shao, J., Zhuang, Y., Xiao, J.: Natural language video localization with learnable moment proposals. arXiv preprint arXiv:2109.10678 (2021)

31. Van den Oord, A., Li, Y., Vinyals, O.: Representation learning with contrastive predictive coding. arXiv e-prints (2018). arXiv-1807

32. Misra, I., van der Maaten, L.: Self-supervised learning of pretext-invariant representations. In: Proceedings of the IEEE/CVF Conference on Computer Vision and Pattern Recognition, pp. 6707–6717 (2020)

33. Wu, Z., Xiong, Y., Yu, S.X., Lin, D.: Unsupervised feature learning via non-parametric instance discrimination. In: Proceedings of the IEEE Conference on Computer Vision and Pattern Recognition, pp. 3733–3742 (2018)

34. Chen, T., Kornblith, S., Norouzi, M., Hinton, G.: A simple framework for contrastive learning of visual representations. In: International Conference on Machine Learning, pp. 1597–1607. PMLR (2020)

35. Grill, J.B., et al.: Bootstrap your own latent-a new approach to self-supervised learning. Adv. Neural. Inf. Process. Syst. **33**, 21271–21284 (2020)

36. Gupta, T., Vahdat, A., Chechik, G., Yang, X., Kautz, J., Hoiem, D.: Contrastive learning for weakly supervised phrase grounding. In: Vedaldi, A., Bischof, H., Brox, T., Frahm, J.-M. (eds.) ECCV 2020. LNCS, vol. 12348, pp. 752–768. Springer, Cham (2020). https://doi.org/10.1007/978-3-030-58580-8_44

37. Tran, D., Bourdev, L., Fergus, R., Torresani, L., Paluri, M.: Learning spatiotemporal features with 3d convolutional networks. In: Proceedings of the IEEE International Conference on Computer Vision (ICCV), pp. 4489–4497 (2015)

38. Carreira, J., Zisserman, A.: Quo vadis, action recognition? a new model and the kinetics dataset. In: proceedings of the IEEE Conference on Computer Vision and Pattern Recognition (CVPR), pp. 6299–6308 (2017)

39. Schuster, M., Paliwal, K.K.: Bidirectional recurrent neural networks. IEEE Trans. Signal Process. **45**, 2673–2681 (1997)

40. Pennington, J., Socher, R., Manning, C.D.: GloVe: global vectors for word representation. In: Proceedings of the 2014 Conference on Empirical Methods in Natural Language Processing (EMNLP), pp. 1532–1543 (2014)

41. Yu, A.W., et al.: QANet: combining local convolution with global self-attention for reading comprehension. arXiv preprint arXiv:1804.09541 (2018)

42. Miech, A., Alayrac, J.B., Smaira, L., Laptev, I., Sivic, J., Zisserman, A.: End-to-end learning of visual representations from uncurated instructional videos. In: Proceedings of the IEEE/CVF Conference on Computer Vision and Pattern Recognition, pp. 9879–9889 (2020)

43. Vaswani, A., et al.: Attention is all you need. In: Advances in Neural Information Processing Systems 30 (2017)

44. Sigurdsson, G.A., Varol, G., Wang, X., Farhadi, A., Laptev, I., Gupta, A.: Hollywood in homes: crowdsourcing data collection for activity understanding. In: Leibe, B., Matas, J., Sebe, N., Welling, M. (eds.) ECCV 2016. LNCS, vol. 9905, pp. 510–526. Springer, Cham (2016). https://doi.org/10.1007/978-3-319-46448-0_31

45. Chen, Y.W., Tsai, Y.H., Yang, M.H.: End-to-end multi-modal video temporal grounding. In: Advances in Neural Information Processing Systems 34 (2021)

46. Rodriguez, C., Marrese-Taylor, E., Saleh, F.S., Li, H., Gould, S.: Proposal-free temporal moment localization of a natural-language query in video using guided attention. In: Proceedings of the IEEE/CVF Winter Conference on Applications of Computer Vision, pp. 2464–2473 (2020)

47. Liu, D., et al.: Context-aware biaffine localizing network for temporal sentence grounding. In: Proceedings of the IEEE/CVF Conference on Computer Vision and Pattern Recognition, pp. 11235–11244 (2021)

48. Zhao, Y., Zhao, Z., Zhang, Z., Lin, Z.: Cascaded prediction network via segment tree for temporal video grounding. In: Proceedings of the IEEE/CVF Conference on Computer Vision and Pattern Recognition, pp. 4197–4206 (2021)

49. Zhou, H., Zhang, C., Luo, Y., Chen, Y., Hu, C.: Embracing uncertainty: decoupling and de-bias for robust temporal grounding. In: Proceedings of the IEEE/CVF Conference on Computer Vision and Pattern Recognition, pp. 8445–8454 (2021)

Two-Stage Multimodality Fusion for High-Performance Text-Based Visual Question Answering

Bingjia Li[1], Jie Wang[2], Minyi Zhao[1], and Shuigeng Zhou[1(✉)]

[1] Shanghai Key Lab of Intelligent Information Processing, and School of Computer Science, Fudan University, Shanghai 200438, China
{bjli20,zhaomy20,sgzhou}@fudan.edu.cn
[2] ByteDance, Shanghai, China
wangjie.bernard@bytedance.com

Abstract. Text-based visual question answering (TextVQA) is to answer a text-related question by reading texts in a given image, which needs to jointly reason over three modalities—question, visual objects and scene texts in images. Most existing works leverage graph or sophisticated attention mechanisms to enhance the interaction between scene texts and visual objects. In this paper, observing that compared with visual objects, the question and scene text modalities are more important in TextVQA while both layouts and visual appearances of scene texts are also useful, we propose a two-stage multimodality fusion based method for high-performance TextVQA, which first semantically combines the question and OCR tokens to understand texts better and then integrates the combined results into visual features as additional information. Furthermore, to alleviate the redundancy and noise in the recognized scene texts, we develop a denoising module with contrastive loss to make our model focus on the relevant texts and thus obtain more robust features. Experiments on the TextVQA and ST-VQA datasets show that our method achieves competitive performance without any large-scale pre-training used in recent works, and outperforms the state-of-the-art methods after being pre-trained.

Keywords: TextVQA · Scene text recognition · Multimodal information fusion · Contrastive learning

1 Introduction

Nowadays, numerous methods [2, 15, 24, 30] have been proposed to solve the task of visual question answering (VQA) [3], which is to answer questions about images. However, these methods fail in answering text-related questions as they usually focus on objects and scenes while ignoring texts in the images. Actually, text matters in real life as it appears ubiquitously in practical images like advertisements, conveying valuable information that is essential for scene understanding and reasoning. Thus, text-based visual question answering (TextVQA) [5, 23, 28] is gaining

L. Wang et al. (Eds.): ACCV 2022, LNCS 13844, pp. 658–674, 2023.
https://doi.org/10.1007/978-3-031-26316-3_39

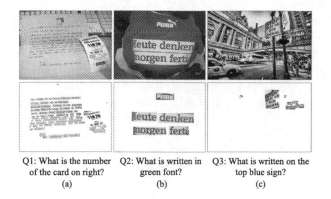

Q1: What is the number   Q2: What is written in   Q3: What is written on the
of the card on right?   green font?   top blue sign?
        (a)   (b)   (c)

Fig. 1. Some examples of TextVQA. The 1st, 2nd and 3rd rows show the original images, all the scene text areas extracted by an OCR system in each image, and the questions, respectively. All these questions can be correctly answered with only these cropped scene texts in the images. To correctly answer these questions, the model requires to first understand the semantics of the questions and scene texts, and then use some additional scene text information: (a) layout information for Q1, (b) visual appearance for Q2, (c) both layout and visual appearance for Q3.

popularity in recent years as an extension of VQA to answer text-related questions by reading and understanding scene texts in images.

Motivation. In general, the TextVQA task requires the model to read scene texts in images by an OCR system and jointly reason over three modalities - question, visual objects and scene texts. In our points of view, TextVQA research faces two major technical challenges as follows:

1) *How to effectively exploit multimodal information?* Most existing works [9, 11,14,19] use an object detector to extract global visual objects and utilize graph or complicated attention mechanisms to enhance the interaction between OCR tokens and visual objects. However, Wang *et al.* [31] pointed out that the accuracy of M4C [12] is almost unaffected after discarding the visual object modality, because the information of visual objects is not utilized well. Then, do global visual objects matter in TextVQA? As we all know, there are three modalities available for TextVQA: question, visual objects and scene texts in each image. And each piece of scene texts contains textual content, layouts and visual appearances (*font, color, background etc.*). Our preliminary study shows that compared to visual objects, scene texts are more important, and for high performance TextVQA, both layouts and visual appearances are indispensable. Concretely, we found that over 70% questions can be answered by using only the scene text areas of the images, and 60% questions can still be correctly answered even after discarding the visual appearances of scene texts while keeping only the layout information. As layouts can provide contextual information, which are not only critical to answer position-related questions, but also helpful for text understanding. Figure 1 shows some examples of TextVQA. We can see that to answer Q1 (Fig. 1(a)), layouts of scene texts are required, and visual appearances of scene texts are critical to

Q2 (Fig. 1(b)). While to correctly answer Q3 (Fig. 1(c)), besides layouts, the blue background of the text *"Park ave"* is also indispensable, without which the answer may become *"STOP"* mistakenly.

2) *How to select proper scene texts to answer the question?* Existing methods try to use as many OCR tokens as possible in the image to provide enough semantic information and make sure that the correct answer texts are included in the input sequence. Although more OCR tokens bring more contextual information, which does help improve the performance, they also introduce noise inevitably. Note that not all scene texts are relevant to the question. For example, in Fig. 1(a), only texts containing numbers (e.g. *"11870"* or *"15"*) need to be considered as the question is about number. Considering that too many unrelated OCR tokens may confuse the model, especially in text-intensive scenarios. An ideal solution is to select the texts most relevant to the question when semantic relationship is not enough to support question answering.

Solution and Contributions. In this paper, we pay more attention to the question and scene text modalities. On the one hand, to address the first challenge mentioned above, we propose a two-stage multimodality fusion module to take full advantage of the textual content, layouts and visual appearances of scene texts. In the first fusion stage, our model tries to understand the question and scene texts by combining them and the layouts of scene texts as contextual information with the help of LayoutLM [35]. After the textual and contextual features interaction, visual features are then included in the second fusion stage to handle questions that need the help of visual clues from scene texts. On the other hand, in order to handle the second challenge above, i.e., reducing redundancy and noise in the recognized scene texts, we develop a denoising module that masks irrelevant OCR tokens and uses contrastive loss to integrate the features of positive samples. In such a way, our model is able to focus on relevant texts and obtain more robust features.

In summary, the contributions of this paper are:

- Observing that the question and scene text modalities are of first importance in TextVQA, while both layouts and visual appearances of scene texts are useful, we propose a two-stage multimodality fusion based method to take full advantage of these information to boost TextVQA.
- We develop a denoising module with contrastive loss as an auxiliary task to reduce the redundancy and noise of recognized scene texts and thus make the model focus on the relevant texts and get more robust features.
- We validate the effectiveness and superiority of our method on the TextVQA and ST-VQA benchmarks. Experimental results show that our method achieves competitive results without any large-scale pre-training used in recent works, and outperforms the state-of-the-art methods after being pre-trained.

2   Related Work

TextVQA aims to answer text-related questions by first reading scene texts in images and then reasoning over three modalities—question, visual objects and scene texts. As a pioneer work, Singh *et al.* [28] proposed the first dataset TextVQA with a framework LoRRA by extending the VQA model Pythia [13] with an OCR attention branch. Later, several other datasets were built with texts of different scenarios, *e.g.* ST-VQA [5] in daily natural scenes, OCR-VQA [23] of book covers, STE-VQA [32] with bilingual texts and M4-ViteVQA [39] in video text understanding.

Recent works [4,9–12,14,19,21,29,31,36–38,40] have tried to improve the performance of TextVQA by various network architectures, more powerful OCR systems or large-scale datasets. Among them, M4C [12] utilizes multimodal transformers to fuse all modalities with a dynamic pointer network supporting multi-step answer decoding, which is the basis of most later works. With M4C, SA-M4C [14] proposes a spatiality-aware self-attention layer and handles different spatial relationships by different attention heads. Similarly, some other works [9,10,19,38,40] leverage graph or complicated attention mechanisms to emphasize the relationships between objects and OCR tokens, but the performance improvement is mainly gained by stronger OCR systems. TAP [36] is the first work to introduce pre-training to this task and pre-trains the model with three auxiliary tasks. With the help of the Microsoft-OCR system and the proposed large-scale dataset OCR-CC, it significantly boosts the TextVQA performance. LOGOS [21] enhances the model's understanding ability with two grounding tasks to better localize the key information of the image. LaTr [4] bases its architecture on T5 [25] and applies the pre-training strategy on large-scale scanned documents.

Though some latest works emphasize the significance of question and scene texts, yet none of them take full advantage of these two modalities with layouts and visual appearances of scene texts simultaneously. In this paper, we propose a two-stage multimodality fusion based method to comprehensively exploit such information. In addition, we also develop a denoising module with contrastive loss to reduce the redundancy and noise of recognized scene texts, which makes the model focus on the relevant texts. Our experiments verify the effectiveness and advantage of the proposed method.

3   Methodology

3.1   Overview

Figure 2 shows the architecture of our method, which mainly consists of three components: multimodal feature extraction, two-stage multimodality fusion and denoising. Besides, an optional pre-training component is considered. Given a sample X with an image I and a text-related question Q, we first extract features of the question, scene text and visual object modalities. These features are then progressively fused and reasoned with our two-stage multimodality fusion module, where the first stage focuses on textual and layout information from Q and scene texts in I, and the second stage includes visual appearances of scene texts and utilizes global visual objects as auxiliary information. The denoising module first masks the input OCR tokens, and then utilizes the masked result

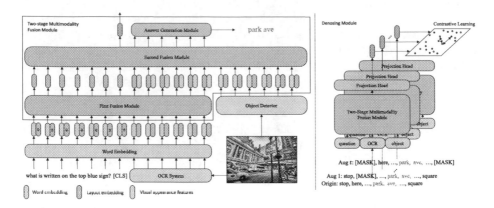

Fig. 2. The architecture of our method. After features of different modalities are extracted, they are fused and reasoned progressively with the two-stage fusion module. The output features of scene texts will be used for further answer decoding with the help of the denoising module, which masks the input and uses contrastive loss as an auxiliary task.

and a contrastive loss as an auxiliary task to make the model focus on the relevant texts. Optionally, our model can be further pre-trained on the question and scene text modalities to boost the performance. Note that in our method, the denoising module is used only in the fine-tuning stage.

3.2 Multimodal Features

OCR Features. After extracting OCR tokens in image I with an OCR system [7], previous works [12,19,36] obtain multiple features of OCR tokens with various pre-trained models and add them together before fusion. Unlike them, here we categorize the features into three types: layout embeddings, visual appearance features and word embeddings. Let $O = \{O_i\}_{i=1}^N$ be the OCR tokens after tokenization, where N is the length of the sequence. The layout embedding $x_i^{ocr,l}$ of the i-th OCR token O_i indicates its size and 2-D spatial position, which is defined as follows:

$$x_i^{ocr,l} = E_x(x_i^0) + E_y(y_i^0) + E_x(x_i^1) + E_y(y_i^1) + E_w(w_i) + E_h(h_i) \qquad (1)$$

where E_x, E_y, E_w and E_h are learnable embedding layers, (x_i^0, y_i^0) denotes the coordinates of the upper left corner, (x_i^1, y_i^1) denotes the coordinates of the bottom right corner, and w_i and h_i correspond to the width and height of the detected bounding box. All the coordinates have been scaled to 0–1000. For each OCR bounding box, we use Faster R-CNN [26] to extract the visual appearance feature $x_i^{ocr,v}$. As for texts, the word embedding is represented as $x_i^{ocr,t} = E_t(O_i)$, where E_t is a learnable embedding layer. The final representations of the detected OCR tokens are then defined as $X^{ocr,l} = \{x_i^{ocr,l}\}_{i=1}^N$, $X^{ocr,v} = \{x_i^{ocr,v}\}_{i=1}^N$ and $X^{ocr,t} = \{x_i^{ocr,t}\}_{i=1}^N$.

Question Features. Let $Q = \{Q_i\}_{i=1}^L$ be the sequence of question tokens after tokenization, where L is the sequence length. For the i-th token Q_i, the word embedding is represented as $x_i^{q,t} = E_t(Q_i)$, which shares the same embedding layer as the OCR word embedding. And the final representation is defined as $X^{q,t} = \{x_i^{q,t}\}_{i=1}^L$.

Object Features. As we have mentioned above, the visual object modality is not the key factor in the TextVQA task, so we just use it as a supplementary. To unify the inputs, we use the same Faster R-CNN to detect visual objects $V = \{V_i\}_{i=1}^M$ in image I, where M is the number of objects, and then extract features of each object V_i as $x_i^{obj,v}$. Similar to the OCR embedding, we obtain the layout embedding $x_i^{obj,l} = E_x'(x_i^{0'}) + E_y'(y_i^{0'}) + E_x'(x_i^{1'}) + E_y'(y_i^{1'}) + E_w'(w_i') + E_h'(h_i')$. As the features of visual objects are applied only in the second fusion stage, here we just sum them up as $x_i^{obj} = W_1 x_i^{obj,v} + x_i^{obj,l}$, where W_1 is a linear layer to control the dimension. The final object representation is defined as $X^{obj} = \{x_i^{obj}\}_{i=1}^M$.

3.3   Two-Stage Multimodality Fusion

After the features of different modalities are extracted, a common routine is to add the unimodal features together and fuse them with a multimodal transformer, which is not effective enough in our work. To take full advantage of the text, layout and visual appearance information, we propose two-stage multimodality fusion.

In the **first fusion stage**, the model focuses on understanding the texts, including the question and scene texts. We believe that there is a semantic connection between the scene texts and the question. Most questions are semantically closely related to the texts in the image, so texts can provide valuable clue for answering the question, which constitutes the basis of the TextVQA task. So we put these two modalities at the highest priority and jointly understand them. Besides, previous works [34, 35] on document understanding have shown the value of layout information, which provides contextual information. Similarly in natural scenes, a specific OCR token's 2-D position and positional relationship with its contextual tokens help us to understand the OCR token.

Here, we base the first stage fusion on LayoutLM [35]. LayoutLM is a BERT-like model that incorporates the visually rich layout information and align it with the input texts. With both word embeddings and layout embeddings as input, it is pre-trained with the masked visual language model (MVLM) on the document dataset IIT-CDIP Test Collection 1.0 [18], which can also bring more knowledge to our model. As there is a second fusion stage, we use only the first 6 layers of LayoutLM with the weights from HuggingFace [33] as initialization. To jointly reason over the question and OCR tokens, as shown in Fig. 3, we concatenate the features of the question and OCR tokens. The input of OCR tokens is defined as $X^{ocr,t} + X^{ocr,l}$ and we add a special [CLS] token as the beginning of OCR tokens to represent the whole texts, which will be used in the denoising module. As to the question, because there is no layout embedding, we

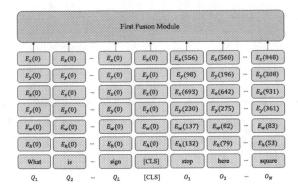

Fig. 3. The input of the first multimodality fusion stage. We concatenate the features of the question and OCR tokens together, including both word embeddings and layout embeddings. All the coordinates for the question are set to zero.

set all the coordinates as zero. Finally, we input the unified features into the fusion module and obtain

$$[X^{q'}; X^{ocr'}] = \Phi([X^{q,t}; X^{ocr,t} + X^{ocr,l}]) \tag{2}$$

where $X^{q'}$ and $X^{ocr'}$ are the semantically enhanced features for the question and OCR tokens respectively, $\Phi(\cdot)$ is the first fusion module LayoutLM and $[;]$ is a concatenation operation.

In the **second fusion stage**, the visual appearance features of scene texts are introduced as additional information to help handle questions that require visual clues from scene texts. Here, we combine them with the semantic features obtained in the first fusion stage. In addition, we also use the visual objects as an aid and then fuse the three modalities with a multimodal transformer. The output features are represented as

$$[X^{q''}; X^{ocr''}; X^{obj'}] = \Psi([X^{q'}; X^{ocr'} + W_2 X^{ocr,v}; X^{obj}]) \tag{3}$$

where $X^{q''}$, $X^{ocr''}$ and $X^{obj'}$ are the outputs for the question, scene texts and visual objects, respectively, $\Psi(\cdot)$ is the second fusion module and W_2 is a linear layer that projects $X^{ocr,v}$ to the same dimension as $X^{ocr'}$. Finally, $X^{ocr''}$ will be used in the denoising module and for further answer decoding.

3.4 Denoising Module

More scene texts bring more noise, considering only the semantic relationship between the question and OCR tokens is not enough to search for the relevant texts for answering the question. To relieve this issue, we design a denoising module, which augments the samples by randomly masking the input OCR tokens to reduce modality interaction and understanding difficulty, and utilizes contrastive learning to make the model focus on the relevant texts. In the following, we introduce this module in detail.

Masking Strategy. The text of correct answer provides important information, without which the text-related question about the image cannot be answered, so we avoid masking the tokens appearing in the answer. For the remaining OCR tokens, we randomly mask them with a certain probability. It is worth mentioning that masking the OCR tokens by totally deleting them may lead to confusion. For example, for some questions like *"What is the second word on the page?"*. If the first word is masked and we simply mask the final feature of it, the original second word will no longer be the correct answer as it has become the first one. Therefore, we replace the masked OCR tokens with a special [MASK] token while keeping the corresponding layout and visual appearance features. In this way, our model gets the information that there exists a word but does not know what the word is, which effectively reduces the interaction and understanding difficulty. Besides, in order to enlarge the differences between the original sample and the augmented ones, we use **whole word masking**. For example, in our model the input OCR word *"vegeburger"* is tokenized into three subtokens *"ve"*, *"##ge"* and *"##burger"* and the other features will be duplicated for each token. Once the word is chosen, we mask these subtokens of the word separately.

Contrastive Learning. With this masking strategy, for a batch of l samples, we augment each sample t times with a probability p. Given a pair of question and answer (a training sample), these augmented samples can be seen as positive samples of the original one while the rest are negative samples. As mentioned above, for the i-th sample X_i in the batch we can get the output features $X_i^{ocr''}$ of scene texts with the two-stage fusion module. We take the feature of the first [CLS] token as the representation of the whole sequence and project it to $z_i \in \mathbb{R}^d$ with a contrastive projection head, which is composed of two linear layers. Similarly, we can get the contrastive output of all the augmented samples using the same module. Then, we utilize a contrastive loss to constrain the distance between positive samples in the latent space as follows:

$$\mathcal{L}_{cont} = \sum_{i \in B} \mathcal{L}_i^{cont} = -\sum_{i \in B} \log \frac{\exp(z_i \cdot z_{j(i)}/\tau)}{\sum\limits_{a \in A(i)} \exp(z_i \cdot z_a/\tau)} \tag{4}$$

where $B = \{1, ..., l * (t + 1)\}$ is the index of samples in the batch after augmentation, '$\cdot$' represents the dot production, $j(i)$ denotes the indexes of positive samples of the i-th sample, $A(i) = B \backslash i$, and $\tau \in \mathbb{R}^+$ is a scalar temperature parameter. As only the correct answer will never be masked, it plays a key role in the loss, and the model is then forced to pay more attention on it.

Training. For the main task, we use teacher-forcing technique [17] and multi-label binary cross-entropy loss $\mathcal{L}_{bce}$ to train the model, which is defined as follows:

$$\mathcal{L}_{bce} = -y_{gt}\log(y_{pred}) - (1 - y_{gt})\log(1 - y_{pred})$$
$$y_{pred} = \frac{1}{1 + \exp(-f(X^{ocr''}))} \tag{5}$$

where y_{gt} is the ground-truth target, y_{pred} is the final prediction of our model and $f(\cdot)$ is the answer generation module following previous works of the TextVQA task. Finally, the total loss $\mathcal{L}$ is defined as a linear combination of the two losses above as follows:

$$\mathcal{L} = \lambda_1 * \mathcal{L}_{bce} + \lambda_2 * \mathcal{L}_{cont} \tag{6}$$

where λ_1 and λ_2 are hyper-parameters to trade-off the two losses.

3.5   Pre-training

So far, we have introduced the main components of our method. Recent works [4, 36] tend to take advantage of pre-training to improve the performance. As an option, our model also can be pretrained to achieve higher performance. Here, we conduct the masked language modeling (MLM) task on our model. Unlike TAP [36], we directly per-train our model on the question and scene text modalities with both layout and visual appearance features, but do not introduce additional texts to enhance the question modality.

LayoutLM has been pre-trained with both texts and layout information of documents. Considering the domain gap between documents and daily natural scenes, and the fact that natural scenes contain more visual information, we further pre-train our model to align all these features. In the pre-training stage, we randomly mask each text token of both question and scene texts with a probability 15%. The masked tokens are replaced by a special [MASK] with 80% probability, a random token with 10% probability, and remain unchanged with 10% probability. Note that we only mask the input tokens while keeping the corresponding 2-D position embeddings and visual appearance features as we believe that they provide additional contextual information. Tokens with close spatial relationship or sharing similar visual appearance tend to be more related in semantics. Then, the model is required to recover the masked word W_{mask} with two fully-connected layers. In our experiments, we find that the image-text matching (ITM) task used in TAP brings no performance improvement, so we do not use it. Finally, we pre-train our model with cross-entropy loss and use our denoising module in the fine-tuning stage.

4   Experiments

4.1   Datasets and Evaluation Metrics

TextVQA is the first proposed large-scale dataset for the TextVQA task, which contains a total of 45,336 questions about 28,408 images sampled from the Open Image dataset [16]. All the questions are related to the texts in the images, and each of them has 10 answers provided by 10 different annotators.

ST-VQA is similar to TextVQA, it consists of 23,038 images collected from more diverse sources. There are in total 31,791 questions while each question has up to two answers. All the questions can only be answered based on the texts that appear in the images. We report our results on the open dictionary task (task 3), which contains 18,921 training images and 2,971 test images. Following previous works [12], we split the training images into a training set with 17,028 images and a validation set with 1,893 images.

TextCaps and **OCR-CC** are introduced additionally during the pre-training stage. TextCaps [27] reuses images from the TextVQA dataset and attaches 145,329 captions to them, while OCR-CC was proposed along with TAP [36], which contains 1.367 million text-related image-caption pairs. These captions play the same role as the questions to increase data in the pre-training stage.

Evaluation Metrics. For the TextVQA dataset, we report the VQA accuracy [3] measured via the soft voting of the 10 answers. For the ST-VQA dataset, besides VQA accuracy we also report the Average Normalized Levenshtein Similarity (ANLS), which is defined as $1 - d_L(a_{pred}, a_{gt})/\max(|a_{pred}|, |a_{gt}|)$, where a_{pred} and a_{gt} are the predicted and ground-truth answers, while d_L refers to the edit distance. The final result is the average of all scores with those below the threshold 0.5 being truncated to 0.

4.2   Implementation Details

We implement our method based on the code of TAP [36]. The maximum input sequence lengths of question tokens, OCR tokens and object numbers are set to $L = 20$, $N = 300$, and $M = 100$ respectively. For our two-stage fusion module, the base model contains a LayoutLM with 6 layers and a multimodal transformer with 4 layers and 12 attention heads, while '†' refers to a larger model with 8 and 12 layers adopted in LayoutLM and the multimodal transformer respectively. The LayoutLM is initialized with weights from HuggingFace [33] and the multimodal transformer is initialized from scratch. The dimension in the joint embedding space is set to 768. In our denoising module, we use $d = 128$ for the output features of the contrastive projection head. Besides, we set $t = 3$ and $p = 0.15$ to augment each sample. For other hyper-parameters, we use $\tau = 0.07$, $\lambda_1 = 1$ and $\lambda_2 = 1$ by experience.

During the training stage, we adopt AdamW [20] as our optimizer and set the batch size l to 32. The learning rate is 1e-4 for the multimodal transformer and 1e-5 for the pre-trained LayoutLM. The warm-up learning ratio and warm-up iteration are set as 0.2 and 1,000. In the pre-training stage, we pre-train the model for 24,000 iterations when only TextVQA or ST-VQA datasets are used, and 240,000 iterations when TextCaps and OCR-CC datasets are included. In the fine-tuning stage, the model is further trained for another 30,000 iterations.

4.3   Experimental Results

Results on TextVQA. Following previous works [36], we evaluate our method under two different settings: the first is the constrained setting that uses only TextVQA for training and Rosetta [7] for OCR detection, the second is the unconstrained setting. All results are presented in Table 1.

As shown in Table 1, the **top part** of the table reports the results under the constrained setting. Because TAP [36] uses 100 OCR tokens, Latr [4] uses

Table 1. Performance comparison on the TextVQA dataset. The top part reports the results in the constrained setting that only uses TextVQA for training and Rosetta for OCR detection, while the bottom part displays the results in the unconstrained setting. We compare our method with several state-of-the-art methods, and our method clearly outperforms them.

Method	OCR system	Extra data	Val acc.	Test acc.
M4C [12]	Rosetta-en	✗	39.40	39.01
SMA [9]	Rosetta-en	✗	40.05	40.66
CRN [19]	Rosetta-en	✗	40.39	40.96
LaAP-Net [11]	Rosetta-en	✗	40.68	40.54
BOV [37]	Rosetta-en	✗	40.90	41.23
TAP [36]	Rosetta-en	✗	44.06	-
LaTr [4]	Rosetta-en	✗	44.06	-
Ours (100 tokens)	Rosetta-en	✗	44.20	-
Ours	Rosetta-en	✗	**44.75**	-
M4C [12]	Rosetta-en	ST-VQA	40.55	40.46
LaAP-Net [11]	Rosetta-en	ST-VQA	41.02	40.54
SA-M4C [14]	Google-OCR	ST-VQA	45.40	44.6
SMA [9]	SBD-Trans OCR	ST-VQA	-	45.51
BOV [37]	SBD-Trans OCR	ST-VQA	46.24	46.96
TAP [36]	Microsoft-OCR	✗	49.91	49.71
TAP [36]	Microsoft-OCR	ST-VQA	50.57	50.71
LOGOS [21]	Microsoft-OCR	ST-VQA	51.53	51.08
TAP [36]	Microsoft-OCR	ST-VQA, TextCaps, OCR-CC	54.71	53.97
Ours	Microsoft-OCR	✗	53.33	52.35
Ours	Microsoft-OCR	ST-VQA	54.33	54.47
Ours[†]	Microsoft-OCR	ST-VQA, TextCaps, OCR-CC	**55.96**	**55.33**

512 OCR tokens and previous works tend to use 50 OCR tokens, to make a fair comparison, we conduct experiments on different numbers of OCR tokens. As can be seen, when using only 100 OCR tokens, our method has already lifted the accuracy achieved by TAP and LaTr from 44.06% to 44.20%. When using 300 OCR tokens, our method achieves the state-of-the-art accuracy of 44.75%.

The **bottom part** displays results in the unconstrained setting and we also list the OCR system and extra data used by different methods. Following TAP [36], we use Microsoft-OCR to detect scene texts in the images and gradually expand the training data. As we can see, **(1)** when switching to the Microsoft-OCR, without any extra data our method achieves 53.33% and 52.35% on the validation and test set, improving TAP by **+3.42%** and **+2.64%** respectively in the same setting. **(2)** When adding ST-VQA as another training dataset like previous works, our method improves the test accuracy from 51.08% of LOGOS [21] to 54.47% (**+3.39%**), which has already surpassed the final result of TAP after pre-trained on larger datasets (53.97%). Besides, unlike

Table 2. Performance comparison on the ST-VQA dataset.

Method	Val acc.	Val ANLS	Test ANLS
M4C [12]	38.05	0.472	0.462
SA-M4C [14]	42.23	0.512	0.504
SMA [9]	-	-	0.466
CRN [19]	-	-	0.483
LaAP-Net [11]	39.74	0.497	0.485
BOV [37]	40.18	0.500	0.472
LOGOS [21]	48.63	0.581	0.579
TAP [36]	50.83	0.598	0.597
Ours	50.49	0.598	0.587
Ours†	**55.51**	**0.646**	**0.634**

Table 3. Ablation study results on the TextVQA dataset. "TMFM" and "DM" refer to the two-stage multimodality fusion module and denoising module.

Configuration	TMFM	DM	Pre-train	Val acc.
(1) TextVQA	✓	✗	✗	50.71
(2) TextVQA	✓	✓	✗	51.32
(3) w/ ST-VQA	✓	✗	✗	51.49
(4) w/ ST-VQA	✓	✓	✗	52.56
(5) w/ MLM	✓	✗	✓	52.82
(6) w/ MLM, ITM	✓	✗	✓	52.47
(7) w/ MLM	✓	✓	✓	53.33

these methods, our accuracy on the test set is better than that on the validation set (54.33%), demonstrating that our method has better generalization ability. **(3)** After introducing image caption datasets when pre-training, our method achieves a final result of 55.33%, increasing the accuracy of TAP by **+1.36%**.

Results on ST-VQA. Table 2 presents the results on the ST-VQA dataset in the unconstrained setting. The base model is pre-trained and fine-tuned on the training set of ST-VQA and † refers to the large model that uses TextVQA, ST-VQA, TextCaps and OCR-CC in pre-training. Unlike the TextVQA dataset, in ST-VQA all the answers are texts in the images. Our method works better on ST-VQA as we focus more on scene texts. As can be seen, without any extra data, our method achieves the accuracy of 50.49% and the ANLS score of 0.598, which is nearly at the same level with the large-scale pre-trained TAP. When pre-trained on larger datasets, the final accuracy and ANLS score of our method reach 55.51% and 0.646, i.e., **+4.68%** and **0.048** higher than that of TAP.

4.4 Ablation Study

Here, we conduct ablation study to demonstrate the effectiveness of each component in our method. All the experiments are done on the TextVQA dataset.

Overall Results. As shown in Table 3, we first report the overall ablation results of different components in our method. Only by our two-stage multimodality fusion module but **without any pre-training**, the accuracy reaches 50.71%, which outperforms TAP after being pre-trained (49.91%), while our denoising module increases the performance to 51.32% (+0.61%). Especially, when introducing ST-VQA as an additional training dataset, the improvement gained by our denoising module increases to +1.07% (Row 3 and Row 4), showing that our denoising module works better with more data. One possible reason is that all the answers of the ST-VQA dataset are texts in the images without any external vocabulary, which is beneficial to the denoising module. Finally, our method achieves a competitive result 52.56% without pre-training. On the other hand, Row 5 and Row 6 show the results of our model with pre-training on

Table 4. Ablation study on the two-stage multimodality fusion module.

Fusion module	$X^{ocr,l}$	$X^{ocr,v}$	X^{obj}	Val acc.
One-stage fusion	✓	✓	✓	49.41
Two-stage fusion	✗	✗	✗	45.12
	✓	✗	✗	49.24
	✓	✓	✗	50.50
	✓	✓	✓	50.71

Table 5. Ablation study on augmentation times t and masking probability p of the denoising module.

	$p=0.15$	$p=0.3$	$p=0.5$
$t=1$	50.53	51.04	50.41
$t=3$	51.32	51.16	50.36

TextVQA dataset. As can be seen, the MLM task makes a positive impact while introducing the ITM task causes a decrease from 52.82% to 52.47% (-0.35%). This is possibly because that the semantic relationship between scene texts and the image is not so strong (different from the fact that image caption is usually closely related to the image). At last, Row 7 shows the final result of our model trained on the TextVQA dataset. Our denoising module still works well after the model being pre-trained, increasing the accuracy from 52.82% to 53.33%.

Effect of Two-Stage Multimodality Fusion Module. We conduct further ablation study on our two-stage multimodality fusion module to verify its effectiveness and show the influence of each feature. First, we test the one-stage fusion paradigm. In this setting, features of different modalities are extracted respectively and unimodal features are added together before fusion. Here, we use BERT [8] to extract the feature of the question following previous works while keeping LayoutLM to extract the semantic features of OCR tokens. The semantic and visual features of OCR tokens are then added together before interacting with the other modalities. As shown in Table 4, the final result is 49.41% while our two-stage multimodality fusion module improves the accuracy to 50.71 ($+1.30\%$), which verifies our expectation that jointly considering the question and scene texts are beneficial to the understanding of both modalities.

Then, we display the importance of various features. As we have mentioned in the beginning that texts are the basis, by removing the object features and layout/visual appearance features of scene texts in our two-stage fusion module, the model can still obtain an accuracy of 45.12%. Adding layout features in the first stage to help text understanding brings an increase of $+4.12\%$, and introducing visual appearance features of scene texts based on the former setting achieves additional improvement of $+1.26\%$, which shows the importance of both layout and visual appearance of scene texts. Finally, we add object features as auxiliary information. The gap of the accuracy between our model with or without object features is only about 0.2%, which is consistent with our claim that global visual objects are not the first important role in the TextVQA task.

Effect of Denoising Module. As shown in Table 5, we conduct ablation study on the hyper-parameters in our denoising module by setting different values for augmentation times and masking probability. Augmenting only one time

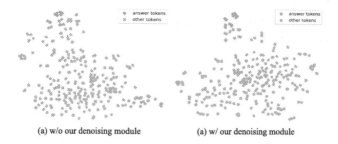

(a) w/o our denoising module   (a) w/ our denoising module

Fig. 4. tSNE results of the output features of OCR tokens in a text-rich sample. (a) Without our denoising module, there are more noise points around the answer tokens. (b) With our denoising module, fewer unrelated tokens are around the correct answers.

makes no obvious improvement, which may because of the strong randomness. Increasing augmentation times is helpful while increasing masking probability leads to accuracy decrease. At last, we choose to augment each sample 3 times with the masking probability being 15%.

In order to show the superiority of our masking strategy, we further test some other methods. Masking without excluding the correct answers obtains an accuracy of 50.74% (−0.58%) while deleting the whole features achieves 50.13% (−0.19%), which shows that our masking strategy is the best choice.

As using only data augmentation can also boost performance, we conduct experiment by removing our contrastive loss, and the accuracy decreases to 51.08%, which shows the effectiveness of contrastive learning. Besides, we compare the visualization results of models trained with and without our denoising module. Figure 4 shows the visualization results of the output features of OCR tokens by tSNE [22]. As seen, when trained without our denoising module, there are more noise points around the correct answer. While with our denoising module, as tokens in the correct answer are not masked and fewer tokens are considered each time, there are fewer noise points around the answer tokens.

Moreover, our proposed denoising module can also work well with previous works. Here we conduct experiments with the common baseline model M4C [12] on the TextVQA dataset. As the input features of M4C are different from ours, here we randomly mask the semantic features of FastText [6] and PHOC [1]. Our experimental results show that the denoising module lifts the accuracy of M4C from 45.55% to 46.24%, which verifies the effectiveness and flexibility of it.

5   Conclusion

In this paper, we observe that compared to visual objects, question and scene text modalities are more important in the TextVQA task while both layout and visual appearance are useful. Based on this observation, we propose a two-stage multimodality fusion based method to boost TextVQA. Besides, in order to alleviate the redundancy and noise of recognized scene texts, we develop a denoising module that utilize contrastive loss to make the model focus on the relevant texts. Extensive experiments on two benchmarks are conducted, which verify the effectiveness and superiority of the proposed method.

Acknowledgments. This work was supported in part by a ByteDance Research Collaboration Project.

References

1. Almazán, J., Gordo, A., Fornés, A., Valveny, E.: Word spotting and recognition with embedded attributes. IEEE Trans. Pattern Anal. Mach. Intell. **36**(12), 2552–2566 (2014)
2. Anderson, P., et al.: Bottom-up and top-down attention for image captioning and visual question answering. In: Proceedings of the IEEE Conference on Computer Vision and Pattern Recognition, pp. 6077–6086 (2018)
3. Antol, S., et al.: VQA: visual question answering. In: Proceedings of the IEEE International Conference on Computer Vision, pp. 2425–2433 (2015)
4. Biten, A.F., Litman, R., Xie, Y., Appalaraju, S., Manmatha, R.: LATR: layout-aware transformer for scene-text VQA. arXiv preprint arXiv:2112.12494 (2021)
5. Biten, A.F., et al.: Scene text visual question answering. In: Proceedings of the IEEE/CVF International Conference on Computer Vision, pp. 4291–4301 (2019)
6. Bojanowski, P., Grave, E., Joulin, A., Mikolov, T.: Enriching word vectors with subword information. Trans. Assoc. Comput. Linguist. **5**, 135–146 (2017)
7. Borisyuk, F., Gordo, A., Sivakumar, V.: Rosetta: large scale system for text detection and recognition in images. In: Proceedings of the 24th ACM SIGKDD International Conference on Knowledge Discovery and Data Mining, pp. 71–79 (2018)
8. Devlin, J., Chang, M.W., Lee, K., Toutanova, K.: Bert: pre-training of deep bidirectional transformers for language understanding. arXiv preprint arXiv:1810.04805 (2018)
9. Gao, C., et al.: Structured multimodal attentions for TextVQA. IEEE Trans. Pattern Anal. Mach. Intell. **44**(12), 9603–9614 (2021)
10. Gao, D., Li, K., Wang, R., Shan, S., Chen, X.: Multi-modal graph neural network for joint reasoning on vision and scene text. In: Proceedings of the IEEE/CVF Conference on Computer Vision and Pattern Recognition, pp. 12746–12756 (2020)
11. Han, W., Huang, H., Han, T.: Finding the evidence: localization-aware answer prediction for text visual question answering. arXiv preprint arXiv:2010.02582 (2020)
12. Hu, R., Singh, A., Darrell, T., Rohrbach, M.: Iterative answer prediction with pointer-augmented multimodal transformers for TextVQA. In: Proceedings of the IEEE/CVF Conference on Computer Vision and Pattern Recognition, pp. 9992–10002 (2020)
13. Jiang, Y., Natarajan, V., Chen, X., Rohrbach, M., Batra, D., Parikh, D.: Pythia v0.1: the winning entry to the VQA challenge 2018. arXiv preprint arXiv:1807.09956 (2018)
14. Kant, Y., et al.: Spatially aware multimodal transformers for TextVQA. In: Vedaldi, A., Bischof, H., Brox, T., Frahm, J.-M. (eds.) ECCV 2020. LNCS, vol. 12354, pp. 715–732. Springer, Cham (2020). https://doi.org/10.1007/978-3-030-58545-7_41
15. Kim, W., Son, B., Kim, I.: VILT: vision-and-language transformer without convolution or region supervision. In: International Conference on Machine Learning, pp. 5583–5594. PMLR (2021)

16. Krasin, I., et al.: Openimages: a public dataset for large-scale multi-label and multi-class image classification, vol. 2, no. 3, p. 18 (2017). Dataset. https://github.com/openimages

17. Lamb, A.M., Alias Parth Goyal, A.G., Zhang, Y., Zhang, S., Courville, A.C., Bengio, Y.: Professor forcing: a new algorithm for training recurrent networks. In: Advances in Neural Information Processing Systems, vol. 29 (2016)

18. Lewis, D., Agam, G., Argamon, S., Frieder, O., Grossman, D., Heard, J.: Building a test collection for complex document information processing. In: Proceedings of the 29th Annual International ACM SIGIR Conference on Research and Development in Information Retrieval, pp. 665–666 (2006)

19. Liu, F., Xu, G., Wu, Q., Du, Q., Jia, W., Tan, M.: Cascade reasoning network for text-based visual question answering. In: Proceedings of the 28th ACM International Conference on Multimedia, pp. 4060–4069 (2020)

20. Loshchilov, I., Hutter, F.: Decoupled weight decay regularization. arXiv preprint arXiv:1711.05101 (2017)

21. Lu, X., Fan, Z., Wang, Y., Oh, J., Rosé, C.P.: Localize, group, and select: boosting text-VQA by scene text modeling. In: Proceedings of the IEEE/CVF International Conference on Computer Vision, pp. 2631–2639 (2021)

22. Van der Maaten, L., Hinton, G.: Visualizing data using t-SNE. J. Mach. Learn. Res. 9(11) (2008)

23. Mishra, A., Shekhar, S., Singh, A.K., Chakraborty, A.: OCR-VQA: visual question answering by reading text in images. In: 2019 International Conference on Document Analysis and Recognition (ICDAR), pp. 947–952. IEEE (2019)

24. Niu, Y., Tang, K., Zhang, H., Lu, Z., Hua, X.S., Wen, J.R.: Counterfactual VQA: a cause-effect look at language bias. In: Proceedings of the IEEE/CVF Conference on Computer Vision and Pattern Recognition, pp. 12700–12710 (2021)

25. Raffel, C., et al.: Exploring the limits of transfer learning with a unified text-to-text transformer. arXiv preprint arXiv:1910.10683 (2019)

26. Ren, S., He, K., Girshick, R., Sun, J.: Faster R-CNN: towards real-time object detection with region proposal networks. In: Advances in Neural Information Processing Systems, vol. 28 (2015)

27. Sidorov, O., Hu, R., Rohrbach, M., Singh, A.: TextCaps: a dataset for image captioning with reading comprehension. In: Vedaldi, A., Bischof, H., Brox, T., Frahm, J.-M. (eds.) ECCV 2020. LNCS, vol. 12347, pp. 742–758. Springer, Cham (2020). https://doi.org/10.1007/978-3-030-58536-5_44

28. Singh, A., et al.: Towards VQA models that can read. In: Proceedings of the IEEE/CVF Conference on Computer Vision and Pattern Recognition, pp. 8317–8326 (2019)

29. Singh, A., Pang, G., Toh, M., Huang, J., Galuba, W., Hassner, T.: TextOCR: towards large-scale end-to-end reasoning for arbitrary-shaped scene text. In: Proceedings of the IEEE/CVF Conference on Computer Vision and Pattern Recognition, pp. 8802–8812 (2021)

30. Su, W., et al.: VL-BERT: pre-training of generic visual-linguistic representations. In: International Conference on Learning Representations (2019)

31. Wang, Q., Xiao, L., Lu, Y., Jin, Y., He, H.: Towards reasoning ability in scene text visual question answering. In: Proceedings of the 29th ACM International Conference on Multimedia, pp. 2281–2289 (2021)

32. Wang, X., et al.: On the general value of evidence, and bilingual scene-text visual question answering. In: Proceedings of the IEEE/CVF Conference on Computer Vision and Pattern Recognition, pp. 10126–10135 (2020)

33. Wolf, T., et al.: Huggingface's transformers: state-of-the-art natural language processing. arXiv preprint arXiv:1910.03771 (2019)
34. Xu, Y., et al.: LayoutLMv2: multi-modal pre-training for visually-rich document understanding. arXiv preprint arXiv:2012.14740 (2020)
35. Xu, Y., Li, M., Cui, L., Huang, S., Wei, F., Zhou, M.: LayoutLM: pre-training of text and layout for document image understanding. In: Proceedings of the 26th ACM SIGKDD International Conference on Knowledge Discovery and Data Mining, pp. 1192–1200 (2020)
36. Yang, Z., et al.: Tap: text-aware pre-training for text-VQA and text-caption. In: Proceedings of the IEEE/CVF Conference on Computer Vision and Pattern Recognition, pp. 8751–8761 (2021)
37. Zeng, G., Zhang, Y., Zhou, Y., Yang, X.: Beyond OCR+ VQA: involving OCR into the flow for robust and accurate TextVQA. In: Proceedings of the 29th ACM International Conference on Multimedia, pp. 376–385 (2021)
38. Zhang, X., Yang, Q.: Position-augmented transformers with entity-aligned mesh for TextVQA. In: Proceedings of the 29th ACM International Conference on Multimedia, pp. 2519–2528 (2021)
39. Zhao, M., et al.: Towards video text visual question answering: benchmark and baseline. In: Thirty-Sixth Conference on Neural Information Processing Systems Datasets and Benchmarks Track (2022)
40. Zhu, Q., Gao, C., Wang, P., Wu, Q.: Simple is not easy: a simple strong baseline for TextVQA and TextCaps. In: Proceedings of the AAAI Conference on Artificial Intelligence, vol. 35, pp. 3608–3615 (2021)

Bright as the Sun: In-depth Analysis of Imagination-Driven Image Captioning

Huyen Thi Thanh Tran[1(✉)] and Takayuki Okatani[1,2]

[1] RIKEN Center for AIP, Tokyo, Japan
{tran,okatani}@vision.is.tohoku.ac.jp
[2] Graduate School of Information Sciences, Tohoku University, Sendai, Japan

Abstract. Existing studies on image captioning mainly focus on generating "literal" captions based on visual entities in images and their basic properties such as colors and spatial relationships. However, to describe images, humans use not only literal descriptions but also "imagination-driven" descriptions that characterize visual entities by some different entities; they are often more vivid, precise, and visually comprehensible by readers/hearers. Nonetheless, none of the existing studies seriously consider captions of this type. This study presents the first comprehensive analysis of the generation and evaluation of imagination-driven captions. Specifically, we first analyze imagination-driven captions in existing image captioning datasets. Then, we present the comprehensive categorizations of imagination-driven captions and their usage, discussing the (potential) issues with the current image captioning models to generate such captions. Next, compiling these captions extracted from the existing datasets and synthesizing fake captions, we create a dataset named *IdC-I* and *-II*. Using this dataset, we examine nine existing metrics of image captioning about how accurately they can evaluate imagination-driven caption generation. Last, we propose a baseline model for imagination-driven captioning. It has a built-in mechanism to select which to generate between literal and imagination-driven captions, which existing image captioning models can not do. Experimental results demonstrate that our model performs better than six existing models, especially for imagination-driven caption generation. Dataset and code will be publicly available at: https://github.com/ TranHuyen1191/Imagination-driven-Image-Captioning.

Keywords: Image captioning · Imagination-driven image captioning

1 Introduction

Image captioning is the task of automatically generating a description (also known as caption) in natural language for a given image. It is one of the fundamental tasks of computer vision, which has broad applicability in various areas

Supplementary Information The online version contains supplementary material available at https://doi.org/10.1007/978-3-031-26316-3_40.

GT: "a sign on a pole near a park"
SAT: "a sign that is on the side of a road"
(a)

GT: "an assortment of cupcakes made to look like sheep"
SAT: "a close up of cupcakes on a table"
(b)

GT: "the bright colors are fun they remind me of the sun"
SAT: "a close up of a glass of orange juice"
(c)

GT: "man in a black suit and straw hat holds a tennis racket as if it were a guitar"
SAT: "a man in a suit and hat sitting on a bench"
(d)

Fig. 1. Examples of image captioning. *GT* denotes human-generated descriptions: (a): "Literal" description; (b) (c) & (d): "Imagination-driven" descriptions. SAT denotes captions generated by Show-Attend-Tell [5] that is trained on the *MS COCO* dataset [6]. (a) & (b) are from *MS COCO* [6], (c) & (d) are from *ArtEmis* [7] and *Flickr30K* [8], respectively.

of biomedicine, commerce, and web searching [1]. For instance, image captioning can help visually-impaired people understand the content of images and, to some extent, form similar images in their minds.

Existing studies on image captioning mainly focus on generating "literal" captions, which are based on visual entities in images and their basic properties (e.g., colors, spatial relationships) [2–4]. As an example in Fig. 1(a), the caption *"a sign that is on the side of a road"*, which is generated by an existing model of Show-Attend-Tell (SAT) [5], is based on three visual entities of *sign*, *side*, and *road*, and their relationships of *on* and *of*.

To describe images, humans actually use not only literal descriptions but also "imagination-driven" descriptions. Unlike literal descriptions, which directly describe visual entities, imagination-driven descriptions characterize visual entities by other entities, which we call *imaginary entities* in this paper. Imaginary entities are usually not presented in images, but typically share one or several common properties with visual entities. Examples of imagination-driven descriptions are shown in Fig. 1(b), (c), and (d). When observing the image in Fig. 1(b), rather than only thinking about the cupcakes, *a visual entity*, the annotator associates them with sheep, *an imaginary entity*, since they are similar in shape and color. Similarly, in the example of Fig. 1(c), the bright colors in the image evoke the Sun in the annotator's mind; the activity of the man in the image of Fig. 1(d) makes the annotator liken a tennis racket to a guitar.

We tend to generate imagination-driven descriptions for images that are capable of spontaneously stimulating our imagination to produce similar images in our minds. It is hard, possibly impossible, for us to give concise and understandable literal descriptions of such images. This is because of the limited expression power of literal language; humans can perceive many more colors and shapes than concepts or words we have in language to describe them [9]. In addition, forcibly generated literal descriptions for such images are usually unnatural and complicated, making them visually incomprehensible for humans. For instance, it is quite challenging to create literal descriptions for the example images in Fig. 1(b) (c) and (d). Instead, the annotators provide imagination-driven descriptions, which are vivid and easy to be visualized by readers/hearers.

In this paper, we study the generation and evaluation of such imagination-driven captions, which have been overlooked in previous studies. Although a lot of efforts have been made to provide various datasets [6,8,10–12] and methods [13–16] for image captioning, there has been no prior study seriously considering imagination-driven captions. There are three main challenges to enable the generation of imagination-driven captions. The first challenge is the lack of datasets fit for the study. The second challenge is the absence of proper methods for generating such captions. The existing image captioning models are not designed to properly handle imagination-driven captions, as discussed later. The third challenge is the lack of the methods that can evaluate imagination-driven caption generation. Existing metrics for image captioning does not perform well, as we will show later.

Towards conquering the above challenges, in this paper, we first analyse four existing datasets, showing the statistics of imagination-driven captions in them. We then show the comprehensive categorization of imagination-driven captions and their usages. We also discuss the issues with existing image captioning models to generate such captions. By collecting imagination-driven captions from the four source datasets and synthesizing fake captions, we create a dataset, named *IdC-I* and *-II*. Using this dataset, we assess the accuracy of nine state-of-the-art evaluation metrics. Experimental results show that UMIC [17] shows high accuracy in evaluating imagination-driven captions of natural images. Meanwhile, for art images, all the considered metrics are not very effective.

Finally, we propose a novel image captioning method that can adequately handle imagination-driven caption generation. We design it to address a particular difficulty that existing models have. It is that they do not distinguish between visual and imaginary entities, leading to the inability to generate good imagination-driven captions and select the appropriate caption type fit for the input image. In the above example of Fig. 1(b), *sheep* in the imagination-driven caption does not appear as real entities in the image. Thus, the models need to be able to differentiate between real and imaginary sheep. Moreover, they must adequately judge which type of captions to generate for a given image; literal captions are sufficient for some images, and imagination-driven ones are better or necessary for others. To address these, our model generates literal and imagination-driven captions separately and selects the one that best fits the input image. For the latter, we build a scorer that scores the quality of the generated imagination-driven caption based on the content of the images. The scorer is built based on CLIP [18], which is a pre-trained vision-language model. Experimental results demonstrate the effectiveness of the proposed model over six existing methods developed for standard image-captioning.

In summary, the main contributions of this work are as follows:

- We analyze the existing image captioning datasets and present the comprehensive categorization of imagination-driven captions and their usages.
- We introduce a dataset for the generation and evaluation of imagination driven image captioning.

– Using this dataset, we examine nine state-of-the-art metrics of image captioning to measure the accuracy of imagination-driven caption generation.
– We propose a new image captioning model and experimentally examine its
effectiveness.

2 Related Work

Image Captioning Datasets. In the literature, a lot of efforts have been made
to build image captioning datasets [6–8,10,11,19]. Some of them are domain-
generic datasets with images of various scenes and objects such as *MS COCO* [6],
Flickr30K [8], and *CC12M* [19]. Because of the broad coverage of scenes and
objects, these datasets are usually considered as standard benchmarks to build
and evaluate image captioning methods [20].

Besides, there are some domain-specific datasets that are constructed for several specific tasks [1]. For instance, the *CUB* dataset consists of 117,880 captions
of bird images [21]. Considering linguistic aspects, the authors in [22] focus on
captions including negations. In [23], an attempt has been made to build a corpus of commonly used phrases that are repeated almost verbatim in captions
of different images. To the best of our knowledge, there has been no previous
research constructing datasets specific to imagination-driven caption generation.

Image Captioning Models. Most existing image captioning models are based
on an encoder-decoder paradigm. In this paradigm, an image encoder is used to
project images to visual representations, which are then fed into a text generator to generate captions [1]. Many models use CNN as an image encoder [5,13–
15]. However, CNN usually results in loss of granularity [1]. To address this
problem, an attention mechanism over visual regions is exploited [16,24–26].
Typically, Faster R-CNN [27] is used to extract bounding boxes of concrete
objects, whose representations are then fed to a text generator to output captions. Recently, thanks to computational efficiency and scalability, transformer
architectures based on self-attention mechanisms [28] are also adopted as image
encoders [29,30].

Regarding text generators, LSTM [31] has become a predominant architecture for a long time due to its ability to learn dependencies in captions. However,
it faces issues about training speed and the ability to learn long-term dependencies [20]. Recent studies employ transformer architectures [28] as an alternative,
since it can better learn long-term dependencies [24,25,32]. Besides, to enrich
generated captions, some studies adopt graphs to detect spatial relationships of
visual entities [3,25]. Existing image captioning studies mainly focus on detecting
visual entities and their spatial relationships. So far, there has been no previous study that aims at generating imagination-driven captions, where forming
imaginary entities is also of crucial importance.

3 Analyzing Imagination-Driven Captions

We first analyze imagination-driven captions to answer the following three questions: 1) how frequently imagination-driven captions are used in existing image captioning datasets; 2) how many types exist, and 3) when humans (annotators) use such captions. For this purpose, we first examine imagination-driven captions in four image captioning datasets. We then classify imagination-driven captions in terms of their types and usages. We finally introduce datasets, named *IdC-I* and *IdC-II*, for the study of imagination-driven caption generation, which are the collection of extracted captions from the above datasets and synthesized fake captions to assess evaluation metrics of image captioning.

3.1 Analysis of Existing Datasets

To answer the above questions, we consider the following existing image captioning datasets: *MS COCO* [6], *Flickr30K* [8], *VizWiz* [11], and *ArtEmis* [7]. We adopt a filtering strategy to extract imagination-driven captions from them. *MS COCO* and *Flickr30K* are domain-generic datasets while *VizWiz* and *ArtEmis* are domain-specific datasets. *MS COCO* contains 995,684 captions from 164,062 images, which were gathered by searching for 80 object categories and different scene types on Flickr [33]. *Flickr30K* consists of 158,920 captions from 31,784 images of everyday activities and scenes. *VizWiz* is constructed to study image captioning for people who are blind. It includes 195,905 captions from 39,181 images taken by blind photographers in their daily lives. *ArtEmis* containing 454,684 captions from 80,031 art images is created to investigate affective human experiences evoked by artworks.

As in the examples in Fig. 1(b) (c) and (d), it can be seen that the annotators use the phrases of *look like, remind me,* and *as if* to associate visual entities (i.e., *cupcakes, colors,* and *tennis racket*) with imaginary entities (i.e., *sheep, Sun,* and *guitar*). Taking advantage of this feature, a list of 34 keywords as given in Table 1 is exploited to automatically extract imagination-driven captions from the source datasets. A caption will be considered imagination-driven if it includes at least one of these keywords.

Table 2 shows the statistics of imagination-driven captions extracted by the above procedure; *ratioImg* (*ratioCapt*) is defined as the ratio of the extracted images (captions) over the total number of images (captions) in each source dataset. It can be seen that imagination-driven captions account for only a small portion of captions (i.e., <1%) in *MS COCO*, *Flickr30K*, and *VizWiz*. There are multiple possible reasons. One is that these datasets mainly include natural images with typical objects/activities that do not trigger human imagination. In addition, the annotators of these datasets are requested to be as objective as possible to provide captions [10,11]; in other words, they tend to avoid imagination-driven captions that are likely to be subjective. Thus, note that just because their percentages are small does not mean that they are unimportant.

Meanwhile, for *ArtEmis*, 70.86% of the included images are described by imagination-driven captions. Also, imagination-driven captions account for

Table 1. Keywords used to extract imagination-driven captions.

Keywords						
'looks like'	'look like'	'look as'	'looks as'	'looks likely'	'look likely'	'is likely'
'are likely'	'is like'	'are like'	'looks almost like'	'look almost like'	'shaped like'	'shapes like'
'shape like'	'is almost as'	'are almost as'	'seems to be'	'seem to be'	'seems like'	'thinks of'
'think of'	'as if'	'as though'	'seems as'	'seem as'	'seem like'	'calm like'
'feels like'	'feel like'	'resemble'	'resembling'	'reminds me'	'remind me'	

Table 2. Statistics of imagination-driven captions extracted from source datasets.

Source dataset	MS COCO	Flickr30K	VizWiz	ArtEmis
Number of extracted images	1489	577	1133	56,707
Number of extracted captions	1699	595	1160	100,393
ratioImg(%)	1.21	1.82	3.63	70.86
ratioCapt(%)	0.28	0.37	0.74	22.08

22.08% of all the captions. This is because this dataset is created to investigate affective human experiences; the included images (i.e., artworks) are mostly abstract and have a tendency to evoke human emotion as well as human imagination.

3.2   Classifying Imagination-Driven Captions and Their Usages

Through the analyses of the above datasets and others, we found that imagination driven captions can be categorized into two types, which we name object-based and action-based. We also found that there are three scenarios of their usage. We show their details below, along with the challenges for image captioning models to generate each type in each scenario.

Object-Based Caption Type. Object-based captions are created based on similarities of characteristics between visual and imaginary objects. Imaginary objects in these captions are either "common" or "proper". Figures 2(a), (b), and (c) show three examples of this caption type. In the first example, the clocks in the image evoke cats in the annotator's mind, *a common object/animal*. Interestingly, the clocks do not have identical shapes and colors to real cats; they are modified/deformed based on the designer's imagination and creativity. This raises the first challenge for image captioning models: how they can connect visual entities that are the products of human imagination and creativity with the right imaginary entities. On the other hand, the cake in Fig. 2(b) looks so similar to a real dog that they are indistinguishable even from humans without considering the context of the image. Thus, the second challenge is how models distinguish real objects and their "look-alike" objects, generating good captions. In the third example, the vanity style makes the annotator imagine a Victorian house, *a proper object*, which refers to a popular architectural revival style during the reign of Queen Victoria (1837–1901). To generate such captions, the model

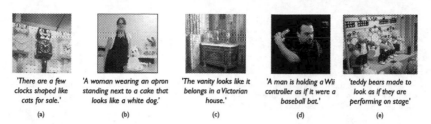

'There are a few
clocks shaped like
cats for sale.'

(a)

'A woman wearing an apron
standing next to a cake that
looks like a white dog.'

(b)

'The vanity looks like it
belongs in a Victorian
house.'

(c)

'A man is holding a Wii
controller as if it were a
baseball bat.'

(d)

'teddy bears made to
look as if they are
performing on stage'

(e)

Fig. 2. Examples of imagination-driven caption types. (a), (b) & (c): Object-based captions; (d) & (e): Action-based captions. Examples are from *MS COCO* [6].

LC: 'a red and white fire hydrant
out on a road',
IdC: 'A white and red fire hydrant
made to look like a dalmatian dog.'

(a)

LC: 'A giraffe standing next to
a fence a green field.'
IdC: 'A giraffe who looks like he
needs friends.'

(b)

LC: 'A man on a cellphone
near an arched bridge.'
IdC: 'Reminds me of my trip to Venice. Going
under the bridge in a boat is so romantic.'

(c)

LC: 'A man is trimming something
round with some scissors.'
IdC: 'A man trimming what seems to
be a piece of cheese or bread'

(d)

Fig. 3. Examples of imagination-driven description usages. *LC* and *IdC* denote literal and imagination-driven descriptions. Examples are from *MS COCO* [6].

needs to retain special knowledge they may not be able to acquire through learning using generic image captioning datasets.

Action-Based Caption Type. Unlike object-based captions, imaginary entities in action-based captions are triggered by the poses of visual humans/animals or the contexts of the images. Figures 2(d) and (e) show two examples of this caption type. When observing the image in Fig. 2(d), the man's pose makes the annotator think about a baseball bat. It is challenging for models to detect poses and link them to actions as humans do regardless of irrelevant objects in images. In the second example of Fig. 2(e), based on the context of the image (i.e., microphone, stage, audience, etc.), teddy bears are personified as humans performing on stage. It will be hard for existing captioning models to understand contexts to generate such personification captions.

Usage Scenarios. There are three typical usage scenarios for imagination-driven descriptions. The first scenario is when objects/actions in an image remind us of imaginary entities due to their similarities in key characteristics. The second scenario is when annotators want to express their emotions when observing the images or the expressions of humans/animals in images. Finally, the third scenario is when annotators attempt to use their imagination to describe a visual entity that they cannot accurately recognize.

To illustrate these three usage scenarios, Fig. 3 shows four examples of the pairs of literal and imagination-driven descriptions. In the first example, to depict the color of the fire hydrant, the first annotator uses a literal adjective phrase

Fig. 4. Example of real and fake captions in *IdC-II*. The bold texts indicate *subjects* while the underlined texts indicate *predicates*.

of *red and white*. Meanwhile, the second annotator utilizes the imagination-driven phrase of *look like a dalmatian dog*. Compared to the literal description, the imagination-driven description is more vivid and easier to be visualized by humans.

In the second example, unlike the literal description of *"A giraffe standing next to a fence a green field"*, which describes the action and position of the giraffe, the imagination-driven description, *"A giraffe who looks like he need friends"*, emphasizes the affective expression of the giraffe. Similarly, the feeling of the annotator when observing the image is conveyed through the imagination-driven description in the third example, *"Reminds me of my trip to Venice. Going under the bridge in a boat is so romantic."*. In the final example, while the literal description uses *something* to explain the unrecognizable object in the image, the imagination-driven caption associates the object to *"a piece of cheese or bread"*.

These examples illustrate that imagination-driven captions can be used in various scenarios to make descriptions more vivid and effectively express the feelings of annotators and the expressions of visual entities in images. However, imagination-driven captions are often unnecessary or inadequate for some types of images that do not evoke human imagination. How to select which to generate between literal and imagination-driven captions for individual images is challenging for existing models.

3.3   Dataset: *IdC-I* and *IdC-II*

In this section, we present a dataset that will be useful for studying imagination driven image captioning, named *IdC-I* and *-II*. *IdC-I* contains all the imagination driven captions extracted from the existing datasets as above, which we call "real" captions in what follows. *IdC-II* contains *IdC-I* and additionally "fake" captions we create by combining irrelevant visual and imaginary entities. Using *IdC-II*, we can assess existing metrics for image captioning; how accurately they can evaluate imagination-driven captions. This is done by checking if each metric yields higher scores with input images for their real captions than fake captions.

To create fake captions, we first split the sentence of each real caption into two parts: the set of words before keywords, called *subject*, and the set of the remaining words, called *predicate*. Then, we select the pairs of real captions satisfying: 1) they are of different images, and 2) the number of overlapping words of the two subjects is highest. Next, through an analysis of failures of image captioning models, we exploit three typical error types to create fake captions, namely subject failures (SF), predicate failures (PF), and grammatical mistakes (GM). The first type refers to failures at detecting visual objects that are usually included in subjects. For the second type, predicates are not aligned with subjects considering image content. Related to grammatical mistakes, we generate incomplete captions by randomly excluding some ending words in the real captions. Figure 4 shows an example of fake captions corresponding to the three error types. A good metric should give the highest score to real captions and lower scores to fake captions.

4   A Method for Generating both Types of Captions

4.1   Overview of the Proposed Method

As discussed earlier, existing image captioning models cannot generate imagination driven captions properly. This is because it requires higher-level skills than literal caption generation. For instance, it requires the ability to interpret a visual entity in two ways (i.e., literal and imaginary) and then link them to generate a meaningful description. An image captioning model may not be able to acquire such an ability, if not impossible, through pure learning using generic image-captioning datasets.

To cope with this difficulty, we build a model that internally generates literal and imagination-driven captions for the input image and then selects the one best fit for the image. Specifically, it has two text generators, a literal caption generator (LCGen) and an imagination-driven caption generator (IdCGen); each is designed to generate one caption type. Figure 5(b) depicts the model's architecture, which consists of four main modules: an image encoder, LCGen, IdCGen, and a selector. Given an image, the image encoder extracts visual representation $\mathbf{V}$ from the input image, which is then fed into LCGen and IdCGen to produce two captions. To select one of them, the selector scores the imagination-driven caption using a CLIP-based scorer, named *CScorer*. If the score is higher than a threshold θ, the model selects the imagination-driven caption. Otherwise, it selects the literal caption. The image encoder is designed based on Vision Transformer (ViT) [29] while the text generators are built based on the transformer decoder architecture [28]. We will explain the image encoder, the text generator, and the scorer in this order.

4.2   Image Encoder

Figure 5(a) shows the image encoder architecture. First, an image with the resolution of $H \times W$ is spatially divided into patches with a fixed resolution of $P \times P$. Consequently, there are totally $N = HW/P^2$ patches. Next, a linear projection is applied to generate the patch embeddings $\mathbf{X}^p$. The embeddings are then concatenated with a learnable embedding $\tilde{\mathbf{x}}_0$. To retain positional information, the concatenated embeddings are added to positional encodings $\mathbf{P}^{image}$.

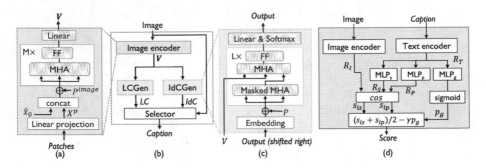

Fig. 5. (a) Image encoder architecture, (b) General architecture of the proposed model, (c) Text generator architecture, (d) CScorer architecture.

The result is then fed into a transformer encoder [28]. The transformer encoder is constructed from a stack of M identical layers, each contains two sub-layers of multi-head self-attention (MHA) and position-wise fully connected feed forward (FF) in sequence. The FF sub-layer is composed of two linear transformations with a GELU non-linearity in between. Note that each sub-layer is preceded by layer normalization and followed by a residual connection [34]. Layer normalization and a linear transformation are then employed to map the output of the transformer encoder to the visual representation $\mathbf{V}$.

4.3   Text Generator

Figure 5(c) shows the text generator architecture. Given the visual representation $\mathbf{V}$, the output is a sequence of T tokens one at a time. Similar to existing studies on text generation [35,36], the proposed model does also work in an auto-regressive manner [24]. To retain positional information, the token embeddings are added to positional encodings $\mathbf{P}$. The result is then fed into a stack of L identical layers, each includes three alternating sub-layers of masked MHA, MHA, and FF. Each sub-layer is in between layer normalization and a residual connection [34]. To map the result to the token probability, we use layer normalization, a linear transformation, and a softmax transformation. The loop to generate tokens ends when either the number of the output tokens reaches the maximum length T or a special token $<eos>$ (end of sequence) is produced.

4.4   CLIP-Based Scorer

Figure 5(d) illustrates the architecture of the scorer. By using the image encoder and the text encoder of CLIP, the image and the caption are projected to the corresponding representations of R_I and R_T. With the input of R_T, three multi-layer perceptrons (MLP) modules compute the grammatical penalty p_g, the subject representation R_S, and the predicate representation R_P. Each MLP module includes two linear transformations with a GELU non-linearity in between. The image-caption alignment score is then computed as the average of the two cosine similarity values: $s_{is} = cos(R_I, R_S)$ and $s_{ip} = cos(R_I, R_P)$. The final score is calculated by $s = (s_{is} + s_{ip})/2 - \gamma p_g$.

Similar to [18], we train CScorer by using the multi-class N-pair loss [37]. However, because each image is paired with not only one real caption but also three fake captions, an asymmetric loss is adopted instead of a symmetric loss. The loss function is $\mathcal{L} = \mathcal{L}_{is}^{\mathcal{CE}}(s_{is}) + \alpha \mathcal{L}_{ip}^{\mathcal{CE}}(s_{ip}) + \beta \mathcal{L}_s^{\mathcal{CE}}(s) + \mu \mathcal{L}_g^{\mathcal{BCE}}(p_g)$, where $\mathcal{L}^{\mathcal{CE}}$ and $\mathcal{L}^{\mathcal{BCE}}$ denote cross entropy loss and binary cross entropy loss, respectively. Note that image-caption pairs are used to calculate $\mathcal{L}_{is}$, $\mathcal{L}_{ip}$, and $\mathcal{L}_s$, while $\mathcal{L}_g$ requires only captions.

5 Experiments

In this section, we show the results of two experiments. The first one assesses the accuracy of nine existing metrics in evaluating imagination-driven image captions. The second experiment evaluates the effectiveness of the image captioning model presented in Sect. 4. To indicate the source dataset of captions, we will use names with the prefix of the source dataset. For instance, *MS COCO-IdC-I* denotes the set of captions in *IdC-I* that originate from the *MS COCO* dataset.

5.1 Evaluation of Metrics

Experimental Settings. We use *IdC-II* to evaluate the accuracy of nine existing metrics, namely TIGEr [38], VIFIDEL$_{no\text{-}refs}$ (nrVIFIDEL) [39], VIFIDEL [39], BERTScore (BERTS) [40], ViLBERTScore (ViLBERTS) [41], UMIC [17], SMURF [42], CLIPScore (CLIPS) [43], and RefCLIPScore (RefCLIPS) [43]. Among these metrics, nrVIFIDEL, UMIC, and CLIPS are unreferenced metrics [17], which do not require human-generated annotations, whereas the others are referenced metrics.

As mentioned in Sect. 3.3, *IdC-II* includes caption tuples, each consisting of one real caption and three fake captions. For each tuple, a good metric should give the highest score to the only real caption and lower scores for the fake captions. Based on this idea, we regard the scoring of a tuple as accurate if and only if the real caption has the highest score. We use the ratio of the number of accurately scored caption tuples over the total number of caption tuples as the evaluation measure of the metrics.

For the three sets of *MS COCO-IdC-II*, *VizWiz-IdC-II*, and *Flickr30K-IdC-II*, we evaluate only the three unreferenced metrics, due to the deficient numbers of human-generated annotations per image. For *ArtEmis-IdC-II*, to enable the evaluation of all the metrics, we use only 2497 images that have at least four human-generated annotations per image.

Results. Table 3 shows the accuracy of the metrics on *IdC-II*. For *MS COCO-IdC-II*, *VizWiz-IdC-II*, and *Flickr30K-IdC-II*, which include natural images, UMIC generally achieves the highest accuracy. In terms of average accuracy,

Table 3. Accuracy of the metrics on *IdC-II*.

Metric		*MS COCO-IdC-II*			*VizWiz-IdC-II*			*Flickr30K-IdC-II*		
		nrVIFIDEL	UMIC	CLIPS	nrVIFIDEL	UMIC	CLIPS	nrVIFIDEL	UMIC	CLIPS
Error type	SF	0.87	**0.98**	0.97	0.65	0.86	**0.87**	0.63	**0.97**	0.97
	PF	0.67	**0.91**	0.89	0.65	0.81	**0.88**	0.61	0.89	**0.89**
	GM	0.47	**0.83**	0.80	0.35	**0.69**	0.68	0.37	**0.78**	0.64
Average		0.67	**0.91**	0.88	0.55	0.78	**0.81**	0.54	**0.88**	0.83

Metric		*ArtEmis-IdC-II*								
		TIGEr	nrVIFIDEL	VIFIDEL	BERTS	ViLBERTS	UMIC	SMURF	CLIPS	refCLIPS
Error type	SF	0.68	0.60	0.71	0.78	0.77	0.72	0.55	0.78	**0.78**
	PF	0.66	0.59	0.63	0.70	0.62	0.70	0.64	0.77	**0.77**
	GM	0.54	0.33	0.26	**0.66**	0.28	0.60	0.57	0.55	0.55
Average		0.63	0.51	0.54	**0.71**	0.56	0.68	0.59	0.70	0.70

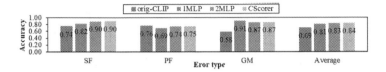

Fig. 6. Accuracy of CScorer and baselines on *ArtEmis-IdC-II* test set.

the results of UMIC and CLIPS are rather high (i.e., ≥ 0.78) while that of nfV-IFIDEL is quite low (i.e., ≤ 0.67). When comparing the three error types, the accuracies are highest for the subject failure type, followed by the predicate failure type. With the accuracies higher than 0.81, UMIC and CLIPS can be used to evaluate the degree of image-text alignment of imagination-driven captions. Meanwhile, for grammatical mistakes, the accuracies range from 0.64 to 0.83, suggesting that there is still room for improvement in evaluating this error type.

Regarding *ArtEmis-IdC-II*, which includes art images, all the nine considered metrics are not very effective. Their average accuracies are from 0.51 to 0.71. In particular, all of them show poor performance for grammatical mistakes (i.e., ≤ 0.66). These results imply that it is essential to build dedicated metrics for evaluating imagination-driven captions for art images.

5.2　Evaluation of Image Captioning Models

Experimental Settings. In this experiment, we use the *ArtEmis* dataset, which has a sufficient number of imagination-driven captions for training and testing. *ArtEmis* consists of *ArtEmis-IdC-I* and the set of literal captions named *ArtEmis-LC*. Regarding the data splitting, *ArtEmis* is divided into the training, validation, and test sets, including 68,028 images, 6000 images, and 5497 images. To make a comprehensive evaluation of the models, we additionally consider two subsets of the test set: 1) a set of all the literal captions with 4019 images (*ArtEmis-LC-TS*) and 2) a set of all the imagination-driven captions from 2497 images (*ArtEmis-IdC-I-TS*). By using the same data split, *ArtEmis-IdC-II*, which is used to train CScorer, is also divided into the training, validation, and test sets (Fig. 6).

Table 4. Performances of the models. The bold number indicate the highest performance.

Models	Whole test set						ArtEmis-LC-TS						ArtEmis-IdC-I-TS					
	B1	B2	B3	B4	M	R	B1	B2	B3	B4	M	R	B1	B2	B3	B4	M	R
NN	40.8	17.6	7.6	3.4	10.9	22.8	40.3	16.8	6.9	2.9	10.7	22.7	38.6	16.2	7.0	3.2	10.6	21.3
SAT	57.5	34.4	20.5	12.5	15.2	31.5	58.4	32.8	17.3	8.6	14.7	31.5	54.2	31.9	19.0	12.1	14.9	30.1
$\mathcal{M}^2$	57.1	33.4	19.6	11.7	14.8	31.3	57.1	31.9	17.3	8.9	14.5	31.6	51.0	28.7	16.2	9.4	13.6	28.0
CLIPCap	47.8	27.2	15.4	9.3	**16.0**	25.6	47.8	24.8	11.7	5.6	15.0	24.3	44.1	24.9	14.6	9.2	14.7	23.7
Oscar	44.1	24.0	13.0	7.2	15.5	28.2	54.0	29.0	14.9	7.4	14.5	30.3	50.6	28.6	16.7	10.3	14.5	28.8
OFA	59.9	35.1	18.8	9.9	15.8	32.3	59.4	34.6	18.1	9.3	**15.7**	32.7	53.4	28.8	14.4	6.8	13.4	27.6
1GEN	60.1	37.6	**22.8**	**14.1**	15.4	32.6	**61.2**	36.3	**19.9**	**10.2**	15.3	32.9	53.6	32.0	18.7	11.5	14.2	29.6
2GEN	**61.8**	**38.1**	22.5	13.6	15.9	**33.2**	61.2	**36.6**	19.9	10.1	15.3	**33.4**	62.6	43.0	29.1	19.9	19.1	37.6

(a)

GT-1: "the person looks like a ghost or zombie from a horror film"
GT-2: "the man looks like he has blood on his face and it looks scary"
SAT: "the colors are very dark and gloomy"
OFA: "the blue and green colors make me feel sad"
LCGen: "the colors are very sad and the face is sad"
IdCGen: "the man looks like a zombie is bleeding to be a lot of a scary"

(b)

GT-1: "it reminds me of an english garden full of flowers in bloom very cheerful"
GT-2: "portrait of a beautiful colorful field of flowers of all kinds"
SAT: "the bright colors of the flowers make me feel happy"
OFA: "the bright colors of the flowers make me feel happy"
LCGen: "the flowers are very colorful and the colors are beautiful"
IdCGen: "the bright colors and flowers remind me of a garden"

Fig. 7. Examples of human-generated descriptions (*GT-1* and *GT-2*) and captions generated by SAT, OFA, LCGen and IdCGen.

Regarding the image encoder in the proposed model, we use the pre-trained image encoder of CLIP with the ViT-B/16 backbone [18]. For LCGen and IdC-Gen, the number of layers and heads are set to 8. The maximum caption length is set to $K = 65$. The loss is calculated by the average loss over LCGen and IdCGen using cross-entropy loss functions. For CScorer, we use the pre-trained CLIP with the ResNet-50x16 backbone [18]; θ and γ are set to 0.5 and 0.2. The parameters of the loss function are set to $\alpha = \beta = \mu = 1$. The optimizer is Adam with (0.9,0.999) [44]. To assess the effectiveness of using the two text generators (called *2GEN*), we also evaluate the case of using only one text generator for both the caption types, called *1GEN*.

Our model is compared with six reference models of Nearest-Neighbor (NN) [7], Show-Attend-Tell (SAT) [5], Meshed-Memory transformer ($\mathcal{M}^2$) [26], Oscar [45], CLIPCap [46], and OFA [47]. Among the considered models, Oscar, CLIP-Cap, OFA, and our model are based on large-scale vision-language pre-training (VLP). For the remaining models, they are based on basic backbones: ResNet-34 pre-trained on ImageNet for NN and SAT, Faster R-CNN [27] pretrained on Visual Genome for $\mathcal{M}^2$. For a fair comparison, we use beam search for the reference models with the beam size of 2 since our model generates two captions simultaneously from LCGen and IdCGen. We use six commonly used evaluation metrics of BLEU 1-4 (*B1-B4*) [48], ROUGE-L (*R*) [49], and *METEOR* (*M*) [50].

Results. By using *ArtEmis-IdC-II*, we evaluate the accuracy of CScorer and three baselines, namely 2MLP, 1MLP, and orig-CLIP. These baselines excludes one, two, or three MLP modules, respectively, from CScorer. Figure 7 shows the accuracy of CScorer and baselines. It can be seen that CScorer is the most effective scorer with high and stable accuracies (i.e., ≥0.75). These results also demonstrate the effectiveness of adding MLP modules in increasing the accuracy of CScorer, compared to orig-CLIP.

Table 4 shows the obtained results of the models. Among the reference models, SAT and OFA usually perform best. Comparing all the models, we see that 1GEN and 2GEN generally achieve the highest performance for the whole test set and *ArtEmis-LC-TS*. Also, the gap between them is small. However, for *ArtEmis-IdC-TS*, 2GEN outperforms all the other models with a significant margin, suggesting that 2GEN can effectively generate literal and imagination-driven captions. This result also implies the advantage of separating the generation of these two caption types.

Figure 7 shows two examples of captions generated by SAT, OFA, and LCGen/IdCGen of our model[1]. Compared with SAT and OFA, our model generally produces captions closer to human-generated descriptions. Particularly, our model accurately detects visual entities and associates them with imaginary entities similar to the annotators. As an example in Fig. 7(a), LCGen and IdCGen detect *face* and *man* as visual entities; *man* is linked with *a zombie is bleeding* in the imagination-driven caption generated by IdCGen. SAT and OFA generate only literal captions about the image's colors; no visual or imaginary object is included in these captions. In Fig. 7(b), although all the captions include *flowers* as a visual entity, only IdCGen successfully associates *flowers* with *a garden* as in the first annotator's description.

In summary, it can be observed that the proposed model is more effective than the existing models in generating both literal and imagination-driven captions. Also, it is suggested to learn these two caption types separately.

6   Summary and Conclusion

In this paper, we have shed light on the previously overlooked problem of generating imagination-driven captions. Specifically, we have analyzed existing datasets, classified imagination-driven captions, and discussed their usage. In addition, we have introduced the dataset for generating and evaluating imagination-driven image captioning methods. By using this dataset, we have assessed the nine existing evaluation metrics. Also, we have proposed a model capable of generating literal and imagination-driven captions. By separately learning the two caption types, our model is experimentally found to be more effective than the six existing models, especially for generating imagination-driven captions. We hope this study will be the groundwork for future studies on imagination-driven captions.

Acknowledgements. This work was supported by JST [Moonshot Research and Development], Grant Number [JPMJMS2032] and by JSPS KAKENHI Grant Number 20H05952 and 19H01110.

---

[1] See the captions generated by the other models and more examples in the supplementary material.

References

1. Hossain, M.Z., Sohel, F., Shiratuddin, M.F., Laga, H.: A comprehensive survey of deep learning for image captioning. ACM Comput. Surv. **51**, 1–36 (2019)
2. Nguyen, K., Tripathi, S., Du, B., Guha, T., Nguyen, T.Q.: In defense of scene graphs for image captioning. In: Proceedings of the IEEE/CVF International Conference on Computer Vision, pp. 1407–1416 (2021)
3. Yao, T., Pan, Y., Li, Y., Mei, T.: Exploring visual relationship for image captioning. In: Proceedings of the European Conference on Computer Vision, pp. 684–699 (2018)
4. Gu, J., Joty, S., Cai, J., Zhao, H., Yang, X., Wang, G.: Unpaired image captioning via scene graph alignments. In: Proceedings of the IEEE/CVF International Conference on Computer Vision, pp. 10323–10332 (2019)
5. Xu, K., et al.: Show, attend and tell: neural image caption generation with visual attention. In: International Conference on Machine Learning, pp. 2048–2057 (2015)
6. Chen, X., et al.: Microsoft coco captions: data collection and evaluation server. arXiv preprint arXiv:1504.00325 (2015)
7. Achlioptas, P., Ovsjanikov, M., Haydarov, K., Elhoseiny, M., Guibas, L.J.: Artemis: affective language for visual art. In: Proceedings of the IEEE/CVF Conference on Computer Vision and Pattern Recognition, pp. 11569–11579 (2021)
8. Young, P., Lai, A., Hodosh, M., Hockenmaier, J.: From image descriptions to visual denotations: new similarity metrics for semantic inference over event descriptions. Trans. Assoc. Comput. Linguist. **2**, 67–78 (2014)
9. Carston, R.: Figurative language, mental imagery, and pragmatics. Metaphor. Symb. **33**, 198–217 (2018)
10. Hodosh, M., Young, P., Hockenmaier, J.: Framing image description as a ranking task: data, models and evaluation metrics. J. Artif. Intell. Res. **47**, 853–899 (2013)
11. Gurari, D., Zhao, Y., Zhang, M., Bhattacharya, N.: Captioning images taken by people who are blind. In: Vedaldi, A., Bischof, H., Brox, T., Frahm, J.-M. (eds.) ECCV 2020. LNCS, vol. 12362, pp. 417–434. Springer, Cham (2020). https://doi.org/10.1007/978-3-030-58520-4_25
12. Sharma, P., Ding, N., Goodman, S., Soricut, R.: Conceptual captions: a cleaned, hypernymed, image alt-text dataset for automatic image captioning. In: Proceedings of Annual Meeting of the Association for Computational Linguistics (ACL) (2018)
13. Vinyals, O., Toshev, A., Bengio, S., Erhan, D.: Show and tell: a neural image caption generator. In: Proceedings of the IEEE Conference on Computer Vision and Pattern Recognition, pp. 3156–3164 (2015)
14. Chen, F., Ji, R., Sun, X., Wu, Y., Su, J.: GroupCap: group-based image captioning with structured relevance and diversity constraints. In: Proceedings of the IEEE Conference on Computer Vision and Pattern Recognition, pp. 1345–1353 (2018)
15. Mao, J., Xu, W., Yang, Y., Wang, J., Huang, Z., Yuille, A.: Deep captioning with multimodal recurrent neural networks (M-RNN). In: The International Conference on Learning Representations (ICLR) (2015)
16. Liao, W., Rosenhahn, B., Shuai, L., Ying Yang, M.: Natural language guided visual relationship detection. In: Proceedings of the IEEE/CVF Conference on Computer Vision and Pattern Recognition Workshops (2019)
17. Lee, H., Yoon, S., Dernoncourt, F., Bui, T., Jung, K.: UMIC: an unreferenced metric for image captioning via contrastive learning. In: Proceedings of the 59th Annual Meeting of the Association for Computational Linguistics and the 11th

International Joint Conference on Natural Language Processing (Volume 2: Short Papers), pp. 220–226 (2021)

18. Radford, A., et al.: Learning transferable visual models from natural language supervision. In: International Conference on Machine Learning, pp. 8748–8763 (2021)

19. Changpinyo, S., Sharma, P., Ding, N., Soricut, R.: Conceptual 12M: pushing web-scale image-text pre-training to recognize long-tail visual concepts. In: Proceedings of the IEEE/CVF Conference on Computer Vision and Pattern Recognition, pp. 3558–3568 (2021)

20. Stefanini, M., Cornia, M., Baraldi, L., Cascianelli, S., Fiameni, G., Cucchiara, R.: From show to tell: a survey on deep learning-based image captioning. IEEE Trans. Pattern Anal. Mach. Intell. **45**(1), 539–559 (2022)

21. Reed, S., Akata, Z., Lee, H., Schiele, B.: Learning deep representations of fine-grained visual descriptions. In: Proceedings of the IEEE Conference on Computer Vision and Pattern Recognition, pp. 49–58 (2016)

22. van Miltenburg, C., Vallejo, R.M., Elliott, D.: Pragmatic factors in image description: the case of negations. In: Proceedings of the Workshop on Vision and Language, pp. 54–59 (2016)

23. Chen, J., Kuznetsova, P., Warren, D., Choi, Y.: Déja image-captions: a corpus of expressive descriptions in repetition. In: Proceedings of the Conference of the North American Chapter of the Association for Computational Linguistics: Human Language Technologies, pp. 504–514 (2015)

24. Xu, G., Niu, S., Tan, M., Luo, Y., Du, Q., Wu, Q.: Towards accurate text-based image captioning with content diversity exploration. In: Proceedings of the IEEE/CVF Conference on Computer Vision and Pattern Recognition, pp. 12637–12646 (2021)

25. Chen, H., Wang, Y., Yang, X., Li, J.: Captioning transformer with scene graph guiding. In: IEEE International Conference on Image Processing (ICIP), pp. 2538–2542 (2021)

26. Cornia, M., Stefanini, M., Baraldi, L., Cucchiara, R.: Meshed-memory transformer for image captioning. In: Proceedings of the IEEE/CVF Conference on Computer Vision and Pattern Recognition, pp. 10578–10587 (2020)

27. Girshick, R.: Fast R-CNN. In: Proceedings of the IEEE International Conference on Computer Vision, pp. 1440–1448 (2015)

28. Vaswani, A., et al.: Attention is all you need. In: Advances in Neural Information Processing Systems, vol. 30 (2017)

29. Dosovitskiy, A., et al.: An image is worth 16x16 words: transformers for image recognition at scale. In: The International Conference on Learning Representations (ICLR) (2021)

30. Shen, S., et al.: How much can clip benefit vision-and-language tasks? arXiv preprint arXiv:2107.06383 (2021)

31. Hochreiter, S., Schmidhuber, J.: Long short-term memory. Neural Comput. **9**, 1735–1780 (1997)

32. Khan, S., Naseer, M., Hayat, M., Zamir, S.W., Khan, F.S., Shah, M.: Transformers in vision: a survey. ACM Comput. Surv. **54**, 1–41 (2021)

33. Lin, T.-Y., et al.: Microsoft COCO: common objects in context. In: Fleet, D., Pajdla, T., Schiele, B., Tuytelaars, T. (eds.) ECCV 2014. LNCS, vol. 8693, pp. 740–755. Springer, Cham (2014). https://doi.org/10.1007/978-3-319-10602-1_48

34. He, K., Zhang, X., Ren, S., Sun, J.: Deep residual learning for image recognition. In: Proceedings of the IEEE Conference on Computer Vision and Pattern Recognition, pp. 770–778 (2016)

35. Aneja, J., Deshpande, A., Schwing, A.G.: Convolutional image captioning. In: Proceedings of the IEEE Conference on Computer Vision and Pattern Recognition, pp. 5561–5570 (2018)
36. Li, Z., Tran, Q., Mai, L., Lin, Z., Yuille, A.L.: Context-aware group captioning via self-attention and contrastive features. In: Proceedings of the IEEE/CVF Conference on Computer Vision and Pattern Recognition, pp. 3440–3450 (2020)
37. Sohn, K.: Improved deep metric learning with multi-class N-pair loss objective. In: Advances in Neural Information Processing Systems, vol. 29 (2016)
38. Jiang, M., et al.: TIGEr: text-to-image grounding for image caption evaluation. In: Proceedings of the 2019 Conference on Empirical Methods in Natural Language Processing and the 9th International Joint Conference on Natural Language Processing (EMNLP-IJCNLP), pp. 2141–2152 (2019)
39. Madhyastha, P.S., Wang, J., Specia, L.: VIFIDEL: evaluating the visual fidelity of image descriptions. In: Proceedings of the 57th Annual Meeting of the Association for Computational Linguistics, pp. 6539–6550 (2019)
40. Zhang, T., Kishore, V., Wu, F., Weinberger, K.Q., Artzi, Y.: BERTScore: evaluating text generation with BERT. In: International Conference on Learning Representations (2020)
41. Lee, H., Yoon, S., Dernoncourt, F., Kim, D.S., Bui, T., Jung, K.: ViLBERTScore: evaluating image caption using vision-and-language BERT. In: Proceedings of the First Workshop on Evaluation and Comparison of NLP Systems, pp. 34–39 (2020)
42. Feinglass, J., Yang, Y.: SMURF: semantic and linguistic understanding fusion for caption evaluation via typicality analysis. In: Proceedings of the 59th Annual Meeting of the Association for Computational Linguistics and the 11th International Joint Conference on Natural Language Processing (Volume 1: Long Papers), pp. 2250–2260 (2021)
43. Hessel, J., Holtzman, A., Forbes, M., Le Bras, R., Choi, Y.: Clipscore: a reference-free evaluation metric for image captioning. In: Proceedings of the 2021 Conference on Empirical Methods in Natural Language Processing, pp. 7514–7528 (2021)
44. Kingma, D.P., Ba, J.: Adam: a method for stochastic optimization. arXiv preprint arXiv:1412.6980 (2014)
45. Li, X., et al.: OSCAR: object-semantics aligned pre-training for vision-language tasks. In: Vedaldi, A., Bischof, H., Brox, T., Frahm, J.-M. (eds.) ECCV 2020. LNCS, vol. 12375, pp. 121–137. Springer, Cham (2020). https://doi.org/10.1007/978-3-030-58577-8_8
46. Mokady, R., Hertz, A., Bermano, A.H.: Clipcap: clip prefix for image captioning. arXiv preprint arXiv:2111.09734 (2021)
47. Wang, P., et al.: OFA: unifying architectures, tasks, and modalities through a simple sequence-to-sequence learning framework. In: The International Conference on Machine Learning (ICML) (2022)
48. Papineni, K., Roukos, S., Ward, T., Zhu, W.J.: Bleu: a method for automatic evaluation of machine translation. In: Proceedings of the 40th Annual Meeting of the Association for Computational Linguistics, pp. 311–318 (2002)
49. Rouge, L.C.: A package for automatic evaluation of summaries. In: Proceedings of Annual Meeting of the Association for Computational Linguistics (ACL), Workshop on Text Summarization, Spain (2004)
50. Denkowski, M., Lavie, A.: Meteor universal: language specific translation evaluation for any target language. In: Proceedings of the Workshop on Statistical Machine Translation, pp. 376–380 (2014)

Heterogeneous Interactive Learning Network for Unsupervised Cross-Modal Retrieval

Yuanchao Zheng◯ and Xiaowei Zhang(✉)◯

Qingdao University, Qingdao, China
xiaowei19870119@sina.com

Abstract. Cross-modal hashing has received a lot of attention because of its unique characteristic of low storage cost and high retrieval efficiency. However, these existing cross-modal retrieval approaches often fail to align effectively semantic information due to information asymmetry between image and text modality. To address this issue, we propose Heterogeneous Interactive Learning Network (HILN) for unsupervised cross-modal retrieval to alleviate the problem of the heterogeneous semantic gap. Specifically, we introduce a multi-head self-attention mechanism to capture the global dependencies of semantic features within the modality. Moreover, since the semantic relations among object entities from different modalities exist consistency, we perform heterogeneous feature fusion through the heterogeneous feature interaction module, especially through the cross attention in it to learn the interaction between different modal features. Finally, to further maintain semantic consistency, we introduce adversarial loss into network learning to generate more robust hash codes. Extensive experiments demonstrate that the proposed HILN improves the accuracy of $T \rightarrow I$ and $I \rightarrow T$ cross-modal retrieval tasks by 7.6% and 5.5% over the best competitor DGCPN on the NUS-WIDE dataset, respectively. Code is available at https://github.com/Z000204/HILN.

Keywords: Cross-modal hashing · Heterogeneous interactive · Adversarial loss

1 Introduction

With the explosive growth of data, cross-modal hashing retrieval has attracted more and more attention. Cross-modal hashing (CMH) as a hot topic is to map data of different modalities to the common binary hash space for matching, which improves the efficiency of retrieval and storage consumption [7,19]. CMH is divided into unsupervised and supervised methods, depending on whether label information is used. At present, supervised hashing methods have achieved good performance [1,5,9] due to a large amount of hand-labeled prior knowledge. However, these methods based on supervised learning require a lot of manual annotations and are often not suitable for the real world. Recently, more and more attention has been paid to unsupervised cross-modal hashing [12,16,18,21],

© The Author(s), under exclusive license to Springer Nature Switzerland AG 2023
L. Wang et al. (Eds.): ACCV 2022, LNCS 13844, pp. 692–707, 2023.
https://doi.org/10.1007/978-3-031-26316-3_41

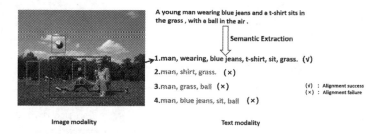

Fig. 1. An exemplar of semantic consistency within image-text pair. The red font in text and red bounding box represents aligned semantic information between image and text modality, the green ones denote semantic alignment failure. (Color figure online)

which can reduce the dependence on data annotations in the training process and has achieved significant progress.

However, due to the lack of artificial prior knowledge (label annotations), unsupervised cross-modal hashing methods face the problem of the heterogeneous semantic gap between image-text pairs, leading to the failure of cross-modal retrieval. How to model the semantic similarity between image modality and text modality becomes the key to improving the performance of cross-modal retrieval. Many recent unsupervised cross-modal hashing methods [12,16,21] proposed to align local semantic entities for building the semantic similarity. However, these methods do not take into account the information asymmetry between image-text pairs, where the local semantic entities between image and text modality often are unequal for cross-modal retrieval. As shown in Fig. 1, there are many "person" in image modality to respond to the "man" in-text modality, while "A man sits in the grass" in image modality is corresponding to "A young man wearing blue jeans and a t-shirt sits in the grass" with the similarity of semantic relations. This means that it often does not unique or even unequal entities to associate local semantics between image-text pairs, while there is a consistent semantic relation for cross-modal retrieval.

Based on the above analysis, we propose Heterogeneous Interactive Learning Network (HILN) for unsupervised cross-modal retrieval from the view of the similarity of semantic relations. First, we introduce a multi-head self-attention mechanism to capture the global dependencies of semantic features within the modality for modeling semantic relations among object entities. Secondly, we perform heterogeneous feature fusion through the heterogeneous feature interaction module, especially through cross attention to learn the interaction between different modal features. Finally, to further maintain semantic consistency, we introduce adversarial loss into network learning to generate more robust hash codes. Our contributions can be summarized as follows:

- We propose a novel end-to-end cross-modal hashing method, named Heterogeneous Interactive Learning Network (HILN) for unsupervised cross-modal retrieval, which models global semantic consistency between image and text modality from the view of similarity of semantic relations to generate high-quality hash codes.

- We introduce a multi-head self-attention mechanism to capture the global dependencies of semantic features within a single modality for modeling semantic relations among object entities.
- HILN performs heterogeneous feature fusion through the heterogeneous feature interaction module, especially through the cross attention in it to learn the interaction between different modal features, to better align the semantic relation between modalities. Finally, to further maintain semantic consistency, we introduce adversarial loss into network learning to generate more robust hash codes.
- Extensive experiments demonstrate that the proposed HILN improves the accuracy of $T \rightarrow I$ and $I \rightarrow T$ cross-modal retrieval tasks by 7.6% and 5.5% over the best competitor DGCPN [22] on the NUS-WIDE dataset, respectively.

In the following sections, we present related work (Sect. 2), the proposed method (Sect. 3), analytical experiments (Sect. 4), and conclusion (Sect. 5).

2   Related Work

2.1   Cross-Modal Hashing

Cross-modal hash retrieval methods can be broadly divided into two categories: supervised methods and unsupervised methods. The former is to explore semantic information from manual labels to bridge the semantic gap between different modalities for the generation of the hash codes, such as DCMH [9], DADH [1], AGAH [5].

Compared with the supervised methods, unsupervised methods mainly use co-occurrence information between images and texts to maximize the relationship between similar data between modalities and bridge the heterogeneous semantic gap between different modalities. Based on this, several unsupervised approaches have been proposed. DBRC [6] generates hash codes by reconstructing the original data from binary representation. To preserve the similarity between the original data, DJSRH [16] suggests a joint semantic similarity matrix and the use of hash codes to rebuild the similarity values of features. Next, JDSH [12] uses the characteristics of the data distribution to generate a better cross-modal similarity matrix to supervise the generation of hash codes based on DJSRH. At the same time, DSAH [21] builds on this by introducing semantic alignment loss to enhance the interaction between different modalities.

Although these approaches have achieved promising performance, they do not adequately align the global semantic relations between image and text modality, and even the performance imbalance of cross-modal retrieval between retrieving text from image and retrieving the image from text.

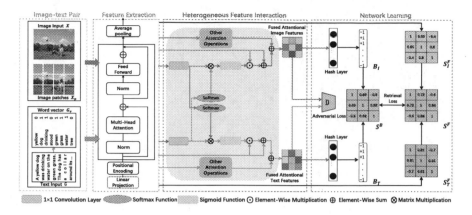

Fig. 2. The structure of our approach. It consists of an image-text pair, feature exaction, heterogeneous feature interaction, and Network learning. The purple arrows represent upsampling, and the red arrows represent dimensionality reduction and full connected layer. (Color figure online)

2.2 Attention Mechanism

Neural network-based attention mechanism has achieved great success and is widely used in a variety of tasks, such as natural language processing [11,20] and computer vision [4,27]. The self-attention mechanism [17] is a variant of the attention mechanism, which can capture long-distance dependencies. A Structured Self-attentive Sentence Embedding [11] uses self-attention for sentence embedding to enhance the semantics of the sentence. Context-aware Self-Attention Networks for Natural Language Processing [20] contextualize the representations with global information. Vision Transformer [4] represents the use of a transformer with the self-attention mechanism for image, capturing the global semantic information of the image, enhancing the ability of image feature characterization. In this paper, we utilize a multi-head self-attention mechanism to capture the global dependencies of semantic features within the single modality and introduce channel attention, spatial attention, and cross-attention to enhance the context-aware similarity of semantic relations between image-text pairs.

2.3 Generative Adversarial Network

Generative Adversarial Network, capable of modeling the distribution of data, have now achieved great success. It has been widely used in cross-modal hash retrieval tasks [2,24,25]. Among them, SCH-GAN [25] makes the generative model learn to fit the correlation distribution of unlabeled data, and tries to select samples from the unlabeled data of one modality to get the query of another modality, so as to better reflect the data of the unlabeled data distributed. MGAH [24] utilizes the ability of generative adversarial network for unsupervised representation learning to fully mine the underlying popular relationship of multimedia data, thereby improving retrieval performance. SAALDH [2] utilizes self-attention mechanism to enhance hash expression and utilizes adversarial loss to further maintain hash code consistency.

In this paper, we introduce adversarial loss into network learning, which utilizes an adversarial loss to model the feature distribution of image and text data on the basis of obtaining different modal attention enhancements to generate more robust hash codes.

3   Proposed Approach

3.1   Probelm Definition

Assume that we have M image-text pairs which can be denoted as $O = \{X_i, G_i\}_{i=1}^{M}$. X_i and G_i represent the i-th image and the i-th text in the instance respectively. The structure of our method is shown in Fig. 2. Given $\boldsymbol{F}^I$ and $\boldsymbol{F}^T$, our approach aims to learn two effective hash functions and simultaneously generates hash codes $\boldsymbol{B}_I \in \{-1, +1\}^{M \times K}$ and $\boldsymbol{B}_T \in \{-1, +1\}^{M \times K}$ for image and text modalities respectively, where K is the length of the hash codes.

3.2   Feature Extraction

Multi-head Self-attention. Like vision Transformer [4], we use the multi-head self-attention module to model the long-range relationships within the modalities.

Specifically, give the $1D$ embedding sequence $O \in \mathbb{R}^{L \times C}$ as input through the transformer-based encoder to learn feature representations, in which L is the length of the sequence, C is the numbers of the channel. Transformer encoder consists of J layers of multi-head self-attention (MSA) and Multilayer Perceptron (MLP) blocks. Layernorm is applied before every block and MLP. In each layer j, the query, key, and value computed from the input $O^{j-1} \in \mathbb{R}^{L \times C}$ with the corresponding weights are used as input for the self-attention as:

$$query = O^{j-1}\boldsymbol{W}_q, key = O^{j-1}\boldsymbol{W}_k, value = O^{j-1}\boldsymbol{W}_v, \qquad (1)$$

where $\boldsymbol{W}_q, \boldsymbol{W}_k, \boldsymbol{W}_v \in \mathbb{R}^{C \times d}$ are the learnable weight parameter. Self-attention (SA) is then formulated as:

$$SA\left(O^{j-1}\right) = O^{j-1} + softmax\left(\frac{O^{j-1}\boldsymbol{W}_q\left(O\boldsymbol{W}_k\right)^{\mathsf{T}}}{\sqrt{d}}\right)\left(O^{j-1}\boldsymbol{W}_v\right). \qquad (2)$$

MSA is an extension of SA in which contains n separate SA operations and projects their concatenated outputs.

$$MSA(O^{j-1}) = \left[SA_1\left(O^{j-1}\right); \cdots; SA_n\left(O^{j-1}\right)\right]\boldsymbol{W}_{mas}, \qquad (3)$$

where $\boldsymbol{W}_{mas} \in \mathbb{R}^{nd \times C}$ and d is set to C/n. The output of the MSA is then fed into the MLP and added to the MSA result by a residual.

$$O^j = MSA\left(O^{j-1}\right) + MLP\left(MSA\left(O^{j-1}\right)\right) \in \mathbb{R}^{L \times C}, \qquad (4)$$

which $j = 1, 2, \cdot, \cdot, \cdot, J$ represents the transformer layers.

Image Modality. We denoted $X \in \mathbb{R}^{H \times W \times C}$ to be a raw input image from the image dataset, where (H, W) is the resolution of the image and C is the number of channels. We reshape the image into a sequence of flattened $2D$ patches $X_p \in \mathbb{R}^{N \times P^2 \times C}$, where $N = (H \times W/P^2)$ is the number of image patches and (P, P) is the resolution of each patch. Then we generate an attentional image features $X_p^J \in \mathbb{R}^{M \times 1024}$ by the MSA and MLP:

$$X_p^J = MSA(X_p) + MLP(MSA(X_p)). \tag{5}$$

Finally, we obtain the final feature $\boldsymbol{Z}_a$ by pooling the aggregated features $\boldsymbol{X}_p^J$ on average.

Text Modality. Unlike most existing cross-modal hashing methods that only use bag-of-words as input and fully connected layers as the encoder, we expect the encoder can enhance the global connections of the text features. The word vector of the bag-of-words $G_v \in \mathbb{R}^d$ (where d is the dimension of word vector) is then turned into an attentional image features $\boldsymbol{G}_v^J \in \mathbb{R}^{M \times 1024}$ by the MSA and MLP:

$$G_v^J = MSA(G_v) + MLP(MSA(G_v)), \tag{6}$$

where G_v^J is the output of the last layer of the transformer and 1024 is the number of the channel. Finally, we obtain the final feature $\boldsymbol{E}_a$ by pooling the aggregated features $\boldsymbol{G}_v^J$ on average.

3.3   Heterogeneous Feature Interaction

Since the semantic relations among object entities from different modalities exist consistency, inspired by [23], as illustrated in Fig. 3, we exploit a heterogeneous feature interaction (HFI) module to perform heterogeneous feature fusion, especially through the cross attention in it to learn the interaction between different modal features, so as to better align the semantic relation between modalities. Specifically, we introduce channel attention, spatial attention, and cross-attention to align the semantic relations between the different modalities. We set the expanded attentional feature maps $\boldsymbol{Z}_a$ of the image and the attentional feature maps $\boldsymbol{E}_a$ of the text as Z and E, with attentional feature shapes of $H \times W \times C$ and $H \times W \times C$.

Spatial Attention. In this work, we use spatial attention to enhance the context-aware similarity of semantic relations of images and text by learning the global contextual information from the images and text. Spatial attention is used in space only with the self-attention mechanism. And spatial attention is computed separately on the attentional image modality and the attentional text modality.

Specifically, suppose the input attentional features are $\boldsymbol{Z} \in \mathbb{R}^{H \times W \times C}$, we first apply two separate convolution layers with 1×1 kernels on $\boldsymbol{Z}$ to generate query $\boldsymbol{Q}$ and $\boldsymbol{V}$ respectively, where $\boldsymbol{Q}, \boldsymbol{V} \in \mathbb{R}^{H \times W \times C'}$ and $C' = \frac{1}{2}C$ is the reduced channel number. Then a average pooling with aggregate the feature expressions

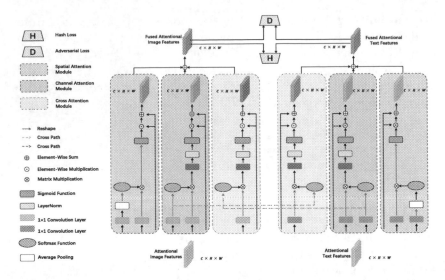

Fig. 3. The proposed heterogeneous feature interaction module, which includes spatial attention, channel attention, and cross-attention. The output of the heterogeneous feature interaction is computed simultaneously for adversarial loss and hashing loss.

is applied to the features Q to generate features $\bar{Q} \in \mathbb{R}^{1 \times 1 \times C'}$. Next the feature is reshaped to $\hat{Q} \in \mathbb{R}^{1 \times C'}$. The features V is reshaped to $\bar{V} \in \mathbb{R}^{HW \times C'}$. We can generate a spatial attention map $A_s \in \mathbb{R}^{HW \times 1}$ via softmax operations and matrix multiplication as:

$$A_s = softmax(\hat{Q})\bar{V} \in \mathbb{R}^{HW \times 1}. \tag{7}$$

Then the attention map A_s is reshaped to $A_s^s \in \mathbb{R}^{H \times W \times 1}$ and the sigmoid function is used to keep all parameters between 0 and 1 to get $\hat{A}_s^s$:

$$\hat{A}_s^s = sigmoid(A_s^s), \in \mathbb{R}^{H \times W \times 1}. \tag{8}$$

Finally, combining the $\hat{A}_s^s$ and Z by element-wise multiplication and Z by element-wise sum to get the final fused image features:

$$Z_s^s = \alpha \hat{A}_s^s \cdot Z + Z \in \mathbb{R}^{H \times W \times C}, \tag{9}$$

where α is a scalar parameter.

Channel Attention. The per-channel mapping of high-level semantic features is typically responsive to a specific target category. Similarly, processing features across all channels will hinder representation capabilities. Therefore, we use channel attention to selectively enhance the features within each modality. We can compute channel attention map A_c and the fused channel-features Z_c^c in a similar manner. Likewise, channel attention is a self-attention mechanism used only on channels. Notice the process, compare with the spatial attention without average pooling and add a layer norm, 1×1 convolution layer after matric multiplication.

Cross Attention. To enhance the context-aware similarity of semantic relations between different modalities, we use the cross-attention to learn such mutual information from the image branch and text branch.

Specifically, we use $\boldsymbol{Z} \in \mathbb{R}^{H \times W \times C}$ and $\boldsymbol{E} \in \mathbb{R}^{H \times W \times C}$ to denote attentional image features and attentional text features respectively. Taking the attentional branch for example, we first apply a 1×1 convolution layers with a reshape operation on $\boldsymbol{E}$ to generate query $\boldsymbol{Q}'$, where $\boldsymbol{Q}' \in \mathbb{R}^{H \times W \times 1}$ and reshape it to $\hat{\boldsymbol{Q}}' \in \mathbb{R}^{HW \times 1 \times 1}$. Then we compute the cross-attention from the image branch by performing similar operations as channel attention and the cross-attention computed from the image branch is encoded into the text value $\boldsymbol{V}'$ as,

$$\boldsymbol{A}_c = softmax(\hat{\boldsymbol{Q}}')\boldsymbol{V}' \in \mathbb{R}^{1 \times 1 \times C'}. \tag{10}$$

Then the attention map $\boldsymbol{A}_c$ is applied to 1×1 convolution layer with a reshape operation, and layernorm gets $\boldsymbol{A}_c^c \in \mathbb{R}^{1 \times 1 \times C}$ to improve model training speed and accuracy. The next is to use sigmoid function to keep all parameters between 0 and 1 to get $\hat{\boldsymbol{A}}_c^c$:

$$\hat{\boldsymbol{A}}_c^c = sigmoid(\boldsymbol{A}_c^c), \in \mathbb{R}^{1 \times 1 \times C}. \tag{11}$$

Combining the $\hat{\boldsymbol{A}}_c^c$ and $\boldsymbol{E}$ by element-wise multiplication and $\boldsymbol{E}$ by element-wise sum to get the final fused image features $\bar{\boldsymbol{Z}}_c^c$:

$$\bar{\boldsymbol{Z}}_c^c = \eta \hat{\boldsymbol{A}}_c^c \cdot \boldsymbol{E} + \boldsymbol{E} \in \mathbb{R}^{H \times W \times C}, \tag{12}$$

where η is a scalar parameter.

Finally, the spatial fused features $\boldsymbol{Z}_s^s$ and channel fused features $\boldsymbol{Z}_c^c$, cross-attentional features $\bar{\boldsymbol{Z}}_c^c$ are simply combined with an element-wise sum, generating the fused features $\boldsymbol{F}^I$ for image modality.

$$\boldsymbol{F}^I = \boldsymbol{Z}_s^s + \boldsymbol{Z}_c^c + \bar{\boldsymbol{Z}}_c^c. \tag{13}$$

In the same way, the fused features $\boldsymbol{F}^T$ for text modality is got.

$$\boldsymbol{F}^T = \boldsymbol{E}_s^s + \boldsymbol{E}_c^c + \bar{\boldsymbol{E}}_c^c. \tag{14}$$

3.4   Network Learning and Optimization

Adversarial Loss. To maintain modality invariance and semantic consistency across modalities, we introduce adversarial loss to align the semantic relations across modalities, and inspired by [5], we design a discriminator an as a classifier to discriminate the modalities to which the unknown relations belong. In this process, the semantic relations captured by self-attention in one modality and the semantic relations after attentional enhancement are treated as true semantic relations, while the semantic relations acquired in the other modality are treated as false semantic relations.

As the discriminator struggles to discriminate unknown relations, the captured semantic relations $\boldsymbol{F}^I$ and $\boldsymbol{F}^T$ struggle to confuse the discriminator. We define the adversarial loss as $\mathcal{L}_{adv}$. The adversarial loss in semantic relation learning $\mathcal{L}_{adv}$ can be formulated as follows:

$$\mathcal{L}_{adv} = -\frac{1}{n}\sum log\left(D_F\left(\boldsymbol{F}^I;\theta_{D_F}\right)\right) - \frac{1}{n}\sum log\left(1 - D_F\left(\boldsymbol{F}^T;\theta_{D_F}\right)\right) \quad (15)$$

Hash Loss. In order to better achieve a balanced state between modalities in the generated similarity matrix, we obtain fusion features $\boldsymbol{F}^I$ and $\boldsymbol{F}^T$ with semantic relations enhancement through the HFI module. After normalizing $\boldsymbol{F}^I, \boldsymbol{F}^T$ to $\boldsymbol{F}_I, \boldsymbol{F}_T$ which have unit L_2 norm each row, we can calculate the cosine similarity matrices $\boldsymbol{S}_I^F \in [-1,1]^{M \times M}$ on $\boldsymbol{F}_I$ and $\boldsymbol{S}_T^F \in [-1,1]^{M \times M}$ on $\boldsymbol{F}_T$ to describe the original neighborhood structure for the input image modality and text modality, respectively. We calculate the similarity matrix $\boldsymbol{S}^F \in [-1,1]^{M \times M}$ between the modal features of two images and text. Meanwhile, we further integrate it with image matrix $\boldsymbol{S}_I^F$ and text matrix $\boldsymbol{S}_T^F$ generated similarity matrix $\boldsymbol{S}$, so as to further bridge the heterogeneous semantic gap between modalities.

$$\boldsymbol{S} = \lambda\boldsymbol{S}_I^F + \beta\boldsymbol{S}_T^F + \omega\boldsymbol{S}^F = \{\boldsymbol{S}_{ij}\}_{i,j=1}^M,$$
$$s.t. \lambda, \beta, \omega \geq 0, \lambda + \beta + \omega = 1, \boldsymbol{S}_{ij} \in [-1, +1] \quad (16)$$

where S_{ij} represents the pairwise similarity of an image text data pair. λ, β, ω are the trade-off parameters that balance the similarity information from different modalities.

To obtain a high-level semantic description of the image and text modalities, we learn to obtain the hidden states using two functions $\boldsymbol{H}_I = f_I(\boldsymbol{F}^I; \theta_I)$ and $\boldsymbol{H}_T = f_T(\boldsymbol{F}^T; \theta_T)$ respectively, where θ_I and θ_T are two learnable parameters. In order to get the hash codes of the image and text, we adopt the sign function.

$$\boldsymbol{B}_* = sign(\boldsymbol{H}_*), * \in \{I, T\}. \quad (17)$$

Inspired by DSAH [21], although the ways and contents of obtaining matrix $\boldsymbol{S}$ in our method are different, in order to bridge the modal gap between different modalities, we adopt hash loss, including intra-modal loss, inter-modal loss, and symmetric loss function, just like them.

$$\mathcal{L}_{hl} = \min_{\boldsymbol{B}_I, \boldsymbol{B}_T} \sum \left\|1 - \boldsymbol{S}^B\right\|^2 + \min_{\boldsymbol{B}_I, \boldsymbol{B}_T} \sum \left\|\gamma\boldsymbol{S}_*^F - \boldsymbol{S}_*^B\right\|^2$$
$$+ \min_{\boldsymbol{B}_I, \boldsymbol{B}_T} \sum \left\|\gamma\boldsymbol{S} - \boldsymbol{S}_*^B\right\|^2 * \in \{I, T\}, \quad (18)$$

where after normalizing $\boldsymbol{B}^I, \boldsymbol{B}^T$ to $\boldsymbol{B}_I, \boldsymbol{B}_T$ which have unit L_2 norm each row, we can calculate the cosine similarity matrices $\boldsymbol{S}_I^B \in [-1,1]^{M \times M}$ on $\boldsymbol{B}^I$ and $\boldsymbol{S}_T^B \in [-1,1]^{M \times M}$ on $\boldsymbol{F}^T$ to describe the original neighborhood structure for the input images modality and texts modality respectively. And we calculate the cosine distances between hash codes of image and text to generate similarity matrices $\boldsymbol{S}^B = \boldsymbol{B}_I\boldsymbol{B}_T^\mathsf{T}$. And γ is a trade-off parameter.

We combine the losses of different modules into our final objective function, as shown below:

$$\mathcal{L}_{total} = \mathcal{L}_{adv} + \mathcal{L}_{hl}, \tag{19}$$

where $\mathcal{L}_{adv}$ and $\mathcal{L}_{hl}$ include the adversarial loss, hash loss. The hash loss includes inter-modal loss and symmetric loss. Since the sign function has the problem of gradient zero for any non-zero input to human, we substitute the tanh function for the sign function:

$$B_* = tanh(\eta H_*), * \in \{I, T\}, \tag{20}$$

where $\eta > 0$ is a scaling parameter, when $\lim_{\eta \to \infty} tanh(\eta H_*) = sign(H_*)$.

4   Experiments and Evaluations

4.1   Datasets and Evaluation

We conducted experiments on three public cross-modal retrieval datasets, including Wiki [14], MIRFlickr-25K [8] and NUS-WIDE [3], to verify the validity of our method. In our experiments, we adopt $mAP@50$ to evaluate the retrieval performance.

4.2   Implementation Details

In this section, we present the implementation details of our HILN in the experiments. We implement the method HILN by pytorch [13], and workstation configured with NVIDIA RTX 3090 GPU. We set the dimension of the feature representation extracted by the transformer to 1024. After the heterogeneous feature interaction, the feature representation obtained by dimensionality reduction is 1024 and set the dimension of the hash layer to be consistent with the length of the hash code.

Moreover, we train the proposed HILN in a mini-batch way and set the batch size as 32. For other comparison methods, we set the optimal experimental parameter configuration provided by their authors for training. For fairness, for all methods, we use the same dataset for performance comparison. The weight decay rate is 0.0005 and the momentum is set to 0.9. When training on the Wiki dataset, the learning rate of the network is set to 0.01, $\lambda = 0.4$, $\beta = 0.4$, $\omega = 0.2$ when training on the MIRFlickr-25k and NUS-WIDE datasets, the learning rate of the network is set to 0.001, $\lambda = 0.45$, $\beta = 0.45$, $\omega = 0.1$. The training epochs on the Wiki, MIRFLickr-25K, and NUS-WIDE datasets are set to 150, 100, and 80, respectively.

4.3   Performance

We compare our HILN with several representative deep unsupervised cross-modal hashing retrieval methods including DBRC [6], UDCMH [18],DJSRH [16],

Table 1. The mAP@50 values of Wiki, MIRFlickr-25k, and NUS-WIDE at various code lengths. Bold data represent the best performance among all contrasting methods.

Task	Method	Wiki				MIRFlickr-25k				NUS-WIDE			
		16bits	32bits	64bits	128bits	16bits	32bits	64bits	128bits	16bits	32bits	64bits	128bits
I→T	DBRC [6]	0.253	0.265	0.269	0.288	0.617	0.619	0.620	0.621	0.424	0.459	0.447	0.447
	UDCMH [18]	0.309	0.318	0.329	0.346	0.689	0.698	0.714	0.717	0.511	0.519	0.524	0.558
	DJSRH [16]	0.388	0.403	0.412	0.421	0.810	0.843	0.862	0.876	0.724	0.773	0.798	0.817
	JDSH [12]	0.346	0.431	0.433	0.442	0.832	0.853	0.882	0.892	0.736	0.793	0.832	0.835
	DSAH [21]	0.416	0.430	0.438	0.445	0.863	0.877	0.895	0.903	0.775	0.805	0.818	0.827
	AGSH [15]	0.397	0.434	0.446	-	0.679	0.691	0.698	-	0.543	0.552	0.562	-
	KDCMH [10]	-	-	-	-	0.713	0.716	0.724	0.728	0.615	0.628	0.637	0.642
	DGCPN [22]	0.420	0.438	0.440	0.448	0.875	0.891	0.908	0.918	0.788	0.820	0.826	0.833
	AGCH [26]	0.408	0.425	0.433	0.450	0.865	0.887	0.892	0.912	**0.809**	0.830	0.831	0.852
	OURS	**0.446**	**0.453**	**0.467**	**0.472**	**0.878**	**0.908**	**0.932**	**0.944**	0.793	**0.830**	**0.862**	**0.879**
T→I	DBRC [6]	0.573	0.588	0.598	0.599	0.618	0.626	0.626	0.628	0.455	0.459	0.468	0.473
	UDCMH [18]	0.622	0.633	0.645	0.658	0.692	0.704	0.718	0.733	0.637	0.653	0.695	0.716
	DJSRH [16]	0.611	0.635	0.646	0.658	0.786	0.822	0.835	0.847	0.712	0.744	0.771	0.789
	JDSH [12]	0.630	0.631	0.647	0.651	0.825	0.864	0.878	0.880	0.721	0.785	0.794	0.804
	DSAH [21]	0.644	0.650	0.660	0.662	0.846	0.860	0.881	0.882	0.770	0.790	0.804	0.815
	AGSH [15]	0.431	0.443	0.453	-	0.674	0.689	0.693	-	0.543	0.567	0.570	-
	KDCMH [10]	-	-	-	-	0.711	0.715	0.731	0.733	0.623	0.636	0.647	0.651
	DGCPN [22]	**0.644**	0.651	0.660	**0.662**	0.859	0.876	0.890	0.905	0.783	0.802	0.812	0.817
	AGCH [26]	0.627	0.640	0.648	0.658	0.829	0.849	0.852	0.880	0.769	0.780	0.798	0.802
	OURS	0.643	**0.651**	**0.660**	0.661	**0.877**	**0.908**	**0.932**	**0.944**	**0.793**	**0.831**	**0.862**	**0.879**

JDSH [12], DSAH [21], AGSH [15], KDCMH [10], AGCH [26] DGCPN [22]. Table 1 shows the *mAP*@50 values of HILN and other comparison methods on MIRFlickr-25k, NUS-WIDE, and Wiki in two cross-modal retrieval tasks for four lengths of hash codes. And Fig. 4 shows the precision@top-K curves on three datasets at 128 bits among five comparison methods. $I \rightarrow T$ means that the query is image and text modality is the database. $T \rightarrow I$ is the opposite. It can be seen that HILN is significantly better than the latest unsupervised cross-modal hashing methods. Specifically, compared to DGCPN [22], for the Wiki, as shown in the results, we achieve boosts of 5.3% in average *mAP*@50 for different hash code lengths in the $I \rightarrow T$ task. Moreover, HILN achieves boosts of 1.9% and 3.7% in average *mAP*@50 with different hash code lengths in $I \rightarrow T$ task and $T \rightarrow I$ task on MIRFlickr-25k respectively, and achieves boosts of 2.9% and 4.6% in two retrieval tasks on NUS-WIDE.

The main reason for the performance improvement is the HFI proposed by HILN. It also ensures that the performance of image retrieval for text and text retrieval for images is essentially the same.

4.4   Ablation Study

We design several variants to validate the impact of our proposed modules and to demonstrate the superiority of the original HILN.

Table 2. The mAP@50 results at 128 bits of ablation study about MSA on MIRFlickr-25k and NUS-WIDE. Bold data represent the best performance.

Model	Configuration	MIRFlickr		NUS-WIDE	
		I→T	T→I	I→T	T→I
Baseline	-	0.903	0.882	0.827	0.815
HILN-1	Baseline+IMSA	0.914	0.892	0.842	0.827
HILN-2	Baseline+TMSA	0.910	0.888	0.838	0.823
HILN-3	Baseline+I,TMSA	**0.919**	**0.896**	**0.848**	**0.832**

Table 3. The mAP@50 results at 128 bits of ablation study about HFI on MIRFlickr-25k and NUS-WIDE. Bold data represent the best performance of the HILN method.

Model	Configuration	MIRFlickr		NUS-WIDE	
		I→T	T→I	I→T	T→I
HFI-1	-	0.919	0.896	0.848	0.832
HFI-2	HFI-1+ATT	0.939	0.939	0.871	0.871
HFI-3	HFI-1+$\mathcal{L}_{adv}$	0.931	0.928	0.862	0.853
HFI-4	HFI-2+ $\mathcal{L}_{adv}$	**0.944**	**0.944**	**0.879**	**0.879**

The Effectiveness of Multi-head Self-attention. As shown in Table 2, several variants we designed to verify the effectiveness of the multi-head self-attention module. HILN-1 extracts global semantic information from images using only the multi-head self-attention mechanism. HILN-2 extracts global semantic information from text using only the multi-head self-attention mechanism. HILN-3 extracts global semantic information from images and texts using the multi-head self-attention mechanism.

From the results of Baseline, HILN-1, HILN-2, and HILN-3, we find the effectiveness of multi-head self-attention(MSA). HILN-1 improves $mAP@50$ by 1.1% and 1.5% over Baseline for the $T \rightarrow I$ task on both datasets MIRFlickr and NUS-WIDE, respectively. HILN-1 improves $mAP@50$ by 1.2% and 1.8% over Baseline for the $I \rightarrow T$ task on both datasets MIRFlickr and NUS-WIDE, respectively. We find that the reason for the improved performance of HILN-1 is due to the extraction of global semantic information of the image using the MSA. HILN-2 improves $mAP@50$ by 0.7% and 1% over Baseline for the $T \rightarrow I$ task on both datasets MIRFlickr and NUS-WIDE, respectively. HILN-2 improves $mAP@50$ by 0.7% and 1.3% over Baseline for the $I \rightarrow T$ task on both datasets MIRFlickr and NUS-WIDE, respectively. We find that the reason for the improved performance of HILN-2 is due to the extraction of global semantic information of the text using the MSA. From the $mAP@50$ results of HILN-1, HILN-2, and HILN-3, we find that the MSA can effectively capture the global dependencies of semantic features within the modality for modeling semantic relations among object entities. These results suggest that the MSA is more effective for image modalities and performs better if used simultaneously.

The Effectiveness of Heterogeneous Feature Interaction and Adversarial Loss. Meanwhile, as shown in Table 3, there are several variants we designed to verify the effectiveness of the heterogeneous feature interaction and adversarial loss. HFI-1 stands for only using the MSA mechanism to capture the global semantic information of image and text modalities. HFI-2 builds on HFI-1 by adding only the heterogeneous feature interaction module. HFI-3 adds only the adversarial loss to HFI-1. HFI-4 builds on HFI-2 by adding only the adversarial loss.

Table 4. The mAP@50 results at 128 bits of ablation study about three of attention on MIRFlickr-25k and NUS-WIDE.

Model	Configuration	MIRFlickr-25k		NUS-WIDE	
		I→T	T→I	I→T	T→I
HFI-a	MSA+$\mathcal{L}_{adv}$	0.919	0.896	0.848	0.832
HFI-b	Baseline+Spatial attention	0.927	0.907	0.861	0.849
HFI-c	Baseline+Spatial+Channel Attention	0.935	0.918	0.869	0.856
HFI-d	Baseline+Spatial+Channel+Cross Attention	**0.944**	**0.944**	**0.879**	**0.879**

HFI-2 improves $mAP@50$ by 4.8% and 4.7% over HFI-1 for the $T \rightarrow I$ task on both datasets, MIRFlickr and NUS-WIDE, respectively. HFI-2 improves $mAP@50$ by 2.2% and 2.7% over HFI-1 for the $I \rightarrow T$ task on both datasets, MIRFlickr and NUS-WIDE, respectively. The performance improvement of HFI-2 attributes the attention from the heterogeneous feature interaction module, which effectively aligns the global semantic similarity relations between different modalities so that global semantic relations between modalities are consistent. HFI-3 improves $mAP@50$ by 3.2% and 2.5% over HFI-1 for the $T \rightarrow I$ task on both datasets, MIRFlickr and NUS-WIDE, respectively. HFI-3 improves $mAP@50$ by 1.3% and 1.7% over HFI-1 for the $I \rightarrow T$ task on both datasets, MIRFlickr and NUS-WIDE, respectively. And from the $mAP@50$ results of HFI-2 and HFI-4, the performance improvement of HFI-2 comes mainly from the adversarial loss. Thus, it is shown that adversarial loss effectively maintains semantic consistency.

Comparing HFI-2 and HFI-3, HFI-2 makes the performance of $I \rightarrow T$ and $T \rightarrow I$ tasks comparable, while the performance of HFI-3 is stronger for $I \rightarrow T$ tasks than for $T \rightarrow I$ tasks. Meanwhile, the performance of HFI-2 is better than that of HFI-3. From the above comparative analysis, it can be found that if the heterogeneous feature interaction or the adversarial loss is used alone, the effect is not as good as the effect of using both simultaneously. Therefore, the adversarial loss we introduce is effective for cross-modal hashing.

The Effectiveness of Spatial Attention, Channel Attention, and Cross Attention. As shown in Table 4, several variants we designed to verify the effectiveness of spatial attention, channel attention, and cross attention. HFI-a represents the use of MSA and adversarial loss. HFI-b builds on HFI-a by adding

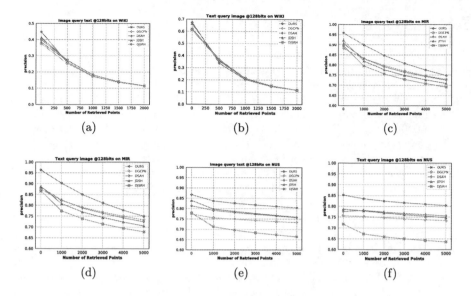

Fig. 4. Precision@top-K curves on three datasets at 128 bits.

only spatial attention. HFI-c builds on HFI-b by adding only the channel attention. HFI-d builds on HFI-c by adding only the cross attention. From the results of HFI-a, HFI-b, HFI-c, and HFI-d, we find the effectiveness of spatial attention, channel attention, and cross attention. In particular, when three kinds of attention are used simultaneously, the retrieval effect will be better.

5 Conclusion

In this paper, we propose a novel unsupervised cross-modal hashing method called Heterogeneous Interactive Learning Network (HILN) for unsupervised cross-modal retrieval. To bridge the heterogeneous semantic gap between different modalities, we introduce a multi-head self-attention mechanism to capture the global dependencies of semantic features within the modality for modeling semantic relations among object entities. Meanwhile, we exploit a heterogeneous feature interaction module for feature fusion to align the semantic relationships between different modalities. Moreover, we introduce adversarial loss into network learning to further maintain semantic consistency. Extensive experiments have shown the effectiveness of our method.

Acknowledgments. This work was supported in part by the National Natural Science Foundation of China (Grant No. 61902204), in part by the Natural Science Foundation of Shandong Province of China (Grant No. ZR2019BF028).

References

1. Bai, C., Zeng, C., Ma, Q., Zhang, J., Chen, S.: Deep adversarial discrete hashing for cross-modal retrieval. In: Proceedings of the 2020 International Conference on Multimedia Retrieval, pp. 525–531 (2020)
2. Chen, S., Wu, S., Wang, L., Yu, Z.: Self-attention and adversary learning deep hashing network for cross-modal retrieval. Comput. Electr. Eng. **93**, 107262 (2021)
3. Chua, T.S., Tang, J., Hong, R., Li, H., Luo, Z., Zheng, Y.: NUS-WIDE: a real-world web image database from National University of Singapore. In: Proceedings of the ACM International Conference on Image and Video Retrieval, pp. 1–9 (2009)
4. Dosovitskiy, A., et al.: An image is worth 16x16 words: transformers for image recognition at scale. arXiv preprint arXiv:2010.11929 (2020)
5. Gu, W., Gu, X., Gu, J., Li, B., Xiong, Z., Wang, W.: Adversary guided asymmetric hashing for cross-modal retrieval. In: Proceedings of the 2019 on International Conference on Multimedia Retrieval, pp. 159–167 (2019)
6. Hu, D., Nie, F., Li, X.: Deep binary reconstruction for cross-modal hashing. IEEE Trans. Multimed. **21**(4), 973–985 (2018)
7. Hu, H., Xie, L., Hong, R., Tian, Q.: Creating something from nothing: unsupervised knowledge distillation for cross-modal hashing. In: Proceedings of the IEEE/CVF Conference on Computer Vision and Pattern Recognition, pp. 3123–3132 (2020)
8. Huiskes, M.J., Lew, M.S.: The MIR Flickr retrieval evaluation. In: Proceedings of the 1st ACM International Conference on Multimedia Information Retrieval, pp. 39–43 (2008)
9. Jiang, Q.Y., Li, W.J.: Deep cross-modal hashing. In: Proceedings of the IEEE Conference on Computer Vision and Pattern Recognition, pp. 3232–3240 (2017)
10. Li, M., Wang, H.: Unsupervised deep cross-modal hashing by knowledge distillation for large-scale cross-modal retrieval. In: Proceedings of the 2021 International Conference on Multimedia Retrieval, pp. 183–191 (2021)
11. Lin, Z., et al.: A structured self-attentive sentence embedding. arXiv preprint arXiv:1703.03130 (2017)
12. Liu, S., Qian, S., Guan, Y., Zhan, J., Ying, L.: Joint-modal distribution-based similarity hashing for large-scale unsupervised deep cross-modal retrieval. In: Proceedings of the 43rd International ACM SIGIR Conference on Research and Development in Information Retrieval, pp. 1379–1388 (2020)
13. Paszke, A., et al.: Pytorch: an imperative style, high-performance deep learning library. In: Advances in Neural Information Processing Systems, vol. 32 (2019)
14. Pereira, J.C., et al.: On the role of correlation and abstraction in cross-modal multimedia retrieval. IEEE Trans. Pattern Anal. Mach. Intell. **36**(3), 521–535 (2013)
15. Shen, X., Zhang, H., Li, L., Liu, L.: Attention-guided semantic hashing for unsupervised cross-modal retrieval. In: 2021 IEEE International Conference on Multimedia and Expo (ICME), pp. 1–6. IEEE (2021)
16. Su, S., Zhong, Z., Zhang, C.: Deep joint-semantics reconstructing hashing for large-scale unsupervised cross-modal retrieval. In: Proceedings of the IEEE/CVF International Conference on Computer Vision, pp. 3027–3035 (2019)
17. Vaswani, A., et al.: Attention is all you need. In: Advances in Neural Information Processing Systems, pp. 5998–6008 (2017)
18. Wu, G., et al.: Unsupervised deep hashing via binary latent factor models for large-scale cross-modal retrieval. In: IJCAI, pp. 2854–2860 (2018)
19. Yan, C., Bai, X., Wang, S., Zhou, J., Hancock, E.R.: Cross-modal hashing with semantic deep embedding. Neurocomputing **337**, 58–66 (2019)

20. Yang, B., Wang, L., Wong, D.F., Shi, S., Tu, Z.: Context-aware self-attention networks for natural language processing. Neurocomputing **458**, 157–169 (2021)
21. Yang, D., Wu, D., Zhang, W., Zhang, H., Li, B., Wang, W.: Deep semantic-alignment hashing for unsupervised cross-modal retrieval. In: Proceedings of the 2020 International Conference on Multimedia Retrieval, pp. 44–52 (2020)
22. Yu, J., Zhou, H., Zhan, Y., Tao, D.: Deep graph-neighbor coherence preserving network for unsupervised cross-modal hashing. In: Proceedings of the AAAI Conference on Artificial Intelligence, pp. 4626–4634. AAAI (2021)
23. Yu, Y., Xiong, Y., Huang, W., Scott, M.R.: Deformable Siamese attention networks for visual object tracking. In: Proceedings of the IEEE/CVF Conference on Computer Vision and Pattern Recognition, pp. 6728–6737 (2020)
24. Zhang, J., Peng, Y.: Multi-pathway generative adversarial hashing for unsupervised cross-modal retrieval. IEEE Trans. Multimed. **22**(1), 174–187 (2019)
25. Zhang, J., Peng, Y., Yuan, M.: SCH-GAN: semi-supervised cross-modal hashing by generative adversarial network. IEEE Trans. Cybern. **50**(2), 489–502 (2018)
26. Zhang, P.F., Li, Y., Huang, Z., Xu, X.S.: Aggregation-based graph convolutional hashing for unsupervised cross-modal retrieval. IEEE Trans. Multimed. **24**, 466–479 (2021)
27. Zhu, L., Tian, G., Wang, B., Wang, W., Zhang, D., Li, C.: Multi-attention based semantic deep hashing for cross-modal retrieval. Appl. Intell. **51**, 1–13 (2021)

Biometrics

GaitStrip: Gait Recognition via Effective Strip-Based Feature Representations and Multi-level Framework

Ming Wang[1], Beibei Lin[2], Xianda Guo[3], Lincheng Li[4], Zheng Zhu[3], Jiande Sun[5], Shunli Zhang[1(✉)], Yu Liu[1], and Xin Yu[6]

[1] Beijing Jiaotong University, Beijing, China
slzhang@bjtu.edu.cn
[2] National University of Singapore, Singapore, Singapore
[3] PhiGent Robotics, Beijing, China
[4] NetEase Fuxi AI Lab, Beijing, China
[5] Shandong Normal University, Jinan, China
[6] University of Technology Sydney, Ultimo, Australia

Abstract. Many gait recognition methods first partition the human gait into N-parts and then combine them to establish part-based feature representations. Their gait recognition performance is often affected by partitioning strategies, which are empirically chosen in different datasets. However, we observe that strips as the basic component of parts are agnostic against different partitioning strategies. Motivated by this observation, we present a strip-based multi-level gait recognition network, named GaitStrip, to extract comprehensive gait information at different levels. To be specific, our high-level branch explores the context of gait sequences and our low-level one focuses on detailed posture changes. We introduce a novel StriP-Based feature extractor (SPB) to learn the strip-based feature representations by directly taking each strip of the human body as the basic unit. Moreover, we propose a novel multi-branch structure, called Enhanced Convolution Module (ECM), to extract different representations of gaits. ECM consists of the Spatial-Temporal feature extractor (ST), the Frame-Level feature extractor (FL) and SPB, and has two obvious advantages: First, each branch focuses on a specific representation, which can be used to improve the robustness of the network. Specifically, ST aims to extract spatial-temporal features of gait sequences, while FL is used to generate the feature representation of each frame. Second, the parameters of the ECM can be reduced in test by introducing a structural re-parameterization technique. Extensive experimental results demonstrate that our GaitStrip achieves state-of-the-art performance in both normal walking and complex conditions. The source code is published at https://github.com/M-Candy77/GaitStrip.

1 Introduction

Gait recognition is one of the most popular biometric techniques. Since it can be used in a long-distance condition and cannot be disguised, gait recognition

L. Wang et al. (Eds.): ACCV 2022, LNCS 13844, pp. 711–727, 2023.
https://doi.org/10.1007/978-3-031-26316-3_42

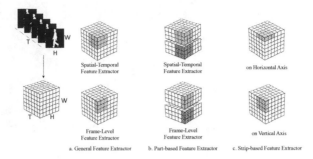

Fig. 1. Visualization of feature extractors of different methods.

is widely applied in video surveillance and access control systems. However, this technology has experienced a huge challenge due to the complexity of the external environment, such as cross-view, speed changes, bad weathers and variations in appearances [4,13,14,35,37].

Recently, many Convolutional Neural Networks (CNNs) based gait recognition frameworks have been proposed to generate discriminative feature representations [1,2,7,15–19,21,22,26–28,33,36,40,44,45]. As shown in Fig. 1(a), some researchers extract gait features directly from the whole gait sequence, which captures global context information of gait sequences [2,27,40]. As those methods take the human gait as a unit to extract features, some local gait changes that are important for gait recognition might not be fully captured, which may affect the recognition performance. On the other hand, some other researchers [7,44] propose part-based feature representation to represent the human gait, which is shown in Fig. 1(b). They first partition the human gait into N-parts and then extract the detailed information of each part. Although carefully choosing the number of partitions in different convolutional layers can achieve appealing performance, it is unclear how to build an accurate part-based model on new datasets, which limits the generalization of the methods.

According to these findings, we argue that the part-based feature representation is not a general feature representation for gait recognition. Hence, we question *whether there is a gait descriptor that is insensitive to various partitions?* Through carefully analysis of recent part-based methods, we find that strips are the minimal effective representation elements for gaits instead of parts. Using strips, we will be able to circumvent the handcrafted partition in part-based methods. As shown in Fig. 1(c), the strip can be considered as an extreme form of the part-based representation, thus it is not necessary to manually determine the reasonable number of the parts. Motivated by this observation, we propose a new gait recognition network, called GaitStrip, to learn more discriminative feature representations based on strips. Specifically, GaitStrip is implemented under a multi-level framework to improve the representation capability. The multi-level framework includes two branches, i.e., the low-level branch and the high-level one. In particular, the high-level branch extracts the global context information from low-resolution gait images, while the low-level one captures more details from high-resolution images.

Furthermore, we introduce Enhanced Convolution Module (ECM), as a multi-branch block, to our GaitStrip. ECM includes three branches, i.e., the StriP-Based feature extractor (SPB), the Spatial-Temporal feature extractor (ST) and the Frame-Level feature extractor (FL), where each branch corresponds to a specific representation. Specifically, SPB is designed to generate strip-based feature representations by taking each strip of the human body as a basic unit, ST aims to extract spatial-temporal information of a gait sequence, and FL is used to extract each frame's spatial features. On the other hand, we introduce a structural re-parameterization technique to reduce the parameters of the ECM module in test [6]. Specifically, the parameters of SPB, ST and FL can be merged into a single $3 \times 3 \times 3$ convolution.

After feature extraction, we obtain an effective feature representation by using temporal aggregation and spatial mapping operations. The temporal aggregation ensembles temporal information of a variable-length gait sequence [20]. The spatial mapping first partitions the feature maps into multiple horizontal vectors and aggregates each vector by Generalized-Mean (GeM) pooling operations [25] for better representation. Extensive experiments on widely-used gait recognition benchmarks demonstrate that our GaitStrip outperforms the state-of-the-arts significantly.

The main contributions of the proposed method are three-fold, shown as follows:

- Based on the observation that the strip-based method can achieve more effective gait representations than part-based partitioning, we propose a multi-level gait recognition framework with strip to extract more comprehensive gait features, in which the high-level representation contains the context information while the low-level representation extracts local details of gait sequences.
- We develop an effective enhanced convolution module including three branches, which can not only take the advantage of both frame-level and spatial-temporal features but also use SPB to enhance the representation ability. Furthermore, we use the structural re-parameterization technique to reduce the parameters for high efficiency in test.
- We compare the proposed method with several state-of-the-art methods on two public datasets, CASIA-B and OUMVLP. The experimental results demonstrate that the performance of the proposed method achieves superior performance to these approaches.

2   Related Work

2.1   Gait Recognition

Existing gait recognition methods can be divided into two types, *i.e.* , global-based and local-based.

The global-based methods usually take the human gait as a sample to generate global feature representations [27,34,40]. For instance, Shiraga et al. [27] first calculate the Gait Energy Image (GEI) by using the mean function to compress the temporal information of gait sequences, and then utilize 2D CNNs to extract gait features. However, the generation of the GEI causes the loss of temporal information, which may degrade the representation ability. Thus, some other researchers [2,3,10,43] use 2D CNNs to extract each frame's feature before building the template. On the other hand, some researchers [20,30,32] extract spatial-temporal information from gait sequences for representation. Recently, 3D CNN has been used in gait recognition to learn the spatial-temporal representation of the entire gait sequence. For example, Lin et al. [20] use 3D CNNs to extract spatial-temporal information, and employ temporal aggregation to integrate temporal information, addressing the variable-length issue of video sequences.

The local-based methods usually take the part of the human gait as input to establish the part-based feature representations [7,44]. For example, Fan et al. [7] propose a focal convolution layer to extract part-based gait features. The focal convolutional layer first splits the feature maps into several local parts and then uses a shared convolution to extract each part's feature. Zhang et al. [44] first partition the human gait into four parts and then use 2D CNN to obtain feature representations of each part. However, these local-based methods need to predefine the number of partitions for specific datasets, which limit the generalization ability.

2.2   Strip-Based Modeling

Recently many strip-based modeling methods have been proposed in the visual field. For example, Ding et al. [5] propose a novel block, called Asymmetric Convolution Block (ACB), to generate discriminative feature representations. They use 1D asymmetric forms (e.g. 3×1 Conv and 1×3 Conv) to improve the feature representation ability of the standard square-kernel convolution (3×3 Conv). Note that the asymmetric convolutions can exploit the information of the horizontal and vertical strips. In particular, the asymmetric convolutions can be fused into the original square-kernel convolution. Huang et al. [12] propose the CCNet network to capture global contextual information. CCNet which is built with Criss-Cross Attention blocks models the relationships of horizontal and vertical strips.

However, the aforementioned methods only focus on the spatial strip-based information, which do not capture the temporal changes of each strip. Therefore, in this paper, we propose a novel strip-based feature extractor, which can be used to establish each strip's spatial-temporal information. In particular, as far as we know, GaitStrip is the first network which models strip-based feature representations in gait recognition.

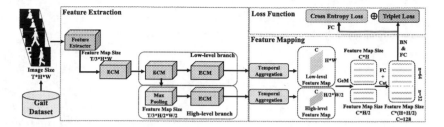

Fig. 2. Overview of the entire gait recognition framework.

3 Proposed Method

In this section, we first overview the whole gait recognition framework. Then, we describe the enhanced convolution module, the multiple-level structure and feature mapping in detail. Finally, we introduce the strategies of training and test.

3.1 Overview

The proposed gait recognition framework, GaitStrip, which includes the feature extraction stage and feature mapping stage is shown in Fig. 2. The GaitStrip is constructed based on 3D convolutions, which can effectively extract spatial-temporal information of gait sequences. During the feature extraction stage, a novel enhanced convolution module which uses both frame-level feature extractor and strip-based feature extractor to improve the representation ability of the traditional spatial-temporal feature extractor is proposed. Then, we design the multi-level framework which includes both the high-level and the low-level branches. During the feature mapping stage, the temporal aggregation operation is introduced to integrate the temporal information of feature maps [20]. Then, the feature maps are partitioned into multiple horizontal vectors and the information is aggregated by Generalized-Mean (GeM) pooling [25]. Finally, a combined loss function consisting of both cross-entropy loss and triplet loss is employed to train the proposed network.

3.2 Enhanced Convolution Module

Recently, many excellent feature extractors have been proposed to extract robust gait features, which can be divided into two types. One is the frame-level feature extractor which extracts gait features of each frame [2,3,7], and the other one is the spatial-temporal feature extractor which generates spatial-temporal feature representations of a gait sequence [20,30,32].

Assume that the feature map $X_{in} \in \mathbb{R}^{C_{in} \times T_{in} \times H_{in} \times W_{in}}$ is the input of a convolution operation, where C_{in} is the number of channels, T_{in} is the length of gait sequences and (H_{in}, W_{in}) is the image size of each frame. The frame-level and spatial-temporal feature extractors can be designed as

$$X_{FL} = c^{1 \times 3 \times 3}(X_{in}), \tag{1}$$

$$X_{ST} = c^{3 \times 3 \times 3}(X_{in}), \tag{2}$$

where $c^{a \times b \times c}(\cdot)$ represents the 3D convolution with kernel size (a, b, c). $X_{FL} \in \mathbb{R}^{C_{out} \times T_{in} \times H_{in} \times W_{in}}$ and $X_{ST} \in \mathbb{R}^{C_{out} \times T_{in} \times H_{in} \times W_{in}}$ are the output of the frame-level and spatial-temporal feature extractors, respectively.

The frame-level features ignore the temporal information of the gait sequence, while the spatial-temporal features focus on the spatial-temporal changes, which may not pay enough attention to the detailed information of each frame. Thus, we propose a combined framework which takes advantage of frame-level and spatial-temporal information as our backbone. The combined structure includes two branches, i.e. the spatial-temporal feature extractor and frame-level feature extractor, which can be designed as

$$X_{STFL} = X_{FL} + X_{ST}. \tag{3}$$

To further improve the global representation and address the inflexibility issue in the part-based representation, we present a StriP-Based feature extractor (SPB) which extracts strip-based features on horizontal axis and vertical axis, respectively. The strip-based feature extractor on horizontal axis first splits the human body into multiple horizontal strips and then applies convolution to extract spatial-temporal information of each horizontal strip. This extractor can be denoted as

$$X_{SPB-H} = c^{3 \times 1 \times 3}(X_{in}), \tag{4}$$

where $c^{3 \times 1 \times 3}(\cdot)$ denotes the 3D convolution with kernel size $(3, 1, 3)$. $X_{SPB-H} \in \mathbb{R}^{C_{out} \times T_{in} \times H_{in} \times W_{in}}$ is the output of this extractor.

Similarly, the strip-based feature extractor is used for the vertical strip's spatial-temporal extraction, represented as

$$X_{SPB-V} = c^{3 \times 3 \times 1}(X_{in}), \tag{5}$$

where $c^{3 \times 3 \times 1}(\cdot)$ denotes the 3D convolution with kernel size $(3, 3, 1)$. $X_{SPB-V} \in \mathbb{R}^{C_{out} \times T_{in} \times H_{in} \times W_{in}}$ is the output of this extractor. Finally, by combining the horizontal-based and vertical-based feature extractors, the strip-based feature extractor can obtain the following feature maps

$$X_{SPB} = X_{SPB-H} + X_{SPB-V}. \tag{6}$$

The proposed SPB can be used to enhance the feature representation ability of the traditional feature extractor. By combining SPB with aforementioned spatial-temporal feature extractor and frame-level feature extractor, as shown in Fig. 3, the ECM module can be obtained as follows

$$X_{ECM} = X_{ST} + X_{FL} + X_{SPB}. \tag{7}$$

Thus more comprehensive feature representations can be achieved.

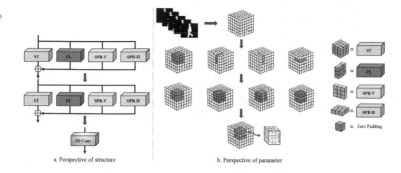

a. Perspective of structure   b. Perspective of parameter

Fig. 3. Overview of the enhanced convolution module. ST represents the Spatial-Temporal feature extractor, FL represents the Frame-Level feature extractor, SPB-V represents StriP-Based feature extractor in vertical and SPB-H represents StriP-Based feature extractor in horizontal.

3.3 Structural Re-parameterization

To reduce the parameters of the proposed ECM, we introduce the structural re-parameterization [6] method to ensemble different convolutions during the test stage. As shown in Eq. 7, the ECM block includes four convolutions: $c^{3 \times 3 \times 3}(\cdot)$, $c^{1 \times 3 \times 3}(\cdot)$, $c^{3 \times 1 \times 3}(\cdot)$ and $c^{3 \times 3 \times 1}(\cdot)$. During the test stage, these convolutions can be integrated into a single 3D convolution $c_{emb}^{3 \times 3 \times 3}(\cdot)$, which can be designed as

$$c_{emb}^{3 \times 3 \times 3} = c^{3 \times 3 \times 3} + c_t^{3 \times 3 \times 3} + c_h^{3 \times 3 \times 3} + c_w^{3 \times 3 \times 3}, \qquad (8)$$

where $c_t^{3 \times 3 \times 3}$, $c_h^{3 \times 3 \times 3}$ and $c_w^{3 \times 3 \times 3}$ are zero-padding expansions of $c^{1 \times 3 \times 3}$, $c^{3 \times 1 \times 3}$ and $c^{3 \times 3 \times 1}$, respectively, to make the kernels maintain the same dimensions. According to Eq. 8, the ECM in the test stage can be designed as

$$X_{ECM} = c_{emb}^{3 \times 3 \times 3}(X_{in}). \qquad (9)$$

Note that although four convolutions are employed to improve the representation ability in the training stage, only a single convolution is required in the test stage, which does not increase the parameter number and not degrade the inference running efficiency.

3.4 Multi-level Framework

To further improve the representation ability, we design the multi-level framework based on the proposed ECM block for both high-level and low-level feature extraction. The low-level branch directly extracts features from the large-size feature maps, which focuses on details of the human body. By contrast, the high-level one which works on down-sampled feature maps based on max pooling can extract more abstract information.

3.5   Temporal Aggregation and Spatial Mapping

To adaptively aggregate the temporal information of variable-length gait sequences, we introduce the temporal aggregation [20]. Assuming that the feature map $X_{out} \in \mathbb{R}^{C_f \times T_f \times H_f \times W_f}$ is the output of the feature extraction module, the temporal aggregation operation can be represented as

$$Y_{ta} = F_{Max}^{1 \times T_f \times 1 \times 1}(X_{out}), \tag{10}$$

where $Y_{ta} \in \mathbb{R}^{C_f \times 1 \times H_f \times W_f}$ is the output of the temporal aggregation module.

For the spatial mapping, we generate multiple horizontal feature representations and then combine them to improve the spatial representation ability [2,7,20,24,39]. The spatial mapping can be represented as

$$Y_{out} = F_s(F_{GeM}^{1 \times 1 \times 1 \times W_f}(Y_{ta})), \tag{11}$$

where $Y_{out} \in \mathbb{R}^{C_f \times 1 \times H_f \times 1}$ is the output of the spatial mapping. $F_{GeM}(\cdot)$ means the Generalized-Mean (GeM) pooling operation [25]. $F_s(\cdot)$ denotes the multiple separate fully connected (FC) layers. After spatial mapping, we obtain the final feature representation Y by concatenating the high-level and low-level feature maps in horizontal axis.

3.6   Loss Function

To train the proposed network, we employ the combined loss function which consists of the triplet loss and cross entropy loss. Besides the traditional cross entropy loss used for classification, the triplet loss is also introduced to make the samples from the same ID as close as possible while those from different IDs have larger distance in the feature space. The combined loss function is calculated with the obtained the output of spatial mapping, which is represented as

$$L_{combined} = L_{tri} + L_{cse}, \tag{12}$$

where the L_{tri} and L_{cse} denote the triplet loss and cross entropy loss, respectively. L_{tri} is defined as

$$L_{tri} = \max(d(r, s) - d(r, t) + m, 0) \tag{13}$$

where r and s are samples of the same category, while r and t are samples from different categories. $d(\cdot)$ represents the Euclidean distance between the two samples and m is the margin of the triplet loss.

3.7   Training and Test Details

Training. In this paper, we introduce a combined loss function consisting of cross-entropy loss and triplet loss [2,7,31,38,41,42] to train the proposed Gait-Strip. Specifically, the feature representation Y is fed into the triplet loss function for calculation [2], and input into the cross-entropy loss function through an FC layer. The Batch ALL (BA) [8] is used as the sampling strategy. The number of samples of each batch is $P \times K$, which contains P classes and each class corresponds to K samples. Considering the memory limit, the length of gait sequences is set to T in the training stage.

Test. During the test stage, we input the whole sequences into the GaitStrip to produce the feature representation Y. After that, Y is flattened into a vector to represent the corresponding sample. In general, the gait datasets are usually divided into two sets, *i.e.* , the gallery set and the probe set. The feature vectors from the gallery set are taken as the standard view to be retrieved, while those from the probe set are used to evaluate the performance. Specifically, we calculate the Euclidean distance between the feature vectors in the probe set and all feature vectors in the gallery set. The class label of the gallery sample with the smallest distance will be assigned to the probe sample.

4   Experiments

4.1   Datasets and Evaluation Protocol

CASIA-B. The CASIA-B dataset [37] is one of the largest gait datasets for evaluation. It includes 124 subjects, each of which contains 10 groups of gait sequences (six groups of normal walking (NM) #01-#06, two groups of walking with a bag (BG) #01-#02 and two groups of walking with a coat (CL) #01-#02). Each group contains 11 view angles (0°–180°) and the sampling interval is 18°. Hence, the CASIA-B dataset contains 124 (subject) × 10 (groups) × 11 (view angle) = 13,640 gait sequences. The dataset is divided into two subsets, the training set and the test set. We use the protocol [2] for evaluation, which includes three different settings, i.e., Small-sample Training(ST), Medium-sample Training (MT) and Large-sample Training (LT). In the three settings, 24, 62 and 74 subjects are used to form the training set, respectively, and the rest are used for test. During the test stage, four groups of sequences (NM#01-#04) are used as the gallery set and the rest (NM#05-#06, BG#01-#02 and CL#01-#02) are taken as the probe set.

OUMVLP. The OUMVLP dataset [29] is one of the most popular gait datasets, which includes 10,307 subjects. Each subject was collected in two groups of video sequences (Seq#00 and Seq#01), each of which contains 14 view angles (0°, 15°, ..., 75°, 90°, 180°, 195°, ..., 255°, 270°). During the test phase, the sequences in Seq#01 are used as the gallery set and the sequences in Seq#00 are taken as the probe set.

4.2   Implementation Details

Gait sequences are preprocessed and normalized into the same size 64 × 44 on both datasets [2]. In CASIA-B, GaitStrip has four blocks, where the last three blocks are built with the proposed ECM. The channel number of the four blocks

Table 1. Rank-1 accuracy (%) on CASIA-B under all view angles, different settings and conditions, excluding identical-view case.

Gallery NM#1-4			0°–180°											
Probe			0°	18°	36°	54°	72°	90°	108°	126°	144°	162°	180°	Mean
ST (24)	NM#5-6	GaitSet [2]	64.6	83.3	90.4	86.5	80.2	75.5	80.3	86.0	87.1	81.4	59.6	79.5
		MT3D [20]	71.9	83.9	90.9	90.1	81.1	75.6	82.1	89.0	91.1	86.3	69.2	82.8
		GaitGL [23]	77.0	87.8	93.9	92.7	83.9	78.7	84.7	91.5	92.5	89.3	74.4	86.0
		Ours	**79.6**	**89.5**	**95.6**	**94.3**	**86.4**	**82.0**	**86.6**	**93.0**	**93.6**	**90.1**	**75.1**	**87.8**
	BG#1-2	GaitSet [2]	55.8	70.5	76.9	75.5	69.7	63.4	68.0	75.8	76.2	70.7	52.5	68.6
		MT3D [20]	64.5	76.7	82.8	82.8	73.2	66.9	74.0	81.9	84.8	80.2	63.0	74.0
		GaitGL [23]	68.1	81.2	87.7	84.9	76.3	70.5	76.1	84.5	87.0	83.6	65.0	78.6
		Ours	**71.4**	**82.6**	**90.4**	**88.1**	**77.9**	**73.6**	**79.8**	**86.4**	**89.1**	**86.3**	**71.3**	**81.5**
	CL#1-2	GaitSet [2]	29.4	43.1	49.5	48.7	42.3	40.3	44.9	47.4	43.0	35.7	25.6	40.9
		MT3D [20]	46.6	61.6	66.5	63.3	57.4	52.1	58.1	58.9	58.5	57.4	41.9	56.6
		GaitGL [23]	46.9	58.7	66.6	65.4	58.3	54.1	59.5	62.7	61.3	57.1	40.6	57.4
		Ours	**54.3**	**67.8**	**75.0**	**71.6**	**66.2**	**59.7**	**65.5**	**70.5**	**69.6**	**63.6**	**46.6**	**64.6**
MT (62)	NM#5-6	GaitSet [2]	86.8	95.2	98.0	94.5	91.5	89.1	91.1	95.0	97.4	93.7	80.2	92.0
		MT3D [20]	91.9	96.4	98.5	95.7	93.8	90.8	93.9	97.3	97.9	95.0	86.8	94.4
		GaitGL [23]	93.9	97.6	**98.8**	97.3	95.2	92.7	95.6	98.1	98.5	96.5	91.2	95.9
		Ours	**94.0**	**98.0**	98.7	**97.8**	**95.6**	**93.0**	**96.1**	**98.2**	**98.6**	**97.0**	**92.6**	**96.3**
	BG#1-2	GaitSet [2]	79.9	89.8	91.2	86.7	81.6	76.7	81.0	88.2	90.3	88.5	73.0	84.3
		MT3D [20]	86.7	92.9	94.9	92.8	88.5	82.5	87.5	92.5	95.3	92.9	81.2	89.8
		GaitGL [23]	88.5	95.1	95.9	94.2	91.5	85.4	89.0	95.4	97.4	94.3	86.3	92.1
		Ours	**88.8**	**95.2**	**96.8**	**95.5**	**92.7**	**87.4**	**90.7**	**95.7**	**97.6**	**95.3**	**87.0**	**93.0**
	CL#1-2	GaitSet [2]	52.0	66.0	72.8	69.3	63.1	61.2	63.5	66.5	67.5	60.0	45.9	62.5
		MT3D [20]	67.5	81.0	85.0	80.6	75.9	69.8	76.8	81.0	80.8	73.8	59.0	75.6
		GaitGL [23]	70.7	83.2	87.1	84.7	78.2	71.3	78.0	83.7	83.6	77.1	63.1	78.3
		Ours	**69.2**	**86.7**	**90.0**	**88.3**	**83.6**	**75.8**	**82.3**	**88.1**	**88.1**	**81.7**	**65.7**	**81.8**
LT (74)	NM#5-6	GaitSet [2]	90.8	97.9	99.4	96.9	93.6	91.7	95.0	97.8	98.9	96.8	85.8	95.0
		GaitPart [7]	94.1	98.6	99.3	98.5	94.0	92.3	95.9	98.4	99.2	97.8	90.4	96.2
		MT3D [20]	95.7	98.2	99.0	97.5	95.1	93.9	96.1	98.6	99.2	98.2	92.0	96.7
		GaitGL [23]	96.0	98.3	**99.0**	97.9	**96.9**	**95.4**	97.0	98.9	**99.3**	98.8	94.0	97.4
		Ours	**96.0**	**98.4**	98.8	**97.9**	96.6	95.3	**97.5**	**98.9**	99.1	**99.0**	**96.3**	**97.6**
	BG#1-2	GaitSet [2]	83.8	91.2	91.8	88.8	83.3	81.0	84.1	90.0	92.2	94.4	79.0	87.2
		GaitPart [7]	89.1	94.8	96.7	95.1	88.3	84.9	89.0	93.5	96.1	93.8	85.8	91.5
		MT3D [20]	91.0	95.4	97.5	94.2	92.3	86.9	91.2	95.6	97.3	96.4	86.6	93.0
		GaitGL [23]	92.6	96.6	96.8	95.5	93.5	89.3	92.2	96.5	98.2	96.9	**91.5**	94.5
		Ours	**92.8**	**96.6**	**97.2**	**96.5**	**95.2**	**90.5**	**93.5**	**97.5**	**98.3**	**97.6**	91.4	**95.2**
	CL#1-2	GaitSet [2]	61.4	75.4	80.7	77.3	72.1	70.1	71.5	73.5	73.5	68.4	50.0	70.4
		GaitPart [7]	70.7	85.5	86.9	83.3	77.1	72.5	76.9	82.2	83.8	80.2	66.5	78.7
		MT3D [20]	76.0	87.6	89.8	85.0	81.2	75.7	81.0	84.5	85.4	82.2	68.1	81.5
		GaitGL [23]	76.6	90.0	90.3	87.1	84.5	79.0	84.1	87.0	87.3	84.4	69.5	83.6
		Ours	**79.9**	**92.3**	**93.4**	**89.2**	**86.0**	**80.0**	**86.0**	**88.5**	**91.7**	**87.5**	**73.5**	**86.2**

is set to 32, 64, 128 and 128, respectively. In OUMVLP, we use five blocks to construct the proposed GaitStrip and the last two blocks are implemented by the ECM module. The channel number of the five blocks is set to 64, 128, 196, 256 and 256, respectively. The margin of the triplet loss is set to 0.2 and Adam is selected as the optimizer. During the training stage, the parameters P and K are both set to 8. And the length of sequences T is set to 30. The learning rate is set to 1e-4 and reset to 1e-5 in the last 10K iterations. For the settings ST, MT and LT on CASIA-B dataset, the iteration number is set to 60K, 70K and 80K, respectively. On OUMVLP dataset, the parameter $P \times K$ is set to 32×8. The iteration number is set to 210K. The learning rate is initialized to 1e-4 and reset to 1e-5 after 150K iterations.

4.3   Comparison with the State-of-the-Art

Evaluation on CASIA-B. We compare the proposed method with several gait recognition approaches including GaitSet [2], GaitPart [7], MT3D [20] and GaitGL [23] on the CASIA-B dataset. The experimental results are shown in Table 1. It can be observed that the proposed method achieves the highest average accuracy under all settings (ST, MT and LT) and conditions (NM, BG and CL). Furthermore, we explore the performance of the proposed method under different settings and conditions in details.

Evaluation Under Various Settings (ST, MT and LT). We observe that our method achieves high performance under all three settings (ST, MT and LT) and exceeds the best result reported before. We display the complete experimental results under these three settings in Table 1. The recognition accuracy of GaitGL under ST MT and ST settings in NM condition is 86.0%, 95.9% and 97.4%, respectively. For the proposed method, the gait recognition accuracy is 87.8%, 96.3% and 97.6%, respectively. Furthermore, our method obtains significant improvement comparing with other methods in all three settings.

Evaluation Under Various Conditions (NM, BG and CL). It can be seen that when the external environment changes and more challenges exist, the accuracy decreases heavily. Under the LT setting, the accuracy of GaitGL in NM, BG and CL conditions is 97.4%, 94.5% and 83.6%, respectively. Comparing with GaitGL, our results are 0.2%, 0.7% and 2.6% higher, respectively. Under ST and MT settings, we can also observe that the proposed method owns the best performance. In the ST setting, our method outperforms GaitGL by 1.8%, 2.9% and 7.2% under NM, BG and CL, respectively. In the MT setting, the accuracy of the proposed method is 96.3%, 93.0% and 81.8%, which exceeds GaitGL by 0.4%, 0.9% and 3.5%, respectively.

Evaluation on Specific Angles (0°, 90°, 180°). The proposed method shows significant improvement in some extreme view angles (0°, 90° and 180°). For example, the average accuracy of MT3D in the setting LT and NM is 96.7%, but the accuracy corresponding to the three specific view angles are 95.7%, 93.9% and 92.0%, respectively. For the proposed method, the accuracy in the setting LT and NM is 97.6%, which outperforms MT3D by 0.9%. And the accuracy corresponding to the specific view angles (0°, 90° and 180°) are 96.0%, 95.3% and 96.3%, respectively, which outperforms MT3D by 0.3%, 1.4% and 4.3%, respectively. The main reason may be that the proposed SPB module extracts the feature of each strip, making the proposed ECM obtain more effective feature representation in the specific view angles.

Evaluation on OUMVLP. Compared with the CASIA-B, the OUMVLP dataset contains more subjects. Hereby, we compare GaitStrip with several famous gait recognition methods, including GEINet [27], GaitSet [2], GaitPart [7], GLN [9], SRN+CB [10], GaitGL [23] and 3D Local [11] on this dataset. The experimental results are shown in Table 2 which indicates that the proposed method achieves the optimal performance in all conditions. For example, the accuracy of GaitGL with invalid probe is 89.7%. For the proposed method,

Table 2. Rank-1 accuracy (%) on OUMVLP dataset under different view angles, excluding identical-view cases.

Method	Probe view														Mean
	0°	15°	30°	45°	60°	75°	90°	180°	195°	210°	225°	240°	255°	270°	
GEINet [27]	24.9	40.7	51.6	55.1	49.8	51.1	46.4	29.2	40.7	50.5	53.3	48.4	48.6	43.5	45.3
GaitSet [2]	84.5	93.3	96.7	96.6	93.5	95.3	94.2	87.0	92.5	96.0	96.0	93.0	94.3	92.7	93.3
GaitPart [7]	88.0	94.7	97.7	97.6	95.5	96.6	96.2	90.6	94.2	97.2	97.1	95.1	96.0	95.0	95.1
GLN [9]	89.3	95.8	97.9	97.8	96.0	96.7	96.1	90.7	95.3	97.7	97.5	95.7	96.2	95.3	95.6
SRN+CB [10]	91.2	96.5	98.3	98.4	96.3	97.3	96.8	92.3	96.3	98.1	98.1	96.0	97.0	96.2	96.4
GaitGL [23]	90.5	96.1	98.0	98.1	97.0	97.6	97.1	94.2	94.9	97.4	97.4	95.7	96.5	95.7	96.2
3D Local [11]	–	–	–	–	–	–	–	–	–	–	–	–	–	–	96.5
Ours	**92.8**	**97.0**	**98.4**	**98.5**	**97.6**	**98.2**	**97.8**	**96.0**	**96.2**	**97.8**	**97.9**	**96.6**	**97.3**	**96.7**	**97.0**

Table 3. Rank-1 accuracy (%) of different ECM blocks.

ST	FL	SPB	NM	BG	CL
✓			97.4	94.9	85.3
✓		✓	97.4	95.2	85.5
	✓		96.2	92.9	78.5
	✓	✓	97.2	94.9	85.2
✓	✓		97.4	95.2	85.9
✓	✓	✓	**97.6**	**95.2**	**86.2**

the accuracy in the same conditions is 90.5%, which outperforms GaitGL by 0.8%. The accuracy of GaitGL excluding invalid probe sequences is 96.2%, while the accuracy of the proposed method is 97.0%.

4.4 Ablation Study

In this paper, to obtain effective feature representation, we propose the GaitStrip with ECM block, SPB feature extractor and multi-level framework. Therefore, we design several ablation studies to explore the contribution of the key components.

Analysis of the SPB module. We propose the novel SPB extractor to extract more discriminative gait features. To explore the contribution of the SPB, we first design three groups of comparative experiments, i.e., only using the ST to compare with the combination of ST and SPB, only using the FL to compare with the combination of FL and SPB, and comparing the combination of ST and FL to the combination of ST, FL and SPB. The experimental results are shown in Table 3. We can find that the performance of the modules with SPB is improved compared with that without SPB. The accuracy of methods with and without SPB in NM condition is very close, but the methods with SPB in CL condition perform better. Specifically, the accuracy in CL condition by using FL is 78.5%, while the accuracy in CL condition with the combination of FL and SPB is 85.2%, which increases by 6.7%. In the CL condition, the accuracy with the combination of FL, ST and SPB is 86.2%, which increases 0.3% compared with that with only the combination of

Table 4. Rank-1 accuracy (%) of different levels.

Multi-level structure		NM	BG	CL
High-level	Low-level			
✓		97.3	94.4	83.4
	✓	97.2	94.4	84.4
✓	✓	**97.6**	**95.2**	**86.2**

Table 5. The accuracy (%) of different strip-based modeling on the CASIA-B dataset.

Method	NM	BG	CL
baseline+ECM	**97.6**	**95.2**	**86.2**
baseline+ACB	96.1	92.8	79.4
baseline+CCA	80.4	75.1	67.6

FL and ST. Hence, the SPB can help to extract more comprehensive gait features, which plays an important role in recognition improvement.

Analysis of the ECM Block. In this paper, we propose the ECM to generate the discriminative feature representations by taking full advantage of the frame-level and strip-based information. The ECM consists of the ST, FL and SPB. To explore the advantage of the combination of the ST, FL and SPB in robust feature extraction, we design ablation experiments by using only one or two modules. The results of the ablation experiments are shown in Table 3. In NM condition, the accuracy of the combination of ST and SPB is 97.4%, the accuracy of the combination of FL and SPB is 97.2%, and the combination of the ST, FL and SPB is 97.6%, which increases by 0.2% and 0.4%, respectively, compared with the other two modules. The accuracy of the study shows that the combination of the ST, FL and SPB can obtain better accuracy in NM, BG and CL conditions than using only one or two of the modules.

Analysis of Multi-level Framework. The proposed GaitStrip works with multiple levels. To investigate the contribution of the low-level and high-level branches, we design the comparison methods with only one branch. The experimental results are shown in Table 4, from which we can observe that the accuracy of the methods with only high-level or low-level branch is 97.3% and 97.2%, respectively, while the accuracy with both levels is 97.6%, which achieves 0.3% and 0.4% improvement, respectively, demonstrating that the multi-level structure can effectively enhance the representation ability and then improve the recognition performance.

4.5   Comparison with Other Strip-Based Modeling

In Sect. 2.2, we introduce two different modules to model the strip-based information. To analyze their performance, we design some experiments by using the Asymmetric Convolution Block (ACB) or Criss-Cross Attention Block to replace the ECM module. All experiments are built with the LT setting on CASIA-B. The experimental results are shown in Table 5. It can be observed that the proposed ECM achieves better performance than other strip-based modelings. This may be because our ECM utilizes the spatial-temporal information of each strip, improving the feature representation ability. The accuracy of the ECM method in NM, BG and CL is 97.6%, 95.2% and 86.2% respectively, which exceeds the ACB method by 1.5%, 2.4% and 6.8%. The accuracy of the CCA method in NM, BG and CL is 80.4%, 75.1% and 67.7% respectively, which is inferior to our method as well. By comparing with other strip-based methods, we can note that the proposed method can better exploit the spatial-temporal representation, especially in some complex conditions, which achieves significant improvement.

4.6   Computational Analysis

In the inference stage, the proposed ECM can be embedded into a standard 3D convolution, which reduces parameters and inference time. The computational analysis is shown in Table 6. It can be observed that the average accuracy of using ECM is 93.0%, outperforming the accuracy of using ST by 0.5%. However, the parameters of both modules are equal.

Table 6. The accuracy (%), inference time (second/sequence) and parameters (M) of different methods on CASIA-B dataset

	Re-param	ST	ST+FL	ECM
Accuracy	–	92.5	92.8	**93.0**
Inference time	×	0.025	0.027	0.035
Parameters	×	3.87	4.33	5.25
Accuracy	–	92.5	92.8	**93.0**
Inference time	✓	0.025	0.025	0.025
Parameters	✓	3.87	3.87	3.87

5   Conclusion

In this paper, we propose a novel gait recognition network GaitStrip with ECM block and multi-level framework. On the one hand, the proposed ECM which aggregates spatial-temporal, frame-level and strip-based information can generate more comprehensive feature representations. Moreover, the spatial-temporal, frame-level and strip-based feature extractors can be embedded into a common 3D convolution in the inference stage, which does not introduce additional parameters. On the other hand, the multi-level structure containing both low-level and high-level branches can ensemble global semantic and local detailed information. The experiment results verify that the proposed GaitStrip achieves appealing performance in normal environment as well as complex conditions.

Acknowledgements. This work was supported by the National Natural Science Foundation of China (61976017 and 61601021), the Beijing Natural Science Foundation (4202056), the Fundamental Research Funds for the Central Universities (2022JBMC013) and the Australian Research Council (DP220100800, DE230100477). The support and resources from the Center for High Performance Computing at Beijing Jiaotong University (http://hpc.bjtu.edu.cn) are gratefully acknowledged.

References

1. Chai, T., Mei, X., Li, A., Wang, Y.: Silhouette-based view-embeddings for gait recognition under multiple views. In: ICIP (2021)
2. Chao, H., He, Y., Zhang, J., Feng, J.: GaitSet: regarding gait as a set for cross-view gait recognition. In: AAAI (2019)
3. Chao, H., Wang, K., He, Y., Zhang, J., Feng, J.: GaitSet: cross-view gait recognition through utilizing gait as a deep set. TPAMI **44**(7), 3467–3478 (2021)
4. Connor, P., Ross, A.: Biometric recognition by gait: a survey of modalities and features. In: CVIU (2018)
5. Ding, X., Guo, Y., Ding, G., Han, J.: ACNet: strengthening the kernel skeletons for powerful CNN via asymmetric convolution blocks. In: ICCV (2019)
6. Ding, X., Zhang, X., Ma, N., Han, J., Ding, G., Sun, J.: RepVGG: making VGG-style convnets great again. In: CVPR (2021)
7. Fan, C., et al.: GaitPart: temporal part-based model for gait recognition. In: CVPR (2020)
8. Hermans, A., Beyer, L., Leibe, B.: In defense of the triplet loss for person re-identification. arXiv preprint arXiv:1703.07737 (2017)
9. Hou, S., Cao, C., Liu, X., Huang, Y.: Gait lateral network: learning discriminative and compact representations for gait recognition. In: Vedaldi, A., Bischof, H., Brox, T., Frahm, J.-M. (eds.) ECCV 2020. LNCS, vol. 12354, pp. 382–398. Springer, Cham (2020). https://doi.org/10.1007/978-3-030-58545-7_22
10. Hou, S., Liu, X., Cao, C., Huang, Y.: Set residual network for silhouette-based gait recognition. TBIOM **3**(3), 384–393 (2021)
11. Huang, Z., et al.: 3D local convolutional neural networks for gait recognition. In: ICCV (2021)
12. Huang, Z., Wang, X., Huang, L., Huang, C., Wei, Y., Liu, W.: CCNet: criss-cross attention for semantic segmentation. In: ICCV (2019)
13. Jin, Y., Sharma, A., Tan, R.T.: DC-ShadowNet: single-image hard and soft shadow removal using unsupervised domain-classifier guided network. In: ICCV (2021)
14. Jin, Y., Yang, W., Tan, R.T.: Unsupervised night image enhancement: when layer decomposition meets light-effects suppression. arXiv preprint arXiv:2207.10564 (2022)
15. Li, S., Liu, W., Ma, H.: Attentive spatial-temporal summary networks for feature learning in irregular gait recognition. TMM **21**(9), 2361–2375 (2019)
16. Li, X., Makihara, Y., Xu, C., Yagi, Y., Ren, M.: Gait recognition invariant to carried objects using alpha blending generative adversarial networks. PR **105**, 107376 (2020)
17. Li, X., Makihara, Y., Xu, C., Yagi, Y., Ren, M.: Gait recognition via semi-supervised disentangled representation learning to identity and covariate features. In: CVPR (2020)
18. Li, X., Makihara, Y., Xu, C., Yagi, Y., Yu, S., Ren, M.: End-to-end model-based gait recognition. In: ACCV (2020)

19. Lin, B., Liu, Y., Zhang, S.: GaitMask: mask-based model for gait recognition. In: BMVC (2021)
20. Lin, B., Zhang, S., Bao, F.: Gait recognition with multiple-temporal-scale 3D convolutional neural network. In: ACM MM (2020)
21. Lin, B., Zhang, S., Liu, Y., Qin, S.: Multi-scale temporal information extractor for gait recognition. In: ICIP (2021)
22. Lin, B., Zhang, S., Wang, M., Li, L., Yu, X.: GaitGL: learning discriminative global-local feature representations for gait recognition. arXiv2208 (2022)
23. Lin, B., Zhang, S., Yu, X.: Gait recognition via effective global-local feature representation and local temporal aggregation. In: ICCV (2021)
24. Liu, J., et al.: Leaping from 2D detection to efficient 6DoF object pose estimation. In: Bartoli, A., Fusiello, A. (eds.) ECCV 2020. LNCS, vol. 12536, pp. 707–714. Springer, Cham (2020). https://doi.org/10.1007/978-3-030-66096-3_47
25. Radenović, F., Tolias, G., Chum, O.: Fine-tuning CNN image retrieval with no human annotation. TPAMI **41**(7), 1655–1668 (2018)
26. Shen, C., Lin, B., Zhang, S., Huang, G.Q., Yu, S., Yu, X.: Gait recognition with mask-based regularization. arXiv preprint arXiv:2203.04038 (2022)
27. Shiraga, K., Makihara, Y., Muramatsu, D., Echigo, T., Yagi, Y.: GEINet: view-invariant gait recognition using a convolutional neural network. In: ICB (2016)
28. Song, C., Huang, Y., Huang, Y., Jia, N., Wang, L.: GaitNet: an end-to-end network for gait based human identification. PR **96**, 106988 (2019)
29. Takemura, N., Makihara, Y., Muramatsu, D., Echigo, T., Yagi, Y.: Multi-view large population gait dataset and its performance evaluation for cross-view gait recognition. IPSJ Trans. Comput. Vis. Appl. **10**(1), 1–14 (2018). https://doi.org/10.1186/s41074-018-0039-6
30. Thapar, D., Jaswal, G., Nigam, A., Arora, C.: Gait metric learning Siamese network exploiting dual of spatio-temporal 3D-CNN intra and LSTM based inter gait-cycle-segment features. PRL **125**, 646–653 (2019)
31. Tian, Y., Yu, X., Fan, B., Wu, F., Heijnen, H., Balntas, V.: SOSNet: second order similarity regularization for local descriptor learning. In: CVPR (2019)
32. Wolf, T., Babaee, M., Rigoll, G.: Multi-view gait recognition using 3D convolutional neural networks. In: ICIP (2016)
33. Wu, H., Tian, J., Fu, Y., Li, B., Li, X.: Condition-aware comparison scheme for gait recognition. TIP **30**, 2734–2744 (2020)
34. Wu, Z., Huang, Y., Wang, L., Wang, X., Tan, T.: A comprehensive study on cross-view gait based human identification with deep CNNs. TPAMI **39**(2), 209–226 (2016)
35. Yeoh, T., Aguirre, H.E., Tanaka, K.: Clothing-invariant gait recognition using convolutional neural network. In: ISPACS (2016)
36. Yu, S., et al.: HID 2021: competition on human identification at a distance 2021. In: IJCB (2021)
37. Yu, S., Tan, D., Tan, T.: A framework for evaluating the effect of view angle, clothing and carrying condition on gait recognition. In: ICPR (2006)
38. Yu, X., et al.: Unsupervised extraction of local image descriptors via relative distance ranking loss. In: ICCV Workshops (2019)
39. Yu, X., Zhuang, Z., Koniusz, P., Li, H.: 6DoF object pose estimation via differentiable proxy voting loss. In: BMVC (2020)
40. Zhang, C., Liu, W., Ma, H., Fu, H.: Siamese neural network based gait recognition for human identification. In: ICASSP (2016)

41. Zhang, J., et al.: Gigapixel whole-slide images classification using locally supervised learning. In: Wang, L., Dou, Q., Fletcher, P.T., Speidel, S., Li, S. (eds.) Medical Image Computing and Computer Assisted Intervention–MICCAI 2022. Lecture Notes in Computer Science, vol. 13432, pp. 192–201. Springer, Cham (2022)
42. Zhang, X., et al.: Sleep stage classification based on multi-level feature learning and recurrent neural networks via wearable device. Comput. Biol. Med. **103**, 71–81 (2018)
43. Zhang, Y., Huang, Y., Wang, L., Yu, S.: A comprehensive study on gait biometrics using a joint CNN-based method. PR **93**, 228–236 (2019)
44. Zhang, Y., Huang, Y., Yu, S., Wang, L.: Cross-view gait recognition by discriminative feature learning. TIP **29**, 1001–1015 (2019)
45. Zhu, Z., et al.: Gait recognition in the wild: a benchmark. In: ICCV (2021)

Soft Label Mining and Average Expression Anchoring for Facial Expression Recognition

Haipeng Ming, Wenhuan Lu, and Wei Zhang$^{(\boxtimes)}$

Tianjin University, Tianjin, China
{minghaipeng,wenhuan,tjuzhangwei}@tju.edu.cn

Abstract. Facial expression recognition (FER) suffers from high inter-class similarity and large intraclass variation, leading to ambiguity or uncertainty and further confusing annotators. They also hinder the network in learning the valuable features of facial expression. Recently, many studies have revealed that the uncertainty or ambiguity is one of the key challenges in FER. In this paper, we propose a new method to address this issue from two aspects: a soft label mining module to convert the original hard labels to soft labels dynamically during training, and an average facial expression anchoring module to separate unique expression features from similarity expression features. The soft label mining module breaks the limits of the categorical model and mitigates the uncertainty or ambiguity. And the average facial expression anchoring module suppresses the high interclass similarity of facial expressions. Our method can train any backbone network for facial expression recognition. The experiments on the popular datasets show that our method achieves state-of-the-art results by 92.82% on RAF-DB and 67.91% on SFEW, and achieves a comparable result of 62.26% on AffectNet. The code is available at https://github.com/HaipengMing/SLM-AEA.

1 Introduction

Facial expression is one of the most natural, powerful, and universal signals for human beings to convey their emotional states and intentions [6, 32]. In the past years, facial expression recognition (FER) has attracted much attention due to its important role in human-computer interaction, health care, and many other applications. Similar to other modalities in affective computing, a facial expression is commonly characterized as one of several discrete affective states(*e.g.*, basic emotions defined by Ekman and Friesen [9, 10]), which is also known as the categorical model. Generally, annotating facial expressions with the categorical model is much easier and cheaper than other models (FACS [11] and dimension models [26]). It's more consistent with people's intuition as well. Existing facial expression datasets are mostly annotated with categorical model, such as Oulu-CASIA [44], SFEW/AFEW [7], FERPlus [1], RAF-DB [16], AffectNet [20] (AffectNet also annotated Valence-Arousal dimensions), etc.

© The Author(s), under exclusive license to Springer Nature Switzerland AG 2023
L. Wang et al. (Eds.): ACCV 2022, LNCS 13844, pp. 728–744, 2023.
https://doi.org/10.1007/978-3-031-26316-3_43

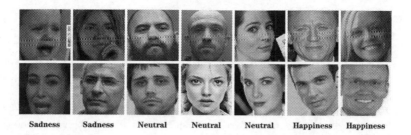

| Sadness | Sadness | Neutral | Neutral | Neutral | Happiness | Happiness |

Fig. 1. An illustration of the uncertainty or ambiguity. These 14 images are from AffectNet and their labels are attached to the bottom of the images. From left to right, sadness, neutral, happiness. But there is no clear boundary between sadness and neutral as well as neutral and happiness.

However, the categorical model is limited in the ability to represent the complexity and subtlety of facial expressions, especially in the wild. As illustrated in Fig. 1, there is not usually a clear boundary between the different expression categories. Even worse, due to the high interclass similarity, annotators may even annotate incorrectly. This means that the quality and consistency of the datasets are difficult to guarantee because of the subjectivity of the annotators. Compared to discrete affective states, the soft label has a greater expressivity, which can describe ambiguity appropriately. High interclass similarity can be described as other expression components, while high intraclass variation can be described as different extensions of the real expression component. However, it is expensive and time-consuming to provide soft labels for large-scale FER datasets. A compromised way is soft label mining. Based on [41], we designed a simple yet efficient *soft label mining module* to mine soft labels from original annotations, *i.e.*, the hard labels. Specifically, we introduce the label smoothing method to initially transfer the hard labels to soft labels according to an artificially set value p, which reflects the confidence level of the original annotations. The initialized soft labels act as targets to train the network parameters. We update them by taking into account both the network predicted distribution and the original distribution(*i.e.*, the hard labels) during training. To make a trade-off between them, we design a ramp function to achieve a balance dynamically. Note that the soft label mining module is designed for training, imposes no additional burden on inference, and adds only a very small additional burden to training. Meanwhile, we also design a novel *average facial expression anchoring module* to suppress the high interclass similarity. Specifically, we take the average feature of a mini-batch as the anchor feature, *i.e.*, the average expression, and introduce a learnable vector with the same size as the attention weight of the anchor feature. The weighted anchor features are further summed with the expression features as the final extracted features. We suppress the uncertainty or ambiguity in FER through these two modules, especially the soft label mining module, which has an excellent and stable performance on different FER datasets. Our method can be used to train any backbone network for facial expression recognition.

Overall, the main contributions can be summarized as follows:

(1) We propose a novel method to address the uncertainty or ambiguity problem in FER by designing a soft label mining module and an average expression anchoring module. In comparison with existing related work, our approach is simpler and more efficient.
(2) Our approach is extensively evaluated on laboratory databases and real-world datasets. Experimental results show that our method achieves the state-of-the-art performance by 92.82% on RAF-DB and 67.91% on SFEW, and achieves a comparable result of 62.26% on AffectNet.

2   Related Work

2.1   Facial Expression Recognition

Facial expression recognition has been an active topic for many years. Early work focused on using *handcraft features* (*e.g.*, SIFT [22], HOG [5], LBP [27], Gabor Wavelets [2] etc.) to exact the emotion feature and recognize facial expressions. While with the development of deep learning, now the *learning-based* methods have become mainstream, which can be roughly classified into three groups: data-focused [21,42], model-focused [3,25,34,40] and label-focused [4,28,35]. The results observed by Ng *et al.* [21] show that pre-fine-tuning on an additional FER dataset can improve the performance. Zeng *et al.* [42] propose a new model termed IPA2LT to address the inconsistency issue in fusion of different FER datasets. Yang *et al.* [40] and Wang *et al.* [34] introduce adversarial mechanism into FER. Ruan *et al.* [25] propose a novel model to take into account subtle differences between different facial expressions. Recently, some researchers began to consider the uncertainty of annotations in FER datasets. Chen *et al.* [4] introduce label distribution learning and draw on an auxiliary space of FACS or landmarks. Wang *et al.* [35] propose a Self-Cure Network (SCN) to suppress the uncertainty by correcting possible mislabeling. She *et al.* [28] proposed a model named DMUE, which achieved previous leading performance, to mine latent distribution and estimation uncertainty. Zhao *et al.* [45] propose a lightweight FER network considering both model and label.

2.2   Methods for Uncertainty or Ambiguity

Low-quality annotations caused by uncertainty or ambiguity can be considered as label noise, which also appears in other computer vision tasks. Numerous methods have been proposed to resolve this issue. Some methods leverage a small set of clean data [17,33]. Qu *et al.* [23] proposed a label-noise robust network by matching the feature distributions. In the field of FR (Face Recognition), a GCN-based model [43] is proposed to address the large-scale label noise. Recently, label distribution has been proposed to mitigate the adverse impact of label noise by converting logical labels to discretized bivariate Gaussian label distribution with the help of prior knowledge [12,13,31]. In the classification problem, label distribution means soft label. Label enhancement (LE) [38,39] is the universal way to find the latent ground truth. Uncertainty estimation methods can be used to [30,37] address the inconsistent data quality. The uncertainty or ambiguity in FER is intrinsic. Compared to other classification tasks, facial expressions of different classes suffers from high interclass similarity but also high intraclass variation.

3   Method

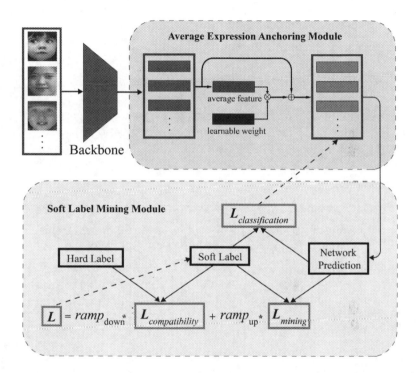

Fig. 2. The overall framework of our method. Face images are first fed into a backbone network for feature extraction. The extracted features are then averaged to obtain the anchor feature, further multiplied element-wise by a learnable weight vector. The obtained results and the extracted initial features are added element-wise as the final extracted features of the network. We view the soft label as the target to train the network. When training, the soft labels are updated dynamically with a newly designed loss function. The solid line in the figure represents forward propagation while the dashed line represents back-propagation.

Notation. Given a FER dataset $\mathcal{X}$, we donate its corresponding hard label space as $\mathcal{H} = \{\boldsymbol{y} : \boldsymbol{y} \in \{0,1\}^{C}, \|\boldsymbol{y}\|_1 = 1\}$, and the soft label space as $\mathcal{S} = \{\boldsymbol{y} : \boldsymbol{y} \in [0,1]^{C}, \|\boldsymbol{y}\|_1 = 1\}$, where C is the number of classes. For the i-th facial image $\boldsymbol{x}_i$ belonging to $\mathcal{X}$, we denote $\hat{\boldsymbol{y}}_i = \{\hat{y}_{i1}, \hat{y}_{i2}, ..., \hat{y}_{iC}\} \in \mathcal{H}$ as its annotated deterministic hard label, which is a one-hot vector. The output vector of backbone network is donated as $\boldsymbol{z}_i = \{z_{i1}, z_{i2}, ..., z_{iC}\}$.

3.1   Soft Label Mining

We first transfer the hard labels to soft labels with the help of label smoothing [15], which is a trick to improve the performance in classification tasks. A hyperparameter p is introduced for initialization. The more uncertain or ambiguous the FER dataset is, the lower the p is set. As for hard labels, p is 1. We take the initialized soft labels as the new target to train the network. While training, we update the soft labels dynamically.

Specifically, in a C-classes classification task, the process of prediction with the model can be formulated as:

$$z_i = f(\boldsymbol{x}_i; \boldsymbol{\theta}) \tag{1}$$

where f is a model and $\boldsymbol{\theta}$ is the set of network parameters. Then the predicted result of the model f will be normalized by softmax function.

$$\bar{z}_i = softmax(\boldsymbol{z}_i) \tag{2}$$

$$\bar{z}_{ij} = \frac{exp(z_{ij})}{\sum_{j=1}^{C} exp(z_{ij})} \tag{3}$$

Obviously, $\bar{z}_i$ belongs to $\mathcal{S}$. It can be viewed as the label distribution generated by model f. Generally, the purpose of training is to minimize the cross-entropy loss function:

$$\mathcal{L} = -\frac{1}{N} \sum_{i=1}^{N} \sum_{j=1}^{C} \hat{y}_{ij} log \bar{z}_{ij} \tag{4}$$

For the i-th image belongs to c-th class,

$$\mathcal{L}_i = -z_{ic} + log(\sum_{j=1}^{C} exp(z_{ij})) \tag{5}$$

Its optimal solution is $z_{ic}^* = \boldsymbol{inf}$, while keeping others small enough. But many ambiguous facial expressions have similar intensity in different emotional components. It is not reasonable to force other components small enough. The idea of label smoothing is to transfer $\hat{\boldsymbol{y}}_i \in \mathcal{H}$ to $\bar{\boldsymbol{y}}_i \in \mathcal{S}$ by changing the construction as:

$$\bar{y}_{ij} = \begin{cases} 1 - \varepsilon & if \ j = c, \\ \varepsilon/(C-1) & otherwise \end{cases} \tag{6}$$

where ε is a small constant. $\bar{\boldsymbol{y}}_i$ is a label distribution. We donate its predecessor before the softmax operator as $\tilde{\boldsymbol{y}}_i$, its components are computed as:

$$\tilde{y}_{ij} = \begin{cases} k & if \ j = c, \\ k - log\frac{(1-\varepsilon)(C-1)}{\varepsilon} & otherwise \end{cases} \tag{7}$$

where k is an arbitrary number. It determines the data scale before softmax. But when ε is fixed, k has no effect on $\bar{y}$, that is, the distribution of $\bar{y}$ in $\mathcal{S}$ is only related to ε. In our experiments, k is set to 5. The $\tilde{y}_i$ is used as the score-form soft labels and updated with a learning rate during training.

Note that we use p ($p = 1-\varepsilon$) instead of ε as the hyperparameter. It is because p responds to the confidence level of the original annotations. Our experimental results show that in the synthetic noise datasets, p should be adjusted to lower as the noise ratio increases.

As mentioned earlier, $\tilde{y}_i$ is used as the score-form soft label. The cross entropy loss function in Eq. 4 now is:

$$\mathcal{L}_{cls} = -\frac{1}{N} \sum_{i=1}^{N} \sum_{j=1}^{C} \bar{y}_{ij} log \bar{z}_{ij} \qquad (8)$$

$$\bar{y}_i = softmax(\tilde{y}_i) \qquad (9)$$

$\mathcal{L}_{cls}$ is the loss function to update the network parameter $\boldsymbol{\theta}$. We update $\tilde{y}_i$ and $\boldsymbol{\theta}$ in two continuous but different backpropagation stages. After $\boldsymbol{\theta}$ has been updated, $\tilde{y}_i$ is regarded as the learnable parameters. Both [35] and [28] show that the network prediction $\bar{z}_i$ of some ambiguous facial expressions is more credible than the annotations during training. So an intuitive approach is to take full advantage of $\bar{z}_i$.

$$\mathcal{L}_m = -\frac{1}{N} \sum_{i=1}^{N} \sum_{j=1}^{C} \bar{z}_{ij} log \bar{y}_{ij} \qquad (10)$$

We mine the soft label by Eq. 10. But certain samples should benefit from the original annotations, so we also utilize another loss function (Eq. 11) to explore the compatibility between the original label and the latent label distribution.

$$\mathcal{L}_{cpt} = -\frac{1}{N} \sum_{i=1}^{N} \sum_{j=1}^{C} \hat{y}_{ij} log \bar{y}_{ij} \qquad (11)$$

We introduce a ramp function [28] to adjust the weight between different loss functions. The overall loss function guiding $\tilde{y}_i$ to update is:

$$\mathcal{L} = ramp_{up}(e) \cdot \mathcal{L}_m + ramp_{down}(e) \cdot \mathcal{L}_{cpt} \qquad (12)$$

$$ramp_{up}(e) = \begin{cases} exp(-(\alpha - \frac{e}{\beta})^2) & e \leq \beta, \\ 1 & e > \beta \end{cases} \qquad (13)$$

$$ramp_{down}(e) = \begin{cases} 1 & e \leq \beta, \\ exp(-(\alpha - \frac{\beta}{e})^2) & e > \beta \end{cases} \qquad (14)$$

where α and β are hyperparameters. α is used to adjust the slope of ramp function, which is fixed at 1 in [28]. β is the epoch threshold. In the training process, the network trusts the original annotations more before β-th epoch, and trusts itself more after β-th epoch.

$$\tilde{\boldsymbol{y}} \leftarrow \tilde{\boldsymbol{y}} - \lambda \frac{\partial \mathcal{L}}{\partial \tilde{\boldsymbol{y}}} \tag{15}$$

The backpropagation of $\mathcal{L}$ takes another learning rate λ to update $\tilde{\boldsymbol{y}}$ as shown in Eq. 15. Note that λ has a strong relationship with k in Eq. 7. k decides the scale of score-form soft label, while λ is the learning rate to update it. In our experiment, k is fixed to 5 and λ is fixed to 200. Different k has different optimal λ, but once a pair of k and λ is determined, there is no need to change it. For different datasets, only the hyperparameter p, α and β need to be adjusted.

3.2 Average Expression Anchoring

To handle the high interclass similarity in FER, we design an average expression anchoring module to separate the common feature and the unique feature. We share a similar mindset with a small part of [29].

In general, the features extracted from backbone networks are processed by a fully-connected layer to get $\boldsymbol{z}_i$:

$$\boldsymbol{z}_i = \boldsymbol{W}^T \boldsymbol{F}_i \tag{16}$$

where $\boldsymbol{F}_i \in \boldsymbol{R}^D$ and belongs to feature space.

The average expression anchoring module is comprised of a learnable attention weight $\tilde{\boldsymbol{W}} \in \boldsymbol{R}^D$, and an element-wise summation operation layer which can be donated as:

$$\tilde{\boldsymbol{F}}_i = \tilde{\boldsymbol{W}} * \bar{\boldsymbol{F}} + \boldsymbol{F}_i \tag{17}$$

$$\bar{\boldsymbol{F}} = \frac{1}{b} \sum_{i=1}^{b} \boldsymbol{F}_i \tag{18}$$

where $*$ means element-wise multiplication and $+$ means element-wise summation, b is the batch size of the train set.

The average expression anchoring module replaces $\boldsymbol{F}_i$ with $\tilde{\boldsymbol{F}}_i$, and we think it addresses the uncertainty or ambiguity problem from two aspects. First, because the expressions of the different categories have very similar information, which disturbs the network to distinguish, it is necessary to separate the unique features from the common features. $\bar{\boldsymbol{F}}$ represents the highly similar information while $\boldsymbol{F}_i$ represents the special information. Experimental results show that the network advantages of the separation. Second, for low-quality samples, anchoring the average expression allows them to take advantage of information from high-quality samples in the same batch, thus improving the overall quality of the dataset.

Note that the improvements of this module depends on "average expression", $i.e.$, $\bar{\boldsymbol{F}}$. That means it will not work if inferencing a single expression rather than a batch of expressions.

3.3   The Overall Framework

As illustrated in Fig. 2, given a FER dataset with hard labels, we first evaluate the confidence level of the dataset annotations by p. Then we fix p and soften the hard labels by label smoothing to get the initialized soft labels. The initialized soft labels are used as the target to train the backbone networks. An average expression anchoring module is introduced at the end of the backbone to mitigate the high interclass similarity in FER, by separating the unique features from common features. The soft labels are further updated dynamically during training by minimizing multiple loss functions. We force them to approximate both the network prediction and the original annotations by $\mathcal{L}_{cls}$ and $\mathcal{L}_m$. We designed different weights to make the network trust original annotations more in the beginning and trust itself more as the training progressed. Not that we use not only traditional CNNs but also Swin-transformers [18] as our backbones to extract facial expression features. The effects of different backbone networks will be compared in Sect. 4.

4   Experiments

4.1   Datasets

Oulu-CASIA [44] contains videos captured in controlled lab conditions. Subjects were asked to pose six basic expressions (happiness, surprise, sadness, anger, disgust, fear). We select the last three frames in each sequence in the condition of the visible light and strong illumination (consisting of 1,440 images in total), to construct the training set and the test set. Similar to [35], we employ the subject-independent tenfold cross-validation protocol for evaluation.

RAF-DB [16] is the real-world facial expression dataset, and contains 15,339 facial images annotated with six basic expressions and neutral expression. Among them, 12,271 images are used for training, and the other 3,068 images for testing. We report the overall sample accuracy of the testing set for measurement.

AffectNet [20] is by far the largest FER dataset collected in the wild. It was annotated with both categorical and Valence-Arousal dimension labels. It contains more than 100,000 images from the Internet by querying 1,250 expression-related keywords in three search engines, of which 450,000 images are manually annotated with seven expression classes the same as RAF-DB and the extra contempt expression. Among them, 280K images are used for training and the remaining 4K images for testing. The overall sample accuracy is used for measurement.

SFEW [7] is created by selecting static frames from Acted Facial Expressions in the Wild (AFEW). The images in SFEW are labeled with six basic expressions and neutral expression same as RAF-DB. We use 958 images for training and 436 images for testing. The overall sample accuracy is used for measurement.

4.2 Implementation Details

We take ResNet18 [14] pre-trained on MS-Celeb-1M as the default backbone network with the standard routine for a fair comparison. We also aligned the faces of the in-the-wild datasets for pose normalization. The facial expression images are resized to 256×256 pixels and further augmented by random cropping to 224×224 pixels, horizontally flipped, and added Gaussian noise with a probability of 0.5. As mentioned earlier, the parameter k and λ are set to 5 and 200, respectively. The hyperparameter α is set to 1.6. But the other parameters (β and p) need to be adjusted according to the datasets. p reflects the confidence level of the original annotations and β is the epoch that the backbone has learned enough knowledge so that it can trust its prediction. Considering carry out in practice, β is set to 7 and p is set to 0.9 can achieve a performance that greatly exceeds the baseline generally. We use Adam with a weight decay of 10^{-4}. The initial learning rate is 10^{-3}, which is further reduced to 10^{-6} as a cosine function. The training ends at epoch 40. We use the Pytorch toolbox to implement our model on a single Nvidia 2080Ti GPU and train it in an end-to-end manner.

4.3 Ablation Studies

We conduct ablation experiments to observe the effect of key parameters and different modules on the final performance. We choose RAF-DB and AffectNet as the benchmark since they are two of the popular largest real-world FER datasets.

Influence of Different Modules. In order to evaluate the contribution of different modules, we conduct an ablation study to investigate the soft label mining module and the average expression anchoring module on RAF-DB and AffectNet. Considering some related works did not pre-train the network on large scale face recognition datasets, we also investigate the effect of pertaining on MS-Celeb-1M. The experimental results are shown in Table 1. Some observations can be concluded. First, the soft label mining module improves the performance by 1.82% on RAF-DB and 3.86% on AffectNet without pertaining on MS-Celeb-1M, and improves the performance by 1.98% on RAF-DB and 4.31% on AffectNet with pre-training. It plays the most important role in our method due to its outstanding and stable improvements. Second, pre-trained on large scale face recognition datasets and the average expression anchoring module improves the accuracy on RAF-DB observably. But pre-training achieves very little improvement on AffectNet. The average expression anchoring module can even degrade the accuracy on AffectNet slightly. We think it can be explained by the consistency of the datasets. Due to the uncertainty or ambiguity problem in FER, larger scale FER datasets tend to be more inconsistent. At this point, the requirement for the soft mining module is even greater because it breaks the limitations of the categorical model. Third, pre-trained on large scale face recognition datasets and the union of two modules can improve the performance greater. Taking the result of pre-training on MS-Celeb-1M as a baseline, our method improves the accuracy by 5.97% on RAF-DB and 4.74% on AffectNet finally.

Table 1. Accuracy (%) comparison of different modules on RAF-DB and AffectNet. SLM and AEA are abbreviations for the soft label mining module and the average expression anchoring module, respectively. ✓ on "Pre-train" means ResNet18 is pre-trained on MS-Celeb-1M, while × means the initialized parameters of ResNet18 are provided by Pytorch, which is pre-trained on ImageNet.

Pre-train	SLM	AEA	RAF-DB	AffectNet
×	×	×	86.34	59.34
×	✓	×	87.91	61.63
×	×	✓	89.12	58.78
×	✓	✓	91.65	61.82
✓	×	×	87.59	59.44
✓	✓	×	89.32	62.00
✓	×	✓	91.99	59.29
✓	✓	✓	**92.82**	**62.26**

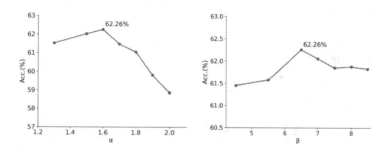

Fig. 3. The accuracy (%) with different α and β on AffectNet.

Influence of α. α determines the slope and the ramp function's initial value. We introduced α in our model to improve the ramp function. This is because the original ramp function has a higher starting point, but we want to force the network to concentrate more on the original labels in the early phases of training. The experimental results show that the introduction of α helps to resolve this problem. Figure 3 shows the influence of α. If α is set too small, the network's predictions at the beginning moments of training will have a greater impact on relabeling. If α is set too large, the slope of the ramp function will be too steep.

Influence of β. β is the epoch that the network starts to mine soft labels. Before β-th epoch, soft label mining depends more on the original hard label. But after β-th epoch, it depends more on the logits of the network. Theoretically, the lower the quality of the dataset, the lower the β. But too small β will harm learning enough useful features. Figure 3 shows the impact of β on AffectNet, and proves this speculation well. Different datasets need different β. As shown in

Table 2, AffectNet has the minimum optimal β and RAF-DB has the maximum optimal β.

Table 2. Optimal β and p for different FER datasets.

	Oulu-CASIA	RAF-DB	AffectNet	SFEW
β	7	8	6.5	7
p	0.95	0.97	0.85	0.91

Optimal p of Different FER Datasets. We conduct plenty of experiments to test the optimal p of different datasets. As shown in Table 2, the result shows that the AffectNet has the minimum optimal p. This means that it may be the most uncertain or ambiguous of the four FER datasets. We think it can be explained by the scale of AffectNet. As early mentioned, larger-scale FER datasets tend to be more inconsistent.

Table 3. Accuracy (%) comparison of different backbone networks on RAF-DB and AffectNet. SLM-AEA is an abbreviation for our method. All the backbone networks are pre-trained on ImageNet for a fair comparison.

Backbone	SLM-AEA	RAF-DB	AffectNet
ResNet18 [14]	✗	86.34	59.34
ResNet18 [14]	✓	91.65	61.82
ResNet50 [14]	✗	85.39	59.04
ResNet50 [14]	✓	89.08	60.95
Swin Transformer [18]	✗	87.61	59.74
Swin Transformer [18]	✓	90.63	62.17

Different Backbone Networks. Our framework can be used to train any backbone networks for facial expression recognition. We conduct experiments not only on CNNs but also on the recently popular transformer. We choose ResNet18, ResNet50, and Swin-transformer as the backbone networks for comparison. For a fair comparison, all the backbones are pre-trained on ImageNet. Pre-training on ImageNet makes the performance slightly inferior compared to pre-training on MS-Celeb-1M. As shown in Table 3, one interesting thing is that ResNet50 does not perform as well as ResNet18 on RAF-DB and AffectNet. In contrast, the performance of the Swin-transformer on AffectNet is a bit surprising. We believe the transformer has even more potential for facial expression recognition. However, on RAF-DB, the performance of the Swin-transformer decreases when our method is implemented.

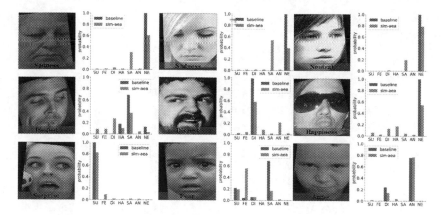

Fig. 4. Nine images and their predicted results from RAF-DB. There is some ambiguity in each image. The orange soft labels were mined by our model (Su: Surprise, Fe: Fear, Di: Disgust, Ha: Happy, Sa: Sad, An: Anger, Ne: Neutral). The blue logits were output from the baseline method. (Color figure online)

4.4   Visualization Analysis

To demonstrate the superiority of our method, we analyze it visually in comparison with the baseline. We selected 9 images from RAF-DB with different degrees of ambiguity. As shown in Fig. 4, the label of each expression is attached to the green rectangle in the lower part of the image. Compared with the baseline, our method mined more latent ground truth. The rationality of the soft labels can be illustrated with the first two expressions labeled as sadness in Fig. 4. In the baseline method, both images are predicted to be "Neutral", but our predictions have a larger weight on both "Neutral" and "Sadness". Although our method predicts only one of them correctly, we think it is closer to the ground truth. The same conclusion can be observed in other images. Another interesting example is the fourth expression, it's hard to determine its real label. And it seems that our model has some "confusion" with it, too. Despite the presence of many uncertain or ambiguous expressions in the FER dataset, we assume that the labels provided by the annotators are correct overall (at least for one component).

4.5   Evaluation on Synthetic Datasets

One of the consequences of ambiguity is that annotators tend to mislabel expressions, which can be viewed as label noise. Label noise is usually classified into two types: symmetric label noise and asymmetric label noise. Asymmetric noise means that each expression can be incorrectly labeled as any other expression uniformly. We synthesize asymmetrical noise like [34], and quantitatively added 10%, 20%, and 30% noise to RAF-DB. We reproduced SCN [34] to compare with our method. To make a fair comparison, we use the same initialization parameters which pre-train on MS-Celeb-1M with the backbone of ResNet18. We report

Table 4. Accuracy (%) comparison with the state-of-the-art method on synthetic noise RAF-DB datasets.

Method	n(%)	RAF-DB/best	RAF-DB/last
Baseline	10	82.88	81.64
SCN	10	86.77	86.40
SLM-AEA	10	**91.59**	**91.59**
Baseline	20	80.84	74.88
SCN	20	85.20	84.78
SLM-AEA	20	**89.86**	**89.57**
Baseline	30	79.19	63.88
SCN	30	82.69	82.34
SLM-AEA	30	**87.74**	**85.95**

Table 5. Comparison with the State-of-the-arts. [+] denotes both AffectNet and RAF-DB are used as the training set. [*] denotes oversampling is used since the train set of AffectNet is imbalanced.

Benchmark dataset	Method	Pre-trained dataset	Acc. (%)
Oulu-CASIA	IPA2LT[+] [42]	-	61.49
Oulu-CASIA	FN2EN [8]	2.6M face images	87.71
Oulu-CASIA	DeRL [40]	BU-4DFE & BP4D	88.00
Oulu-CASIA	DDL [24]	Multi-PIE	88.26
Oulu-CASIA	FDRL (ResNet18) [25]	MS-Celeb-1M	88.26
Oulu-CASIA	**SLM-AEA (ResNet18)**	**MS-Celeb-1M**	**88.61**
RAF-DB	LDL-ALSG[+] [4]	-	85.53
RAF-DB	IPA2LT[+] [42]	-	86.77
RAF-DB	SCN (ResNet18)[+] [35]	MS-Celeb-1M	88.14
RAF-DB	DMUE (ResNet18) [28]	MS-Celeb-1M	88.76
RAF-DB	FDRL (ResNet18) [25]	MS-Celeb-1M	89.47
RAF-DB	**SLM-AEA (ResNet18)**	**MS-Celeb-1M**	**92.82**
AffectNet	IPA2LT[+] [42]	-	55.71
AffectNet	RAN[*] [36]	MS-Celeb-1M	59.50
AffectNet	SCN (ResNet18)[*] [35]	MS-Celeb-1M	60.23
AffectNet	CVT[*] [19]	MS-Celeb-1M	61.70
AffectNet	**SLM-AEA[*] (ResNet18)**	**MS-Celeb-1M**	**62.26**
AffectNet	DMUE (ResNet18)[*] [28]	MS-Celeb-1M	62.84
SFEW	RAN [36]	MS-Celeb-1M	56.40
SFEW	DMUE(ResNet18) [28]	MS-Celeb-1M	57.12
SFEW	IPA2LT[+] [42]	-	58.29
SFEW	DDL [24]	Multi-PIE	59.86
SFEW	FDRL(ResNet18) [25]	MS-Celeb-1M	62.16
SFEW	**SLM-AEA (ResNet18)**	**MS-Celeb-1M**	**67.91**

the mean accuracy of ten experiments. As shown in Table 4, our method outperforms both the baseline and the SCN for different ratios of label noise. That is, our method has better error correction capability. We believe that this error correction capability comes from representing expressions with soft labels rather than hard labels. Note that the value of p should be adjusted according to the synthesized noise ratio. With the noise ratio increasing, p should be tuned down slightly to achieve better performance.

4.6   Comparison with the State-of-the-Arts

We compared our method with the state-of-the-arts as shown in Table 5. IPA2LT aims to address the inconsistency of different datasets. DDL is proposed to disentangle the disturbing factors in facial expression images. DeRL decomposes expressions into expressive components and unexpressed neutral components with the help of GAN. RAN is developed to deal with the occlusion and head pose in FER. LDL-ALSG introduces label distribution learning to FER and uses extra information from related tasks. SCN and DMUE are recently proposed and also committed to addressing the uncertainty or ambiguity problem of FER. [35] is the first paper to propose the uncertainty problem in FER. DMUE (ResNet50-IBN version) achieved previous leading performance. We choose version ResNet18 of the DMUE for a fair comparison. FDRL is also a recently proposed method and achieves leading performance on both the Oulu-CASIA and RAF-DB datasets. CVT introduced transformer into FER and also achieved state-of-the-art performance. Compared with these methods, we definitely achieve better performance and set new records on Oulu-CASIA, RAF-DB, and SFEW. SLM-AEA outperforms these state-of-the-art methods consistently with a huge margin on RAF-DB and SFEW. Although it did not outperform DMUE on AffectNet, a very competitive result was achieved and our method was simpler compared to DMUE.

5   Conclusion

In this paper, we proposed a novel method to address the uncertainty and ambiguity problem. Our method is composed of two main modules: the soft label mining module and the average expression anchoring module. The former is designed to break the limitations of the classification model by dynamically converting hard labels to soft labels, and the latter aims to alleviate the high interclass similarity. The soft label mining module plays the most important role in our method due to its outstanding and stable improvements. Compare with the state-of-the-art methods, our SLE-AEA is more simple yet effective. Experiments on popular benchmarks show that our method is extremely competitive.

Acknowledgements. This work was supported by the Tianjin Municipal Science and Technology Program for New Generation of Artificial Intelligence (19Z-XZNGX00030).

References

1. Barsoum, E., Zhang, C., Ferrer, C.C., Zhang, Z.: Training deep networks for facial expression recognition with crowd-sourced label distribution. In: ACM International Conference on Multimodal Interaction, pp. 279–283 (2016)
2. Bazzo, J.J., Lamar, M.V.: Recognizing facial actions using Gabor wavelets with neutral face average difference. In: International Conference on Automatic Face and Gesture Recognition, pp. 505–510 (2004)
3. Cai, J., Meng, Z., Khan, A.S., Li, Z., O'Reilly, J., Tong, Y.: Island loss for learning discriminative features in facial expression recognition. In: International Conference on Automatic Face and Gesture Recognition, pp. 302–309 (2018)
4. Chen, S., Wang, J., Chen, Y., Shi, Z., Geng, X., Rui, Y.: Label distribution learning on auxiliary label space graphs for facial expression recognition. In: IEEE Conference on Computer Vision and Pattern Recognition, pp. 13984–13993 (2020)
5. Dalal, N., Triggs, B.: Histograms of oriented gradients for human detection. In: IEEE Conference on Computer Vision and Pattern Recognition (2005)
6. Darwin, C., Prodger, P.: The Expression of the Emotions in Man and Animals. Oxford University Press, Oxford (1998)
7. Dhall, A., Goecke, R., Lucey, S., Gedeon, T.: Static facial expression analysis in tough conditions: data, evaluation protocol and benchmark. In: International Conference on Computer Vision, pp. 2106–2112 (2011)
8. Ding, H., Zhou, S.K., Chellappa, L.: FaceNet2ExpNet: regularizing a deep face recognition net for expression recognition. In: International Conference on Automatic Face and Gesture Recognition (2017)
9. Ekman, P.: Strong evidence for universals in facial expressions: a reply to Russell's mistaken critique. Psychol. Bull. **115**(2), 268–287 (1994)
10. Ekman, P., Friesen, W.V.: Constants across cultures in the face and emotion. J. Pers. Soc. Psychol. **17**(2), 124–129 (1971)
11. Ekman, P., Rosenberg, E.: What the Face Reveals: Basic and Applied Studies of Spontaneous Expression Using the Facial Action Coding System (FACS). Oxford University Press, Oxford (1997)
12. Gao, B., Xing, C., Xie, C., Wu, J., Geng, X.: Deep label distribution learning with label ambiguity. IEEE Trans. Image Process. **26**(6), 2825–2838 (2017)
13. Geng, X.: Deep label distribution learning with label ambiguity. IEEE Trans. Knowl. Data Eng. **28**(7), 1734–1748 (2016)
14. He, K., Zhang, X., Ren, S., Sun, J.: Deep residual learning for image recognition. In: IEEE Conference on Computer Vision and Pattern Recognition (2016)
15. He, T., Zhang, Z., Zhang, H., Zhang, Z., Xie, J., Li, M.: Bag of tricks for image classification with convolutional neural networks. In: IEEE Conference on Computer Vision and Pattern Recognition, pp. 558–567 (2019)
16. Li, S., Deng, W., Du, J., Zhang, Z.: Reliable crowd-sourcing and deep locality-preserving learning for expression recognition in the wild. In: IEEE Conference on Computer Vision and Pattern Recognition, pp. 285–2861 (2017)
17. Li, Y., Yang, J., Song, Y., Cao, L., Luo, J., Li, L.: Learning from noisy labels with distillation. In: IEEE Conference on Computer Vision and Pattern Recognition, pp. 1910–1918 (2017)
18. Liu, Z., et al.: Swin transformer: hierarchical vision transformer using shifted windows. In: International Conference on Computer Vision (2021)
19. Ma, F., Sun, B., Li, S.: Robust facial expression recognition with convolutional visual transformers. IEEE Trans. Affect. Comput. (2021)

20. Mollahosseini, A., Hasani, B., Mahoor, M.H., Mahoor, M.H.: AffectNet: a database for facial expression, valence, and arousal computing in the wild. TAC **10**(1), 18–31 (2017)

21. Ng, H.W., Nguyen, V.D., Vonikakis, V., Winkler, S.: Deep learning for emotion recognition on small datasets using transfer learning. In: ACM International Conference on Multimodal Interaction, pp. 443–449 (2015)

22. Ng, P.C., Henikoff, S.: SIFT: predicting amino acid changes that affect protein function. Nucleic Acids Res. **31**(13), 3812–3814 (2003)

23. Qu, Y., Mo, S., Niu, J.: DAT: training deep networks robust to label-noise by matching the feature distributions. In: IEEE Conference on Computer Vision and Pattern Recognition (2021)

24. Ruan, D., Yan, Y., Chen, S., Xue, J.H., HanziWang: deep disturbance-disentangled learning for facial expression recognition. In: ACM International Conference on Multimedia (2020)

25. Ruan, D., Yan, Y., Lai, S., Chai, Z., Shen, C., Wang, H.: Feature decomposition and reconstruction learning for effective facial expression recognition. In: IEEE Conference on Computer Vision and Pattern Recognition, pp. 7660–7669 (2021)

26. Russell, J.A.: A circumplex model of affect. J. Pers. Soc. Psychol. **39**(6), 1161–1178 (1980)

27. Shan, C., Gong, S., McOwan, P.W.: Facial expression recognition based on local binary patterns: a comprehensive study. Image Vis. Comput. **27**(6), 803–816 (2009)

28. She, J., Hu, Y., Shi, H., Wang, J., Shen, Q., Mei, T.: Dive into ambiguity: latent distribution mining and pairwise uncertainty estimation for facial expression recognition. In: IEEE Conference on Computer Vision and Pattern Recognition, pp. 6248–6257 (2021)

29. Shi, J., Zhu, S., Liang, Z.: Learning to amend facial expression representation via de-albino and affinity (2021). arXiv:2103.10189

30. Shu, J., et al.: Meta-weight-net: Learning an explicit mapping for sample weighting. In: Annual Conference on Neural Information Processing Systems (2019)

31. Su, K., Geng, X.: Soft facial landmark detection by label distribution learning. In: AAAI Conference on Artificial Intelligence, pp. 5008–5015 (2019)

32. Tian, Y.I., Kanade, T., Cohn, J.F.: Recognizing action units for facial expression analysis. T-PAMI **23**(2), 97–115 (2001)

33. Veit, A., Alldrin, N., Chechika, G., Krasin, I., Gupta, A., Belongie, S.: Learning from noisy large-scale datasets with minimal supervision. In: IEEE Conference on Computer Vision and Pattern Recognition, pp. 839–847 (2017)

34. Wang, C., Wang, S., Liang, G.: Identity- and pose-robust facial expression recognition through adversarial feature learning. In: ACM International Conference on Multimedia, pp. 238–246 (2019)

35. Wang, K., Peng, X., Yang, J., Lu, S., Qiao, Y.: Suppressing uncertainties for large-scale facial expression recognition. In: IEEE Conference on Computer Vision and Pattern Recognition, pp. 6897–6906 (2020)

36. Wang, K., Peng, X., Yang, J., Meng, D., Qiao, Y.: Region attention networks for pose and occlusion robust facial expression recognition. IEEE Trans. Image Process. **29**, 4057–4069 (2020)

37. Wang, Y., Ma, X., Chen, Z., Luo, Y., Yi, J., Bailey, J.: Symmetric cross entropy for robust learning with noisy labels. In: International Conference on Computer Vision (2019)

38. Xu, N., Liu, Y., Geng, X.: Label enhancement for label distribution learning. IEEE Trans. Knowl. Data Eng. **33**(4), 1632–1643 (2021)

39. Xu, N., Shu, J., Liu, Y., Geng, X.: Variational label enhancement. In: Proceedings of the 37th International Conference on Machine Learning, pp. 10597–10606 (2020)
40. Yang, H., Ciftci, U.A., Yin, L.: Facial expression recognition by de-expression residue learning. In: IEEE Conference on Computer Vision and Pattern Recognition, pp. 2168–2177 (2018)
41. Yi, K., Wu, J.: Probabilistic end-to-end noise correction for learning with noisy labels. In: IEEE Conference on Computer Vision and Pattern Recognition (2019)
42. Zeng, J., Shan, S., Chen, X.: Facial expression recognition with inconsistently annotated datasets. In: European Conference on Computer Vision, pp. 22–237 (2018)
43. Zhang, Y., et al.: Global-local GCN: large-scale label noise cleansing for face recognition. In: IEEE Conference on Computer Vision and Pattern Recognition (2020)
44. Zhao, G., Huang, X., Taini, M., Li, S.Z., äInen, M.P.: Facial expression recognition from near-infrared videos. Image Vis. Comput. **29**(9), 607–619 (2011)
45. Zhao, Z., Liu, Q., Zhou, F.: Robust lightweight facial expression recognition network with label distribution training. In: AAAI Conference on Artificial Intelligence, pp. 3510–3519 (2021)

'Labelling the Gaps': A Weakly Supervised Automatic Eye Gaze Estimation

Shreya Ghosh[1(✉)], Abhinav Dhall[1,2], Munawar Hayat[1], and Jarrod Knibbe[3]

[1] Monash University, Melbourne, Australia
{shreya.ghosh,munawar.hayat}@monash.edu, abhinav@iitrpr.ac.in
[2] IIT Ropar, Rupnaga, India
[3] University of Melbourne, Melbourne, Australia
jarrod.knibbe@unimelb.edu.au

Abstract. Over the past few years, there has been an increasing interest to interpret gaze direction in an unconstrained environment with limited supervision. Owing to data curation and annotation issues, replicating gaze estimation method to other platforms, such as unconstrained outdoor or AR/VR, might lead to significant drop in performance due to insufficient availability of accurately annotated data for model training. In this paper, we explore an interesting yet challenging problem of gaze estimation method with a limited amount of labelled data. The proposed method utilize domain knowledge from the labelled subset with visual features; including identity-specific appearance, gaze trajectory consistency and motion features. Given a gaze trajectory, the method utilizes label information of only the start and the end frames of a gaze sequence. An extension of the proposed method further reduces the requirement of labelled frames to only the start frame with a minor drop in the generated label's quality. We evaluate the proposed method on four benchmark datasets (CAVE, TabletGaze, MPII and Gaze360) as well as web-crawled YouTube videos. Our proposed method reduces the annotation effort to as low as 2.67%, with minimal impact on performance; indicating the potential of our model enabling gaze estimation 'in-the-wild' setup[1] (https://github.com/i-am-shreya/Labelling-the-Gaps).

Keywords: Gaze estimation · Weakly-supervised learning · Neural network

1 Introduction

The 'language of the eyes' provides an insight into a complex mental state such as visual attention [29] and human cognition (emotions, beliefs and desires) [52]. Accurate gaze estimation has wide applications in computer vision-related assistive technologies [16, 33] where the gaze is measured as a line of sight of the pupil

---

Supplementary Information The online version contains supplementary material available at https://doi.org/10.1007/978-3-031-26316-3_44.

in 3D/2D space or 2D screen location [24]. Recent advances in computer vision and deep learning have significantly enhanced the accuracy of gaze estimation [15]. Most promising eye gaze estimation techniques either require specialized hardware (for example Tobii [34]) or use supervised image processing solutions [46,62]. Device and sensor based gaze estimation methods are highly dependent on user assistance, illumination specificity, the high device failure rate in an uncontrolled environment and constraints on the device's working distance. On the other hand, supervised methods require a large amount of labelled data for training. Manual labelling of human gaze information is a complex, noisy, resource-expensive and time-consuming task. To overcome these limitations, weakly-supervised learning provides a promising paradigm since it enables learning from a large amount of readily available non-annotated data. Few works [10,17,26,57] explore in this direction to eliminate the data curation and annotation issue. However, these methods mostly investigate from a spatial analysis perspective. Human eye movement is a spatio-temporal, dynamic process which is either task-driven or involuntary action. Thus, it would be interesting to simplify the ballistic eye movement and curate large-scale training data for gaze representation learning. To this end, we propose a weakly supervised eye gaze estimation framework. Our proposed technique reduces the requirement of a large number of annotated training samples. We show that the technique can also be used to facilitate the annotation process and reduce the bias in the data annotations. The proposed method requires the ground truth labels of start and end frames in a pre-defined gaze trajectory. We further refine this strategy where only the start frame's gaze annotation is required. Our proposed method significantly reduces the annotation effort which could be beneficial for annotating large-scale gaze datasets quickly. Moreover, it can be used in several applications such as immersive, augmented and virtual reality [9,49] (especially in Foveated Rending (FR)), animation industry [18,38] and social robotics [2,58], where unsupervised or weakly supervised calibration is highly desirable. In Foveated Rending (FR), gaze based interaction demands low latency gaze estimation to reduce energy consumption. To achieve this, the virtual environment displays high-quality images only from the user's point of view and blurs the peripheral region. Due to the subsequent delays in the frame-wise gaze estimation pipeline, the usage of FR is quite limited and mostly headpose direction is used to approximate the field of view [9]. Our proposed method has the potential to bridge the gap and reduce energy consumption by interpolating the gaze trajectory of the user. Another potential application includes animation industry [18,38]. Given the start and end Point of Gaze (PoG) of a virtual avatar, our method can easily generate realistic labels for intermediate frames to display realistic facial gestures in the interaction environment [18,38]. Similarly, in social robotics, multi-modal gaze control strategies have been explored for guiding the robot's gaze. For example, an array of microphones has been utilized [2] to guide the gaze direction of a robot named Maggie. The other well-established methods include the usage of infrared laser and multimodal stimuli (e.g., visual, auditory and tactile) for modelling any known gaze trajectories [58]. Our proposed methods could eliminate the aforementioned requirement of specialised hardware or predefined heuristics to navigate the environment. The **main contributions** of the paper are as follows:

1. We propose two weakly supervised neural networks (2-labels, 1-label) for gaze estimation. '2-labels' and '1-label' require the labels of two and one frames in a gaze sequence, respectively.
2. We use task-specific information to bridge the gap between labelled and unlabelled samples. Our proposed method leverages facial appearance, relative motion, trajectory ordering and embedding consistency. This task-specific knowledge bridge the gap between labelled and unlabeled samples via learning.
3. We evaluate the performance of the proposed networks in two settings: 1) *On benchmark datasets* (CAVE, TabletGaze, MPII and Gaze360) and 2) *On unlabelled 'in the wild' YouTube data* where ground truth annotation is not available. Additionally, we perform cross-dataset experiments to validate the generalizability of the framework.
4. We also demonstrate the effectiveness of our proposed techniques by re-learning state-of-the-art eye gaze estimation methods with the labels generated by our method with very few prior annotations. The results indicate comparable performance for state-of-the-art frameworks (for example, 3.8° by pictorial gaze [37] and 4° with our 2-label technique).
5. We also validate our learning based interpolation method on unlabelled YouTube data where ground truth annotation is not available. Our experimental results suggest that this annotation method can be useful for extracting substantial training data for learning gaze estimation models.

2 Related Work

Gaze Estimation. Recent advances in computer vision and deep learning techniques have significantly enhanced the gaze estimation performance [24,37]. A thorough analysis of gaze estimation literature is mentioned in a recent survey [15]. Appearance based gaze estimation methods [30,31,63] learn image to gaze mapping either via support vector regression [46] or deep learning methods [12,23,27,59,60,63]. Among the deep learning based methods, supervised learning methods [23,37,60,63] mostly encode appearance based gaze which require a large amount of annotated data. To overcome the limitation, few works explore gaze estimation with limited supervision such as 'learning-by-synthesis' [48], hierarchical generative models [51], conditional random field [6], unsupervised gaze target discovery [62], unsupervised representation learning [10,57], weakly supervised learning [26], pseudo labelling [17] and few-shot learning [36,56]. Among these studies, the few shot learning approach required very few (≤ 9) calibration samples for gaze inference. Our method requires even less data annotation for gaze estimation (CAVE: 6.56%, TabletGaze: < 1%, MPII: 4.67% and Gaze360: 2.38%).

Gaze Motion. Eye movements are divided into the following categories: *1) Saccade.* Saccades are voluntary eye movements to adjust the PoG in a visual field and it usually lasts for 10 to 100 ms. *2) Smooth Pursuit.* It is an involuntary eye movement that occurs while tracking a moving visual target. *3) Fixations.* Fixations consist of three involuntary eye movements termed as *tremor,*

Fig. 1. Weakly supervised labelling approach illustration. The zig-zag path $\overrightarrow{AD}$ in the left is an example of a human gaze movement. This path can further be broken into several 'gaze trajectories' ($\overrightarrow{AB}, \overrightarrow{BC}$ and $\overrightarrow{CD}$). Given a gaze trajectory $\overrightarrow{AB}$ with gaze annotation for A and B, the objective of this work is to annotate all the intermediate unlabelled frames i.e. P_1, P_2 and P_3. The right side is an example of a gaze trajectory [63].

drift and *microsaccades* [11]. The main objective is to stabilize the PoG on an object. Prior works along this line mainly used velocity based thresholding [25], BLSTM [47], Bayesian framework [42], and hierarchical HMM [67] for classification. Arabadzhiyska et al. [3] model saccade dynamics for gaze-contingent rendering. Our proposed method uses trajectory constrained gaze interpolation using temporal coherency with limited ground truth labels.

Gaze Datasets. In the past decade, several datasets [12–14,21,24,35,46,59,60] have been proposed to estimate gaze accurately. The dataset collection technique has evolved from constrained lab environments [46,61] to unconstrained indoor [12,21,59–61] and outdoor settings [24]. To consider both the constrained and unconstrained settings, we evaluate our weakly supervised framework in CAVE [46], TabletGaze [21], MPII [60] and Gaze360 [24] datasets.

Weakly Supervised Neural Networks. Over the past few years, several promising weakly supervised methods have been proposed which mainly infer on the basis of prior knowledge [8,28], task-specific domain knowledge [4,64,65], representation learning [53,66], loss-imposed learning paradigms [4,41] and combinations of the above [4]. Williams et al. [54] propose a semi-supervised Gaussian process model to predict the gaze. This method simplifies the data collection process as well. Bilen et al. [8] use pre-trained deep CNN for the object detection task. Arandjelovic et al. [4] propose a weakly supervised ranking loss for the place recognition task. Haeusser et al. [19] introduce 'associative learning' paradigm, which allows semi-supervised end-to-end training of any arbitrary network architecture. Unlike these studies, we explore loss-imposed domain knowledge for our framework.

3   Preliminaries

Gaze Trajectory. Human eye movement follows an arbitrary continuous path in three-dimensional space termed as the 'gaze trajectory' [40]. Gaze trajectories generally depend on the person, context and external factors [32]. Eye movements can be divided into three types: fixations, saccades and smooth pursuit [40]. Fixation occurs when the gaze may pause in a specific position voluntarily or involuntarily. Conversely, gaze moves from one to another position for a saccade.

The human gaze consists of a series of fixations and saccades in random order. In this work, we consider a small duration of eye movement from one position to another. Let us assume, the zig-zag path (left of Fig. 1) is an example of the human gaze movement path. This path can be divided into three small sub-paths $(\overrightarrow{AB}, \overrightarrow{BC}$ and $\overrightarrow{CD})$. We conduct our experiments on each such small sub-path. We use the term 'gaze trajectory' to refer to this simplified version of the gaze path (i.e. $\overrightarrow{AB}$). The red points (P_1, P_2 and P_3) in $\overrightarrow{AB}$ are the frames in the A to B sequence. Considering these as discrete, they can be split into 3-point subsets with a constraint: each subset should contain the start and endpoints and the points should maintain the order of trajectory sequence. We term the 3 points set as *3-frame set*. For example, $\{A, P_1, B\}$, $\{A, P_2, B\}$ and $\{A, P_3, B\}$ are three 3-frame sets.

Problem Statement. In the context of a video, the points (A, P_1, P_2, P_3 and B) are frames in a specific trajectory order. Throughout this paper, we term A and B as start and end frames. Given the annotated start and end frames, the gaze trajectory sequence can be divided into small segments. There might be a high and insignificant temporal coherence if the segment duration is too small. On the other hand, a longer duration could affect the learning of meaningful representation due to diversity. We observed that the average difference in large time segment consisting of approx. ~ 80 frames is around $\sim 35°$. On the other hand, the smallest segment has angular difference of $< 1°$. On this front, from a long sequence, we mine 3-frame subsets of gaze trajectory for learning meaningful representation. We work on two experimental settings: (a) **2-labels:** *When the start and end frame annotations are available.* (b) **1-label:** *When only the start frame annotation is available.* Given a gaze trajectory similar to $\overrightarrow{AB}$ with labels for A and B, the objective of this work is to annotate the unlabelled frames i.e. $P_1, P_2, P_3, \dots P_n$ where, n is the number of intermediate frames.

Notations. Suppose that we have a set of N '3-frame set' samples in a dataset $D = \{X_n, Y_{s_n}, Y_{e_n}\}_{n=1}^{N}$, where X_n is a nth 3-frame set consisting of start, middle/unlabelled and end frames $(f_s, f_{ul}, f_e)_n$, Y_{s_n} and Y_{e_n} are the nth start and end frame labels, respectively. Lets assume our model G with learnable parameter θ maps input $X_n \in \mathbb{R}^{3 \times 100 \times 50 \times 3}$ to the relevant label spaces i.e., $Y_{s_n} \in \mathbb{R}^3$, $Y_{ul_n} \in \mathbb{R}^3$ and $Y_{e_n} \in \mathbb{R}^3$. The mapping function is denoted as $G_\theta : X_n \rightarrow \{Y_{s_n}, Y_{ul_n}, Y_{e_n}\}$.

4 Gaze Labelling Framework

4.1 Architectural Overview of '2-Labels'

The overview of the proposed framework is shown in Fig. 2. Given a 3-frame set $X_t : \{f_s, f_{ul}, f_e\}$, we define an encoder E which maps the input X_t to latent space $Z_t : \{Z_s, Z_{ul}, Z_e\}$, where $Z_s \in \mathbb{R}^{2048}, Z_{ul} \in \mathbb{R}^{2048}, Z_e \in \mathbb{R}^{2048}$ (Refer Fig. 2 Left). After $E : X_t \rightarrow Z_t$ mapping, two motion features M_{s_ul} and M_{ul_e} are extracted between start-middle frames and middle-end frames. These motion features are concatenated with the latent embeddings. On top of it, Fully Connected (FC) layers having 512 and 1024 nodes are appended before prediction. Finally, the network predicts Y_s^p, Y_{ul}^p and Y_e^p, where, Y_s^p, Y_{ul}^p and Y_e^p are gaze information corresponding to the input frames $\{f_s, f_{ul}, f_e\}$. The backbone network is not architecture-specific, although we use VGG-16 [45] and Resnet-50 [20] for our experiments. The rationale behind the framework design is explained as follows:

Identity Adaptation. The obvious usefulness of user adaptation of gaze calibration for AR and VR devices motivates us to design an identity specific gaze labelling framework [43]. Thus, we purposefully select identity specific 3-frame sets. At the same time, it is important for the framework to be able to learn the variations across a large number of subjects with a different head pose, gaze direction, appearance, illumination, image resolution and many other configurations. Additionally, the framework should encode rich features relevant to eye region appearance, which is the most important factor for weakly supervised gaze labelling. Similar to recent studies [36,59], we use eye region images as input for gaze inference. The 3-frame set is selected over a small duration temporal window in a video. As there is a subtle change in the appearance of eyes from one frame to another, the latent representation of the image also has minimal change. At a conceptual level, we are motivated by the smoothness constraint in optical flow algorithms. We choose to calculate cosine distance between start and middle, middle and end pair while calculating consistency loss. It helps in preserving the identity specific features across the 3-frame set.

Motion Feature. At first glance, the task of predicting gaze from sparsely labelled data may seem overly challenging. However, given a '3-frame set' sequence of a subject in very small time duration, there will be a high correspondence between frame A, P_i and B, where $i = \{1, 2, 3\}$. Given this constraint, the objective is reduced to modelling the head and eye motion information to bridge the gap. This motivates us to use motion features for sequence modelling. Similar to [7] in pose estimation domain, we encode a weak inter-frame motion by computing the ℓ_1 distance between two consecutive frames in the latent space. We define motion feature by $M_{s\_ul} \oplus M_{ul\_e}$ where $M_{s\_ul} = Z_s - Z_{ul}$ and $M_{ul\_e} = Z_{ul} - Z_e$. M_{s_ul} and M_{ul_e} represent motion feature between start-middle and middle-end frames, respectively. Further, we use this feature to estimate the gaze-direction in a given trajectory. To train the above-mentioned network, we use the following loss functions.

Regression Loss. Corresponding to each 3-frame set X_t, the start and end frames are annotated in the 2-labels setting. These annotations provide strong supervisory information to predict the gaze information of the middle unlabelled frame. It provides information regarding an arbitrary gaze trajectory. The unlabelled middle frame lies in between start and end frames in that specific trajectory. Thus, it belongs to the same distribution of the start and end frames. The regression loss is defined as: $l_{reg} = MSE(Y_s, Y_s^p) + MSE(Y_e, Y_e^p)$ Here, MSE is Mean Squared Error, Y_s, Y_s^p, Y_e and Y_e^p are start label, predicted start label, end label and predicted end label, respectively.

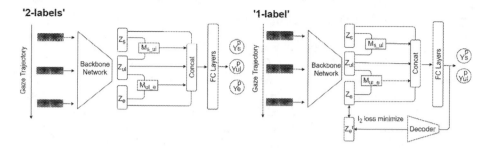

Fig. 2. Overview of '2-labels' and '1-label' network pipelines. The frameworks take 3-frame set as input and learn to interpolate intermediate labels via weak supervision. Refer Sect. 4 for more details.

Consistency Loss. The main aim of the consistency loss is to maintain consistency between latent and label space. Here, latent space is denoted as Z and label space is the output space of the '2-labels' framework. In the latent space, let the distance between Z_s and Z_{ul} be $d^z_{s\_ul}$ and the distance between Z_{ul} and Z_e be $d^z_{ul\_e}$. Similarly in the label space, let the distance between Y^p_s and Y^p_s be $d^y_{s\_ul}$ and the distance between Y^p_s and Y_e be $d^y_{ul\_e}$. According to our hypothesis, the distance from start to unlabelled frame and unlabelled to end frame remains consistent. The loss is defined as follows: $l_{consistency} = \{|d^y_{s\_ul} - d^z_{s\_ul}| + |d^y_{ul\_e} - d^z_{ul\_e}|\}$

In the equation, the cosine distance is considered. The rationale behind the choice is as follows: the cosine distance is applied pairwise to utilize the partial annotations and bridge the gap between labelled and unlabelled frames. The distance has following properties: it leverages the identity specific information across the 3-frame set and it captures the motion-similarity information which indirectly encodes relative ordering of the frames. The angular distance (d_{a_b}) between the frames a and b is defined as follows: $d_{a\_b} = \frac{a}{||a||_2} \cdot \frac{b}{||b||_2}$ where, a and b are latent or label space embeddings of the start, unlabelled and end frames. '(.)' denotes the dot product. According to [5], this term is calculated in two stages. First, the latent embeddings are L_2-normalized (i.e. $\frac{a}{||a||_2}$), which maps the d-dimensional latent embedding to a unit hyper-sphere, where the cosine similarity/distance and dot product are equivalent. As the human gaze follows a spatio-temporal trajectory [40], the 3-frame sets satisfy their relative trajectory ordering as well. The condition is satisfied by the consistency loss as it incorporates their relative position w.r.t. the middle unlabelled frame from label space to the latent space.

Overall Loss Function. The final loss function is defined below:

$$Loss = \lambda_1 l_{reg} + \lambda_2 l_{consistency} \qquad (1)$$

It includes regression and consistency losses. Here, $\lambda_{1,2}$ are the regularization parameters. Further, we reduce the label requirement (by removing the requirement of the end frame's label) in the next framework with minimal impact in the performance.

4.2   Architectural Overview of '1-Label'

The network architecture of 1-label network is shown in Fig. 2 Right. Similar to the 2-labels network, the motion feature and identity specific appearance are leveraged for gaze estimation. The main motivation for moving from 2-labels to 1-label architecture is to leverage rich features with even lesser number of total annotations in a dataset. In '1-label' architecture, we have additional decoder module defined as $D(Z'_e; \phi_D)$ consisting of an FC layer. The decoder D, parameterized by ϕ_D, maps the penultimate layer features (i.e. $Z'_e \in \mathbb{R}^{1024}$) to latent-embeddings $Z^r_e \in R^{2048}$. The rationale behind this remapping is adding constraints which can enhance the performance of the network by encoding meaningful representation from f_e. l_2 loss is computed between Z_e and Z^r_e. The backbone network is not architecture-specific similar to the 2-labels architecture, although we use VGG-16 [45] and Resnet-50 [20] as backbone networks for our experiments. To train this network, we use the following loss functions:

Regression Loss. Similar to 2-labels architecture, we compute the regression loss corresponding to the start label as follows: $l_{reg} = MSE(Y_s, Y^p_s)$ Here, MSE is mean squared error, Y_s and Y^p_s are start and predicted start label, respectively.

Consistency Loss. Similar to 2-labels, in the 1-label technique, we add the consistency loss $l_{consistency}$. The loss is defined as follows: $l_{consistency} = \{|d^y_{s\_ul} - d^z_{s\_ul}|\}$ Here, $d^y_{s\_ul}$ and $d^z_{s\_ul}$ are distance between start and unlabelled frame in label and latent space, respectively.

Similar Distribution. We leverage on the constraint that the gaze information belongs to similar distribution. Given a 3-frame set X_t, the output gaze should belong to a specific distribution and for that we compute Kullback-Leibler divergence (KL). To computes KL divergence loss between Y_{true} and Y_{pred}, the following equation is followed: $loss = Y_{true}(log\frac{Y_{true}}{Y_{pred}})$. The objective function $l_{divergence}$ is the loss term to minimize the divergence between the gaze information of start and unlabeled frames, defined as follows: $l_{divergence} = (KL(Y_s, Y_{ul}))$

Embedding Loss. It is the ℓ_2 loss computed between Z_e and Z^r_e. This pattern is predicting future embeddings from prior knowledge.

Overall Loss Function. The final loss function for '1-label' is an ensemble of regression, consistency, similar distribution and embedding losses, where, $\lambda_1 \cdots \lambda_4$ are the regularization parameters.

$$Loss = \lambda_1 l_{reg} + \lambda_2 l_{consistency} + \lambda_3 l_{divergence} + \lambda_4 l_{embedding} \qquad (2)$$

5   Experiments

On Benchmark Datasets. We validate the proposed methods on 4 benchmark datasets: CAVE, Tabletgaze, MPII and Gaze360. These datasets are collected in fully constraint (CAVE), less constrained (Tabletgaze and MPII) and unconstrained (Gaze360) environments.

Table 1. Comparison of benchmark dataset statistics to show the amount of images in the original data and derived data (3-frames set mined).

Dataset	Original dataset	Derived dataset
CAVE [46]	5,880	3,024
MPII [63]	213,659	32,751
TabletGaze [21]	1,428 min	108,524
Gaze360 [24]	172,000	197,588

Automatic 3-frame Set Mining: The first step to implement our proposed method is 3-frame set mining. For generating 3-frame sets, we perform dataset-specific pre-processing as the datasets are collected in different setups. Please note that we require ground truth labels to define the gaze trajectories, especially for CAVE and MPII datasets as temporal information is not present. Moreover, the annotated subset required for weak supervision is a special subset which contains start and end frame of the trajectories. Additionally, there is no temporal overlap between 3-frame sets for any of the datasets. Table 1 shows the comparison among dataset statistics to show the amount of images in the original data and derived data (3-frames set mined used for weak-supervision). See supplementary material for more details.

On Unlabelled 'in the Wild' YouTube Data. We evaluate our method on an 'in the wild' data i.e. when the expert/ground truth labels are not available. We leverage two eye symmetry property i.e. the change in relative position of the iris is symmetrical while scanning 3D space [10]. We collect approximately 400,000 frames from YouTube videos using this strategy. The details are mentioned in the supplementary material.

Experimental Settings. We define the following terminologies for easy navigation in the upcoming sections. *1) Original Data (OD):* It refers to benchmark dataset's unaltered data; *2) Derived Data (DD):* 3-frame set mined data derived from OD; *3) Original Labels (OL):* Original ground-truth labels provided with OD; and *4) Predicted Labels (PL):* Labels predicted from '2-labels' and '1-label' methods (i.e. Y_{ul}^p). We perform experiments with the following settings: **1) Validation w.r.t. Ground Truth Labels.** First, we applied '3-frame set mining' to obtain the DD (Refer Table 1) from OD. Further, we split the DD into 80%-20% train-test splits without any identity overlap. We evaluate our proposed method with OL. **2) Label Quality Assessment via State-of-the-art Methods' Performance.** We train the state-of-the-art methods [21,24,37] on PL and validate on OL for label quality assessment. **3) Experiments with Different Data Partitions.** We train state-of-the-art models [21,24,37] with different input data settings as follows: a) start and end frames of 3-frame sets, b) 50% of the whole data and c) newly labelled frames. **4) Ablation Studies.** We have conducted extensive ablation studies to show the importance of loss function, motion feature, sequential modelling and regularization parameters. **5) Gaze Labelling 'in the wild'.** We collect 'in the wild' gaze data from YouTube videos having *creative common licence* and compare the label quality with various model based techniques [22,50,55].

Table 2. Comparison of '2-labels' and '1-label' techniques using MAE, CC and Angular Error. Both frameworks with both VGG-16 and ResNet-50 backbone are trained on 80% of the Derived Data (DD, i.e. 3-frame set mined) and validated on 20% of the DD. This 80%-20% partition does not have any identity overlap. Here, MAE: Mean Absolute Error, CC: Correlation Coefficient, TG: TabletGaze.

| Dataset | VGG-16 | | | | ResNet-50 | | | | 2-labels | 1-label |
| | 2-labels | | 1-label | | 2-labels | | 1-label | | Angular error (in °, TG: in cm) | |
	MAE	CC	MAE	CC	MAE	CC	MAE	CC		
CAVE	0.29	0.85	0.53	0.21	0.25	0.90	0.43	0.25	3.05	3.30
MPII	0.43	0.62	0.57	0.25	0.42	0.63	0.57	0.25	5.00	5.40
Gaze360	0.38	0.69	0.45	0.42	0.34	0.71	0.40	0.70	15.00	15.80
TabletGaze	0.49	0.54	0.50	0.58	0.47	0.55	0.49	0.55	2.27	2.61
YouTube 'in the wild'	0.27	0.90	0.36	0.81	0.24	0.92	0.34	0.86	9.41	12.07

Further, we did perform additional experiments to evaluate the generalizibity of the proposed method. See supplementary material for *cross dataset evaluation.*

Evaluation Metrics. For quantitative evaluation, we use Mean Absolute Error (MAE), Correlation Coefficient (CC) and Angular Error (in °). Following each database's evaluation protocol, we follow 'leave-one-person-out' for MPII, cross-validation for CAVE and TabletGaze; and train-val-test partitions for Gaze360 dataset. We randomly split the 3-frame sets into 80%-20% training and testing sets. For a fair comparison between the datasets, we estimate the errors in normalized space i.e. we apply min-max normalization to the labels to map it to a [0–1] range. Additionally, we compute angular error (in °) except for the TabletGaze dataset, for which we compute the error in cm (similar to [21]). To compare with the state-of-the-art methods, we use similar evaluation protocols mentioned in the respective studies. See supplementary material for more details.

Training Details. After the 3-frame set mining, we apply Dlib face detector [44] for eye detection. If face detection (dlib) fails especially for Gaze360 dataset, we use cropped headpose provided with the dataset[1]. Otherwise, we use the resized input image. For the backbone network, we choose VGG-16 [45] and ResNet-50 [20] architectures. For training, we use SGD optimizer with 0.001 learning rate with $1 \times e^{-6}$ decay per epoch. The values of $\lambda_1 \cdots \lambda_4$ are 1. In each case, the models are trained for 1,000 epochs with batch size 32 and early stopping.

6   Results

Gaze Labelling and Estimation Performance Comparison. To show the effectiveness of the proposed method, we evaluate on four benchmark datasets and YouTube data. First, we applied 3-frame set mining to get the derived data (Refer Table 1). Further, we split the derived data randomly into train and test sets (train set: 80% and test set: 20%). Please note that the test set does not have

---

[1] https://github.com/erkil1452/gaze360/tree/master/dataset.

Table 3. Results of the re-trained state-of-the-art methods on MPII and CAVE dataset in terms of angular error in °. OP=Original Predictions is the result as mentioned in the original papers [37].

Dataset	Method	Alexnet	VGG 16	Pictorial Gaze [37]
CAVE	OP	4.20	3.90	3.80
	2-labels	4.10	4.46	4.00
	1-label	4.55	4.84	4.36
MPII	OP	5.70	5.40	4.50
	2-labels	5.90	5.79	4.70
	1-label	6.30	6.20	4.90
Train:YouTube	CAVE	–	–	3.90
	MPII	–	–	4.57

Table 4. Results of the re-trained methods on TabletGaze and Gaze360 dataset in terms of angular error in ° and cm. OP=Original Predictions (based on manual annotation) is the result as mentioned in the [21,24].

TabletGaze	mHOG+SVR [21]
OP	2.50
2-labels	2.70
1-label	3.10
Train:YouTube	2.30

Gaze360	Pinball LSTM [24]
OP	13.50
2-labels	14.40
1-label	17.20
Train:YouTube	12.80

Table 5. Performance comparison for different input data settings with the state-of-the-art methods on MPII, CAVE, TabletGaze (TG) and Gaze360 dataset. NL: Newly Labelled, SF: Start Frame, EF: End Frame.

Dataset	CAVE	TG	Gaze360	MPII
Method	[37]	[21]	[24]	[37]
(SF+EF)	7.14	8.60	22.10	6.10
50% data	5.50	4.50	18.70	5.40
2-labels	4.00	2.70	14.40	4.70
NL frames	4.30	3.00	14.90	4.30
1-label	4.36	3.10	17.20	4.90

identity overlap with training partition. The results are mentioned in Table 2 in terms of MAE, CC and angular error. We use VGG-16 and Resnet-50 as backbone networks to show the impact of different network architectures. Quantitatively, ResNet-50 performs slightly better than the VGG-16. From Table 2, it is also observed that '2-labels' technique is closer to the original label distribution as compared to the '1-label' technique due to the absence of supervisory signal (i.e. absence of end frame label). Due to high-resolution images in CAVE dataset, generated labels' similarity as compared with original labels is high for

Table 6. Ablation study: effect of loss functions for '2-labels' (Eq. 1) & 1-label (Eq. 2). NA: Not Applicable

Loss	MAE (2-labels)	MAE (1-label)
l_{reg}	0.98	0.99
$l_{reg} + l_{consistency}$	0.42	0.76
$l_{reg} + l_{consistency} + l_{divergence}$ (for 1-label)	NA	0.58
$l_{reg} + l_{consistency} + l_{divergence}$ $+l_{embedding}$ (for 1-label)	NA	0.57

'2-labels' setting. Instead of sequential modelling, our loss imposed gaze estimation method improves model performance.

Comparison with State-of-the-Art Methods. We also evaluate the label generation quality of our proposed methods. For this purpose, we conduct experiments by training existing state-of-the-art methods [12,21,24,37] with the labels predicted from our method. The state-of-the-art network's performance is measured by comparing with the original ground truth labels. The performance comparison is mentioned in the Table 3 and 4. For [24,37], we use author's GitHub implementations [1]. It is observed that '2-labels' performs better than '1-label' for all the datasets. For CAVE dataset, 'gazemap' based method [37] for '2-labels' (i.e. 4.08°) performs better than other settings. It is to be noted that the results of re-trained methods are comparable to when they were trained with the original labels. By using less than 5% labelled data (For MPII 4.67% and Gaze360 2.38%), the labels generated by our weakly-supervised method perform favourably when evaluated on state-of-the-art methods [24,37]. This shows the usefulness of our weakly-supervised approach of label generation.

Re-train State-of-the-Art Methods with Subset of Data. To further validate the generated labels, we perform following experiments: we re-train from scratch [21,24,37] using the following labelled sets independently: a) start and end frames of 3-frame sets only, b) 50% of originally labelled training data, c) frames with labels generated with 2-labels method, d) frames with labels generated with 2-labels method apart from start and end frames, and e) frames with labels generated with 1-labels method. The results are shown in Table 5. When we train the networks with start and end frames, the error is high as the start and end frames consist of < 10% of the whole dataset. When we use 50% of the whole labelled data, the results significantly improve. Similarly, for newly labelled frames, the error is less as compared to above two settings. Please note that in the newly labelled case the training is performed on 90–95% of the training data. These results validate the quality of labels generated by our methods.

Ablation Studies
Impact of Loss Function. We progressively integrate different parts of our method. We assess the impact of each loss term mentioned in Eq. 1 and Eq. 2

by considering them one at a time during learning on the MPII dataset. The results are shown in Table 6. For '2-labels' and '1-label', if only regression loss is considered, the error is high for the two techniques (0.98 and 0.99). We argue that this could be due to lack of domain knowledge. Further, consistency loss is added to the network, which reduces the error (0.42 and 0.76) significantly for both settings. On the other hand, KL divergence based loss is added to the '1-label' framework, which again reduces the error significantly from 0.76 to 0.58. Additionally, embedding loss is introduced to consider future frame consistency, though we note that for MPII, the change in MAE is very less (i.e. from 0.58 to 0.57). These experiments clearly establish the individual importance of each of the proposed loss terms in our framework.

Impact of Motion Feature. To judge the impact of the motion feature, we evaluate the performance of '2-labels' on the CAVE dataset without using the motion feature i.e. the latent space features are directly concatenated (Refer Fig. 2). The results suggests that with the motion feature the MAE is reduced from 0.34 to 0.25. Thus, it is important to include this information in the framework.

Sequential Modelling Vs '2-labels'. We incorporate an LSTM module having 3 steps instead of the whole pipeline. The input to the LSTM module is 3-frame set and it estimate the gaze of the unlabelled frame. The MAE is quite high (i.e. 0.95) as compared to our ResNet-50 based '2-labels' framework (i.e. 0.25). The possible reason for this performance enhancement owes to task-relevant losses posed within very small temporal information.

Gaze Labelling 'in the Wild'. The best results on the collected data in terms of MAE and CC are mentioned in the last row of Table 2. The angular errors w.r.t. the SLERP [39] for '2-labels' and '1-label' are 9.41° and 12.07°, although SLERP is a weak baseline to compare eye gaze in an unconstrained environment. When we use the generated labels on YouTube dataset to complement the other dataset, the per-

Table 7. Comparison of model based methods on YouTube data. MAE: Mean Absolute Error, AE: Angular Error.

Method	MAE	AE
2-labels	**0.24**	**9.41**
[50]	0.74	15.30
[55]	0.52	13.90
[22]	0.57	14.01

formance improves for TabletGaze and Gaze360 (see Table 3 and 4). For adapting the SOTA methods (Table 3 and 4), we fine-tune the models following standard protocol [10,56]. The results suggest that the network learns a meaningful representation. Moreover, we compare our method with model based approaches [22,50,55]. The results are depicted in Table 7. From the table, it is observed that our proposed method outperforms model based methods.

Qualitative Analysis. Figure 3 illustrates few examples of gaze trajectories along with the predictions of the proposed method. The trajectories consist of start, unlabelled and end frames and the trajectory length is not limited to 3. During training, the proposed method takes start, end and *one of the unlabelled frame* as input. Please note that the eye patch cropped from the facial images are used as input. In the '2-labels' case, the ground truth labels of

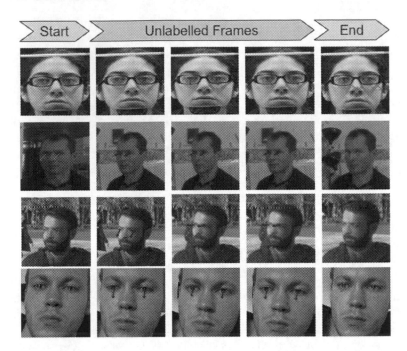

Fig. 3. Few examples of gaze trajectories along with qualitative prediction results. Here, red and green arrow represent predicted and ground truth gaze direction, respectively. (Color figure online)

both start and end frames are provided for weak supervision. In Fig. 3, the green and red arrow indicates ground truth and predicted gaze direction. For better understanding, we plot the gaze direction corresponding to each eye originated from detected pupil center. This qualitative analysis indicates that our weakly supervised method learns to interpolate gaze labels efficiently from terminal frames of a trajectory.

7   Limitations, Conclusion and Future Work

Our study introduces a weakly-supervised approach for generating labels for intermediate frames in a defined gaze trajectory. The proposed frameworks leverage task-specific domain knowledge i.e. trajectory ordering, motion and appearance features etc. With extensive experiments, we show that the labels generated by our methods are comparable to the ground truth labels. Further, we also show that the state-of-the-art existing techniques re-trained using the labels generated by our method give comparable performance. This applies that with just 1%-5.6% labelled data (dependent on the dataset) training can be performed with performance comparable to when 100% training data is available. Further, we also propose a technique to collect and label eye gaze 'in the wild'. The proposed method can be used for other computer vision based applications (e.g.

gaze tracking devices for AR and VR) without the prior need of having to use the whole labelled dataset during training. Main limitations of our study are the fixed gaze trajectory requirement, inclusion of eye-blink and near frontal face. In the future, we will investigate subject specific gaze estimation in challenging situations such as low-resolution images, uneven illumination conditions and extreme head-poses. Although our methods consider gaze annotation in few of the aforementioned diverse conditions, it would be interesting to have more in-depth study in this domain.

References

1. gaze code:https://github.com/swook/GazeML, https://github.com/Erkil1452/gaze360 (-)
2. Alonso-Martín, F., Gorostiza, J.F., Malfaz, M., Salichs, M.A.: User localization during human-robot interaction. Sensors **12**(7), 9913–9935 (2012). https://doi.org/10.3390/s120709913, http://dx.doi.org/10.3390/s120709913
3. Arabadzhiyska, E., Tursun, O.T., Myszkowski, K., Seidel, H.P., Didyk, P.: Saccade landing position prediction for gaze-contingent rendering. ACM Trans. Graphics **36**(4), 1–12 (2017)
4. Arandjelovic, R., Gronat, P., Torii, A., Pajdla, T., Sivic, J.: NetVLAD: CNN architecture for weakly supervised place recognition. IEEE Comput. Vision Pattern Recog. pp. 5297–5307 (2016)
5. Barz, B., Denzler, J.: Deep learning on small datasets without pre-training using cosine loss. In: 2020 IEEE Winter Conference on Applications of Computer Vision (WACV), pp. 1371–1380 (2020)
6. Benfold, B., Reid, I.: Unsupervised learning of a scene-specific coarse gaze estimator. In: IEEE International Conference on Computer Vision, pp. 2344–2351 (2011)
7. Bertasius, G., Feichtenhofer, C., Tran, D., Shi, J., Torresani, L.: Learning temporal pose estimation from sparsely-labeled videos. arXiv preprint arXiv:1906.04016 (2019)
8. Bilen, H., Vedaldi, A.: Weakly supervised deep detection networks. In: Conference on IEEE Computer Vision and Pattern Recognitio, pp. 2846–2854 (2016)
9. Blattgerste, J., Renner, P., Pfeiffer, T.: Advantages of eye-gaze over head-gaze-based selection in virtual and augmented reality under varying field of views. In: Proceedings of the Workshop on Communication by Gaze Interaction, pp. 1–9 (2018)
10. Dubey, N., Ghosh, S., Dhall, A.: Unsupervised learning of eye gaze representation from the web. In: 2019 International Joint Conference on Neural Networks (IJCNN), pp. 1–7. IEEE (2019)
11. Duchowski, A.T., Duchowski, A.T.: Eye tracking Methodology: Theory and Practice. Springer, London (2017). https://doi.org/10.1007/978-1-84628-609-4
12. Fischer, Tobias, Chang, Hyung Jin, Demiris, Yiannis: RT-GENE: real-time eye gaze estimation in natural environments. In: Ferrari, Vittorio, Hebert, Martial, Sminchisescu, Cristian, Weiss, Yair (eds.) ECCV 2018. LNCS, vol. 11214, pp. 339–357. Springer, Cham (2018). https://doi.org/10.1007/978-3-030-01249-6_21
13. Funes Mora, K.A., Monay, F., Odobez, J.M.: Eyediap: a database for the development and evaluation of gaze estimation algorithms from RGB and RGB-d cameras. In: ACM Symposium on Eye Tracking Research and Applications (2014)

14. Garbin, S.J., Shen, Y., Schuetz, I., Cavin, R., Hughes, G., Talathi, S.S.: Openeds: Open eye dataset. arXiv preprint arXiv:1905.03702 (2019)
15. Ghosh, S., Dhall, A., Hayat, M., Knibbe, J., Ji, Q.: Automatic gaze analysis: a survey of deep learning based approaches. arXiv preprint arXiv:2108.05479 (2021)
16. Ghosh, S., Dhall, A., Sharma, G., Gupta, S., Sebe, N.: Speak2label: using domain knowledge for creating a large scale driver gaze zone estimation dataset. arXiv preprint arXiv:2004.05973 (2020)
17. Ghosh, S., Hayat, M., Dhall, A., Knibbe, J.: Mtgls: multi-task gaze estimation with limited supervision. In: Proceedings of the IEEE/CVF Winter Conference on Applications of Computer Vision, pp. 3223–3234 (2022)
18. Gumilar, I., et al.: Connecting the brains via virtual eyes: eye-gaze directions and inter-brain synchrony in VR. In: Extended Abstracts of the 2021 CHI Conference on Human Factors in Computing Systems, pp. 1–7 (2021)
19. Haeusser, P., Mordvintsev, A., Cremers, D.: Learning by association-a versatile semi-supervised training method for neural networks. In: 2017 IEEE Conference on Computer Vision and Pattern Recognition, pp. 89–98 (2017)
20. He, K., Zhang, X., Ren, S., Sun, J.: Deep residual learning for image recognition. In: 2017 IEEE Conference on Computer Vision and Pattern Recognition, pp. 770–778 (2016)
21. Huang, Q., Veeraraghavan, A., Sabharwal, A.: Tabletgaze: dataset and analysis for unconstrained appearance-based gaze estimation in mobile tablets. Mach. Vis. Appl. 28(5–6), 445–461 (2017)
22. Ishikawa, T.: Passive driver gaze tracking with active appearance models (2004)
23. Jyoti, S., Dhall, A.: Automatic eye gaze estimation using geometric & texture-based networks. In: International Conference on Pattern Recognition, pp. 2474–2479. IEEE (2018)
24. Kellnhofer, P., Recasens, A., Stent, S., Matusik, W., Torralba, A.: Gaze360: physically unconstrained gaze estimation in the wild. In: IEEE International Conference on Computer Vision (2019)
25. Komogortsev, O.V., Karpov, A.: Automated classification and scoring of smooth pursuit eye movements in the presence of fixations and saccades. Behav. Res. Methods 45(1), 203–215 (2013)
26. Kothari, R., De Mello, S., Iqbal, U., Byeon, W., Park, S., Kautz, J.: Weakly-supervised physically unconstrained gaze estimation. In: Proceedings of the IEEE/CVF Conference on Computer Vision and Pattern Recognition, pp. 9980–9989 (2021)
27. Krafka, K., Khosla, A., Kellnhofer, P., Kannan, H., Bhandarkar, S., Matusik, W., Torralba, A.: Eye tracking for everyone. In: IEEE Computer Vision and Pattern Recognition, pp. 2176–2184 (2016)
28. Lee, D.H.: Pseudo-label: The simple and efficient semi-supervised learning method for deep neural networks. In: International Conference on Machine Learning Workshop. vol. 3, p. 2 (2013)
29. Liu, H., Heynderickx, I.: Visual attention in objective image quality assessment: based on eye-tracking data. IEEE Trans. Circuits Syst. Video Technol. 21(7), 971–982 (2011)
30. Lu, F., Chen, X., Sato, Y.: Appearance-based gaze estimation via uncalibrated gaze pattern recovery. IEEE Trans. Image Process. 26(4), 1543–1553 (2017)
31. Lu, F., Sugano, Y., Okabe, T., Sato, Y.: Gaze estimation from eye appearance: a head pose-free method via eye image synthesis. IEEE Trans. Image Process. 24(11), 3680–3693 (2015)

32. Majaranta, P.: Gaze Interaction and Applications of Eye Tracking: Advances in Assistive Technologies. IGI Global (2011)
33. Mustafa, A., Kaur, A., Mehta, L., Dhall, A.: Prediction and localization of student engagement in the wild. arXiv preprint arXiv:1804.00858 (2018)
34. Niehorster, D.C., Hessels, R.S., Benjamins, J.S.: Glassesviewer: open-source software for viewing and analyzing data from the TOBII pro glasses 2 eye tracker. Behav. Res. Methods, **52**, 244–1253 (2020)
35. Palmero, C., Sharma, A., Behrendt, K., Krishnakumar, K., Komogortsev, O.V., Talathi, S.S.: Openeds 2020: Open eyes dataset. arXiv preprint arXiv:2005.03876 (2020)
36. Park, S., Mello, S.D., Molchanov, P., Iqbal, U., Hilliges, O., Kautz, J.: Few-shot adaptive gaze estimation. In: IEEE International Conference on Computer Vision, pp. 9368–9377 (2019)
37. Park, Seonwook, Spurr, Adrian, Hilliges, Otmar: Deep pictorial gaze estimation. In: Ferrari, Vittorio, Hebert, Martial, Sminchisescu, Cristian, Weiss, Yair (eds.) ECCV 2018. LNCS, vol. 11217, pp. 741–757. Springer, Cham (2018). https://doi.org/10.1007/978-3-030-01261-8_44
38. Park, Wooyeong, Heo, Jeongyun, Lee, Jiyoon: Talking through the eyes: user experience design for eye gaze redirection in live video conferencing. In: Kurosu, Masaaki (ed.) HCII 2021. LNCS, vol. 12763, pp. 75–88. Springer, Cham (2021). https://doi.org/10.1007/978-3-030-78465-2_7
39. Peters, C., Qureshi, A.: A head movement propensity model for animating gaze shifts and blinks of virtual characters. Comput. Graphics **34**(6), 677–687 (2010)
40. Purves, D., Morgenstern, Y., Wojtach, W.T.: Perception and reality: why a wholly empirical paradigm is needed to understand vision. Front. Syst. Neurosci. **9**, 156 (2015)
41. Sajjadi, M., Javanmardi, M., Tasdizen, T.: Mutual exclusivity loss for semi-supervised deep learning. In: IEEE International Conference on Image Processing, pp. 1908–1912 (2016)
42. Santini, T., Fuhl, W., Kübler, T., Kasneci, E.: Bayesian identification of fixations, saccades, and smooth pursuits. In: Proceedings of the Ninth Biennial ACM Symposium on Eye Tracking Research & Applications, pp. 163–170 (2016)
43. Schmidt, S., Bruder, G., Steinicke, F.: Depth perception and manipulation in projection-based spatial augmented reality. PRESENCE Virtual Augment. Real. **27**(2), 242–256 (2020)
44. Sharma, S., Shanmugasundaram, K., Ramasamy, S.K.: FAREC-CNN based efficient face recognition technique using DLIB. In: International Conference on Advanced Communication Control and Computing Technologies, pp. 192–195 (2016)
45. Simonyan, K., Zisserman, A.: Very deep convolutional networks for large-scale image recognition. In: International Conference on Learning Representations (2014)
46. Smith, B., Yin, Q., Feiner, S., Nayar, S.: Gaze locking: passive eye contact detection for human-object interaction. In: Proceedings of the 26th Annual ACM Symposium on User Interface Software and Technology (2013)
47. Startsev, M., Agtzidis, I., Dorr, M.: 1D CNN with BLSTM for automated classification of fixations, saccades, and smooth pursuits. Behav. Res. Methods **51**(2), 556–572 (2019)
48. Sugano, Y., Matsushita, Y., Sato, Y.: Learning-by-synthesis for appearance-based 3d gaze estimation. In: IEEE Conference on Computer Vision and Pattern Recognition, pp. 1821–1828 (2014)

49. Swaminathan, A., Ramachandran, M.: Enabling augmented reality using eye gaze tracking, US Patent 9,996,150.12 June 2018
50. Valenti, R., Sebe, N., Gevers, T.: Combining head pose and eye location information for gaze estimation. IEEE Trans. Image Process. **21**(2), 802–815 (2011)
51. Wang, K., Zhao, R., Ji, Q.: A hierarchical generative model for eye image synthesis and eye gaze estimation. In: Proceedings of the IEEE Conference on Computer Vision and Pattern Recognition, pp. 440–448 (2018)
52. Wang, W., Shen, J.: Deep visual attention prediction. IEEE Trans. Image Process. **27**(5), 2368–2378 (2017)
53. Weston, Jason, Ratle, Frédéric., Mobahi, Hossein, Collobert, Ronan: Deep learning via semi-supervised embedding. In: Montavon, Grégoire., Orr, Geneviève B.., Müller, Klaus-Robert. (eds.) Neural Networks: Tricks of the Trade. LNCS, vol. 7700, pp. 639–655. Springer, Heidelberg (2012). https://doi.org/10.1007/978-3-642-35289-8_34
54. Williams, O., Blake, A., Cipolla, R.: Sparse and semi-supervised visual mapping with the s3gp. In: 2006 IEEE Computer Society Conference on Computer Vision and Pattern Recognition (CVPR'06)(2006)
55. Yamazoe, H., Utsumi, A., Yonezawa, T., Abe, S.: Remote gaze estimation with a single camera based on facial-feature tracking without special calibration actions. In: Proceedings of the 2008 Symposium on Eye Tracking Research & Applications,. pp. 245–250 (2008)
56. Yu, Y., Liu, G., Odobez, J.: Improving few-shot user-specific gaze adaptation via gaze redirection synthesis. In: IEEE Conference on Computer Vision and Pattern Recognition, pp. 11937–11946 (2019)
57. Yu, Y., Odobez, J.: Unsupervised representation learning for gaze estimation. IEEE Conference on Computer Vision and Pattern Recognition, pp. 1–13 (2020)
58. Zabala, U., Rodriguez, I., Martínez-Otzeta, J.M., Lazkano, E.: Modeling and evaluating beat gestures for social robots. Multim. Tools Appl. **81**, 3421–3438 (2021)
59. Zhang, X., Sugano, Y., Fritz, M., Bulling, A.: It's written all over your face: Full-face appearance-based gaze estimation. In: IEEE Computer Vision and Pattern Recognition Workshop (2017)
60. Zhang, X., Sugano, Y., Fritz, M., Bulling, A.: Mpiigaze: real-world dataset and deep appearance-based gaze estimation. IEEE Trans. Pattern Anal. Mach. Intell. (2017)
61. Zhang, Xucong, Park, Seonwook, Beeler, Thabo, Bradley, Derek, Tang, Siyu, Hilliges, Otmar: ETH-XGaze: a large scale dataset for gaze estimation under extreme head pose and gaze variation. In: Vedaldi, Andrea, Bischof, Horst, Brox, Thomas, Frahm, Jan-Michael. (eds.) ECCV 2020. LNCS, vol. 12350, pp. 365–381. Springer, Cham (2020). https://doi.org/10.1007/978-3-030-58558-7_22
62. Zhang, X., Sugano, Y., Bulling, A.: Everyday eye contact detection using unsupervised gaze target discovery. In: 30th Annual ACM Symposium on User Interface Software and Technology, pp. 193–203 (2017)
63. Zhang, X., Sugano, Y., Fritz, M., Bulling, A.: Appearance-based gaze estimation in the wild. In: 2015 IEEE Conference on Computer Vision and Pattern Recognition (CVPR), pp. 4511–4520 (2015)
64. Zhang, Y., Dong, W., Hu, B.G., Ji, Q.: Weakly-supervised deep convolutional neural network learning for facial action unit intensity estimation. In: IEEE Computer Vision and Pattern Recognition, pp. 2314–2323 (2018)

65. Zhang, Y., Zhao, R., Dong, W., Hu, B.G., Ji, Q.: Bilateral ordinal relevance multi-instance regression for facial action unit intensity estimation. In: Proceedings of the IEEE Conference on Computer Vision and Pattern Recognition, pp. 7034–7043 (2018)

66. Zhao, J., Mathieu, M., Goroshin, R., Lecun, Y.: Stacked what-where auto-encoders. International Conference on Learning Representations Workshop (2015)

67. Zhu, Ye., Yan, Yan, Komogortsev, Oleg: Hierarchical HMM for eye movement classification. In: Bartoli, Adrien, Fusiello, Andrea (eds.) ECCV 2020. LNCS, vol. 12535, pp. 544–554. Springer, Cham (2020). https://doi.org/10.1007/978-3-030-66415-2_35

Author Index

Printed in the United States
by Baker & Taylor Publisher Services